Astrophysical Concepts

Astrophysical Concepts

MARTIN HARWIT

Astronomy Department
Cornell University

JOHN WILEY & SONS
New York · London ·
Sydney · Toronto

Library of Congress Cataloging in Publication Data:

Harwit, Martin, 1931—
Astrophysical concepts.

Bibliography: p.
1. Astrophysics. I. Title.

QB461.H37   523.01   73-3135
ISBN 0–471–35820–7

Printed in the United States of America

10 9 8 7 6 5 4 3 2 1

Preface

My principal aim in writing this book was to present a wide range of astrophysical topics in sufficient depth to give the reader a general quantitative understanding of the subject. The book outlines cosmic events but does not portray them in detail—it provides a series of astrophysical sketches. I think this approach befits the present uncertainties and changing views in astrophysics.

The material is based on notes I prepared for a course aimed at seniors and beginning graduate students in physics and astronomy at Cornell. This course defined the level at which the book is written.

For readers who are versed in physics but are unfamiliar with astronomical terminology, Appendix A is included. It gives a brief background of astronomical concepts and should be read before starting the main text.

The first few chapters outline the scope of modern astrophysics and deal with elementary problems concerning the size and mass of cosmic objects. However, it soon becomes apparent that a broad foundation in physics is needed to proceed. This base is developed in Chapters 4 to 7 by using, as examples, specific astronomical situations. Chapters 8 to 10 enlarge on the topics first outlined in Chapter 1 and show how we can obtain quantitative insights into the structure and evolution of stars, the dynamics of cosmic gases, and the large-scale behavior of the universe. A final chapter discusses life in the universe.

Throughout the book I emphasize astrophysical concepts. This means that objects such as asteroids, stars, supernovae, or quasars are not described in individual chapters or sections. Instead, they are mentioned throughout the text whenever relevant physical principles are discussed. Thus the common features of many astronomical situations are underlined, but there is a partition of information about specific astronomical objects. For example, different aspects of neutron stars and pulsars are discussed in Chapters 5, 6, 8, Appendix A, and elsewhere. To compensate for this treatment a comprehensive index is included.

I have sketched no more than the outlines of several traditional astronomical topics, such as the theories of radiative transfer, stellar atmospheres, and polytropic gas spheres, because a complete presentation would have required extensive mathematical development to be genuinely useful. However, the main physical concepts of these subjects are worked into the text, often as remarks without specific mention. In addition I refer, where appropriate, to other sources that treat these topics in greater detail.

The bibliography is designed for readers who wish to cover any given area in greater depth. I have cited only authors who actively have contributed to a field and whose views bring the reader closer to the subject. Although some of the cited articles are popular, the writing is accurate.

A book that covers a major portion of astrophysics must be guided by the many excellent monographs and review articles that exist today. It is impossible to acknowledge all of them properly and to give credit to the astrophysicists whose viewpoint strongly influenced my writing. I am also grateful for the many suggestions for improvements offered by my colleagues at Cornell and by several generations of students who saw this book evolve from a series of informal lecture notes. Finally, I would like to thank Barbara L. Boettcher for preparing the drawings.

Martin Harwit

Contents

CHAPTER 3.   DYNAMICS AND MASSES OF ASTRONOMICAL BODIES

CHAPTER 4.   RANDOM PROCESSES

CHAPTER 5. PHOTONS AND FAST PARTICLES

CHAPTER 6.   ELECTROMAGNETIC PROCESSES IN SPACE

CHAPTER 7.   QUANTUM PROCESSES IN ASTROPHYSICS

CHAPTER 8.   STARS

CHAPTER 9.   COSMIC GAS AND DUST

CHAPTER 10.   STRUCTURE OF THE UNIVERSE

1

An Approach to Astrophysics

In a sense each of us has been inside a star; in a sense each of us has been in the vast empty spaces between the stars; and—if the universe ever had a beginning—each of us was there!

Every molecule in our bodies contains matter that once was subjected to the tremendous temperatures and pressures at the center of a star. This is where the iron in our red blood cells originated. The oxygen we breathe, the carbon and nitrogen in our tissues, and the calcium in our bones, also were formed through the fusion of smaller atoms at the center of a star.

Terrestrial ores containing uranium, plutonium, lead, and many other massive atoms must have been formed in a supernova explosion—the self-destruction of a star in which a sun's mass is hurled into space at huge velocity. In fact, most of the matter on earth and in our bodies must have gone through such a catastrophic event!

The elements lithium, beryllium, and boron, which we find in traces on earth, seem to have originated through cosmic ray bombardment in interstellar space. At that epoch the earth we now walk on was distributed so tenuously that a gram of soil would have occupied a volume the size of the entire planet.

To account for the deuterium, the heavy hydrogen isotope found on earth, we may have to go back to a cosmic explosion signifying the birth of the entire universe. Some of the helium in a child's balloon would also have been formed at that time.

How do we know all this? And how sure are we of this knowledge?

This book was written to answer such questions and to provide a means for making astrophysical judgments.

We are just beginning a long and exciting journey into the universe. There is much to be learned, much to be discarded, and much to be revised. We have excellent theories, but theories are guides for understanding the truth. They are not truth itself. We must therefore continually revise them if they are to keep leading us in the right direction.

In going through the book, just as in devising new theories, we will find ourselves baffled by choices between the real and the apparent. We will have to learn that it may still be too early to make such choices, that reality in astrophysics has often been short-lived, and that—disturbing though it would be—we may some day have to reconcile ourselves to the realization that our theories had recognized only superficial effects—not the deeper, truly motivating, factors.

We may therefore do well to avoid an immediate preoccupation with astrophysical "reality." We should take the longer view and look closely at those physical concepts likely to play a role in the future evolution of our understanding. We may reason this way:

The development of astrophysics in the last few decades has been revolutionary. We have discarded what had appeared to be our most reliable theories, replaced them, and frequently found even the replacements lacking. The only constant in this revolution has been the pool of astrophysical concepts. It has not substantially changed, and it has provided a continuing source of material for our evolving theories.

This pool contained the neutron stars 35 years before their discovery, and it contained black holes three decades before astronomers started searching for them. The best investment of our efforts may lie in a deeper exploration of these concepts.

In astrophysics we often worry whether we should organize our thinking around individual objects—planets, stars, pulsars, and galaxies—or whether we should divide the subject according to physical principles common to the various astrophysical processes.

Our emphasis on concepts will make the second approach more appropriate. It will, however, also raise some problems: Much of the information about individual types of objects will be distributed throughout the book, and can be gathered only through use of the index. This leads to a certain unevenness in the presentation.

The unevenness is made even more severe by the varied mathematical treatment: No astrophysical picture is complete if we cannot assign a numerical value to its scale. In this book, we will therefore consistently aim at obtaining a rough order of magnitudes characteristic of the different phenomena. In some cases, this aim leads to no mathematical difficulties. In other problems, we will have to go through rather complex mathematical preparations before even the crudest answers emerge. The estimates of the curvature of the universe in Chapter 10 are an example of these more complex approaches.

Given these difficulties, which appear to be partly dictated by the nature of modern astrophysics, let us examine the most effective ways to use this book.

For those who have no previous background in astronomy, Appendix A may provide a good starting point. It briefly describes the astronomical objects we will study and introduces astronomical notation. This notation will be used throughout the book and is generally not defined in other chapters. Those who have previously studied astronomy will be able to start directly with the present chapter that presents the current searches going on in astrophysics—the questions that we seek to answer. Chapters 2 and 3 show that, while some of the rough dimensions of the universe can be measured by conceptually simple means, a deeper familiarity with physics is required to understand the cosmic sources of energy and the nature of cosmic evolution. The physical tools we need are therefore presented in the intermediate Chapters 4 to 7. We then gather these tools to work our way through theories of the synthesis of chemical elements mentioned right at the start of this section, the formation and evolution of stars, the processes that take place in interstellar space, the evolution of the universe, and the astrophysical setting for the origins of life.

This is an exciting, challenging venture; but we have a long way to go. Let us start!

1:1 CHANNELS FOR ASTRONOMICAL INFORMATION

Imagine a planet inhabited by a blind civilization. One day an inventor discovers an instrument sensitive to visible light and this device is found to be useful for many purposes, particularly for astronomy.

Human beings can see light and we would expect to have a big headstart in astronomy compared to any civilization that was just discovering methods for detecting visible radiation. Think then of an even more advanced culture that could detect not only visible light but also all other electromagnetic radiation and that had telescopes and detectors sensitive to *cosmic rays, neutrinos,* and *gravitational waves.* Clearly that civilization's knowledge of astronomy could be far greater than ours.

Four entirely independent channels are known to exist by means of which information can reach us from distant parts of the universe.

(a) Electromagnetic radiation: γ-rays, X-rays, ultraviolet, visible, infrared, and radio waves.

(b) Cosmic ray particles: These comprise high energy electrons, protons, and heavier nuclei as well as the (unstable) neutrons and mesons. Some cosmic ray particles consist of antimatter.

(c) Neutrinos and antineutrinos: There are two different types of neutrinos and antineutrinos; those associated with electrons and others associated with μ-mesons.

(d) Gravitational waves.

1:1

Most of us are familiar with channel *a*, currently the channel through which we obtain the bulk of astronomical information. However, let us briefly describe channels (b), (c), and (d).

(b) There are fundamental differences between cosmic ray particles and the other three information carriers: (i) cosmic ray particles move at very nearly the speed of light, while the others move at precisely the speed of light; (ii) cosmic rays have a positive rest mass; and (iii) when electrically charged the particles can be deflected by cosmic magnetic fields so that the direction from which a cosmic ray particle arrives at the earth often is not readily related to the actual direction of the source.

Cosmic ray astronomy is far more advanced than either neutrino or gravitational wave work. Detectors and detector arrays exist, but the technical difficulties still are great. Nonetheless, through cosmic ray studies we hope to learn a great deal about the chemistry of the universe on a large scale and we hope, eventually, to single out regions of the universe in which as yet unknown, grandiose accelerators produce these highly energetic particles. We do not yet know how or where the cosmic ray particles gain their high energies; we merely make guesses, expressed in the form of different theories on the origin of cosmic rays (Ro64a, Go69, Gu69).

(c) Neutrinos, like photons, have zero rest mass. They have one great advantage in that they can traverse great depths of matter without being absorbed. Neutrino astronomy could give us a direct look at the interior of stars, much as X-rays can be used to examine a metal block for internal flaws or a medical patient for lung ailments. Neutrinos could also convey information about past ages of the universe because, except for a systematic energy loss due to the expansion of the universe, the neutrinos are preserved in almost unmodified form over many aeons.* Much of the history of the universe must be recorded in the ambient neutrino flux, but so far we do not know how to tap this information (We62).

A first serious search for solar neutrinos has been conducted and has shown that there are fewer emitted neutrinos than had been predicted (Da68). This has led to a re-examination of theories on the nuclear reactions taking place in the interior of the sun; but the puzzle persists. The predicted neutrino flux exceeds the observed!

(d) Gravitational waves, when reliably detected, will yield information on the motion of very massive bodies. Despite many attempts, it is not absolutely certain whether gravitational waves have so far actually been detected. But the detectors constructed for this purpose have measured the presence of signals that, thus far, cannot be explained in terms of anything else and actually may represent gravitational waves. We seem therefore to be on the threshold of

---

* One aeon = 10^9 y.

important discoveries that are sure to have an important influence on astronomy (We70).

It is clear that astronomy cannot be complete until techniques are developed to detect all of the four principal means by which information can reach us. Until that time astrophysical theories must remain provisory.

Not only must we be able to detect these information carriers, but we will also have to develop detectors that cover the entire spectral range for each type of carrier. The importance of this is shown by the great contribution made by radio astronomy. Until two or three decades ago, all our astronomical information was obtained in the visible, near infrared, or near ultraviolet regions; no one at that time suspected that a wealth of information was available in the radio spectrum. Yet, today the only map we have of our own Galaxy is a radio astronomical map showing the distribution of interstellar gas in distant spiral arms. All this is hidden in visible light because great clouds of dust obscure the view. Moreover, a large variety of new objects has recently been discovered in the radio spectrum. Many of these are highly powerful sources, but do not appear significant in the visible part of the spectrum.

There can be no doubt that great efforts will be made to advance cosmic ray, neutrino, and gravitational wave astronomy. These developments will presumably involve great technological difficulties, but it does not appear possible that astrophysical knowledge could become complete without them.

Just as we are reaching for our first cosmic neutrino and gravitational wave detections, the possibility of yet a fifth channel for communicating has been suggested. The information carriers would be tachyons; these thus-far hypothetical particles would travel at speeds greater than the speed of light. Whether such particles exist and are detectable is not clear. What is clear, however, is that their existence in measurable quantities would revolutionize astrophysics. We would then be able to obtain a nearly up-to-date picture of the distant regions of the universe, for which the other carriers only provide information some aeons old. A far better understanding of cosmic evolution would thus come about. Moreover, should intelligent life exist elsewhere in the Galaxy or universe, it would certainly make use of tachyons to communicate more quickly. The current theoretical discussions and laboratory experiments on tachyons are therefore potentially of greater interest to astrophysics.

1:2 X-RAY ASTRONOMY: DEVELOPMENT OF A NEW FIELD

The development of a new branch of astronomy often follows a general pattern: Vague theoretical thinking tells us that no new development is to be expected at all. Consequently, it is not until some chance observation focuses attention onto a new area that serious preliminary measurements are undertaken. Many of these initial findings later have to be discarded as techniques improve.

These awkward developmental stages are always exciting; let us outline the evolution of X-ray astronomy, as an example, to convey the sense of advances that should take place in astronomy and astrophysics in the next few years.

Until 1962 only solar X-ray emission had been observed. This flux is so weak that no one expected a large X-ray flux from sources outside the solar system. Then, in June 1962, R. Giacconi, H. Gursky, and F. Paolini of the American Science and Engineering Corporation (ASE) and B. Rossi of M.I.T. (Gi62) flew a set of large area Geiger counters in an Aerobee rocket. The increased area of these counters was designed to permit detection of X-rays scattered by the moon, but originating from the sun. The counters were sensitive in the wavelength region from 2 to 8 Å.

No lunar X-ray flux could be detected. However, a source of X-rays was discovered in a part of the sky not far from the center of the Galaxy and a diffuse background flux of X-ray counts was evident from all portions of the sky. Various arguments showed that this flux probably was not emitted in the outer layers of the earth's atmosphere, and therefore should be cosmic in origin. Later flights by the same group verified their first results.

At this point a team of researchers at the U.S. Naval Research Laboratory became interested. They had experience with solar X-ray observations and were able to construct an X-ray counter some 10 times more sensitive than that flown by Giacconi's group. Instead of the very wide field of view used by that group, the NRL team limited their field of view to 10 degrees of arc so that their map of the sky could show somewhat finer detail (Bo64a).

An extremely powerful source was located in the constellation Scorpius about 20 degrees of arc from the Galactic center. At first this source remained unidentified. Photographic plates showed no unusual objects in that part of the sky. The NRL group also discovered a second source, some eight times weaker than the Scorpio source. This was identified as the Crab Nebula, a remnant of a supernova explosion observed by Chinese astronomers in 1054 A.D. The NRL team, whose members were Bowyer, Byram, Chubb, and Friedman, believed that these two sources accounted for most of the emission observed by Giacconi's group.

Many explanations were advanced about the possible nature of these sources. Arguments were given in favor of emission by a new breed of highly dense stars whose cores consisted of neutrons. Other theories suggested that the emission might come from extremely hot interstellar gas clouds. No decision could be made on the basis of observations because none of the apparatus flown had fine enough angular resolving power. Nor did the NRL team expect to attain such instrumental resolving power for some years to come.

Then, early in 1964, Herbert Friedman at NRL heard that the moon would occult the Crab Nebula some seven weeks later. Here was a great opportunity to test whether at least one cosmic X-ray source was extended or stellar. For, as the edge of the moon passes over a well-defined point source, all the radiation is suddenly cut off. On the other hand, a diffuse source is slowly covered as the moon moves across the celestial sphere; accordingly, the radiation should be cut off gradually.

No other lunar occultation of either the Scorpio source or the Crab Nebula was expected for many years; so the NRL group went into frenzied preparations and seven weeks later a payload was ready. The flight had to be timed to within seconds since the Aerobee rocket to be used only gave 5 minutes of useful observing time at altitude. Two possible flight times were available; one at the beginning of the eclipse, the other at the end. Because of limited flight duration it was not possible to observe both the initial immersion and subsequent egress from behind the moon.

The first flight time was set for 22:42:30 Universal Time on July 7, 1964. That time would allow the group to observe immersion of the central 2 minutes of arc of the Nebula. Launch took place within half a second of the prescribed time. At altitude, an attitude control system oriented the geiger counters. At 160 seconds after launch, the control system locked on the Crab. By 200 seconds a noticeable decrease in flux could be seen and by 330 seconds the X-ray count was down to normal background level. The slow eclipse had shown that the Crab Nebula is an extended source. One could definitely state that at least one of the cosmic X-ray sources was diffuse. Others might be due to stars. But this one was not (Bo64b).

Roughly seven weeks after this NRL flight the ASE-MIT group was also ready to test angular sizes of X-ray sources. Their experiment was more general in that any source could be viewed. Basically it made use of a collimator that had been designed by the Japanese physicist, M. Oda (Od65). This device consisted of two wire grids separated by a distance D that was large compared to the open space between wires, which was slightly less than the wire diameter d.

The principle on which this collimator works is illustrated in Fig. 1.1. When the angular diameter of the source is small compared to d/D, alternating strong and weak signals are detected as the collimator aperture is swept across the source. If $\theta \gg d/D$ virtually no change in signal strength is detected as a function of orientation.

In their first flight the MIT-ASE group found the Scorpio source to have an angular diameter small compared to $1/2°$. Two months later a second flight confirmed that the source diameter was small, in fact, less than $1/8°$. A year and a half later this group found that the source must be far smaller yet, less than $20''$ in diameter. On this flight two collimators with different wire spacing were used. This meant that the transmission peaks for the two collimators coincided only

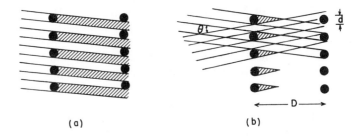

Fig. 1.1 (*a*) For parallel light the front grid casts a sharp shadow on the rear grid. As the collimator is rotated light is alternately transmitted and stopped depending on whether the shadow is cast on the wires of the rear grid or between them. (*b*) For light from a source whose angular dimension $\theta \gg d/D$, the shadow cast by the front grid is washed out. Rotation of the collimator assembly then does not give rise to a strong variation of the transmitted X-ray flux.

for normal incidence and, in this way, yielded an accurate position of the Scorpio source (Gu66). An optical identification was then obtained at the Tokyo Observatory and subsequently confirmed at Mount Palomar (Sa66a). It showed an intense ultraviolet object that flickered on a time scale less than one minute. These are characteristics associated with old novae near their minimum phase.

The brightness and color of neighbouring stars in the vicinity of Sco XR-1 showed that these stars were at a distance of a few hundred light years from the sun, and this gave us a good first estimate of the total energy output of the source. A search on old plates showed that the mean photographic brightness of the object had not changed much since 1896.

At the time of writing these early observations have been largely confirmed. Interestingly, however, the discovery that the Crab Nebula contains a pulsar, sent X-ray astronomers back to data previously collected. Some of these records showed the characteristic 33 millisecond pulsations, and showed that an appreciable fraction of the flux—10 to 15%—comes from a point source—now believed to be a neutron star formed in the supernova explosion (Fr69). Our views of the Crab as a predominantly diffuse X-ray source had to be revised.

Many other Galactic X-ray sources have by now been located and identified; and frequently they have a violet, stellar (pointlike) appearance similar to Sco XR-1. These objects sometimes suddenly increase in brightness by many magnitudes within hours. Others pulsate regularly, somewhat like the Crab Nebula pulsar. The range of X-ray energies at which the observations have by now been carried out, is quite wide too, and both visual and X-ray spectra are available for many sources.

A few extragalactic X-ray sources have also been seen. The first of these was M87, a galaxy known to be a bright radio source (By67). It is a peculiar galaxy consisting of a spherical distribution of stars from which a jet of gas seems to be ejected. The jet is bluish in visible light, and probably gives off light by virtue of highly relativistic electrons spiraling about magnetic lines of force and emitting radiation by the synchrotron mechanism (see Chapter 6)—a mechanism by which highly accelerated energetic particles lose energy in a synchroton.

Theorists have now proposed a variety of explanations for cosmic X-ray sources and also for the continuum X-ray background that appears to permeate the universe. Many experiments are being planned to test these theories. X-ray, visual, infrared, and radio astronomers compare their results to see if a common explanation can be found. Progress is rapid and perhaps in a few years this field will no longer be quite as exciting. But by then another branch of astronomy will have opened up and the excitement will be renewed.

The fundamental nature of astrophysical discoveries being made—or remaining to be made—leaves little room for doubt that a large part of current theory will be drastically revised over the next decades. Much of what is known today must be regarded as tentative and all parts of the field have to be viewed with healthy skepticism.

We expect that much will still be learned using the methods that have been so successful in the past. However, there are parts of astrophysics—notably cosmology—in which the very way in which we think and our whole way of approaching scientific problems may be a hindrance. It is therefore useful to describe the starting point from which we always set out.

1:3 THE APPROPRIATE SET OF PHYSICAL LAWS

Nowadays *astrophysics* and *astronomy* have come to mean almost the same thing. In earlier days it was not clear at all that the study of stars had anything in common with physics. But physical explanations for the observations not only of stars, but of interstellar matter and of processes that take place on the scale of galaxies, have been so successful that we confidently assume all astronomical processes to be subject to physical reasoning.

Several points must, however, be kept in mind. First, the laws of physics that we apply to astrophysical processes are largely based on experiments that we can carry out with equipment in a very confined range of sizes. For example, we measure the speed of light over regions that maximally have dimensions of the order of 10^{14} cm, the size of the inner solar system. Our knowledge of large-scale dynamics also is based on detailed studies of the solar system. We then extrapolate the dynamical laws gained on such a small scale to processes that go on, on a cosmic scale of $\sim 10^{18}$ to 10^{28} cm; but we have no guarantee that this extrapolation is warranted.

It may well be true that these local laws do in fact hold over the entire range of cosmic mass and distance scales; but we only have to recall that the laws of quantum mechanics, which hold on a scale of 10^{-8} cm, are quite different from the laws we would have expected on the basis of classical measurements carried out with objects 1 cm in size.

A second point, similar in vein, is the question of the constancy of the "constants of nature." We do not know, in observing a distant galaxy from which light has traveled many *aeons*, whether the electrons and atomic nuclei carried the same charge in the past as they do now. If the charge was different, then perhaps the energy of the emitted light would be different too, and our interpretation of the observed spectra would have to be changed.

A third point concerns the uniqueness of the universe.

Normal questions of physics are answered by experiment. We alter one feature of our apparatus and note the effect on another. Cosmic questions, however, do not permit this kind of approach. The universe is unique. We cannot alter phenomena on a very large scale, at least not at our current level of technological development; and if we did, it is not clear that we would be able to discern real changes. There would simply be no available apparatus that would not in itself become affected by the experiment—no reference frame against which to detect the change. In short, we may not be asking questions that can be answered in physical terms, because the methods of physics, and more generally of all science, depend on our ability to conduct experiments; truly cosmic problems may just not permit such an approach.

This then is the current situation: We know a great deal about some as yet apparently unrelated astronomical events. We feel that an interconnection must exist, but we are not sure. Not knowing, we divide our knowledge into a number of different "areas": cosmology, galactic structure, stellar evolution, cosmic rays, and so on. We do this with misgivings, but the strategy is to seek a connection by solving individual small problems. All the time we expect to widen the areas of understanding, until some day contact is made between them and a firm bridge of knowledge is established between previously separated domains.

How far will this approach work for us? How soon will the philosophical difficulties connected with the uniqueness of the universe arise? We do not yet know; but we expect to face the problem when we get there.

In the meantime we can address ourselves to a number of concrete problems that, although still unsolved, nevertheless are expected to have solutions that can be reached using the laws of physics as we know them. Among these are questions concerning the origin and evolution of stars, of galaxies, of planetary systems. There are also questions concerning the origins of the various chemical elements;

and perhaps the origin of life itself will become clearly established as astrophysical processes become better understood!

The next few sections will sketch the more important features of these problems.

1:4 THE FORMATION OF STARS

We believe that no observed star has existed forever—since sooner or later the energy supply must run out—and so we must account for the birth of stars. Inasmuch as those stars that we believe to be young are always found close to clouds of interstellar dust and gas, we argue that such clouds of cosmic matter must be contracting slowly, giving rise to more and more compact condensations, some of which eventually collapse down to stellar size.

This picture makes a good deal of sense. Dust grains in interstellar space are very effective in radiating away heat. Whenever a hydrogen atom collides with a dust grain, the grain becomes slightly heated, and this energy is radiated away in the infrared part of the electromagnetic spectrum. We speak about thermal radiative emission in such a case.

The energy that is radiated away by the grains reduces the kinetic energy of the gas, because it is this energy that is transferred to the grains in the collision of atoms with dust. When the gas loses kinetic energy, it falls toward the center of the cloud through gravitational attraction, gains some kinetic energy in falling, and again transfers some of this to a dust grain to repeat the cooling cycle. The atom also transfers some of its centrally directed momentum to the grain thus also causing the grains to drift in toward the center of the contracting cloud. As a result of many such interactions, the cloud as a whole contracts.

Grain radiation is not the only radiative process that rids a *protostar* of energy. As it collapses, the protostellar cloud becomes hotter and hotter, and various molecular and atomic states can be excited through collisions. The excited particles can give off radiation, thus returning to the ground state. The radiation may escape from the cloud into outer space, and this net loss of energy again gives rise to cooling (Fig. 1.2).

Attractive though it may seem, there are difficulties with this picture. First, the protostar cannot just lose *energy* in forming a star. It also must lose *angular momentum*. This comes about because the formation of a star requires the condensation of an initially extended, highly tenuous cloud of gas. But observations have shown that matter close to the Galactic center has higher orbital speed w than matter at great distances w'; hence the contracting cloud will have an initially large angular momentum. We can think of the mean orbital speed of the matter to be about $(w + w')/2$, and can then consider the matter at the cloud's circumference to be rotating about the cloud center with a velocity $(w - w')/2$ (Fig. 1.3).

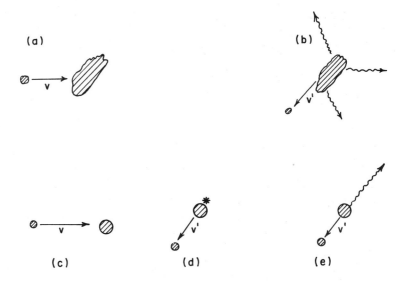

Fig. 1.2 Cooling processes in *protostellar clouds.* (*a*) An atom with velocity *v* hits a grain. Its kinetic energy is $v^2/2$ multiplied by the atomic mass *m.* (*b*) The grain absorbs energy that it radiates away and the atom leaves with reduced velocity *v'* and reduced kinetic energy $mv'^2/2$.

In (*c*), (*d*), and (*e*) an atom collides with another atom or with a molecule. This second particle first goes into an excited (higher) energy state, denoted by an asterisk (*), and then emits radiation, to return to its inital state. The first atom loses kinetic energy in this process, and if the emitted radiation escapes from the cloud, this represents an energy loss for the entire cloud. In this way, the *protostellar matter* slowly contracts to form a star.

In general, the amount of matter needed to form a star from an interstellar cloud with a density 1 atom cm^{-3} requires the collapse of gas from a volume whose initial radius *r* would be order 10^{19} cm. Over such distances, $w - w'$ is observed to be $\sim 3 \times 10^3$ cm sec^{-1} in our part of the Galaxy, so that the angular momentum per unit mass $r(w - w')/2 \sim 10^{22}$ cm^2 sec^{-1}. On the other hand, the observed surface velocities of typical stars indicate that the angular momentum per unit mass is many orders of magnitude lower: 10^{16} to 2×10^{18} cm^2 sec^{-1}! For the sun, it actually is only $\sim 10^{15}$ cm^2 sec^{-1}; but the angular momentum of the solar system taken as a whole corresponds to 10^{17} cm^2 sec^{-1}. In the solar system most of the angular momentum resides in the motion of the planets, and in particular in the motion of Jupiter, orbiting about the sun (see also Fig. 1.9 and section 1:7).

From this it is clear that in becoming a star or a planetary system, the initially

1:4

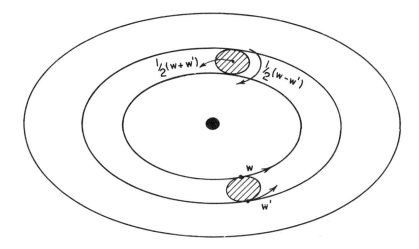

Fig. 1.3 Matter close to the Galactic center has higher orbital speed than at the periphery. $w > w'$. This phenomenon is called *differential rotation*.

contracting cloud of interstellar matter must somehow have contrived to lose almost all of its angular momentum. Only one part in several thousands can remain.

A similar problem concerns the *magnetic field* initially present in the interstellar medium. If this field is predominantly oriented along some given direction, then the final field after contraction of the cloud to form a star would also have that direction; and moreover, the initially very weak, oriented component would become strongly concentrated in the contraction of the protostellar mass. A field, B, initially as weak as 10^{-7} gauss—and this is less than observed values (Ma72)— would become some 10^{16} times stronger, as the protostellar radius decreased from 10^{19} down to 10^{11} cm. B would be proportional to r^{-2} in such a contraction. This comes about because the magnetic lines of force act as if frozen to the gaseous material (section 6:2), and the number of these lines of force threading through the cross-sectional area of the contracting cloud stays constant. Actual fields found on the surfaces of stars normally are only of the order of one gauss, although some peculiar stars have magnetic fields of the order of tens of thousands of gauss; but this is still very different from the 10^9 gauss that would be required if stars contracted without the destruction or loss of magnetic field lines permeating the interstellar material.* How this loss occurs is a major unsolved part of the problem of star formation!

*The very compact white dwarf stars and the even denser neutron stars, respectively, have fields believed to be of order 10^5 and 10^{12} gauss; but these are just the field strengths we would expect if the sun was to shrink to the size of such compact stars.

Let us backtrack for a moment and view the question of star formation in a different way. Early in our argument we made an assumption that need not necessarily be correct: The association of dust clouds with recently formed stars is not absolute proof that stars form from these clouds. Some causal relation presumably exists, but is it impossible that stars just form out of nothing at all, and that a lot of dust gets raised in the process? Such a picture, while unsatisfying because it postulates an apparently unphysical origin, after all at least avoids the angular momentum and magnetic field difficulties.

We should keep this important point in mind: Perhaps stars do form out of "nothing"! For the moment, however, we prefer to work, as far as we are able, within the framework of ordinary physics.

With that decision made, it becomes fruitful to investigate not only whether stars are forming out of dust clouds at the present time, but also whether they might not have been able to form from a simple, dust-free, helium-hydrogen mixture at earlier epochs in the history of the Galaxy. Such an overall approach could eventually lead to greater understanding of galaxy formation and related cosmological questions. For, as we will see, the chemical composition of matter in a galaxy seems to be affected to a large extent by the conversion of hydrogen into helium and helium into heavier elements. Some of these heavier elements are eventually ejected from a star to form part of the interstellar medium. If stars are continually formed from the interstellar medium, then the initial chemical composition of stars, freshly formed at the present time, could be appreciably different from that of stars formed a long time ago when the galaxy was young. Such differences in chemical composition are in fact observed in the spectra of some extremely old stars although, somewhat surprisingly, stars formed at the present time seem to have about the same composition as the sun, which is 5×10^9 y old.

These differences in the surface composition of the very oldest stars do indicate that they were formed out of a chemically different medium. Moreover, the basic physical processes that should occur in the interior of stars—if our theories of stellar evolution (section 1:5, below) are correct—are in good agreement with the observation that stars forming today contain more of the heavier chemical elements than stars formed in the earliest stages of galactic evolution. By studying the evolution of stars once they are formed, we therefore gather evidence about the formation of stars from the interstellar medium and, at the same time, gain insight into the life cycle of galaxies. Only if such views were to lead to incompatible results, would we wish to switch to a theory that required the spontaneous formation of stars out of nothing. But then we would still not be home free. We would still have to explain why stars formed from "nothing" some 10 aeons ago had low metal abundances, while stars formed in the same way, but within the past few million years, have higher metal abundances. We would have to face the somewhat uncomfortable implication that "nothing" had changed.

This section has emphasized the difficulties faced in understanding star formation; but the past 20 years of research have given many new insights into physical processes that could produce the contraction of interstellar clouds, could produce a loss of angular momentum, and could bring about a loss of magnetic field strength. The existence of such theories signifies considerable progress in coming to grips with the basic problems of star formation. What we still need, however, are much more detailed observational data that would verify whether any of these models actually describe stellar birth, or whether star formation depends on quite different sets of processes that we have not yet considered.

1:5 EVOLUTION OF STARS

Granted that we do not know very much about how stars are born, can we say anything about how they evolve after birth? The answer to that is a convincing "Yes," as we will see.

When the absolute brightness of a set of stars is plotted against surface temperature, as measured either by the star's color or from an analysis of the stellar spectrum, we find that only certain portions of such *Hertzsprung-Russell* and *color-magnitude diagrams* are populated appreciably. The concentration of stars in different parts of these diagrams represents the main clue we have to the way stars evolve. The color-magnitude diagram looks a little different when plotted for differing groups of stars. It has one appearance when drawn for the group of stars that lie closest to the sun. It has a different appearance (Fig. 1.4) when plotted for stars found in the loosely agglomerated *galactic clusters* in the Milky Way plane —clusters that must be very young since the stars they contain are too bright to have existed a long time on their limited fund of nuclear energy. It has a different appearance still (again see Fig. 1.4) when drawn for some of the oldest groups of stars in the Galaxy, faint stars that have slowly used up their nuclear fuel over some 10 aeons. They are represented by stars found in *globular clusters*—spherically symmetric aggregates consisting of hundreds of thousands of stars. These clusters are found primarily in the Galactic halo, distributed with no preference for closeness to the *Milky Way plane*.

Although differing in detail, the Hertzsprung-Russell plots have a number of common features, illustrated schematically in Fig. 1.5. While this figure represents no actual group of stars found in practice, it gives the designation of stars found in differing portions of the H-R diagram.

1:5

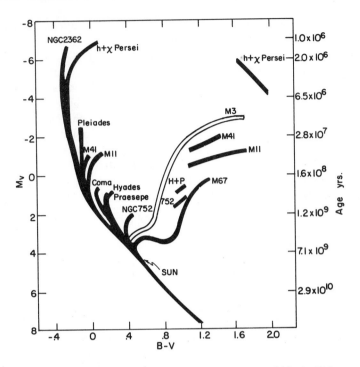

Fig. 1.4 Color-magnitude diagram for galactic clusters within the Galaxy and for the globular cluster M3. These clusters show differing turn-off points from the main sequence. From the theory of nuclear evolution we can determine the age of stars at the various turn-off points. Their ages are shown on the right of the figure (after A. Sandage, Sa57). (With the permission of the University of Chicago Press.)

On the ordinate we plot the logarithm of a star's *luminosity*, choosing the sun's luminosity—the total that the sun radiates away each second—as a convenient unit. The abscissa gives the logarithm of the effective temperature of the star's radiating surface (4:13). Stars on the left of the diagram are hot, having extreme surface temperatures up to nearly 10^5 °K. Stars on the right are cool. Bright stars are found at the top of the diagram and faint stars at the bottom. Running diagonally from top left to bottom right is the *main sequence* along which roughly 99% of the stars fall. All other stars are comparatively rare. Stars on the *subgiant* and *red giant branch* belong to a population that is more or less spherically distributed about the Galactic center in a *halo*. These stars are sometimes called *population II stars*, to distinguish them from *population I stars*, objects that lie in the Milky Way plane and make this portion of the Galaxy appear particularly

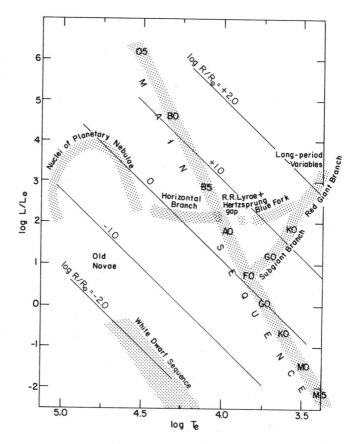

Fig. 1.5 Schematic Hertzsprung-Russell diagram. The lines of constant slope represent stars having identical radii (see section 4:13).

bright. The O and B stars, which are the bluest, brightest main sequence stars, are population I objects. We know that they must be very recently formed stars, judging on the basis of luminosity alone. For, the theory of stellar evolution in its present form specifies that hydrogen to helium conversion provides the prime source of energy for stars on the main sequence; and these stars are so bright that they must convert most of their hydrogen content in a very short time. The energy available per gram of hydrogen is known for this process, and we can compute that the O and B stars must be converting their available hydrogen into helium over a period of a few million years. These stars, then, are thought to be no older than several millions of years. In contrast, population II objects are probably $\sim 10^{10}$ y old, judging by their rate of hydrogen burning and by the fact that the brighter members of this group are just turning into *red giants*—stars that have consumed all the hydrogen in a central core.

The theory of stellar evolution attempts to explain the distribution of stars within the H-R diagram, showing not only why certain regions are populated and others not, but also why some regions—particularly the main sequence— are heavily populated, while stars are very sparse in others.

The calculations are based on nuclear reaction rates, computed for different stages of a star's life. They give the following evolutionary history.

As a star contracts from an initial dust cloud, it becomes very bright but radiates at such low temperatures that the star would be off the scale of our diagram, beyond the right edge. At this stage the star radiates only in the infrared; but the star's life time in this state is so short that such objects have not yet been identified. We only have theoretical grounds for believing that this phase exists; but one of the current aims of infrared astronomy is to actually find and study such stars. During this stage, contraction converts gravitational potential energy into kinetic energy and into radiation that is lost into space. The protostar's surface temperature stays nearly constant during this phase but as the surface area diminishes during contraction, the protostar becomes fainter. It follows a nearly vertical downward path—called the Hayashi track—at the extreme right side of the H-R diagram.

Eventually the star's contraction slows down, potential energy is no longer being lost at a high rate, but the temperature increases and the star crosses the H-R diagram from the right-hand edge, moving over toward the main sequence along a nearly horizontal line. As it reaches the main sequence it stops contracting. At this stage it has reached a compact configuration in which the temperature at the center of the star is high enough to permit hydrogen to helium conversion. In this con- version, 0.7 % of the initial mass can be converted into energy to be radiated away. For each gram of hydrogen $0.007 \, c^2 = 6 \times 10^{18}$ erg of energy can leave the star's surface to travel out into space.

Massive stars are hottest in their central regions, radiate at a greater rate, and convert mass into radiant energy at a correspondingly greater rate also. The B0 stars represent objects some 15 times as massive as the sun. Figure 1.5, which shows relative luminosities, indicates that B0 stars are using their fuel at a rate some 10^4 times faster than the sun does. While the sun will attain a main sequence age amounting to 10 aeons, B0 stars are expected to live only some 10 million years before changing structure. The O5 stars would evolve even faster. We can therefore expect that an old grouping of stars will only contain yellow and red, low mass stars on its main sequence.

Structural changes that result when all the hydrogen at a star's center has been used up should be associated with some change in the star's surface temperature and brightness. In fact, we find that the star moves off the main sequence in the

Hertzsprung-Russell diagram. In the galactic cluster $h + \chi$ Persei (Fig. 1.4) we see evidence for such a move. We see a curling from the main sequence toward the right—toward lower temperatures—and a new grouping in the right hand, upper corner of the plot where bright red stars are to be found. Detailed calculations based on model stars, and on the rates at which the nuclear reactions would proceed in them, indicate that stars just turning off the main sequence in $h + \chi$ Persei cannot be more than two million years old.

In contrast, the galactic cluster M67 shows no main sequence stars bluer than spectral type F. These stars therefore all are relatively small—the most massive are not much more massive than the sun. Stars of this brightness would complete the hydrogen burning in their central regions in some 7×10^9 y or so, and we think that this must represent the present age of M67.

This cluster also has a well-developed giant branch. Evidently stars that leave the main sequence travel out into this branch. Since the actual number of stars lying along the branch is small compared to the number populating the main sequence, we conclude that the stars do not spend much time in the subgiant or red giant stages before going on to some other stage. If they did spend more time as giants, we would expect this portion of the diagram to be densely populated —perhaps just as densely as the main sequence near the *turn-off point*. After all the subgiants and giants, according to the view we have presented, all are former main sequence stars. At an earlier age they lived on the portion of the main sequence just above the present turn-off point.

The H-R diagram therefore turns out to be a very useful tool indeed. It shows us not only where the various stars are found but, by using appropriate theoretical tools, we can estimate how old a given assembly of stars might be, assuming that all the member stars were formed at one and the same time. Further, on the basis of a continuity argument alone, not making use of nuclear burning theories at all, we are able to estimate the relative lengths of time spent by a star in various stages, making use only of the density of stars in differing portions of the H-R diagram.

This may sound simpler than it actually is, because we do after all have to determine the direction along which the stars are moving through the diagram and the sequence in which various stages are reached. For that, we very much depend on nuclear burning theories—theories that tell us in what sequence a star converts various elements found in its interior into other elements, and how much energy can be released in each process.

We note one other feature, that is shown in Fig. 1.5. As a star moves off the main sequence it apparently gets redder and, if anything, it becomes brighter than it was on the main sequence. At lower temperatures, however, an object always emits less radiation per unit area. Correspondingly, the only way that this particular course of evolution can proceed is for the star to be growing in size as

Fig. 1.6   Color-magnitude diagram for the globular cluster M3, with a superposed schematic evolutionary path. (After H. L. Johnson and A. R. Sandage, Jo56, and M. Schwarzschild, Sc70. With the permission of the University of Chicago Press, and with the permission of the Officers and Council of the Royal Astronomical Society.)

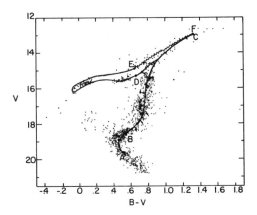

it moves off the main sequence. In comparison to their size on the main sequence, the stars now become giants. Their radii can increase by factors of 10 or 100.

Let us see how such stages evolve.

For this purpose the study of globular clusters is most instructive. These clusters contain what are believed to be some of the oldest stars in the Galaxy. Their faint, red turnoff points indicate that. The stars in each given cluster presumably were formed almost simultaneously; and the turnoff age therefore gives the age of the cluster. Figure 1.6 shows the color-magnitude diagram for the globular cluster M3. It shows population not only of the subgiant and red giant branches, but also of the *horizontal branch*. It also shows the *Hertzsprung gap* in the horizontal branch at a $B - V$ value of ~ 0.3. How do the stars evolve along these branches? In what directions do they go? How do they jump across a gap, so that no stars appear in it?

The line segment *AB* represents the very small change that actually does take place in a star's appearance while it stays on the main sequence. The distance shown is covered in a lifetime of $\sim 10^{10}$ y. The star begins life on the main sequence at *A*, and then slowly moves toward *B*, becoming slightly redder and brighter. The sun, for example, must be undergoing such a change. As we come to understand them better, geological and paleontological records reflecting the last few aeons may attest to similar brightness changes of the sun.

At point *B*, hydrogen at the center of the star is exhausted, and there is hydrogen to helium conversion only in a shell around a depleted *central core*. The star moves from *B* to *C* in a first red giant phase in which this *hydrogen shell burning* is the main source of energy; and the *helium core* slowly increases in mass while the burning shell moves outward converting more hydrogen into helium as it goes (see Fig. 8.8).

1:5

At point C, the massive core has heated up through contraction to a temperature at which helium can be converted into carbon. This occurs very rapidly, in what is called the *helium flash*. In this flash, roughly 3% of the core, whose total mass is $\sim 0.5 M_\odot$, is burned. This is enough to heat the core to the point where it expands against gravitational forces and then keeps burning helium in a convectively churning core mass.

It is interesting that during the flash, the energy conversion rate in the core may equal the total energy being released by all the stars in a cluster. But this energy cannot immediately come to the surface of the star. Instead, it goes into effecting an expansion of the highly gravitationally bound core.

The loop DE, which is associated with the horizontal branch stars, is believed to represent a stage in which the main source of energy comes from helium burning in the core—with hydrogen burning continuing in a shell further out. The evolution from D to E lasts some 10^8 y, while the evolution from B to C may have lasted somewhat longer. Figure 1.6 shows that the lifetime in the stage C to D must be very short, because few members of M3 are shown there.

At the point E, the core has become exhausted of helium and a stage of helium shell burning sets in. Segment EF represents a phase in which the star consists of an inert *carbon-oxygen core*, surrounded by a *helium-burning shell*, a further helium zone in which no nuclear conversion is going on, a hydrogen burning shell, and finally an envelope made up of the original unconverted matter from which the star formed. The chemical makeup of the outermost portions of the stars, the only ones that we actually see, therefore show no evidence of the complicated changes that have occurred in the star's interior, even at this advanced stage of evolution.

During this *second giant red stage EF*, the star follows almost precisely the same path through the H-R diagram that it traced out before in going from C to D and in the top portion of the track from B to C. These three different groups have an outwardly similar appearance.

It is interesting to note that the initial concentration of helium when the star formed can perhaps be determined by the loop from D to E. The theory of stellar evolution predicts that for low initial helium content, a star should spend more time in the upper portion of the loop on the return leg toward E. For about 10% helium abundance, we would expect most of the stars to be concentrated toward the left end of the loop. On the other hand, for a 20 to 30% initial helium abundance, a star starts its horizontal branch traverse by a very rapid movement to the left: thereafter it slowly proceeds to the left and even around the corner. Once it reaches the top part of the loop, the star rapidly moves toward E. The observed distribution of stars in globular cluster horizontal branches is best fitted by stars with this high $\sim 30\%$ helium abundance (Sc70).

The high initial helium abundance, while an interesting result, immediately brings up many new problems. Does such a high abundance represent the initial *primordial matter* from which all matter in the universe has been formed? Or is there a stage of hydrogen to helium conversion that occurs at some earlier epoch, preceding even formation of the earliest stars we know in the Galaxy? If so, is this process part of the galaxy formation stage, or is it a universal phenomenon representing a stage prior to galaxy formation? We will discuss this question a number of times from differing points of view; but we will have to admit that we do not have any clear answers: The problem remains an unresolved question in astrophysics.

One feature of the helium shell burning that has not yet been mentioned is that it does not occur steadily, but rather proceeds in a series of flashes—just as helium core burning proceeds in a flash. In each flash of shell burning the peak luminosity may be of order 10^5 $L_\odot$. There are indications that, at least in certain model stars, the helium shell flash causes the star to move off the red giant branch in a loop that leads to the left and back again in a matter of about 10^3 y. This motion takes the star into a portion of the diagram occupied by population II *cepheid variables*, and it is possible that these pulsating stars should be associated with helium shell flash burning processes. This is a question that needs much further study.

Similarly we also need to know more about the *planetary nebulae*—hot *central stars* surrounded by a shell of ejected material. Since one planetary nebula is known to exist in a globular cluster, it is likely that the mass of at least that one central star is only slightly higher than the turnoff mass from the main sequence in that cluster. We are tempted to associate the planetary nebula stage with some stage following the red giant phase just discussed. It is possible that a star that has evolved to the stage described by the path *EF* in Fig. 1.6 comes to a region of instability where the outer layers are ejected in one or several violent outbursts that separates off an outer shell of the star lying above a point somewhere between the hydrogen and helium burning shells. The central portion of the star would then comprise a relatively low mass carbon and oxygen rich core, which slowly contracts to the *white dwarf* stage shown on the H-R diagram (Fig. 1.5). The central star of the planetary would first appear very hot and bright, as indicated by the loop drawn in Fig. 1.5, but then would cool down toward the portion occupied by white dwarfs. The mass of the white dwarf might at this stage be no more than ~ 0.7 $M_\odot$, the rest of the mass having been ejected during the outbursts that led to the gaseous envelope of the planetary nebula (Sa68b).

Many objections can be raised to this picture of planetary nebula formation. It is not clear that the predicted mass tallies with observed ejected masses. The chemical composition of the ejected mass might also be expected to have a high concentration of heavy elements, or at least of helium, but that does not seem to be consistent with observations. It is clear that this stage of stellar evolution is still very poorly understood; but there is optimism that the problem will be solved in the next few years.

Similar questions can be raised about explosive stars: the *novae* and *supernovae*. It is possible that such explosive events have a close relationship to the emission of neutrinos (Sa69b) in the dense cores of highly evolved stars. The neutrino loss allows the central portions of a star to contract without appreciable temperature increase. The small neutrino interaction cross sections permit ready energy transport away from the collapsing core. This then results in a separation of the rapidly contracting inner core from an outer portion in which neutrino electron scattering occurs with a consequent energy deposit in the outer layers, which eventually leads to violent ejection. For this to happen, even the outer layers first have to collapse down to a density of order 10^{11} gm cm^{-3} where the star's radius would only be 100 km.

The ejected mass then probably becomes the observed material of a *supernova* outburst, while the shrinking core may evolve into a *neutron star*.

To conserve angular momentum, this central star would have to rotate rapidly. It also would have a strong embedded magnetic field, since the original field present in the star could not escape during its collapse. The magnetic field rotating with the star could cause electrons and ions to become accelerated to a very high velocity and could give rise to the strongly polarized *pulsar* radio wave pulsations that are now believed to represent the emission of highly relativistic particles radiating away energy as they corotate with the star. *Cosmic ray particles* may also have their origin in such pulsars.

This is as far as the theory of stellar evolution has come: It promises many detailed insights into the nuclear history of stars and of the universe, and into the major energy sources available in the universe. It may eventually also lead to a better understanding of the formation of extremely energetic cosmic ray particles, and of the ultimate fate of stellar matter in the universe.

Studies of stellar evolution then are much more than just a conjecture about the life and death of stars; they may ultimately provide the detailed physical insights necessary for understanding the more important energy generating processes on a cosmic scale!

1:6 THE ABUNDANCE OF CHEMICAL ELEMENTS IN STARS AND THE SOLAR SYSTEM

Spectroscopic determinations of the abundances of chemical elements in the atmospheres of stars can provide us with information on the chemical composition of the medium from which the star was formed. The theory of stellar structure shows that for most types of star, the outer layers remain unaffected by the nuclear processes that liberate energy at the star's center. Only the lithium, beryllium, and boron content of the stellar atmosphere will not be representative of the protostellar material, because these three elements are readily destroyed in reactions with protons that take place at relatively low temperatures. Any deuterium that may have been present could similarly be destroyed during the early convective contraction that mixes material from the protostar's surface into the hot central portions of the star. This *convection* is absent in the main sequence hydrogen burning phase that produces helium.

There are some kinds of stars, however, that are anomalously overabundant in, say, helium, barium, or carbon, which evidently has been brought to the stellar surface through some kind of convection. In these stars, we think we see the results of the chemical abundance changes produced by nuclear reactions in the stellar interior, and we hope to be able to use the observations to learn more about the nature of these reactions and of the conditions that prevail deep inside the star (Un69).

In the atmosphere of normal stars, we also see strong variations in the abundances of different chemical elements. The stars that we suspect of being oldest show abundances of the elements from carbon to barium that are up to a couple of orders of magnitude smaller than in younger stars like the sun. This low but non-negligible *metal abundance*—in this context the word "metal" denotes any atom heavier than helium—is a real puzzle. Were the younger stars formed from material that considerably differed from primordial matter—possibly pure hydrogen—initially present when the Galaxy was formed? In fact, do these oldest stars represent the first stage of nuclear changes in the Galaxy? Or was there an earlier stage in which helium and the metal elements were formed—a stage that left no apparent survivors?

In Chapter 8 we will see that (a) the sun, (b) a very young B0 star, Tau Scorpii, (c) planetary nebulae, (d) a red giant ε Virginis, and many other "normal" stars all have the same chemical composition, within the limits of observational error. This is significant because the ages of these objects cover the lifetime of the Galaxy since the first known stage of star formation during which the globular cluster red giants are thought to have been formed.

These analyses show that throughout the lifetime of the Galaxy, the interstellar matter has had an almost unchanged composition. This may be due to a recently discovered, apparently continuous, infall of gas from outside the galaxy. The infall rate appears to amount to two solar masses each year, and this is about equal to the star formation rate in the Galaxy (La72). Whether this effect is confirmed or not, it is evident that the admixture of matter exploded out from supernovae or slowly pushed out in planetary nebulae or in the *stellar winds* of stars, has not changed the composition of the medium appreciably. This might be attributed to the very limited mixing of the outer layers of stars with material in the stellar interior, and to the involvement only of stellar surface material in the massive eruption of stars.

A small number of exceptional stars do, however, show a quite different chemical make-up. In these stars we believe that mixing of central material with surface layers has indeed occurred; and the composition of the surface material can be used to analyze the nuclear reactions that must have taken place at the star's center. Just why mixing takes place in these stars, and not in others, is not yet understood. There may be several reasons.

Three types of stars are particularly interesting: the *helium stars*, the *carbon stars*, and the *S stars*.

The helium stars are very hot objects that seem to have had all or very nearly all of their hydrogen converted into helium. In some of these stars, the chemical composition indicates that helium probably was produced by straight fusion of four hydrogen nuclei to form a helium nucleus. In others the catalytic action of carbon, nitrogen, and oxygen helped to form helium from hydrogen, and in the process appears to have converted most of the initially present oxygen and carbon into nitrogen.

The carbon stars, in contrast to helium stars, are cool red giants in which spectra of the radicals CH, C_2, and CN appear together with atomic lines of carbon. Evidently the stellar helium has been burned here to form carbon through fusion of three helium nuclei. Even here a detailed spectral analysis shows that different stars have reached these high carbon contents by differing routes.

The cool stars of spectral type S show great spectral line strength in the elements zirconium, barium, lanthanum, yttrium, strontium, and in such molecules as ZrO, LaO, YO and so on formed from them. These heavy elements seem to have been formed by the growth of nuclei through absorption of neutrons.

These processes are discussed in Chapter 8. They give concrete evidence that nuclear reactions are proceeding in stars. Further study of these unusual stars may provide a detailed tracing of the various evolutionary paths that a star can sometimes follow. Hand in hand with laboratory work on the rates of nuclear reactions that take place at different temperatures, these spectral analyses should go a long way toward explaining the nuclear or chemical history of the Galaxy and of the universe as a whole.

At the other extreme—representing the least evolved matter we see—we have some of the oldest stars in the Galaxy that seem to have metal abundances that are systematically low compared to their hydrogen content, by as much as a couple of orders of magnitude. Table 1.1 shows the ratios of abundances, relative to the same ratios found in the sun.

Although the oldest stars in the Galaxy exhibit a metal deficiency relative to hydrogen, helium does not seem to be deficient. Helium may therefore have been produced in a prestellar stage. Whether that stage was a *protogalactic stage*, a

Table 1.1 Abundances by number density of chemical elements in HD140283 a metal deficient subdwarf, and in HD161817 a horizontal-branch star.[a]

Atomic Number Z	Element	HD140283 ($\log n/n_\odot$ + 2.32)	HD161817 ($\log n/n_\odot$ + 1.11)
1	H	+ 2.32	+ 1.11
6	C	− 0.5	− 0.26
11	Na	− 0.30	− 0.05
12	Mg	+ 0.01	+ 0.18
13	Al	− 0.26	− 0.26
14	Si	+ 0.07	− 0.19
20	Ca	− 0.03	+ 0.09
21	Sc	+ 0.61	+ 0.21
22	Ti	+ 0.05	+ 0.25
23	V		− 0.42
24	Cr	− 0.09	− 0.27
25	Mn	+ 0.35	− 0.43
26	Fe	− 0.16	− 0.10
27	Co	+ 0.02	+ 0.12
28	Ni	− 0.31	+ 0.11
38	Sr	0.00	+ 0.24
39	Y		− 0.19
40	Zr		− 0.13
56	Ba		− 0.05

[a]These are compared to the sun; the average deficiencies have been subtracted to facilitate recognition of possible effects in individual elements. This table is taken from Unsöld (Un69). It is based on data compiled by B. Baschek and by K. Kodaira.

first collapse of all the galactic material into the galactic center, or whether the helium was produced at an early stage in the life of the (here assumed to be) *evolving universe*, is currently under debate. It is possible that the *3° K cosmic background radiation* detected at mm-wavelengths is energy that was produced at such an early epoch.

It is also possible that helium was formed during a very early stage of the universe and that a weak metal abundance was subsequently produced in a second protogalactic nuclear processing stage. Such questions are being actively investigated and an answer may be found in a few years.

In connection with the study of chemical abundances, it is interesting to speculate on the chemical composition of the solar system at its formation (Ca68). Some of the elements have been preserved since that time in a few types of meteorites in apparently unchanged ratios. Surface material on earth cannot readily yield corresponding information because considerable fractionation has taken place on earth. Table 1.2 shows the abundances derived mainly from *carbonaceous chondrites*, meteorites considered to be most representative of the *primeval solar nebula* and hence probably also of the solar surface composition some 4.5 aeons ago. Because not all element abundances can be reliably determined in this way—some volatile elements, for example, may have escaped from the meteorites through diffusion—the table has been augmented using information obtained from solar spectra and from the cosmic rays emitted by the sun. Figure 1.7 shows

Fig. 1.7 Abundances of the nuclides plotted as a function of mass number (after Cameron Ca68).

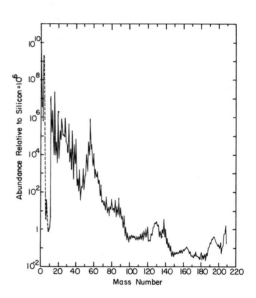

these results graphically on a logarithmic scale. We note that the heaviest elements, which can be readily determined in meteorites, are not easily obtained in spectra of *stellar atmospheres*. The two types of information therefore complement each other and serve also to point out agreement or differences for elements for which direct comparisons are available. A table of stellar atmospheric abundances is given in Chapter 8 (Table 8.3).

1:6

Table 1.2 Abundance, by Mass, of the Chemical Elements in the Solar System. Compilations of Abundance Normalized to Si $= 10^6$ (after Cameron Ca68).[a]

Element	Abundance	Element	Abundance	Element	Abundance
1 H	2.6×10^{10}	29 Cu	919	58 Ce	1.17
2 He	2.1×10^{10}	30 Zn	1500	59 Pr	0.17
3 Li	45	31 Ga	45.5	60 Nd	0.77
4 Be	0.69	32 Ge	126	62 Sm	0.23
5 B	6.2	33 As	7.2	63 Eu	0.091
6 C	1.35×10^7	34 Se	70.1	64 Gd	0.34
7 N	2.44×10^6	35 Br	20.6	65 Tb	0.052
8 O	2.36×10^7	36 Kr	64.4	66 Dy	0.36
9 F	3630	37 Rb	5.95	67 Ho	0.090
10 Ne	2.36×10^6	38 Sr	58.4	68 Er	0.22
11 Na	6.32×10^4	39 Y	4.6	69 Tm	0.035
12 Mg	1.050×10^6	40 Zr	30	70 Yb	0.21
13 Al	8.51×10^4	41 Nb	1.15	71 Lu	0.035
14 Si	1.00×10^6	42 Mo	2.52	72 Hf	0.16
15 P	1.27×10^4	44 Ru	1.6	73 Ta	0.022
16 S	5.06×10^5	45 Rh	0.33	74 W	0.16
17 Cl	1970	46 Pd	1.5	75 Re	0.055
18 Ar	2.28×10^5	47 Ag	0.5	76 Os	0.71
19 K	3240	48 Cd	2.12	77 Ir	0.43
20 Ca	7.36×10^4	49 In	0.217	78 Pt	1.13
21 Sc	33	50 Sn	4.22	79 Au	0.20
22 Ti	2300	51 Sb	0.381	80 Hg	0.75
23 V	900	52 Te	6.76	81 Tl	0.182
24 Cr	1.24×10^4	53 I	1.41	82 Pb	2.90
25 Mn	8800	54 Xe	7.10	83 Bi	0.164
26 Fe	8.90×10^5	55 Cs	0.367	90 Th	0.034
27 Co	2300	56 Ba	4.7	92 U	0.0234
28 Ni	4.57×10^4	57 La	0.36		

[a]This table is intended to be characteristic of primitive solar matter, and is based as much as possible on abundances of Type 1 carbonaceous chondrites, since volatile substances probably escape least from this type of meteorite. Other sources that were used when needed were data from ordinary chondrites, solar atmospheric abundances as obtained from observed spectra, and solar cosmic ray abundance as measured in the vicinity of the earth. Eight elements were interpolated on the basis of nuclear synthesis theory in stars, because insufficient data were available.

1:7 ORIGIN OF THE SOLAR SYSTEM

Sometime around the era when the sun formed, the system of planets became established too.

Was the *solar system* formed after the sun already was several hundred million years old; or were the sun and the *planets* born in one and the same process? Did the solar system form out of a single cloud of matter surrounding the sun; or was there another star involved in the birth of the planets?

The planets occupy orbits that are regularly spaced according to a pattern first noticed by Bode (see Fig. 1.8). Is *Bode's law* just a coincidence of numbers, or does it describe some deeper interrelation among the planets' orbits? In particular, does that interrelationship provide any insights into the early history of the solar system, or is it an arrangement that would hold for any system of bodies orbiting about a central mass, given only enough time for these bodies to reach some state of dynamic equilibrium?

Does planet formation always accompany star formation, or do only a small fraction of the stars have planetary systems? Are stars having the sun's spectral characteristics more prone to have planetary systems than some other stars? And if so, are these planetary systems largely identical with the solar system so that life might be expected to exist there too?

These are only some of the more important questions on a long list of unanswered problems. What may be more distressing is that the tools we have right now probably are not yet the proper ones to yield answers. Instead, the methods that have been of help thus far, and that will be partly described in this section, may do no more than show us what newer lines of approach might be more useful to cope with the problems we have to answer.

To give one example: Newton's laws of motion describe the orbits of the planets about the sun; they also describe the changes in these orbits as the planets interact with each other. This is the subject matter with which *celestial mechanics* concerns itself. From the motion of the moons around individual planets, and from the short-term interactions of neighbouring planets, we can judge the masses of the major bodies that make up the solar system. Knowing these masses, and knowing the instantaneous orbits, we should be able to calculate ahead and see how the solar system will evolve in the future. Not only that, we should also be able to to see how it evolved in the past, what its appearance was like a few hundred years ago, a few million years ago, and possibly a few aeons ago!

This kind of hope was first voiced more than a hundred years ago. At that time, however, it was quite clear that the calculations were so detailed, that it would not be practical to reconstruct the past history of the solar system in this way. The amount of work involved was simply too great.

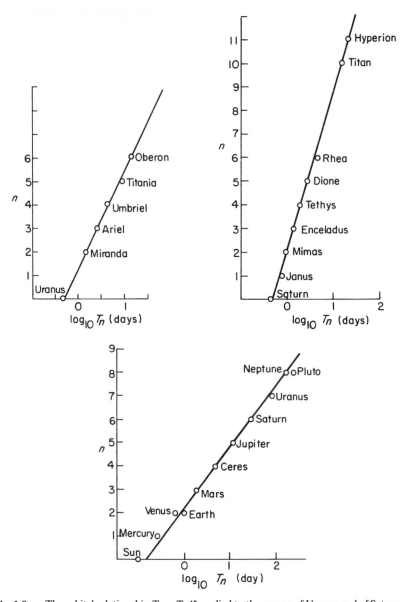

Fig. 1.8   The orbital relationship $T_n = T_0 A^n$ applied to the moons of Uranus and of Saturn and to the solar system. T_n is the orbital period of the nth moon or planet, T_0 is chosen close to the rotational period T_e of the parent planet or the sun, respectively. (After Dermott, De68. With the permission of the Officers and Council of the Royal Astronomical Society.)

There is now a revival of celestial mechanics, based on the prodigious capability of computers in rapid calculations involving large numbers of repeated steps; and the dream of reconstructing a dynamic history of the solar system, by computing backward in time, may not be as remote as it looked half a century ago when the complexity of the calculations had already become abundantly apparent.

Still, this approach just may not be fruitful. It might not work, because there may have been cataclysmic changes in the solar system about which we have no present information. In that case, a set of computations carried out in ignorance of these changes would come out with false results. Sometimes, however, such false results are so incongruous that they induce us to actually search for evidence of abrupt changes, using other tools that might be at our disposal. In this fashion, analyses using a number of different tools and representing a variety of different lines of reasoning, often allow us to step backward into past eras, to partially reconstruct the history of the solar system.

Just what are these other tools, or lines of attack? There are several of them. Some are quite clearly going to be useful, but may be very laborious. Others are simpler but may be more in the nature of apparently relevant questions. Exactly how relevant they really are may however not become obvious until after we have answered them—and we do not yet know just how to get those answers.

To illustrate this latter type of approach, we can give three dynamical arguments.

(a) Dynamical Questions

First, we know that the planets all orbit close to the earth's orbital plane, the *ecliptic.* Only Mercury, the smallest and nearest planet to the sun has an *inclination* as high as 7°. Pluto has a higher inclination yet > 17°; but its inclination, in contrast to that of the other planets, is believed to vary rapidly under the perturbing influence of its far more massive, neighbouring planets. Generally, then, the *orbital angular momentum* axis for all the planets lies along the same direction. The mean angular momentum has a direction nearly normal to the orbital plane of Jupiter, the most massive planet.

Surprisingly this angular momentum has a direction 7° away from the sun's spin axis. The sun's equatorial plane is inclined that strongly relative to the ecliptic.

How can this be? Does it mean that the sun and planets could not have been formed from one and the same rotating mass? Does it mean that some other massive body was present and instrumental in the birth of the planets? On a more detailed basis, could it be that Mercury, whose orbit does have about the same inclination to the ecliptic as the sun's equator was formed later than the more distant planets?

1:7

The number of questions raised by this single consideration is large. It may therefore not be a very productive line of pursuit. Perhaps some future theory involving much more complex arguments will automatically also produce the proper relationships among orbital inclinations as a natural side product; but the side product alone may not be a sufficient clue to the overall structure of that theory, and may not help us very much right now.

The second example involves Bode's law, which we have already mentioned. Recently Dermott (De68) has shown that a slightly new phrasing of this law permits us to include not only planetary orbits around the sun, but also the orbits of *moons* around their parent planets.

Dermott writes the *orbital period* T of the nth body of the orbital system in terms of a basic period T_0, close to the *spin period* T_p of the parent body

$$T = AT_0^n$$

Figure 1.8 shows the results for the planetary system, for the moons of Uranus, and for the moons of Saturn. The moons orbiting other planets show similarly good fits. We have to be careful, however, to note three factors. First, the abscissa is drawn logarithmically, and may more easily mask defects. Secondly, the rotational period of the parent T_p, does not always agree with T_0 chosen for the formula—so that this parameter is fairly arbitrary in determining a good fit. A second arbitrary parameter is A, chosen to give the best fit. Actually A is only partially arbitrary; it is always the square root of a small integral number. Finally, not all the slots corresponding to integral values of n are filled equally. Sometimes there are gaps, sometimes there are two objects for a given value of n. Given so many arbitrary features, is the remaining, apparently good fit to the Bode-type equation really significant?

A very detailed statistical analysis is needed to answer that question. Such analyses are difficult, because it is hard to estimate objectively how many parameters have actually been kept arbitrary, one way or another. How many different types of fit could have been attempted and how many would in fact have given better results, given the same number of free parameters? Data of this kind may, however, prove valuable in the long run, just as Balmer's empirical analysis of the hydrogen spectrum eventually led to a correct understanding of atomic structure.

A third, potentially useful insight still based on dynamical reasoning, concerns the observed *rotation of stars*. Stars with spectral types O and B have rotational velocities of 100 km sec^{-1} and often appreciably more. On the other hand, dwarf G stars like the sun have surface velocities of the order of only 1 km sec^{-1}. Between these extremes a fairly sharp transition in rotational velocity becomes apparent, along the main sequence, around stars of spectral type A5 (Fig. 1.9). We note, however, that the sun has an angular momentum that amounts to only

Fig. 1.9 The angular momentum per unit mass for stars of differing masses. The solar system is shown as S.S. McNally (McNa65) points out that A stars have the highest angular velocities $\sim 10^{-4}$ rad sec^{-1} and suggests that the centrifugal forces become so great that a low mass star cannot hold onto surface material. If such material is shed during star formation, a planetary system could result. (With the permission of the Officers and Council of the Royal Astronomical Society.)

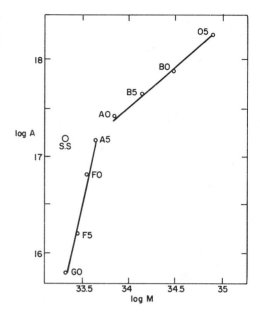

one half of one percent of the total angular momentum of the planets, as they orbit the sun. The angular momentum of the entire solar system, in fact, is equal to that of stars of spectral type A. It is therefore tempting to associate the observed low angular momentum of stars of late spectral type, with the formation of planetary systems. Perhaps all main sequence stars of this spectral type have companions, just as the sun does, and perhaps the formation of such companions is an integral part of the star formation process. (McNa65)

There are indications that a number of stars other than the sun have a planet, or perhaps several planets. The nearby *Barnard's star* has been found to have a *proper motion* in the sky that is only explained by the presence of a planet several times as massive as Jupiter, or to a better fit by the presence of two planets with masses 1.1 and 0.8 times Jupiter's mass (vdKa69). The star itself moves about the common center of mass of the system, and although this motion is very small, careful observations have shown it to be real. Two stars cataloged as 61 Cygni and Lalande 21185 also show evidence of having planets.

Studies such as these are laborious and can only be carried out for the nearest stars. However, the fact that a relatively nearby star also has a planetlike, small companion gives us immediate confidence that the solar system is not a unique, or highly uncommon chance phenomenon. Rather, we infer that planetary systems must be quite common, and that the formation of such systems must have an explanation that does not involve events of very low probability. This is an important conclusion, because it rules out a number of theories that, for example, might have required close passage of a star by the sun. Such a close passage simply is too improbable to account for the formation of two planetary systems as close to each other as the sun and Barnard's star.

In case we were to wonder whether Barnard's star might have passed near the sun and induced planet formation in both systems, we need only note that the sun has a velocity of 20 km sec^{-1} relative to nearby stars. The sun would have been extremely distant from its present neighbors 5×10^9 y ago.

Dynamical studies, of course, yield only a small fraction of the knowledge we have gathered about the history of the solar system. We can obtain much more complete information in a number of other ways.

(b) Radioactive Dating

Radioactive dating, for example, permits us to estimate that at least some rocks on the earth must have solidified some 3.5 æ ago (Ba71a), and that meteorites falling onto the earth from interplanetary space have ages of 4 to 4.4 æ. The earth as a whole seems to be about 4.5 æ old. Lunar surface samples brought back to earth indicate ages in excess of 3.5 æ so that the moon and earth may have comparable ages.

The age of solidified rocks can be determined by the ratio of radioactive *parent* and *decay products* found in a sample. For example, the uranium isotope U^{238} decays into lead Pb^{206}, emitting eight alpha particles in the process. The half-life for this decay is 4.5 æ. If the rock is not porous, the alpha particles become trapped as helium atoms after combining with some of the electrons that are released as the nuclear charge diminishes in the alpha decay. By measuring the ratio of U^{238} to Pb^{206} and helium present in the rock, some estimate of the age can be obtained. This estimate must of course take into account that other radioactive decay may be going on simultaneously. The uranium isotope U^{235}, for example, decays into lead Pb^{207} with the release of seven alpha particles in a half-life of 0.7 æ; Thorium Th^{232} decays into lead Pb^{208} and six helium atoms in 13.9 æ, rubidium Rb^{87} turns into strontium Sr^{87} in 46×10^9 y, and potassium K^{40} turns into argon Ar^{40} in 1.25 æ (Wh64).

A complete age determination usually involves several of these decays. Only if all the dates obtained agree can we feel safe in setting an age for a studied sample.

Studies of this kind indicate that the earth and the meteorites solidified about 4.5×10^9 y ago, within a time of about 10^8 y. Other theories, involving nuclear processes going on in stars, predict the abundance ratios for the various isotopes of given elements at the time that matter is ejected from a star. The currently found abundance ratios for some of the radioactive isotopes present within the solar system, therefore, also can be used to fix the time of formation of these elements in the interior of an earlier star. Somewhat surprisingly that time is only of the order 6×10^9 y, so that the sun must have formed within $\sim 10^9$ y after the heavier elements found in the solar system were formed, perhaps in the explosion of an earlier generation star. In fact, the whole process of star and planet formation probably took place within a time span of only one aeon.

(c) Abundance of the Isotopes of Elements

The relative abundance of different elements can also yield information of a different type: Deuterium, lithium, beryllium, and boron are found in appreciable quantities, both in the meteorites and on the earth. On the sun, however, they are nearly absent, because they are destroyed by *thermonuclear reactions* even at relatively low temperatures. Such elements can however be produced when cosmic rays bombard heavy nuclei, causing them to break up (the *spallation process*). This process releases neutrons that, if slowed down by colliding with hydrogen atoms, can produce the observed ratios of lithium isotopes Li^6/Li^7 and boron B^{10}/B^{11}. It has therefore been argued that the early solar system may have contained much of its mass in the form of icy spheres. If these spheres were about 10 m in diameter, cosmic ray particles produced in the sun could have caused spallation, while the hydrogen contained in the postulated ice spheres would have slowed down the neutrons released in spallation (Fo62). This would have permitted the slowed neutrons to react with Li^6 to form tritium, and with B^{10} to form Li^7, thus leading to an overall enrichment of Li^7 relative to Li^6. This somewhat involved process is believed necessary, because spallation tends to produce equal amounts of Li^6 and Li^7, while on earth the abundance of Li^7 is much greater than Li^6.

In modified versions of this theory, larger spheres containing less H_2O ice are found to give similar results, and only a moderate fraction of the earth's material need have initially been bombarded by *solar cosmic rays*. Larger preplanetary masses could therefore also have contributed to the proper Li^6/Li^7 ratios, and the 10 m dimensions need not have been universal. It is interesting that these bodies of frozen gases are reminiscent of the structure of comets.

(d) Comets and the Chemical Makeup of Planets

Perhaps at an early stage the solar system consisted of comet-sized objects. Comets are objects that contain frozen gases like ammonia and probably water. They contain large amounts of hydrogen, trapped in these molecules, or larger mother molecules that can break up into NH_3, OH, CO_2, and CH on exposure to solar radiation. There may be large amounts of frozen hydrogen gas present too. Some comets approach the sun from distances as great as 10^{18} cm. They appear to be bound members of the solar system that have been at great distances from the sun most of their life and are approaching now after an absence of a hundred million years or perhaps considerably more. In these comets we may be seeing the primordial matter from which planets formed. The comets apparently were pushed out to large distances from the sun early in the formation of the solar system, and have been orbiting there ever since. They may represent deep-frozen samples of matter preserved from the early solar system and, therefore, are extremely interesting objects to study if the history of the solar system is to be reconstructed. Thus far, unfortunately, we can study comets only as they fall apart on approaching the sun when solar heating evaporates some of the frozen gases and liberates solid materials that were held together by the ices. Some of these solid particles later strike the earth's atmosphere and heat or burn up because of their high initial velocity relative to the earth. This heating and burning gives rise to emission of light whose spectrum can be analyzed for the presence of various elements. From this and from the spectra of gases released by the comet on approaching the sun, we can make crude chemical analyses of the comets' contents. We find that, in addition to large amounts of hydrogen, they contain those elements that also are abundant in the outer planets. This estimate still is rough; but a refined chemical analysis may have to await the launching of a space probe that could actually obtain a sample of comet material and examine it.

The gravitational influence of Jupiter is so great that it can significantly alter the orbits of at least some comets, and bring them appreciably closer to the sun. These comets are captured from the highly elliptical orbits that have taken them to the most distant parts of the solar system and places them into relatively small, short-period orbits with *aphelion points* near Jupiter's orbit.

The continual heating by solar irradiation can then evaporate most of the short-period comet's gases. The comet nucleus itself is too small to hold on to these gases through gravitation and soon the entire comet disintegrates. If it has a solid core, only that core would remain after a few thousand years. It is possible that at least some asteroids—bodies whose sizes largely range from a few kilometers down to fractions of kilometers—are the remnants of earlier comets. They certainly have orbits very similar to the short-period comets and might, therefore, have a common origin. The largest asteroids, however, are more than 100 kilometers in diameter, much larger than observed comets seen in the past, and it is likely that these larger asteroids do not represent comet remnants.

Let us now return to our discussion of the planets. The differences in *density* and *chemical composition* of the various planets may provide evidence about how they were formed. The inner planets are much more dense than the outer ones. They also contain silicates and iron, which solidify at relatively high temperatures and hence could have solidified close to the sun. They contain lesser amounts of hydrogen, because hydrogen is readily evaporated from a small planet close to the sun where temperatures are high. Because of this evaporation, the atmospheres of the inner planets as seen today may be quite different from the atmospheres as they were during early times in the solar system. The earth's atmosphere in particular is thought to have been reducing—meaning that hydrogen was prevalent and that oxygen was tied up in molecules and unavailable for combining with other elements. Now of course the atmosphere is definitely *oxidizing* with its 20% abundance of free oxygen gas. We can see from Table 1.3 that the major planets are less dense, but more massive than the inner planets. They contain a large fraction of their mass in the form of hydrogen and are able to retain it because of the low temperatures determined by their relatively large distances from the sun and because of their stronger gravitational fields.

This distribution by volatility suggests that elements with low vapor pressures were able to solidify at small distances from the sun in the early life of a gaseous *protoplanetary nebula* surrounding the sun. Initially the size of condensations may have been no bigger than dust grains, but these grains could have aggregated by successive collisions, some of which would have vaporized both colliding grains, while others would have caused the grains to stick. The vaporization and sticking both would act to narrow the velocity ranges of successively condensing dust grains until they were generally able to clump together. As such clumps grew to 1000-kilometer proportions, they could start sweeping up a wider range of matter through their gravitational attraction and in this way increase their capture radius. Larger bodies, capable of holding on to an atmosphere, also could slow down infalling particles before impact and outgoing ejecta after impact and, thus, increase the capture efficiency for colliding matter. Eventually the large bodies therefore would grow rapidly at the expense of small ones and the formation of a few large bodies rather than multitudes of small ones would be assured.

The natural abundance of elements as found in the sun appears to be mirrored in the composition of the solar system as a whole, if much of the hydrogen and helium is taken to be contained in the larger planets; this certainly seems to be true. These planets no doubt contain iron and silicates too, but the hydrogen is much more obvious, because its natural abundance is so great.

Thus far we have seen that we know very little about the development of the solar system. Besides the few factors mentioned, there are many other indicators that may lead to a better understanding of early solar history. For example, much information comes from the study of meteorites. Many *stony meteorites* show an abundance of millimeter-sized spheres held together by a matrix of silicates. Were these spherical *chondrules* present already in the early solar nebula and do they therefore contain information that could be used to infer primitive conditions? *Iron meteorites* again show crystalline structure that can only form under very high pressure conditions. Does this mean that these meteorites originated in the interior of a large planet that once broke up? Could this exploded planet have given rise

Table 1.3 Characteristics of the Planets

Planet	Mercury	Venus	Earth	Mars	Jupiter	Saturn	Uranus	Neptune	Pluto
Orbital semi-major axis	0.387	0.723	1.000	1.524	5.203	9.54	19.2	30.1	39.4 AU
Sidereal Period	0.241	0.615	1.000	1.881	11.86	29.46	84.02	164.8	248 y
Eccentricity	0.206	0.007	0.017	0.093	0.048	0.056	0.047	0.009	0.250
Inclination	7°0'	3°24'	0°00'	1°51'	1°18'	2°29'	0°46'	1°46'	17°10'
Equatorial radius	2420	6050	6378	3380	70,850	60,000	25,400	24,750	3000 (?) km
Mass	3×10^{26}	4.9×10^{27}	5.98×10^{27}	6.7×10^{26}	1.9×10^{30}	5.7×10^{29}	8.67×10^{28}	1.2×10^{29}	5×10^{27} g
Density	5.4	5.1	5.5	3.97	1.36	0.70	1.3	1.7	(?)g
Number of known moons	0	0	1	2	12	10	5	2	0
Brightness, m_v at maximum angle from the sun	− 0.2	− 4.2		− 2.0	− 2.5	+ 0.70	+ 5.5	+ 7.9	+ 14.9
Typical surface magnetic field	(?)	$< 5 \times 10^{-2}$	0.5	$< 10^{-3}$	5 (?)				(?) gauss
Surface gravity	360	870	982	376	2350	905	~ 830	1100	(?)cm sec^{-2}
Main atmospheric constituents		CO_2, CO H_2O	N_2 O_2	CO_2 CO, H_2O	H_2, CH_4 NH_3, He	H_2, He CH_4 prob NH_3, H_2O	H_2, He CH_4 prob NH_3, H_2O	H_2, He CH_4 prob NH_3, H_2O	
Rotation period	58.6d	243d	24h	24h	10h	10h	10.8h	15.8h	6d 9h
Oblateness	0.0	0.0	0.0034	0.005	0.06	0.1	0.03	0.025	(?)
Equator centrifugal repulsion	0	0	3.4	1.7	225	176	62	28	(?)cm sec^{-2}
Subsolar surface temperature	620	250 (cloud top)	295	270	140 ± 10	138 ± 6	125 ± 15	134 ± 18	(?) °K
					(at ~ 1 cm radio wavelengths, whole disk)				
Albedo—fraction of light that is reflected	0.06	0.85	0.4	0.15	0.58	0.57	0.8	0.71	0.15

1.7

to the asteroids which, according to Bode's law, occupy a region that should contain a planet? Or, on the other hand, can the high pressure conditions needed for the crystalline structure be provided by shocks that would naturally occur when meteorites collide from time to time?

We also may find out more about the structure of the solar system, by studies of the magnetic fields of planets, from the interaction of planets with their moons, and by studies of the cloud of *zodiacal dust* that orbits the sun and perhaps represents comet debris or debris from the intercollision of asteroids or smaller bodies. Detailed chemical and nucleochemical studies also bring in new insight at a rapid rate. But each new method appears to raise more questions than it answers. Evidently we still are very far from our goal. Perhaps it will take a few decades before we actually start to piece together a consistent history of the solar system.

We may be able to learn a great deal from missions to planets, in which a vehicle carrying scientific apparatus carries out detailed observations from a circumplanetary orbit. The Mariner 9 mission to Mars has given us a wealth of information gathered by such means. Vehicles that land on planetary surfaces, or analyze samples of the atmosphere on their way to the surface, may give information of even greater significance about a planet's surface chemistry and about the existence of life.

1:8 FORMATION AND EVOLUTION OF GALAXIES AND CLUSTERS

Matter is inhomogeneously distributed throughout the universe. Much of the universe is empty; but embedded in these empty spaces we find concentrations of gases, concentrations of stars, and concentrations of galaxies, in a hierarchical sequence that has rich variety. There are neutral and ionized clouds of gas, groups of stars, clusters of stars, dwarf and giant galaxies, and groupings of galaxies as in Fig. 1.10, and clusters of galaxies. Evidently a number of different physical processes compete in determining the characteristics and makeup of these various types of aggregates. The physical processes which lead to the substantial differences that distinguish globular clusters from galaxies, and individual galaxies from clusters of galaxies, are completely unknown. We simply do not know why there should be distinctions as we switch from a scale of 10^{20} cm to 10^{23} cm and lastly to 10^{25} cm. These are the respective dimensions of *globular clusters*, *galaxies*, and *clusters of galaxies*.

Larger yet, by a factor of $\sim 10^3$, is the scale of the entire *universe* whose horizon is some 10^{28} cm distant.

Before we make too much of these size differences, let us first make sure that they really exist—that they are not just some subtle feature ingrained in our

Fig. 1.10   The group of galaxies VV282 from the catalog compiled by Vorontsov-Vel'iaminov. The drawing, based on a photograph obtained by H. Arp, was published by E. M. Burbidge and W. L. W. Sargent (Bu71a).

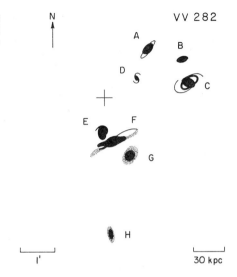

particular way of making observations. For example, we could conceive of very small galaxies whose characteristics are similar to those of the largest globular clusters. However, since these galaxies would be very faint, we would not see them at large distances, and would not know of their existence.

To circumvent this problem, we can undertake an exhaustive study of all discoverable clusters and groupings in the immediate neighbourhood of our own galaxy, where we might be able to determine the properties of galaxies and clusters in some depth. Specifically we would like to know the answers to such questions as:

(a) Are there small galaxies whose sizes are intermediate between those of globular clusters and of large galaxies like the *Andromeda Nebula,* our nearest spiral galactic neighbor in space?

(b) If yes, do such galaxies show the same kind of structure as a globular cluster, or are they more like galaxies in that they actually contain a number of subordinate globular clusters as part of their makeup? Furthermore, do they contain interstellar gas like the larger galaxies do?

(c) Do globular clusters sometimes behave like independent galaxies, in the sense that they are not gravitationally bound to any galaxy, but exist in space by themselves? And would such clusters have their own quota of interstellar gas?

Questions of this kind fortunately can be answered. For the Galaxy and the Andromeda nebula do not form a pair of galaxies completely isolated from all others. A careful search over the past decades has shown that there is a *Local Group* of objects containing some 21 thus-far identified members (Table 1.4). There are certain to be several others obscured by the absorbing matter within the Galaxy, and probably many more that simply have been too faint to be observed by present day techniques.

Not listed in Table 1.4 are two strongly obscured objects, Mafei 1 and Mafei 2, named after their discoverer. Although Mafei 1 may be a giant elliptical galaxy only 1 Mpc distant (Sp71) and Mafei 2 may be a nearby galaxy too, it is not yet clear whether these sources really are part of the local group. This question should soon be settled.

For the moment let us describe the minor members that we are sure belong to the Local Group.

The group contains a number of *dwarf spheroidal systems.* These are very small galaxies, devoid of gas and dust, looking very much like extremely large globular clusters, but with very low surface brightness. One of these systems, Fornax, contains five apparently normal globular clusters and, therefore, must be considered to be more like a galaxy than like a cluster of stars.

In Table 1.4 the quantities b and a present the minor and major diameters of the objects, and show that these systems are appreciably nonspherical. Among the Galaxy's companions only Leo appears round.

As will be discussed in Chapter 3, a loosely bound group of stars, such as any one of these spheroidal systems, cannot come too close to a massive, gravitationally attracting center, before being pulled apart by the difference in the gravitational forces acting on its near and far sides. From this we can conclude that none of these objects can ever have come very close either to the Galaxy or to the *Andromeda Nebula, M31.* Otherwise, on a near passage, the closer portions of the dwarf galaxy would have tended to fall into the larger object more rapidly than the distant portions, and the dwarf system would have become disrupted, unable to hold itself together by its gravitational self-attraction.

Although they are quite near, it is not clear whether the dwarf systems are gravitationally bound to our Galaxy; there is too little information available on their velocity relative to the Galaxy. If the dwarf systems are not tied to the Galaxy, then they should be more or less uniformly distributed throughout the Local Group, and there could be some 200 of them. We would only see the nearest members, because they are too faint to be seen far away. It may, however, be that all these objects are bound to either the Galaxy or to M31 and, in that case, the total number would be smaller. We would then suspect that such systems are formed at the edge of a galaxy, in some protogalactic stage, and that no close approach to the center ever occurred. Interestingly, the colors of the stars in the dwarf systems are rather different—and their H-R diagrams differ strongly—from H-R diagrams for components of the Galaxy. This indicates a different

1:8

Table 1.4 Known Members of the Local Group (After van den Bergh vdBe 68, 72).

Name	R.A. 1950 (h m)	Dec. (°)	Type	$M/M_\odot$	M_v	Distance	$(1-b/a)$	radius pc
M31 = NGC 224	00 40.0	+41 00	Sb I-II	3.1×10^{11}	−21.1	690 Kpc		
Galaxy	17 42.5	−28 59	Sb or Sc	1.3×10^{11}	−20?	—		
M33 = NGC 598	01 31.1	+30 24	Sc II-III	3.9×10^{10}	−18.9			
LMC	05 24	−69 50	Ir or SBc III-IV	6×10^9	−18.5	50		
SMC	00 51	−73 10	Ir or Ir IV-V	1.5×10^9	−16.8	60		
NGC 205	00 37.6	+41 25	E6p		−16.4			
M32 = NGC 221	00 40.0	+40 36	E2		−16.4			
NGC 6822	19 42.1	−14 53	Ir IV-V	1.4×10^9	−15.7			
NGC 185	00 36.1	+48 04	dE0		−15.2			
NGC 147	00 30.4	+48 14	dE4		−14.9			
IC 1613	01 02.3	+01 51	IrV	3.9×10^8	−14.8			
Fornax	02 37.5	−34 44	Spheroidal		−13.6	~180	0.35	900
Sculptor	00 57.5	−33 58	Spheroidal		−11.7	~84	0.35	300
Leo I	10 05.8	+12 33	Spheroidal		−11.0	~220	0.31	200
Leo II	11 10.8	+22 26	Spheroidal		−9.4	~220	0.01	200
Ursa Minor	15 08.2	+67 18	Spheroidal		−8.8	~67	0.55	200
Draco	17 19.4	+57 58	Spheroidal		−8.6	~67	0.29	130
And I	00 43.0	+37 44	Spheroidal		−11	40 Kpc from M31		
And II	01 13.5	+33 09	Spheroidal		~−11	125 Kpc from M31		
And III	00 32.6	+36 14	Spheroidal		~−11	60 Kpc from M31		
And IV	00 39.8	+40 18	?			10 Kpc from M31		

helium or metal abundance in the dwarf systems. That view is also supported by studies of individual variable stars in these objects.

In these systems that apparently have always been well isolated from the Galaxy itself, we therefore seem to have the interesting possibility of studying the evolution of stars having a different initial chemical composition from that found in most of the stars in the Galaxy.

These miniature galaxies may therefore be valuable as test samples for studies of the theory of production of heavy elements in stars and as indicators of the original composition of the material from which our Galaxy was formed. Contamination by the Galaxy, at least on a heavy scale, may be ruled out by the tidal argument on the closest distance of approach, and by the fact that these objects are without interstellar gas of their own. This lack of gas would make it difficult for them to trap gases ejected by our Galaxy during any violent outbursts in the past.

There also seems to exist abundant evidence that the stars, at least in our Galaxy and in M31, have an increasingly great metal abundance as the center of the galaxy is approached. The nuclear region appears to be particularly metal rich, and this seems to indicate that the evolution of chemical elements is somehow speeded up in these regions and is not uniform throughout the galaxy.

Further interesting differences between the galaxies in the Local Group are the differences in gas abundance. Although the dwarf spheroidal systems have no apparent gas content, the *Magellanic clouds* have rich gas abundances of about 9% (LMC) and 30% (SMC). The Galaxy and M31 only have a gas content of 2% and 1%, respectively, by mass.

Related to the question of Local Group membership is the existence of globular clusters whose velocities are so great that even though the clusters are near to the Galaxy, they cannot be physically bound to it. Tidal considerations also show that some of these globular clusters could never have been close to the Galactic center and hence again there is an isolation that might indicate quite separate chemical evolution, independent of the evolution of the Galaxy.

These isolated systems, and also the various chemically distinct stellar populations in the Galaxy itself, should provide useful tests of the theory of buildup of chemical elements. At the same time, they should provide greater insight into the initial chemical composition of the Galaxy and of separate major events that led to chemical differentiation, both within the Galaxy, and between the Galaxy and its chemically isolated companions.

More than that, however, we see that the Local Group already is beginning to throw light on some of the questions posed at the beginning of this section. We see that there is no absolutely sharp cutoff distinguishing galaxies from globular clusters. We sense that there are differences; but the criteria we choose do not always provide a clear distinction in borderline cases.

Let us ask next how galaxy formation might take place. Why are there even difficulties in coming up with explanations for the phenomenon (Re68a, Re70)? We know galaxies exist in all parts of the universe. At the largest distances accessible with modern techniques, galaxies can still be found in apparently unchanged form and undiminished numbers. Why are they there?

No sure answer is known to any of these questions, and we should perhaps start by pointing out why: The normal approach of astrophysicists has been to postulate that galaxies are condensations formed out of previously tenuous material. There is a parallel between this line of thinking and the approach taken in theories of star formation. In both cases a compact configuration is to be attained, starting from an initially dispersed gas. In both cases there are difficulties.

For galaxy formation, the prime difficulty is the cosmic expansion. To understand galaxy formation we would like to think of gravitation, or some other force, acting to concentrate matter into a small volume. This action, however, is always countered by the continuing recession of any two or more elements of matter, as the expansion of the universe forces them further apart. There results a tug-of-war between cosmic expansion and gravitation. Gravitation evidently wins but we do not observe the process anywhere and so we cannot be certain.

Let us now re-examine our main concept: How sure are we that condensation actually is the process through which the galaxies are formed? Not very! There is an alternate approach first seriously documented by the Soviet astronomer Ambartsumian. He noted that there are many *pairs of galaxies* observed in the sky and that the difference in radial *recession velocity* sometimes is so great that the galaxies should be escaping from one another. Their mutual gravitational attraction is too weak to hold them together, unless there is some large unobserved mass that makes the total aggregate much more massive than we infer from the brightness of individual galaxies alone. Sometimes these pairs of galaxies are still connected by a faint *intergalactic bridge*—evidence of a prior genetic relation.

Ambartsumian argued that such galaxies might be very young, perhaps recently formed in a violent explosion. There is no doubt that peculiar galaxies, quite irregular in shape, exist in abundance. About 3 % of all galaxies exhibit large scale irregularities that could become smoothed out on a time scale τ of one galactic rotation, or several hundred million years. If $\tau \sim 3 \times 10^8$ y, and the apparent age of the universe is 10^{10} y, then the 3 % occurrence of disruptions is consistent with the idea that each galaxy goes through such a violent stage once in its life. Ambartsumian suggested that this stage represents the birth of galaxies. A large number of peculiar galaxies exhibiting behavior of this kind was collected by Vorontsov-Vel'iaminov in a catalog of *peculiar galaxies.*

Many of these objects have been studied in detail (Ar71).* *Rotation curves* for some of the galaxies that occur in chains, apparently about to come apart, have shown that the member galaxies seem quite normal, but that they may have formed recently, since the total gravitational mass of the galaxies is not sufficient to keep them together. They should drift apart over a period of a few hundred million years—1 % of the age of the universe!

The discovery of *quasars* also was interpreted by some astrophysicists as possible evidence in favor of present day galaxy formation. The quasars are so bright that they probably are consuming their energy resources at a very rapid rate. We think they cannot shine for longer than a few million years with their present luminosity. Whether the quasars are formed from the intergalactic medium through contraction, or whether they are sources forming galaxies out of nothing, is still a matter of speculation.

From the point of view of the steady state cosmology, the idea of continuing galaxy formation is attractive. In this type of cosmology we have to explain the formation of galaxies under present day conditions. This means that we should be able to observe evidence of galaxy formation in our immediate neighbourhood —within perhaps tens of megaparsecs, if the universe is homogeneous. The predictions of this cosmology would therefore be subject to straightforward tests, if we only knew how to recognize newly forming galaxies. Unfortunately we do not!

In evolving universes, galaxy formation may also be going on at the present epoch; but we believe that most of the galaxies were formed at a much earlier stage, and that contemporary galaxy formation could only represent the closing stages of a much more vigorous period.

For evolving cosmological models, galaxy formation from an intergalactic medium could be most easily envisioned in a *Lemaître universe*. This is a *cosmological model* in which a rapid expansion following the birth of the cosmos is followed by a slowing down, and a quiescent stage of zero expansion. Thereafter expansion picks up again. During the quiescent period the universe is roughly described by a static state in which galaxies might well be able to form, since there is no cosmic expansion to interfere with the condensation process. If instabilities leading to clumping of the extragalactic gas can occur at all, this is the kind of universe in which the instabilities would be most likely to set in.

Whether this is a sufficient reason to favor Lemaître models is, of course, not clear. An extreme tendency toward instability can lead to the formation of quite small condensations, rather than objects of the size of galaxies. The formation process in Lemaître models will therefore have to be theoretically analyzed not only with a view toward understanding how condensations can arise at all, but also with the idea of determining whether condensations of the right-size range are formed. Thus far this problem has not yet been successfully tackled. While galaxy formation then constitutes a serious problem in astrophysics, it does have the great promise of leading to greater insights into processes on a very large scale in cosmology. If we came to understand how galaxies are formed, we would also understand a great deal more about the origin—provided there ever was

1:8

one—of the universe and about the dynamic laws that hold on a scale some 10^{10} times larger than the scale of the solar system on which our current laws of dynamics are tested.

1:9 PROBLEMS OF LIFE

One of the most fascinating problems of astrophysical science concerns the origin of life. Since physical and chemical methods have consistently shown themselves able to clarify biological problems, there now is great confidence that the origins of life and the conditions under which life can originate will some day be understood.

To some extent we are hindered by not knowing how wide a range of phenomena is to be included when we talk about life. The definition of life itself is not yet finally agreed on. Is a virus alive? Or is virus formation just a matter of the reproduction of rather complex forms, just like crystal formation is a reproduction of a complex form? To what extent are natural *mutation* and eventual death requisite features of living matter? Somewhere a line between animate and inanimate matter must be drawn, and we do not yet know just how to do that.

Even when we understand how to define life and living matter, we still will have to investigate whether entirely different physical or chemical bases of life are not possible, and whether life on quite different scales might not be able to proceed in the universe.

In the last section we gave evidence showing that galaxies may be formed on a continuing basis, through explosive division of older galaxies, into two or more parts. If hydrogen were continually gathered up into existing galaxies, then quite conceivably galaxies would be able to grow, to reproduce by binary fission like bacteria do, to metabolize hydrogen into helium inside stars, or to metabolize matter completely through gravitational collapse of stars. Would this make galaxies alive? If not, this picture at least shows that life in the universe might take on quite distinct forms and occur on scales vastly different from those we now recognize.

Even in the more restricted problem of life as we know it on earth, we are faced with formidable difficulties. We know of millions of different forms of life on our planet. We also know that species die out and that new, quite different, species are born. Why? Do conditions on earth change sufficiently, so that the habitat becomes too unfriendly for one kind of life and more hospitable to another? We think so.

The primitive earth, as it was when first formed from a nebula surrounding the sun, had an atmospheric composition quite different from what it is today. Its hydrogen content was far greater. The form that life took at that time must have been entirely *anaerobic*. As the atmosphere slowly became rich in oxygen, and life changed to take advantage of oxygen as a source of energy, some anaerobes remained and sought refuge where oxygen could not penetrate and where competition from the *aerobes*, or oxygen-metabolizing organisms, was not severe (Op61a, b; Sh66).

One of the interesting problems of astrophysics, then, is to try to understand the chemistry of the primitive earth. By noting the overall composition of solar surface material and the chemical composition of the atmospheres of other planets where conditions may have always remained stable, we may be able to understand what changes have taken place on earth. As already mentioned, the chemistry of comets may also help to produce an understanding of the initial conditions that existed on the young earth.

Is life, even as we know it, abundant in the universe? The probabilities involved in an answer to that question are still thoroughly speculative. If we estimated conservatively, we might suggest that life exists only on planets around stars having the same general characteristics as the sun; and even then, we might need to postulate the existence of a planet just at the distance where water neither freezes nor boils. Unfortunately, we do not know enough about the formation of planets to be able to estimate the likelihood for the occurrence of such a combination of star and planet. Once this kind of information is available, however, we will still face the problem of estimating the likelihood of life spontaneously catching on, on such a planet.

Increasingly sophisticated laboratory tests now are possible. They seek to establish the kind of lifelike molecule that could occur under conditions assumed to have held on the primitive earth. In time, these experiments might go far enough to synthesize lifelike primitive organisms. Once that is done, the probability of the formation of life could be estimated more realistically.

There are other possibilities too. Perhaps life is more in the nature of an infection that, having started in a given planetary system, is then able to spread from one system to another, either through natural causes, or through the intervention of intelligent beings who would like to see life propagate over wide regions.

If this second situation is true, then life would have had to be formed only once, and from then on no further spontaneous formation would have been necessary. The study of the spontaneous origin of life on a primitive earth could then lead to considerable error if extrapolated to deal with the probability for the occurrence of life elsewhere.

The assumption of intelligent life existing in other regions of the Galaxy or universe is of course fascinating. Can we contact such life? How would we communicate? If such an intelligent civilization, far more advanced than ours, exists, is it trying to communicate with us? Is there some unique best way of communicating, which a better understanding of physics and astrophysics will some day provide? Do we have to communicate by means of electromagnetic signals, or are there perhaps faster than light particles—tachyons—that we will discover later on and that would almost certainly be used by an intelligent civilization bent on saving time?

If other civilizations exist, should we visit them, or is that even possible outside our solar system? After all, the purpose of a visit is to see, talk, and touch; all of that could be done by improved communication techniques provided only that the distant civilization is able, and also willing, to communicate. There are relatively few things that cannot be settled that way, although without some actual exchange of mass, we probably could not decide whether a given civilization was made of matter or antimatter.

There are many fundamental questions of life on which astrophysics can throw new light and the interest of astrophysicists in biological problems is bound to increase in the next few years.

1:10 UNOBSERVED ASTRONOMICAL OBJECTS

In Appendix A we list a wide variety of astronomical objects; and we might think that we know enough to form a world picture with some reasonable assurance of—at least—being sensible.

To avoid this trap of complacency, we should complete our list of astronomical objects by citing those that have not yet been observed. We might think that this would be difficult; but it is not. To illustrate this, we first restrict ourselves to photographic observations of diffuse objects. An extension to other techniques will become obvious later.

We produce a plot comparing the absolute photographic magnitude and the logarithm of the diameter of different objects (Fig. 1.11). This was first conceived by Halton Arp (Ar65).

We note that all objects normally discovered on photographic plates have to lie on a strip between the two slanted lines on the Arp plot. Objects lying to the left or above this strip appear stellar. But since there are about 10^{11} stars that can appear on photographs taken of the Galaxy, abnormal or highly compact objects with a stellar appearance could never be separated from bona fide stars, without an inordinate amount of labor.

To detect something unusual about objects falling into this upper region on the chart, some other peculiar earmark must be found. For example, quasars lie above the strip. They were first discovered by virtue of their radio emission and only later identified as distant objects by means of individually obtained spectra.

To the right and below the strip, the surface brightness of a diffuse object is so low that the foreground glow emitted by the night sky outshines the object,

Fig. 1.11 Diagram showing the diameter-brightness strip onto which extended objects observed through the atmosphere tend to fall. Objects in the upper left-hand corner are very compact, and are not readily distinguished from ordinary stars. In the lower right-hand corner, atmospheric night sky emission interferes with observations. The upper and lower crosses, respectively, represent the quasars 3C273 and 3C48. Their diameters are quite uncertain. The upper and lower filled circles represent the Fornax and Draco galaxies—minor members of the Local Group of galaxies (Based on a drawing by Arp Ar65. With the permission of the University of Chicago Press.)

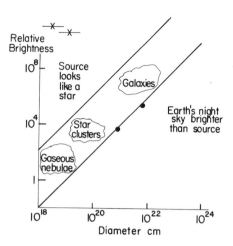

making it undetectable. Exceptions to this are the Local Group minor galaxies Fornax and Draco in which individual stars can be counted. If these objects were more distant, individual stars would not be detected and the objects would not be visible.

We note that the strip of observable objects covers only a small portion of the available area on the plot. This means that we have not yet had the opportunity to see many different varieties of objects that probably occur in nature. For it would be too much of a coincidence to expect all classes of objects in the universe to fit neatly into a pattern defined by our own instrumental capabilities—and to fall onto the strip of observables in the Arp diagram.

By taking instruments above the atmosphere, in rockets and satellites, we are able to get above much of the night sky emission. The demarcation line on the right can therefore be moved downward and further to the right. It also is possible to obtain a higher image resolving power above the atmosphere, since the main limitation to the resolving power on the ground is due to atmospheric scintillation. This feature would move the line on the left of the strip upward and to the left. The combined effect is to widen the strip altogether and to allow us to identify a larger variety of objects than is accessible from the ground. This is one reason for launching an observatory into an orbit high above the atmosphere, and we may expect new discoveries in astronomy when such observations become possible.

Of course, not all objects emit visible radiation, and so we cannot expect to find out all there is to know in astronomy simply by making visual observations. Table 1.5 gives 10 different observable entities, chosen to be roughly representative of the various kinds of astronomical objects we know. We note that only three

Table 1.5 Comparison of Objects Seen in the Visual and Radio parts of the Electromagnetic Spectrum.[a]

[a]We find that many of the objects are more readily apparent observed with one technique than with the other, even though some emission may be present in both parts of the spectrum.

	VISIBLE	RADIO	CORRELATION	NEW
1. COMETS	✓	X	0	
2. PLANETS	✓	✓	1	
3. STARS	✓	X	0	
4. IONIZED REGIONS	✓	✓	1	
5. INTERSTELLAR MASERS	X	✓	0	□
6. GLOBULAR CLUSTERS	✓	X	0	
7. GALAXIES	✓	✓	1	
8. PULSARS	X	✓	0	□
9. QUASARS	X	✓	0	□
10. BACKGROUND	X	✓	0	□
			3/10	4/10

objects could have been discovered equally well through visual or through radio observations. Four of the objects emit only radio waves, or are most easily distinguished as unusual objects through their radio emission. *Quasars*, for example, do emit in the visual domain; but they look like ordinary stellar objects until we painstakingly observe their spectra. As stated above, any stellar object is easily lost among the 10^{11} ordinary stars in our galaxy; and it is therefore only through their radio emission that *quasistellar objects* were discovered as early as they were.

We note from the highly subjective Fig. 1.12 that most of our knowledge about the universe still comes from visual observations mainly because more observations have been carried out in the visible than in other parts of the spectrum.

A few more decades of radio observations may however reverse this state of affairs and, certainly, continued infrared, X-ray, and γ-ray observations will also follow that trend.

Table 1.5 shows vividly that observations made with new techniques help to discover new phenomena. They do not merely tell us something additional about phenomena we have already observed. We can therefore expect that an entirely new set of astronomical objects will become uncovered as we perfect observations throughout

(a) The entire electromagnetic spectrum, going all the way from the lowest frequencies in the hundred kilocycle radio band up to the highest energy gamma rays.

(b) The entire modulation frequency spectrum, going up to megacycle frequencies: Pulsars would never have been discovered had it not been for electronic innovations that permitted observations of intensity changes over millisecond

Fig. 1.12 Subjective drawing indicating the amount of information that has been gained through observations in the various portions of the electromagnetic spectrum. We could equally well have plotted "total time spent making observations throughout the history of astronomy", and come up with a similar plot. The ordinate has a quite arbitrary scale, probably more nearly logarithmic than linear. The peak *V* represents observations in the visual part of the spectrum.

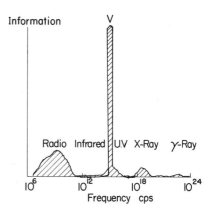

time intervals. Using photographic plates, where exposure times of the order of an hour are representative, one could not expect to discover objects with periodic brightness undulations much shorter than an hour. At the other extreme, analysis of old photographic plates cannot yield discernible variations for phenomena whose period is much longer than a few decades.

(c) The entire spatial frequency domain. As already indicated, many observing techniques are good for stellar or at least highly compact objects, but are not capable of detecting a uniform background. Other techniques permit background measurements, but not the observation of faint compact objects. This is particularly true of infrared observations. Until we have observed the entire range of possible angular sizes, from the lowest limits of angular resolution, all the way up to a uniform background, there must remain potentially interesting unobserved astronomical objects.

(d) The entire set of communication channels: electromagnetic and gravitational radiation, cosmic rays, neutrinos, and, if they exist, tachyons. These channels again can be expected to exhibit the existence of new phenomena, in a universe rich far beyond our most adventurous speculation.

(e) All of the above should lead to progress. However, there are many possible astronomical objects that simply are beyond observation through any currently planned telescopes. For example, if 10 percent of the mass of our galaxy consisted of snowballs (fist-sized chunks of frozen water freely floating through interstellar space) we would never know it. The amount of light scattered from these objects would be too low to make them detectable. They would not be able to penetrate the solar system to become visible as meteors because the sun would evaporate them away long before they ever approached the earth's orbit. Snowballs would therefore have to remain undetected until spaceships traveled beyond the solar system. After that they could become a major nuisance, since a spaceship moving nearly at the speed of light could be completely destroyed on colliding with one of these miniature icebergs. Similarly black dwarf stars, or unaccompanied planets, would have been difficult to detect thus far.

When we look at the unfinished work implied by the points (a) to (e), we must be prepared to accept the thought that astronomers may have seen no more than a few percent by number of all observable important phenomena characterizing the universe. From this point of view, it almost seems premature to construct sophisticated cosmological theories and cosmic models.

On the other hand, these theories and models often suggest novel observations that produce new results. We should therefore think of astrophysical theory not so much as a structure that summarizes all we know about the universe. Rather, it is a continually changing pattern of thought that permits us to grope in the right directions.

1:10

2

The Cosmic Distance Scale

2:1 SIZE OF THE SOLAR SYSTEM

A first requirement for the establishment of a cosmic distance scale is the correct measurement of distances within the solar system. The basic step in this procedure is the measurement of the distance to Venus. The most precise way of obtaining this distance is through the use of radar techniques.

A radar pulse is sent out in the direction of Venus, and the time between its transmission and reception is measured. Since time measurements can be made with great accuracy, the distance to Venus and the dimensions of its orbit can be established within a kilometer.

Once the distance to Venus is known at closest approach a, and most distant separation b, and these measurements are repeated over a number of years, the diameter and eccentricity of both the earth's and Venus's orbit can be computed. The mean distance from the earth to the sun is then directly available as the mean value of $(a + b)/2$ (Fig. 2.1). This distance is called the *astronomical unit*. A check on the earth-Venus distance is obtained from trajectories of space vehicles sent to Venus.

2:2 TRIGONOMETRIC PARALLAX

When observations are made from opposite extremes in the earth's orbit about the sun, a nearby star will appear displaced relative to more distant stars in the same part of the sky. The *parallax, p*, is defined as half the apparent angular dis-

2:2

placement measured in this way. The distance d to the star is then

$$d = \frac{\text{astronomical unit}}{\tan p} \qquad (2-1)$$

or

$$d = 1.5 \times 10^{13}(\tan p)^{-1} \text{ cm}$$

A star whose parallax is one second of arc is at a distance of 3×10^{18} cm, since $\tan 1'' = 5 \times 10^{-6}$. This distance forms a convenient astrophysical unit length, and is called the parsec, pc. 1 pc $= 3 \times 10^{18}$ cm.

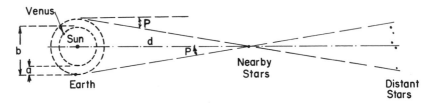

Fig. 2.1   Measurement of the astronomical unit and trigonometric parallax.

The trigonometric parallax can be reliably determined out to distances of about 50 pc, where the parallax is 0.02 sec of arc.

2:3 SPECTROSCOPIC PARALLAX

Once the distance to nearby stars has been determined, we can correlate absolute brightness with spectral type. Bright stars of recognizable spectral type then become distance indicators across large distances where only the brightest stars are individually recognized and trigonometric parallax cannot be used.

2:4 THE MOVING CLUSTER METHOD

As explained below this method has thus far been used only for the Hyades star cluster, whose members are close together and move through space as a group. Three measurements need to be made:

(a) The *radial velocity v*, of the cluster stars along the line of sight is measured from spectral displacement—Doppler shifts.

2:4

(b) and (c) The *proper motion* (apparent motion perpendicular to the line of sight) of individual stars in the cluster is measured by intercomparing their positions on photographic plates taken many decades apart. This motion has the form of an apparent contraction of the cluster angular diameter θ. Its rate of decrease, $\dot{\theta}$, can be computed and gives the fractional rate of increase of the cluster's distance from the sun (Fig. 2.2).

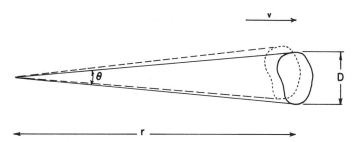

Fig. 2.2 The Hyades cluster at distance r recedes from us with a radial velocity component v. If the cluster diameter D remains constant, θ must decrease. Equation 2–2 shows how r can be determined from a measurement of v, θ, and $\dot{\theta}$, the rate of change of θ with time.

The quantities θ, $\dot{\theta}$, and v can be related to the cluster's distance r, since $\theta = D/r$ and the time derivative is $\dot{\theta} = -D\dot{r}/r^2$ for a constant cluster diameter D. Since $\dot{r} = v$,

$$\dot{\theta} = -\frac{v\theta}{r} \qquad r = -\frac{v\theta}{\dot{\theta}} \tag{2-2}$$

The distance r to the Hyades cluster is therefore available in terms of three directly measured quantities. This simple scheme is restricted to use with the Hyades stars because no other cluster is near enough to yield accurate proper motions. Clearly use of the method depends on the dynamical stability of the Hyades group. If the stars were not gravitationally bound but, say, were expanding away from a common origin, the method would yield a false measure of the distance, because the time derivative $\dot{D}$ would then contribute to $\dot{\theta}$.

2:5 METHOD OF WILSON AND BAPPU

The H and K lines of singly ionized calcium appear in the spectra of most late-type stars. The shape of these lines is rather complex. First, there is a broad absorption line. In its center we see a thinner central emission feature and superposed on this is an even finer central dark absoprtion band. The broad absorption labeled H_1 and K_1, respectively, is due to cool gas in the outer atmosphere of a star. The lines are broad because the absorption of calcium is so strong that it extinguishes radiation even in the wings of the spectral lines where the absoprtion coefficient is relatively low.

The emission lines H_2 and K_2 are due to re-emission higher in the atmosphere. The emission from these lines can be absorbed a second time—and produces fine lines repsectively labeled H_3 and K_3. This absoprtion line is produced by cool gas lying even higher in the star's atmosphere than the atoms responsible for the emission lines H_2 and K_2.

Bappu and Wilson (Wi57) noticed that there exists a correlation of the width of the H_2 and K_2 components with the stellar brightness:

$$\frac{dM_v}{d \log (W_2)} = \text{constant} \tag{2-3}$$

W_2 is the line width measured in cycles per second. There is no known explanation for this linear relationship between the visual magnitude and the logarithm of the line width; it simply is empirical.

Using the brightness of the sun as one data point and the brightness of four Hyades stars as the other, the slope of the line can be determined. Hence, the measured brightness of a star and its H_2 and K_2 line widths establish its distance. When properly used, the uncertainty of the distance measured by this method is of the order of 10%. Some care must, however, be taken since the Wilson–Bappu relationship does not appear to hold for stars of all spectral types. Figure 2.3 shows a calibration of the line width W_2 in terms of stars whose trigonometric parallax is known. The relation (2–3) is illustrated by this plot. This is not a fundamental method for calibrating distances. It does, however, serve as a useful cross check, and gives us additional confidence in our other methods of judging the distances of stars.

2:6 SUPERPOSITION OF MAIN SEQUENCES

This method is based on the assumption that main sequence stars have identical properties in all galactic clusters. This means that the slope of the main sequence is the same for all such clusters and, moreover, it requires that main sequence stars of a given spectral type or color have the same absolute brightness in all clusters (see Fig. 1.4). On this assumption we can compare the brightness of the main sequence of the Hyades and any other galactic cluster. The vertical shift necessary to bring the two main sequences into superposition gives the relative distances of the clusters.

2:6

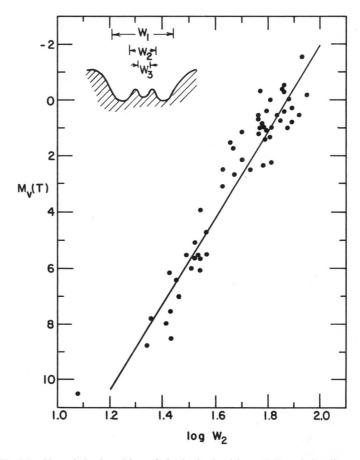

Fig. 2.3 Plot of the logarithm of the ionized calcium. CaII emission line width W_2 against absolute magnitudes derived from trigonometric parallaxes. The insert explains the labeling of line widths W_1, W_2, and W_3. (After Wilson and Bappu, Wi57. With the permission of the University of Chicago Press.)

PROBLEM 2-1. If the shift in apparent magnitudes is $\Delta m = m_{GC} - m_{Hya}$ show that the relative distances are

$$\Delta m = 5 \log \frac{r_{GC}}{r_{Hya}} + A' \qquad (2\text{--}4)$$

where A' is a correction for the difference in interstellar reddening of the galactic cluster, GC, and the Hyades. The derivation is analogous to the work leading to equation A–2.

The factor A' can be determined through use of stellar line spectra as explained in section A:6.

To obtain the distance to the globular clusters we can proceed on one of three different assumptions:

(a) The Hertzsprung–Russell diagram of the globular cluster has a segment that runs essentially parallel to the galactic cluster main sequence. We can assume that this segment coincides with the main sequence of the Hyades cluster. The distance of the globular cluster can then be calculated in terms of equation 2–4.

(b) Alternately we can assume that the segment coincides with the main sequence defined by a group of dwarf stars in the sun's immediate neighborhood. The distance to these dwarf stars is determined by trigonometric parallax.

(c) Finally we can assume that the mean absolute magnitude for short-period variables (RR Lyrae variables) is the same in globular clusters and in the solar neighborhood (see section 2:7, below).

None of these three choices is safe in itself. However, when applied to the globular cluster M3, the third entry in the Messier catalog, all three methods give distance values in fair agreement with each other. This verifies that the main sequence of different groupings of stars coincide reasonably well, and can be used as distance indicators.

2:7 BRIGHTNESS OF RR LYRAE VARIABLES

We find that the apparent brightness of all RR Lyrae variables in a given globular cluster is the same regardless of the variable's period. Since these stars are intrinsically bright, and since their short period makes them stand out, they serve as ideal distance indicators. We assume that the absolute brightness of these stars is the same not only within a given cluster, but also elsewhere. The relative distance of two clusters can then be determined by the inverse square law corrected for interstellar extinction (equation 2–4).

2:8 BRIGHTNESS OF CEPHEID VARIABLES

At the turn of the century, cepheid variables in the Magellanic clouds were found to have periods that are a function of brightness. The Magellanic clouds are dwarf companions to the Galaxy. They are small galaxies in their own right and are compact enough so that all their stars can be taken to be at essentially the same distance from the sun. By intercomparing the brightness of cepheids in the Magellanic clouds and in globular clusters, one was able to obtain relative distances to these objects.

However, there was a pitfall in this comparison. The cepheids in the Clouds are population I stars, normally found in the disk of a galaxy. Globular clusters, on the other hand, belong to the halo component that is more or less spherically distributed about the center of a galaxy. Generally, *population I* consists of bright early stars, and dwarf late stars that lie on the main sequence. *Population II* is characterized by a large number of late-type giants.

In 1952 Baade analyzed the brightness of cepheids in M31, comparing population I with population II regions. He found that population I cepheids were about 1.5 magnitudes brighter than population II cepheids. The distance modulus of M31 had previously been derived by comparison of these brighter cepheids with type II cepheids in clusters within our own Galaxy. The distance to M31 had therefore been erroneously underestimated by a factor of two. Baade's measurements showed that this distance and, in fact, the distance to all galaxies had to be doubled.

2:9 THE BRIGHTNESS OF NOVAE AND H$_{II}$ REGIONS

Novae have an absolute brightness that is related to the decay rate of the brightness after an outburst. The great intrinsic brightness of a nova makes it a very useful distance indicator for nearby galaxies.

The diameters of bright H$_{II}$ (ionized hydrogen) regions also form good yardsticks by which to judge the distances of such galaxies.

2:10 DISTANCE–RED-SHIFT RELATION

The distances of various galaxies can be compared by making a comparison of bright objects within the galaxies. Suitable candidates are O stars, novae, cepheid variables, and H$_{II}$ regions. These individual objects can be detected out to distances about as far as the nearer Virgo cluster galaxies. By comparing the distance estimated from the apparent brightness of such stars and the sizes of H$_{II}$ regions, it is possible to show that the spectral red shift of light from these galaxies is linearly related to distance: $\Delta\lambda/\lambda \propto r$.

We can also compare the brightness of individual galaxies to estimate relative distances. Here we must be careful to compare galaxies of the same general type. To minimize errors due to statistical variation in brightness, we sometimes compare not the brightest, but rather, say, the 10th brightest galaxy in two different clusters. By this device we hope to avoid selecting galaxies that are unusually

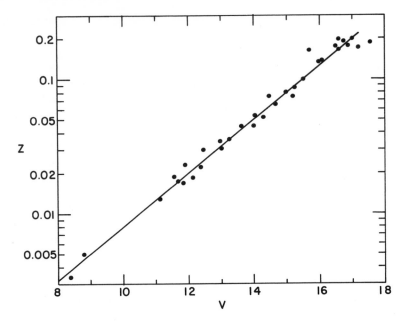

Fig. 2.4 Red-shift–magnitude diagram for the brightest members of 38 clusters of galaxies. After J. V. Peach (Pe69) based on data obtained by A. Sandage. z is the red shift $\Delta\lambda/\lambda$. V is the visual magnitude.

bright. Figure 2.4, however, is a plot of the brightest cluster galaxies as a function of red shift.

The data show a linear distance–red-shift relation. It is not clear how far this linearity persists, but for many cosmological purposes we use the red shift as a reliable indicator of a galaxy's distance. This procedure may, however, not be appropriate for quasars.

We should still note that distance measurements are not easy and that errors cannot always be avoided. In 1958, Sandage (Sa58) discovered that previous observers had mistaken ionized hydrogen regions for bright stars. This had led them to underestimate the distance to galaxies by a factor of ~ 3 beyond the error previously unearthed by Baade. Within a space of five years the dimensions of the universe therefore had to be revised upward by a total factor of ~ 6. It is not unlikely that, from time to time, similar errors may lead to further revisions of the cosmic distance scale. However, Fig. 2.5 shows that we can frequently check astronomical distances by several different methods, and eventually we should be able to derive a reliable distance scale. At present, a red-shift velocity of 75 km sec^{-1} is estimated to indicate that a galaxy is at a distance of 1 Mpc.

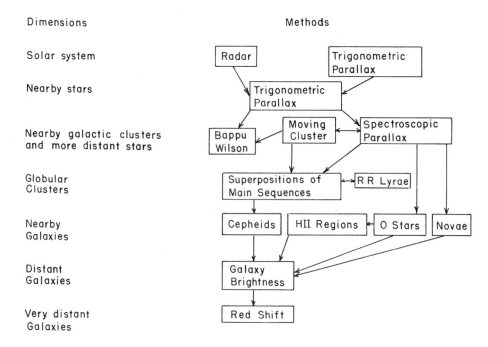

Fig. 2.5 Flowchart of distance indicators.

The velocity-distance proportionality constant—the Hubble constant H—has a value $H = 75$ km sec^{-1} Mpc^{-1}.

Once we know the distance to the various galaxies, we can estimate typical intergalactic distances and typical number densities of galaxies for cosmological purposes. The variation of number density with distance, or more accurately, with spectral red shift, can in principle be used to determine the geometric properties of the universe. By such means we may hope to determine whether the universe is open or closed, and whether it is finite or infinitely large. We will return to such questions in Chapter 10, but a simple argument based on Euclidean geometry is given in the next section.

2:11 SEELIGER'S THEOREM AND NUMBER COUNTS IN COSMOLOGY

If a set of emitting objects is homogeneously distributed in space, then N_m/N_{m-1}, the ratio of objects whose apparent magnitude is less than m to those whose apparent magnitude is less than $m - 1$ is 3.98. This is called Seeliger's theorem. Let us see how this result is obtained.

2:11

Let $m - 1$ be the apparent magnitude of a given star at distance r_1 (see Fig 2.6). Then the distance r_0 at which its apparent magnitude would be m is $r_0 = (2.512)^{1/2}r_1$. At that distance its apparent brightness is reduced by $(r_0/r_1)^2 = 2.512$. All this follows directly from our definition of the magnitude scale in section A:5.

Fig. 2.6

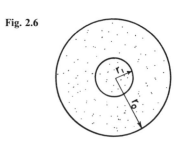

If stars are uniformly distributed in space and have a fixed brightness, they will appear brighter than apparent magnitude m out to a distance r_0, but brighter than $m - 1$ only out to a distance r_1. The ratio of the number of stars brighter than a certain magnitude, N_m/N_{m-1}, is proportional to the volume occupied.

$$\frac{N_m}{N_{m-1}} = \frac{r_0^3}{r_1^3} = (2.512)^{3/2} = 3.98 \tag{2-5}$$

Since this is true for stars of any given brightness, it will also be true for any homogeneous distribution of stars, regardless of their luminosities. Equation 2–5 states that the flux obtained from a source is proportional to r^{-2}, while the number of sources observed down to a given flux limit is proportional to r^3. Hence the number of sources observed brighter than a certain strength (flux density) $S(v)$ at a given spectral frequency v is

$$N \propto S(v)^{-3/2} \quad \text{since} \quad N \propto r^3 \text{ and } S(v) \propto r^{-2} \tag{2-6}$$

This proportionality, which already was of interest in classical stellar astronomy, has become even more important in modern cosmology, where it is usually found in a somewhat different form. If we take the logarithm of both sides, we find

$$\log N \propto -\tfrac{3}{2} \log S(v) \tag{2-7}$$

In radioastronomy a comparison of $\log N$ and $\log S$, often called the $\log N$–$\log S$ plot, means this: If the logarithm of the number of sources brighter than a given level is plotted against the logarithm of the brightness, at the spectral frequency at which the instrument operates, then the slope of the plot should be constant,

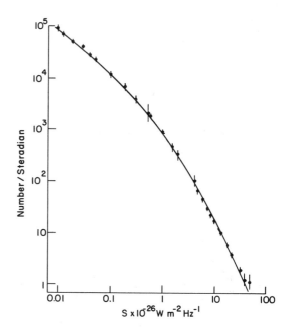

Fig. 2.7   Plot of log N against log S, where N is the number of sources per unit solid angle with flux density, measured in watts per square meter per Hertz, greater than S. The measurements were made by Pooley and Ryle (Po68) and have been discussed in the context of other studies by Ryle (Ry68).

with a value of $-3/2$, provided (a) the sources are homogeneously distributed in space, (b) space is Euclidean, and (c) we compensate for any cosmic red shift in apparent brightness. This latter requirement comes about because observations are made at one given frequency v. If, say, the source intrinsically is very bright at high frequencies, then red shift to lower frequencies would make it look deceptively bright. A correction for the spectral shape for radio sources is therefore required. A correction for the red shift is required, in any case, because a cosmically red shifted source already appears weaker just from the time dilatation effect, that is, from the increased apparent spacing between the emission times of radio frequency photons (Ke68). How these corrections are to be made is discussed in section 10:6.

   If these red-shift effects are ignored, say at small cosmic distances, or else corrected for on a statistical basis, then the homogeneity and Euclidean geometry required by a steady state universe predict that it show a log N–log S slope of

− 1.5. Actual observations indicate a slope not precisely of this value, but near enough to keep the theory within contention.

If true deviations from this slope eventually become established, we will have evidence to show either that space is non-Euclidean, or that the radio sources are inhomogeneously distributed. Either of these consequences would imply an evolving cosmos (see Chapter 10). Figure 2.7 shows the current state of observational results. The slope at high flux densities is − 1.85; at low densities, it is − 0.8. These results are partly dependent on the frequency at which observations are carried out. At 1400 MHz, a slope of − 1.5 appears to hold for flux densities $S \geq 0.5 \times 10^{-26} \, \mathrm{Wm}^{-2} \, \mathrm{Hz}^{-1}$, except that for the very few strong sources a steeper slope exists. This may just be a local inhomogeneity characteristic of the Galaxy's neighborhood (Br72).

PROBLEMS DEALING WITH THE SIZE OF ASTRONOMICAL OBJECTS

The methods here described are not those normally used by astronomers. However, they allow us to obtain insight into the dimensions of planetary and stellar systems without recourse to the more sophisticated methods covered in this chapter. The first six steps were already known to Newton (Ne00).

2–2.   The distance, R, to Venus can be accurately obtained by triangulation when Venus is at its point of closest approach. Two observers separated by 10^4 km along a line perpendicular to the direction of Venus find the position of Venus on the star background to differ by 49″ of arc. Calculate the distance of Venus at closest approach.

2–3.   At this distance, the angular diameter of Venus is 64″, while at greatest separation its angular diameter is 10″. Assuming both the earth's and Venus's orbit to be circular, compute the two orbital radii. Assume the orbits to be concentric.

2–4.   The mean angular diameter of Saturn at smallest separation is about 1.24 times as great as at the largest separation. (These mean angular diameters have to be averaged over many orbital revolutions, because Saturn's orbit about the sun is appreciably eccentric). Calculate the semimajor axis, a, of Saturn's orbit about the sun.

2–5.   Both the sun and the moon subtend an angular diameter of $1/2°$ at the earth. The lunar disk at full moon is only about 2×10^{-6} as bright as the sun's disk. Knowing that the moon is much nearer the earth than the sun, compute the reflection coefficient K of the lunar surface, assuming that the light is reflected isotropically, into 2π sterad. Show that this reflection coefficient is appreciably lower than that of terrestrial surface material (which is estimated to have a mean reflection coefficient of order 0.3). Actually the moon scatters light mainly in the backward direction, so the result obtained here gives an artificially elevated value for K.

2–6. Assume that Saturn has an angular diameter of $\sim 17''$ at the sun. Let its distance from both the earth and the sun be considered to be 9.5 AU. If the light received from Saturn is 0.86×10^{-11} that received from the sun, compute the reflection coefficient of Saturn's surface. Note that Saturn is known to shine primarily by reflection, since its moons cast a shadow on the surface when they pass between the planet and the sun.

2–7. Saturn appears to emit 0.86×10^{-11} as much light as the sun. How far would the sun have to be removed from the earth to appear with a luminosity identical to that of Saturn, that is, to appear like a first magnitude star?

2–8. Assuming the sun to be a typical star, we conclude that the nearest stars are of the order of 5.2×10^{18} cm distant. We further assume that this is the characteristic distance between stars in the disk of the Andromeda spiral galaxy M31. We note that M31 appears to be a system viewed more or less perpendicular to the disk containing the spiral arms. Other spiral galaxies viewed in profile indicate that the thickness of the disk is about $0.003L$, where L is the diameter of the galaxy. In terms of the distance D of M31, show that the flux received would be

$$\sim \left(\frac{0.003 \, SL^3}{D^2} \right) \times \frac{\pi}{4} \frac{10^{-18}}{5.2}$$

where S is the flux we would expect to receive from the sun if it were 5.2×10^{18} cm from the earth.

2–9. If the brighter regions of M31 subtend an angular diameter of $3°$ at the earth, and if the galaxy is a 5th magnitude object, calculate the distance of the galaxy. Show that its diameter is ~ 6 kpc. (Note that the actual diameter of M31 is about half an order of magnitude larger.)

2–10. Find the distance of the smallest resolved galaxies, on the assumption that all spiral galaxies are of the size of M31. The smallest resolved objects for currently available telescopes are of the order of $2''$ of arc in diameter.

2–11. We note that the light from distant galaxies appears red shifted in proportion to their distance as judged by their angular diameters. If the smallest resolved objects are red shifted by 30% of the spectral frequency, that is, $\Delta v/v \sim 0.3$, calculate at what distance the linear distance–red-shift law would require galaxies to attain the speed of light. This distance is sometimes called the effective radius of the universe.

2–12.   Olbers' Paradox: Let there be *n* stars per unit volume throughout the universe.

(a) What is the number of stars seen at distances *r* to *r* + *dr* within a solid angle Ω?

(b) How much light from these stars is incident on unit area at the observer's position, assuming each star to be as bright as the sun?

(c) Integrating out to *r* = ∞ how much light is incident on unit detector area at the observer?

This problem will be discussed at length in Chapter 10.

ANSWERS TO SELECTED PROBLEMS

2–2. $R = 4.2 \times 10^7$ km.

2–3. $R_e = 1.5 \times 10^8$ km, $R_v = 1.1 \times 10^8$ km.

2–4. $(a + 1)/(a - 1) = 1.24$. Hence $a = 9.5$ AU.

2–5.   If $L_\odot$ is the solar luminosity, *r* is the radius of the moon, and *R* is the distance of the moon—and the earth—from the sun, then $S = (\pi r^2/4\pi R^2) L_\odot$ is the radiation accepted by the moon. This light is spread into 2π solid angle so that, at the distance *D* of the earth, the flux per unit area is $(K \cdot S)/2\pi D^2$, which has to be compared with $L_\odot/4\pi R^2$ coming directly from the sun.

$$\therefore \frac{Kr^2}{2D^2} = 2 \times 10^{-6} \quad \text{and} \quad K \sim 0.2.$$

2–6.   Saturn's diameter is $2r \sim 7.8 \times 10^{-4}$ AU

$$\therefore \frac{\dfrac{\pi r^2 L_\odot}{4\pi(9.5)^2} \cdot \dfrac{K}{2\pi(9.5)^2}}{\dfrac{L_\odot}{4\pi(1)^2}} = 0.86 \times 10^{-11}$$

Hence $K \sim 0.90$.

2–7.   The distance at which the sun would appear to be a first magnitude star is $r = 5.2 \times 10^{18}$ cm

2–8.   If $L_\odot$ is the sun's luminosity, and *D* is the distance, the flux from the galaxy is

$$\frac{\text{(volume of galaxy) (number density) } L_\odot}{4\pi D^2}$$

$$\sim \frac{\pi/4\,(L^2)\,(.003L)\cdot\left(\dfrac{1}{5.2\times10^{18}}\right)^3\cdot L_\odot}{4\pi D^2}$$

$$= \frac{\pi/4\,(L^2)(.003L)}{D^2}\cdot\frac{S}{(5.2\times10^{18})}.$$

2–9. Comparing the magnitude of M31 to a first magnitude star, and taking $\theta = 3/57 = L/D$ and we see from Problem 2–8 that $D \sim 0.1$ Mpc, $L \sim 6$ kpc.

2–10. Distance $= 2 \times 10^{27}$ cm

2–11. Distance $= 7 \times 10^{27}$ cm.

2–12. (a) $\Omega\, nr^2\, dr.$

(b) $\Omega nr^2\, dr\,\dfrac{L_\odot}{4\pi r^2} = \dfrac{\Omega n}{4\pi} L_\odot\, dr.$

(c) The integral (b) would diverge if distant stars were not eclipsed by nearer stars. When eclipses are taken into account, the flux at the observer is finite and is equal to the flux at the surface of the sun.

3

Dynamics and Masses
of Astronomical Bodies

The motion of astronomical bodies was first correctly analyzed by Newton, in the second half of the 17th century. He saw that a variety of apparently unrelated observations all had common features and should form part of a single theory of gravitational interaction. To formulate the theory, he had to invent mathematical techniques that described the observations and showed their interrelationship. His struggles with the mathematical problems are recorded in his book *Principia Mathematica* (NeOO).

The intervening three centuries since Newton's discoveries have allowed his mathematical formulation to be streamlined, so that it can now be presented in brief form; but the underlying astrophysics remains unchanged.

The aim of this chapter will be to show how astronomical observations lead to the conclusions reached by Newton (1642–1727). We will then show the importance of Newtonian dynamics in determining the masses of all astronomical objects. It is interesting that a correct evaluation of these masses was not obtained until more than a century after Newton's work. We will discuss the gravitational interaction of matter with antimatter and finally mention some of the limitations of Newton's work.

3:1 UNIVERSAL GRAVITATIONAL ATTRACTION

A number of astronomical observations and experimental results were known to Newton when he first tried to understand the dynamics of bodies. Many of the experimental results dealing with the motion of falling bodies had been found by Galileo (1564–1642) (GaOO). The astronomical observations, which treated the motions of planets, had been gathered over many years by Tycho Brahe (1546–1601). Johannes Kepler (1571–1630) had then analyzed these data and summarized them in terms of three empirical laws. Newton postulated that the work of Kepler and of Galileo was related. We will not retrace his reasoning here, but rather will outline the evidence with some of the advantages of three centuries of hindsight.

We know from experiments with sets of identical springs and sets of identical masses that a single mass accelerated through the release of, say, two stretched springs, mounted side by side, is accelerated at twice the rate experienced by the same mass when impelled by one spring only (Fig. 3.1). Of course, the springs have to be stretched to the same length. Measurements of this kind lead us to assert that an acceleration always is associated with a force, and is directly proportional to it.

$$F \propto \ddot{r} \qquad (3\text{--}1)$$

This is a brief way of stating Newton's first and second laws.

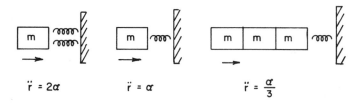

Fig. 3.1 Definition of inertial mass.

In a related experiment three interconnected masses accelerated by releasing a single spring would be accelerated at only one third the rate experienced by one mass acted on by the same spring. This second type of measurement shows that the acceleration produced is inversely proportional to the mass of the impelled body:

$$\ddot{r} \propto \frac{1}{m} \qquad (3\text{--}2)$$

Combining relations (3–1) and (3–2) we obtain the proportionality relation

$$\ddot{r} \propto F/m \qquad (3\text{--}3)$$

The acceleration produced on a body is proportional to the force acting on it and inversely proportional to its mass. When the impelling force is zero, the body remains unaccelerated; its velocity stays constant and may be zero.

We can go one step further and say that the force is equal to the mass times the acceleration. This defines the unit of force in terms of the other two quantities:

$$F = m\ddot{r} \qquad (3\text{--}4)$$

3:1

With these ideas in mind we can draw a significant conclusion from Galileo's experiments that showed that two bodies placed at identical points near the earth fall (are accelerated) at equal rates, even though their masses may be quite different. This independence of mass, interpreted in terms of the proportionality relation (3–3), shows that the accelerating force is proportional to the mass of the falling body. We will need to make use of this point in the arguments that follow.

We can now consider Galileo's work that concerned itself with projectiles. A projectile fired at a given angle falls to earth at a greater distance if its initial velocity is large. We can ask what would happen if the initial velocity was increased indefinitely. The projectile would keep falling to earth at larger and larger distances and, neglecting atmospheric effects, it could presumably circle the earth if given enough initial velocity. If the projectile still retained its original velocity on returning to its initial position, the circling motion would continue. The projectile would orbit the earth much like the moon.

Newton already knew a number of facts about the motion of the moon and he performed calculations to show that the moon behaves in every way just like a projectile placed into an orbit around the earth.

In addition to the experiments of Galileo, Newton also was aware of the observational results summarized by Kepler. There were three principal observations that are summarized in *Kepler's laws*:

(i) The orbits along which planets move about the sun are ellipses.

(ii) The area swept out by the radius vector joining sun and planet is the same in equal time intervals. This means that the angular velocity about the sun is small when the planet is distant, and is large when the planet is close to the sun. The moon shows the same behavior as it orbits about the earth.

(iii) The period a planet requires to describe a complete elliptical orbit about the sun is related to the length of the semimajor axis of the ellipse: The square of the period P is proportional to the cube of the semimajor axis a (Figure 3.2). This law also describes the motion of satellites (moons) about their parent planets.

Newton therefore had three pieces of information:

(i) He knew that projectiles fall because they are gravitationally attracted toward the earth.

(ii) He knew that there are certain similarities between the motions of projectiles and the motion of the moon about the earth.

(iii) He knew that the motion of the moon is similar to that of Jupiter's and Saturn's satellites and that those motions are governed by the same laws that described the motions of planets about the sun.

These ideas led him to attempt an explanation of all these phenomena in terms of accelerations produced by gravitational attraction.

3:1

He already suspected that in the interaction of two bodies equal but oppositely directed forces act on both bodies (Newton's third law). The fact that a planet is attracted by the sun, but can also attract a satellite by gravitational means, indicates that there is no real difference between the *attracting* and the *falling* body. If the force acting on one of Galileo's falling bodies was proportional to its own mass—as stated above—then the force must also be proportional to the mass of the earth. The gravitational force of attraction between two bodies must then be proportional to the product of their masses m_a and m_b:

$$F \propto m_a m_b \tag{3-5}$$

Since the acceleration of distant planets is smaller than that of planets lying close to the sun, this force must also be inversely dependent on the distance between the bodies. Similarly, the distance and orbital period of the moon show it to have an acceleration toward the earth, much smaller than that of objects at the earth's surface. Quantitatively it indicates that $F \propto r^{-2}$. In any case, F must drop faster than $F \propto r^{-1}$, because otherwise the effects of distant stars would influence a planet's orbital motion more strongly than the sun.* As seen from Problems 2–2 and 2–7, Newton knew the distances to other stars, and knew that there were a large number of stars surrounding the sun. As a reasonable choice of distance dependence he tried an inverse square relationship. We will show in the next section that a force law of the form

$$F \propto m_a m_b r^{-2} \tag{3-6}$$

allows us to derive Kepler's laws of motions. To turn this proportionality relation into the form of an equation, we write

$$F = \frac{m_a m_b}{r^2} G \tag{3-7}$$

where the proportionality constant G is the *gravitational constant*. This constant must be experimentally determined.

3:2 ELLIPSES AND CONIC SECTIONS

Since the planets are known to describe elliptical orbits about the sun, it is convenient to start the discussion of their motions by defining a set of parameters in terms of which the elliptical paths can be described.

---

* If *differential* acceleration of the sun and earth is considered, the effect of the distant stars is not so striking. However, with a r^{-1} force, the sun would still rob the earth of its moon.

Fig. 3.2   The terminology of conic sections:

f, f' The two foci of the ellipse.
a Semimajor axis.
b Semiminor axis.
e Eccentricity: the focus of the ellipse is displaced from the center by a distance ae.
q Distance of the pericenter; we can see that $q = a(1 - e)$.
Q Distance of the apocenter; $Q = a(1 + e)$.
θ True anomaly, the angle between the radius vector r, and the major axis as seen from the focus f.
r The radius vector from the focus f.
r' The radius vector from the focus f'.

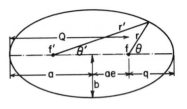

We can define an ellipse as the set of all points, the sum of whose distances from the two foci is constant:

$$r + r' = \text{constant}$$

Since the ellipse is symmetrical about the two foci, we can see from Fig. 3.2 that this constant must have the value $2a$:

$$r + r' = 2a \tag{3–8}$$

Hence $b = \sqrt{a^2 - a^2 e^2}$ by the theorem of Pythagoras. The figure also shows that

$$r \sin \theta = r' \sin \theta' \tag{3–9}$$

and that

$$r \cos \theta - r' \cos \theta' = -2ae \tag{3–10}$$

These two equations, respectively, represent the laws of sines and of cosines for plane triangles. Squaring (3–9) and (3–10) and adding these expressions gives

$$r^2 + 4aer \cos \theta + 4a^2 e^2 = r'^2 \tag{3–11}$$

Substituting from (3–8) then gives

$$r = \frac{a(1 - e^2)}{1 + e \cos \theta} \tag{3–12}$$

an equation that we will need below. Actually equation (3–12) is more general than shown here; it describes any conic section. When the eccentricity is $0 < e < 1$, the figure described is an ellipse. If $e = 0$, we retrieve the expression for a circle of radius a. If $e = 1$, a becomes infinite, the product $a(1 - e^2)$ can remain finite, and the equation describes a parabola. If $e > 1$, equation (3–12) describes a hyperbola.

3:2

3:3 CENTRAL FORCE

From Newton's laws and Kepler's second law, a simple but important deduction can be drawn at once. Kepler's law can be stated in vector form as

$$\mathbf{r} \wedge \dot{\mathbf{r}} = 2 A \mathbf{n} \tag{3-13}$$

Here $\mathbf{r}$ is the radius vector from the sun to the planet, $\dot{\mathbf{r}}$ is the planet's velocity with respect to the sun. A is a constant and the symbol $\wedge$ stands for the *vector*, or *cross*, product. The product of r, $\dot{r}$, and the sine of the angle between these two vectors is twice the area swept out by the radius vector in unit time. $\mathbf{n}$ is a unit vector whose direction is normal to the plane in which the planet moves.

We see that the time derivative of equation 3–13 is

$$\frac{d}{dt}(\mathbf{r} \wedge \dot{\mathbf{r}}) = \mathbf{r} \wedge \ddot{\mathbf{r}} = 0 \tag{3-14}$$

since both A and $\mathbf{n}$ are constant. Multiplying this expression by the mass of the planet m, and using equation 3–14 we find that

$$\mathbf{F} \wedge \mathbf{r} = 0 \tag{3-15}$$

Since neither the force nor the radius vector vanishes in elliptical motion, it is clear that the force and radius vectors must be colinear. Whatever the nature of the force acting on the planet may be, it is clear that this force acts along the radius vector: Such a force is called a *central force*. A planet is pulled toward the sun at all times; and the components of a binary star always are mutually attracted.

3:4 TWO-BODY PROBLEM WITH ATTRACTIVE FORCE

Define a coordinate system whose origin lies at the *center of mass* of bodies a and b. The positions and masses of the bodies are related (Fig. 3.3) by

$$\mathbf{r}_a = -\frac{m_b}{m_a}\mathbf{r}_b \tag{3-16}$$

Fig. 3.3   Center of mass (CM) of two bodies a and b.

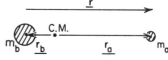

3:4

We know that in planetary motion we are dealing with a central attractive force, and that the force should decrease more rapidly than the inverse first power of the distance between attracting bodies. We postulate that the attractive force is an inverse square law force. If this postulate is correct, we should obtain the correct laws of motion as given by Kepler's laws. We will show below that this indeed is true.

For a central force decreasing as the square of the distance between two attracting bodies, we write the force F_a on body a due to body b as

$$\mathbf{F}_a = m_a \ddot{\mathbf{r}}_a = -\frac{m_a m_b G}{r^3} \mathbf{r} \tag{3-17}$$

where m_a and m_b are the masses of the two bodies and G is sometimes called *Newton's* gravitational constant. It is a universal constant whose value will be discussed below. From the definition of $\mathbf{r}$ and the center of mass, we have

$$\mathbf{r} = \mathbf{r}_a - \mathbf{r}_b = \left(1 + \frac{m_a}{m_b} \right) \mathbf{r}_a \tag{3-18}$$

Combining (3–17) and (3–18) we have

$$\ddot{\mathbf{r}}_a = -\frac{GM}{r^3} \mathbf{r}_a \qquad M \equiv m_a + m_b \tag{3-19}$$

where M is the total mass of the two bodies. Subtracting a similar expression for r_b we obtain

$$\ddot{\mathbf{r}} = -\frac{GM}{r^3} \mathbf{r} \tag{3-20}$$

We see that the acceleration of each body relative to the other is influenced only by the total mass of the system and the separation of the bodies. If equation 3–20 is multiplied by a mass term μ, we obtain a force term that is a function only of r, M, μ, and the gravitational constant:

$$\mathbf{F}(\mu, M, r) = -\frac{GM\mu}{r^3} \mathbf{r} = \frac{-Gm_a m_b \mathbf{r}}{r^3} \tag{3-21}$$

If this force is to be equal to the force acting between the two masses, we must satisfy equation 3–7 which means that

$$\mu = \frac{m_a m_b}{m_a + m_b} \tag{3-22}$$

μ is called the *reduced mass*.

3:4

The equation of motion (3–20) taken together with equation 3–21, shows that the orbit of each mass about the other is equivalent to the orbit of a mass μ about a mass M that is fixed—or moves in unaccelerated motion. There is a great advantage to this reformulation. Newton's laws of motion only hold when referred to certain reference frames, for example, a stationary coordinate system, or one in uniform unaccelerated motion (see also sections 3:8 and 5:1). It was for this reason that the motion of each mass a and b was initially referred to the center of mass. This procedure, however, required us to keep separate accounts of the time evolution of $\mathbf{r}_a$ and $\mathbf{r}_b$. The separation $\mathbf{r}$ was only determined subsequently by adding r_a and r_b. This two-step procedure is avoided if equations 3–20 and 3–21 are used, since $\mathbf{r}$ can then be determined directly.

3:5 KEPLER'S LAWS

Consider a polar coordinate system with unit vectors $\boldsymbol{\varepsilon}_r$ and $\boldsymbol{\varepsilon}_\theta$ (Fig. 3.4) A particle is placed at position $\mathbf{r} = r\boldsymbol{\varepsilon}_r$. Since the rate of change of the unit vectors can be expressed as

$$\dot{\boldsymbol{\varepsilon}}_r = \dot{\theta}\boldsymbol{\varepsilon}_\theta \tag{3–23}$$

defining the rate of change (rotation) of the radial direction, and

$$\dot{\boldsymbol{\varepsilon}}_\theta = -\dot{\theta}\boldsymbol{\varepsilon}_r \tag{3–24}$$

giving the rate of change for the tangential direction, we can write the first and second time derivatives of r as

$$\dot{\mathbf{r}} = \dot{r}\boldsymbol{\varepsilon}_r + r\dot{\theta}\boldsymbol{\varepsilon}_\theta \tag{3–25}$$

$$\ddot{\mathbf{r}} = (\ddot{r} - r\dot{\theta}^2)\boldsymbol{\varepsilon}_r + (2\dot{r}\dot{\theta} + r\ddot{\theta})\boldsymbol{\varepsilon}_\theta \tag{3–26}$$

From equations 3–20 and 3–26 we obtain two separate equations, respectively, for the components along and perpendicular to the radius vector

$$\ddot{r} = -\frac{GM}{r^2} + r\dot{\theta}^2 \tag{3–27}$$

and

$$2\dot{r}\dot{\theta} + r\ddot{\theta} = 0 = 2\dot{r}r\dot{\theta} + r^2\ddot{\theta} \tag{3–28}$$

Equation 3–28 integrates to

$$r^2\dot{\theta} = h \tag{3–29}$$

where h is a constant that is twice the area swept out by the radius vector per unit time. This relationship has a superficial resemblance to the law of conservation

Fig. 3.4   Vector components of the velocity $\dot{r}$.

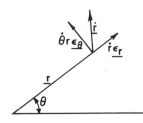

of angular momentum (per unit mass). But that law would involve the distances r_a and r_b, instead of r. Equation 3–29 does state Kepler's second law, however, and that is satisfactory.

Combining equations 3–27 and 3–29 we have

$$\ddot{r} - \frac{h^2}{r^3} + \frac{MG}{r^2} = 0 \tag{3–30}$$

PROBLEM 3–1. Choose a substitution of variables

$$y = r^{-1}, \qquad \dot{\theta}\frac{d}{d\theta} = \frac{d}{dt} \tag{3–31}$$

to rewrite equation 3–30 in the form

$$\frac{d^2y}{d\theta^2} + y = \frac{MG}{h^2} \tag{3–32}$$

Show that this has the solution

$$y = B\cos(\theta - \theta_0) + \frac{MG}{h^2} \tag{3–33}$$

This leads to

$$r = \frac{1}{B\cos(\theta - \theta_0) + \dfrac{MG}{h^2}} \tag{3–34}$$

This is the expression for a conic section (see equation 3–12). It therefore represents a generalization of Kepler's first law. Gravitationally attracted bodies move along conic sections that, in the case of planets, are ellipses. If we set

$$a(1 - e^2) = \frac{h^2}{MG} \tag{3–35}$$

and

$$e = \frac{Bh^2}{MG} \tag{3-36}$$

The minimum value of r occurs for $\theta = \theta_0$.

Let r_m be a relative maximum or minimum distance between the two bodies. Then the entire velocity at separation r_m must be transverse to the radius vector and by equation 3–29

$$\frac{(r_m\dot\theta)^2}{2} = \frac{h^2}{2r_m^2} \tag{3-37}$$

is the kinetic energy per unit mass. The *total energy* per unit mass is the sum of *kinetic* and *potential energy* per unit mass.

$$\mathscr{E} = \frac{h^2}{2r_m^2} - \frac{MG}{r_m} \tag{3-38}$$

Solving for r_m we have

$$r_m = \left(\frac{MG}{h^2} \pm \sqrt{\frac{M^2G^2}{h^4} + \frac{2\mathscr{E}}{h^2}} \right)^{-1} \tag{3-39}$$

Hence the quantity B in equation 3–34 has the value

$$B = + \sqrt{\frac{M^2G^2}{h^4} + \frac{2\mathscr{E}}{h^2}} \tag{3-40}$$

the sign being determined by the condition that the minimum r value occur at $\theta - \theta_0 = 0$.

Equations 3–12 and 3–35 show that the minimum value of r is

$$q = \frac{h^2}{MG(1 + e)} \tag{3-41}$$

Substituting this into equation 3–38 we then have an expression for the energy in terms of the semimajor axis a,

$$\mathscr{E} = (e^2 - 1)\frac{M^2G^2}{2h^2} = -\frac{MG}{2a} \tag{3-42}$$

where we have made use of expression 3–35. Since the total energy per unit mass is the sum of kinetic and potential energy, also per unit mass, we see that

$$\mathscr{E} = \frac{v^2}{2} - \frac{MG}{r} \tag{3-43}$$

and from (3–42) we obtain the orbital speed as

$$v^2 = MG\left(\frac{2}{r} - \frac{1}{a}\right)$$
(3–44)

We now can make a number of useful statements:

(i) If S is the area swept out by the radius vector

$$\frac{dS}{dt} = \frac{1}{2}h, \quad S - S_0 = \frac{1}{2}ht$$
(3–45)

For an ellipse, the total area is

$$S - S_0 = \pi ab = \pi a^2 (1 - e^2)^{1/2}$$

so that from equation 3–35 the period of the orbit is
(3–46)

$$P = \frac{2}{h}\pi a^2 (1 - e^2)^{1/2} = \frac{2\pi a^{3/2}}{\sqrt{MG}}$$
(3–47)

Equation 3–47 is a statement of Kepler's third law.

(ii) If the eccentricity is $e = 1$, the total energy is zero, by equation 3–42 and the motion is parabolic. Astronomical observations have shown that some comets approaching the sun from very large distances have orbits that are practically parabolic, although they may be slightly elliptical or slightly hyperbolic. At best, these comets therefore are only loosely bound to the sun. A small gravitational perturbation by a passing star evidently can make the total energy of some of these comets slightly positive, and they escape from the solar system to wander about in interstellar space.

We should still note that one of the big advances brought about by Newton's theory was the realization that both cometary and planetary orbits could be understood in terms of one and the same theory of gravitation. Prior to that no such connection was known.

(iii) If the eccentricity $e > 1$, the total energy is positive, and the motion of the two masses is unbound. After one near approach the bodies recede from each other indefinitely.

(iv) If the eccentricity is zero, the motion is circular with some radius R and the energy obtained from equation 3–42 is $- MG/2R$ per unit mass. Equation 3–44 then states that v^2 equals MG/R or that the gravitational attractive force MG/R^2 must equal v^2/R, which sometimes is called the *centrifugal force*,—a fictitious force that is supposed to "keep the orbiting mass at constant radius R despite the attractive pull of M."

Thus far we have shown that the motion of one mass about another describes a conic section. In addition, we can show that the orbit of each mass about th common center of mass is a conic section also. Equation 3–19 can be rewritten as

$$\ddot{\mathbf{r}}_a = - \frac{GM}{(1 + m_a/m_b)^3} \frac{\mathbf{r}_a}{r_a^3} \tag{3–19a}$$

This is of the same form as equation 3–20 and we can, therefore, readily obtain equations similar in form to expressions 3–27, 3–28, 3–29, and finally 3–34. This argument also holds true if we were to talk about the vector $\mathbf{r}_b$ instead of $\mathbf{r}_a$. Hence both masses m_a and m_b are orbiting about the center of mass along paths that describe conic sections.

Let us still see how we can determine the masses of the components of a spectroscopic binary. This is the most important means we have for determining stellar masses. For such binaries we can measure the radial velocities of both stars throughout their orbits (Fig 3.5).

It is relatively easy to determine the period of such a binary by looking at the repeating shifts of the superposed spectral lines. Equation 3–47 then gives the ratio (a^3/M) of the semimajor orbital axis cubed and the sum of the masses. If the binary, in addition, is an eclipsing binary, so that the line of sight is known to lie close to the orbital plane, then the semimajor axes of the orbits of the two components about the common center of mass can be found; and this gives the individual component masses if use is made of component equations derived from (3–19a).

For a few visual binaries that are close enough to permit accurate observations, the motion of the individual components relative to distant background stars again permits computation of the individual semimajor axes, provided the trigonometric parallax also is known. The orbital period then allows us to compute the individual masses through Kepler's third law and equation 3–19a.

We note that expressions such as 3–35, 3–36, 3–44, and 3–47, which connect measurable orbital characteristics to M and G always depend on the product MG and, hence, permit a determination neither of the system's total mass, nor

Fig. 3.5 Binary star orbits and the individual semimajor axes for the two stars orbiting their common center of mass.

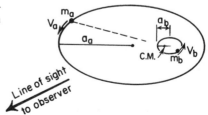

of the gravitational constant. For a long time this presented a serious difficulty. However:

PROBLEM 3–2.   Show how a rough measure of G can be obtained from falling mass experiments when the known size of the earth and some estimate of its density are used to determine the earth's mass. In section 3:6 we show how G was eventually measured by Cavendish. Note that for an accurate determination of the earth's density, G has to be accurately known.

3:6 DETERMINATION OF THE GRAVITATIONAL CONSTANT

Henry Cavendish (1731–1810) an English chemist, discovered a means of measuring the gravitational constant G, late in the 18 century, more than a 100 years after Newton had first shown how the motion of the planets depends on the mass the sun. Until Cavendish performed his experiment, the absolute masses of celestial objects could not be accurately determined; there were only relative values of, say, planetary masses as judged by orbits of their moons.

In the Cavendish experiment a torsion balance is used. Typically such a device may consist of a fine quartz fiber to which a rod bearing masses m_1 and m_2 is attached, as shown in Figure 3.6a. Each mass is at some distance L from the fiber. We can calibrate the balance by noting the torsion that can be induced in the fiber when a small measurable torque is applied to the system. We can apply this torque by having a spring with known force constant exert a horizontal force at the position of m_2.

If masses M_1 and M_2, respectively, are placed at a small horizontal distance from masses m_1 and m_2, we may observe a twist of the fiber, in the sense shown in Figure 3.6b. We can determine the distances r_1 and r_2, respectively, between m_1 and M_1, and m_2 and M_2 to find the horizontal forces acting on the ends of the bar and, hence, establish the torque acting on the quartz fiber. That torque N is

$$N = L\left(\frac{m_1 M_1 G}{r_1^2} + \frac{m_2 M_2 G}{r_2^2}\right) \tag{3–48}$$

From the measured deflection of the masses we can determine the value of N, in terms of the calibration previously obtained on the twisted quartz wire. Since we can measure N, L, r_1, r_2, and the masses of all the different bodies, we now have the values of all quantities in (3–48) except for G, which can then be directly determined from the equation. That value is $G = 6.7 \times 10^{-8}$ dyn cm^2 g^{-2}.

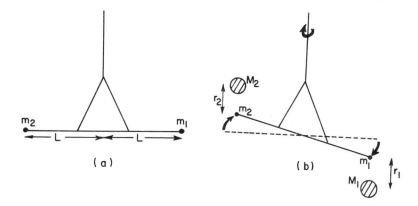

Fig. 3.6 The Cavendish experiment to determine the gravitational constant, G.

Once the value of G is known, the mass of the sun is readily determined using Kepler's third law, equation 3–47. That law actually involves the total mass M of the sun and planet, but by performing the calculations for a number of different planets we can verify that the mass of the sun is very nearly equal to M and that the mass of the planets $\sim 0.0013\, M_\odot$ can be neglected to a good approximation. The approximate mass of the earth can be derived in a similar way, making use of the known orbit of the moon.

3:7 THE CONCEPT OF MASS

If we examine what we have said about the measurement of masses, we find that there really are two quite distinct ways of determining the mass of a body. (i) We can measure its acceleration in response to a measured force (equation 3–4), or (ii) we can measure the force acting on the body when a given mass is placed at a specified distance—this is what we do when we weigh the body with a spring balance.

The first of these is a dynamic measurement, the second can be static. The mass of a body measured in the first way is called its *inertial mass*, while the mass measured by means of the second method is called the *gravitational mass*.

Suppose now that we take a steel ball whose gravitational mass is m_1. We take a wooden ball that is slightly too heavy, and slowly file away excess material until its gravitational mass is also equal to m_1. If the two balls now are placed on a pan balance, they should leave the balance arm in a horizontal position, because the earth attracts both masses equally.

3:7

The question now is whether the inertial mass of these two bodies is always the same. Will the wooden ball be accelerated at the same rate as the steel ball in response to a given force? Until we try the experiment we cannot be sure.

This question intrigued the Hungarian baron Roland von Eötvös around the turn of the century. He suspended two weights of different composition but identical weight on a torsion balance, with the horizontal bar along the East-West direction (Fig. 3.7a). As the earth rotated, two forces acted on each mass, (i) a gravitational attraction that is equal, since the weights of the masses are equal and (ii) a *centrifugal force* due to the earth's rotation. If the centrifugal force on mass *A* was greater or less than on mass *B*, this would indicate that their inertial masses differed. The bar would rotate until the torsional force in the suspending wire compensated for the inequality in the centrifugal forces.* Eötvös never observed such a rotation of the bar and concluded that the inertial and gravitational masses of the bodies were identical to within one part in about 10^8.

Fig. 3.7 Eötvös and Dicke experiments.

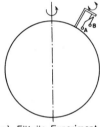

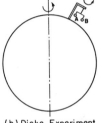

(a) Eötvös Experiment (b) Dicke Experiment

This experiment now has been refined by R. H. Dicke and his co-workers. They suspended their weights with the bar in a North-South direction. As the earth turns about its axis, mass *A* might be attracted more or less strongly toward the sun than mass *B*. One should then observe a diurnal effect with the balance arm first swinging in one direction, then in the other. No such effect was observed and the conclusion is that the gravitational and inertial masses are identical to within about one part in 10^{11}. The advantage of this kind of experiment is that it is dynamic, not static, and has a definite expected periodicity for which we can look (Ro64b).

We may now ask whether the gravitational mass of matter is the same as that of antimatter. If there existed galaxies composed of antimatter, would they attract or repel a galaxy consisting of matter? L. I. Schiff (Sc58a) has given a tentative

---

* This experiment works best, roughly halfway between the equator and poles, say, Budapest or Princeton.

answer to such questions. He points out that many atomic nuclei emit virtual positron electron pairs. This means that part of the time a fraction of the total nuclear energy is to be found in the form of an electron and a matched positron. Such a pair of particles is continually formed and reassimilated and is never actually emitted.

Schiff computes that if the positrons had a negative gravitational mass, then the ratio of inertial to gravitational mass would be affected for a number of substances for which the electron-positron virtual pair formation is a major effect. The ratio of the two kinds of masses would then be different from unity by about 1, 2, and 4 parts in 10^7 for aluminum, copper, and platinum, respectively. Experiments with such substances have been performed, and inequalities of this size are ruled out by the Eötvös and Dicke experiments. It follows that matter and antimatter ought to have gravitational masses of the same sign and that galaxies interacting gravitationally with antigalaxies cannot be distinguished on dynamical grounds.

Note that we have really only shown that the inertial and gravitational mass have the same sign for positrons as for electrons. But we actually know from dynamical experiments in magnetic fields that the inertial mass of the positron equals that of the electron, so that our previous conclusion should follow at once: Matter and antimatter both have positive mass.

There is one difficulty with this argument. As Schiff himself recognized, we are not absolutely certain that virtual electron-positron pairs behave exactly like real pairs. Could it be that the gravitational mass of a real positron differs from that of a virtual positron? Unfortunately, we will not be absolutely sure until we make a direct measurement of the positron's motion in a gravitational field.

3:8 INERTIAL FRAMES OF REFERENCE—THE EQUIVALENCE PRINCIPLE

We noted earlier that Newton's laws of motion held only when the motion is described in coordinates that refer to a frame of reference that is either fixed, or is moving at constant velocity with respect to the distant galaxies. Such a reference frame is known as an *inertial coordinate system.*

Several perplexing questions arise when we try to understand the significance of these frames of reference. They can be described by some simple experiments.

(1) Suppose that a man were blindfolded and placed on a merry-go-round. He could determine quite accurately whether he was being spun around because he would be able to feel the centrifugal force acting on him when the merry-go-round was moving. If he adjusted the mechanism until he felt no centrifugal force

he would find, on taking off his blindfold, that the merry-go-round was stationary with respect to the distant galaxies.

(2) A blindfolded man placed in a rocket in interstellar space could adjust his controls until he felt no forces on himself. On taking a closer look, he would find that he had adjusted the engine to give zero thrust. He might find that he was moving at constant velocity with respect to the distant galaxies. Alternately, however, he might find that he had strayed into the vicinity of a star and was freely falling toward it! Einstein first postulated that freely falling, nonrotating coordinate frames are fully equivalent to Newton's inertial frames that move at constant velocity with respect to the distant galaxies. All laws of physics have precisely the same form in both types of frames. This *equivalence principle* will prove to be very useful in sections 3:9 and 5:13.

When we talk about a motion with respect to the distant galaxies, we really mean a motion with respect to the mean velocity of all galaxies at very large distances. Galaxies are receding in all directions but, as far as we can tell, there always exists a local frame of reference in which the motions of distant galaxies appear symmetrical, no matter which direction we look.

This suggests that perhaps the local frame of zero acceleration is determined by the distribution of the galaxies in the universe. Just how this determination comes about is a basic unanswered question of the theory of gravitation. The thought that the overall distribution of mass within the universe should determine a local inertial framework is due to E. Mach and is sometimes called *Mach's principle*. There are many related questions all of which are involved in the same basic thought: "Is the inertial mass of a body determined by the distribution of matter in the universe? Is the gravitational constant determined by the distribution of the distant galaxies? As a result, would the value of the gravitational constant change with time, as the galaxies recede from each other? Are the atomic constants of physics related to the large-scale structure of the universe?" As yet there are no answers to these fundamental questions.

3:9 GRAVITATIONAL RED-SHIFT AND TIME DILATION

Einstein's *principle of relativity* (5:1) states that mass and energy are related in such a way that any stationary mass m has an equivalent energy mc^2 associated with it (Ei07). Einstein showed that the separate laws of conservation of mass and of energy merged into a more general *conservation of mass-energy*. This predicts a gravitational red shift for radiation emitted at the surface of a star. Consider two particles, an electron and a positron, at rest at a very great distance from a star. The rest mass of each particle is m_0. If the particles fall in toward the star's surface each one acquires a total mass-energy

$$E \equiv m_r c^2 = \left(m_0 c^2 + \frac{m_0 MG}{r} \right) = m_0 c^2 \left(1 + \frac{MG}{rc^2} \right) \tag{3–49}$$

at distance r from the star. The second term in the parantheses represents the conversion of potential into kinetic energy. Now let the two particles be deflected without loss of energy or momentum, so that they collide head-on, and annihilate. Two photons, each with frequency

$$\nu_r = \frac{m_r c^2}{h} \tag{3–50}$$

will be formed in this process. These photons are now permitted to escape from r, but through reflections from stationary mirrors—which produce no frequency shifts—we can make them collide again at a large distance from the star.

In this collision they can form an electron-positron pair. If energy is conserved, then the frequency ν_0 at a large distance from the star must again be

$$\nu_0 = \frac{m_0 c^2}{h} \tag{3–51}$$

Otherwise there would be either too much or too little energy to recreate a positron-electron pair at rest. Hence

$$\nu_0 = \frac{\nu_r}{1 + MG/rc^2} \tag{3–52}$$

The frequency at a large distance from the star is less than the emitted frequency. For a star like the sun $M \sim 2 \times 10^{33}$ g, the radius $R \sim 7 \times 10^{10}$ cm, and $MG/Rc^2 \sim 2 \times 10^{-6}$ at the sun's surface. For a neutron star whose mass would be about the same, but whose radius would be 10^5 times less, the fractional frequency shift

$$\frac{\Delta\nu}{\nu_r} = \frac{\nu_0 - \nu_r}{\nu_r} = -\frac{MG}{rc^2}\left[1 + \frac{MG}{rc^2} \right]^{-1} \tag{3–53}$$

becomes comparable to unity, and the frequency shift $\Delta\nu$ becomes comparable to the frequency itself!

We will see in the next section that the frequency of electromagnetic waves can give a very accurate measure of time and can therefore be used as a clock. Such a clock, placed in a strong gravitational field, would therefore run more slowly. Quite generally, the rate at which a clock runs is determined by the potential $\mathbb{V}(r)$ at the position r, of the clock. The period P of this clock measured by an observer outside the potential field, that is by an observer located at $\mathbb{V} = 0$, appears to be

$$P_0 = P_r\left(1 - \frac{\mathbb{V}}{c^2}\right) \tag{3-54}$$

In section 3:11 we outline an experiment that has been proposed to measure this *time dilation,* which leads to a delay in the arrival, at the earth, of pulsar pulses that have passed close to the sun.

3:10 MEASURES OF TIME

In describing the orbital motions of planets about the sun, we have obtained expressions for position as a function of time. But how is this parameter, time, actually measured?

There are a number of ways (see D. H. Sadler Sa68a) of measuring time and it is interesting to see how these methods interrelate—some rather basic questions of physics are involved. Let us first describe some imaginary clocks that could be constructed. They may not be practical but they should work in principle.

First Clock.

Take an amount of tritium H^3 that beta decays into the helium isotope He^3. If the tritium is kept at a temperature around $10°K$, the helium will diffuse out as it is formed. We weigh the tritium. When the mass has dropped to half its initial value, we say that a time of one unit, NT, has elapsed. We could set up a clock that struck each time the remaining mass was reduced by a factor of two.

Second Clock.

Take a quantity of the cesium isotope Cs^{133}. It has a transition between two hyperfine levels of the ground state. We measure the frequency of the radiation (radio wave) emitted in this transition. The period of this electromagnetic wave can serve as a unit of time, AT.

Third Clock.

We set up a telescope that is always pointed at the local zenith. Each time a given distant galaxy reappears exactly in the center of the telescope's field of view, we say that one unit of time, UT, has elapsed.

Fourth Clock.

We note the plane described by Jupiter as it rotates about the sun. We mark the instant that the earth crosses this plane in its motion about the sun. The earth crosses the plane twice per orbit; once it does so from North to South, and the next time from South to North. If we define the interval between successive N to S crossings as one unit of time, ET, we have still another means of measuring time.

We call these measures — NT, AT, UT and ET — *nuclear time, atomic time, universal time,* and *ephemeris time.* As we have chosen to define them here, the units of time, respectively, correspond to ~ 12 years, $(9, 192, 631, 770)^{-1}$ seconds, 1 1 day, and 1 year.

The basic differences between the clocks are these: The first clock uses beta decay, a weak interaction, as its basic mechanism. The second clock uses an electromagnetic process to measure time. The third clock uses the earth's rotation to measure time; this is an inertial process. Finally, the fourth clock makes use of a gravitational force to measure time.

Since each of these clocks depends on quite different physical processes, we worry that they might not measure the same "kinds" of time. There is no reason, for example, why atomic time and ephemeris time as defined above should describe intervals having a constant ratio. At present the ratio of these time units is about $3 \times 10^{17} : 1$. Will this ratio be the same some 10^9 y from now? Or does the strength of the gravitational field, or of the weak interactions, change after years in such a way that one of these clocks becomes accelerated relative to the other?

We can test this question experimentally and such tests have, in fact, been proposed. Their results would be of great importance to cosmology. For, in order to understand the nuclear history of the universe and the formation of chemical elements, we have to know how nuclear reaction rates in stars may have been affected by the overall evolution of the universe over the past aeons. This will appear more clearly after the synthesis of nuclei in stars has been discussed in Chapter 8.

The important point to realize is that we have enumerated four quite different ways of defining time.* The last two are related if the general theory of relativity holds true; and their interrelationship becomes a test for theories of gravitation.

In practice, the intercomparison of these clocks is difficult. Planetary perturbations make the earth's orbit about the sun irregular. Earthquakes and other

---

* Actually, there are five ways. Nuclear β and α decay rates are based on weak and strong nuclear forces, respectively. We will show in Chapter 10 that these two kinds of clock apparently have run at identical rates over the past few æons.

disturbances affect the rotation rate of the earth. An incomplete understanding of these effects makes it hard to compare the UT and ET rates with time measured by atomic clocks. Eventually, however, such practical difficulties should be overcome and an intercomparison of time scales will be possible.

3:11   USES OF PULSAR TIME

Many pulsars emit signals with a periodicity that appears to vary by less than one part in 10^8 over an interval of a year. These signals can therefore be used to define a time scale. The mechanism of the clock is not yet understood, but is thought to involve the rotational period of neutron stars. In any case, its regularity allows us to put it to scientific use. For many purposes we do not need to know how a clock works as long as its accuracy can be verified.

Counselman and Shapiro have listed a number of interesting gravitational effects that can be studied using pulsars (Co68).

(a) The orbit of the earth could be determined more precisely than it is known now. The pulsar emission would act like a "one-way" radar. Counting pulse rates from different pulsars would allow us to measure the instantaneous velocity of the earth, relative to some arbitrarily defined inertial frame. Integrating these velocities over a series of time intervals would yield the earth's position as a function of time, that is, the shape of the orbit and its orientation.

Such measurements can also yield data on the positions and masses of the outermost planets. Their motions affect the position of the solar system's *barycenter* and hence also the orbit of the earth. The periodicity of these effects is determined by the orbital periods of the planets and we should find corresponding periodic variations in the pulse counts (As71).

(b) A pulsar located near the ecliptic plane will appear close to the sun once each year. When the light pulses pass very close to the limb of the sun, they should be slowed down because all clocks are slowed by the presence of a strong gravitational field and because the speed of light measured locally at the sun would still appear to be c. The arrival time of pulses at the earth will therefore be delayed by an amount of order 100 μsec, depending only on how close to the sun the radiation passes. By keeping track of the arrival times we can compute the delay and see whether the measured delay agrees with the predictions of relativity theory. To do this, we have to first correct for the time delay due to the relatively high index of refraction of the solar corona. This is possible because the delay due to refraction is proportional to v^{-2}, while the gravitational delay is independent of frequency v. Several pulsars pass within $1°$ of the sun and should be suitable objects for such tests.

(c) Since pulsars are located within the galaxy, the shear motion of stars in the galaxy would yield an acceleration relative to the sun that could be detected by keeping track of pulse arrival times. The differential rotation of the galaxy could therefore be mapped very accurately.

3:12 GALACTIC ROTATION

The mass of the Galaxy is not distributed evenly. Much of it is concentrated near the nucleus. For this reason, stars near the Galactic center tend to have angular velocities $\dot{\theta}(r)$ appreciably larger than stars at greater distances, that is, $d\dot{\theta}/dr < 0$. Suppose for simplicity that all stars have idealized circular orbits about the Galactic center. Let the sun be at distance r_s from the center. Relative to the sun, matter at Galactic longitude l, and at distance r from the center C has an approach velocity $v(r, l)$ along the line of sight (Fig. 3.8).

Fig. 3.8 Notation used for discussion of differential rotation.

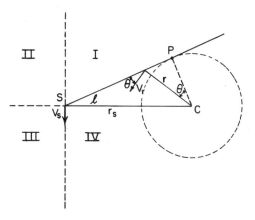

$$v(r,l) = [- r_s\dot{\theta}(r_s)\sin l + r\dot{\theta}(r)\cos\theta] = [\dot{\theta}(r) - \dot{\theta}(r_s)]\, r_s \sin l \qquad (3\text{–}55)$$

where the simple form of the expression on the extreme right is due to the fact that $r\cos\theta = r_s \sin l$, as evident from Fig. 3.8. We note from (3–55) and from the fact that $d\dot{\theta}/dr < 0$, that $v(r,l)$ is positive in the quadrants I and III so that stars and gas in these directions should appear to approach, and their spectra should be blue-shifted. In quadrants II and IV stellar spectra should appear red-shifted.

This, in fact, is what is observed. In 1927 the Dutch astronomer Oort was able to use this evidence to prove that stars in our Galaxy are in *differential rotation* about the Galactic center (Oo27a,b).

At any given Galactic longitude l, the highest velocity should be observed at point P, where the line of sight is tangential. By noting the maximum velocity at any given elongation l, we can construct a model of the Galaxy giving both its mass distribution and distance of the sun from the Galctic center. Present results yield $r_s \sim 9.5$ kpc ± 1.5 kpc (vdBe68).

Differential rotation tends to shear aggregates of gas and dust as they orbit the Galactic center. For some time this effect was considered responsible for the appearance of spiral arms in some galaxies. However, more recently, Lin (Li67) has suggested that the spiral structure represents a local increase in density and that this enhanced density spiral travels around the galaxy like a wave, at a "pattern" velocity different from that of the speed of the stars involved. For the Galaxy this speed is about 13.5 km sec^{-1} times the distance from the center measured in kiloparsecs. At our distance from the center, this would be 135 km sec^{-1}, while the Galactic rotation (velocity of the stars) is ~ 250 km sec^{-1}. In contrast to this, the stars in the bar of a barred spiral galaxy do appear to move with the pattern, that is, similar to a solid body (see Problem 3–13).

3:13 SCATTERING IN AN INVERSE SQUARE LAW FIELD

When a meteorite approaches the earth, its orbit can become appreciably changed. Similarly, a comet passing close to Jupiter can be given enough energy to escape the solar system. In both cases the smaller object is scattered or deflected by the larger body. For a particle initially approaching from direction $\theta_\infty - \theta_0$ (Fig. 3.9), the orbital equation is given by equations 3–34 and 3–40.

$$\frac{1}{r} = \frac{MG}{h^2}\left[1 + \sqrt{1 + \frac{2\mathscr{E}h^2}{M^2 G^2}}\cos(\theta - \theta_0)\right] \tag{3–56}$$

At large distances from the scatterer, the asymptotic motion is along directions (see equation 3–42)

$$\cos(\theta_\infty - \theta_0) = -\left(1 + \frac{2\mathscr{E}h^2}{M^2 G^2}\right)^{-1/2} = -\frac{1}{e}, \quad r \to \infty \tag{3–57}$$

This has solutions for two values of $|\theta_\infty - \theta_0|$, one corresponding to the incoming and the other to the scattered asymptotic direction. The angle through which the object is deflected is $\textcircled{H} = 2(\theta_\infty - \theta_0) - \pi$. We see that

$$\sin\frac{\textcircled{H}}{2} = -\cos(\theta_\infty - \theta_0) = \frac{1}{e} \tag{3–58}$$

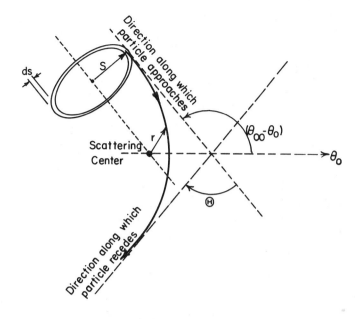

Fig. 3.9 Scattering in an attractive inverse square law field.

We note that h, twice the area swept out per unit time, is

$$h = v_0 s$$

where s is the *impact parameter* (Fig. 3.9) and v_0 is the *approach velocity* of the scattered particle at a large distance, $r \to \infty$.

$$\text{Since } \mathscr{E} = \frac{v_0^2}{2}, \quad h^2 = 2\mathscr{E}s^2 \tag{3-59}$$

and

$$\sin \frac{\text{\textcircled{H}}}{2} = \left[1 + \left(\frac{2s}{MG} \right)^2 \right]^{-1/2} \tag{3-60}$$

This leads to

$$\cot \frac{\text{\textcircled{H}}}{2} = \frac{2\mathscr{E}s}{MG} \tag{3-61}$$

If any object having an impact parameter between s and $s + ds$ is scattered into an angle between $\text{\textcircled{H}}$ and $\text{\textcircled{H}} + d\text{\textcircled{H}}$, we say that the *differential cross section*

$\sigma(\Theta)$ for scattering is given by

$$2\pi s\, ds \equiv -\sigma(\Theta)\, d\Omega = -2\pi\,\sigma(\Theta)\sin\Theta\, d\Theta \qquad (3\text{–}62)$$

In this equation the expression on the left represents the area of a ring through which all particles approaching from a given direction have to flow if they are to be scattered into the solid angle $d\Omega$ enclosed between two cones having half angles Θ and $\Theta + d\Theta$, respectively. The expression on the right gives the solid angle between these two cones multiplied by the differential cross section. The differential cross section is therefore just a parameter that assures conservation of scattered particles. The negative sign appears because an increase in the impact parameter s results in a decreasing scattering angle Θ. The differential cross section is proportional to the probability for scattering into an angle between Θ and $\Theta + d\Theta$, since $2\pi s\, ds$ (see equation 3–62) is the probability for encounter at impact parameter values between s and $s + ds$.
We now can rewrite (3–62) in the form

$$\sigma(\Theta) = \frac{s\, ds}{\sin\Theta\, d\Theta} \qquad (3\text{–}63)$$

which, together with expressions 3–60 for s yields

$$\sigma(\Theta) = \frac{1}{4}\left(\frac{MG}{2\mathscr{E}}\right)^2 \frac{1}{\sin^4(\Theta/2)} \qquad (3\text{–}64)$$

3:14   STELLAR DRAG

If a high velocity star moves through a surrounding field of low velocity stars, it experiences a drag because it is slightly deflected in each *distant encounter*. We can compute this drag in an elementary way through the use of the scattering theory derived above.

First, we note that the star's velocity loss along the initial direction of approach to the scattering star is

$$\Delta v = v_0\,(1 - \cos\Theta)$$

where v_0 is the approach velocity relative to the scattering center at large distances. This is not an overall velocity loss; just a decrease in the component along the direction of approach. The change in momentum is $\mu\Delta v$ where, again, μ is the reduced mass. The force on the high velocity star, opposite to its initial direction of motion, therefore, is

$$F = \sum_i \frac{\mu_i \Delta v_i}{\Delta t} \tag{3-65}$$

where Δt is the time during which a change Δv_i takes place and the summation is taken over all stars i that are being encountered during this time interval. This summation can be replaced by considering a gas of stars with number density n. In terms of the probability or cross section for scattering into angle Ⓗ, at any given encounter, the force becomes

$$F = 2\pi \mu v_0^2 n \int_{Ⓗ_{\max}}^{Ⓗ_{\min}} (1 - \cos Ⓗ) \, \sigma(Ⓗ) \sin Ⓗ \, dⓀ \tag{3-66}$$

This assumes that all the deflections due to interactions with individual stars are small, and that the forces along the direction of motion add linearly. Using equation 3-62 we have

$$F = 2\pi \mu v_0^2 n \int_{s_{\min}}^{s_{\max}} (1 - \cos Ⓗ) \, s \, ds \tag{3-67}$$

Instead of integrating for all stars, we integrate the impact parameter s over all possible values for a single star and then multiply by the stellar density n. This is an equivalent procedure because the probability of encountering a star at impact parameter s is proportional to s and to n. The extra factor v_0 that appears in the expression takes account of the increasing number of encounters, per unit time, at large velocities. If we set $\theta_0 \equiv 0$ and $\theta_\infty \equiv \theta$, then

$$- \cos Ⓗ = \cos 2\theta \equiv \frac{1 - \tan^2\theta}{1 + \tan^2\theta} \tag{3-68}$$

but (see Fig. 3.9),

$$\cot \frac{Ⓗ}{2} \equiv \tan \theta = \frac{2\mathscr{E}s}{MG} = \frac{s v_0^2}{MG} \equiv \alpha s \tag{3-69}$$

so that

$$F = 2\pi \mu v_0^2 n \int s \frac{2}{1 + \tan^2\theta} \, ds \tag{3-70}$$

and

$$F = 4\pi \mu v_0^2 n \int \frac{s \, ds}{1 + \alpha^2 s^2} \tag{3-71}$$

$$F = \frac{4\pi \mu v_0^2 n}{2\alpha^2} \ln(1 + \alpha^2 s^2) \Bigg]_{s_{\min}}^{s_{\max}} \tag{3-72}$$

We define a slowing down time, or *relaxation time*, τ

$$\tau \equiv \frac{\mu v_0}{F} \tag{3-73}$$

In this calculation we have assumed that the star is moving through an assembly of stationary "field" stars. As long as the random motion of these stars is low compared to v_0, equation 3–72 holds quite well. However, when the random stellar velocities approach v_0, the particle can alternately be accelerated or slowed down by collisions and the above derivation no longer holds. For the sun, moving with a velocity of $v_0 = 20 \text{ km sec}^{-1}$ through the ambient star field, $\alpha = 3 \times 10^{-14} \text{ cm}^{-1}$, $n \sim 10^{-56} \text{ cm}^{-3}$, and $\mu \sim 10^{33}$ g. If $s_{max} \sim 10^{19}$ cm, roughly the mean separation of stars,* and s_{min} is much smaller, then $F \sim 10^{18}$ dyn. However, even with this large force, $\tau \sim 10^{21}$ sec, that is, far greater than the estimated age of the universe. This large value of τ is disconcerting because it is symptomatic of a general problem in stellar dynamics. We find such aggregates as globular clusters to be in configurations close to those we would expect in thermodynamic equilibrium. This would mean that the stars must interact quite strongly to transfer energy to each other; and yet the above mechanism will not accomplish this at anywhere near a satisfactory rate! Neither will other mechanisms of the same general class. The interaction of these stars must be dominated by some other process that we do not yet understand. We will discuss this again in section 3:16. However, we might note that interaction of stars with gas clouds or clouds of stars produces a larger effect than that of individual stars encounters (Sp51a). If the mass of the cloud is $M \sim 10^6 \ M_\odot$ and $n \sim 10^{-65} \text{ cm}^{-3}$, F increases by 10^3, and τ decreases by 10^3. Here s_{max} might be chosen $\sim 10^{22}$ cm.

Collisions need not always act to slow particles down. When stars in the plane of the Galaxy interact with the much more massive clouds of gas, they can actually become accelerated to high velocities. In Table A.6 we showed that, relative to the sun, older stars have higher root mean squared random velocities than younger stars. This may be due to collisions with such clouds. As will be shown in Chapter 4 an assembly of bodies tends to arrange itself in such a way that translational energies are equal (equipartition of energy). The massive clouds, therefore, tend to pass some of their energy on to the less massive stars and, in so doing, accelerate them to velocities higher than v_c, the velocity of the clouds.

A quite different class of problems in which the above calculations are useful deals with charged particles. The inverse square law electrostatic forces allow us to derive equations quite similar to (3–72) and (3–73) and we can compute the

---

* When s_{max} is much larger than the mean separation, encounters begin to overlap in time, and (3–72) becomes an overestimate of F because the effects of individual stars will tend to cancel through symmetry in their distribution (Ch43).

$$\therefore \ s_{max} \sim n^{-1/3} \tag{3-74}$$

electrostatic drag on fast electrons traveling through the interstellar medium and on charged interstellar or interplanetary dust grains moving at typical velocities of ~ 10 km sec^{-1} through a partially ionized medium. This effect plays a major role especially in the dynamics of dust grains moving through the interstellar medium. In section 6:16 we will also see that the distant collisions of electrons and ions are described by equations like (3–67) and that the opacity or emissivity of an ionized plasma can be computed making use of these equations. The radio emission from hot ionized interstellar gas can then be directly related to the plasma density, or rather to the collision frequency in a line-of-sight column through the cloud.

3:15 VIRIAL THEOREM

The theorem we will prove here again is statistical. It describes the overall dynamic behavior of a large assembly of bodies, rather than the precise behavior of any given individual body belonging to the assembly.

Consider a system of masses m_j at positions $\mathbf{r}_j$. Let the force on m_j be $\mathbf{F}_j$. We now write the identity

$$\frac{d}{dt}\sum_j \mathbf{p}_j \cdot \mathbf{r}_j = \sum_j \mathbf{p}_j \cdot \dot{\mathbf{r}}_j + \sum_j \dot{\mathbf{p}}_j \cdot \mathbf{r}_j \tag{3–75}$$

$$= 2\mathbb{T} + \sum_j \mathbf{F}_j \cdot \mathbf{r}_j \tag{3–76}$$

where $\mathbb{T}$ is the kinetic energy of the entire system and the time derivative of the momentum $\dot{\mathbf{p}}_j$ is equal to the force $\mathbf{F}_j$. For the moment we do not identify the left side of the equation with any physically interesting quantity. Taking the time average of both sides, we obtain

$$\frac{1}{\tau}\int_0^\tau \frac{d}{dt}\sum_j \mathbf{p}_j \cdot \mathbf{r}_j \, dt = \langle 2\mathbb{T} + \sum_j \mathbf{F}_j \cdot \mathbf{r}_j \rangle \tag{3–77}$$

where the brackets denote a time average. A particularly interesting situation concerns a bound system in which each member of the assembly remains a member for all time. In this situation all the $\mathbf{r}_j$ values must remain finite since no particle escapes from the system, and all $\mathbf{p}_j$ values must remain finite because the total energy of the system is finite.

Since $\sum_j \mathbf{p}_j \cdot \mathbf{r}_j$ remains finite, the integral of its derivative must also remain finite for all time. This means that the left side of equation 3–77 consists of a finite quantity divided by τ that can be made arbitrarily large—if we average over a very large or infinite period. The left side of equation 3–77 therefore approaches zero and we can set

$$\langle 2\mathbb{T} \rangle + \langle \sum_j \mathbf{F}_j \cdot \mathbf{r}_j \rangle = 0 \tag{3-78}$$

If the force is derivable from a potential, this equation becomes

$$\langle 2\mathbb{T} \rangle - \langle \sum_j \nabla\mathbb{V}(r_j) \cdot \mathbf{r}_j \rangle = 0 \tag{3-79}$$

where $\mathbb{V}(r_j)$ is the potential energy of mass m_j at position r_j. The force in such a situation is a function of position only and can be written as the negative gradient ∇ of the potential energy:

$$\mathbf{F}_j = -\nabla\mathbb{V}(r_j) \tag{3-80}$$

If the potential is proportional to r^n, the gradient lies along the radial direction and

$$\sum_j \nabla\mathbb{V}(r_j) \cdot \mathbf{r}_j = \sum_j \frac{\partial\mathbb{V}(r_j)}{\partial r_j} r_j \tag{3-81}$$

Calling the total potential energy of the entire assembly $\mathbb{V}$, we obtain

$$\langle \mathbb{T} \rangle = \frac{n}{2}\langle \mathbb{V} \rangle \qquad \mathbb{V} \equiv \sum_j \mathbb{V}(r_j) \qquad -2 < n \tag{3-82}$$

This relation runs into difficulty for $n < -2$, since the total energy $\langle \mathbb{T} \rangle + \langle \mathbb{V} \rangle$ would then be positive, indicating that the system would no longer be bound. For an inverse square law force, as in gravitation or electrostatics, the potential goes as the inverse first power, $n = -1$, and

$$\langle \mathbb{T} \rangle = -\frac{1}{2}\langle \mathbb{V} \rangle \tag{3-83}$$

This theorem is of great importance and finds many applications in astrophysics. For example, it provides the only current estimate for the mass of clusters of galaxies. That estimate is obtained by observing the spread in radial *Doppler velocities* among different galaxies in the cluster, which gives the mean kinetic energy per unit mass. Equation 3–83 then yields the mean potential energy per unit mass, and if a typical cluster diameter is known from the cluster's distance and from the angle it subtends in the sky, we can obtain a rough estimate of the total cluster mass on the assumption that

$$\frac{\mathbb{V}}{M} \sim \frac{MG}{R} \tag{3-84}$$

Here M is the cluster mass and R is some weighted cluster radius, somewhat smaller than the observed radius of the cluster. The estimated cluster mass would normally be in error (that is, too high) by a factor less than ~ 2 if the actually observed cluster radius is used in equation 3–84.

An interesting problem occurs when we measure the masses of clusters of galaxies making use of the virial theorem. The masses of individual galaxies within the cluster can be determined by the method of Problem 3–9. From these we can compute the potential energy of the entire cluster, if the cluster dimensions are computed from the apparent diameter and red-shift distance. An independent estimate of the potential energy, however, is obtained from (3–83) if the random velocities of the individual galaxies are taken to compute $\mathbb{T}$. To do this, we note the variations in red shifts from galaxy to galaxy and estimate the actual random velocities. Strangely the results of using (3–83) always give values of $\langle \mathbb{T} \rangle$ and, hence, $\langle \mathbb{V} \rangle$ that are about an order of magnitude higher than the total potential energy computed on the basis of individual galactic masses. We conclude that either (i) there is a lot of undetected matter in clusters, or (ii) all the clusters are breaking up, or (iii) we do not understand dynamics on such a large scale. For example, we might ask whether galaxies in a cluster could not participate in the overall cosmic expansion. Could this account for the apparent disruption of the clusters? The answer seems to depend on factors we do not yet know. If the bulk of the mass-energy in the universe is in the form of matter, cosmic expansion seems to play a minor role; but if gravitational or electromagnetic radiation and neutrinos have an energy density exceeding that of matter, the cosmic expansion can make significant contributions (No71). Problem 4–5 and the discussion following it treat the cluster problem from an observational viewpoint.

3:16 STABILITY AGAINST TIDAL DISRUPTION

When a swarm of gravitationally bound particles, having a total mass m, approaches too close to a massive object M, the swarm tends to be torn apart. The same thing can happen to a solid body held together by gravitational forces, when it approaches a much more massive object.

The reason for this is quite simple. If we consider that the center of mass of the swarm is at a distance r from the mass M, and is falling straight toward it, then its acceleration toward M is $-MG/r^2$. Let r' be the swarm radius. A particle P_0 (Fig. 3.10), at the surface of the swarm nearest to M, would be accelerated at a rate $-MG/(r-r')^2$ towards M, if it were not for the fact that a gravitational attraction from the center of the swarm tends to accelerate it away from M at a rate mG/r'^2. In order for the particle to be pulled steadily away from the swarm, we must have the condition

$$MG\left(-\frac{1}{r^2} + \frac{1}{(r-r')^2} \right) > \frac{mG}{r'^2}$$

(3–85)

3:16

Fig. 3.10 A swarm of gravitationally bound particles—stars, atoms, molecules—can be disrupted through close encounter with a massive object M.

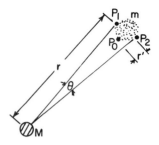

Expanding the expression on the left and keeping only terms down to first order in r', we obtain

$$\frac{2M}{r^3} > \frac{m}{r'^3} \tag{3-86}$$

Similarly for a swarm in a perfectly circular orbit about M, disruption occurs when

$$\frac{3M}{r^3} > \frac{m}{r'^3} \tag{3-87}$$

PROBLEM 3-3. Derive the result (3–87). In doing this, it is helpful to think of the swarm as moving without rotation about its center, and to consider its center of mass as having a *centrifugal repulsion*

$$F_c = r\,\dot{\theta}^2 \tag{3-88}$$

away from M. This is different from the "repulsion" $(r - r')\,\dot{\theta}^2$ at P_0.

The precise ratios of the masses M and m will therefore vary with differing orbits and the rotation of m will also play a role in determining its stability. What is important to note, however, is that the density of the swarm is more important a consideration than its actual mass, or size, taken individually.

There is a second effect that also plays an important role. Again, consider a direct infall. Here points P_1 and P_2 (Fig. 3.10) would be accelerated radially toward M and would tend to converge. The effective acceleration of P_1 and P_2 relative to each other would be roughly

$$2\,\frac{MG}{r^2}\,\frac{r'}{r} = \frac{2MGr'}{r^3}$$

due to this effect taken by itself. This is important whenever it is larger than the acceleration mG/r'^2 due to the mass of the swarm itself, that is, when (3–86) holds. There exists a lateral compression, therefore, that accompanies the tidal disruption and tends to concentrate the swarm, while the tidal forces attempt to tear it apart. What actually happens under these combined effects will be better understood in terms of the Liouville theorem in Chapter 4.

It is now worth noting that there are two particular conditions where this kind of tidal disruption seems to play a leading role. First, comets that approach too close to the sun or even too close to the massive planet Jupiter have been observed to break up into two or more fragments, and the general nature of the tidal theory seems to be borne out.

Equally interesting is the effect that tidal disruption seems to have on globular clusters. Von Hoerner (vH057) has examined the orbits of these clusters statistically, and finds that they have orbits that draw them very close to the Galactic center. The massive Galactic nucleus then seems able to rob such clusters of loosely bound outer members. At the center of the cluster, where the density is largest, disruptive effects are then relatively small, while at the periphery, where m/r^3 is low, stars can be pulled away more readily.

We can now also see why the interaction of stars within a globular cluster may only play a limited role in determining the ultimate velocity distribution of stars in the cluster. The treatment of section 3:14, and the very long star encounter relaxtion time τ predicted by equation 3–73 may not give a true picture of the actual evolution of clusters into the well-defined, compact, spherical aggregates we observe. Interaction with the Galactic nucleus must have an appreciable, perhaps even a dominant, influence on this distribution. We will touch on this problem again in section 4:21.

In section 1:8 we had said that some of the dwarf galaxies of the Local Group can never have been very close to either the Galaxy or $M31$. We can conclude this directly from criteria like (3–86) and (3–87).

PROBLEMS

3–4. The orbital period for the earth moving about the sun is given by equation 3–47. The distance of the sun can be obtained most accurately by the radar method described in section 2:1. Averaged over the earth's eccentric orbit it has a mean value of 1.5×10^{13} cm. Assuming the earth's mass, $m_E \ll M_\odot$, show that the sun's mass $M_\odot = 2.0 \times 10^{33}$ g.

3–5. A radar signal reflected from the moon returns 2.56 sec after transmission. The speed of light is 3.00×10^{10} cm sec^{-1}. Assume the period of the moon to be roughly 27.3 days. Find the mass of the earth assuming the moon's mass is small compared to the earth.

Note: In this way we can determine the mass of any planet with a moon. When a planet has no moon, its mass is determined by the perturbations it produces on the orbits of nearby planets. Such a calculation is quite time-consuming, but introduces no essentially new physical concepts. The calculations proceed within the framework of Newtonian dynamics.

3–6. Since the moon and the earth revolve about a common center of mass, the apparent motion of Mars has a periodicity of one month superposed on its normal orbit. The distance of the moon is $D \sim 3.8 \times 10^5$ km. The distance of Mars at closest approach is $L \sim 5.6 \times 10^7$ km. The apparent displacement of Mars over a period of a half month is ~ 34 sec of arc. What is the mass of the moon?

3–7. A meteor approaches the earth with a speed v_0, when it is at a very large distance from the earth. Show that the meteor will strike the earth, at least at grazing incidence if its impact parameter s is given by

$$s \lessgtr [R^2 + 2MGR\, v_0^{-2}]^{1/2}$$

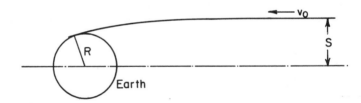

Fig. 3.11 Impact of a meteorite or a cloud of meteors on the earth's atmosphere.

3–8. If a cloud of meteors approaches the earth at relative speed v_0, show that the rate of mass capture is $\pi(R^2 v_0 + 2MGR/v_0)\,\rho$, where ρ is the mass density of the cloud. Both Problems 3–7 and 3–8 neglect the sun's influence on the meteors. This problem might be more appropriately done after reading section 4:5.

3–9. A disk-shaped rotating galaxy is seen edge on. By Doppler-shift spectroscopic measurements we can determine the speed V with which the stars near the edge of the galaxy rotate about its center. Show that the mass of the galaxy in terms of the observed velocity is $\sim V^2 R/G$. State the assumptions made. R is the radius of the galaxy.

3–10. In the vicinity of young clusters we occasionally see *runaway stars,* O or B stars that evidently were part of the cluster until recently but are receding rapidly. Blaauw (Bl61) has suggested that the runaways initially may have been part of binaries in which the companion exploded as a supernova, leaving only part of its mass behind. Suppose that the initial motion was circular, with initial orbital velocity v for the surviving star. If the initial mass of the companion was M, and the final mass after the explosion is only $M/10$, what will be the final velocity V of the runaway star at large distance from the explosion? The mass of the surviving star is m. Refer v and V to the system's center of mass.

3–11. A gravitationally bound body spins rapidly (but not at relativistic velocities). At what rotational velocity will it break up if its mass is m and radius is r? Assume the body remains spherical until breakup—even though this assumption will not normally hold.

3–12. Observations on the compact radio source 3C279, which is occulted by the sun once a year, show that radio waves are bent as they pass very close to the sun (Hi71). Show that this bending is a consequence of the equivalence principle.

3–13. In the barred spiral galaxy NGC 7479, Doppler-shift velocities indicate that the bar rotates as a solid body (Bu60), that is, with constant angular velocity ω along its entire length. Show that such a motion can take place when the distribution of mass is actually spherical (but only the bar consists of luminous stars) and when the mass $M(r)$ enclosed by a sphere of radius r increases sufficiently rapidly with increasing distance r from the galaxy's center (Fig. 1.10). Show that $dM(r)/dr$ is proportional to r^2 and to ω^2, in that case. Barred spirals may, however, be set up in ways altogether different from this process. We do not yet know! Aarseth (Aa60, 61) has discussed the stability of an actual cylindrical bar of stars.

ANSWERS TO SELECTED PROBLEMS

3–1.
$$\dot{r} = -\frac{1}{y^2}\theta\frac{dy}{d\theta} = -h\frac{dy}{d\theta}.$$

$$\dot{r} = -h\theta\frac{d^2y}{d\theta^2} = \frac{-h^2}{r^2}\frac{d^2y}{d\theta^2}.$$

Substituting in (3–30), we see that

$$\frac{d^2y}{d\theta^2} + y = \frac{MG}{h^2}.$$

Substitution of $y = B\cos(\theta - \theta_0) + MG/h^2$ satisfies the equation.

3–2. $m\ddot{r} = GmM_E/(R_E + H)^2$ at a height $H \ll R_E$.
If we take $M_E = \rho_E(4/3)\,\pi R_E^3$, where the symbols represent the earth's mass, density, and radius, we can estimate G from the measured acceleration, $G \sim g\,[\rho_E(4\pi/3)\,R_E]^{-1}$.

3–3. At the center of mass of the swarm, the centrifugal and gravitational forces are equal: $(r\dot{\theta})^2 = GM/r$. A particle, p, at the swarm's near surface, will experience a centrifugal acceleration away from M, smaller than that of the swarm's center by MGr'/r^3. It will also experience a stronger gravitational acceleration towards M, by

$$\frac{MG}{r^2}\left[-1 + \frac{r^2}{(r - r')^2}\right].$$

For disruption to occur, these accelerations must be stronger than mG/r'^2. Expanding this inequality gives

$$\frac{3M}{r^3} > \frac{m}{r'^3}. \tag{3–87}$$

This solution assumes no rotation of the swarm.

3–6. Let m be the lunar mass and M the terrestrial mass. The distance R of the earth from the center of mass is then given by

$$RM = (D - R)\,m$$

The apparent displacement of Mars is $2R/L$, where L is the distance to Mars. Hence $2R = 1.7 \times 10^{-4}\,L$. $R = 4.8 \times 10^3$ km and with $M = 6.0 \times 10^{27}$ obtained in Problem 3.2 we can now evaluate m as $\sim 7.4 \times 10^{25}$ g.

3–7. Call V the velocity the meteor has at grazing incidence, that is, when it hits the earth tangentially. Then this velocity is perpendicular to the radius vector R: We can therefore write conservation of angular momentum as

$$sv_0 = RV$$

Conservation of energy per unit meteor mass is

$$\frac{v^2}{2} = \frac{V^2}{2} - \frac{MG}{R}$$

Eliminating V from the equation one obtains the expression

$$s = \left(R^2 + \frac{2MGR}{v^2}\right)^{1/2}$$

Clearly all meteors with impact parameter less than s also can hit the earth. This gives rise to the desired expression.

3–8. The number of meteors hitting the earth per second is given by the density of meteors in space, times the volume of the cylinder of radius impact parameter *s* swept up in unit time:

$$\pi s^2 \cdot v_0 \cdot \rho$$

s is given in Problem 3–7.

3–9. Assume circular motion. The mass of the galaxy *M* acting on a star at its periphery is then given by the relation between kinetic and potential energy per unit mass of the star:

$$\frac{V^2}{2} = \frac{MG}{2R}$$

3–10. This problem is somewhat complex. Initially the linear momentum of each star about the center of mass is *mv*. If the star *M* explodes and leaves a remnant of mass $M/10$, the two remaining stars will move with a momentum $0.9mv$ relative to the initial center of mass. This gives rise to a kinetic energy of translation of the center of mass, and additional kinetic energy of rotation of these stars about their new center of mass. The new potential binding energy is only one tenth the initial binding energy; but if $m \gg M/10$, much of this decreased potential energy just goes into translational motion of the system. In that case the two surviving stars remain bound and $V \sim 0.9\,mv/(m + M/10)$. If $M/10$ is still large, the decrease in potential energy permits *m* to escape, and much of *V* then represents a true velocity of separation of the surviving stars.

3–11. The centrifugal force > gravity: $r\omega^2 > mG/r^2$; $\omega > \sqrt{mG/r^3}$.

3–12. Suppose an observer falls toward the sun in a space ship. Light rays passing by the sun enter the window of his cabin. The equivalence principle states that he should see the light moving in a straight line. But since he is falling toward the sun, this means that the rays must actually be following a parabolic path relative to a stationary observer.

3–13. For a body rotating as a solid with constant angular velocity, ω,

$$\frac{dv}{dr} = \frac{V}{r} = \omega\,.$$

But by (3–44)

$$v(r) = \left(\frac{M(r)\,G}{r}\right)^{1/2}.$$

Here $M(r)$ is the mass enclosed by the circle of radius *r*:

$$M(r) = \frac{\omega^2 r^3}{G}\,, \qquad \frac{dM(r)}{dr} = \frac{3\omega^2 r^2}{G}$$

4

Random Processes

4:1 RANDOM EVENTS

If a bottle of ether is opened at one end of a room, we can soon smell the vapors at the other end. But the ether molecules have not traversed the room in a straight line, nor in a single bound. They have undergone myriad collisions with air molecules, bouncing first one way, then another in a random walk that takes some molecules back into the bottle from which they came, others through a crack in the door, and others yet into the vicinity of an observer's nose where they can be inhaled to give the sensation of smell.

In general, molecules diffuse through their surroundings by means of two processes: (i) individual collision with other atoms and molecules and (ii) turbulent and convective bulk motions that involve the transport of entire pockets of gas. These, too, are the mechanisms that act to mix the constituents of stellar and planetary atmospheres. Both processes give rise to random motions that can best be statistically described.

In an entirely different context, think of a broadband amplifier whose input terminals are not connected to any signal source. On displaying the output on an oscilloscope, we would find that the trace contains nothing but spikes, some large, others smaller, looking much like blades of grass on a dense lawn. An exact description of this pattern would be laborious; but a statistical summary in terms of mean height and mean spacing of spikes can be provided with ease and may in many situations present all the information actually needed.

The spikes are the noise inherent in any electrical measurement. If we are to detect, say, a radio-astronomical signal fed into the amplifier, we must be able to distinguish the signal from the noise. That can only be done if the statistics of the noise are properly understood.

Again, consider a third situation, a star embedded in a dense cloud of gas. Light emitted at the surface of the star has to penetrate through the cloud if it is to reach clear surroundings and travel on through space. An individual photon may be absorbed, re-emitted, absorbed again, and re-emitted many times in succession. The direction in which the photon is emitted may bear no relation at all to the direction in which it was traveling just before absorption. The photon may then travel about the cloud in short, randomly directed steps until it eventually reaches the edge of the cloud and escapes. This *random walk* can be described statistically. We can estimate the total distance covered by the photon before final escape and, at any given time in its travel, we can predict the approximate distance of the photon from the star.

These three physically distinct situations can all be treated from a single mathematical point of view. In its simplest form each problem can be reduced to a random walk. We picture a man taking a sequence of steps. He may choose to take a step forward, or a step backward; but, for simplicity, we will assume that his step size remains constant. If the direction of each step is randomly determined, say by toss of a coin, the man will execute a random walk. The toss of the coin might tell him that his first step should be backward, the next forward, the next forward again, backward, backward, forward, and so on. After 10 steps, how far will the man have moved from his initial position? How far will he be after 312 steps or after 10,000,000? We cannot give an exact answer, but we can readily evaluate the probability of terminating at any given distance from the starting point.

4:2 RANDOM WALK

Consider a starting position at some zero point. We toss a coin that tells the man to move forward or backward. He ends up at either the $+1$ or the -1 position (Fig. 4.1). If he ends up in the $+1$ position, the next toss of the coin will take him to the $+2$ or the 0 position, depending on whether the toss tells him to move forward or back. Similarly from the -1 position he could move to 0 or -2.

There exist two possible ways of arriving back at the zero position, and only one possible way of getting to the -2 or to the $+2$ position. Since all of these sequences are equally probable, there is a probability of $\frac{1}{4}$ that the man ends up in the $+2$ position, a probability of $\frac{1}{4}$ that he ends up at -2, and a probability of $\frac{1}{2}$ that he ends up in the zero position after two steps. The zero position is more probable because there are two distinct ways of reaching this position, while there is only one way to get to the $+2$ or -2 positions when only two steps are allowed.

Figure 4.1 shows the number of ways $p(m, n)$ of ending up at a distance of m steps from the starting point, if the man executes a total of n steps. We will call m the deviation from the starting position. We will call $p(m, n)$ the *relative probability*

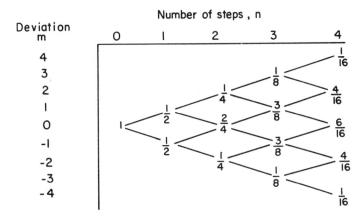

Fig. 4.1   Probability $P(m, n)$ of terminating at position m after n steps.

of terminating at distance m. The *absolute probability* $P(m, n)$ of terminating at position m, after n steps is

$$P(m, n) = \frac{p(m, n)}{\sum_k p(k, n)} = \frac{\text{Number of paths leading to position } m}{\text{Sum of all distinct paths leading to any position, } k} \quad (4\text{–}1)$$

Figure 4.1 shows that the quantities $p(m, n)$ have a binomial distribution; they are the same numbers that appear as coefficients in the expansion

$$\left[\frac{1}{x} + x\right]^n = x^n + nx^{n-2} + \frac{n(n-1)}{2!}x^{n-4} + \frac{n!\, x^{n-2r}}{(n-r)!\, r!} \cdots + \frac{1}{x^n} \quad (4\text{–}2)$$

Knowing this, we can easily evaluate the sum of coefficients in the series, $\sum_k p(k, n)$. It is the sum of the coefficients in the binomial expansion and can be obtained by setting $x = 1$ on the right side of equation 4–2.

Substituting $x = 1$ on the left side of (4–2) shows that the sum of terms must have the value 2^n:

$$\sum_{k=-n}^{n} p(k, n) = 2^n \quad (4\text{–}3)$$

and

$$P(m, n) = \frac{p(m, n)}{2^n} \quad (4\text{–}4)$$

4:2

We note also that if the exponent of a given term in equation 4–2 represents the deviation m, in Fig. 4.1, then the coefficient of that term represents the relative probability $p(m, n)$. In that sense we can rewrite (4–2) as

$$\left(\frac{1}{x} + x \right)^n = \sum_{k=-n}^{n} p(k, n) \, x^k \tag{4–5}$$

Every second term of this series has a coefficient zero. We now wish to determine the mean deviation from the zero position after a random walk of n steps. By this we mean the sum of distances reached in any of the 2^n possible paths that we could take, all divided by 2^n. Since there are $p(k, n)$ ways of reaching the distance k, the numerator of this expression is $\sum_k k p(k, n)$ and we see that the *mean deviation*, $\langle k \rangle$, is

$$\langle k \rangle \equiv 2^{-n} \sum_{k=-n}^{n} k p(k, n) \;=\; \frac{\text{sum of all possible terminal distances after } n \text{ steps}}{\text{number of all possible paths using } n \text{ steps}}$$

$$\tag{4–6}$$

We notice from Fig. 4.1 and from the binomial distribution (4–2) that the relative probability $p(k, n)$ of having a deviation k equals the relative probability of having deviation $-k$: $p(-k, n) = p(k, n)$. Since the summation in (4–6) is carried out over values from $-n$ to n, there will be an exact cancellation of pairs involving $k = m$ and $-m$, and the only uncancelled term is the one having $k = 0$. This shows that the value of $\langle k \rangle$ must be zero also. The mean deviation from the starting position must be zero, no matter how many steps we take.

This does not mean that the absolute value of the deviation is zero. Far from it! But there are equally many ways of ending up at a positive as at a negative distance and the average position is right at the starting point itself.

This much is evident from symmetry. However, we usually need to know something about the actual distance reached after n steps. For example, we want to know the actual distance from a star that a photon has traveled after n absorptions and re-emissions in a surrounding cloud. A useful measure of such distances is the *root mean square deviation* Δ

$$\Delta \equiv \langle k^2 \rangle^{1/2} = \left[\frac{\displaystyle\sum_{k=-n}^{n} k^2 p(k, n)}{\displaystyle\sum_{k=-n}^{n} p(k, n)} \right]^{1/2} = \left[\frac{\text{sum of (distances)}^2}{\text{sum of all possible paths}} \right]^{1/2} \tag{4–7}$$

4:2

This is obtained by first taking the mean of the deviation squared $\langle k^2 \rangle$, and then taking the root of this mean value to obtain a deviation in terms of a number of unit length steps. If we did not take the square root, the quantity obtained would have to be measured in units of $(\text{step})^2$; this is an area, rather than a length or distance. To evaluate the sum

$$\sum_{k=-n}^{n} k^2 p(k, n) \tag{4-8}$$

we can employ a simple technique. We substitute the quantity $x = e^y$ in equation 4–5 and differentiate twice in succession with respect to y. In the limit of small y values, we then obtain

$$\sum_{k=-n}^{n} k^2 p(k, n) = \lim_{y \to 0} \frac{d^2}{dy^2} (e^{-y} + e^y)^n$$

$$= n(n-1)(e^{-y} + e^y)^{n-2}(e^y - e^{-y})^2 + n(e^{-y} + e^y)^n]_{y=0} = n2^n \tag{4-9}$$

In summary, we can write

$$\sum_{k=-n}^{n} k^2 p(k, n) = n2^n \tag{4-10}$$

Equations 4–3 and 4–10 can now be substituted into (4–7) to obtain a root mean square deviation

$$\Delta = n^{1/2} \tag{4-11}$$

After n steps of unit length the absolute value of the distance from the starting position is therefore approximately $n^{1/2}$ units.

The following four problems widen the applications of the random walk concept.

PROBLEM 4-1. For a one-dimensional random walk, involving steps of unequal length, prove that the mean position after a given number of steps is zero, the starting position.

Note that for a finite number of differing step lengths, this walk can be reduced to a succession of random walks, each walk having only one step length.

PROBLEM 4-2. Prove that the root mean square deviation for a walk involving the sum of different numbers n_i of steps of length λ_i, is

$$\Delta = N^{1/2} \lambda_{\text{rms}} \tag{4-12}$$

4:2

where $N = \sum_i n_i$ and λ_{rms} is the root mean square value of the step length

$$\lambda_{\rm rms} = \left[\frac{\sum\limits_i n_i \lambda_i^2}{N} \right]^{1/2} \tag{4–13}$$

PROBLEM 4–3. Show that the root mean square deviation in a three-dimensional walk with step lengths L_0 is $s^{1/2} L_0$ after s steps. To show this, take the three Cartesian components of the $i^{\rm th}$ step (see Fig. 4.2) as

$$L_0 \cos \theta_i, \qquad L_0 \sin \theta_i \cos \phi_i, \qquad L_0 \sin \theta_i \sin \phi_i \tag{4–14}$$

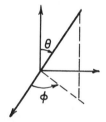

Fig. 4.2 Polar coordinate system used in describing the three-dimensional random walk.

The mean square deviations along the three coordinates are, respectively,

$$\Delta_z^2 = \sum_{i=1}^{s} L_0^2 \cos^2 \theta_i \qquad \Delta_x^2 = \sum_{i=1}^{s} L_0^2 \sin^2 \theta_i \cos^2 \phi_i$$

$$\Delta_y^2 = \sum_{i=1}^{s} L_0^2 \sin^2 \theta_i \sin^2 \phi_i \tag{4–15}$$

These components can be added by the Pythagorean theorem to give the overall mean square deviation as

$$\Delta^2 = s L_0^2 \tag{4–16}$$

PROBLEM 4–4. A hot star is surrounded by a cloud of hydrogen that is partly ionized, partly neutral. Radiation emitted by the star at the wavelength of the Lyman-α line can be absorbed and re-emitted by the neutral atoms. Let the mean path traveled by a photon, between emission and absorption have length L. Let the radius of the cloud be R. About how many absorption and re-emission processes are needed before the photon finally escapes from the cloud? We make use of this result in section 9:6.

4:2

The random walk concept provides an essential basis for all radiative transfer computations. We will tackle such problems later, in discussing the means by which energy can be transported from the center of a star, where it is initially released, to the surface layers, and then through the star's atmosphere out into space. In the general theory of radiative transfer, the *opacity* of the material is inversely proportional to the step length we assumed for the random walk above. The added complication that arises in most practical problems is that the mean energy per photon becomes less and less as energy is transported outward from the center of a star. Energy initially found in hard gamma rays eventually leaves the stellar surface as visible and infrared radiation. One gamma photon released in a nuclear reaction at the center of the star provides enough energy for about a million photons emitted at the stellar surface. The walk from the center of a star, therefore, involves not a single photon alone, but also all its many descendants.

4:3 DISTRIBUTION FUNCTIONS, PROBABILITIES, AND MEAN VALUES

In section 4:2 we calculated the mean deviation and root mean square deviation after a number of steps in a random walk. Often we are interested in computing mean values for functions of the deviation; and for distributions other than binomial distributions there is also a procedure for obtaining such values.

Suppose a variable x can take on a set of discrete values x_i. Let the absolute probability of finding the value x_i in any given measurement be $P(x_i)$. If we pick a function $F(x)$ that depends only on the variable x, we can then compute the mean value that we would obtain for $F(x)$ if we were to make a large number of measurements. This mean is obtained by multiplying $F(x_i)$ by the probability $P(x_i)$ that the variable x_i will be encountered in any given measurement; summation over all i values then yields the mean value $\langle F(x) \rangle$

$$\langle F(x) \rangle = \sum_i P(x_i) F(x_i) \qquad (4-17)$$

Sometimes the absolute probability is not immediately available but the relative probability $p(x_i)$ is known. We then have the choice of computing $P(x_i)$ as in equation 4–1, or else we can proceed directly to write

$$\langle F(x) \rangle = \frac{\sum_i p(x_i) F(x_i)}{\sum_i p(x_i)} \qquad (4-18)$$

where the denominator gives the normalization that is always needed when relative probabilities are used.

4:3

If x can take on a continuum of values within a certain range, the integral expressions corresponding to equations 4–17 and 4–18 are

$$\langle F(x) \rangle = \int P(x) \, F(x) \, dx = \frac{\int p(x) \, F(x) \, dx}{\int p(x) \, dx} \tag{4–19}$$

where the integrals are taken over the range of the variable for which a mean value $\langle F(x) \rangle$ is of interest. In some situations this range is $-\infty < x < \infty$.

We note that the expressions 4–6 and 4–7 already have the general form required by equations 4–17 to 4–19. Basically, in equation 4–6 the function $F(x)$ is just x itself, while in (4–7) it is x^2. We have merely substituted a new symbol x, for the values previously denoted by the position symbol k.

4:4 PROJECTED LENGTH OF RANDOMLY ORIENTED RODS

Let a system be viewed along a direction defining the axis of polar coordinates (θ, ϕ) (Fig. 4.3). A rod of length L has some arbitrary orientation θ with respect to the axis, and its projected length transverse to the line of sight is $L \sin \theta$, independent of ϕ, $0 \leq \phi < 2\pi$.

Fig. 4.3 Polar coordinate system for discussion of projected lengths.

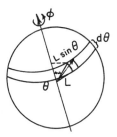

We wish to determine the mean value of the observed length, the average being taken over all possible orientations of the rod: The probability of finding the rod with an orientation that lies within an increment $d\theta$ at angle θ is proportional to the area that the strip $d\theta$ defines on the surface of a sphere of unit radius. The normalized probability $P(\theta)$ is

$$P(\theta) \, d\theta = \frac{1}{2\pi} \int p(\theta, \phi) \, d\theta \, d\phi = \sin \theta \, d\theta \tag{4–20}$$

We see that this is a properly normalized probability since

$$\int_0^{\pi/2} P(\theta) \, d\theta = -\cos \theta \Big|_0^{\pi/2} = 1 \tag{4–21}$$

that is, the probability of finding the rod with *some* orientation between 0 and $\pi/2$ is unity.* The probability of finding the rod with projected length $L \sin \theta$ is therefore $\sin \theta$, and the mean value of the projected length averaged over all position angles is

$$\frac{\int_0^{\pi/2} P(\theta) L \sin \theta \, d\theta}{\int_0^{\pi/2} P(\theta) \, d\theta} = \int_0^{\pi/2} L \sin^2 \theta \, d\theta = \frac{\pi}{4} L \qquad (4\text{--}22)$$

Here, the integral in the numerator is a summation over the lengths obtained over all orientations, and the integral in the denominator assures an average value by dividing the numerator by the whole range of probabilities. This division is not strictly necessary because we already have normalized correctly. However, had we, for example, wished to find the mean projected lengths only for those rods having inclinations to the polar axis in the range $0 < \theta \leq \pi/4$, the limits of integration both in the numerator and denominator would be 0 and $\pi/4$, and the integral in the denominator would no longer be trivial. Reversing the problem, we can ask for the actual value of a length S when only the random projected lengths can be observed to have mean value D. Then

$$S = \frac{4 \langle D \rangle}{\pi} \qquad (4\text{--}23)$$

by simple inversion of the argument developed in (4–22). We can ask a slightly different question, "Given a particular observed value of D, what is the mean of all the values that S could have?" To answer this, we average $D \sin \theta$ for an isotropic distribution and find $\langle S \rangle = \pi D/2$. This approach is useful in obtaining the mean distance between members of double galaxies, when only the projected distance is known from observations.

Similarly we can use our approach to decide whether elliptical galaxies are prolate—cigar-shaped—or oblate—disk-shaped. To make such an analysis, we do have to assume that all elliptical galaxies have roughly the same shape; so that according to this view, the globular galaxies would just be ordinary ellipticals viewed along their symmetry axis.

PROBLEM 4–5. When a series of double galaxies is observed, the total mass of each pair can be statistically determined by measuring the projected separation between the galaxies and the projected radial component of their motions about

---

* The limits of integration are $0 \leq \phi < 2\pi$, $0 \leq \theta \leq \pi/2$, since a rod with orientation (θ, ϕ) is equivalent to one with orientation $(-\theta, \phi + \pi)$.

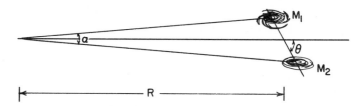

Fig. 4.4 Diagram to illustrate estimation of the total mass found in binary galaxies.

each other. If R is the distance to the pair as determined by their mean red shift, and α is the angular separation, then we can obtain the projected separation d_p. The difference between the two red shifts gives the projected orbital velocity component v_p. Assuming that the galaxies move in circular orbits about each other, and that one of the galaxies is much more massive, show that the mass of the pair is statistically given by

$$M_{\text{pair}} = \frac{\langle v^2 \rangle}{G \langle 1/r \rangle} = \frac{3\pi}{2} \frac{\langle v_p^2 \rangle}{G \langle 1/d_p \rangle} \qquad (4\text{–}24)$$

It may first be useful to show that $\langle v^2 \rangle = 3 \langle v_p^2 \rangle$ and that $\langle 1/r \rangle = 2 \, \pi \langle 1/d_p \rangle$. Note that $\langle r \rangle \neq \langle 1/r \rangle^{-1}$. Actually the projection angle in this case is not independent for r and v, and alternate forms of (4–24) can be presented which take this correlation into account.

When we talk about clusters of galaxies, the same considerations apply, because the virial theorem (3–83) again sets the mean potential energy equal to twice the (negative of the) kinetic energy; and the mass of the entire cluster is then substituted on the left side of equation 4–24. The right side gives the mean squared velocities of the cluster galaxies and their mean reciprocal distances from the cluster center. As discussed in 3:15, when the cluster mass is estimated in this way, it always turns out to be some 10 times greater than the sum of the masses of the individual galaxies determined as in Problem 3–9. We will return to this puzzle in Chapter 10; but even there we will not be able to resolve the difficulty.

Salpeter and Bahcall (Sa69a) have used (4–24) to obtain an upper limit on the mass of the QSO B264 that appears to lie in a cluster of galaxies. The upper limit is $5 \times 10^{13} \, M_\odot$. Their estimate essentially says that the QSO must be less massive than the total mass of the cluster. Since nothing at all was previously known about QSO masses, even this very high upper limit is interesting.

4:4

4:5 THE MOTION OF MOLECULES

An assembly of molecules surrounding an interstellar dust grain exerts pressure on the grain's surface. This pressure arises because the molecules are moving randomly and sometimes collide with the dust. A molecule initially moving toward the grain is deflected at the grain's surface and recedes following the collision. Since the particle's velocity is changed, its momentum **p** also is altered. For a brief interval the surface, therefore, exerts a force on the molecule, because, by definition, a force is required to produce the change of momentum. This follows from Newton's law, equation 3–4, which can be rewritten as

$$\mathbf{F} = m\ddot{\mathbf{r}} = \dot{\mathbf{p}} \qquad (3\text{--}4)$$

If the grain exerts a force on a molecule during a given time interval τ, the molecule too must be reacting on the grain in that time. The sum of all the forces exerted by all the individual molecules impinging on unit grain area at any given time then constitutes the pressure—or force per unit area—acting on the dust.

To calculate the pressure we must first decide how many molecules hit a grain per unit time. Figure 4.5 shows a spherical polar coordinate system by means of which we can label the direction from which the particles initially approach. That direction is given by angles (θ, ϕ). If there are $n(\theta, \phi, v)$ molecules per unit volume coming from an increment of solid angle $d\Omega = \sin \theta \, d\theta \, d\phi$ about the direction (θ, ϕ) with a speed v to $v + dv$, then the number of particles incident on unit surface area in unit time is

$$\iiint v \cos \theta \, n(\theta, \phi, v) \sin \theta \, d\theta \, d\phi \, dv \qquad (4\text{--}25)$$

The factor $\cos \theta$ has to be included because the volume of an inclined cylinder that contains all the incident particles is the product of the base area and the height (Figure 4.6).

Expression 4–25 is proportional to v since particles with larger speeds can reach the impact area from greater distances in any given time interval.

If we assume that each molecule is reflected specularly—as from a mirror—then the angle of incidence is equal to the angle of reflection from the surface, and the total change in momentum for a reflected particle is

$$\Delta p = -2p \cos \theta \qquad (4\text{--}26)$$

Only the momentum component normal to the surface changes in such a reflection and this gives rise to the factor $\cos \theta$. We can now compute the pressure

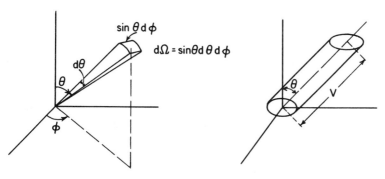

Fig. 4.5 Spherical polar coordinates for computing of pressure.

Fig. 4.6 Inclined cylindrical element containing all molecules striking the surface from direction θ, with speed v, in unit time interval.

that is just (the negative of) the total change of momentum suffered by all molecules incident on unit area in unit time.

$$P = \int_0^{2\pi} d\phi \int_0^\infty dv \int_0^{\pi/2} d\theta (2p\cos\theta)\, v \cos\theta\, n(\theta, \phi, v) \sin\theta \qquad (4\text{--}27)$$

In an *isotropic gas* the number of molecules arriving from unit solid angle is independent of θ and ϕ and we can write

$$n(\theta, \phi, v)\, dv = \frac{n(v)}{4\pi} dv \qquad (4\text{--}28)$$

Here $n(v)$ is the number density of molecules with speeds in the range v to $v + dv$ and the factor $1/4\pi$ is a normalization constant that arises because 4π steradians are needed to describe all possible approach directions.

Expression 4–28 allows us to separate out a velocity dependent part of the integrand in (4–27). It is independent of the direction coordinates θ and ϕ. If $v \ll c$—as in a nonrelativistic gas—then $p = mv$, where m is the mass of a molecule. We can then write

$$\int_0^\infty n(v)\, v^2\, dv = n \langle v^2 \rangle \qquad (4\text{--}29)$$

where n is the number density of particles per unit volume regardless of speed and direction, and $\langle v^2 \rangle$ is the mean squared value of the velocity. Equation 4–29 simply is a definition of the mean squared velocity.

The other part of the integral in (4–27) now can be written as

$$\frac{1}{2\pi} \int_0^{2\pi} \int_0^{\pi/2} \cos^2 \theta \sin \theta \, d\theta \, d\phi = \frac{1}{3} = \langle \cos^2\theta \rangle \qquad (4\text{–}30)$$

This integral defines the mean value of $\cos^2 \theta$ averaged over a hemisphere $0 \le \theta \le \pi/2$. This is the hemisphere from which all particles striking the wall must approach. Because of symmetry about $\theta = \pi/2$, the mean squared value of the cosine function actually is $1/3$ even if we integrate over all possible directions, rather than just one hemisphere.

Substituting equations 4–29 and 4–30 into 4–27 we can rewrite the expression for pressure as

$$P = \frac{nm\langle v^2 \rangle}{3} \qquad (4\text{–}31)$$

Writing the product of the pressure P with the volume V that encloses N particles of the assembly, we then have the expression

$$PV = \frac{Nm\langle v^2 \rangle}{3} = N \text{Ⓗ} \qquad (4\text{–}32)$$

where $N = nV$ and $\text{Ⓗ} = m\langle v^2 \rangle/3$.

PROBLEM 4–6. The random velocity of galaxies is thought to amount to $v \sim 100$ km sec^{-1}. Their number density is $n \sim 10^{-1}$ Mpc^{-3}. Typical galaxies have a mass of 3×10^{44} g. What is the cosmic pressure due to galaxies? This pressure will be considered in Chapter 10, where the cosmic pressure needs to be known to determine the dynamics of the universe in its expansion or contraction.

PROBLEM 4–7. The number density of stars in the sun's vicinity is $n \sim 10^{-57}$ cm^{-3}. The sun's velocity relative to these stars is $v \sim 2 \times 10^6$ cm sec^{-1} and we can take the cross section for collision with another star to be $\sigma \sim 5 \times 10^{22}$ cm^2. In the Jeans theory of the birth of the solar system, such an encounter is considered responsible for the formation of the planets. How probable is it that the sun would have formed planets in $P = 5 \times 10^9$ y? How many planetary systems would we expect altogether in the Galaxy if there are 10^{11} stars and if the sun is representative?

4:5

4:6 IDEAL GAS LAW

Tenuous gases obey a simple law at temperatures far above the temperature of condensation. This law relates the temperature of a gas to its pressure and density. Since it becomes exact only at extremely high temperatures and low densities, it represents an idealization that a real gas can only approach; and we speak of the *ideal gas law*. In practice, deviations from ideal behavior are small for a large variety of gases in many different situations, and the law is therefore very useful.

To understand this law, we must first know what is to be meant by *temperature*. We can easily "feel" whether a body is hot or cold; but it is not simple to describe this feeling in terms of a measurable physical quantity. One way of prescribing temperatures is in terms of a device, for example, an ordinary mercury bulb thermometer. When the thermometer is dipped into a bowl of water that feels hot, the mercury expands out of the bulb and rises in the capillary tube. When the thermometer is placed into a cold bowl of water the mercury contracts. We can attach an arbitrary scale to the capillary portion of the thermometer and take readings to obtain the temperature in terms of the location of the mercury meniscus in the capillary. To show just how arbitrary such a scale may be, we need only recall that there are at least five different temperature scales in common use in the Western world.

Choosing a given mercury thermometer as a standard, we can make observations of the behavior of gases and eventually arrive at a relation between the density, pressure, and temperature of a given gas. This relation is called an *equation of state*. It has the functional form

$$F(T, P, \rho) = 0 \tag{4-33}$$

The density is sometimes expressed in terms of its reciprocal, the volume per unit mass, or more often in terms of the *molar volume*, or volume per mole of gas. The *mole* is a quantity of matter represented by $\mathcal{N} = 6.02 \times 10^{23}$ molecules. $\mathcal{N}$ is called *Avogadro's number*. Avogadro's number is the number of atoms of the carbon isotope C^{12} weighing exactly 12 grams—one gram-atomic-weight of C^{12}.

Writing the molar volume as $\mathscr{V}$, we obtain the ideal gas law as

$$P\mathscr{V} = RT \tag{4-34}$$

where R is a constant called the *gas constant*. At constant pressure the volume of a given amount of gas increases linearly with temperature. At fixed volume the pressure rises linearly with temperature. Some gases, notably helium, behave very nearly like an ideal gas and can, therefore, be used to define a *gas thermometer* temperature scale. The important point to realize is that, in any event, temperature has to be defined operationally in terms of a convenient device.

We note the similarity between equations 4–32 and 4–34. When N in equation 4–32 is chosen to be Avogadro's number $\mathcal{N}$, we find that

$$\frac{RT}{\mathcal{N}} = \textcircled{H} = \frac{m\langle v^2 \rangle}{3} = mc_s^2 \qquad (4\text{--}35)$$

Here we have introduced a new symbol c_s, standing for the *speed of sound* in an ideal gas, mainly brcause it is interesting to note that sound and pressure waves are propagated at a speed equal to the root mean square velocity component of the gas molecules along the direction of propagation. In solids and liquids incompressibility—stiffness—dominates over molecular motion in propagating pressure, and the speed of sound is appreciably higher than the molecular speeds (Mo68).

We can define a new constant $k = R/\mathcal{N}$, called Boltzmann's constant. Equation 4–35 then becomes

$$\frac{3}{2}kT = \frac{m\langle v^2 \rangle}{2} \qquad (4\text{--}36)$$

The right side of equation 4–36 is the mean kinetic energy per particle in the assembly, and the temperature is therefore nothing other than an index of the mean kinetic energy. In a hot gas the molecules move at high velocity; in a cooler gas they move more slowly. The Boltzmann constant k has to be experimentally determined by direct or indirect measurement of the kinetic energy of molecules in a gas at a given temperature: $k = 1.38 \times 10^{-16}\, \text{erg}\,{}^\circ\text{K}^{-1}$.

Equation 4–32 can now be rewritten as

$$P\mathcal{V} = \mathcal{N}kT \qquad \text{or} \qquad P = nkT \qquad (4\text{--}37)$$

This is straightforward as long as we deal with one particular kind of gas or one given type of molecule. But what happens if the gas consists of a mixture of different molecules? The kinetic theory developed thus far predicts that the total pressure should still be determined by the total number density of molecules as prescribed by equation 4–37. If there are j different kinds of molecules present in thermal equilibrium, each with number density n_i, the complete relation would read

$$P = \sum_{i=1}^{j} P_i = \sum_{i=1}^{j} n_i kT = nkT \qquad (4\text{--}38)$$

where P_i is the *partial pressure* exerted by molecules of type i alone. Equation 4–38 expresses *Dalton's law* of partial pressures: The total pressure of an ideal gas is the sum of the partial pressures of the various constituents.

PROBLEM 4–8. Interstellar atomic hydrogen is often found in neutral, H I clouds whose temperature is 100°K. What is the root mean squared velocity at which the hydrogen atoms travel? If the number density $n = 1 \text{ cm}^{-3}$, what is the pressure in interstellar space?

PROBLEM 4–9. These clouds also contain dust grains that might characteristically have diameters 5×10^{-5} cm and unit density. Treating the dust as though it were an ideal gas, what would be the random motion of dust grains?

PROBLEM 4–10. If the gas had systematic velocity v relative to the dust grains, how much momentum would be transferred to each dust grain per unit time, and what is the acceleration? Assume that the gas density $n = 1 \text{ cm}^{-3}, v = 10^6$ cm sec^{-1}, and that the gas atoms stick to the grain in each collision.

PROBLEM 4–11. What would be the rate of mass gain for this grain? How soon would its mass increase by 1 %?

PROBLEM 4–12. In an ionized hydrogen (H II) region, protons and electrons are dissociated. If the temperature of this interstellar gas is 10^4°K, calculate electron and proton velocities.

4:7 RADIATION KINETICS

Electromagnetic radiation is transmitted in the form of photons—discrete quanta having momentum p and energy $\mathscr{E}$. The experimentally determined relationship between the spectral frequency v—color of the radiation—and the energy and momentum is

$$p = \frac{hv}{c} \tag{4-39}$$

$$\mathscr{E} = hv \tag{4-40}$$

where h is Planck's constant, c is the speed of light, and v is the spectral frequency.

We can substitute expression 4–39 into the pressure equation 4–27, replacing v by c, and neglecting the integration over velocity since all photons have the same speed c. Expression 4–27 then reads

$$P(v)\,dv = \int_0^{2\pi} d\phi \int_0^{\pi/2} d\theta \, \frac{2hv}{c} \cos\theta \, c \cos\theta \, n(\theta, \phi, v) \sin\theta \, dv \tag{4-41}$$

4:7

The two factors c cancel, and hv can be replaced by $\mathscr{E}$. For an isotropic radiation field, $n(\theta, \phi, v) = n(v)/4\pi$, and use of equation 4–30 leads to

$$P(v) = \frac{n(v)\,\varepsilon}{3} = \frac{hvn(v)}{3} \tag{4–42}$$

If there are quanta of j different spectral frequencies present, expression 4–42 becomes

$$P = \frac{U}{3} \tag{4–43}$$

where U is the total energy density summed over all spectral frequencies:

$$U = \sum_{i=1}^{j} n_i h v_i \tag{4–44}$$

PROBLEM 4–13. The radiation energy density in the universe is of the order of 6×10^{-13} erg cm^{-3}, if primarily a $3°$K radiation field (4:13) is considered to exist on a cosmic scale. What is the pressure due to this field and how does it compare to the galactic pressure of Problem 4–6?

PROBLEM 4–14. The radiation energy incident from the sun on unit area per unit time is 1.37×10^6 erg cm^{-2} sec^{-1}, at the earth. This quantity is called the *solar constant*. Find the radiative repulsive force on a 10^{-2} cm diameter black (totally absorbing) grain, at the distance of the earth.

PROBLEM 4–15. A 10^{-4} cm radius grain absorbs $\frac{1}{3}$ of solar radiation incident on its surface and scatters the remainder isotropically. Calculate the ratio of gravitational attraction to radiative repulsion from the sun, assuming that the grain has density 6 g cm^{-3}. Show that this ratio is constant as a function of distance from the sun.

PROBLEM 4–16. If the repulsive force of radiation for a grain is $\frac{1}{3}$ the attraction to the sun due to gravitation, we can define an "effective" gravitational constant $G_{\mathrm{eff}} = \frac{2}{3}G$, where G is the gravitational constant. This will characterize the motion of the grain. What is the orbital period of such a grain moving in the earth's orbit? How does its orbital velocity compare to that of the earth?

4:8 ISOTHERMAL DISTRIBUTIONS

We say that a gas is *isothermal* if its temperature is the same throughout the volume it occupies. Consider an isothermal spherically symmetric gas configuration in space. There is a decreasing gas density and pressure at increasing central distance r. The pressure change dP between positions r (Fig. 4.7) and $r + dr$ is given by the gravitational force acting on matter between r and $r + dr$:

$$dP = -dr\,\rho(r)\,\nabla\mathbb{V}(r) \tag{4-45}$$

Here $\rho(r) = n(r)\,m$ and $\mathbb{V}(r)$ is the gravitational potential due to mass enclosed by the sphere r. For an ideal gas (see equation 4–38) $P/\rho = kT/m$. Dividing this expression into equation 4–45 we have

$$\frac{dP}{P} = -\frac{m}{kT}\nabla\mathbb{V}(r)\,dr$$

which integrates to

$$P = P_0 e^{-m\mathbb{V}(r)/kT} \tag{4-46}$$

Reapplying the ideal gas law, we can also obtain the forms

$$n = n_0 e^{-m\mathbb{V}(r)/kT} \qquad \text{or} \qquad \rho = \rho_0 e^{-m\mathbb{V}(r)/kT} \tag{4-47}$$

The exponential term appearing in equations 4–46 and 4–47 is called the *Boltzmann factor*. It plays an important role throughout the theory of statistical thermodynamics and, as we will see in section 4:21, gives a useful starting point for describing the distributions of stars in globular clusters, and molecules in protostars.

Fig. 4.7 Pressure-distance relation for a spherically symmetric configuration.

4:9 ATMOSPHERIC DENSITY

Using equation 4–47, we can readily find the density distribution in the atmosphere of a star, planet, or satellite. In what follows we will keep referring to the parent body as a planet, but the theory holds equally well for a star, moon, or any other massive body.

The gravitational potential at any location in the atmosphere is given by

$$\mathbb{V}(r) = -\frac{MG}{r} \tag{4-48}$$

4:9

where r is the distance measured from the center of the planet and M is its mass. Expression 4–48 also assumes that the atmosphere is tenuous so that M can be assumed to be constant and independent of r. Let R be the planet's radius, and consider a point at height x above the surface. The difference between the potential at height x and at the surface is

$$\mathbb{V}(R + x) - \mathbb{V}(R) = -\frac{MG}{R + x} + \frac{MG}{R} = \frac{MGx}{R^2}, \quad x \ll R \qquad (4\text{–}49)$$

Equation 4–47 then becomes

$$n = n_0 e^{-(mMG/kTR^2)x} = n_0 e^{-mgx/kT} \qquad (4\text{–}50)$$

where n_0 now represents the density at the surface and $MG/R^2 \equiv g$ is the *surface gravity* of the planet. It is clear that the atmospheric density decreases exponentially with height. Moreover, we can define a *scale height*

$$\Delta = \frac{kTR^2}{mMG} = \frac{kT}{mg} \qquad (4\text{–}51)$$

We see that the density at height $x + \Delta$ is reduced by a factor e below the value at height x. It is worthwhile noting that the scale height is small for low temperature gases composed of heavy molecules—m large—and for dense parent bodies—large M, small R.

PROBLEM 4–17. Show that an atmosphere consisting of a combination of gases has a variety of scale heights, one for each gas component present. Show that the total pressure is

$$P = \sum_i P_i = \sum_i P_{i0} e^{-(m_i gx/kT)} \qquad (4\text{–}52)$$

and that the total density is

$$\rho = \sum_i n_i m_i = \sum_i n_{i0} m_i e^{-(m_i gx/kT)} \qquad (4\text{–}53)$$

where the subscript 0 denotes a value at the base of the atmosphere. Assume that there is no convection in the atmosphere. (Convection normally requires bulk motion of entire volumes of gas and gives rise to winds that do not allow complete separation of different gaseous constituents. The concept of scale height then cannot be applied.

The earth's atmosphere exhibits some of these features. At the low densities found in the upper atmosphere, there is some separation of gases with different scale heights. Helium, for example, appears in appreciable concentrations only at high altitudes. In the lower atmosphere three features complicate any analysis. There are winds, temperature gradients, and atmospheric water vapor. The vapor is near the condensing point and a local atmospheric temperature drop can give rise to condensation and a decrease in pressure. This gives rise to winds. More important, the lower atmosphere is not isothermal and does not behave in the simple way described here.

PROBLEM 4-18. The mass of the atmosphere is negligible compared to the mass of the earth. If the gravitational attraction at the surface of the earth is 980 dynes g^{-1}, calculate the scale height of the atmosphere's main constituent, molecular nitrogen N$_2$.

4:10 PARTICLE ENERGY DISTRIBUTION IN AN ATMOSPHERE

The exponential decline of particle density with height is an important clue to the velocity distribution of particles. We note that molecules at a height x_1, having an upward directed velocity component $v_x = (2gh)^{1/2}$, have enough energy to reach a height $x_1 + h$. Whether a given molecule with this instantaneous velocity actually reaches height $x_1 + h$ cannot be predicted. The molecule might collide with another one, and lose most of its energy. However, as long as thermal equilibrium exists, and the gas temperature remains stable, we can be sure that, for every molecule that loses energy through a collision, there will be a *restituting collision* at some nearby point in which some other molecule gains a similar amount of energy. This concept, sometimes referred to as *detailed balancing*, allows us to neglect the effect of collisions in the remainder of our argument.

Since the temperature is the same at all levels of an isothermal atmosphere, the velocity distribution must also be the same everywhere, and only the number of particles changes with altitude. The ratio of the particle densities at heights $x_1 + h$ and x_1, (see equation 4–53) is $\exp(-mgh/kT)$. Since the particles encountered at height $x_1 + h$ have all come up from the lower height x_1, to which they will eventually return—fall back—we can be certain that the fraction of particles passing through a plane at height x_1 and having speeds greater than $v_h = (2gh)^{1/2}$ also is $\exp(-mgh/kT)$, because this is the fraction of particles that has enough energy to reach altitude $x_1 + h$. We can therefore express the number density of particles with vertical velocity v_x greater than v_h as

$$\frac{N(v_x > v_h)}{N(v_x > 0)} = e^{-mgh/kT} = e^{-mv_h^2/2kT} \tag{4-54}$$

Note that N is not a density, it is a number crossing unit area in unit time interval.

Consider an isotropic velocity distribution $f(v)$ that is normalized by the integral

$$\iiint_{-\infty}^{\infty} f(v_x, v_y, v_z)\, dv_x\, dv_y\, dv_z = 1 \tag{4–55}$$

As a trial solution for the function f, we can use an exponential v_x dependence, like that given by equation 4–54. The isotropy requirement then demands a similar dependence on v_y and v_z, and equation 4–55 gives the full function as

$$f(v_x, v_y, v_z) = \left(\frac{m}{2\pi kT}\right)^{3/2} e^{-(m/2kT)(v_x^2 + v_y^2 + v_z^2)} \tag{4–56}$$

where the coefficient is a normalization factor required by (4–55). This function is separable in the variables v_x, v_y, and v_z. To test whether it also obeys equation 4–54 we note that

$$\frac{N(v_x > v_h)}{N(v_x > 0)} = \frac{\displaystyle\int_{v_h}^{\infty} v_x e^{-(m/2kT)v_x^2}\, dv_x}{\displaystyle\int_0^{\infty} v_x e^{-(m/2kT)v_x^2}\, dv_x} = e^{-mv_h^2/2kT} \tag{4–57}$$

The quantity v_x in the integrand plays the same role here as in equation 4–27. It takes into account that the higher velocity particles can reach a given surface from a larger distance, and from a larger volume, in unit time. We can write the distribution (4–56) in terms of the speed

$$v = (v_x^2 + v_y^2 + v_z^2)^{1/2} \tag{4–58}$$

We then obtain

$$f(v) = \left(\frac{m}{2\pi kT}\right)^{3/2} e^{-mv^2/2kT} \tag{4–59}$$

PROBLEM 4–19. Satisfy yourself that the normalization condition for $f(v)$ is

$$4\pi \int_0^{\infty} f(v)\, v^2\, dv = 1 \tag{4–60}$$

Show also that, in terms of momentum, the distribution function is

$$f(p) = \frac{1}{(2\pi mkT)^{3/2}} e^{-p^2/2mkT} \tag{4–61}$$

and

$$\int_0^\infty 4\pi f(p)\, p^2\, dp = 1 \tag{4-62}$$

Note that equations 4–56, 4–59, and 4–61 all are independent of the gravitational potential initially postulated. The equations derived here therefore have much wider applicability than just to the gravitational problem. We will discuss this further in section 4:15.

The velocity and momentum distribution functions (4–59) and (4–61) are called Maxwell-Boltzmann distributions, after L. Boltzmann and J. C. Maxwell who were two of the founders of classical kinetic theory. The momentum distribution is plotted in Fig. 4.8. These distribution functions have extremely wide application.

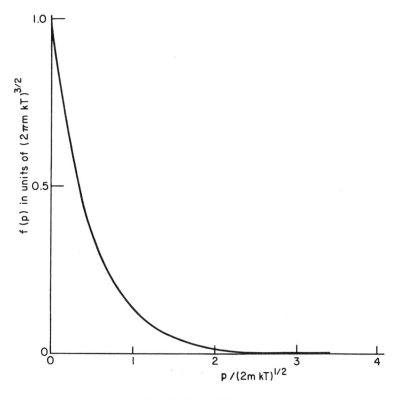

Fig. 4.8 Maxwell-Boltzmann momentum distribution.

4:10

PROBLEM 4-20. If the moon had an atmosphere consisting of gases at $300°K$, calculate the mass of the lightest gas molecules for which $3kT/2 < MmG/R$. m is the mass of the molecule, M and R are the mass and radius of the moon, respectively, 7.3×10^{25} g and 1.7×10^8 cm. Note that the quantity on the left is related to the escape velocity at the moon. What is this velocity? Actually heavier molecules than those with mass m, calculated above, can escape from the moon, because (a) in a Maxwell-Boltzmann distribution, gases have many molecules with speeds larger than the mean speed and (b) because the sunward side of the moon reaches temperatures near $400°K$ during the two weeks it is exposed to the sun.

Despite their great usefulness there are a number of important situations in which we cannot use Maxwell–Boltzmann statistics. These effects can appear at high densities in the center of stars, or at high temperatures in the radiation emitted by stars. They are quantum effects that have no classical basis. The next few sections will concern themselves with such situations.

4:11 PHASE SPACE

The quantum effects that lead to deviations from classical statistical behavior always involve particles that are identical to each other. We might deal with electrons that have almost identical positions and momenta, and spin; or we might have photons with identical frequency, position, direction of propagation, and polarization.

For the case of electrons an important restriction comes into play. The *Pauli exclusion principle* forbids any two electrons from having identical properties. *Neutrons, protons, neutrinos,* and, in fact, all particles with odd half-integral spin $(\frac{1}{2}, \frac{3}{2}, \ldots)$ also obey this principle. *Photons* and *pions,* on the other hand, have integral or zero spin, and any number of these particles can have identical momenta, positions, and spins. The first group of particles—those that obey the injunction of the Pauli principle—are called *Fermi–Dirac particles* or *Fermions;* the others are called *Bosons* and their behavior is governed by the *Bose–Einstein statistics.*

Thus far we have not stated what we mean by "identical." Clearly we could always imagine an infinitesimal difference in the momenta of two particles, or in their positions. Should such particles still be termed identical, or should they not? The question is essentially answered by *Heisenberg's uncertainty principle,* which denies the possibility of distinguishing two particles if the difference in the momentum δp, multiplied by the difference in position δr, is less than h. This comes about because the uncertainty in the simultaneous measurement of momentum and position components for any one particle is at least

$$\Delta p_x \Delta x \sim \hbar, \qquad \Delta x \equiv \langle (x - \langle x \rangle)^2 \rangle^{1/2} = \langle x^2 - \langle x \rangle^2 \rangle^{1/2}$$

$$\Delta p_x = \langle p_x^2 - \langle p_x \rangle^2 \rangle^{1/2} \qquad\qquad (4\text{--}63)$$

where $\hbar = h/2\pi$, and h is *Planck's constant.* If two particles are to be distinguishable, then $\delta p_x \delta x$ should be somewhat greater than $h = 6.626 \times 10^{-27}$ erg sec.

Quantum mechanically one can show that two particles are to be considered identical if their momenta and positions are identical within values

$$\delta p_x \delta x = h, \qquad \delta p_y \delta y = h, \qquad \delta p_z \delta z = h \qquad (4\text{--}64)$$

and if their spins are identical.

In this description each particle is characterized by a position (x, y, z, p_x, p_y, p_z) in a six-dimensional phase space. It occupies a six-dimensional *phase cell* whose volume is (Figs. 4.9 and 4.10):

$$\delta x \delta y \delta z \delta p_x \delta p_y \delta p_z = h^3 \qquad\qquad (4\text{--}64a)$$

Particles within one phase cell are identical—physically indistinguishable—while those outside can be distinguished. Since δx is the dimension of the phase cell, it must be at least twice as large as Δx, the root mean square deviation from the central position. The same relation holds between δp_x and Δp_x. That is why the right side of equation 4–63 involves $\hbar$ while equation 4–64 contains the larger value, h (Fig. 4.9).

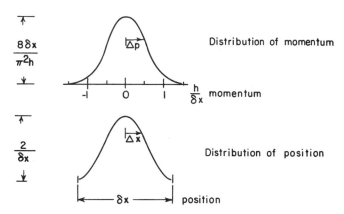

Fig. 4.9 Relation between phase cell dimensions, distribution of positions and momenta, and uncertainties in these variables. Only the simplest of a large family of distribution functions corresponding to different energies are shown.

Fig. 4.10   Phase space is a six-dimensional hypothetical space, having three momentum and three spatial dimensions. Projected onto the $p_x - x$ plane, individual cells always present an area h. Although their shapes may be quite arbitrary, as shown, it is often useful to think of them as square or rectangular since that makes computations simpler. In section 4:14 (Fig. 4.13) we will see how an initially rectangular cell becomes distorted.

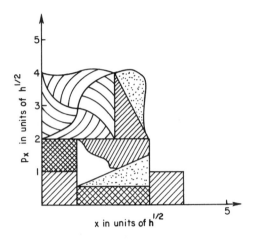

We can now ask how many electrons could be fitted into a box with volume V? The answer depends on how high a particle momentum we wish to consider. If momenta up to some value p_m are to be permitted, the available volume in phase space is $2 \cdot 4\pi/3 \cdot p_m^3 \cdot V$. The factor 2 accounts for the different spin polarization, since electrons whose spins differ, always can be distinguished and therefore must belong to different phase cells. Hence the number of available phase cells is $8\pi/3 \cdot p_m^3 \cdot V/h^3$, and that also is the maximum number of electrons that could occupy the box.

In general, the number of phase cells with momenta in a range p to $p + dp$ is

$$Z(p)\,dp = 2V\,\frac{4\pi p^2\,dp}{h^3} \qquad (4\text{–}65)$$

At the center of a star, ionized matter is sometimes packed so closely that all the lowest electron states are filled. Further contraction of the star can then force the electrons to assume much higher momenta than the value $(3kTm)^{1/2}$ normally found in tenuous gases. Such a closely packed gas of Fermions is said to be *degenerate.* We will study this form of matter in section 4:14 and in Chapter 8, where very dense cores of stars are discussed.

Sometimes we may prefer to talk about *frequency space* instead of *momentum space.* Defining the particle frequency v, by $v \equiv pc/h$, we obtain the number of phase cells with frequencies between v and $v + dv$ as

$$Z(v)\,dv = 2\left[\frac{4\pi v^2\,dv}{c^3}\right]V \qquad (4\text{–}65a)$$

4:11

4:12 ANGULAR DIAMETERS OF STARS

The fact that two photons sometimes occupy the same phase cell allows us to measure the angular diameter of stars. The idea is this: Two photon counters are placed a distance D apart, transverse to the direction of the star. If D is small enough, we have the possibility that one photon from a cell will hit one detector, while the other photon hits the other detector, the simultaneous arrival being detected by a coincidence counter. Let the diameter of the star be d and its distance R. The angle it subtends is $\theta = d/R$. The photon pair impinging on either detector has a distribution in momentum, along the direction of D, amounting to $\Delta p_D = p\theta = (h\nu/c)\,\theta$ where ν is the frequency of radiation to which the detector is sensitive. But the nonzero value of Δp_D makes it necessary that D itself be small so that photons reaching either detector may be in the same phase cell. That is, it is necessary that

$$D\Delta p_D \lesssim h$$

or $$\frac{Dh\nu}{c}\theta \lesssim h \qquad (4\text{–}66)$$

or $$D\theta \lesssim \lambda, \quad \lambda = c/\nu$$

λ is the *wavelength* of the radiation. By increasing D a decreasing coincidence rate is observed, and for values of D at which coincidences no longer occur the angular diameter is $\theta \lesssim \lambda/D$. The stellar *angular* diameter (Figure 4.11) is

$$\theta \sim d/R \sim \lambda/D \qquad (4\text{–}67)$$

in such observations. This technique was first discovered by R. Hanbury Brown and R. Q. Twiss (Ha54). A second method of measuring angular diameters of stars makes use of the Michelson stellar interferometer that essentially depends on the same type of phenomenon in that the normal diffraction peak width again is given by uncertainty principle considerations.

Fig. 4.11 The Hanbury-Brown-Twiss interferometer.

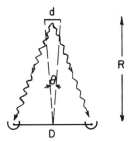

4:13 THE SPECTRUM OF LIGHT INSIDE AND OUTSIDE A HOT BODY

Any opaque body is permeated by a radiation bath. Atoms, molecules, or ions and electrons are continually absorbing and re-emitting quanta of light. From time to time a photon approaches the edge of the body and escapes. This diffusion of photons from the interior of the hot body out to its boundary, and the subsequent escape into empty space is an important process in stars. Energy generated at the center of the star slowly diffuses outward and escapes. The escaping radiation gives the star its luminous appearance.

To understand this phenomenon in some detail, we want to describe the radiation to be expected inside a body. The spectrum of the radiation is going to be different for different temperatures, and we would like to deduce that spectrum and obtain a complete description of the radiation bath.

We first consider a photon gas embedded in material at temperature T. The radiation is in thermal equilibrium with the material if there is ample opportunity for the photons to interact with the atoms through scattering or absorption and re-emission. Two factors have to be considered:

(a) Photons are Bose–Einstein particles and can aggregate in single phase cells.

(b) If the frequency of the photons aggregating in a phase cell is v, and if there are n photons in the cell, we can consider the assembly of photons in this phase cell to be in an energy state with energy $(n + \frac{1}{2}) hv$. We sometimes speak of a *quantum oscillator* in the nth state when we describe this phenomenon. Even when a phase cell is completely empty, in the ground state, a residual ground state energy of $hv/2$ is present. This energy normally cannot be observed, because it does not participate in the absorption and emission processes. [However, there exist (Fr46) processes involving the surfaces of deformable bodies in which this ground state energy can give rise to surface tension effects.]

We can compute the probability of finding a quantum oscillator in the nth excited state. The relative probability of that state is given by the Boltzmann factor $e^{-(n+1/2)hv/kT}$. The absolute probability is given by dividing the relative probability by the sum of all the relative probabilities:

$$P(v, T) = \frac{e^{-(n+1/2)hv/kT}}{\sum\limits_{n} e^{-(n+1/2)hv/kT}} = \frac{e^{-(nhv/kT)}}{\sum\limits_{n} e^{-nhv/kT}} \tag{4–68}$$

In these terms we can give the average energy $\langle \mathscr{E} \rangle$, per phase cell, of all phase cells corresponding to frequency v. We sum the energies of all the oscillators

4:13

and divide by the total number of oscillators. Choosing $x \equiv hv/kT$, we obtain

$$\langle \mathscr{E} \rangle = \sum_n \left(n + \frac{1}{2} \right) hv e^{-nhv/kT} \left[\sum_n e^{-nhv/kT} \right]^{-1}$$

$$= \frac{kT(xe^{-x} + 2xe^{-2x} + 3xe^{-3x} + \ldots)}{1 + e^{-x} + e^{-2x} + e^{-3x} + \ldots} + \frac{hv}{2} \qquad (4\text{-}69)$$

This allows us to evaluate the denominator of equation 4–69 as $(1 - e^{-x})^{-1}$. To evaluate the numerator, we can make use of the same binomial expansion formula twice in succession.

$$kT\{x(e^{-x} + e^{-2x} + e^{-3x} + \ldots)$$
$$+ x(e^{-2x} + e^{-3x} + \ldots)$$
$$+ x(e^{-3x} + \ldots)$$
$$+ \ldots \ldots \ldots \ldots \ldots \ldots \ldots \ldots \ldots \ldots)\}$$

$$= kT\left\{ \frac{xe^{-x}}{1 - e^{-x}} + \frac{xe^{-2x}}{1 - e^{-x}} + \frac{xe^{-3x}}{1 - e^{-x}} + \ldots \right\}$$

$$= kT \frac{xe^{-x}}{(1 - e^{-x})^2}$$

In these terms

$$\langle \mathscr{E} \rangle = \frac{kTxe^{-x}}{1 - e^{-x}} + \frac{hv}{2} = \frac{kTx}{(e^x - 1)} + \frac{hv}{2} = \frac{hv}{(e^{hv/kT} - 1)} + \frac{hv}{2} \qquad (4\text{-}70)$$

Knowing the number of phase cells per unit volume, $8\pi v^2 \, dv/c^3$, and the mean energy per phase cell, we can write the energy density of photons as a function of frequency and temperature. This is the *blackbody radiation* spectrum:

$$\rho(v, T) \, dv = \frac{8\pi v^2 \, dv}{c^3} \left(\frac{hv}{e^{hv/kT} - 1} + \frac{hv}{2} \right) \qquad (4\text{-}71)$$

As previously stated, the term $hv/2$ is not observable in terms of photon absorption or emission. We will therefore neglect it from now on, and concentrate on the remainder of the expression, which can give rise to observable astronomical signals. Integrating equation 4–71 over all frequencies from zero to infinity, we can obtain the total energy density. The total photon density is similarly obtained:

$$\rho(T) = \frac{8}{15} \frac{\pi^5}{c^3} \frac{k^4}{h^3} T^4 = aT^4 = U = 7.6 \times 10^{-15} T^4 \, \text{erg cm}^{-3}$$

(4-72)

$$n(T) = \frac{8\pi}{c^3} \int_0^\infty \frac{v^2 \, dv}{e^{hv/kT} - 1} = 20T^3 \, \text{photons cm}^{-3}$$

PROBLEM 4-21. Note that all this is strictly true only if the index of refraction, n, in the medium is $n = 1$. For general values of n, show that

$$\rho(T) = n^3 aT^4$$

This is more generally the case inside a star. Show also what happens if the index of refraction is frequency dependent—which it always is.

Equation 4-72 is a well-known definite integral. We note that the second term in parentheses in equation 4-71 would give rise to an infinite zero point energy as written there. The coefficient of the T^4 term is sometimes abbreviated (see equation 4-72) by the symbol a. We can also define another useful constant $\sigma = ac/4$, *the Stefan–Boltzmann constant*. This constant allows us to write the energy emitted per unit area of a hot blackbody in unit time, as

$$W = \sigma T^4 \qquad (4\text{-}73)$$

To see this, we can think of photons that escape from the surface as being representative of the density of photons immediately within the surface of the body. Only those photons can be considered with velocities directed outward through the surface. So only one half of the photons come into consideration. These photons have an average velocity component normal to the surface equal to $c\langle \cos \theta \rangle$ where θ is the angle of emission with respect to the direction normal to the surface. We therefore have to evaluate $\langle \cos \theta \rangle$, averaged over all possible angles. This is

$$\langle \cos \theta \rangle = \frac{1}{2\pi} \int_0^{2\pi} \int_0^{\pi/2} \cos \theta \sin \theta \, d\theta \, d\phi = \left. \frac{\cos^2 \theta}{2} \right|_0^{\pi/2} = \frac{1}{2}$$

(4-74)

$$\therefore \, c\langle \cos \theta \rangle = \frac{c}{2}$$

But since only half the photons are outward directed, the total flux is $1/2 \cdot c/2 \cdot aT^4 = acT^4/4$, as previously stated.

The spectrum of most stars is closely approximated by a blackbody spectrum, with individual spectral emission and absorption lines superposed. To the extent that the blackbody approximation holds, it is possible to ascertain the temperature of the star's photosphere where most of the light is emitted. Using two different wide band filters, say the B and V filters often used in observations, we can determine the ratio of intensities in these spectral ranges. This ratio can be uniquely related to the temperature. The temperature derived in this way is called the *color temperature*. A useful formula is (Al63):

$$T_c = \frac{7300}{(B - V) + 0.73} \tag{4–75}$$

PROBLEM 4–22. Using the effective wavelengths given in Table A.1, compare the ratio of blue and visual intensities predicted, respectively, by equations 4–71 and 4–75 for a star at temperature 6000°K (spectral class G) and one at 10,000°K (spectral class A).

Another means of defining temperature involves the luminosity of the star. Since the total power emitted per unit area is a function of temperature alone, we can calculate an *effective temperature T_e* for the star if both its luminosity and surface area can be determined:

$$L = \sigma T_e^4 4\pi R^2 \tag{4–76}$$

If the distance of the star is known from observations of the kind described in Chapter 2, the stellar radius can be obtained using the Michelson or Hanbury Brown–Twiss interferometers discussed in section 4:12. From (4–76) it is readily seen that

$$\log \frac{L}{L_\odot} = 4 \log \frac{T_e}{T_{e\odot}} + 2 \log \frac{R}{R_\odot} \tag{4–77}$$

where $T_{e\odot} \sim 5800°K$ and $R_\odot = 6.96 \times 10^{10}$ cm, are the solar values. $T_{e\odot}$ is uncertain by as much as $\sim 50°K$ because the ultraviolet and infrared components of $L_\odot$ are still not well known. When the Hertzsprung–Russell diagram is plotted in terms of the logarithm of luminosity and effective temperature, as in Fig. 1.5, stars with identical radii lie on lines of constant slope, as required by equation 4–77.

It is worth mentioning two typical astrophysical situations in which temperature is a useful concept.

(a) Temperatures in the Solar System

The temperature of a black interplanetary object is determined by the energy equilibrium equation

$$\frac{L_\odot}{4\pi R^2}\,\pi r^2 = \sigma T^4 4\pi r^2$$

where $L_\odot$ is the solar luminosity, R is the distance from the sun, and r is the radius of the object. If the mean efficiency for absorption (in the visible) is ε_a and the mean efficiency of reradiation (at infrared wavelengths) is ε_r, we have

$$T = \left(\frac{\varepsilon_a}{\varepsilon_r}\,\frac{L_\odot}{16\pi\sigma R^2}\right)^{1/4} \tag{4–78}$$

We note that:

(i) At the earth's distance

$$T \sim \left(\frac{\varepsilon_a}{\varepsilon_r}\right)^{1/4}\left(\frac{4\times 10^{33}}{16\pi(5.7\times 10^{-5})2.3\times 10^{26}}\right)^{1/4} \sim 282\left(\frac{\varepsilon_a}{\varepsilon_r}\right)^{1/4}\,{}^\circ K$$

(ii) A gray body ($\varepsilon_a = \varepsilon_r$) has the same temperature as a black one.
(iii) For increasing distance from the sun $T \propto R^{-1/2}$.
(iv) If the thermal conductivity of the body is small and its rotation slow, as for the moon, the subsolar point assumes a temperature

$$T \sim \left(\frac{\varepsilon_a}{\varepsilon_r}\,\frac{L_\odot}{R^2 4\pi\sigma}\right)^{1/4} \tag{4–79}$$

which is $(4)^{1/4} \sim 1.4$ higher than the temperature of an equivalent rotating body.

(b) Radio Astronomical Temperatures

Some characteristics of radio astronomical measurements can be understood in terms of temperatures. At very low frequencies, $v \ll kT/h$, we find that the energy density in a source can be written (equation 4–71) as

$$\rho(v) = \frac{8\pi kTv^2}{c^3} = \frac{8\pi kT}{c\lambda^2} \tag{4–80}$$

where $\lambda \equiv c/v$ is the wavelength. The flux emanating per solid angle and area normal to the surface is called the *intensity* $I(v)$, at frequency v. This is the *surface brightness* of the source, and

$$I(v) = \frac{c\rho(v)}{4\pi} = \frac{2v^2 kT}{c^2} = \frac{2kT}{\lambda^2} \qquad (4\text{–}81)$$

If a flux $I(v)$ is measured in an observation then, regardless of whether the source is thermal, we can pretend that a temperature parameter can be assigned to the observation. This is called the *brightness temperature* T_b and is defined at frequency v as

$$T_b(v) \equiv \frac{I(v)\, c^2}{2kv^2} = \frac{I(v)}{2k}\lambda^2 \qquad (4\text{–}82)$$

T_b then is the temperature of an ideal blackbody whose radiant energy in the particular energy range v to $v + dv$ is the same as that of the observed source. A related concept is that of *antenna temperature*—which has nothing to do with the temperature that the antenna actually assumes under ambient climatic conditions. To examine this concept we must first consider some practical properties of antennas. In general, an antenna absorbs differing amounts of power depending on the direction of the source. If we draw a *directional diagram* of an antenna, it usually has the shape of Figure 4.12. The response $A(\theta, \phi)$ of the antenna is called its *effective area*. The power absorbed is

$$P \equiv \frac{1}{2} \int A(\theta, \phi)\, I(v, \theta, \phi)\, dv\, d\Omega$$

and for a small source,

$$P(v, \theta, \phi)\, dv = \frac{1}{2} F(v)\, A(\theta, \phi)\, dv, \qquad F(v) = \int I(v)\, d\Omega \qquad (4\text{–}83)$$

Fig. 4.12 Directional diagram of an antenna, showing a main lobe and a set of sidelobes. The angle θ is the beamwidth (see text).

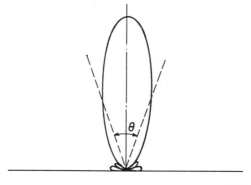

Here $F(v)$ is the flux density at the antenna, and the factor $\frac{1}{2}$ comes about because the antenna only accepts one component of polarization. Now if A is independent of the angle ϕ, and a diagram like Fig. 4.12 is drawn, $A(\theta)$ normally has a very large value in one particular direction, $\theta = 0$, and the large lobe around this direction is called the *main lobe*. The smaller lobes in the diagram are called side lobes. *Back lobes* can also occur. A well-designed radio telescope has a narrow main lobe for greatest positional accuracy, and minimized sidelobes to minimize the confusion produced by sources outside the desired field of view (Sh60).

We can define a mean value of the effective area of the antenna taken over all directions as

$$\langle A \rangle \equiv \frac{1}{4\pi} \int A(\theta, \phi) \, d\Omega \qquad (4\text{--}84)$$

and the *gain* of the antenna is the dimensionless quantity

$$G(\theta, \phi) = \frac{A(\theta, \phi)}{\langle A \rangle} \qquad (4\text{--}85)$$

which gives the ratio of the effective area in a given direction to the mean effective area. The function G has a maximum value in the direction $\theta = \phi = 0$ in a properly designed instrument. The *beamwidth* is the angle θ between points in the directional diagram at which $A = A(\theta, \phi)/2$.

In these terms, we can now return to the concept of *antenna temperature* T_a. If a source has directional and spectral brightness $I(v, \theta, \phi)$, then a radio telescope with effective area $A(\theta, \phi)$ receives an amount of power

$$P(v) = \frac{1}{2} \int A(\theta, \phi) \, I(v, \theta, \phi) \, dv \, d\Omega \qquad (4\text{--}86)$$

On the other hand, we could replace the antenna by a resistor at temperature T connected to the receiver. Such a resistor can be shown experimentally and theoretically to produce *thermal noise* power in an amount

$$P = kT\Delta v \qquad (4\text{--}87)$$

where Δv is the receiver bandwidth. We can therefore define an antenna temperature T_a, so that

$$T_a = \frac{1}{k\Delta v} \cdot \frac{1}{2} \int A(\theta, \phi) \, I(v, \theta, \phi) \, dv \, d\Omega \qquad (4\text{--}88)$$

This equation is useful for practical reasons. It is relatively easy to compare the power received from a celestial source to that received from a resistor switched to the receiver input in place of the antenna. The noise in (4–87) is sometimes called *Johnson noise* or *Nyquist noise*. Johnson (Jo28) and Nyquist (Ny28), respectively, supplied the experimental data and theoretical explanation leading to (4–87).

4:14 BOLTZMANN EQUATION AND LIOUVILLE'S THEOREM

Define a function $f(\mathbf{r}, \mathbf{p}, t)$, the density of particles in phase space, so that the number of particles in volume element $d\mathbf{r}$ at position $\mathbf{r}$, having moments that lie in some momentum-space volume $d\mathbf{p}$, is $f(\mathbf{r}, \mathbf{p}, t)\, d\mathbf{r}\, d\mathbf{p}$. We ask how the function f evolves with time. Since each particle in the assembly can be described in terms of three momentum and three spatial coordinates, the general form of the equation reads

$$\frac{\partial f}{\partial t} + \sum_i \frac{\partial f}{\partial \mathbf{r}_i} \frac{d\mathbf{r}_i}{dt} + \sum_i \frac{\partial f}{\partial \mathbf{p}_i} \frac{d\mathbf{p}_i}{dt} = \frac{df}{dt}\bigg|_{\text{collisions}} \tag{4–89}$$

The left side of this equation gives the time rate of change of particles in the volume element $d\mathbf{r}\, d\mathbf{p}$ as a function of the coordinates $\mathbf{r}_i, \mathbf{p}_i, i = 1, 2, 3, \dots, n$, for an n particle assembly. As the particles move, the surface enclosing them in phase space becomes distorted and the expression gives the rate of change of density through this distortion and through any other effects. The right side gives the loss or gain of particles through collisions. Equation 4–89 is called the *Boltzmann equation*.

To see how the evolution proceeds in the case of a collisionless situation, in which the right side of equation 4–89 is zero, we draw a simple two-dimensional picture. In Figure 4.13 we have an assembly of particles initially confined between positions $\mathbf{r}_1$ and $\mathbf{r}_2$ and between momentum values $\mathbf{p}_a$ and $\mathbf{p}_b$. Some time later, the momentum values are unchanged, but the particles have moved so that the higher momentum particles are now at positions between $\mathbf{r}'_1$ and $\mathbf{r}'_2$ while the lower momentum particles are at positions between $\mathbf{r}''_1$ and $\mathbf{r}''_2$. However, since the base and height of the enclosing area has not changed, the number density of particles per unit area has remained constant.

A similar argument holds when forces are applied to the particles. In that case the momenta of particles are not constant and the parallelepiped in Fig. 4.13 will also be displaced in a vertical direction. However, a similar argument can then be applied to show that the area covered by the particles still remains constant and the density of particles in this two-dimensional situation is unchanged. This is particularly easy to see if the force is the same on all particles. In that case, $d\mathbf{p}/dt$ is uniform and the difference in values $\mathbf{p}_a - \mathbf{p}_b$ is maintained constant. When different forces are exerted on differing constitutents of the gas, the area occupied

Fig. 4.13 Evolution of a collisionless assembly of particles in two-dimensional phase space.

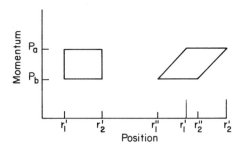

by each constitutent remains constant. These results also hold when there are gradients in the force fields.

A further extension of the argument can be applied to the full six-dimensional situation. Unless there are some means for creating or destroying particles in the assembly (such as through collisions or through, say, particle-antiparticle pair formation) the density of particles will be constant along the trajectory in the six-dimensional space.

This is the sense of *Liouville's theorem*: The six-dimensional space density of particles in an assembly remains unchanged unless collisions occur:

$$df/dt = 0 \qquad\qquad (4\text{–}90)$$

This theorem has interesting applications to cosmic ray particles. Many of these particles are so energetic that they must be able to escape from the magnetic fields resident in our Galaxy. Their density in space outside the Galaxy must therefore be the same as the density that we measure in the vicinity of the earth, provided only that they have had enough time since their creation to travel distances comparable to the dimensions of the universe. This follows at once if we consider an assembly of energetic particles passing by the earth. These particles move on throughout the Galaxy, guided by magnetic field lines (see section 6:6 for further discussion of this topic). Eventually they escape the Galaxy, because the Galaxy's magnetic field is not strong enough to keep them trapped. As these particles go out into intergalactic space their density in phase space must still be constant. If the arrival rate of cosmic ray particles at the earth could be shown to be constant in time this would mean that the spatial density of cosmic rays in extragalactic space is the same as that measured at the earth. This argument need not be true for low energy particles if these particles can remain bottled up in local magnetic fields within our Galaxy. However, if these magnetic fields allow a small leakage of low energy particles into extragalactic space, Liouville's theorem again demands that the density of the particles eventually become uniform throughout these accessible spaces. A local probe of cosmic ray particle intensities can, for this reason, potentially be useful in giving information on particle densities throughout the universe. On the other hand, we should not be overly optimistic. The recent discovery that pulsars in our Galaxy may be responsible for most of the cosmic ray particles we see, suggests that local measurements at the earth may not have a direct bearing on the density of particles in extragalactic space, because there may not have been enough time to produce a homogeneous distribution of cosmic ray particles in extragalactic regions.

Another interesting application of Liouville's theorem concerns the use of optical telescopes in concentrating light beams onto small detectors. In many applications we could obtain very high instrumental sensitivity if light from some cosmic source could be concentrated onto the smallest possible detector. Let the solid angle subtended by the astronomical object be Ω and the telescope area be A. Then Liouville's theorem states that the smallest detector area onto which the light could be focused is

$$ a = \frac{A\Omega}{4\pi} \tag{4–91} $$

and that is only possible if light can be made to impinge on the detector from all sides. Usually we are able to make light fall onto the detector only from some smaller solid angle $\Omega' < 4\pi$ so that the minimum area of the detector becomes

$$ a = A\Omega/\Omega' \tag{4–92} $$

A violation of this situation would also imply a violation of the second law of thermodynamics that states that heat cannot freely flow from a cold to a hot object. For the density of photons in phase space would then be greater at the detector than at the source. That would imply that the radiation temperature at the source was lower than at the detector and that radiation was actually flowing from a cooler to a hotter object.

Finally we should still mention the problem discussed in section 3:16, where a swarm of particles moves through a gravitational field. There we were concerned with tidal disruption of globular clusters, but noted that while the clusters became extended along a direction pointing toward the Galactic center, the gravitational forces also tended to produce a compression lateral to that direction. This compression produces some additional transverse velocities so that the overall evolutionary pattern becomes quite complex. Liouville's theorem, however, gives us at least one solid guide toward understanding the overall development. It tells us that whatever detailed dynamical arguments we apply— such as those of section 3:16—the results must always agree at least with Liouville's requirement of a constant phase space density.

4:14

4:15 FERMI-DIRAC STATISTICS

In a *Fermi–Dirac assembly* a phase cell can contain only one particle or none. For any given assembly there exists a *Fermi energy*, $\mathscr{E}_F$, up to which all states are filled at zero temperature. At $T > 0$ excitation to a higher state of energy $\mathscr{E}$ can take place. The relative probabilities of being at energy $\mathscr{E}$ and αkT are, respectively,

$$e^{-(\mathscr{E} - \alpha kT)/kT} \quad \text{and } 1 \tag{4–93}$$

The relative probability of occupancy of a state of energy $\mathscr{E}$, in an assembly at temperature T, therefore, is

$$\frac{e^{\alpha - \mathscr{E}/kT}}{1 + e^{\alpha - (\mathscr{E}/kT)}} = \frac{1}{1 + e^{(\mathscr{E}/kT) - \alpha}} \tag{4–94}$$

Here we have not specified the energy αkT, but we can see that at very low temperatures, $T \sim 0$, αkT must approach $\mathscr{E}_F$ because the Fermi function

$$F(\mathscr{E}) = \left[1 + e^{(\mathscr{E} - \mathscr{E}_F)/kT}\right]^{-1} \tag{4–95}$$

has the form shown in Fig. 4.14.

Fig. 4.14 The Fermi function $F(\mathscr{E})$.

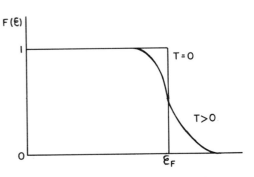

We define the Fermi energy $\mathscr{E}_F$ as that energy for which $F(\mathscr{E}) = \frac{1}{2}$. Note that for $T = 0$, the exponent in (4–95) has a large absolute value, whenever $\mathscr{E} - \mathscr{E}_F \neq 0$, so that

$$F(\mathscr{E}) = 1 \quad \text{for} \quad \mathscr{E} < \mathscr{E}_F$$
$$F(\mathscr{E}) = 0 \quad \text{for} \quad \mathscr{E} > \mathscr{E}_F \tag{4–96}$$

This gives rise to the step function in Figure 4.14. Whenever all the available energy levels are filled—which means whenever $T = 0$—we say that the gas of

4:15

Fermions is completely *degenerate*. When $T > 0$, the step is seen to roll off more gently. The product of this probability $F(\mathscr{E})$ and $\mathscr{E}$ gives a mean value for the energy contained in all phase cells corresponding to an energy $\mathscr{E}$. Filled as well as empty cells have to be considered to obtain this value.

$$\langle\mathscr{E}\rangle = \frac{\mathscr{E}}{1 + e^{(\mathscr{E}/kT) - \alpha}} \tag{4-97}$$

We know that the number of states in the momentum range

$$p \quad \text{to} \quad p + dp \quad \text{is} \quad Z(p)\,dp = \frac{8\pi p^2 V\,dp}{h^3}$$

But

$$\mathscr{E} = \frac{p^2}{2m}, \quad d\mathscr{E} = \frac{p}{m}\,dp$$

$$\therefore Z(\mathscr{E})\,d\mathscr{E} = \frac{8\pi}{h^3}\,V\sqrt{2m\mathscr{E}}\,m\,d\mathscr{E} = \frac{4\pi V}{h^3}(2m)^{3/2}\,\mathscr{E}^{1/2}\,d\mathscr{E} \tag{4-98}$$

The overall mean energy of the particles integrated over all values $\mathscr{E}$ is therefore

$$\langle\mathscr{E}\rangle = \frac{\displaystyle\int_0^\infty Z(\mathscr{E})\langle\mathscr{E}\rangle d\mathscr{E}}{\displaystyle\int_0^\infty Z(\mathscr{E})\,d\mathscr{E}} = \frac{\displaystyle\int_0^\infty \frac{\mathscr{E}^{3/2}\,d\mathscr{E}}{1 + e^{\mathscr{E}/kT - \alpha}}}{\displaystyle\int_0^\infty \mathscr{E}^{1/2}\,d\mathscr{E}} \tag{4-99}$$

Again, setting αkT equal to the Fermi energy, $\mathscr{E}_F$, for an assembly of particles at temperature $T = 0$, we obtain

$$\langle\mathscr{E}\rangle_{T=0} = \int_0^{\mathscr{E}_F} \mathscr{E}^{3/2}\,d\mathscr{E}\left[\int_0^{\mathscr{E}_F} \mathscr{E}^{1/2}\,d\mathscr{E}\right]^{-1}$$

$$\langle\mathscr{E}\rangle_{T=0} = \frac{3}{5}\mathscr{E}_F \tag{4-100}$$

One can show that for $T > 0$, $\mathscr{E}_F < \mathscr{E}_{F_0}$. The Fermi energy drops slightly. For $\mathscr{E} - \mathscr{E}_F \gg kT$, that is, in the limit of large particle energies, we have

$$F(\mathscr{E}) \sim e^{-(\mathscr{E} - \mathscr{E}_F)/kT}$$

which approaches a Boltzmann distribution for very energetic Fermions.

At the center of stars degenerate conditions often exist. This is true mainly for electrons since at a given energy $\mathscr{E}$, $p = \sqrt{2\mathscr{E}m}$ is less for electrons than for protons, by a factor $\sqrt{m_p/m_e}$. The lower energy electron states therefore become fully occupied—degenerate—much more readily than proton states.

PROBLEM 4–23. Suppose the universe is filled with completely degenerate neutrinos up to an energy Φ_v at a neutrino temperature $T_v = 0$. Show that ρ_v, the mass density (energy density divided by c^2) of neutrinos is

$$\rho_v = \frac{\pi \Phi_v^4}{h^3 c^5} \tag{4-101}$$

Note that neutrinos exist in only one spin state, not two (Wa67).

To see why electrons and protons, which are actually Fermions, appear to have the characteristics of Maxwell–Boltzmann particles in many astrophysical situations we note the following. We can derive the velocity distribution for classical particles in a way similar to the derivation of the Fermi–Dirac distribution. Assume that particles can occupy arbitrary positions in momentum and extension space. This is equivalent to saying that the phase cells are infinitessimally small. We can obtain such a system by pretending that Planck's constant goes to zero as a limit: $h \to 0$. This makes $\mathscr{E}_F = 0$, since arbitrarily many particles can have zero and near-zero energies, and the probabilities in (4–93) become $e^{-\mathscr{E}/kT}$ and 1. We now write the number of particles in the assembly, having momenta near p:

$$n(p)\, dp \propto \frac{8\pi p^2 V}{h^3}\, dp\, e^{-p^2/2mkT} \tag{4-102}$$

Integrating over all p values,

$$n = C \int_0^\infty \frac{8\pi V}{h^3} p^2 e^{-p^2/2mkT}\, dp \tag{4-103}$$

where C is the proportionality constant. This is an error function integral whose value is the total number of particles

$$n = C \frac{8\pi V}{h^3} \left(\frac{1}{4} \sqrt{\pi (2mkT)^3} \right)$$

Hence

$$C = \frac{nh^3}{2V(2\pi mkT)^{3/2}} \tag{4-104}$$

so that

$$n(p) = \frac{4\pi n p^2 e^{-p^2/2mkT}}{(2\pi mkT)^{3/2}}$$

a result obtained earlier.

The Maxwell–Boltzmann statistics apply in all problems dealing with the motion of particles in the atmospheres of stars and planets, with nondegenerate matter in the interior of stars, and with situations involving gas and the random motion of dust grains in interplanetary and interstellar space. These statistics also apply in some problems of stellar dynamics in which the stars can be thought of as members of an interacting assembly. Galaxies moving within a cluster are also believed to obey the M–B statistics. The formulas developed in the next sections therefore have wide applications in astrophysics.

4:16 THE SAHA EQUATION

In a very hot gas, we often find multiply ionized states of any given atom in a state of thermal equilibrium. The ratio of the densities of the different states of ionization is then given by the Saha equation

$$\frac{n_{r+1} n_e}{n_r} = \frac{g_{r+1} g_e}{g_r} \frac{(2\pi mkT)^{3/2}}{h^3} \exp\left(-\chi_r/kT\right) \tag{4–105}$$

In this equation n_r is the number density of atoms in the rth state of ionization and n_{r+1} is the number density of atoms in the next higher state of ionization. The electron density is n_e. The corresponding factors g_{r+1}, g_r, and g_e on the right side of the equation are statistical weights to be assigned to the different states of ionization and to the electron. The electron has two states of polarization and so $g_e = 2$; but the other two weighting factors are more complicated. The quantity χ_r is the energy required to ionize the atom from the rth state into the $(r + 1)^{\text{st}}$ state of ionization.

The significance of this equation can now be discussed in rough qualitative terms. We can imagine that thermal equilibrium in a hot gas requires the balancing of a large variety of processes. Not only is there the possibility of ionization through intercollisions of particles or through the absoprtion of radiation, but ions can also recombine with electrons to form species in the next lower state of ionization. The recombination rate will be proportional to the product $n_{r+1} n_e$ given on the left side of the equation. If, for example, the bulk of the ionization proceeds radiatively, the ionization rate would be proportional to n_r and to some function of the equilibrium temperature T, which determines the photon density.

At the same time we would also expect the ratio of number densities n_{r+1}/n_r to be determined by an exponential Boltzmann factor involving the difference in energy χ_r between these states. This is a form similar to that encountered in our previous discussion. Finally, the weighting factors are present because they define the number of different ways that the various configurations can be filled. We have made no attempt here to derive (4–105)—only to make it plausible.

The Saha equation as presented here is simplified in two respects. First, no account has been taken of degeneracy, and so the form given here holds only for dilute gases. In a dense configuration we must also take into account that two electrons cannot occupy the same position in space if their momenta and polarization states are identical. This leads to greater complexity in the final expression obtained. Second, we have ignored the fact that each state of ionization may have many excitation levels. The relation between the number densities of the ith excitation state of the rth level of ionization and the jth state of the $r + 1^{st}$ level then becomes

$$\frac{n_{r+1,j}n_e}{n_{r,i}} = \frac{g_{r+1,j}g_e}{g_{r,i}} \frac{[2\pi mkT]^{3/2}}{h^3} \exp - \left[\frac{\chi_r + \mathscr{E}_{r+1,j} - \mathscr{E}_{r,i}}{kT} \right] \quad (4\text{–}106)$$

We will need this equation to discuss reactions inside stars.

PROBLEM 4–24. In the solar corona, collisional excitation of atoms predominates over other processes of excitation. Among the identified spectral lines are those of CaXIII (12 times ionized calcium) and CaXV (14 times ionized). The ionization potentials of these ions are 655 and 814 ev, respectively. The lines from CaXIII are considerably stronger than those of CaXV. This fact alone can tell us very roughly what the temperature of the corona is. What is it?

4:17   MEAN VALUES

Once the energy, frequency, or momentum distribution of particles in an assembly are known, mean values of various functions of these parameters can be computed. For particles obeying Maxwell–Boltzmann statistics, the mean value of a function $F(p)$ is

$$\langle F(p) \rangle = \frac{\displaystyle\int_0^\infty Z(p)\, F(p)\, e^{-p^2/2mkT}\, dp}{\displaystyle\int_0^\infty Z(p)\, e^{-p^2/2mkT}\, dp} \quad (4\text{–}107)$$

4:17

This equation has exactly the form of equation 4–19. The integrand in the denominator is the probability of finding a particle with momentum p

PROBLEM 4-25. Two frequently used quantities are $\langle p \rangle$ and $\langle p^2 \rangle$. Show that

$$\langle p \rangle = \frac{\displaystyle\int_0^\infty p^3 e^{-p^2/2mkT}\,dp}{\displaystyle\int_0^\infty p^2 e^{-p^2/2mkT}\,dp} = \frac{\frac{1}{2}(2mkT)^2\,\Gamma(2)}{\frac{1}{2}(2mkT)^{3/2}\,\Gamma(\frac{3}{2})} = \sqrt{\frac{8mkT}{\pi}} \qquad (4\text{–}108)$$

Note that this is the mean magnitude of the momentum. The mean momentum $\langle \mathbf{p} \rangle$ is zero because momenta along different directions cancel. Show also that

$$\langle p^2 \rangle = \frac{\displaystyle\int_0^\infty p^4 e^{-p^2/2mkT}\,dp}{\displaystyle\int_0^\infty p^2 e^{-p^2/2mkT}\,dp} = 3mkT \qquad (4\text{–}109)$$

PROBLEM 4-26. In section 6:16 we will make use of the quantity $\langle 1/v \rangle$. Show that

$$\left\langle \frac{1}{v} \right\rangle = \sqrt{\frac{2m}{\pi kT}} \qquad (4\text{–}110)$$

PROBLEM 4-27. In astronomical spectroscopy we can only measure velocities of atoms along a line of sight when observing the shape of a spectral line. To determine the temperature of a gas, whose mean squared random velocity $\langle v_r^2 \rangle$ along the line of sight is known, we therefore have to know how $\langle v_r^2 \rangle$ and T are related. For a Maxwell–Boltzmann distribution show that

$$\langle v_r^2 \rangle = kT/m$$

These integrals all have the form

$$\int_0^\infty x^{n-1} e^{-ax^2}\,dx = \frac{\Gamma(n/2)}{2a^{n/2}} \qquad (4\text{–}111)$$

where $\Gamma(n) = (n-1)\,!$ and $\Gamma(\frac{1}{2}) = \sqrt{\pi}$.

The analogous integrals required for computing mean values for energies or momenta for Fermions or Bosons involve the Fermi-Dirac or Bose–Einstein distribution functions.

4:18 THE FIRST LAW OF THERMODYNAMICS

The first law of thermodynamics expresses the conservation of energy. If a gas is heated, the supplied energy can act in one of two ways. It can raise the gas temperature, or it can perform work by expanding the gas against an externally applied pressure. Symbolically we write*

$$dQ = dU + Pd\mathscr{V} \tag{4-112}$$

where all quantities are normalized to one mole of matter and where the left-hand side gives the amount of heat dQ supplied to the system; dU is the change in internal energy and $Pd\mathscr{V}$ is the work performed. The nature of this last term is easily understood if we recall that *work* is involved in any displacement D against a force F. If the change in volume $d\mathscr{V}$ involves, say, the displacement of a piston of area A, then the force involved is $F = PA$ and the distance the piston moves is $D = d\mathscr{V}/A$

The *internal energy* U of the gas is the sum of the kinetic energy of translation, as the molecules shoot around; the kinetic and potential energy involved in the vibrations of atoms within a molecule; the energy of excited electronic states; and the kinetic energy of molecular rotation.

Q is the *heat content* of the system. The heat Q that must be supplied to give rise to a one degree change in temperature is called the *heat capacity* of the system. Clearly the heat capacity depends on the amount of work that is done. If no work is involved—which means that the system is kept at constant volume—all the heat goes into increasing the internal energy and

$$c_v = \left[\frac{dQ}{dT}\right]_\mathscr{V} = \frac{dU}{dT} \tag{4-113}$$

The subscript $\mathscr{V}$ denotes constant volume.

Sometimes we need to know the heat capacity under constant pressure conditions. For an ideal gas, this relation is quite simple. In differential form, the ideal gas law (4-34) reads

$$Pd\mathscr{V} + \mathscr{V}dP = RdT \tag{4-114}$$

so that the first law becomes

$$dQ = \left(\frac{dU}{dT} + R\right)dT - \mathscr{V}dP \tag{4-115}$$

* dQ is not an *exact differential*. This means that the change of heat, dQ depends on how the change is attained. For example, it can depend on whether we first raise the internal energy by dU, and then do work $Pd\mathscr{V}$, or vice versa.

4:18

For constant pressure

$$c_p = \frac{dU}{dT} + R \qquad (4\text{-}116)$$

For an ideal gas we therefore have the important relation

$$c_p - c_v = R = \mathcal{N}k \qquad (4\text{-}117)$$

This follows from (4–113) and (4–116). $\mathcal{N}$ is Avogadro's number and the heat capacities are figured for one mole of gas.

We have already stated that the internal energy involves the translation, vibration, electronic excitation, and rotational energy of the molecules. We can ask ourselves how these energies are distributed in a typical molecule. We know that the probability of exciting any classical particle to an energy $\mathscr{E}$ is proportional to the Boltzmann factor $e^{-\mathscr{E}/kT}$. This is true whether $\mathscr{E}$ is a vibrational, electronic, rotational, or translational energy. For an assembly of classical particles, then, the mean internal energy per molecule depends only on the number of ways that energy can be excited, that is, the number of *degrees of freedom* multiplied by $kT/2$. This factor $kT/2$ is consistent with our previous finding: That the total translational energy, including three degrees of freedom, is $(\frac{3}{2})\mathcal{N}kT$ per mole. Each translational degree of freedom, therefore, has energy $\frac{1}{2}kT$ and each other available degree of freedom in thermal equilibrium will also be excited to this mean energy. This is called the *equipartition principle*.

PROBLEM 4–28. Show that an interstellar grain in thermal equilibrium with gas at $T \sim 100°$K rotates rapidly. If its radius is $a \sim 10^{-5}$ cm, and the density is $\rho \sim 1$, show that the angular velocity is $\omega \sim 10^{5.5}$ rad sec^{-1}.

The equipartition principle is a part of classical physics. It does not quite agree with observations; the actual values can be explained more easily by quantum mechanical arguments. The difference between classical and quantum theory hinges to a large extent on what is meant by "available" degrees of freedom. The electronically excited states of atoms and molecules normally are not populated at low temperatures. Hence, at temperatures of the order of several hundred degrees Kelvin, no contribution to the heat capacity is made by such states. Even the vibrational states make a relatively small contribution to the heat capacity because vibrational energies usually are large compared to rotational energies. Aside from the translational contribution, it is therefore the low energy rotational states that make a major contribution to the internal energy and the specific heat at constant volume.

The rotational position of a diatomic molecule can be given in terms of two coordinates θ and ϕ. It therefore has two degrees of freedom. A polyatomic molecule having three or more atoms in any configuration excepting a linear one requires three coordinates for a complete description and therefore has three degrees of freedom. A diatomic or linear molecule makes a contribution of kT to the heat capacity and a nonlinear molecule contributes $3kT/2$. Even these relatively simple rules hold only at low temperatures. At higher temperatures the situation is complicated because a quantum mechanical weighting function has to be introduced to take into account the number of identical (degenerate) states that a particle can have at higher rotational energies.

We will be interested in the heat capacity of interstellar gases where temperatures are low and many of the above mentioned difficulties do not arise. Let us define the ratio of heat capacities at constant pressure and volume as $c_p/c_v \equiv \gamma$. Then by (4–115) we have

$$\gamma = \frac{c_v + \mathcal{N}k}{c_v} \tag{4-118}$$

For monatomic gases we simply deal with the translational internal energy and $c_v = 3k\mathcal{N}/2$; $\gamma = \frac{5}{3}$. For diatomic molecules two rotational degrees of freedom are available in addition to the three translational degrees of freedom, so that $c_v = 5k\mathcal{N}/2$ and $\gamma = \frac{7}{5}$.

:19 ISOTHERMAL AND ADIABATIC PROCESSES

The contraction of a cool interstellar gas cloud or, equally well, the expansion of a hot ionized gas cloud can proceed in a variety of ways. Some cosmic processes involving the dynamics of gases can occur quite slowly at constant temperature. These are called *isothermal processes*. The internal energy does not change and the heat put into the system equals work done by it. Another type of process that describes many fast evolving systems, is the *adiabatic process* in which there is neither heat flow into the gas nor heat flowing out, $dQ = 0$.

$$dQ = c_v dT + Pd\mathcal{V} = 0 \tag{4-119}$$

For an ideal gas

$$c_v dT + \frac{RT}{\mathcal{V}} d\mathcal{V} = 0 = c_v \frac{dT}{T} + (c_p - c_v)\frac{d\mathcal{V}}{\mathcal{V}} \tag{4-120}$$

Integrating, we have

$$\log T + (\gamma - 1) \log \mathcal{V} = \text{constant} \tag{4-121}$$

or

$$T\mathcal{V}^{\gamma-1} = \text{constant} \tag{4-122}$$

$$P\mathcal{V}^{\gamma} = \text{constant} \tag{4-123}$$

and

$$P^{(1-\gamma)}T^{\gamma} = \text{constant} \qquad (4\text{--}124)$$

These are the adiabatic relations for an ideal gas. They govern the behavior, for example, of interstellar gases suddenly compressed by a shock front heading out from a newly formed 0-star or from an exploding supernova. We will study these phenomena in Chapter 9.

For electromagnetic radiation the internal energy per unit volume is

$$U = aT^4 \qquad (4\text{--}125)$$

This is just the energy density. The pressure therefore has one-third this value (4–43), and for volume V we can describe an adiabatic process by

$$dQ = dU + P\,dV = 4aT^3V\,dT + \tfrac{4}{3}aT^4\,dV = 3V\,dP + 4P\,dV = 0 \qquad (4\text{--}126)$$

$$P \propto V^{-4/3}, \qquad \gamma = \tfrac{4}{3} \qquad (4\text{--}127)$$

Since $c_p = \infty$ in this case, γ is defined by the form of (4–123), and not by c_p/c_v. In the interstellar medium where, to a good approximation, we deal only with monatomic or ionized particles—particles having no internal degrees of freedom—and with radiation, γ varies from $\tfrac{4}{3}$ to $\tfrac{5}{3}$ depending on whether gas particles or radiation dominate the pressure.

4:20 FORMATION OF CONDENSATIONS AND THE STABILITY OF THE INTERSTELLAR MEDIUM

We think that the stars were formed from gases that originally permeated the whole galaxy, and that galaxies were formed from a medium that initially was more or less uniformly distributed throughout the universe.

There is strong evidence that star formation is going on at the present time. Many stars are in a stage that can only persist for a few million years because the stellar luminosity—energy output—is so great that these stars soon would deplete their available energy and evolve into objects with entirely different appearances. These bright stars are generally found in the vicinity of unusually high dust and gas concentrations and the belief is that the stars were formed from this dense gas.

We now ask how an interstellar gas cloud could collapse to form a star? To answer this question we can study the stability of gases under various conditions. The results indicate that the stability is dependent on the ratio of heat capacities γ, and that collapse of a gas cloud might be explained in these terms. Whether such views have any bearing on the real course of events in star formation is not yet understood.

Consider an assembly of molecules. Their kinetic energy $\mathbb{T}$ per mole is

$$\mathbb{T} = \tfrac{3}{2}(c_p - c_v)\,T \tag{4-128}$$

or

$$\mathbb{T} = \tfrac{3}{2}(\gamma - 1)\,c_v T \tag{4-129}$$

The internal energy is

$$U = c_v T \tag{4-130}$$

Hence

$$\mathbb{T} = \tfrac{3}{2}(\gamma - 1)\,U \tag{4-131}$$

By the virial theorem we then have (3–83)

$$3(\gamma - 1)\,U + \mathbb{V} = 0 \tag{4-132}$$

as long as inverse square law forces predominate among particles. This means that the equation holds true both when gravitational forces are important and where charged particle interactions dominate the behavior on a small scale (see section 4:21 below). In some situations it can even hold when light pressure from surrounding stars acts on particles of the assembly.

If the total energy per mole is

$$\mathscr{E} = U + \mathbb{V} \tag{4-133}$$

we have from equation 4–132 that

$$\mathscr{E} = -(3\gamma - 4)\,U = \frac{3\gamma - 4}{3(\gamma - 1)}\,\mathbb{V} \tag{4-134}$$

Three results are apparent (Ch 39)*:

(a) If $\gamma = \tfrac{4}{3}$, $\mathscr{E}$ is always zero independent of the configuration. Expansion and contraction are possible and the configuration is unstable. This case corresponds to a photon gas (4–127). In its early stages, a planetary nebula has radiation dominated pressure acting to produce its expansion (Ka68). It should therefore be only marginally stable.

(b) For $\gamma = 1$, $\mathbb{V}$ always is zero for any $\mathscr{E}$ value and again no stable configuration exists.

(c) For $\gamma > \tfrac{4}{3}$, equation 4–134 shows that $\mathscr{E}$ always is negative and the system is bound. If the system contracts and the potential energy changes by $\Delta\mathbb{V}$ then, we see that

$$\Delta\mathscr{E} = +\frac{(3\gamma - 4)}{3(\gamma - 1)}\,\Delta\mathbb{V} = -(3\gamma - 4)\,\Delta U \tag{4-135}$$

4:20

An amount of energy $-\Delta\mathscr{E}$ is lost by radiation

$$-\Delta\mathscr{E} = -\frac{3\gamma - 4}{3(\gamma - 1)}\Delta\mathbb{V} \tag{4-136}$$

while the internal energy increases by

$$\Delta U = -\frac{1}{3(\gamma - 1)}\Delta\mathbb{V} \tag{4-137}$$

through a rise in temperature.

As the protostar contracts to form a star it therefore becomes hotter and hotter.

Two comments are necessary:

(a) When theories of the kind developed here are applied on a cosmic scale, say, to formation of galaxies or clusters of galaxies, we run into difficulties in defining the potential $\mathbb{V}$. The zero level of the potential can no longer be defined using Newtonian theory alone, and some more comprehensive approach such as that of general relativity should be used. This complicates the treatment of the problem considerably, and no properly worked out theory exists.

(b) In practice, star formation probably takes place in strong interaction with the surrounding medium. This is indicated by the formation of stars in regions where other stars have just formed and where pressures on the surrounding medium are setting in. There are theories that describe the formation of stars through the compression of cool gas clouds either by surrounding hot ionized regions, or through the compression by starlight emitted from nearby hot stars. The stability of an isolated medium, as treated above, may therefore not be strictly relevant to the discussion. Nevertheless, it is clear that very stable gas configurations will resist applied compressive forces, while intrinsically unstable gases will readily respond to such pressures. As already indicated in section 1:4, magnetic fields, rotational motion about the Galactic center, and rotation about the protostar center, endows protostellar matter with stability against collapse. Random gas motions and radiation energies therefore are not the sole contributors to U in equations 4–131 to 4–134.

4:21 IONIZED GASES AND ASSEMBLIES OF STARS

The behavior of large clusters of stars or galaxies can be described statistically much as we describe the behavior of gases. There are many striking similarities particularly in the physics of ionized gases (plasma) and aggregates of stars. These similarities come about because Newton's gravitational attraction can be written in a form similar to Coulomb's electrostatic force:

Newton's force   Coulomb's force

$$\frac{(iG^{1/2}m_1)(iG^{1/2}m_2)}{r^2} \qquad \frac{Q_1Q_2}{r^2} \qquad\qquad (4\text{--}138)$$

Here the gravitational analogue to electrostatic charge is the product of mass, the square root of the gravitational constant, and the imaginary number, i. The correspondence can be extended to include fields, potentials, potential energies, and other physical parameters. The primary difference between gravitational processes and electrostatic interactions lies in the fact that electric charges can be both positive and negative while the sign of the gravitational analogue to charge is always the same—mass is always positive.

We will first derive some properties of assemblies of gravitationally interacting particles and then make a comparison to plasma behavior. If we take a spherical distribution of particles—a set of stars or galaxies—the force acting on unit mass at distance r from the center is

$$F(r) = -\frac{1}{r^2}\int_0^r 4\pi G r^2 \rho(r)\, dr \qquad\qquad (4\text{--}139)$$

This means that

$$\frac{d}{dr} r^2 F(r) = -4\pi G r^2 \rho(r) \qquad\qquad (4\text{--}140)$$

and setting the force equal to a potential gradient

$$F(r) = -\nabla \mathbb{V}(r) = -\frac{d}{dr}\mathbb{V}(r) \qquad\qquad (4\text{--}141)$$

we have

$$\frac{1}{r^2}\frac{d}{dr} r^2 \frac{d}{dr}\mathbb{V}(r) = +4\pi G \rho(r) \qquad\qquad (4\text{--}142)$$

which is *Poisson's equation*. Substituting from equation 4–47 we have the *Poisson–Boltzmann equation* for a gas at temperature T:

$$\frac{1}{r^2}\frac{d}{dr} r^2 \frac{d\mathbb{V}}{dr} = 4\pi \rho_0 G e^{-[m\mathbb{V}(r)/kT]} \qquad\qquad (4\text{--}143)$$

It is worthwhile noting that the potential appearing in the exponent on the right of this equation really was obtained through the integration of $\nabla \mathbb{V}(r)$ in equation 4–46. The behavior of an assembly of stars or galaxies would, therefore, be no different if some constant potential, present throughout the universe, were added to $\mathbb{V}(r)$. This is essentially the point that was already raised in section 4:20 in connection with the stability of uniform distributions of gas.

PROBLEM 4–29. Show that the substitutions

$$\frac{m\mathbb{V}(r)}{kT} \equiv \psi \quad \text{and} \quad r \equiv \left(\frac{kT}{4\pi\rho_0 mG} \right)^{1/2} \xi \tag{4–144}$$

turn equation 4–143 into

$$\frac{1}{\xi^2} \frac{d}{d\xi} \left(\xi^2 \frac{d\psi}{d\xi} \right) = e^{-\psi} \tag{4–145}$$

We now have to decide on the boundary conditions that have to be imposed on this differential equation. At the center of the cluster there are no forces and the first derivative of $\mathbb{V}$ or ψ must be zero. Since the potential can have an arbitrary additive constant, we can choose the potential to be zero at the center. In terms of the new variables these two conditions are

$$\psi = 0 \quad \text{and} \quad \frac{d\psi}{d\xi} = 0 \quad \text{at} \quad \xi = 0 \tag{4–146}$$

Taken together with equation 4–145 they lead to a solution that has no closed form (Ch 43).

PROBLEM 4–30. Show that in the limit of very small and of large ξ values the respective solutions are (Ch 39)

$$\psi \sim \tfrac{1}{6}\xi^2 - \tfrac{1}{120}\xi^4 + \tfrac{1}{1890}\xi^6 + \cdots \quad , \quad \xi \ll 1 \tag{4–147}$$

$$\psi \sim \log\left(\frac{\xi^2}{2} \right), \quad \xi^2 > 2 \tag{4–148}$$

This can be verified by substitution in equations 4–145 and 4–146. From this, the radial density and mass distributions can be found (Ch 43). The density distribution is plotted in Fig. 4.15.

One difficulty with this plot and with the asymptotic solution (4–148) is that the density $\rho_0 e^{-\psi}$ is proportional to ξ^{-2}. This causes the total mass integrated to large distances to become infinite. We therefore need a cut-off mechanism that will restrict the radius of a cluster of stars to a finite value. We already men-

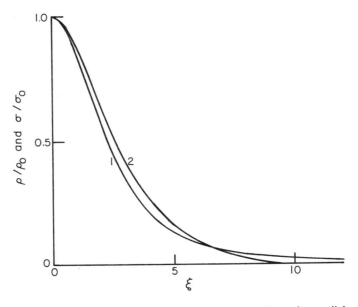

Fig. 4.15 Plot of density (1) and areal density (2) against radial distance from the center of a distribution. For a cluster, σ/σ_0 represents the star density drop with radial distance, as measured directly on a photographic plate. ρ_0 and σ represent values at the center. $\rho/\rho_0 = \exp(-\psi)$, $\xi = (4\pi\rho_0 mG/kT)^{1/2} r$. (From *Principles of Stellar Dynamics* by S. Chandrasekhar. Reprinted through permission of the publisher, Dover Publications, Inc. New York.)

tioned the interaction with the galactic nucleus as representing one such mechanism (section 3:16).

It is interesting to compare these results to those obtained for a plasma in which both positive and negatively charged particles are present. The derivation of the Poisson–Boltzmann equation was in no way based on particle charges. It concerned itself only with an inverse square law force and a uniform mass distribution that could equally well have been a charge distribution. Using the density for an assembly of dissimilar particles (see equation 4–53), the Poisson–Boltzmann equation can be written as

$$\frac{1}{r^2}\frac{d}{dr}r^2\frac{d\mathbb{V}}{dr} = -4\pi \sum_i n_{i0}q_i e^{-q_i\mathbb{V}/kT} \tag{4–149}$$

for plasma. Here q_i is the charge of particles of type i. If we restrict ourselves to large interparticle distances—a condition that holds in intergalactic, interstellar, and interplanetary space—then

$$q_i \mathbb{V} \ll kT \tag{4-150}$$

and we can use the Taylor expansion

$$e^{-q_i \mathbb{V}/kT} = 1 - \frac{q_i \mathbb{V}}{kT} + \frac{1}{2}\left(\frac{q_i \mathbb{V}}{kT}\right)^2 + \cdots \tag{4-151}$$

Neglecting quadratic and higher terms, the charge density on the right of (4-149) then becomes

$$\rho = \sum_i n_{i0} q_i - \frac{\mathbb{V}}{kT} \sum_i n_{i0} q_i^2 \tag{4-152}$$

The first term vanishes because of charge neutrality for the bulk of the plasma. Note that this term does not vanish in the gravitational case. There it is dominant. The second term can be written as

$$\rho = -\frac{\mathbb{V} e^2}{kT} \sum_i n_{i0} Z_i^2 \qquad \text{where} \qquad q_i = e Z_i \tag{4-153}$$

Substituting in (4-149) we have

$$\frac{1}{r^2}\frac{d}{dr}\left(r^2 \frac{d\mathbb{V}}{dr}\right) = \frac{4\pi e^2 \mathbb{V}}{kT} \sum_i n_{i0} Z_i^2 \tag{4-154}$$

or

$$\frac{d}{dr}\left(r^2 \frac{d\mathbb{V}}{dr}\right) = r^2 \mathbb{V}\left(\frac{1}{L^2}\right) \tag{4-155}$$

where

$$\frac{1}{L^2} = \frac{4\pi e^2}{kT} \sum_i n_{i0} Z_i^2 \tag{4-156}$$

L has the dimension of a length. It is called the *Debye shielding length* and is a distance over which a charged particle embedded in a plasma can exert an appreciable electrostatic field. Beyond that distance its electrostatic influence rapidly diminishes.

One reason why the shielding length is of interest in astrophysics is because it points out the impossibility of maintaining an electric field over any large scale. A field cannot be influential over distances much larger than L. For interstellar gas clouds with $n_{i0} = 10^{-3}$ cm^{-3}, $Z_i = 1$, and $T = 100°$K, L turns out to be about 20 meters. This is completely negligible compared to typical interstellar distances ~ 1 pc.

Electrostatic forces may be important in large-scale processes, but only when they appear in conjunction with large-scale magnetic fields, that can prevent the flow of charged particles along the electric field lines and therefore prevent the charge separation required for electrostatic shielding. The behavior of plasmas in the presence of magnetic fields is treated in the theory of magnetohydrodynamic processes (Co57), (Sp62). We will briefly touch these processes in Chapter 6.

ANSWERS TO SELECTED PROBLEMS

4–2.   Suppose we take n_i steps of length λ_i. The mean square deviation then is $n_i \lambda_i^2$; and a similar result holds for all step sizes. Hence the final mean square deviation is $\sum_i n_i \lambda_i^2 = N \langle \lambda^2 \rangle$.

4–4.   For escape, the deviation has to be $\sim n^{1/2}$ steps of length L. Hence $n \sim R^2/L^2$.

4–5.   For a given value of r, changing θ values lead to

$$\frac{1}{r} = \frac{\sin \theta}{R\alpha}, \quad \left\langle \frac{1}{R\alpha} \right\rangle = \frac{1}{r} \left\langle \frac{1}{\sin \theta} \right\rangle = \frac{\pi}{2r}.$$

Hence

$$\frac{1}{r} = \frac{2}{\pi} \left\langle \frac{1}{R\alpha} \right\rangle$$

and

$$\text{total mass} = \frac{\langle v^2 \rangle}{G \langle 1/r \rangle} = \frac{3 \langle v_p^2 \rangle \frac{\pi}{2}}{\left\langle \frac{1}{R\alpha} \right\rangle}$$

4–6. $\dfrac{nm \langle v^2 \rangle}{3} \sim 3 \times 10^{-17}$ dynes cm^{-2}.

4–7.   The collision probability per star pair in unit time is

$$\text{Probability} = nv\sigma \text{ sec}^{-1}.$$

In P sec the probability is $nv\sigma P$ per star pair or about 1.5×10^{-11} that the sun would have formed a planetary system in the time available. If there are 10^{11} stars altogether, 1.5 pairs, or 3 solar systems would have formed in this way in 5×10^9 y. Since the sun and Barnard's star both have companions of the general size range of Jupiter, and since these two systems, as stated in Chapter 1, are unlikely to have had a common origin, the density of planetary systems in the vicinity of the sun already appears too great to be accounted for by the Jeans hypothesis.

4-8. $P \sim 10^{-14}$ dynes cm^{-2}.

4-9. $v \sim 0.8$ cm sec^{-1} $M = 4\pi\rho r^3/3 \sim 7 \times 10^{-14}$ g.

4-10. Momentum transfer rate is $\pi r^2 n v^2 m = dp/dt = 3.1 \times 10^{-21}$ g cm sec^{-1}. The mass of the grains is given in Problem 4-9 and, hence, $dv/dt = 4.8 \times 10^{-8}$ cm sec^{-2}. This gives the initial acceleration. However, as the grain gains velocity and mass, the acceleration decreases toward a value of zero, which is reached when the grain reaches velocity v.

4-11. The mass gain $dM/dt = \pi r^2 n V m = 3 \times 10^{-27}$ g sec^{-1}. At this rate the grain would gain 1% of its mass in 2×10^{11} sec.

4-13. $P \sim 2 \times 10^{-13}$ dyn cm^{-2}.

4-14. If n is the number of photons passing through unit area per unit time, the pressure is

$$P = \frac{\int n(v)\, hv\, dv}{c}.$$

that is, the energy incident on unit area per unit time, divided by the speed of light. The force on the grain is its area, multiplied by P.

$$P = \frac{1.37 \times 10^6}{3 \times 10^{10}} = 4.6 \times 10^{-5} \text{ dynes cm}^{-2}$$

$$F = 3.6 \times 10^{-9} \text{ dynes}$$

4-15. For isotropic scattering one averages the function $(1 - \cos\theta)$ over all solid angle increments to get the mean momentum transfer. Thus

$$P = \mathscr{E}/3c + \frac{2\mathscr{E}/3c}{4\pi} \int\int (1 - \cos\theta)\sin\theta\, d\phi\, d\theta$$

with θ chosen as zero in the forward scattering direction, and $\mathscr{E}$ the energy incident on the grain per unit time. Thus $P = \mathscr{E}/c$, and the grain has a force of 1.44×10^{-12} dynes acting on it at the earth's distance from the sun. The gravitational force is mMG/R^2; here m is the particle mass, M is the solar mass, and R is the distance of the sun from the earth:

$$m = \rho 4\pi s^3/3 = 2.5 \times 10^{-11} \text{ g}$$

the gravitational force is 1.5×10^{-11} dynes and the radiative repulsion is about 10% of the gravitational attraction.

Finally, the ratio of gravitational to radiative force remains constant because the angle subtended by the grain, as seen from the sun, diminishes as R^2—in the same way as the gravitational force.

4–16. From the equation for the period of a grain in elliptic motion

$$\tau = 2\pi a^{3/2}/(MG_{\text{eff}})^{1/2}$$

We note that the period depends on the square root of the effective gravitational constant. The period of the grain will be $(\frac{3}{2})^{1/2}$ years, that is, 1.22 y. The orbital velocity will be $v_E/1.22$, where v_E is the earth's orbital velocity; that speed will be roughly 24 km/sec. The collision velocity of the earth with such a grain would therefore be ~ 6 km sec^{-1}.

4–18. The thinness of the atmospheric layer implies that g is constant throughout. Hence the scale height h is determined by the equation $ghm/kT = 1$ and

$$h \sim 10^6 \text{ cm.}$$

4–20. $m \sim 2 \times 10^{-24}$ g

and $v \sim 2.3$ km sec^{-1}.

4–23. $Z(p)\,dp = \dfrac{4\pi p^2\,dp}{h^3} \cdot V, \qquad p = \dfrac{\mathscr{E}}{c}, \qquad dp = \dfrac{d\mathscr{E}}{c}$

$$\therefore Z(\mathscr{E})\,d\mathscr{E} = \frac{4\pi \mathscr{E}^2\,d\mathscr{E}}{h^3 c^3}\, V.$$

The total energy density is

$$\frac{\mathscr{E}}{V} = \int_0^{\Phi_\nu} \mathscr{E} Z(\mathscr{E})\,d\mathscr{E} = \frac{\pi \Phi_\nu^4}{h^3 c^3}$$

and the mass density $\rho_\nu = \dfrac{\mathscr{E}}{c^2 V} = \dfrac{\pi \Phi_\nu^4}{h^3 c^5}$

4–24. $kT \sim 655$ ev $\sim 10^{-9}$ erg

$$T \sim \frac{10^{-9} \text{ erg}}{1.4 \times 10^{-16} \text{ erg}/^\circ} \sim 7 \times 10^6 \,^\circ\text{K}$$

We reason that $kT \sim$ excitation energy; the higher ionized state gives rise to a weak line because T is not sufficiently high to lead to frequent ionization to this level. That is, for the higher ionized state $kT <$ ionization energy. For the lower ionized states kT is probably more comparable to the ionization energy. Actually T is $\sim 1.5 \times 10^6\,^\circ$K in the corona.

5

Photons and Fast Particles

5:1 THE RELATIVITY PRINCIPLE

When we discussed Newton's laws of motion, we were careful to note that they only held under restricted conditions. All motions had to be described with respect to *inertial frames of reference*—frames at rest or moved at constant velocity with respect to the mean motion of ambient galaxies.

Under these restricted conditions not only Newton's laws but all other laws of physics are obeyed. This general statement—first formulated by Einstein—is called the *principle of relativity*. It implies that an observer cannot determine the absolute motion of his inertial frame of reference—only its motion relative to some other frame. The principle also has many other important consequences which, taken together, form the basis of the theory of *special relativity* (Ei05a). As already mentioned in section 3:8, Einstein also widened the concept of the inertial frame beyond Newton's scope of a frame moving at constant velocity with respect to fixed stars. He showed that we can include coordinate frames fixed in any freely falling, non-rotating bodies. Such local inertial frames may accelerate with respect to frames that are far from any massive objects but are fully equivalent to them as far as the principle of relativity is concerned. Finally, Einstein postulated that the speed of light is the same in all reference frames, whether they move or are stationary. This actually is a consequence of the relativity principle. If this speed were not the same, an observer could determine whether he was at rest or moving.

We should note that the relativity principle is founded on observations. It could not have been predicted from logic alone.

In recent centuries, long before Einstein's birth, there has always been an awareness that some sort of relativity principle might exist. In Galileo's time, when the speed of light was believed to be infinite, the statement of the principle was almost exactly the same as Einstein's. At that time the velocity of light was believed to be infinite as measured in any reference frame. The instantaneous transmission of signals and messages over large distances seemed possible. Since any velocity added to an infinite velocity still gave infinite speed, it was clear that no matter how an observer moved he would always see light traveling at the same, infinite, speed; and similarly all other laws of physics seemed identical for Newton's inertial frames.

Then in 1666 Roemer discovered that the speed of light is finite, though large. This tended to detract from the *Galilean relativity principle* because it seemed that an observer moving into the direction of a light source would see the light wave moving faster than an observer moving away from the source. But at the end of the 19th century Michelson and others discovered that the speed of light was identical in all the moving reference frames they were able to check. Independently Einstein postulated a new principle of relativity similar to Galileo's except that the speed of light now was finite and equal in all reference frames. To some extent this concept had already been present in Maxwell's theory of electromagnetism, but the required constancy of the speed of light was considered a weakness of the theory, not a strength.

As we will see, Einstein's relativity principle also led him to conclude that no physical object could travel at a speed in excess of light, $c = 2.998 \times 10^{10}$ cm sec^{-1}. The concept of an infinite velocity simply had no correspondence in physical moving objects.

The theory of relativity has the task of formulating the laws of physics in such a way that physical processes can be accurately described in any moving coordinate system. This study conveniently divides into two parts. The first theory is more restricted. It deals with physical processes as viewed from inertial reference frames and specifically excludes any consideration of gravity. This is *special relativity*. The second, more general theory incorporates not only special relativity but also the study of gravitational fields and arbitrarily accelerated motions. It is therefore called the *general theory of relativity*. We will not discuss this theory here, partly because it is complicated, and partly because in its present form its application to astrophysics is not entirely established. There may exist rival theories of similar scope that would give greater insight into astrophysical process. One such theory is being advocated by Dicke (Di67a).

5:2   RELATIVISTIC TERMINOLOGY

Suppose a physical process occurs in a system at rest with respect to some inertial reference frame K'. An observer in some other inertial frame K views this process. If K and K' are moving at large volocities V, relative to each other, the observer will see the processes distorted both in space and in time. But the special theory of relativity will allow him to reconstruct the proper scale of events as they occur in system K. This is a very useful property of relativity theory. We will find many applications of it in astrophysics where high velocities are often encountered. The special theory, however, goes beyond this limited function of reconstructing clear pictures from apparently distorted observations. It gives new insight into the interrelation between time and space, and between momentum, energy, and mass; it justifies the impossibility of exceeding the speed of light, and yields many other new results.

To make full use of this theory, we will need to take a few preparatory steps. We must define new concepts and formulate them in mathematical terms.

(a) To the extent that it is valid, the special theory abolishes an absolute standard for a state of rest. It states that there is no way of defining zero speed in an absolute way. Bodies may be at rest—but only *relative* to some other body or frame of reference.

We know that this statement need not be quite true: A preferred natural state of rest does exist for any locality in the universe. It is the state of rest relative to the mean motion of ambient galaxies. This fact was not known at the time relativity theory was established. It tends to weaken the statement we formulated above, and allows us to state only that an absolute standard of rest is inconsequential to special relativity. The theory draws no distinction between absolute rest and constant velocity.

(b) In relativity we will talk about *events* that have to be described both by a place and a time of occurrence. We need four coordinates to define an event—three space coordinates and a time coordinate.

Correspondingly there exists a hypothetical four-dimensional space having spatial and time coordinates. In this space, events are represented by *world points* (x, y, z, t). Any physical process can be described as a sequence of events and can be represented as a grouping or continuum of world points in the four-dimensional space-time representation. Each physical particle can be represented by a *world line* in this four-dimensional plot.

(c) Two distinct events labeled a and b are separated by an *interval*, s_{ab}, of length

$$s_{ab}^2 = -[(x_a - x_b)^2 + (y_a - y_b)^2 + (z_a - z_b)^2 - c^2(t_a - t_b)^2] \qquad (5\text{–}1)$$

This suggests that we could define a new coordinate $\tau = ict$, where i is the imaginary number, to obtain (5–1) in the form

$$s_{ab}^2 = -[(x_a - x_b)^2 + (y_a - y_b)^2 + (z_a - z_b)^2 + (\tau_a - \tau_b)^2] \qquad (5\text{–}2)$$

5:2

This form brings out a symmetry between time and space coordinates. Equation 5–2 is just the Pythagorean expression for the separation of two points in a four-dimensional flat space. Such a space is also called a *Euclidean space*, and the particular four-dimensional space described in (5–1) is known as a *Minkowski space* (Mi08). Equation 5–2 helps to point out some of the properties of space and time coordinates. The time coordinate in the formulation (5–2) is an imaginary quantity, while the spatial coordinates are real. Unfortunately, the substitution $\tau = ict$ is not very useful. Special relativity, in its full form, deals with quantities that are best described in tensor notation. But that notation cannot be properly used if time is taken to be an imaginary quantity. Rather, as we will see, x, y, z, and ct should be considered to be components of a four vector in a space that is said to have *signature* $(+ + + -)$, meaning that the Pythagorean expression for the square of the interval between events is the sum of the squares of the spatial components of the separation, with the square of the time increment subtracted.

(d) We can formulate equation 5–1 in differential form

$$ds^2 = -(dx^2 + dy^2 + dz^2 - c^2 dt^2) \qquad (5\text{–}3)$$

where ds is called the *line element*.

(e) The interval between two events is said to be *timelike* if $s_{ab}^2 > 0$ and *spacelike* if $s_{ab}^2 < 0$. When s_{ab}^2 just equals zero, we see from either equation 5–1 or 5–3, that

$$v_x^2 + v_y^2 + v_z^2 \equiv v^2 = c^2 \qquad (5\text{–}4)$$

The surface containing all intervals $s_{ab} = 0$, or line elements $ds = 0$ is called the *light cone*. It contains all trajectories going through a point (x, y, z, t) with the speed of light. A two-dimensional projection of this cone is shown in Fig. 5.1. This means that we choose coordinates $y = z = 0$ and the projection of the surface

$$x^2 + y^2 + z^2 = c^2 t^2 \qquad (5\text{–}5)$$

now becomes $x = \pm ct$ with slope

$$\frac{dt}{dx} = \pm \frac{1}{c} \qquad (5\text{–}6)$$

All lines representing physical particles must indicate velocities $v < c$ and, therefore, are contained in that part of the light cone containing the t-axis. The lower half of the diagram represents the past. The upper half contains all world points lying in the future. The two parts of the diagram containing the x-axis are absolutely inaccessible in the sense that velocities greater than the speed of light would be required to reach them.

It is interesting that the concept of absolute past and future depends on the

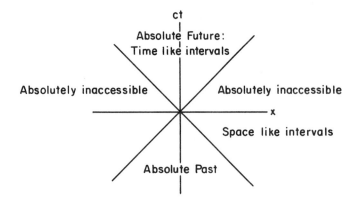

Fig. 5.1 World diagram to show relation between different kinds of events.

fact that the speed of light cannot be exceeded. If it could, we would be able to travel to a sufficiently distant point and "catch up" with light that had been emitted, say, in the supernova of 1054 A.D. With a sufficiently good telescope we could then "look back" and see the star just prior to explosion. The event could thus be brought into our "present," but it would still be inaccessible to us in the sense that we would not be able to influence the event in any way. This problem is looked at more thoroughly in section 5:12.

(f) The time read on a clock moving with the reference frame of an observer is called the *proper time* for that frame; and the length of an object measured in that frame is called the *proper length.*

5:3 RELATIVE MOTION

Consider two inertial frames of reference K and K', whose axes x, y, z and x', y', z' are parallel (Figure 5.2). Relative to K, K' moves with velocity V along the x-axis. An event has coordinates (x, y, z, t) as measured by an observer at rest in system K, and coordinates (x', y', z', t') as measured by an observer at rest in K'.

At some time $t = t' = 0$, let the origins of the two reference frames coincide. The subsequent motion will not affect the identity of the y and z components: $y' = y$ and $z' = z$; but t and x will be related to t' and x' through a more complicated set of relations, the *Lorentz transformations*, which read (Lo04):

$$x = \frac{x' + Vt'}{\sqrt{1 - V^2/c^2}}, \quad y = y', \quad z = z', \quad t = \frac{t' + V(x'/c^2)}{\sqrt{1 - V^2/c^2}} \quad (5\text{–}7)$$

5:3

Fig. 5.2   Notation for moving coordinate frames.

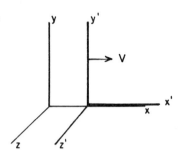

or

$$x_1 = (x_1' + \beta x_4')\gamma(V), \quad x_2 = x_2', \quad x_3 = x_3', \quad x_4 = (x_4' + \beta x_1')\gamma(V) \quad (5\text{--}7a)$$

if we set $x = x_1$, $y = x_2$, $z = x_3$, $ct = x_4$, $\beta = V/c$, and $\gamma(V) = (1 - \beta^2)^{-1}$. The second formulation shows the symmetry between space and time coordinates. These equations can be derived and follow directly from the principle of relativity and the constancy of the speed of light. Here we will only show that the equations are consistent with some of the predictions of the principle.

For example, since the speed of light is the same in systems K and K', we would expect that a light wave emitted at $t = t' = 0$—that is, when the origins of the coordinate systems coincide—would propagate spherically in both systems.

PROBLEM 5–1. Equation 5–5 describes the propagation of a spherical wavefront in the coordinate system K. Show that according to (5–7), the corresponding equation describing the propagation of the wavefront in K' would be

$$x'^2 + y'^2 + z'^2 = c^2 t'^2$$

so that this wave appears spherical too.

Another consequence of the relativity principle is that formulas expressing x', y', z', and t' in terms of x, y, z, and t can be easily obtained by changing V to $-V$.

PROBLEM 5–2. Show that this procedure is valid by actually solving equations 5–7 for x', y', z', and t'.

We also want to examine whether the speed of light will always appear to be c, viewed from any reference frame. We can answer this question by discussing how velocities transform according to equations 5–7. Let us write the expressions

5:3

in differential form:

$$dx = (dx' + V\,dt')\,\gamma(V), \qquad dy = dy', \qquad dz = dz', \qquad dt = \left(dt' + \frac{V}{c^2}dx'\right)\gamma(V)$$

$$(5\text{--}8)$$

where

$$\gamma(V) \equiv \frac{1}{\sqrt{1 - V^2/c^2}}$$

This allows us to write the derivatives

$$v_x = \frac{dx}{dt} = \frac{dx' + V\,dt'}{dt' + \dfrac{V}{c^2}dx'} = \frac{v_x' + V}{1 + v_x'\dfrac{V}{c^2}}$$

$$v_y = \frac{dy}{dt} = \frac{v_y'}{\left(1 + v_x'\dfrac{V}{c^2}\right)\gamma(V)} \qquad (5\text{--}9)$$

$$v_z = \frac{dz}{dt} = \frac{v_z'}{\left(1 + v_x'\dfrac{V}{c^2}\right)\gamma(V)}$$

These equations prescribe the *composition* (addition) *of velocities*. If $v_z' = v_y' = 0$, and we write $v_x' = v'$, then equations 5–9 show that $v_y = v_z = 0$ and $v_x = v$ where

$$v = \frac{v' + V}{1 + \dfrac{v'V}{c^2}} \qquad (5\text{--}10)$$

We interpret this equation in the following way: When all motions are along the x-axis, a velocity measured as having a value v' in reference frame K', will appear to have velocity v in a frame K. v, v' and V are related by equation 5–10. V is the velocity of K' relative to K (Fig. 5.1).

Three cases are of interest

(a) If $v' = V = c$, then substitution shows that $v = c$.

(b) If $v' < c$, $V = c$ or if $v' = c$, $V < c$ then $v = c$. This also can be shown by substitution in equation 5–10. It means that the speed of light is constant and has a value c in all inertial frames of reference.

(c) Finally, if $v' < c$ and $V < c$, then $v < c$.

PROBLEM 5–3. Show that the result (c) is always true by writing $v' = (1 - \delta)\,c$, $V = (1 - \Delta)\,c$ where $0 < \delta, \Delta < 1$.

PROBLEM 5-4. If the speed of light is infinite, Galilean relativity results. Give the transformations equivalent to (5-7) to (5-9) and obtain the law of composition of velocities. These expressions should be consistent with Newtonian physics.

Expression 5-10 is interesting because it also shows that a particle traveling at a speed less than the speed of light can never be accelerated to a speed equaling c. To see that, suppose that the particle initially was moving with velocity V. It is now given an extra velocity v' that also is less than c. From (c), above, we see that the resultant velocity is always less than c. We can keep adding small increments to the particle's velocity, but to no avail. It will always remain at a speed less than the speed of light. This situation characterizes the highly energetic cosmic ray particles. They move very close to the speed of light. When accelerated, they move a little faster, but never faster than c!

The Lorentz transformation leaves the interval s between two events invariant, but this is done at the expense of changes in the apparent time and spatial separations of events. The time separation is affected in the following way: If a clock is at rest at position $x = 0$ in K, then the proper time for K is given by t and the time measured by an observer O' at rest in the K' frame is

$$ t' = \left(t - V\frac{x}{c^2} \right) \gamma(V) \bigg|_{x=0} = t \, \gamma(V) \tag{5-11}$$

Actually, we are not interested in an absolute time, only in time intervals $\Delta t = t_1 - t_2$ and $\Delta t' = t'_1 - t'_2$. Hence the equations 5-11 reduce to

$$ \Delta t' = \frac{\Delta t}{\sqrt{1 - V^2/c^2}} \equiv \Delta t \, \gamma(V) \tag{5-12}$$

A time interval measured as Δt in the K frame would appear longer, $\Delta t' > \Delta t$, in the K' frame. To the observer O', the clock appears to be going slower. We speak of a *time dilation* in moving reference frames. We note that the choice of $x = 0$ was not necessary. The relation between Δt and $\Delta t'$ is independent of the choice of position, x.

In Problem 5-9 we will see that this time dilation can prolong the decay time of fast-moving, unstable, cosmic ray particles by many orders of magnitude! The time dilation is a dominant effect for the decay of such particles.

We can similarly derive the change in spatial separation between simultaneously observed events. If the positions of two points at rest in the K system are x_a and x_b as measured by an observer O at rest in K, the *proper length* of a line joining the two points is $\Delta x = x_b - x_a$. O', the observer at rest in K' measures the separation of the two points at some given time t'. We use the equations

$$x_a = (x'_a + Vt')\gamma(V) \qquad x_b = (x'_b + Vt')\gamma(V) \tag{5–13}$$

where t' is the same in both expressions since O' sees both points simultaneously. The spatial separation observed from the K' frame, then, is

$$\Delta x' = x'_a - x'_b = (x_a - x_b)\sqrt{1 - \frac{V^2}{c^2}} = \frac{\Delta x}{\gamma(V)} \tag{5–14}$$

Since the square root term is always less than unity, this means that the length measured by O' is shorter than the proper length. We call this the *Lorentz contraction* (Lo04).

The Lorentz contraction is found only along the direction of motion, while the transverse dimensions y and z remain unaffected according to equations 5–7. This could at first sight lead us to believe that a moving sphere should appear flattened into an oblate ellipsoid, and that a cube would appear distorted in some way dependent on its orientation with respect to the moving axes.

This view was held for more than half a century after the discovery of the special relativistic transformations by Lorentz and Einstein. But in 1959 Terrell (Te59) suggested that a sphere should always appear spherical, a cube cubical, and so on. The Lorentz transformations, while producing some distortions, primarily act to change the apparent orientation of the object by effectively rotating it.

To see how this comes about, suppose that a cube is moving with velocity V along the x direction. This motion is relative to an observer who looks at the cube in a direction transverse to its motion.

Fig. 5.3 The sides of a rapidly moving cube.

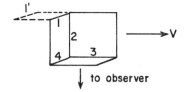

We will be interested in the apparent length of the edges 1, 2, and 3. Let the length of each edge be L, as measured by an observer at rest with respect to the cube and let edge 1 be perpendicular both to the direction of motion and the direction of the observer (Figure 5.3).

When the observer sees both edges 1 and 4 simultaneously—as he would when taking a photographic picture—he does not observe photons that were simultaneously emitted at these two edges. The light reaching him from edge 1 was emitted a time L/c earlier than light arriving from edge 4. But at that earlier

5:3

time, edge 1 occupied position 1'. A photograph will therefore show a view of the cube with the far edge occupying position 1' and the near edge occupying position 4. The projected length of side 2 is the projected distance between 1' and 4, namely Lv/c.

This factor does not enter in discussing the length of edge 4, since the ends of these edges simultaneously emit those light rays that later are simultaneously observed. Side 4 is perpendicular to the direction of motion and its length is left unchanged by the Lorentz transformation; the Lorentz transformations also leave sides 1 and 2 unchanged. But side 3 is shortened by a factor $\sqrt{1 - (V^2/c^2)}$ (see equation 5–14).

A photograph will show sides 1, 2, and 3 having lengths L, LV/c, and $L/\gamma(V)$, respectively. If we define an angle ϕ by $V/c = \sin \phi$, then it is easy to see that these sides have apparent lengths, L, $L \sin \phi$, and $L \cos \phi$. The cube appears rotated by an angle ϕ!

While this is true for a small distant cube at its point of nearest approach, there are added distortions if the same cube is seen, say, earlier in its trajectory: Light arriving from the nearest edge is then emitted later—when the cube is closer—than light from edge 4, the trailing edge. The near edge therefore appears disproportionately long. In general, the cube appears both distorted and rotated (Ma72a)*.

5:4 FOUR-VECTORS

Let us now turn to the interrelationship between the world diagrams of two observers O and O' moving with inertial frames K and K'. As in Fig. 5.2 we will take K' to be moving in the direction of K's positive x-axis with velocity V. The origin of coordinates will then have components $y = y' = 0$ and $z = z' = 0$ at all times.

Let us also choose the origins of K and K' to coincide at some time $t = t' = 0$. This means that $x = x' = 0$ at that time. As seen by O, the origin of K' then has the world line t', shown in Fig. 5.4. The line passes through the origin and has a slope

$$\frac{c\,dt}{dx} = \frac{c}{V} \qquad (5\text{–}15)$$

That t' actually is the time axis for O' follows from the first and last equations of (5–7), for $x' = 0$. Again, if we set $t' = 0$ in these two equations, we see that the slope of the x' axis in O's world diagram must be $cdt/dx = V/c$. The angle ψ between the ct and ct' therefore equals the angle between the x and x' axes:

$$\psi = \tan^{-1}\frac{V}{c} \qquad (5\text{–}16)$$

Fig. 5.4 Minkowski diagram showing characteristics of a moving inertial coordinate system K' as seen by another inertial observer.

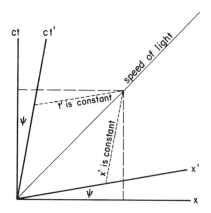

Different observers O and O′ have different spaces described by a fixed proper time. Events that appear simultaneous in one frame no longer appear simultaneous in others. This concept is foreign to Galilean relativity, where the speed of light is infinite and where the slope V/c would be zero. In Galilean relativity, the x and x', and the t and t' axes would therefore coincide. The light cone bisects both the spaces K and K' in this diagram, often called the *Minkowski diagram* (Ro68)*.

A vector in the four-dimensional spaces K and K' appears equally long to both observers. If the vector joins events $(0,0,0,0)$ and (x_1,y_1,z_1,t_1) as seen by O, it will join $(0,0,0,0)$ and (x'_1,y'_1,z'_1,t'_1) as seen by O′. But since we can always choose the x direction to coincide with the direction of motion, we can again set $y = y'$, $z = z'$, so that the lengths squared of the two vectors become

$$L^2 = -\{x^2 + y^2 + z^2 - c^2t^2\}$$
$$= -\{[(x' + Vt')^2 - (ct' + Vx'/c)^2]\gamma^2(V) + y'^2 + z'^2\} \qquad (5\text{--}17)$$
$$= -\{x'^2 + y'^2 + z'^2 - c^2t'^2\} = L'^2$$

$$L = L' \qquad (5\text{--}18)$$

The vector therefore has the same length, judged by either observer. Such a vector with components x, y, z, ct is called a *four-vector*. Four-vectors play a particularly important role in special relativity; first, because the theory's natural setting is a *four-space*; and, second, because the length of the vectors is *invariant* with respect to coordinate transformations: This means that one observer measures exactly the same vector magnitude as any other. But since relativity postulates that the laws of physics are invariant in all inertial frames, these invariant lengths assume a special significance in the formulation of the laws of physics.

We note that the length L specified here corresponds to the interval s defined in equation 5–1. The interval therefore is an invariant. If two events 1 and 2 occur in one and the same place for an observer O', we see that $s_{12}^2 = c^2 t_{12}^2 - l_{12}^2$ $= c^2 t_{12}'^2 > 0$. The square of the interval, s_{12}^2, is positive since the elapsed time t_{12}' is a real quantity. l_{12} is the spatial separation in O's frame. We see that if an interval between events is timelike, there exists a frame in which the events occur in the same place. If the interval is spacelike we can similarly show that a frame exists in which the two events are simultaneous.

The general transformation of a four-vector with components A_1, A_2, A_3, and A_4 reads

$$A_1 = [A_1' + \beta A_4'] \gamma(V), \qquad A_2 = A_2', \qquad A_3 = A_3'$$

$$A_4 = [A_4' + \beta A_1'] \gamma(V)$$

(5–19)

We will find for example, that the *four-momentum* with components $(p_x, p_y, p_z, \mathscr{E}/c)$ transforms as a four-vector. So also does a four-vector (A_x, A_y, A_z, ϕ) having the electromagnetic vector and scalar potentials as components (section 6:13). There are many other such four-vectors comprising useful physical quantities and, in fact, all physically significant quantities must—according to special relativity—mold themselves to conform to this four-dimensional point of view.

5:5 ABERRATION OF LIGHT

Next we will want to use the Lorentz transformations to see how the measurement of angles depends on the relative motion of an observer. We will find here that the measurement of an angle—or rather the sine or cosine of an angle—does not at all involve the measurements of two lengths. Rather it requires the simultaneous measurement of two velocities. This comes about because a distant observer must make his angular measurements using light signals received from an object, and the law of composition of velocities determines the angles these light rays subtend at the observer.

Suppose a particle has a velocity vector that lies in the xy plane. The velocity v has components $v_x = v \cos \theta$ and $v_y = v \sin \theta$ along the x and y axes of the reference frame K. Viewed by an observer at rest in the frame K', the velocity components are $v_y' = v' \sin \theta'$ and $v_x' = v' \cos \theta'$. The velocity transformation equations then allow us to write

$$\tan \theta = \frac{v_y}{v_x} = \frac{v' \sin \theta'}{[v' \cos \theta' + V] \gamma(V)}$$

(5–20)

When we deal with a light ray, $v' = c = v$; and the angles subtended by the light ray transform as

$$\tan \theta = \frac{\sin \theta'}{[V/c + \cos \theta'] \, \gamma(V)} \tag{5-21}$$

for, v_x and v_y alone lead to

$$\sin \theta = \frac{\sin \theta'}{\gamma(V)[1 + \beta \cos \theta']}, \qquad \cos \theta = \frac{\cos \theta' + \beta}{1 + \beta \cos \theta'} \tag{5-22}$$

When $V \ll c$ and the terms in β^2 are negligible, the sine equation becomes

$$[\sin \theta + \beta \sin \theta \cos \theta'] = \sin \theta' \tag{5-23}$$

And the *aberration angle* $\Delta\theta = \theta' - \theta$ is

$$\Delta\theta \sim \beta \sin \theta', \qquad \text{for } \Delta\theta \ll 1, \quad \beta \ll 1 \tag{5-24}$$

We note that since the light travels in a direction opposite to that in which the telescope moves, $\sin \theta'$ has a negative value, that is, $\theta' < \theta$, as shown in Fig. 5.5. This angle is of great practical importance in observational astronomy. For, if a star's position (direction in the sky) is measured at different times of year it will be found to undergo a small annual motion.

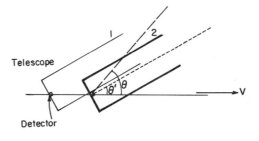

Fig. 5.5   Aberration of light. A telescope (stationary in some coordinate frame K') moves with velocity V, relative to a star. When the starlight enters, the telescope is in position 1. By the time the light has traveled the length of the instrument, the telescope has moved to position 2. All this time the telescope is pointed into direction θ', with respect to the x'-axis. But an observer whose telescope was at rest with respect to the star's reference frame K would have to point it at angle θ relative to x. Aberration is also found in Newtonian physics for a finite speed of light c. This is already indicated by the magnitude of the effect, which is of order V/c, while purely relativistic corrections always are of order V^2/c^2 and, therefore, are smaller. However, a relativistic correction for the Newtonian aberration is needed to obtain the correct value of the aberration angle.

PROBLEM 5-5.   The orbital velocity of the earth about the sun is 30 km/sec, which means that the velocity of the earth changes by 60 km/sec over a six-month interval. Taking $V \sim 60$ km/sec show that

5:5

$$\Delta\theta \sim \frac{60}{3 \times 10^5} \sim 40'' \text{ of arc}$$

This is an easily measured angle in precision astronomy.

The aberration of light was first noticed by James Bradley in 1728. This represented the first verification of the earth's annual motion about the sun and the first conclusive proof that Copernicus had been right in the hypothesis he had advanced nearly two centuries earlier, in 1543

5:6 MOMENTUM, MASS, AND ENERGY

The velocities we have talked about thus far are three-dimensional velocities. In relativity, however, the proper form to use is a four-vector because, as emphasized in section 5:4, four-vectors have *invariant magnitude*. We define a four-velocity with components

$$u_1 = \frac{dx}{ds}, \qquad u_2 = \frac{dy}{ds}, \qquad u_3 = \frac{dz}{ds}, \qquad u_4 = c\frac{dt}{ds} \qquad (5\text{--}25)$$

Since

$$ds = \sqrt{c^2dt^2 - dx^2 - dy^2 - dz^2} = \frac{cdt}{\gamma(v)} \qquad (5\text{--}26)$$

the equations 5–26 become

$$u_1 = \frac{dx}{dt} \; \frac{1}{c\sqrt{1 - \dfrac{v^2}{c^2}}} = \frac{v_x\gamma(v)}{c} \qquad (5\text{--}27)$$

$$u_2 = v_y\gamma(v)/c \qquad u_3 = v_z\gamma(v)/c \qquad u_4 = \gamma(v)$$

or

$$u_i = [\gamma(v)\,dx_i/dt]\,c^{-1} \qquad (5\text{--}28)$$

If we use the notation introduced in section 5:3, we see that u is a dimensionless quantity that does not have the units of velocity: cm sec^{-1}. We can also obtain the square of the magnitude of u:

$$u_1^2 + u_2^2 + u_3^2 - u_4^2 = -1 \qquad (5\text{--}29)$$

This is an invariance property, which we will need below.

Frequently we are interested in a particle's *momentum*. This can be written in the form

$$\mathbf{p} = m_0 \mathbf{v} \gamma(v) \tag{5-30}$$

and involves three components that correspond to the quantities

$$p_1 = m_0 c u, \qquad p_2 = m_0 c u_2, \qquad p_3 = m_0 c u_3 \tag{5-31}$$

m_0 is the particle's mass measured at rest—the *rest mass*. The momentum therefore accounts for the first three components of a four-vector whose fourth component has the form $m_0 c u_4$. In relativity the fourth component is associated with the particle energy $\mathscr{E}$ in the following way

$$p_4 = \frac{\mathscr{E}}{c} = m_0 c u_4 \tag{5-32}$$

The complete relativistic momentum four-vector then has components

$$(p_x, p_y, p_z, \mathscr{E}/c) \tag{5-33}$$

It is clear that in the limit $v \ll c$, the first three components give the classical momentum $\mathbf{p} = m_0 \mathbf{v}$. However, the energy takes on a new form. Written explicitly, the energy equation is

$$\mathscr{E} = \frac{m_0 c^2}{\sqrt{1 - \dfrac{v^2}{c^2}}} = m_0 c^2 \gamma(v) \tag{5-34}$$

At zero velocity this reduces to

$$\mathscr{E} = m_0 c^2 \tag{5-35}$$

an expression that states that mass and energy are equivalent (Ei05b)*. It is precisely this equivalence that allows stars to radiate because the nuclear reactions that ultimately give rise to stellar radiation always involve a mass loss with the simultaneous liberation of energy in the form of photons or neutrinos. As the star radiates it conserves mass-energy by becoming less massive.

For small velocities equation 5–34 can be approximated by the expansion

$$\mathscr{E} = m_0 c^2 + \tfrac{1}{2} m_0 v^2 + \ldots \tag{5-36}$$

where the second term represents kinetic energy. The next higher term would be of order $m_0 v^4/c^4$. In mechanical or chemical processes m_0 remains essentially constant and we normally see changes only in the $mv^2/2$ term. This is why that term has classically been so important even though it is far smaller than the energy contained in a particle's mass.

Equation 5–34 shows that $\mathscr{E} \to \infty$ as $v \to c$, which means that an infinite amount of work would be required to accelerate a particle to the speed of light. As with all special relativistic effects, this statement is valid in inertial frames but need not be true for others. There need therefore be no conflict with the observations that distant galaxies travel at nearly the speed of light and that some may pass across the cosmic horizon when their speed, relative to our galaxy, exceeds the speed of light. Since these distant galaxies are at rest in reference frames that are accelerated relative to ours, special relativity need not hold, and we can make no general statements about speed limitations unless we talk in terms of a less specialized theory, such as the general theory of relativity.

Two important relations should still be stated. First, equations 5–30 and 5–34 show that

$$\mathbf{p} = \frac{\mathscr{E}}{c^2}\mathbf{v} \tag{5–37}$$

Second, writing the four-vector components as

$$p_i = m_0 c u_i \tag{5–38}$$

we obtain the square of the magnitude of the four-momentum vector

$$-(p_1^2 + p_2^2 + p_3^3 - p_4^2) = -p^2 + \frac{\mathscr{E}^2}{c^2} = m_0^2 c^2 \tag{5–39}$$

which again is invariant. Equations 5–39 can be rewritten as

$$\mathscr{E}^2 = p^2 c^2 + m_0^2 c^4 \tag{5–40}$$

Since photons have zero rest mass, and velocity c, (5–37) and (5–40) become

$$p = \frac{\mathscr{E}}{c} \tag{5–41}$$

The relations (5–37) and (5–40) are of particular importance in cosmic ray physics, where particle energies may be as high as $\sim 10^{20}$ ev. The rest mass of a proton with this energy is only 931 Mev, so that the total energy of a cosmic ray particle may be $\sim 10^{11}$ times its rest mass energy. This feature allows a cosmic ray primary, incident on the top layers of the earth's atmosphere, to undergo collisions that produce millions of shower particles whose total rest mass exceeds that of the primary proton by many orders of magnitude. The classical concept of conservation of mass is violated here, but the more encompassing principle of conservation of mass-energy permits this kind of process to take place!

5:7 THE DOPPLER EFFECT

Since energy is the fourth component of a four-vector $(\mathbf{p}, \mathscr{E}/c)$ it transforms as (see equations 5–19 and 5–33)

$$\mathscr{E} = \gamma(V)\left[\mathscr{E}' + Vp'_x\right] \qquad (5\text{–}42)$$

when the relative motion is along the x-direction.

If we wish to see how photon energies will transform, we note for (5–41) that for a ray directed at an angle θ' with respect to the x'-axis

$$\mathscr{E} = \frac{\mathscr{E}' + (\mathscr{E}'V/c)\cos\theta'}{\sqrt{1 - \dfrac{V^2}{c^2}}} = \mathscr{E}'\left[1 + \beta\cos\theta'\right]\gamma(V) \qquad (5\text{–}43)$$

The angle θ' is that shown in Fig. 5.5, but we have to recall that the direction of the photon's travel is opposite to the viewing direction. We know from quantum mechanics that the energy $\mathscr{E}$ is equal to the radiation frequency v, multiplied by Planck's constant, h—a universal constant. Using this in equation 5–43 gives,

$$v = v'(1 + \beta\cos\theta')\gamma(V) \qquad (5\text{–}44)$$

which gives the Doppler shift in frequency for radiation emitted by a moving

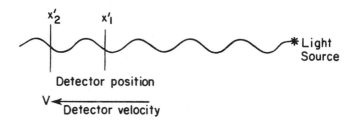

Fig. 5.6   The Doppler effect. The detector moves at velocity V, relative to the source. It starts measuring the frequency of radiation at time t'_1 and finishes at time t'_2. During this interval it is receding from the source, moving from position x'_1 to x'_2. A wave that would just have reached x'_1 by time t'_2 is therefore not counted, nor are any waves lying between x'_1 and x'_2, at time t'_2. The detector therefore senses a lower frequency v'. This explanation basically gives the first order Doppler shift proportional to V/c which is also present classically. The correct relativistic expression contains an additional factor $(\sqrt{1 - V^2/c^2})^{-1}$ as shown in equation (5–44).

source. In contrast to the classical situation, we see that there is a red shift even when the motion of the source is purely transverse ($\cos \theta = 0, \cos \theta' = -V/c$). This corresponds to a time dilation—a frequency decrease. When the source radiates in a direction opposite to its direction of motion, $\beta \cos \theta' < 0$, and $v < v'\gamma(V)$ when it radiates in the forward direction, $v > v'$.

In quasars we frequently see emission lines at some red shifted frequency v_0 and a series of absorption lines, for the same transition at higher frequencies $v_0 + \Delta_1, v_0 + \Delta_2, \ldots$ Here Δ is a shift corresponding to a few hundred or thousand km sec^{-1}. These are likely to be absorption lines due to cooler clouds along the line of sight and moving at different velocities. Some of these clouds may have been ejected from the quasar.

Line of sight components of stellar velocities within the Galaxy are generally determined by the Doppler shift observed in the spectra. Quasi-stellar objects have large frequency shifts sometimes attributed to an explosion of these objects out of our galaxy; but most astronomers prefer to think of the QSO red shift largely in terms of the cosmological red shift of galaxies at large distances. That may or may not be purely a Doppler shift phenomenon.

5:8 POYNTING-ROBERTSON DRAG ON A GRAIN

Consider a grain of dust orbiting about the sun in interplanetary space. It absorbs sunlight, and re-emits this energy isotropically. We can view this two-step process from two differing viewpoints.

(a) Seen from the sun, the particle absorbs light coming radially from the sun and re-emits it isotropically in its own rest frame. A re-emitted photon carries off angular momentum proportional (i) to its equivalent mass hv/c^2, (ii) to the velocity of the grain $R\dot\theta$, and (iii) to the grain's distance from the sun R. Considering only terms linear in V/c, and neglecting any higher terms, we see that the grain loses orbital angular momentum L about the sun at a rate

$$dL = \frac{hv}{c^2} \dot\theta R^2 \qquad (5\text{--}45)$$

$$\frac{1}{L} dL = \frac{hv}{mc^2} \qquad (5\text{--}46)$$

for each photon whose energy is absorbed and re-emitted, or isotropically scattered in the grain's rest-frame. m is the grain's mass.

(b) Seen from the grain, radiation from the sun comes in at an aberrated angle θ' from the direction of motion, instead of at $\theta' = 270°$ (see equation 5–22).

5:8

$$\cos\theta = \frac{\cos\theta' + \dfrac{V}{c}}{1 + \dfrac{V}{c}\cos\theta'} = 0, \quad \cos\theta' = -\frac{V}{c} \tag{5-47}$$

Here V is $\dot{\theta}R$, the grain's orbital velocity. Hence the photon imparts an angular momentum $pR\cos\theta' = -(h\nu/c^2)R^2\dot{\theta}$ to the grain.

For a grain with cross section σ_g

$$\frac{dL}{dt} = -\frac{L_\odot}{4\pi R^2}\frac{\sigma_g}{mc^2}L \tag{5-48}$$

where $L_\odot$ is the solar luminosity.

Either way the grain velocity decreases on just absorbing the light. From the first viewpoint, because the grain has gained mass which it then loses on re-emission; from the second, because of the momentum transfer.

PROBLEM 5-6. A grain having $m \sim 10^{-11}$ g, $\sigma_g \sim 10^{-8}$ cm^2 circles the sun at 1 AU. Calculate the length of time needed for it to fall into the sun, that is, to reach the solar surface, assuming that the motion is approximated by circular orbits throughout the whole time.

PROBLEM 5-7. Suppose one part in 10^8 of the sun's luminosity is absorbed, or scattered isotropically by grains circling the sun. What is the total mass of such matter falling into the sun each second?

5:9 MOTION THROUGH THE COSMIC MICROWAVE BACKGROUND RADIATION

We can derive the apparent angular distribution of light emitted isotropically in the reference frame of a moving object. Let the object be at rest in the K' system. Then the intensity $I(\theta')$ has the same value I', for all directions θ' (Figure 5.7). The energy radiated per unit time into an annular solid angle $2\pi\sin\theta'd\theta'$ is $2\pi I'\sin\theta'd\theta'$.

In the K reference frame the intensity distribution is $I(\theta)$ and we would like to find the relation between $I(\theta)$ and I'. We know that the total energy received is proportional to the observing time, the specific intensity $I(\theta)$ and the solid angle Ω

$$\mathscr{E} \propto \Delta t \cdot I(\theta)\cdot\Omega \tag{5-49}$$

Fig. 5.7   Distribution of radiation, viewed in spherical polar coordinates.

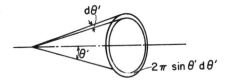

But $\mathscr{E}$ and Δt both transform as the fourth component of a four-vector, so that their ratio is invariant and, hence, $I(\theta) \cdot \Omega$ must be invariant also. This is very important! It means that the total observed power radiated by a source is the same for any set of observers in inertial frames. We will make use of this fact in section 6:19 to compute the total power emitted by a relativistic electron spiraling in a magnetic field. From equation 5–22 we then have

$$\frac{I(\theta)}{I'} = \frac{\sin \theta' \, d\theta'}{\sin \theta \, d\theta} = [(1 + \beta \cos \theta') \gamma(V)]^2 \qquad (5\text{–}50)$$

which is the square of the Doppler shift (5–44).

We see that an isotropically radiating, fast-moving body appears to radiate the bulk of its energy into the forward direction, ($\beta \cos \theta' \sim 1$) and only a small amount in the backward direction ($\beta \cos \theta' - 1$).

Current observations indicate that the universe is bathed by an isotropic bath of microwave radiation. It is interesting that the presence of such a radiation field should allow us to determine an absolute rest frame on the basis of a local measurement. Such a frame would in no way violate the validity of special relativity which, as stated earlier, does not distinguish between different inertial frames. Rather, the establishment of an absolute rest frame would emphasize the fact that special relativity is really only meant to deal with small-scale phenomena and that phenomena on larger scales allow us to determine a preferred frame of reference in which cosmic processes look isotropic.

At each location in an isotropic universe (and we will be discussing such cosmic models in section 10:5) there must exist one frame of reference from which the universe looks the same in all directions. This isotropy of course need hold only on a large scale, beyond the scale of clumping of the nearer galaxies. In the immediate vicinity of any observer, anisotropies on the scale of clusters of galaxies would always persist. The interesting feature of the radiation bath is that we should be able to determine the velocity of the earth relative to any coordinate system in which the cosmic flux appears isotropic. The earth might be expected to move at a speed of the order of 300 km sec^{-1} with respect to such a coordinate system, because the sun's motion about the center of the Galaxy has a velocity of the order of 250 km sec^{-1} and, in addition, galaxies are known to have apparently random velocities of the order of 100 km sec^{-1} relative to one another. These velocities would sum, vectorially, to something of the order of 300 km sec^{-1}.

The relativity principle requires that a body in thermal equilibrium in one inertial frame of reference also be in thermal equilibrium in all others. A blackbody radiator will therefore appear black in all inertial frames. If the isotropic microwave component of the cosmic flux has a blackbody spectrum (4–71) (Pe65)*:

$$I(v) = \frac{2hv^3}{c^2} \left[\frac{1}{e^{hv/kT} - 1} \right] \tag{5–51}$$

then the Doppler shift will transform the observed radiation to higher frequencies in the direction into which the earth is moving. This effect would make the factor v^3 in equation 5–51 systematically larger without changing the spectral shape of the curve that is given by the expression in brackets. An observer would therefore decide that he was seeing a blackbody spectrum at a temperature that was increased in the same amount as the Doppler increase in frequency, since the ratio v/T would then remain unchanged.

An observer moving relative to the cosmic radiation bath will then see a hotter blackbody flux in the direction into which he is moving than in the direction from which he came; and in fact the temperature of the observed flux will increase slowly, as a function of angle, starting from the trailing direction and reaching a maximum in the direction of motion. At each angle with respect to this direction of motion, the observer will see a blackbody spectrum but with a temperature dependent on the angle, as in (5–44).

The difference between blackbody flux in the forward and backward direction should be most apparent near the peak of the spectrum. For the 3°K cosmic background temperature currently postulated, this frequency corresponds to a wavelength in the submillimeter range.

Conklin (Co69) observed the background at a radio frequency of 8000 MHz. His instrument limited his measurements to values of declination $\delta \sim 32°$, and he found a maximum signal at an hour angle $\alpha = 13^h$. The earth's velocity component in that direction appears to be 160 km sec^{-1}.

PROBLEM 5–8. The Lorentz contraction is an important effect for extreme relativistic cosmic ray particles. For a proton with energy 10^{20} ev, for example, the disk of the Galaxy would appear extremely thin. If the width of the disk is of the order of 100 pc in the frame of an observer at rest in the Galaxy, show that this width would appear to be $\sim 3 \times 10^9$ cm, comparable to the earth's circumference, to an observer moving with the cosmic ray proton.

PROBLEM 5–9. The time dilation factor similarly is important at cosmic ray energies. Consider the decay time of a neutron that has an energy comparable to the 10^{20} ev energies observed for protons. How far could such a neutron move across the Galaxy before it beta decayed? In the rest frame of the neutron this decay time is of the order of 10^3 sec; but in the framework of an observer at rest in the Galaxy it would be much longer. Show that the neutron could more than traverse the Galaxy.

PROBLEM 5–10 If an energetic cosmic gamma ray has sufficiently high energy it can collide with a low energy photon and give rise to an electron-positron pair. Because of symmetry considerations, this electron-positron pair has to be moving at a speed equal to the center of momentum of the two photons. The pair formation energy is of the order of 1 Mev. The energy of a typical 3°K cosmic background photon is of the order of 10^{-3} ev. What is the energy of the lowest energy gamma photon that can collide with a background photon to produce an electron-positron pair? Show that in the frame within which the pair is produced at rest, energy conservation gives

$$\frac{h\nu_1\left(1 - \dfrac{v}{c}\right)}{\sqrt{1 - \dfrac{v^2}{c^2}}} + \frac{h\nu_2\left(1 + \dfrac{v}{c}\right)}{\sqrt{1 - \dfrac{v^2}{c^2}}} = 2m_0 c^2. \tag{5–52}$$

and momentum conservation requires the two terms on the left to be equal. These two requirements give

$$(h\nu_1)(h\nu_2) = (m_0 c^2)^2 \tag{5–53}$$

$$h\nu_1 \sim 10^{15}/4 \text{ ev}$$

Interestingly, the cross section for this process is sufficiently high so that no gamma rays with energies in excess of $h\nu_1$ can reach us if they originate beyond 10 kpc—that is, beyond the distance of the Galactic center!

5:10 PARTICLES AT HIGH ENERGIES

Cosmic rays are extremely energetic photons, nuclei, or subatomic particles that traverse the universe. Occasionally such a particle or photon impinges on the earth's atmosphere, or collides with an ordinary interstellar atom. What happens in such interactions?

We have no experimental data on particles whose energies are as high as, say, 10^{20} ev, because our laboratories can only accelerate particles to energies of the order of 3×10^{11} ev. However, the relativity principle permits us some insight even into such interactions. For example, we ask ourselves, how 10^{20} ev protons would interact with low energy photons in interstellar or intergalactic space. Such 3°K blackbody photons have a frequency $v \sim 3 \times 10^{11}$ Hz.

As seen by the proton, these millimeter-wavelength photons appear to be highly energetic gamma rays. This follows because $\gamma(v)$ must be $\sim 10^{11}$ for the proton, since its rest mass is only 9.31×10^8 ev; and by the same token the proton sees the photon Doppler shifted by a factor of 10^{11} (5–34 and 5–44). In the cosmic ray proton's rest frame, the photon appears to have a frequency of $\sim 3 \times 10^{22}$ Hz. That corresponds to a gamma photon with energy ~ 100 Mev, and the proton acts as though it were being bombarded by 100 Mev photons.

This simplifies the problem appreciably. One hundred Mev photons can be produced in the laboratory; and in fact we find that the main effect of photon-proton collisions at this energy is the production of π-mesons, through the interactions

$$\mathscr{P} + \gamma \to \mathscr{P} + \pi^\circ \tag{5–54}$$

$$\mathscr{P} + \gamma \to \mathscr{N} + \pi^+ \tag{5–55}$$

In the first reaction the proton-photon ($\mathscr{P}, \gamma$) collision produces a neutral pion π° and a proton having a changed energy. The second reaction produces a neutron and a positively charged pion.

The cross sections for these interactions can be measured in the laboratory, and the results are then immediately applicable to our initial query. It turns out that the cross sections are so large that the highest energy cosmic ray protons can probably not traverse intergalactic space over distances $\gtrsim 20$ Mpc even if the only extragalactic photons are those found in the 3°K microwave photon flux (Gr66, St68).

For some time this presented a puzzle, because quasi-stellar objects appeared to be the most likely source of energetic cosmic ray protons and QSO's appear to be far more distant than 20 Mpc. However, the suggestion that pulsars may generate sufficiently energetic protons (Gu69) has removed some of these difficulties, since the cosmic ray protons can now be considered to be a more local phenomenon.

Similar relativistic arguments are useful in a variety of other problems involving cosmic ray particles. In Chapter 6, we will discuss the Compton and inverse Compton effects involving collisions of photons and electrons, and we will also consider several other relativistic effects. Frequently a physical problem can be considerably simplified if we choose to view the situation from a favorable inertial frame. The relativity principle shows us how to do that, and gives us many new insights into the symmetries of physical processes.

5:11 HIGH ENERGY COLLISIONS

Consider the elastic collision of a low energy particle with a similar particle initially at rest. If we view this interaction in the resting particle's frame, and both particles have mass m, then the center of mass will move with velocity $v/2$ as shown in Fig. 5.8.

Fig. 5.8   Illustration of elastic collisions for identical particles.

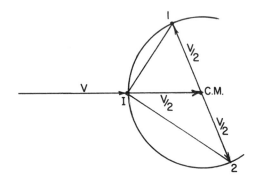

For an initial approach velocity v of the moving particle, conservation of momentum will require that the two particles have velocities $v/2$ relative to the center of mass—after the collision, as well as before. For any time after the collision, a circle can be drawn through the impact point I, and the particle positions 1 and 2 that define a diameter on the circle. This means that the particles always subtend a right angle at the impact point.

So far the treatment has been nonrelativistic. In the relativistic case, the center of mass still lies on a line joining 1 and 2. Effectively, particles 1 and 2 are scattered away from the center of mass in opposing directions. Seen from a rest frame, however, they will appear to be scattered predominantly into the forward direction. This is precisely the same concentration into the forward direction that we saw for the rapidly moving light source that emits radiation isotropically in its own rest frame (5–50).

When a cosmic ray proton collides with the nucleus of a freely moving interstellar atom or with an atom that forms part of an interstellar grain, a fraction of the nucleus can be torn out. This may just be a proton or a neutron, or it could be a more massive fragment, say, a He^3 nucleus. Such *knock-on* particles always come off predominantly in the forward direction, close to the direction along which the primary proton was moving.

Similarly, when a primary arrives at the top of the earth's atmosphere, after its long trek through space, it collides with an atmospheric atom's nucleus, giving rise to energetic secondary fragments, mesons, baryons, and their decay products. These decay into mesons, gammas, electron-positron pairs, or neutrinos, or they may collide with other atoms until a whole shower of particles rains down. Such a *cosmic ray air shower* consisting of electrons, gamma rays, mesons, and other particles, even if initiated at an altitude of 10 or more kilometers, often arrives at ground level confined to a patch no more than some hundred meters in diameter. The forward concentration is so strong that the showers are well confined even though they sometimes consist of as many as 10^9 particles!

This close confinement allows us to deduce the total energy originally carried by the primary; we need only sample the energy incident on a number of rather small detector areas. In fact, much of our information about high energy cosmic ray primaries has come from just such studies made with arrays of cosmic ray shower detectors. These arrays usually sample an area not much more than a few hundred meters in diameter although an array as big as 3.6 km was used by an MIT group. The total energy in the shower can be determined from these samplings, and the time of arrival at each detector shows the direction from which the primary came (Ro64a). Fig. 5.9 shows some of the constituents of cosmic ray air showers.

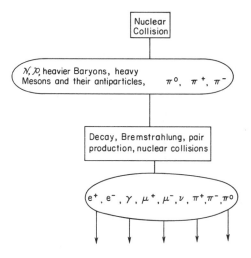

Fig. 5.9 Constituents of a cosmic ray air shower. The primary particle, here shown as a proton, collides with the nucleus of an atmospheric atom, producing a number of secondary particles that suffer nuclear collision, decay, pair production, or *bremsstrahlung*—a process in which a charged particle is slowed down by the emission of a gamma photon. A large succession of such events takes place. By the time the shower arrives at the surface of the earth, most of the charged particles we observe are electrons, positrons, and muons. Although most of the primary nuclei are protons, several percent can be alpha particles (helium nuclei) and about one percent are heavier nuclei. Electrons and positrons also can be primary particles. The air showers are a prime example of the conversion of energy into rest mass. On occasion, the energy of a single primary is sufficient to produce 10^9 shower particles.

5:12 FASTER-THAN-LIGHT PARTICLES

When Einstein first discovered the special relativistic concept he clearly stated that matter could not move at speeds greater than the speed of light. He argued that the relation between rest mass and energy (5–3) already implied that an infinite amount of energy was needed to accelerate matter to the speed of light and that particles with nonzero rest mass could therefore never quite reach even the velocity of light let alone higher velocities.

In recent years, this question has been re-opened by a number of researchers. They have argued that it certainly is not possible to actually reach the speed of light by continuous acceleration, but that this alone does not rule out the existence of faster than light matter created by some other means. They have called particles moving with speeds greater than *c tachyons*, and have examined the possible properties of such entities.

The basic argument in favor of even examining the possibility of tachyon existence is the formal similarity of the Lorentz transformations for velocities greater than and less than the speed of light, and the fact that the transformations taken by themselves say nothing that would rule out tachyon existence.

The similarity of course does not imply that particles and tachyons behave in precisely the same manner. If we look at equation 5–34, we note that a particle moving with speed $V > c$ has an imaginary quantity in the denominator. If the mass of the tachyon is real its energy would therefore be imaginary. In practice, one chooses the mass of the tachyon to be imaginary mainly on the grounds that this is not observationally ruled out. Perhaps this is a somewhat negative approach, but if this assumption is not made, it is more difficult to make headway to the point where predictions on the probable outcome of experiments become possible.

Choosing the mass to be an imaginary number makes the energy $\mathscr{E}$ real. The momentum too is then real as shown by (5–37).

We now combine equations 5–34 and 5–40:

$$\mathscr{E}^2 = m_0^2 c^4 \gamma^2(V) = p^2 c^2 + m_0^2 c^4 \tag{5–56}$$

As V becomes large, $\mathscr{E}$ is seen to become small. In the limit of infinite velocity the energy becomes zero, but the momentum stays finite, asymptotically approaching the value $|m_0 c|$.

Thus far we have only departed from more standard concepts in accepting the possibility of imaginary mass values.

Now, however, we come to an argument that has caused considerable conceptual difficulty in the past few years. This argument considers sequences of events if tachyons can be used as carriers of information. Consider the world lines of two observers O and O′, shown in Fig. 5.10. Let O emit a first tachyon toward O′. The slope of this trajectory is less than that for a photon, in the world diagram. The tachyon enters the "absolutely inaccessible" domain that was shown in Fig. 5.1.

Seen from O′'s vantage point, tachyon 1 arrives from above the x'-axis, apparently moving backward in time. This is bad enough; but let us go a little further.

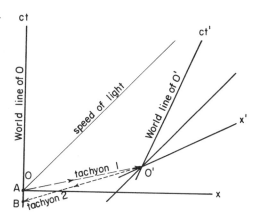

Fig. 5.10 World diagram for tachyon transmission.

O' can also send out tachyons and, in principle, the tachyon 2 he transmits towards O can be a faster moving tachyon than tachyon 1.

We can now set up the following paradox (Th69): Observer O waits until he sees no event at all during a long period of time. He then sends out a tachyon 1 at time A. O' has previous instructions to send out a fast tachyon immediately upon receiving one from O. He does this now, sending out tachyon 2 that arrives at O at time B, during the time interval when O was sure no event had taken place. Cause and effect are thrown into complete confusion! The situation is not helped much even if a tachyon absorbed moving in a negative time direction is really equivalent to a tachyon emitted in the positive time direction.

Should we then disregard tachyons altogether? This seems an unnecessarily harsh edict: The causal argument we have just applied is strongly influenced by traditional special relativity and may be misleadingly restrictive. For example, the concept of simultaneity used in the argument, depends on the idea that no messages propagate faster than the speed of light. We noted in section 5:4 that the Minkowksi diagram would look quite different if the speed of transmission of information was higher than c: The angles ψ in Fig. 5.4 would become smaller, and in the limit of infinite velocity—the Galilean limit—we found $\psi = 0$. For tachyons then, we would expect smaller angles ψ; the angle between the ct'- and x'-axes in Fig. 5.10 should approach a right angle, and the causal paradox should disappear, because as the angle between the tachyon trajectory and the x'-axis increases, it becomes impossible to transmit signals into another observer's past. Arguments of this kind still leave the existence or inadmissibility of tachyons an open question. As we will see in Chapters 10 and 11, the existence of tachyons would have important consequences in cosmology and in rapid communication across large distances within our galaxy or across the Universe.

Preliminary experiments to search for tachyons have been carried out (A168). Thus far none have been detected; but perhaps some day they will be.

5:13 STRONG GRAVITATIONAL FIELDS

As already stated, the introduction of gravitational fields requires a theory more general than the special theory of relativity that restricts itself to inertial frames. For general problems involving gravitation, the general theory of relativity (Ei16) or similar gravitational theories (Di67) have to be used. However, some simple gravitational results can be obtained without such theories if we remember that the set of inertial frames also includes freely falling frames of such small size that the gravitational field can be considered to be locally uniform. We will consider two such local inertial frames in a centrally symmetric gravitational potential Φ.

Consider an observer O' at distance r from the central mass distribution Fig. 5.11. We would like to know the form which the line element ds^2 would take in his frame of reference. We will assume that he was at rest initially, and that he has only just started falling toward the central mass, quite recently. His velocity is therefore still essentially zero, but he is accelerating toward the center. Alternately, we can suppose that O' was initially moving outward from the star but at a speed less than the escape velocity. He only had enough kinetic energy to reach r. Here his velocity reached zero, and he is just beginning his journey back into the center. We see him when his velocity is zero.

Since O' is freely falling, his line element will seem to him to have the form (5–3). In spherical polar coordinates this is

$$ds^2 = c^2 dt'^2 - r'^2(\sin^2\theta' d\phi'^2 + d\theta'^2) - dr'^2 \qquad (5\text{--}57)$$

Fig. 5.11 Freely falling observer near a mass M.

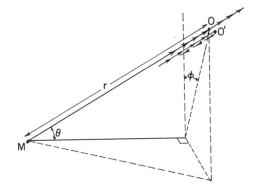

We ask ourselves, however, what ds^2 would seem to be, seen by an observer O far enough away from the mass distribution so that Φ essentially is zero or negligibly small. Φ as used here will be the negative of V in (3–54).

We could of course assume that O gets all his information about O''s system from light signals. But that is not necessary. The physical relationship between O and O' is independent of how the observational information is conveyed.

Let us therefore suppose that O has taken a trip to find out for himself. We can suppose that he was near the central mass distribution, that he is now on his way out, and that he is in unpowered motion, freely falling radially outward, with just exactly enough energy to escape to infinity.

O goes through the radial distance r, close to O', just as O' passes through zero velocity and begins his infall. Since both observers are in inertial frames, the Lorentz transformations can be used to determine what O''s line element would seem like to O. Once again, the components perpendicular to the direction of relative motion should be identical, and

$$r^2(d\theta^2 + \sin^2\theta\, d\phi^2) = r'^2(d\theta'^2 + \sin^2\theta'\, d\phi^2) \tag{5–58}$$

The radial components, however, will appear changed because of the relative motion. If the gravitational potential is weak, the velocity of O relative to O' is immediately obtained from the fact that O just barely has enough kinetic energy to go to infinity, so that, equating kinetic and potential energy, for unit mass

$$\tfrac{1}{2}V^2 = \Phi \tag{5–59}$$

Hence equations 5–12 and 5–14 would now read

$$\Delta t = \frac{\Delta t'}{\sqrt{1 - (2\Phi/c^2)}} \qquad \Delta x = \Delta x'\sqrt{1 - (2\Phi/c^2)} \tag{5–60}$$

This also is the correct form for strong potentials Φ, where the classical concept of kinetic energy no longer has a clear meaning. We can therefore write that the line element (5–57) has the form

$$ds^2 = (c^2 - 2\Phi)\, dt^2 - r^2(\sin^2\theta\, d\phi^2 + d\theta^2) - \frac{dr^2}{1 - (2\Phi/c^2)} \tag{5–61}$$

as seen by O. This represents a translation of the clock rate and scale length in O''s frame as seen from O's coordinate system. O notes this down, and is able to convey these impressions when he reaches infinity. He has been traveling in an inertial frame all this time, and his results are therefore not suspect.

The line element, or *metric*, depicted in (5–61) is called the *Schwarzschild line element.* When the potential is generated by a mass M, we can rewrite (5–61) as

$$ds^2 = \left(c^2 - \frac{2MG}{r} \right) dt^2 - \left(1 - \frac{2MG}{rc^2} \right)^{-1} dr^2 - r^2 (\sin^2 \theta \, d\phi^2 + d\theta^2)$$

$$(5\text{–}62)$$

We see that something odd must take place at the *Schwarzschild radius*

$$r_s = \frac{2MG}{c^2} \tag{5–63}$$

Here, according to (5–60) the clocks would appear to run infinitely slowly. A message emitted at some time, t_0, would not arrive at larger radial distances until an infinite time later. In fact signals emitted at $r < r_s$ never come out. A massive object that was completely enclosed in r_s could therefore not radiate out into the rest of the universe and would appear invisible. Such objects have been called *black holes.* They are only detectable through the gravitational and electromagnetic field they set up, but not through emitted radiation (Ru71a). A star could be orbiting about a black hole companion.

For an ordinary stellar mass, collapsed through a neutron star stage, $r_s \sim 3 \times 10^5$ cm. This is not much smaller than the radius of the neutron star itself. For an object as massive as a galaxy $r_s \sim 10^{16}$ cm. This is somewhat smaller than the radii of quasi-stellar objects are thought to be—provided QSO's are at cosmologically large distances.

We will mention black holes again later. However, for the moment, it is still worth discussing two matters.

First, there cannot be a preponderant number of black holes within the Galaxy, because we can account for roughly 60% of the Galactic mass in terms of ordinary stars and interstellar matter observed by visual and radio observers. Only some 40% of the mass appears unobserved, and some of this is likely to be in the form of nonluminous, cool stars, interstellar molecular hydrogen, and so on. We cannot rule out, however, that something of the order of one-third the Galactic mass might exist in the form of black holes.

Secondly, space travelers must be careful. Once they enter a black hole they can never return. The interior of such an object is as isolated from us as a separate universe!

ANSWERS TO SELECTED PROBLEMS

5–3.   By (5–10)

$$v = \frac{c(2 - \delta - \Delta)}{(2 - \delta - \Delta + \Delta\delta)} < c.$$

For opposing velocities we obtain an expression of form

$$-c < v = \frac{(\Delta - \delta)\,c}{\delta + \Delta - \delta\Delta} = \frac{(\Delta^2 - \delta^2)\,c}{\Delta^2 + \delta^2 + \delta\Delta(2 - \delta + \Delta)} < c.$$

5–4. $x = x' + Vt'$, $y = y'$, $z = z'$, $t = t'$, and since $dx = dx' + V\,dt'$, $dt = dt'$, $v_x = v_x' + V$, $v_y = v_y'$, and $v_z = v_z'$.

5–6. $\dfrac{dL}{dt} = -\dfrac{L}{R^2} \cdot \left(\dfrac{L_\odot}{4\pi}\right) \cdot \dfrac{\sigma_g}{(mc^2)}.$

Since $L = mvR$ and $v^2/R = GM_\odot/R^2$,

$$\frac{dL}{L} = \frac{dR}{2R}.$$

$$t = \int_{R_\odot}^{1AU} R\,dR \left\{ \frac{2\pi}{L_\odot} \cdot \frac{mc^2}{\sigma_g} \right\} = 5 \times 10^3 \text{ y.}$$

5–7. From Problem 5–6, for each grain i

$$m_i c^2 = t_i \frac{\sigma_g}{\pi(R_I^2 - R_\odot^2)} L_\odot$$

where R_I is the grain's initial position.
For all grains $M_{TOT}c^2 \approx t_{TOT} \cdot (10^{-8}\,L_\odot)$.

$\therefore$ Mass/sec falling into sun is $\dfrac{(10^{-8})(4 \times 10^{33} \text{ erg sec}^{-1})}{9 \times 10^{20} \text{ cm}^2 \text{ sec}^{-2}}$

$$= 4.5 \times 10^4 \text{ g.}$$

5–8. $\mathscr{E} = \gamma(V)\,m_0 c^2$; since $\mathscr{E} = 10^{20}$ ev, $\gamma(V) = 10^{11}$ and $\Delta x' = \Delta x/\gamma(V) = 3 \times 10^{20}/10^{11} \sim 3 \times 10^9$ cm.

5–9. $\Delta t' = \gamma(V)\,\Delta t \sim 10^{14}$ sec. At $v \sim c$, the distance traveled is 3×10^{24} cm $= 1$ Mpc.

6

Electromagnetic Processes in Space

6:1 COULOMB'S LAW AND DIELECTRIC DISPLACEMENT

In earlier chapters we noted the similarities between Coulomb's law for the attraction of charged particles and Newton's law for the attraction of masses. Both are inverse square law forces. Coulomb's law states that the attraction between two charges q_1 and q_2 is

$$\mathbf{F} = \left(\frac{q_1 q_2}{r^3} \right) \mathbf{r} \qquad (6\text{--}1)$$

The charges q can be either positive or negative. In the general case where a large number of separate charges exert a force on a given charge q, this force is the vector sum of a whole series of terms of the form of equation 6–1.

$$\mathbf{F} = q \sum_i \left[\frac{q_i}{r_i^3} \right] \mathbf{r}_i \qquad (6\text{--}2)$$

and we can define an *electric field* $\mathbf{E}$

$$\mathbf{E} = \frac{\mathbf{F}}{q} \qquad (6\text{--}3)$$

which can be considered the seat of the force. All this assumes that the charges q_i and q are at rest in a vacuum. If the charges q_i are moving, the charge q will experience an additional force that is magnetic in character, and if the charges are not in vacuum but in a polarizable dielectric material, the material will adjust itself to cancel out some of the force. The actual force acting on q then becomes less than that given in equation 6–2.

To specify this situation completely, we therefore define one more vector quantity, the *dielectric displacement* **D**, which is independent of the properties of the material in which the charges are imbedded. **D** is strictly a geometric quantity and specifies the field that would be obtained if all charges were in a vacuum. In the presence of a uniform dielectric, equation 6–2 becomes

$$\mathbf{F} = \frac{q}{\varepsilon} \sum_i \frac{q_i}{r_i^3} \mathbf{r}_i \qquad (6-4)$$

Equation 6–3 still holds true since it, in fact, is the definition of electric fields; but the dielectric displacement now becomes

$$\mathbf{D} = \varepsilon \mathbf{E} \qquad (6-5)$$

which is seen to be independent of ε, and dependent only on the position and magnitudes of the charges, that is, on the quantities q_i and $\mathbf{r}_i$. ε is the *dielectric constant* that, for most real materials, can be taken to be independent of **E** for field strengths below a critical value.

If we draw a spherical surface around a charge q_1, the displacement produced by this charge at the surface obeys the relation

$$\mathbf{D} = \left(\frac{q_1}{r^3} \right) \mathbf{r} = \left(\frac{4\pi q_1}{4\pi r^3} \right) \mathbf{r} \qquad (6-6)$$

so that

$$\mathbf{D} \cdot \mathbf{n} = \left(\frac{4\pi q_1}{rA} \right) \mathbf{r} \cdot \mathbf{n} \qquad (6-7)$$

where **n** is the normal to the surface at point **r** and A is the total area of the enclosing surface. The dots denote a *scalar product*. If a large number of charges is involved, or if the charge distribution becomes continuous, a more general form of expression (6–7) is applicable:

$$\int \mathbf{D} \cdot d\mathbf{s} = \int 4\pi\rho \, dV = \int \nabla \cdot \mathbf{D} \, dV \qquad (6-8)$$

where the last equality is obtained from *Gauss's theorem* on vector integration that states that for an arbitrary vector **X**

$$\int \mathbf{X} \cdot d\mathbf{s} = \int \nabla \cdot \mathbf{X} \, dV \qquad (6-9)$$

where the integral on the left is a *surface integral* and $d\mathbf{s}$ is an element of the surface over which the integration takes place. $\nabla \cdot$ is the *divergence operator*.

One may wonder why we emphasize the relation between **D** and **E** in such

6:1

detail when we have set out to discuss electromagnetic processes in space. We might expect that the emptiness of the cosmos would assure that **D** and **E** are always identical. In actual fact this is not quite true, and much of our knowledge of the contents of interstellar space depends on small differences between **E** and **D**. We define one more quantity that will be useful later. It is the *polarization field* **P** that is a measure of the difference between the displacement and electric fields:

$$\mathbf{P} = \frac{[\mathbf{D} - \mathbf{E}]}{4\pi} = \frac{(\varepsilon - 1)\,\mathbf{E}}{4\pi} \qquad (6\text{--}10)$$

$4\pi\mathbf{P}$ is the field set up through the rearrangement of charges in the polarizable material. It tends to oppose the externally applied field, reducing its value from **D** to **E**. The factor 4π introduced here is a matter of convention and has the following significance: At a plane boundary, with charge density σ per unit area, **D** just equals $4\pi\sigma$. The polarization field, instead, will depend on σ', the induced charge density per unit area. Now, **P** is the electric dipole moment per unit volume. If this volume contains n dipoles having charge q and separation d, $\mathbf{P} = nqd$. The charge density σ' then is nqd also, because we can visualize a cube of unit volume, made up of d^{-1} dipole layers each of thickness d and containing nd dipoles. This makes P numerically equal to σ'—no factor of 4π occurs!

Thus far we have acted as though static fields perhaps were important on a scale of cosmic dimensions. In general, this is probably not true, because in a near vacuum electric charges generally can quickly rearrange themselves into a configuration where all electric fields are neutralized, that is, into a charge neutralized configuration where any small volume element basically contains the same number of positive and negative charges. The dimensions of such volumes are given by the Debye shielding length discussed previously in Chapter 4. There is one exception to this general rule, and it is important! We will show in the next section that electric charges generally are tied to magnetic field lines in space; and if an electric field is applied perpendicular to the direction of a cosmic magnetic field, the charges cannot flow across the magnetic field lines to neutralize the electric field. In such a situation large-scale electric fields may persist.

6:2 COSMIC MAGNETIC FIELDS

An electric charge q traveling through a cosmic magnetic field experiences a force **F** called the *Lorentz force*:

$$\mathbf{F} = \frac{q\mathbf{v} \wedge \mathbf{B}}{c} \qquad (6\text{--}11)$$

6:2

where **v** is the velocity of the charge, **B** is the *magnetic field* and c is the speed of light. The *cross product*, in equation 6–11 shows that the force, and hence the acceleration experienced by the charge, is perpendicular to both the velocity and the direction of the magnetic field. The charge therefore spirals (see Fig. 6.1)

Fig. 6.1 Diagram to illustrate spiral motion in a magnetic field.

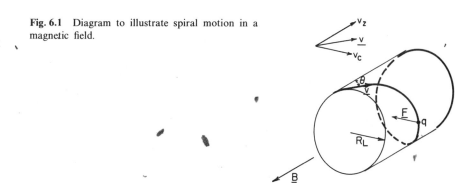

along the magnetic field lines without changing energy. (To do work on the particle one would require a force that has some component along the direction of motion). In a constant magnetic field, the particle describes a helical motion with constant pitch. The velocity component v_z along the direction of the field, is a constant of the motion, and the circular velocity v_c about the field lines then defines a *pitch angle* θ so that

$$\tan \theta = \frac{v_c}{v_z} \qquad (6-12)$$

The *gyroradius* or *Larmor radius*, R_L, of this motion is easily obtained by setting the magnetic force equal to the centrifugal force acting on the particle. If the particle has transverse momentum p_c and *gyrofrequency* $\omega_c = v_c/R_L$, the force has magnitude

$$\dot{p} = \frac{p_c v_c}{R_L} = \frac{qBv_c}{c}, \qquad R_L = \frac{p_c c}{qB} \qquad \text{and} \qquad \omega_c = \frac{v_c}{R_L} = \frac{qBv_c}{p_c c} \qquad (6-13)$$

The gyrofrequency is sometimes also called the *cyclotron frequency*.

PROBLEM 6–1. Show that the Larmor radius of a proton, moving at 10 km sec^{-1} through a field of 10^{-6} gauss is small compared to interstellar and even interplanetary distances.

Because the Larmor radius is small compared to the expected dimensions of interstellar and interplanetary fields, we have a situation in which charged particles, moving with thermal velocities characteristic of cosmic gases, are effectively tied to the magnetic field lines. They can move along the field lines but cannot cross them any appreciable distance. We say that the particles are "frozen" to the field and the motion of such particle-field combinations is called *frozen-in flow*. *Magnetohydrodynamics* is the subject that deals with problems arising from such flow (Co57).

We notice that the only way a particle can escape from being frozen to the lines of force is through an encounter with another particle. Each particle then assumes a completely new orbit. If such collisions are sufficiently frequent, the particles can diffuse across magnetic fields.

Inasmuch as cosmic magnetic fields have their origins in the organized motion of particles, the frozen-in flow not only is due to the presence of a magnetic field, but also maintains the field which causes it. This self-consistent motion of charges is not an obvious result, but magnetohydrodynamics shows that it is real. The collisional processes just mentioned, therefore, conspire not only to prevent freezing-in, but also as a consequence tend to destroy the magnetic fields that are maintained by the frozen in flow. For this reason, frozen-in fields cannot be maintained in dense gases where collisions are frequent. These collisions of charges with surrounding particles provide an electrical resistance that dissipates particle motions and the energy resident in the magnetic field. Frozen-in flow therefore has a short life in a dissipative medium (Sp62).

Magnetohydrodynamics also tells us that the presence of a force, such as a gravitational or electrostatic force, acting normal to the magnetic field, can produce a drifting motion in which charges move in directions perpendicular both to the applied force and to the magnetic field direction. Such particle drifts occur in the *Van Allen belts* of charged particles that constitute part of the earth's *magnetosphere*. These drifts, however, do not directly act to dissipate cosmic magnetic fields, unless the drifting particles suffer collisions.

6:3 OHM'S LAW AND DISSIPATION

A current generally consists of two types of terms. The first of these expresses the actual flow of charge in response to an applied electric field. The second term corresponds to a virtual current representing a change in the applied field. This change gives rise to a magnetic field (see equation 6:5) just as a moving charge would. It is genuinely important! One writes

$$\mathbf{j} = \rho\mathbf{v} + \frac{1}{4\pi} \frac{\partial \mathbf{D}}{\partial t} \qquad (6\text{–}14)$$

6:3

The value of the velocity **v** in this equation is determined by two competing effects. The applied electric field seeks to continuously accelerate the charge, while distant collisions with other electric charges continually seek to slow the particle down. The resistivity of the medium is a measure of this slowing down. Its reciprocal is the conductivity σ. In terms of **E** and σ, equation 6–14 can be written as

$$\mathbf{j} = \sigma\mathbf{E} + \frac{1}{4\pi}\frac{\partial\mathbf{D}}{\partial t} \tag{6–15}$$

In general, the conductivity depends on the density of the gas, its temperature, the state of ionization, and, hence, also on the chemical composition. Distant collisions provide the main process that slows down the motion of charged particles in space, and they determine the value of σ. These collisions were discussed in sections 3:13 and 3:14 and will be fully treated in 6:16.

6:4 MAGNETIC ACCELERATION OF PARTICLES

One of Faraday's contributions to electromagnetism was his discovery that a time-varying magnetic field gives rise to electric currents in a conducting medium that encircles the field. The plane in which this current flows is perpendicular to the direction of the time-varying field component. In integral form *Faraday's law* is expressed as

$$\frac{1}{c}\frac{\partial}{\partial t}\int\mathbf{B}\cdot d\mathbf{s} = -\oint\mathbf{E}\cdot d\mathbf{l} \tag{6–16}$$

where the integral on the left is a surface integral over the area enclosed by the loop through which the current is flowing (Fig. 6.2a). The integral on the right is a line integral taken over that loop and the current observed by Faraday has been replaced by the electric field that gives rise to it in accordance with equation 6–15. We note now that, if any region of interstellar space should suddenly be subjected to a rising magnetic field, electric charges would experience an effective electrical field **E** proportional to the time rate of change of **B**. In the laboratory this effect is used to elevate charges to very high energies. The first device that successfully accomplished this acceleration is the betatron constructed by D. W. Kerst in 1940. Since that time, it has been suggested by various astrophysicists that the betatron process might also be active in interstellar space in accelerating charged particles to the very high energies observed in cosmic rays.

A rapid rise in magnetic field strength could be produced by the compression of a cosmic cloud in a direction perpendicular to its magnetic field. Such a compression can occur in the collision of interstellar clouds, either with one another, or with high velocity gases ejected from exploding supernovae. This process may

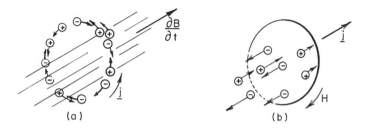

Fig. 6.2   Illustrations for Faraday's and Ampère's Laws. (*a*) Faraday's law states that the current in a conducting loop, and the associated electric field are determined by the rate of change in the number of magnetic *lines of force* enclosed by the loop (see equations 6–15 and 6–16). The number of lines of force crossing unit area is proportional to the magnetic field, *B*.

(*b*) Ampère's law states that the magnetic field, integrated along a loop enclosing a current, is determined by the total current crossing the enclosed area (see equation 6–17).

produce low energy cosmic rays, sometimes called *suprathermal particles*. It is not powerful enough to produce extremely energetic particles. More effective mechanisms are discussed in section 6:6 below.

PROBLEM 6–2.   Suppose the magnetic field in a region of space increases from 10^{-6} to 10^{-5} gauss over a period of 10^7 y. To what energy would electrons and protons be accelerated, if they move perpendicular to the field and suffered no collisions? How does the final energy depend on the initial energy? To do this, it is useful first to derive the energy-field relationship $d\mathscr{E}/\mathscr{E} = dB/B$ that follows from (6–13) and (6–16).

6:5 AMPÈRE'S LAW AND THE RELATION BETWEEN COSMIC CURRENTS AND MAGNETIC FIELDS

In section 6:2 we had noted that cosmic magnetic fields exist by virtue of the gyrating electric charges that are frozen to the field. This idea is more precisely expressed by *Ampère's law* that states that a current produces an encircling magnetic field (Fig. 6.2*b*):

$$\frac{4\pi}{c} \int \mathbf{j} \cdot d\mathbf{s} = \oint \mathbf{H} \cdot d\mathbf{l} \qquad (6\text{–}17)$$

6:5

Here again, the left side of the equation is a surface integral taken over the entire surface encircled by the magnetic field in the line integral on the right.

We believe that cosmic magnetic clouds are configurations in which equation 6–17 is obeyed in every locale. The shapes of the magnetic fields and currents are therefore likely to be quite complicated. One can think of initial configurations called *"force-free" magnetic fields* in which the magnetic fields and the flow of charges are so arranged that no forces result to destroy the configuration. These force-free configurations may well represent the structure of cosmic magnetic fields. Equation 6–11 tells us that a "force-free" structure must have $\mathbf{j} \wedge \mathbf{B} = 0$ everywhere.

6:6 MAGNETIC MIRRORS, MAGNETIC BOTTLES, AND COSMIC RAY PARTICLES

In section 6:4 we described the acceleration of charged particles by a betatron process. Another scheme for magnetically accelerating cosmic ray particles was suggested by Fermi. In the Fermi mechanism the cosmic ray particles are thought to travel between cosmic gas clouds. Each cloud has an embedded magnetic field. When a particle approaches the cloud and enters its field perpendicular to the field direction, it is turned back by virtue of the magnetic force given by equation 6–11. For, after traveling in a semicircle, the particle once again finds itself at the edge of the cloud and headed into the direction from which it came. As shown in Fig. 6.3—and as explained below—a similar reflection occurs for particles approaching along the lines of force.

If the particle impinges on a cloud that is receding from it, the particle's momentum after encounter is smaller than before. If the particle impinges on an approaching cloud, its final momentum is greater than before the collision. In general the probability for collision is greater for an approaching than for a

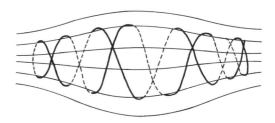

Fig. 6.3 Charged particle trajectories in a magnetic bottle. Light lines denote magnetic lines of force. A high density of field lines indicates a strong magnetic field.

6:6

receding cloud. (This corresponds to normal experience. On a highway we pass more cars going in the opposite direction than going along the direction in which we are traveling.)

Statistically, therefore, particles will derive increased momentum from encounters with clouds and can be accelerated to high energies. The process is similar to the acceleration of a ping-pong ball between two slowly approaching paddles. After the ball has made many bounces off each paddle, it is going far faster than either of these reflecting surfaces.

A number of variants on the Fermi process have been suggested to explain the acceleration of cosmic ray particles. However, none of these suffice to circumvent the following difficulties.

The cosmic ray particle eventually must suffer some destructive collision due to one of several competing processes, such as inelastic impact on another particle, loss from the Galaxy and, hence, loss from contact with the accelerating clouds, and so on. Because of this, the number of accelerating reflections between destructive processes is limited. To reach truly high energies, therefore, cosmic ray particles must be injected into the accelerating fields with rather high initial energies. It is possible that energetic enough particles are provided in supernova explosions. It is also possible that QSO's are capable of producing the high energy cosmic rays in processes such as those described above, and that the highest energy cosmic rays reach our galaxy from these distant objects.

In any case, the Fermi mechanism is no longer thought to be the primary process producing cosmic rays. First, we know that the most energetic protons, those having energies of order 10^{20} ev, have a gyroradius comparable to the galactic radius. These particles could therefore not be retained in the Galaxy for any length of time, certainly not long enough to go through the final stages of acceleration to the observed energies. Such particles must therefore be extragalactic or else produced in very intensely magnetized regions within the Milky Way.

Second, we know that the heavy nuclei, which form an abundant part of the cosmic ray flux, would suffer destructive collisions during the long stay in interstellar space required by the rather slow Fermi acceleration mechanism. Yet we find iron nuclei to be abundant at least up to energies of the order of 10^{12} to 10^{13} ev. Above these energies we have little information about the chemical abundance. Again, we are driven toward a mechanism that could energize these particles rapidly.

Currently, pulsars and perhaps rapidly rotating white dwarfs are considered to be likely sources of cosmic rays. Further observations may lead to a clarification of these hypotheses. We can also hope that the study of solar flares, which are responsible for the solar cosmic ray component, will give us a better understanding of at least one mechanism for accelerating these energetic particles.

Generally, a charged particle moves along the lines of force of a magnetic field, spiraling as it goes. The pitch angle is given by equation 6–12. If the particle encounters a region of the magnetic field where the lines are more compressed, it experiences an increase in the field strength and by Faraday's law (6–16) its circular velocity v_c increases. However, since the field itself is not doing any work on the charge, the increase in v_c must be bought at the expense of kinetic energy initially resident in the longitudinal motion, that is, at the expense of a reduction in v_z. When the particle has advanced into the intense magnetic field to such a depth that all its kinetic energy is spent in circular motion, the pitch angle θ becomes $\pi/2$; the particle is reflected and spirals back out of the field.

As the particle first spirals into the intensifying magnetic field, its angular momentum about the axis of symmetry of the motion is conserved. Hence the *magnetic moment* **M**:

$$\mathbf{M} = \frac{\mathbf{j} \wedge \mathbf{r}}{2c}, \quad \mathbf{j} = q\mathbf{v} \tag{6–18}$$

also is conserved. Substituting the Larmor radius for r (equation 6–13) we find

$$M = \frac{v_c p_c}{2B} \tag{6–19}$$

along the direction of the magnetic field. From this it follows that the transverse kinetic energy is directly proportional to the field B. If a particle has an initial pitch angle θ in a field B, it can therefore penetrate the field until it reaches a region where the field is B_0 and $\sin \theta = 1$:

$$B_0 = \frac{B}{\sin^2\theta} \tag{6–20}$$

Here it is reflected and spirals back out of the intense magnetic field.

A *magnetic bottle* consists of two such *magnetic mirrors* between which a particle is reflected going back and forth without possibility of escape. The Fermi mechanism ping-pong acceleration could involve a (shrinking) magnetic bottle in which the two magnetic mirrors approach.

We sometimes characterize cosmic ray particles by a *magnetic rigidity* BR_L that equals pc/q for motion strictly perpendicular to the field (equation 6–13). The rigidity has dimensions of energy per charge.

PROBLEM 6–3. Consider an interstellar cloud moving with velocity V. It acts as a magnetic mirror so that a particle suffers a change in speed $\Delta V = 2V$, added to its own initial velocity, in any reflection off the cloud. With two approaching clouds, a succession of collisions can occur. Using the law of composition of velocities, compute how many collisions a proton with initial energy $\mathscr{E}$ would need in order to double its energy. Let $V = 7 \text{ km sec}^{-1}$, typical of interstellar cloud velocities, let the distance between approaching clouds (magnetic mirrors) be of the order of 10^{17} cm, and let $\mathscr{E} = 10^{10}$ ev. How long would it take to double the particle's energy? Is this time appreciably different for protons and electrons?

PROBLEM 6-4. Pulsars are now believed to be the source of at least some cosmic ray particles. The particles are thought to be accelerated by magnetic lines of force that co-rotate with the central neutron star. Suppose that the field velocity is simply ωr, where ω is the star's angular velocity and r is the radial distance from the star. Moreover, consider the particles to be dragged along, frozen to the magnetic field lines. What is the energy of the particles, then, as a function of radial distance, if special relativistic physics is approximately valid in this problem? Beyond what radial distance can the particles and magnetic field not co-rotate?

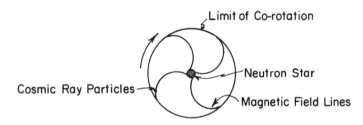

Fig. 6.4   Cosmic ray acceleration near a neutron star.

6:7 MAXWELL'S EQUATIONS

Four equations of electromagnetism allow us to derive all classical electromagnetic effects. They are

$$\nabla \cdot \mathbf{D} = 4\pi\rho \text{ (see equation 6–8)} \tag{6–21}$$

$$\nabla \wedge \mathbf{E} = -\frac{1}{c}\frac{\partial \mathbf{B}}{\partial t} \text{ (equivalent to equation 6–16)} \tag{6–22}$$

$$\nabla \wedge \mathbf{H} = \frac{4\pi}{c}\mathbf{j} \text{ (equivalent to equation 6–17)} \tag{6–23}$$

and finally

$$\nabla \cdot \mathbf{B} = 0 \tag{6–24}$$

6:7

This last equation states that there exist no *magnetic monopoles* (charges) analogous to electric charges. Only magnetic dipoles and higher multipole configurations occur in nature. Compare equation 6–24 to 6–21 in this respect.

Despite this statement, a search for magnetic monopoles has gone on ever since Dirac (Di31) pointed out that quantization of the electron's charge could be understood if a few or even only one such magnetic monopole existed in nature. Thus far no such *Dirac monopoles* have been found.

The four Maxwell equations generally must be supplemented by four auxiliary expressions.

$$\mathbf{D} = \varepsilon\mathbf{E} \quad \text{(equation 6–5)}$$

$$\mathbf{B} = \mu\mathbf{H} \tag{6–25}$$

$$\mathbf{j} = \sigma\mathbf{E} + \frac{1}{4\pi}\frac{\partial\mathbf{D}}{\partial t} \quad \text{(equation 6–15)}$$

$$\nabla \cdot \mathbf{j} = 0 \tag{6–26}$$

Equation 6–25 expresses a relation between the magnetic vectors **B** and **H** that is similar to that between **D** and **E**. Equation 6–26 states that currents in the sense defined by equation 6–15 are continuous, having no sources or sinks. The *magnetic permeability* μ can have values greater than or less than unity, depending on whether the medium is *paramagnetic or diamagnetic*. In most cosmic gases $\mu = 1$, for all practical purposes, but (see section 9:8) paramagnetic grains in interstellar space may be responsible for an observed slight polarization of starlight.

6:8 THE WAVE EQUATION

From equations 6–22 and 6–23 and from the relations (6–15) and (6–25), we can obtain the expression

$$\nabla \wedge (\nabla \wedge \mathbf{E}) = -\frac{1}{c}\frac{\partial}{\partial t}\nabla \wedge \mathbf{B} \tag{6–27}$$

$$= \frac{-4\pi\mu}{c^2}\frac{\partial}{\partial t}\left(\sigma\mathbf{E} + \frac{\varepsilon}{4\pi}\frac{\partial\mathbf{E}}{\partial t}\right)$$

provided the dielectric constant ε and permeability μ do not vary with time, and μ is scalar. Actually both μ and ε are tensor quantities, but they frequently act like scalars. Let us use the identity

$$\nabla \wedge (\nabla \wedge \mathbf{E}) = \nabla(\nabla \cdot \mathbf{E}) - \nabla^2\mathbf{E} \tag{6–28}$$

6:8

and consider only regions in which the space charge is neutral; then $\nabla \cdot \mathbf{E} = 0$ and

$$\nabla^2 \mathbf{E} = \frac{\mu\varepsilon}{c^2} \frac{\partial^2 \mathbf{E}}{\partial t^2} + \frac{4\pi\mu\sigma}{c^2} \frac{\partial \mathbf{E}}{\partial t} \tag{6-29}$$

In a nonconducting medium $\sigma = 0$ so that

$$\nabla^2 \mathbf{E} - \frac{\mu\varepsilon}{c^2} \frac{\partial^2 \mathbf{E}}{\partial t^2} = 0 \tag{6-30}$$

which is the equation for waves propagating with speed

$$V = \frac{c}{\sqrt{\mu\varepsilon}} \tag{6-31}$$

PROBLEM 6-5. Derive a similar expression for the magnetic field:

$$\nabla^2 \mathbf{H} - \frac{\mu\varepsilon}{c^2} \frac{\partial^2 \mathbf{H}}{\partial t} = 0 \tag{6-32}$$

paying particular attention to the limitations imposed on ε, μ, and σ in arriving at this result.

The operator ∇^2 is called the *Laplacian*, sometimes written as Δ. If ε and $\mu = 1$, we can define another operator $\square = \nabla^2 - (1/c^2)(\partial^2/\partial t^2)$, called the *d'Alembertian*.

We note, in passing, that there is no conflict between our previous suggestion (section 6:3) that the conductivity in space is large, while simultaneously setting $\sigma = 0$ in equations 6–30 and 6–32. The conductivity is a frequency dependent quantity. At optical and even at radio frequencies σ usually is very low. Certainly, at optical frequencies the wavelength of the electromagnetic wave is short compared to the distance between charges, and the wave effectively propagates through a vacuum. At the longer radio wavelengths a transition occurs: Charges in the medium can respond to the electric and magnetic fields of a propagated wave, and σ becomes finite. When the second term on the right of equation 6–29 dominates, the expression assumes the form of a diffusion equation and the wave is *damped*.

We note that the propagated waves are *transverse* (Fig. 6.5). If the direction of propagation for a plane wave is the x direction, symmetry dictates that all partial derivatives with respect to y and z are zero. The divergence relations give

$$\frac{\partial E_x}{\partial x} = 0 \quad \text{and} \quad \frac{\partial H_x}{\partial x} = 0 \tag{6-33}$$

and the curl equations, (6–22) and (6–23), give

$$\frac{\partial E_y}{\partial x} = -\frac{1}{c}\frac{\partial H_z}{\partial t}, \qquad \frac{\partial H_y}{\partial x} = \frac{1}{c}\frac{\partial E_z}{\partial t}$$

$$\frac{\partial E_z}{\partial x} = \frac{1}{c}\frac{\partial H_y}{\partial t}, \qquad \frac{\partial H_z}{\partial x} = -\frac{1}{c}\frac{\partial E_y}{\partial t} \qquad (6\text{–}34)$$

$$0 = \frac{\partial H_x}{\partial t}, \qquad 0 = \frac{\partial E_x}{\partial t}$$

If **n** is the unit vector along the direction of propagation, we see that (6–33) and (6–34) are satisfied by an expression of the form

$$\mathbf{H} = \mathbf{n} \wedge \mathbf{E} \qquad (6\text{–}35)$$

so that the **E** and **H** fields always are perpendicular and the solution of the *wave equation* (6–30) has the form

$$f_i = A\cos(2\pi\nu t - kx), \qquad i = y, z \qquad (6\text{–}36)$$

where $k = \sqrt{\mu\varepsilon}\,(\omega/c)$ and $\omega = 2\pi\nu$. ν is called the *frequency* and ω the *angular frequency* of the wave. k is the *wave number*—the number of waves per unit length along the direction of propagation. f_i represents electric and magnetic field components.

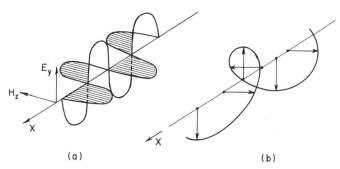

(a)   (b)

Fig. 6.5 Electromagnetic waves. (*a*) Wave propagating along the *x* direction with electric field plane polarized along the *y* direction.

(*b*) Circularly polarized wave propagating along the *x* direction. For simplicity, only the electric field direction is shown here. The direction of the **E** vector rotates about the *x*-axis. The sense of rotation shown here is said to be left-handed circularly polarized (LHP). Any electromagnetic wave can be constructed from a suitable super-position of left and right-handed circularly polarized waves. Plane polarized waves, for example, are obtained by superposing LHP and RHP waves of the same amplitude. Their relative phase determines the plane of the **E** vector (see Fig. 6.7).

6:9 PHASE AND GROUP VELOCITY

We write the equations for propagation of two waves f^- and f^+ that have angular frequencies $\omega - \Delta\omega$ and $\omega + \Delta\omega$, respectively:

$$f^- = A\cos\left[(\omega - \Delta\omega)\,t - (k - \Delta k)\,x\right]$$
$$f^+ = A\cos\left[(\omega + \Delta\omega)\,t - (k + \Delta k)\,x\right]$$

The superposition of these waves gives

$$f = f^- + f^+ = A\{\cos\left[(\omega - \Delta\omega)\,t - (k - \Delta k)\,x\right]$$
$$+ \cos\left[(\omega + \Delta\omega)\,t - (k + \Delta k)\,x\right]\} \tag{6-37}$$
$$f = 2A\cos(\omega t - kx)\cos\left[(\Delta\omega)\,t - (\Delta k)\,x\right]$$

This means that there is a *carrier wave* frequency represented by $\cos(\omega t - kx)$ that is *amplitude modulated* by a wave $\cos(t\Delta\omega - x\Delta k)$. The carrier wave velocity is called the *phase velocity* (6–31), (6–36):

$$V = \frac{\omega}{k} \tag{6-38}$$

while the velocity of the modulation is called the *group velocity*:

$$U = \frac{\partial\omega}{\partial k} \tag{6-39}$$

We will note later that U is the physically more interesting quantity. It represents the speed at which *information* can be conveyed or energy transported. As long as the medium is purely dispersive, that is, $\omega = \omega(k)$, there is no difficulty in defining U. But if the conductivity σ becomes appreciable, one has absorption, the quantity A becomes complex, and U no longer has a clear physical meaning. For long wavelength cosmic radio waves, this kind of absorption prevents transit through the earth's ionosphere. These long waves must then be observed from rockets or satellites. At even longer wavelengths the interstellar medium absorbs, and such waves are not transmitted at all. We will return to this problem in section 6:11.

6:10 ENERGY DENSITY, PRESSURE, AND THE POYNTING VECTOR

The scalar product of equation 6–22 with $\mathbf{H}$, subtracted from the product of (6–23) with $\mathbf{E}$ is

$$\frac{1}{c}\mathbf{H}\cdot\frac{\partial\mathbf{B}}{\partial t} + \frac{\mathbf{E}}{c}\cdot\frac{\partial\mathbf{D}}{\partial t} + \frac{4\pi\sigma\mathbf{E}\cdot\mathbf{E}}{c} = -(H\cdot\nabla\wedge\mathbf{E}) + (\mathbf{E}\cdot\nabla\wedge\mathbf{H}) \tag{6-40}$$

Using the vector identity

$$\nabla \cdot (\mathbf{A} \wedge \mathbf{B}) = \mathbf{B} \cdot \nabla \wedge \mathbf{A} - \mathbf{A} \cdot \nabla \wedge \mathbf{B} \tag{6-41}$$

we find that

$$\frac{1}{8\pi} \frac{\partial}{\partial t} (\varepsilon E^2 + \mu H^2) = -\sigma E^2 - \nabla \cdot \mathbf{S} \tag{6-42}$$

where $\mathbf{S} = (c/4\pi)\, \mathbf{E} \wedge \mathbf{H}$. $\mathbf{S}$ is called the *Poynting vector*. If we apply Gauss's theorem (6–9) relating volume and surface integrals, (6–42) can be written as

$$\frac{\partial}{\partial t} \int \frac{\varepsilon E^2 + \mu H^2}{8\pi} dV = - \int \sigma E^2 dV - \oint \mathbf{S} \cdot d\mathbf{s} \tag{6-44}$$

Here the first term on the right is equivalent to the rate of change of kinetic energy of moving charges. It involves the scalar product of the force on the particles and their velocity since σE represents a current, that is, the motion of charged particles:

$$\int \sigma E^2 dV \rightarrow \sum e\mathbf{v} \cdot \mathbf{E} = \sum \mathbf{v} \cdot \dot{\mathbf{p}} \tag{6-45}$$

This is the time derivative of the kinetic energy summed over all particles. The other two terms in (6–44) represent the flow of electromagnetic energy. The term on the left of equation 6–44 is the rate of change of energy in the volume; $(\varepsilon E^2 + \mu H^2)/8\pi$ is the energy density of the fields. The second term on the right represents the flow of energy through the surface and S therefore is the *electromagnetic flux density*. Equation 6–44 states that the rate of change of energy in a volume equals the rate of change of the kinetic energy of charges plus the rate at which energy is radiated away.

Previously we found that the pressure P due to randomly oriented electromagnetic waves is just $1/3$ the numerical value of the energy density. In section 4:7 we determined this on kinetic grounds:

$$P = \frac{1}{3} \frac{1}{8\pi} (\varepsilon E^2 + \mu H^2) \tag{6-46}$$

The case of static fields is similar except that a magnetic pressure can now exist without an accompanying electric pressure; the conductivity σ is high and a current $\sigma \mathbf{E}$ maintains the magnetic field. There exists a kinetic pressure due to the flow of charges, and this will depend on $\sigma \mathbf{E}$. The situation is further complicated since the magnetic pressure actually is a tensor quantity that depends on the orientation of the fields. For a magnetic field there always exists a tension along the lines of force and an outward pressure perpendicular to the lines of force.

We can see this in the following way. The magnetic energy density in a cube of unit dimension is $\mu H^2/8\pi$. If the cube is compressed an amount dl along a direction parallel to the field lines, the field strength remains constant, but the volume decreases by dl. Since the energy density remains constant while the volume decreases, the total energy in the volume decreases by $(\mu H^2/8\pi)\,dl$, and this amount of energy is given off in the contraction. That means that the amount of work done to compress the cube is $-(\mu H^2/8\pi)\,dl$, and indicates that there is a pressure $-\mu H^2/8\pi$ along the field lines.

If the cube is compressed along a direction transverse to the field lines, the number of lines of force in the volume does not change, and a compression Δl increases the field strength to $H/(1 - \Delta l)$. The energy density now becomes $\sim (\mu H^2/8\pi)(1 + 2\Delta l)$, and because of the decrease in volume $(1 - \Delta l)$, the total energy change on compression is $\sim (\mu H^2/8\pi)\,\Delta l$. In this case an amount of work $(\mu H^2/8\pi)\,\Delta l$ must be done to compress the cube, and the pressure resisting compression is $\mu H^2/8\pi$. For a volume containing randomly directed bundles of field lines, the net effect of averaging over two transverse and one longitudinal direction is an overall outward pressure $P = (\mu H^2/8\pi)/3$.

This is the reason why difficulties arise in the problem of star formation in the presence of fields (section 4:18). It is relatively simple to see how matter can contract along the lines of force in that situation, but it is more difficult to understand how condensation takes place perpendicular to the direction of the field because the gases are frozen to the field lines and the pressure of the magnetic field attempts to resist any contraction. To see how severe this problem is, we note that the transverse pressure is $H^2/8\pi$.

PROBLEM 6-6. The transverse pressure of a static magnetic field is $P_s = H^2/8\pi$; the magnetic part of the radiation pressure (6–46) is $P_r = \frac{1}{3}H^2/8\pi$. What is the significance of the factor $\frac{1}{3}$?

Initially a typical field strength might be 10^{-6} gauss, so that $P_{\text{initial}} \sim 10^{-13}$ dyn cm^{-2}. As a protostar contracts from $\sim 10^{18}$ cm down to 10^{11} cm, conservation of the number of field lines requires that $H \propto r^{-2}$, so that $H^2 \propto r^{-4}$ and we would end up with a protostar having 10^8 gauss magnetic fields and 10^{15} dyn cm^{-2} magnetic pressures. The gravitational forces are far too weak to produce such high fields. We conclude that somehow we are looking at the problem in the wrong way. Stars manage to form despite these difficulties.

6:11 PROPAGATION OF WAVES THROUGH A TENUOUS IONIZED MEDIUM

Consider an ionized medium without electric or magnetic fields. Let this medium be tenuous, so that collisions between ions and electrons are rare. Then, for small departures from equilibrium, electric fields resident in the electromagnetic wave accelerate the electrons in the medium relative to the more massive ions:

$$m\ddot{\mathbf{r}} = e\mathbf{E}(\mathbf{r}, t) \tag{6-47}$$

Here e and m are the charge and mass of the electron, and $\mathbf{E}$ is the field associated with the wave. Let the wave have the form

$$\mathbf{E}(r, t) = E_0(\mathbf{r})\, e^{i\omega t} \quad \text{(Real part)} \tag{6-48}$$

where only the real part will be considered. The displacement of the electron from its equilibrium position then is

$$\mathbf{r} = -\frac{e}{m\omega^2}\mathbf{E} \tag{6-49}$$

This satisfies both equations 6–47 and 6–48. The displacement of the electrons effectively sets up a large number of dipoles that, as discussed in section 6:1, give rise to a polarization field $\mathbf{P}$. If n is the number density of electrons, the polarization field is to be expressed as the sum of the individual dipole fields produced by the passing wave

$$\mathbf{P} = ne\mathbf{r} = -\frac{ne^2}{m\omega^2}\mathbf{E} \tag{6-50}$$

The definition of the polarization field, equation 6–10, then tells us that the dielectric constant of the medium must be

$$\varepsilon = 1 - \frac{4\pi ne^2}{m\omega^2} \tag{6-51}$$

Since the propagation phase velocity is inversely proportional to the refractive index at frequency ω, $n_\omega = \varepsilon^{1/2}$ (we can set $\mu = 1$ in all problems dealing with cosmic wave propagation), the *phase* velocity in a plasma will be greater than the speed of light! But no information and no energy is transmitted at this velocity. Therefore no violation of special relativity is involved. The more significant *group velocity* is always less than c.

If a wave propagates along the x-direction through the cosmic medium, the transverse $\mathbf{E}$ and $\mathbf{B}$ field components have the form (6–36):

$$f = f_0 \cos(kx \pm \omega t) \tag{6-52}$$

and

$$\omega^2 = \frac{k^2 c^2}{\varepsilon} = \frac{k^2 c^2}{1 - (4\pi ne^2/m\omega^2)} \tag{6-53}$$

6:11

where equation 6–51 has been invoked with $\mu = 1$. This can be written as

$$\omega^2 = k^2 c^2 + \frac{4\pi n e^2}{m} \equiv k^2 c^2 + \omega_p^2 \qquad (6\text{–}54)$$

where

$$\omega_p \equiv \left(\frac{4\pi n e^2}{m} \right)^{1/2} \sim 5.6 \times 10^4 n^{1/2} \text{Hz} \qquad (6\text{–}55)$$

is called the *plasma frequency*. It is related to the Debye length L (see equation 4–156) by $(mL^2/kT)^{1/2} = \omega_p^{-1}$. ω_p^{-1} is the time for an electron to cross a Debye length at a velocity $(kT/m)^{1/2}$.

If $\omega < \omega_p$, k becomes imaginary and the wave will not propagate through the medium.

In radio astronomy—as mentioned in section 6:9—observations at low frequencies cannot be carried out from below the ionosphere. Radio waves cannot be transmitted at frequencies below the ionospheric plasma frequency. That frequency varies since the electron density is not uniform. Typically, however, the cut-off is at frequencies of a few megahertz.

When $\omega > \omega_p$, propagation can take place. The group velocity of the wave is

$$U = \frac{d\omega}{dk} = \frac{c}{\sqrt{1 + \omega_p^2/c^2 k^2}} \qquad (6\text{–}56)$$

The velocity of propagation therefore is frequency dependent. A situation in which this phenomenon is important concerns the propagation of pulses emitted by a *pulsar* (He68b)*. If the emitted pulse contains a range of frequency components, the arrival time of these frequencies at the earth will be delayed more at lower frequencies.

We can write (6–56) as

$$U = \frac{c}{\sqrt{1 + \omega_p^2/(\omega^2 - \omega_p^2)}} \qquad (6\text{–}57)$$

The arrival time of a pulse that has traveled a distance D is D/U and the frequency dependence of the arrival time is

$$\frac{d(D/U)}{d\omega} \sim -\frac{D}{c} \frac{\omega_p^2}{\omega^3} \qquad \text{for} \qquad \omega \gg \omega_p \qquad (6\text{–}58)$$

Observations of pulsars show that the pulse arrival time is frequency dependent and the observed delays in arrival time actually take the form (6–58). We therefore conclude that the time delay takes place in a medium whose plasma frequency is less than the radiation frequency. Equation 6–55 therefore puts an upper limit on the number density of electrons in the dispersing medium. More important, however, is the conclusion that the frequency dependence of the time delay is directly proportional to Dn, the total number of electrons per unit cross-sectional area along the line of sight to the emitting object. This useful relation follows directly from (6–55) and (6–58). The integrated electron number density along the line of sight, S, is called the *dispersion measure* $\mathscr{D}$:

$$\mathscr{D} \equiv \int_0^D n(s)\,ds = D\langle n \rangle \qquad (6\text{–}59)$$

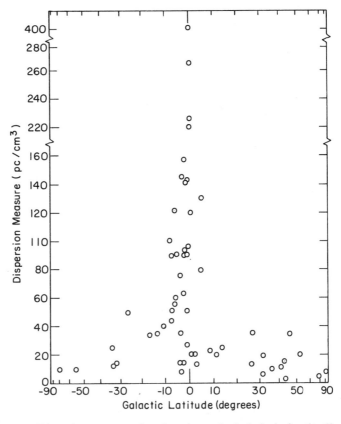

Fig. 6.6 Dispersion measure plotted against galactic latitude for the 63 pulsars known in early 1972 (Te72).

If the mean number density of electrons in the interstellar medium is known, the dispersion relation (6–58) can give us the pulsar distance. Conversely, if D is known from other sources, a mean value of n along the line of sight is obtained. The mean value, estimated in this way, would include a contribution due to any electrons surrounding the emitting region and part of the emitting object, as well as true interstellar electrons. The dispersion measure within the pulsar would not be distance dependent, while the dispersion due to the interstellar medium would be. On this basis we distinguish the two contributions; and we find that the pulsar itself contributes negligibly. The dispersion measure along lines of sight leading to sources at known distances gives a mean electron density of about $\langle n \rangle = 0.03$ cm^{-3}. This value varies from source to source, depending on the number of bright, hot, ionizing stars along the line of sight. Using a mean density of electrons in interstellar space, amounting to ~ 0.03 cm^{-3}, we find that the distribution of the nearer pulsars fits the distances of the nearer spiral arms in the Galaxy. (Da69) These pulsars also show a tendency to cluster close to the Galactic plane (Fig. 6.6).

Throughout this section we have assumed that the collision frequency v_c between ions and electrons is low. However, when v_c becomes high, energy losses through dissipation no longer can be neglected. This problem is treated in section 6:16.

6:12 FARADAY ROTATION

Information about electron number densities in the cosmic medium can also be obtained from the *Faraday rotation* of a wave's plane of polarization. To understand this effect, consider an electron moving in a plane perpendicular to the direction of a magnetic field, **B**. It will be deflected by a force (6–11)

$$\mathbf{F} = \frac{e\mathbf{v} \wedge \mathbf{B}}{c}$$

If the electron is also under the influence of an electromagnetic wave, it will experience a force due to the wave's E field. Finally the gyrations under the combined influence of these fields must be balanced by an outward directed centrifugal force. The relation between these three forces is given by

$$e\mathbf{E} \pm \frac{eB\omega\mathbf{r}}{c} = -m\omega^2\mathbf{r} \tag{6–60}$$

where **E** is the component of the field vector perpendicular to the magnetic field, and the second term on the left has a negative sign when the electron rotates counterclockwise viewed along the direction of the **B** field. This is the motion induced by an electromagnetic wave with a right-handed circular polarization

RHP propagating parallel to **B**. A left-handed circular polarization LHP gives rise to a force $+ eB\omega r/c$ directed along the direction of displacement from the electron's equilibrium position. Note, however, that the value of e is negative for an electron. Solving for **r** gives

$$\mathbf{r} = -\frac{e}{m}\left(\frac{1}{\omega^2 \pm \dfrac{eB\omega}{mc}}\right)\mathbf{E} \tag{6–61}$$

The dielectric polarization, as in (6–50) becomes $\mathbf{P} = ne\mathbf{r}$, giving rise to a dielectric constant

$$\varepsilon = 1 - \frac{4\pi ne^2}{m\omega(\omega \pm \omega_c)}, \qquad \omega_c \equiv \frac{eB}{mc} \tag{6–62}$$

Here ω_c is the gyro- or cyclotron frequency (6–13). Since the index of refraction $\varepsilon^{1/2}$ is not the same, it follows that the left- and right-handed polarized radiation will travel at different velocities through an ionized medium in a longitudinal magnetic field.

If a wave initially is plane polarized with a given direction of polarization, the polarization angle can be expressed as a superposition of two circularly polarized waves of given phase, say θ_0, and equal amplitude. As the waves propagate the phase relationship will change, because one wave lags behind the other, and the direction of polarization therefore rotates. Sometimes the **E** vectors will be in phase; at other times they will be out of phase.

Fig. 6.7 Addition of circularly polarized waves, to give plane polarized radiation. The phase at time $t = 0$ is θ_0.

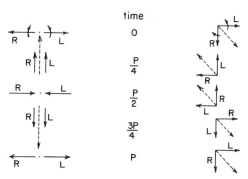

Figure 6.7 shows two sets of superposed, opposite circularly polarized waves. The one on the left has $\theta_0 = 180°$. The one on the right has $\theta_0 = 90°$. The **E** vectors and their sums are shown at different times during the waves' period P. The sum of the vectors is indicated by the dashed line. We can see that the direction of the plane polarized wave is given by an angle equaling half the phase lag.

However, the initial direction of one of the **E** vectors must also be specified. In Fig. 6.7, for example, we took the left-handed polarized **E** vector to point to the right at $t = 0$.

Turning now to the velocity of propagation and refractive index, we find the difference Δn between indices n_L and n_R to be

$$n_L^2 - n_R^2 = \varepsilon_L - \varepsilon_R = 2n_\omega \, \Delta n \qquad (6\text{--}63)$$

We write

$$n_\omega \sim 1 - \frac{4\pi n e^2}{2m\omega^2} \qquad (6\text{--}64)$$

where (6–64) is obtained from equation 6–51 provided $n_\omega - 1 \ll 1$. Substituting the dielectric constants from (6–62) into (6–63) we obtain

$$\Delta n = \frac{\dfrac{4\pi n e^2 (2\omega_c)}{m\omega(\omega^2 - \omega_c^2)}}{2\left(1 - \dfrac{2\pi n e^2}{m\omega^2}\right)} \qquad (6\text{--}65)$$

If $\omega \gg \omega_c$ and $ne^2/m\omega^2 \ll 1$,

$$\Delta n = \frac{4\pi n e^2 \omega_c}{m\omega^3} \qquad (6\text{--}66)$$

This distance lag per unit time is $c \, \Delta n/n_\omega^2 \sim c \, \Delta n$. The phase lag of the LHP relative to the RHP wave therefore becomes $\omega \, \Delta n$, and the plane of polarization rotates through half this angle in unit time:

$$\Delta \theta \sim \frac{\omega \, \Delta n}{2} \qquad (6\text{--}67)$$

The difference in velocity of propagation, and hence the rate at which the polarization vector rotates, is therefore proportional to the number density n and to B. For a given velocity difference, the phase rotates at a rate inversely proportional to the wavelength λ, because the distance one wave has to lag behind the other becomes greater for larger wavelengths. On the other hand, the velocity difference between the waves is proportional to ω^{-3}, according to (6–66), and therefore is proportional to λ^3. Hence the angle $\theta(D)$ through which the plane of polarization is rotated over distance D is proportional to λ^2. In observing distant radio sources emitting polarized radiation, we can determine the angle θ as a function of wavelength. This gives a value for the product of electron density n and the magnetic field component along the viewing direction (provided the path length is known). More correctly, since the rotation depends on the presence of both a properly oriented magnetic field and the local particle density at the field's position, the rotation actually gives a value of the product of particle density and magnetic field, integrated along the line of sight.

If, as is sometimes supposed, the particles and fields actually do not occupy the same positions in space, but are physically separated from one another, then the Faraday rotation only produces a lower limit to the field strength and particle density. Nonetheless, since we know relatively little about the interstellar medium, even this much information is of current astrophysical interest.

In the case of pulsars, the dispersion measure tells us the mean number density of electrons along the line of sight (6:11). The Faraday rotation can then be used to estimate the mean component of the magnetic field strength along the line of sight. This procedure has been followed in obtaining the local Galactic magnetic field (see Fig. 9.9). Since the field direction changes along this path, only a statistical estimate of actual field strength is obtained in this way.

PROBLEM 6–7.   Suppose the field strength is B everywhere, that it varies randomly in direction from region to region, but that its direction is constant over any region of length L. If the source distance is NL, and the electron number density is n, show by a random walk procedure that

$$\theta \sim \sqrt{NL}\left(\frac{2\pi n e^3 B}{m^2 c^2 \omega^2}\right)$$

For simplicity assume that B always points directly toward or away from the observer.

6:13 LIGHT EMISSION BY SLOWLY MOVING CHARGES

When an electric charge is set into accelerated motion, it can emit radiation. If this motion is induced by an incident electromagnetic wave, we may find that the charge—or group of charged particles—absorbs or scatters the radiant energy. To see this, consider the current associated with the accelerated charge. This current will induce a magnetic field at some distance from the position of the charges, but the magnetic field strength variations will normally be somewhat out of phase with the variations in the current. This is due to the time delay involved in transmitting the information about the current strength from one position to another. That information can only be transmitted at the speed of light. For the moment we will regard the charges and currents as sources of electric and magnetic fields. If we use the Maxwell equations 6–22 and 6–23 for empty space where $\mathbf{E} = \mathbf{D}$ and $\mathbf{H} = \mathbf{B}$, and we now write $\mathbf{j}_c$ to symbolize the conduction current $\sigma\mathbf{E}$,

$$\nabla \wedge \mathbf{H} = \frac{4\pi}{c} \mathbf{j}_c + \frac{1}{c} \frac{\partial \mathbf{E}}{\partial t} \tag{6-68}$$

$$\nabla \wedge \mathbf{E} = -\frac{1}{c} \frac{\partial \mathbf{H}}{\partial t} \tag{6-69}$$

Now consider a *vector potential* $\mathbf{A}$ as giving rise to the magnetic field, while a *scalar potential* ϕ, together with $\mathbf{A}$, gives rise to the electric field; then we can write

$$\mathbf{H} = \nabla \wedge \mathbf{A} \tag{6-70}$$

and

$$\mathbf{E} = -\nabla\phi - \frac{1}{c} \frac{\partial \mathbf{A}}{\partial t} \tag{6-71}$$

which are consistent with the Maxwell equations, above. Separable equations, each depending on only one of these potentials, can then be obtained provided that

$$\nabla \cdot \mathbf{A} + \frac{1}{c} \frac{\partial \phi}{\partial t} = 0 \tag{6-72}$$

holds. Equation 6–72 is called the Lorentz condition.

PROBLEM 6-8. Check the validity of this statement by direct substitution of equations 6–70, 6–71, and 6–72 into the Maxwell equations. In this way obtain

$$\nabla^2 \mathbf{A} - \frac{1}{c^2} \frac{\partial^2 \mathbf{A}}{\partial t^2} = -\frac{4\pi}{c} \mathbf{j}_c \tag{6-73}$$

and

$$\nabla^2 \phi - \frac{1}{c^2} \frac{\partial^2 \phi}{\partial t^2} = -4\pi\rho \tag{6-74}$$

In empty space, equations 6–73 and 6–74 have the right side equal to zero; the right side is nonzero only at the actual location of charges and currents. Furthermore, in a static case where the time derivative vanishes, ϕ obeys the Poisson equation (4–142) that we used earlier in discussing plasmas. When we solve the Poisson equation, the potential is expressed in terms of an integral over the volume distributed charges, divided by the distance of the charges from the point at which the potential is evaluated. In view of this, we write the potential as

$$\phi(\mathbf{R}_0, t) = \frac{1}{R_0} \int \rho\left(t - \frac{R_0}{c} + \frac{\mathbf{r} \cdot \mathbf{n}}{c} \right) dV \tag{6–75}$$

where R_0 is the distance from the *center of charge*, and $\mathbf{r} \cdot \mathbf{n}$ is the projected distance of a point in the charge distribution, measured from the center of charge, along the direction joining the point $\mathbf{R}$ to the center of charge (see Fig. 6.8). $\mathbf{n}$ is a unit vector along that direction. Equation 6–75 tells us that the potential at any given time is determined by the charge distribution at a time $R/c = (R_0 - \mathbf{n} \cdot \mathbf{r})/c$ earlier. The similarity between equations 6–73 and 6–74 suggests that we can also write

$$\mathbf{A}(\mathbf{R}_0, t) = \frac{1}{cR_0} \int \mathbf{j}_c\left(t - \frac{R_0}{c} + \frac{\mathbf{r} \cdot \mathbf{n}}{c} \right) dV \tag{6–76}$$

We note that a plane wave in vacuum obeys the relation (6–35)

$$\mathbf{H} = \mathbf{n} \wedge \mathbf{E} \tag{6–35}$$

and that (6–71) therefore leads to

$$\mathbf{H} = \frac{1}{c} \dot{\mathbf{A}} \wedge \mathbf{n} \tag{6–77}$$

as long as the magnetic field strength is measured at a large distance from the charge distribution so that $\nabla\phi$ can be neglected.

It is now a simple matter to determine the energy radiated away by the moving charges. The Poynting vector is immediately obtained from equations 6–35 and 6–43.

$$\mathbf{S} = \frac{c}{4\pi} H^2 \mathbf{n} \tag{6–78}$$

Fig. 6.8   Diagram to illustrate radiation by a dipole, see equations 6–85 and 6–86.

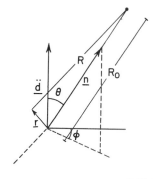

For the simple case of radiation by a *dipole*—that is, two slightly separated dissimilar charges—we can take the integral over the current distribution in (6–76) to be just the rate of change of the dipole moment

$$\mathbf{A} = \frac{1}{cR_0} \dot{\mathbf{d}} \qquad (6\text{–}79)$$

where

$$\dot{\mathbf{d}} = \frac{d}{dt} \sum e\mathbf{r} \qquad (6\text{–}80)$$

Here $\mathbf{d} = \sum e\mathbf{r}$ is the dipole moment of the charge distribution; the time derivative refers to a time $t' = t - (R_0/c)$, and the dimension of the dipole must be small compared to the radiated wavelength λ; for then

$$\frac{\mathbf{r} \cdot \mathbf{n}}{c} \ll \frac{\lambda}{c} = P \qquad (6\text{–}81)$$

and neglect of the term $\mathbf{r} \cdot \mathbf{n}/c$ in equation (6–76) involves a neglect only of a time increment small compared to the *period of oscillation P*. We now see from (6–77) and (6–35) that the field strengths at a distance R_0 from the dipole are

$$\mathbf{H} = \frac{1}{c^2 R_0} \ddot{\mathbf{d}} \wedge \mathbf{n} \qquad (6\text{–}82)$$

$$\mathbf{E} = \frac{1}{c^2 R_0} (\ddot{\mathbf{d}} \wedge \mathbf{n}) \wedge \mathbf{n} \qquad (6\text{–}83)$$

and the *intensity* of radiation dI radiated into a solid angle $d\Omega$ is given by the Poynting vector integrated over that angle:

$$dI = \frac{1}{4\pi c^3} (\ddot{\mathbf{d}} \wedge \mathbf{n})^2 \, d\Omega \qquad (6\text{–}84)$$

Integrating over all angles $d\Omega = \sin\theta \, d\theta \, d\phi$

$$I = \int\int \frac{\ddot{d}^2}{4\pi c^3} \sin^3\theta \, d\theta \, d\phi \qquad (6\text{–}85)$$

$$= \frac{2}{3c^3} \ddot{d}^2 \qquad (6\text{–}86)$$

For two opposite charges e and $-e$, separated by a distance $\mathbf{r}$, the dipole moment is

$$\mathbf{d} = e\mathbf{r} \qquad (6\text{–}87)$$

6:13

and the *total radiated energy* per second is

$$I = \frac{2e^2\ddot{r}^2}{3c^3} \tag{6-88}$$

PROBLEM 6-9. A magnetic dipole can be considered to consist of two fictitious magnetic charges Q and $-Q$ separated by a distance **a**. The magnetic dipole moment would then be $\mathbf{M} = Q\mathbf{a}$. (a) Show that the magnetic field along the axis of this configuration is $\mathbf{H} = 2\mathbf{a}Q/r^3$. (b) At the surface of a pulsar $H \sim 10^{12}$ gauss, $r \sim 10^6$ cm, $\omega \sim 10^2$. By analogy to equation 6-88 show, for **a** perpendicular to the axis of the star's rotation, that the intensity of radiation is (Pa68)

$$I = \frac{2}{3c^3}\ddot{\mathbf{M}}^2 \sim 10^{36} \text{ erg sec}^{-1} \tag{6-89}$$

We should still note that a system of charged particles all of which have the same charge-to-mass ratio cannot radiate as a dipole. The center of charge and the center of mass coincide for such a system; and if the center of mass $\sum m\mathbf{r}$ remains stationary, the derivatives d all vanish:

$$\mathbf{d} = \sum e\mathbf{r} = \sum \frac{e}{m} m\mathbf{r} = 0 \tag{6-90}$$

For such an assembly of charges we can still obtain *electric quadrupole radiation*, or radiation generated by higher electric or magnetic *multipole processes*. These processes depend on the inclusion of terms in $\mathbf{r} \cdot \mathbf{n}/c$ that we had previously neglected. The current j_c is now expressed as an expansion in $\mathbf{r} \cdot \mathbf{n}/c$:

$$j_c\left(t' + \frac{\mathbf{r} \cdot \mathbf{n}}{c}\right) = j_c(t') + \frac{\partial}{\partial t}\left(\frac{\mathbf{r} \cdot \mathbf{n}}{c}\right)j_c(t') + \dots \tag{6-91}$$

where again $t' = t - R_0/c$. If we only retain the first two terms of the expansion and sum over all charges, equation 6-76 yields

$$\mathbf{A} = \frac{\sum e\mathbf{v}}{cR_0} + \frac{1}{c^2 R_0}\frac{\partial}{\partial t}\sum e\mathbf{v}(\mathbf{r} \cdot \mathbf{n}) \tag{6-92}$$

where the first term again is produced only by a time varying dipole moment, and we now understand that **v** and **r** values are measured for time t', although all primes have been dropped for ease in writing. One can show (see La51) that this leads to

$$\mathbf{A} = \frac{\dot{\mathbf{d}}}{cR_0} + \frac{1}{6c^2 R_0}\frac{\partial^2}{\partial t^2}\mathbf{D} + \frac{1}{cR_0}(\dot{\mathbf{M}} \wedge \mathbf{n}) \tag{6-93}$$

where

$$\mathbf{M} = \frac{1}{2c} \sum e\mathbf{r} \wedge \mathbf{v} \quad \text{and} \quad \mathbf{D} = \sum e(3\mathbf{r}(\mathbf{n} \cdot \mathbf{r}) - \mathbf{n}r^2) \quad (6\text{--}94)$$

are the *magnetic dipole moment* and *electric quadrupole moment*, respectively. Note that the magnetic dipole term also vanishes when the *charge-to-mass ratio* is the same for all particles. This comes about because angular momentum is proportional to $\mathbf{M}$ and conservation of angular momentum implies $\dot{\mathbf{M}} = 0$. The second term on the right of equation 6–93 is called the *electric quadrupole term*.

The higher multipole terms are small compared to dipole radiation terms, since they effectively involve an expansion in v/c and, as (6–81) shows, this is a small quantity when the dimensions of the system are small compared to the wavelength.

The considerations presented here in classical terms also apply in the *quantum theory of radiation*. Instead of talking about the intensity of radiation given off by a moving system of charges, we then talk about the probability for emission of radiation. Again, in those situations in which the charge-to-mass ratio does not vanish, the emission probability is normally much higher for electric dipole radiation than for multipole radiation. In those systems having the same e/m ratio for all constituent particles, electric dipole radiation is *"forbidden"* by the quantum mechanical *selection rules*. For example, there may well be large masses of interstellar or intergalactic molecular hydrogen H_2 in the Universe. But the presence of this gas has never been satisfactorily established by observations of its infrared spectrum, primarily because the symmetry of the hydrogen molecule forces us to look for lines that are emitted or absorbed only through the very weak electric quadrupole process. (The ultraviolet transitions break this symmetry and have, in fact, been observed, Ca70a.)

This incidentally brings up one last important point: that *emission* of radiation is just the reverse of *absorption* and the probability for absorption in an atomic system is identical to the probability for *induced emission* (see section 7:10). Induced emission is the process in which an atom or molecule emits radiation in response to stimulation by a light wave that has exactly the same frequency as the wave that the atom can emit. We then find that the stimulated and stimulating radiation have exactly the same characteristics, that is, the photons all belong to the same phase cell. This induced emission is different from the quantum mechanical *spontaneous emission* that has no direct analogue in terms of absorption. The spontaneous emission corresponds to emission by an unperturbed atom or molecule, giving off radiation on its own without any external influence.

It is also interesting that the general approach to radiative processes presented here is relevant to *gravitational radiation*. As discussed in earlier sections, both gravitational forces and electrostatic forces drop as the square of the distance separating the masses or charges. This allows us to use a formalism somewhat similar to electromagnetic theory in dealing with gravitational radiation. One immediate consequence of such considerations is a statement about the strength of the expected gravitational radiation. Since the ratio of inertial to gravitational mass is constant for all matter, gravitational dipole radiation is not permitted. The much weaker quadrupole radiation is the first allowed multipole emission process. Furthermore, the magnitude of the expected radiation at any given multipole level will also be considerably smaller, simply because the ratio of gravitational mass to inertial mass is much smaller than the ratio of electric charge to mass. The ratio of intensities can therefore be expected to differ by factors of order $e^2/m^2G \sim 10^{42}$ if the electron charge-to-mass ratio is used. It is clear then that gravitational radiation can only be expected for large masses and also when the accelerations involved are very large. Such situations would require very compact massive systems to exist in the Universe. We are currently searching to see whether such objects exist. Ordinary binary stars are not massive or compact enough to yield measurable amounts of gravitational radiation, but gravitational radiation may considerably affect the orbits of some compact binaries over periods of 10^{10} y (Fa71).

PROBLEM 6–10. Using equation 6–77 together with expression 6–93 for the quadrupole moment, to obtain a Poynting vector of the form (6–78), show that the radiation intensity for quadrupole radiation is proportional to $(\dddot{D})^2$ and c^{-5}. The dots indicate the third derivative with respect to time.

The actual intensity for gravitational quadrupole radiation (La51) is

$$I = \frac{G}{45c^5} \dddot{D}^2 \tag{6–95}$$

where D is a tensor having the form of (6–94) but with mass replacing the electric charge e. The quadrupole moment is proportional to the ellipticity $\notin \sim \frac{1}{2}[1 - (a_{max}^2/a_{min}^2)]$ and for an ellipsoid rotating with the mass symmetry axis perpendicular to the axis of rotation:

$$I \sim \frac{GM^2a^4\notin^2\omega^6}{c^5} \tag{6–96}$$

(Ch70). Here we assume that $(\notin - 1) \ll 1$.

For a pulsar with $a \sim 10^6$ cm, M $\sim 10^{33}$ g, $\notin \sim 10^{-5}$, and $\omega \sim 10^2$ we see that the intensity of gravitational radiation is 3×10^{32} erg sec^{-1}. This is smaller than the magnetic dipole radiation; but very early in the pulsar's career, when it spins with a period of the order of one millisecond, the ω^6 dependence and a possibly increased $\not\!\ell$ value allows the gravitational radiation to equal or dominate the magnetic dipole radiation.

In addition to pulsars, supernovae, QSO's and galactic nuclei may also be sources of gravitational radiation.

6:14 LIGHT SCATTERING BY UNBOUND CHARGES

When a plane polarized electromagnetic wave moving along the z-direction is incident on a charged particle having mass m and charge e, the particle is subjected to an electric field of form

$$\mathbf{E} = \mathbf{E}_0 \cos{(\mathbf{k} \cdot \mathbf{r} - \omega t + \alpha)} \tag{6–97}$$

If the field is weak enough so that the velocity imparted to the charge is always small—$v \ll c$—then the force $e\mathbf{E}$ is always large compared to the force $e v \wedge \mathbf{H}/c$ acting on the particle. This is evident from (6–35). The acceleration experienced by the particle is given by

$$m\ddot{\mathbf{r}} = e\mathbf{E} \tag{6–98}$$

and the dipole moment produced by the displacement of the charge, $\mathbf{d} = e\mathbf{r}$, has a second time-derivative

$$\ddot{\mathbf{d}} = \frac{e^2}{m}\,\mathbf{E} \tag{6–99}$$

We now see that equation 6–84 predicts a scattered light intensity per solid angle along direction $\mathbf{n}$:

$$dI = \frac{e^4}{4\pi m^2 c^3}\,(\mathbf{E} \wedge \mathbf{n})^2\, d\Omega \tag{6–100}$$

We speak of a *differential scattering cross section*

$$d\sigma(\theta, \phi) = \frac{dI(\theta, \phi)}{S} = \left[\frac{e^2}{mc^2}\right]^2 \sin^2\theta\, d\Omega \tag{6–101}$$

where θ is the angle between the scattering direction $\mathbf{n}$ and the electric field $\mathbf{E}$ of the incident wave. S is given by (6–43). We note:

(a) That the frequency of the radiation is not changed by scattering.
(b) That the angular distribution of scattered light is not dependent on the frequency.

6:14

Fig. 6.9 Direction of incident and scattered waves (see equations 6–101 and 6–105).

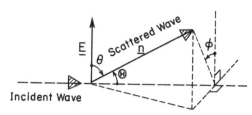

(c) That the total cross section is not frequency dependent. The total cross section is obtained by integrating $d\sigma(\theta, \phi)$ over all angles θ, ϕ:

$$\sigma = \int_0^\pi \int_0^{2\pi} \sigma(\theta, \phi) \sin\theta \, d\phi \, d\theta$$

For electrons

$$\sigma_e = \frac{8\pi}{3}\left(\frac{e^2}{mc^2}\right)^2 = 6.65 \times 10^{-25} \text{ cm}^2 \qquad (6\text{--}102)$$

This is called the *Thomson scattering* cross section.

(d) The differential scattering cross section is symmetrical in θ about $\theta = \pi/2$.

(e) σ is a factor of $(m_p/m_e)^2 \sim 10^6$ times less for protons than for electrons.

Let us still consider the influence that the polarization of the **E** vector has on the actual angular distribution of radiation. If the initial wave incident on the particle is unpolarized, we obtain a scattering cross section, independent of ϕ, but dependent on the polar angle Ⓗ included between the directions of the incident and scattered waves (Fig. 6.9). In this context we note that the angle θ has to be considered a function of the angles ϕ and Ⓗ, and that

$$\cos\theta = \sin Ⓗ \cos\phi \qquad (6\text{--}103)$$

For a given angle Ⓗ

$$\therefore \langle\sin^2\theta\rangle = 1 - \sin^2Ⓗ\langle\cos^2\phi\rangle = 1 - \frac{\sin^2Ⓗ}{2} = \frac{1}{2}(1 + \cos^2Ⓗ) \qquad (6\text{--}104)$$

where we have made use of the fact that $\langle\cos^2\phi\rangle = 1/2$ when the average is taken over all angles ϕ. For unpolarized radiation we can therefore write

$$d\sigma = \frac{1}{2}\left(\frac{e^2}{mc^2}\right)^2 (1 + \cos^2Ⓗ) \, d\Omega \qquad (6\text{--}105)$$

We therefore have the important result that

(f) For unpolarized radiation the cross section has peak values in the forward and backward directions, that is, most of the light is scattered along the direction in which the wave was moving initially—or backward into the direction from which the wave came.

PROBLEM 6–11.   Show that there is a force component

$$F(\text{\textcircled{H}}) = (1 - \cos \text{\textcircled{H}})\, d\sigma\, \frac{\mathbf{S}}{c}$$

acting on the scattering charge along the direction of propagation. Show that when this is averaged over all values $\text{\textcircled{H}}$, one obtains a total force $\mathbf{F}$ along the direction of incidence

$$\mathbf{F} = \frac{2}{3}\left(\frac{e^2}{mc^2}\right)^2 E^2 = \frac{\sigma \mathbf{S}}{c} \qquad (6\text{–}106)$$

In the vicinity of bright hot stars, this can be the dominant force acting on electrons. Much of the visible light reaching us from the solar corona also seems to be due to scattering by electrons. However, the zodiacal glow, that is, the diffuse scattered sunlight in the ecliptic plane, is due to radiation scattered off small solid grains circling the sun in the orbital plane of the planets. This glow extends into the corona and weakly contributes to its brightness.

We can now also consider scattering by a *harmonically bound charge* that would normally oscillate at a natural frequency ω_0. The electric field attempts to force the oscillator to vibrate at a frequency ω instead. The equation of motion for this *forced oscillation* is

$$\ddot{\mathbf{r}} + \omega_0^2 \mathbf{r} = \frac{e\mathbf{E}}{m} \qquad (6\text{–}107)$$

If $\mathbf{E} = 0$, we obtain oscillation at frequency ω_0. If $\mathbf{E}$ has the form (6–97), equation 6–107 has the solution

$$\mathbf{r} = \frac{e\mathbf{E}}{m}\,\frac{1}{(\omega_0^2 - \omega^2)} \qquad (6\text{–}108)$$

and

$$\ddot{\mathbf{d}} = \frac{e^2}{m}\,\mathbf{E}\left(\frac{1}{1 - (\omega_0^2/\omega^2)}\right) \qquad (6\text{–}109)$$

It is clear then (see equation 6–99) that the scattering cross section is

$$\sigma = \frac{\sigma_{\text{Thomson}}}{(1 - \omega_0^2/\omega^2)^2} \qquad (6\text{–}110)$$

When $\omega \gg \omega_0$, the electron acts as though it were free and we again have $\sigma = \sigma_{\text{Thomson}}$. If $\omega_0 \gg \omega$ we obtain

$$\sigma = \frac{8\pi}{3} \frac{e^4}{m^2 c^4} \frac{\omega^4}{\omega_0^4} \tag{6-111}$$

called the *Rayleigh scattering cross section*. Rayleigh scattering is responsible for the scattering of visible light in the daytime sky. The electrons are strongly bound to their parent molecules so that ω_0 is large compared to the frequency of visible light ω. For red light $(\omega/\omega_0)^4$ is smaller than for blue light by a factor close to $(2)^4 = 16$. Hence blue light is scattered most strongly. Red light therefore passes more easily straight through the atmosphere without deflection, while blue light is scattered out of a straight path, and the sky appears blue when we look away from the sun.

There is one more case of scattering that is interesting in astronomy. This is the scattering by fine dust grains. For spherical grains with refractive index n, the cross section for scattering can be shown to be

$$\sigma = 24\pi^3 \left[\frac{n^2 - 1}{n^2 + 2} \right]^2 \frac{V^2}{\lambda^4} \tag{6-112}$$

if the radius a of the sphere is much smaller than the wavelength λ. V is the volume $(4\pi/3)a^3$. We note that the factor λ^{-4} is reminiscent of Rayleigh scattering and, of course, the two types of scattering are related. The differential cross section has exactly the same angular dependence as found for Thomson or Rayleigh scattering (see 6–105). This kind of scattering may be approximately characteristic of the zodiacal (interplanetary) grains mentioned above, and of the interstellar grains discussed in the next section.

6:15 EXTINCTION BY INTERSTELLAR GRAINS

Interstellar grains absorb and scatter radiation so that starlight does not reach an observer directly. We talk about *extinction* by the grains (Gr68). The term extinction refers to the fractional amount of light prevented from reaching us. It is a useful concept when we do not know how much of the radiation is scattered and how much is absorbed. The scattered radiation sometimes can be observed in *reflection nebulas*—clouds of dust grains illuminated by a bright star. These clouds show spectra remarkably similar to those of the illuminating star; and it therefore appears that the scattered portion of the radiation is very much like scattering off snow. The particles are basically white or gray in this sense. On the other hand, when we see starlight that has passed through a cloud, we find an amount of extinction that to first approximation is inversely proportional to

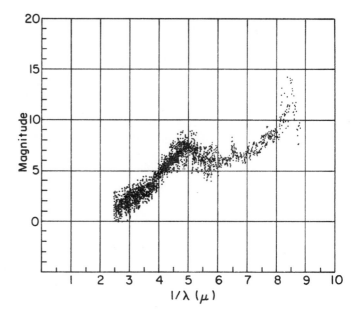

Fig. 6.10 Interstellar extinction curve showing magnitudes of extinction as a function of reciprocal wavelength. The data were obtained from observations of the stars ζ-Persei and ε-Persei, and have been normalized to an extinction difference $E (B - V)$ of one magnitude and $V \sim 0$. The curve would therefore roughly characterize extinction over a path length of order ~ 1 kpc through the galactic plane. (After Stecher St69. With the permission of the University of Chicago Press.)

the wavelength, λ. The data are shown in Fig. 6.10. It may be that λ^{-4} scattering is the predominant scattering process; the size of grains may be roughly comparable to the wavelength of the radiation, and the size distribution may be such that the apparent scattering cross section integrated over all different particle sizes gives an overall mean cross section proportional to λ^{-1}. These considerations all are very uncertain!

From current observations it is not clear what kind of material the grains represent. Some astrophysicists believe that graphite particles are a plausible constituent. Others argue in favor of ice or silicates. Some recent infrared spectral observations indicate a weak absorption band at 3μ wavelength that corresponds to the position at which ice crystals would be expected to absorb. There is a possibility that grains consist of an ice mantle deposited on a graphite core: The graphite could condense into grains in the dense atmosphere of a *carbon star*; water would then be frozen out on these grains in the outer, cooler portions of the star's atmosphere, as the grains streamed outward into interstellar space. There is a strong feeling that interstellar grains have to be formed in some relatively dense regions of the universe, such as in the atmospheres of stars, or in dense clouds, because the general interstellar medium is so tenuous that collisions among molecules are too rare to permit growth of material into micron-sized grains. This is discussed further in section 9:4.

One more factor might be mentioned. Thus far we have only talked about spherical grains that are purely dielectric. It is, however, also possible for grains to have a metallic character; they can then absorb and emit radiation—they do not merely scatter. For metallic grains the dielectric constant has an imaginary component and we talk about a *complex refractive index m*. In these terms, for grains small compared to the wavelength, absorption dominates over scattering, and the extinction is given (Gr68) by

$$\not{E} = 6\pi N \left(\frac{1 - m^2}{m^2 + 2} \right) \frac{V}{\lambda} \quad \text{(IP)} \qquad (6\text{--}113)$$

where $\not{E}$ is the total light extinguished for unit incident energy, and the symbol (IP) means that the imaginary part of the expression in parentheses should be used.

For particles with dimensions comparable to the wavelength of the extinguished radiation, the expressions become quite complicated even for spherical particles; and for nonspherical grains the theory of extinction is extremely laborious.

We might be tempted to attribute interstellar extinction to metallic absorption alone, because then the $1/\lambda$ relation might be directly obtained. However, matters are not that simple. The refractive indices m and n are wavelength dependent, and that wavelength dependence is determined by the chemical makeup of the grains. With all these free parameters, a $1/\lambda$ dependence can therefore be fitted with relative ease—and this is corroborated by the number of theoretical models that astronomers have produced to describe interstellar grains. As in other parts of astronomy, a wealth of models reflects a large degree of uncertainty.

One final comment concerns the overall polarization of scattered, as distinguished from re-emitted, light. Thermally re-emitted radiation from *interstellar* or *circumstellar dust*, should be unpolarized, but scattered radiation by any one of the three processes mentioned above should lead to a polarization that can be shown to have the Ⓗ-dependence

$$P = \frac{\sin^2 Ⓗ}{1 + \cos^2 Ⓗ} \qquad (6\text{--}114)$$

In addition, polarization can also be caused by elongated grains. The light that reaches us from distant stars located close to the plane of the Galaxy shows polarization believed due to this process. How small grains could be aligned to give consistently polarized radiation is discussed in section 9:8. The interaction of grains with radiation is a complicated subject. Detailed discussions can be found in references (Gr68) and (vdHu57).

6:16 ABSORPTION AND EMISSION OF RADIATION BY A PLASMA

In section 6:11, we treated the propagation of radiation through a tenuous ionized medium. Sometimes, however, the plasma encountered in astrophysical situations is dense in that collisions between ions and electrons have a relatively high frequency of occurrence, v_c. In this sense, a medium may be tenuous for high frequency waves, but dense at lower frequencies. This is a quite common situation. Because the nature of the transmission is so different in these two cases, the spectrum of radiation received from a source also may be quite different at high and low frequencies. A relatively abrupt change in spectrum taken together with other data then allows us to determine the collision frequency and, hence, as we will show, also the density of the medium. Radioastronomy therefore provides an extremely useful technique for measuring the density of interstellar ionized gases.

To show how all this comes about, we consider an ionized medium and define the *collision frequency* v_c as the frequency with which an electron successively becomes deflected through a total angle of 90°, usually through a series of small collisions with ions. We choose an angle of 90°, because this is the deflection a particle has to suffer to give up all the directed momentum it had at some previous period, that is, to lose all sense of the direction into which it was accelerated by some previously applied force. We only consider collisions with ions, because we will be interested in the dissipation of energy through collisions. When an electron collides with another electron, the motion is that of a symmetric dipole and no energy is radiated away in such a process. Electron-electron collisions can therefore be neglected.

We now consider a situation in which an electron is accelerated by an applied electric field **E**—in this case the electric field component of an electromagnetic wave. As the particle reaches appreciable velocity induced by the field, it suffers a collision and gives up all its directed momentum. This means that there is an acceleration by the electromagnetic wave traveling through the medium and deceleration through collisions. The net effect is to impose forces on the electron so that

$$m\ddot{\mathbf{r}} = e\mathbf{E}(r, t) - m\dot{\mathbf{r}}v_c \qquad (6\text{–}115)$$

Here m is the reduced electron mass. The second term on the right shows a momentum loss equal to the instantaneous momentum $m\dot{\mathbf{r}}$ of the electron every time there is a collision, or v_c times in unit time interval. This is just what we stated formally in defining v_c; and our problem will be to calculate the actual value v_c has. However, before we do that, we can proceed to solve equation 6–115 and obtain the transmission properties of the plasma in different frequency ranges relative to v_c.

We have already noted that some of the momentum conferred on the electrons by the electromagnetic wave is lost in collisions. This means that the energy transferred from the wave to the particles becomes dissipated, and since the energy of the electromagnetic wave depends on E^2, the square of the wave amplitude, we can expect to find that E will decrease as the wave propagates through the medium. We will therefore make use of a function $\mathbf{E}(r, t)$ of the form

$$\mathbf{E}(r, t) = \mathbf{E}_0 e^{-Kx/2} \cos \omega t \qquad (6\text{--}116)$$

of equation 6–115. Here K is the *absorption coefficient*. The factor 2 in the exponent of this *damping term* is provided so that the energy in the wave, rather than its amplitude, may decay by a factor of $1/e$ in distance $x = 1/K$. Note that the absorption coefficient always has units (length)$^{-1}$.

Because the rate of energy loss from the wave will be determined by the total number of collisions per unit volume, we rewrite equation 6–115 as

$$nm\ddot{\mathbf{r}} + nmv_c\dot{\mathbf{r}} = ne\mathbf{E}_0 e^{-Kx/2} \cos \omega t \qquad (6\text{--}117)$$

where n is the electron density.

If we use a complex field $\mathbf{E}$, instead of a field with real values in equation 6–117, the solution of the differential equation becomes much simpler. However, in order to remember that only the real parts of the equation have physical significance, we add the annotation (RP) in this case:

$$nm\ddot{\mathbf{r}} + nmv_c\dot{\mathbf{r}} = ne\mathbf{E}_0 e^{-(Kx/2)+i\omega t} = ne\mathbf{E} \qquad \text{(RP)} \qquad (6\text{--}118)$$

PROBLEM 6–12. By substitution, show that a particular solution of (6–118) is

$$\mathbf{r} = -\left(\frac{e\mathbf{E}_0}{m\omega}\right) e^{(i\omega t - Kx/2)} \left[\frac{iv_c + \omega}{v_c^2 + \omega^2}\right] \qquad (6\text{--}119)$$

The current due to the n particles per unit volume can now be written as

$$\mathbf{j} = ne\dot{\mathbf{r}} = \frac{ne^2}{m} \left[\frac{v_c - i\omega}{v_c^2 + \omega^2}\right] \mathbf{E} \qquad \text{(RP)} \qquad (6\text{--}120)$$

As in equation 6–50, $ne\mathbf{r}$ is an induced polarization field, so that the second term in the brackets on the right of (6–120) is the induced polarization current

$$\frac{d\mathbf{P}}{dt} = i\omega\mathbf{P} = i\omega\,\frac{\varepsilon - 1}{4\pi}\,\mathbf{E} \qquad (RP) \tag{6–121}$$

Here the imaginary number i enters as a consequence of the assumed field in (6–118), and (6–121) then is a direct consequence of the definition (6–10). The real term in the brackets of (6–120) is just the current $\sigma\mathbf{E}$ due to the flow of charge. We note two features of equation (6–120). The term proportional to v_c on the right represents the dissipation of energy, and will therefore be directly related to the absorption coefficient K. The second term, proportional to $i\omega$, depends on the dielectric constant in the medium, and hence will yield the propagation velocity $c\varepsilon^{-1/2}$ of the wave through the medium. Formally written:

$$\mathbf{j} = \left(\sigma + i\omega\,\frac{\varepsilon - 1}{4\pi}\right)\mathbf{E} \qquad (RP) \tag{6–122}$$

with

$$\varepsilon = 1 - \frac{4\pi e^2 n}{m(\omega^2 + v_c^2)} \qquad \text{and} \qquad \sigma = \frac{e^2 n v_c}{m(\omega^2 + v_c^2)} \tag{6–123}$$

If we write the imaginary and *complex dielectric constants* as

$$\varepsilon_i = -\,\frac{i4\pi\sigma}{\omega} \qquad \text{and} \qquad \varepsilon_c = \varepsilon + \varepsilon_i \tag{6–124}$$

then equation 6–122 can be written in a form characteristic of a pure dielectric. In fact, all of Maxwell's equations take on this form. This can be seen directly by noting that $\mathbf{j}$ appears only in equations 6–15 and 6–23 in the set of Maxwell's differential equations. For a complex field, as it appears in (6–118), a propagating wave will have a form (see equation 6–36)

$$\mathbf{E} = \mathbf{E}_0 \exp i\left[\omega t \pm \frac{\omega\varepsilon_c^{1/2}x}{c}\right] \qquad (RP) \tag{6–125}$$

so that (6–118) will hold if

$$\frac{K}{2} = \frac{i\omega}{c}\,\varepsilon_c^{1/2} \qquad (RP) \tag{6–126}$$

We can always write ε_c in the form

$$\varepsilon_c = (N + iQ)^2 \tag{6–127}$$

where N and Q are real quantities as long as we choose

$$\varepsilon_i = 2NQi = -\left(\frac{4\pi\sigma}{\omega}\right)i \qquad \text{and} \qquad \varepsilon = N^2 - Q^2 \tag{6–128}$$

6:16

We are therefore interested in the quantity

$$\frac{K}{2} = -\frac{\omega Q}{c} = \frac{4\pi\sigma}{2Nc} \tag{6-129}$$

In practice $\omega \gg v_c$, ω_p, in all radio-astronomical situations, so that (6–123) and (6–128) give

$$|\varepsilon| \gg \frac{4\pi\sigma}{\omega} \quad \text{and} \quad N \sim \varepsilon^{1/2} \tag{6-130}$$

With this same approximation we then also obtain

$$K = \frac{4\pi(e^2 n/m\omega^2)\,v_c/c}{\sqrt{1 - 4\pi e^2 n/m\omega^2}} = \frac{v_c(\omega_p^2/\omega^2)/c}{\sqrt{1 - \omega_p^2/\omega^2}} \tag{6-131}$$

where ω_p is the plasma frequency (6–55).

We still need to calculate the collision frequency v_c, but that should be simple because most of the work has already been done. Equation 3–72 gives the force acting on a particle of reduced mass μ deflected in the superposition of inverse square law fields produced by a density of n scattering centers per unit volume. To avoid confusion with the symbol μ used here for magnetic permeability, we will continue to use the symbol m for the electron's reduced mass. In equation (6–115) we had defined the drag force $m\dot{r}v_c$ and we now set this equal to the right side of (3–67), noting that since $\dot{r}$ is the velocity before collision, it plays the same role as v_0 in (3–67):

$$mv_0 v_c = m2\pi n v_0^2 \int_{s_{\min}}^{s_{\max}} s(1 - \cos \textcircled{H})\, ds \tag{6-132}$$

For small deflections $\textcircled{H}$

$$1 - \cos \textcircled{H} \approx 2 \tan^2 \frac{\textcircled{H}}{2} = 2\left[\frac{Ze^2}{v_0^2 sm}\right]^2 \tag{6-133}$$

The second half of this inequality is based on an analogy with equation 3–69 but with Coulomb forces replacing gravitational forces. Z is the typical charge on an ion. We therefore have

$$v_c = 4\pi n \left\langle \frac{1}{v_0} \right\rangle \frac{1}{\langle v_0^2 \rangle} \frac{Z^2 e^4}{m^2} \int_{s_{\min}}^{s_{\max}} s^{-1}\, ds$$

$$= \frac{4\sqrt{2\pi}}{3} n \frac{Z^2 e^4}{\sqrt{m}\,(kT)^{3/2}} \ln \frac{s_{\max}}{s_{\min}} \tag{6-134}$$

6:16

where expressions 4–109 and 4–110 have been used. From (6–131) and (6–134)

$$K(\omega) = \frac{32\pi^{3/2}e^6 n^2 Z^2}{3\sqrt{2}\,c\omega^2 (kTm)^{3/2}} \ln \frac{s_{max}}{s_{min}} \qquad (6\text{–}135)$$

Here the assumption of very many, very weak deflections has been made so that the minimum impact parameter s'_{min} must be large enough to give a potential energy small compared to the kinetic energy. Specifically, for $Z = 1$,

$$s'_{min} \gg \frac{2e^2}{mv_0^2} \qquad (6\text{–}136)$$

In the interstellar medium, s rarely will be less than 10^{-2} cm, while $2e^2 = 5 \times 10^{-19}$; and typically mv^2 is 10^{-12} erg in ionized regions. This shows that the condition (6–136) usually is well satisfied except in very rare chance collisions with a small impact parameter.

A second lower bound is given by the *de Broglie wavelength* of the electron $\lambda_e = h/2\pi mv$. At closer distances than this, the electron no longer behaves like a point charge, and we can use a lower limit $s''_{min} > \lambda_e/2$. There also are two upper bounds we can set. First, we want a collision to appear instantaneous, that is, the time $1/\omega \gg s'_{max}/v_0$; the time during which the electric field changes is long compared to the time in which the electron suffers a collision, or goes through its minimum approach. The second upper limit is $s''_{max} = L$, the Debye length given by equation 4–156. Shielding by nearer particles screens out the effects of charges at distances greater than L. Using the limits s'_{max} and s'_{min} for the ionized interstellar matter, we can then write

$$\frac{s'_{max}}{s'_{min}} = \frac{mv_0^2}{2e^2}\frac{v_0}{\omega} \sim \frac{(2kT)^{3/2}}{2e^2\omega m^{1/2}}$$

The full expression for ionized hydrogen, $Z = 1$, reads

$$K(\omega) = \frac{32\pi^{3/2}}{3\sqrt{2}}\frac{e^6 n^2}{c(mkT)^{3/2}\omega^2}\ln\left[\frac{1.32(kT)^{3/2}}{e^2 m^{1/2}\omega}\right]$$

$$K(\nu) = \frac{8}{3\sqrt{2\pi}}\frac{e^6 n^2}{c(mkT)^{3/2}\nu^2}\ln\left[\frac{1.32(kT)^{3/2}}{2\pi e^2 m^{1/2}\nu}\right] \qquad (6\text{–}137)$$

6:17 RADIATION FROM THERMAL RADIO SOURCES

If we now look at the results obtained in the previous section, we note that we have an absorption coefficient $K(\nu)$ that tells us the amount of absorption obtained

6:17

per unit length of travel through an ionized medium. An electromagnetic wave traveling a distance $D = \int dx$ through the medium will encounter an *optical depth*

$$\tau(v) = \int K(v)\, dx \qquad (6\text{–}138)$$

If the temperature throughout the region is constant, only the density will vary with position x and we find that

$$\tau(v) = F(T, v) \int n^2\, dx \qquad (6\text{–}139)$$

where the function F is $K(v)/n^2$ (see equation 6–137) and the integral

$$\mathscr{E}_m = \int n^2\, dx = \langle n^2 \rangle D \qquad (6\text{–}140)$$

is called the *emission measure*; it is a measure of the amount of absorption and emission expected along D. It is customary in radio-astronomy to express the electron concentration n in terms of cm^{-3} and D, the path length covered, in parsecs. The emission measure then has units cm^{-6} pc.

The emission measure is just a measure that tells how frequently atomic particles approach each other closely along a line of sight through a given region. For this reason, such quantities as the number of atomic recombinations giving rise to a given emission line are also proportional to $\mathscr{E}_m$. Usually the recombination line strength R_1 for any given line v_1 also is a known function of the temperature, so that

$$R(v_1) = F_1(T)\,\mathscr{E}_m \qquad (6\text{–}141)$$

Hence, if we measure both the recombination line strength—possibly in the visible part of the spectrum—and also the radio thermal emission, both the emission measure and the temperature of the region can be determined. For this to be true, radio measurements are best taken at frequencies for which the region is optically thin, so that self-absorption of radiation by the cloud need not be considered.

An interesting feature of the self-absorption by an optically thin cloud is that the brightness should be independent of the frequency v. This comes about because the absorption $K(v)$ is inversely dependent on v^2—if we neglect the weak frequency dependence of the logarithmic term in (6–137). At the same time, the energy density of radio waves corresponding to a blackbody at gas temperature, T, would be

$$\rho(v) \sim \frac{8\pi k T v^2}{c^3} \qquad hv \ll kT \qquad (4\text{–}80)$$

6:17

at very long wavelengths. The product of optical depth or effective emissivity for the gaseous region, and blackbody intensity $I(\nu) = \rho(\nu)\,c/4$, is therefore frequency independent as long as the region is optically thin. At low frequencies, where $\tau(\nu) \gtrsim 1$, this behavior ceases to be true. The effective emissivity then remains close to unity, and the only frequency dependent term is $I(\nu)$. At such frequencies, a thermal source should therefore show a spectrum proportional to ν^2.

For the flat part of the spectrum the product $S(\nu) = \tau(\nu)\,I(\nu)$ is proportional to $T^{-1/2}\mathscr{E}_m$; and this latter product can be immediately determined from a measurement of the surface brightness anywhere in this frequency range. In the steep part of the spectrum, where the region is opaque, the measured surface brightness at frequency ν depends only on T (see equation 4–81). These two sets of observations, taken together, provide data both on the temperature and on the emission measure $\mathscr{E}_m$. Figure 6.11 shows spectra for some very compact ionized hydrogen regions and these show the expected form very clearly.

On this log-log plot the low frequency spectrum has the expected slope of 2. At high frequencies it is flat.

NGC 7027 is a planetary nebula for which Mezger (Me68)—see the data of Fig. 6.11—finds an emission measure of 5.4×10^7 cm^{-6} pc and a temperature of

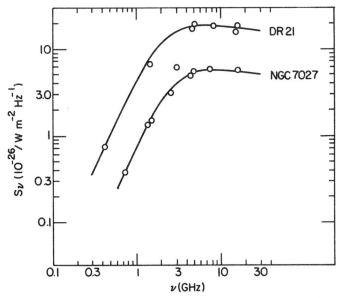

Fig. 6.11 Data obtained by a number of observers on compact H II regions (see text). After P.G. Mezger (Me68). (From *Interstellar Ionized Hydrogen*, Y. Terzian, Ed., 1968, W.A. Benjamin, Inc. Reading, Massachusetts.)

$\sim 1.1 \times 10^{4}$°K. If the object is assumed to have a depth along the line of sight similar to the observed diameter, then an actual density can be computed. Mezger gives the value $n \sim 2.3 \times 10^{4}$ cm^{-3} for this object. From the density and the total volume, we can also obtain the nebular mass, which in this case is roughly 0.25 $M_{\odot}$, or 5×10^{32} g. The observed diameter of NGC 7027 is about 0.1 pc.

Other compact H II regions, which may be found in the plane of the Galaxy in the vicinity of bright young stars, tend to have somewhat lower temperatures, about the same densities, but sometimes much greater masses—up to several stellar masses. These are believed to possibly represent the remains of clouds from which massive stars were formed. When massive protostars light up as they approach the main sequence, they emit an intense ultraviolet flux that heats and ionizes the gas. Such H II regions will be discussed in Chapter 9.

6:18 SYNCHROTRON RADIATION

When a charged particle moves at relativistic velocity across a magnetic field, it describes a spiral motion. The axis of this spiral lies along the direction of the magnetic field and the acceleration experienced by the particle is along directions perpendicular to the field lines. As the particle moves, the direction of the acceleration vector continually changes.

We first consider the motion of a relativistically moving particle orbiting in a plane perpendicular to a magnetic field. This constitutes no restriction in generality, because a constant velocity component along the magnetic field lines would leave the radiation rate unaffected.

If we recall that a force corresponds to a rate of change of momentum, we can use (6–11) to calculate the rate at which the particle is deflected. Consider the direction of motion in Fig. 6.12 to be the x direction, and let the radial direction be the y direction. In a time Δt_0, the momentum change, which is along the y direction, will amount to

$$\Delta p_y = \frac{evB}{c} \Delta t_0 \qquad (6\text{–}142)$$

Since the initial relativistic momentum p_x is

$$p_x = \frac{m_0 v}{\sqrt{1 - v^2/c^2}} \qquad (5\text{–}30)$$

we can see that the angular deflection during time interval Δt_0 is,

$$\delta = \frac{\Delta p_y}{p_x} = \frac{eB}{m_0 c} \sqrt{1 - \frac{v^2}{c^2}} \, \Delta t_0 = \frac{eB \, \Delta t_0}{m_0 c \gamma(v)} \qquad (6\text{–}143)$$

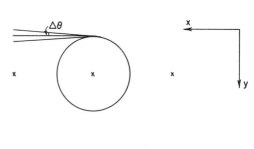

Fig. 6.12 Relativistic charged particle orbiting in a magnetic field. The direction of the field lines is into the paper.

m_0 is the particle's rest mass. From this it follows that the time Δt_1, required for the particle to orbit one radian $[\delta = 1]$ is

$$\Delta t_1 = \frac{m_0 c}{eB} \gamma(v) \qquad (6\text{--}144)$$

The gyrofrequency given in equation 6–13 is the reciprocal of Δt_1; we see that, if we use (5–40) to substitute $m_0 \gamma(v)$ for p_c/v_c in (6–13).

Having obtained the gyrofrequency of the particle, we might think that the problem is completely solved and that the particle will simply radiate energy at that frequency. However, that is not so. The spectrum radiated by the moving charge actually lies at frequencies often many orders of magnitude higher than ω_c. The reason for this is directly related to the strong concentration of emitted radiation into a narrow beam of angular half-width (obtainable from equation 5–50):

$$\Delta\theta \sim \sqrt{1 - \frac{v^2}{c^2}} = \gamma(v)^{-1} \qquad (6\text{--}145)$$

about the forward direction of motion. Because of this, an observer is not properly oriented to receive radiation emitted by the particle except during a very short-time interval

$$\Delta t_2 = 2\Delta\theta\,\Delta t_1 = \frac{2m_0 c}{eB} \qquad (6\text{--}146)$$

of each orbit.

But the radiation emitted during interval Δt actually arrives at the observer over an even smaller time span, because radiation emitted by the particle at the beginning of the interval Δt has a longer distance to travel to the observer than radiation emitted at the end of the interval when the particle is nearer to him. If the particle travels a distance of length L during interval Δt, radiation emitted at the end of the interval will only arrive at a time

$$\Delta t \sim - \left(\frac{L}{c} - \frac{L}{v} \right) \qquad (6\text{–}147)$$

later than radiation emitted at the beginning of the interval. Since

$$L \sim v \, \Delta t_2 \qquad (6\text{–}148)$$

we obtain

$$\Delta t \sim \left(1 - \frac{v}{c} \right) \Delta t_2 \sim \frac{m_0 c}{eB} \left(1 - \frac{v^2}{c^2} \right) \qquad (6\text{–}149)$$

because for highly relativistic particles

$$\left(1 - \frac{v}{c} \right) \sim \frac{1}{2} \left(1 + \frac{v}{c} \right) \left(1 - \frac{v}{c} \right) = \frac{1}{2} \left(1 - \frac{v^2}{c^2} \right) \qquad (6\text{–}150)$$

The radiation frequency corresponding to the reciprocal of this time interval is

$$\omega_m \sim \frac{1}{\Delta t} \sim \frac{eB}{m_0 c} \left(1 - \frac{v^2}{c^2} \right)^{-1} = \gamma^2(v) \, \omega_c = \frac{eB}{m_0 c} \left(\frac{\mathscr{E}}{m_0 c^2} \right)^2 \qquad (6\text{–}151)$$

where $\mathscr{E}$ is the total energy of the particle, $\mathscr{E} \gg m_0 c^2$ and ω_c is the radiation frequency of a non-relativistic particle moving in a magnetic field. We expect to see radiation at frequencies of this order of magnitude from relativistic particles moving in a magnetic field. Since $(1 - v^2/c^2)$ is a very small number, it is clear that ω_m is many orders of magnitude greater than the gyrofrequency,

$$\omega_m \gg \omega_c = \frac{eB}{m_0 c} \qquad (6\text{–}152)$$

Let us summarize what we have done:

(1) First, we computed the orbital frequency of a particle moving in a magnetic field.

(2) Next, we calculated the time in the observer's frame during which the particle was capable of emitting radiation into his direction.

(3) Finally, we computed the length of time elapsing between the arrival of the first and last portion of the electromagnetic wavetrain at the position of the observer. This elapsed time was very small compared to the period of the particle's gyration in the magnetic field and the corresponding frequency $\omega_m \sim 1/\Delta t$ was found to be $(\mathscr{E}/m_0 c^2)^2$ higher than the nonrelativistic gyrofrequency for this field.

6:18

6:19 THE SYNCHROTRON RADIATION SPECTRUM

The actually expected synchrotron radiation spectrum obtained when the above sketched calculations are done rigorously for monoergic electrons is a set of extremely finely spaced lines at high harmonics of the gyrofrequency. The peak of the spectral distribution occurs at a frequency $\omega = 0.5\ \omega_m$. The shape of the spectral function $p(\omega/\omega_m)$ is shown in Fig. 6.13. The maximum value of p is $p(0.5) = 0.10$. Details of the theory are discussed in references (Gi64)* and (Sh60)*.

Fig. 6.13 Envelope of the narrowly spaced lines comprising the synchrotron spectrum of a particle whose frequency $\omega_m = (eB/m_0c)(\mathscr{E}/m_0c^2)^2$. In an actual situation the particle energies vary to some extent and, hence, the finely spaced lines are never seen. One observes a continuum of the shape of the envelope (after I.S. Shklovsky, Sh60).

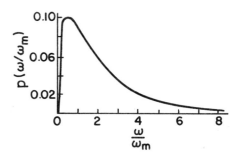

One can show that the energy actually radiated by a particle of energy $\mathscr{E}$ per unit time into unit frequency interval dv is

$$P(v, \mathscr{E})\, dv = P(\omega, \mathscr{E})\, 2\pi\, d\omega = \frac{16e^3 B}{m_0 c^2} p(\omega/\omega_m)\, d\omega \qquad (6\text{–}153)$$

In the limit of very high and very low frequencies the function $p(\omega/\omega_m)$ has asymptotic values

$$p(\omega/\omega_m) = 0.256 \left(\frac{\omega}{\omega_m} \right)^{1/3}, \qquad \omega \ll \omega_m$$

$$p(\omega/\omega_m) = \frac{1}{16} \left(\frac{\pi\omega}{\omega_m} \right)^{1/2} \exp\left(-\frac{2\omega}{3\omega_m} \right), \qquad \omega \gg \omega_m \qquad (6\text{–}154)$$

PROBLEM 6–13. In section (5:9) we saw that the power radiated by a body is independent of an observer's rest frame, as long as both the source and observer move in inertial frames of reference. Make use of this fact to obtain the total power radiated in the form of synchrotron radiation, by computing the emission of the spiraling charge as viewed from an inertial frame moving with the charge's instantaneous velocity. For a charge whose total energy is $\mathscr{E}$, this total power is

$$P(\mathscr{E}) = \frac{2}{3} \frac{e^4 B^2}{m_0^2 c^3} \left(\frac{\mathscr{E}}{m_0 c^2} \right)^2 \qquad (6\text{-}155)$$

for motion perpendicular to the magnetic field. For an electron

$$P(\mathscr{E}) = 1.58 \times 10^{-15} B^2 \left(\frac{\mathscr{E}}{m_0 c^2} \right)^2 \text{ erg sec}^{-1}$$

$$= 1.62 \times 10^{-2} \left(\frac{B^2}{8\pi} \right) \left(\frac{\mathscr{E}}{m_0 c^2} \right)^2 \text{ ev sec}^{-1}$$

Verify that these expressions are at least approximately consistent with expressions 6-151, 6-153, and 6-154 by noting that $P(\omega, \mathscr{E}) \omega_m$ roughly corresponds to $P(\mathscr{E})$, and that equation 6-153 with a numerical integration under the curve in Fig. 6.12 gives the same result.

PROBLEM 6-14. One astronomical object in which synchrotron radiation is important is the Crab Nebula. Take the magnetic field strength in some of the Crab's bright filaments to be of order 10^{-4} gauss, and show that a classically moving electron would radiate at a frequency of about 300 Hz, independent of the energy. On the other hand, if the energy becomes 10^9 ev, some 2×10^3 times the rest-mass energy, the peak radiation will occur at about 600 MHz and if the energy becomes 10^{12} ev, the radiation peaks in the visible part of the spectrum at 6×10^{14} Hz.

It is clear that the exact form of the observed spectrum will depend on the energy spectrum of the radiating particles as well as on the function $P(v, \mathscr{E})$. If we integrate the radiation coming from different distances r along the observed line of sight out to some distance R, the resulting spectral intensity at frequency v will be

$$I_v \, dv = \int_0^{\mathscr{E}_{max}} \int_0^R P(v, \mathscr{E}) \, n(\mathscr{E}, r) \, dr \, d\mathscr{E} \, dv \qquad (6\text{-}156)$$

where $n(\mathscr{E}, r)$ is the number density of particles of energy $\mathscr{E}$ at distance r.

In the frequently encountered situation where $n(\mathscr{E}) \propto \mathscr{E}^{-\gamma}$, that is, where the electrons have an exponential spectrum with constant exponent $-\gamma$, the intensity obeys the proportionality relation $I_v \propto v^{-\alpha}$ where $\alpha = (\gamma - 1)/2$. To show this relationship, we note from Fig. 6.13 and from (6-153) that $P(v, \mathscr{E})$ is equal to $16e^3 B/m_0 c^2$ multiplied by an amplitude 0.1 and 2π, since the bandwidth is $\Delta v = \Delta \omega / 2\pi \sim 3\omega_m/2\pi$.

Let us suppose now that every electron deposits its total radiated power at frequency ω_m. Equation 6–151 shows that $\mathcal{E} \propto \omega_m^{1/2}$ so that $\mathcal{E}^{-\gamma} \propto \omega_m^{-\gamma/2}$, $\Delta\mathcal{E} \propto \Delta\omega/\omega_m^{1/2}$, and the total radiated power in (6–156) obeys the proportionality relation

$$I(\nu)\,\Delta\nu \propto \omega_m \mathcal{E}^{-\gamma}\,\Delta\mathcal{E} \propto \omega_m^{(1-\gamma)/2}\,\Delta\omega \qquad (6\text{–}157)$$

for a constant spectrum along the path of integration. Hence

$$I(\nu) \propto \nu^{-\alpha}, \qquad \alpha = \frac{(\gamma-1)}{2} \qquad (6\text{–}158)$$

α is called the *spectral index* of the source. To obtain this relationship between electron energy and electromagnetic radiation spectra, the source must be optically thin. The optically thick (self-absorbing) sources are discussed below. For a wide variety of nonthermal cosmic sources $0.2 \lesssim \alpha \lesssim 1.2$. For extragalactic objects indices up to $\alpha = 2$ occur. In the frequency range below a few gigahertz, many QSO's have $\alpha < 0.5$; but they often contain optically thick components. Most radio galaxies have $\alpha > 0.5$ (Co72). Spectra of some extragalactic sources appear in Fig. 9.13.

For the Galaxy, confirmation of these general concepts is good. We observe a Galactic cosmic ray electron spectrum with $\gamma \sim 2.6$. This is measured at the earth's position, but the electrons have reached us from great distances. Radio waves also show an overall Galactic spectrum with index $\alpha \sim 0.8$—in agreement with (6–158).

Equation 6–158 is of great importance in astrophysics because it permits us to estimate the relativistic electron energy spectrum by looking at the synchrotron radio emission from a distant region. The total intensity of the radio waves is, however, not only a function of the total number of electrons along a line of sight, it also depends on the magnetic field strength in the region where the relativistic electrons radiate. Proton synchrotron radiation may also be important (Re68b).

PROBLEM 6–15. Show that $I(\nu)$ is proportional to $B^{(\gamma+1)/2}$. For a randomly oriented field, B^2 takes on the mean value of the component of the (magnetic field)^2 perpendicular to the line of sight. Hence

$$I(\nu) \propto B^{(\gamma+1)/2}\nu^{-(\gamma-1)/2}$$

$$\propto B^{\alpha+1}\nu^{-\alpha} \qquad (6\text{–}159)$$

To conclude this discussion, we should still state that synchrotron emission, like any other emission process, also has a corresponding absorption process. Some strong extragalactic radio sources are believed to generate their radiation by means of synchrotron emission. Yet the spectrum of these sources appears black in just those regions where synchrotron radiation would have the highest emissivity. This is interpreted (see section 7:10) as meaning that the sources are opaque to their own radiation. Figure 9.13 shows that the flux for many nonthermal sources is high at low frequencies. On the other hand, at these low frequencies, the flux cannot exceed the flux of a blackbody

$$I(v)\, d\Omega = \frac{2kT}{c^2}\, v^2\, d\Omega \tag{4-81}$$

whose temperature T is determined by the electron energy, $kT \sim \mathscr{E}$. Now, equation 6–151 gives the relation between $\mathscr{E}$, the magnetic field in the source B, and the emitted frequency $v \sim \omega_{max}/2\pi$. Substituting for kT in (4–81), we then have

$$I(v)\, d\Omega = \left(\frac{8\pi v^5 m_0^3 c}{B} \right)^{1/2} d\Omega$$

which expresses the magnetic field strength in the source in terms of the observed flux at frequency v, and the angular size of the source. The low frequency spectrum then is no longer a blackbody spectrum since the energy of electrons decreases at lower radiated frequencies. Effectively the temperature of the electrons, $\mathscr{E}/k$, is frequency dependent. The required data needed to compute the source magnetic field can be gathered with a radio interferometer (Ke71). The peaked spectrum of synchrotron self-absorption characterizes the radio source 3C147 (Fig. 9.13).

6:20 THE COMPTON EFFECT AND INVERSE COMPTON EFFECT

When a high energy photon impinges on a charged particle, it tends to transfer momentum to it, giving it an impulse with a component along the photon's initial direction of propagation. This is an effect that we had neglected in dealing with the low energy Thomson scattering process. Although we talk about *Compton scattering* when we discuss the interaction of highly energetic electromagnetic radiation with charged particles and *Thomson scattering* when lower energies are involved, we must understand that the basic process is exactly the same, and that we are only talking about differences in the mathematical approach convenient for analyzing the most important physical effects in different energy ranges.

Corresponding to the Compton effect, there is an exactly parallel situation, the *inverse Compton effect*, in which a highly energetic particle transfers momentum to a low energy photon and endows it with a large momentum and energy. These processes are exactly alike except that the coordinate frame from which they are viewed differs. To an observer at rest with respect to the high energy particle, the inverse Compton effect will appear to be an ordinary Compton scattering process. To him it would appear that a highly energetic photon was being scattered by a stationary charged particle.

Because of this similarity we will only derive the expressions needed for the Compton effect, and then discuss the inverse effect in terms of a coordinate transformation. We will set down four equations governing the interaction of a photon with a particle. We note here that the effect is more conveniently described in terms of photons than in terms of electromagnetic waves, but again this is only a matter of convenience and does not reflect a physical difference in the radiation involved. The considerations we have to take into account are:

(i) Conservation of mass-energy, given by

$$m_0 c^2 + h\nu = \mathscr{E} + h\nu' \tag{6–160}$$

where ν and ν' are the radiation frequency before and after the collision, m_0 is the rest mass, and $\mathscr{E}$ the relativistic mass energy of the recoil particle (Fig. 6.14).

(ii) The relation of $\mathscr{E}$ to m_0 is (5–34):

$$\mathscr{E} = m_0 c^2 \left(1 - \frac{v^2}{c^2} \right)^{-1/2} \equiv m_0 \gamma(v)\, c^2 \tag{6–161}$$

(iii) Conservation of momentum along the direction of the incoming photon yields

$$\frac{h\nu}{c} = \frac{h\nu'}{c} \cos\theta + m_0 \gamma(v)\, v \cos\phi \tag{6–162}$$

(iv) The corresponding expression for the transverse momentum is

$$0 = \frac{h\nu'}{c} \sin\theta - m_0 \gamma(v)\, v \sin\phi \tag{6–163}$$

We now have four equations in four unknowns, v, v', θ, and ϕ.

PROBLEM 6–16. Show that these four equations can be solved to give the expression

$$\frac{c}{h}\left(\frac{1}{\nu'} - \frac{1}{\nu} \right) = \frac{1 - \cos\theta}{m_0 c} \tag{6–164}$$

By taking the wavelength of radiation to be $\lambda = c/\nu$, $\lambda' = c/\nu'$, we obtain

$$\lambda' - \lambda = 2\lambda_c \sin^2 \frac{\theta}{2} \tag{6–165}$$

Fig. 6.14 Compton scattering.

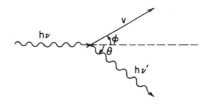

where

$$\lambda_c \equiv \frac{h}{m_0 c} \qquad (6\text{–}166)$$

is called the *Compton wavelength* of the particle. For an electron $\lambda_c = 2.4 \times 10^{-2}$ Å or 2.4×10^{-10} cm. We note that for visible light, the change in wavelength amounts to only ~ 0.05 Å in 5000 Å, a nearly negligible effect. This is why momentum transfer could be neglected in Thomson scattering. However, in the X-ray region, say, at wavelengths of 0.5 Å, we encounter 10% effects; and at higher energies very large shifts can be expected, $(\lambda' - \lambda)/\lambda \gg 1$.

The cross section for Compton scattering must be computed quantum mechanically and turns out to be dependent on the energy of the incoming photon. The expression for this cross section (see Fig. 6.15), known as the *Klein-Nishina formula*, is

$$\sigma_c = 2\pi r_e^2 \left\{ \frac{1+\alpha}{\alpha^2} \left[\frac{2(1+\alpha)}{1+2\alpha} - \frac{1}{\alpha} \ln(1+2\alpha) \right] + \frac{1}{2\alpha} \ln(1+2\alpha) - \frac{1+3\alpha}{(1+2\alpha)^2} \right.$$

$$(6\text{–}167)$$

where r_e is the *classical electron radius* and α is the ratio of photon to electron energy. For an electron

$$r_e \equiv \frac{e^2}{m_0 c^2} = 2.82 \times 10^{-13} \text{ cm}, \qquad \alpha = \frac{h\nu}{m_0 c^2} \qquad (6\text{–}168)$$

In the extreme energy limit this is approximated by

$$\sigma_L = \sigma_e \left\{ 1 - 2\alpha + \frac{26}{5}\alpha^2 + \dots \right\}, \qquad \alpha \ll 1, \qquad \text{low energies} \qquad (6\text{–}169)$$

$$\sigma_H = \frac{3}{8}\sigma_e \frac{1}{\alpha}\left(\ln 2\alpha + \frac{1}{2} \right), \qquad \alpha \gg 1, \qquad \text{high energies} \qquad (6\text{–}170)$$

6:20

where

$$\sigma_e = \frac{8\pi}{3} r_e^2 \qquad (6\text{--}171)$$

is the Thomson scattering cross section.

Fig. 6.15 Comparison of Compton and Thomson scattering cross sections as a function of $\alpha = h\nu/m_0 c^2$ (after Jánossy Já50).

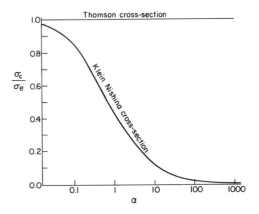

For a proton the cross section would be smaller—inversely proportional to the mass. This means that Compton scattering is primarily an electron scattering phenomenon. Scattering by atoms takes place as though each atom had Z free electrons and the atomic scattering cross section is just Z times greater than that for an individual electron. The atomic binding energy is small compared to the photon energies encountered in Compton scattering and the electrons can be regarded as essentially free.

Let us still turn to the inverse Compton effect. Here we have a highly relativistic electron colliding with a low energy photon and transferring momentum to convert it into a high energy photon. This process can be followed from the point of view of an observer moving with the electron. He will see the incoming radiation blue-shifted (see equation 5–44) to a wavelength

$$\lambda_D = \lambda \sqrt{\frac{c - v}{c + v}} \qquad (6\text{--}172)$$

Still in this frame of reference, the scattered wave will have wavelength (see equation 6–165)

$$\lambda' = 2\lambda_c \sin^2 \frac{\theta}{2} + \lambda_D \qquad (6\text{--}173)$$

since this is a simple Compton process to the observer initially at rest with respect to the electrons. For backscattered radiation $\sin^2\theta/2 = 1$.

Now, when this wave is once again viewed from the stationary reference system—rather than from the viewpoint of the fast electron—a back-scattered photon will be found to have a wavelength

$$\lambda_s \sim \lambda' \sqrt{\frac{c-v}{c+v}} \sim \lambda\left(\frac{c-v}{c+v}\right) + 2\lambda_c\left(\frac{c-v}{c+v}\right)^{1/2} \qquad (6\text{--}174)$$

This is the same transformation as (6–172); a stationary observer also sees back-scattered radiation blue-shifted. Note that we have dropped all directional quantities here and that this expression is therefore only correct in order of magnitude. However, it is quite clear that the wavelength of the photon becomes appreciably shortened in the process and its energy is increased by factors of order

$$\frac{c+v}{c-v} \sim \frac{(1+v/c)^2}{(1-v^2/c^2)} \sim \frac{\mathscr{E}^2}{m_0^2 c^4} \qquad (6\text{--}175)$$

where $\mathscr{E}$ is the initial energy of the particle. As will be discussed in section 9:10, the total power radiated by an electron in inverse Compton scattering is closely related to the power radiated in the form of synchrotron radiation. The total synchrotron emission is proportional to the magnetic field energy density in space, $B^2/8\pi$. The total inverse Compton scattering power loss for electrons is proportional to the electromagnetic radiation energy density in space. The proportionality constant for these two processes is identical.

6:21 SYNCHROTRON EMISSION AND THE INVERSE COMPTON EFFECT ON A COSMIC SCALE

We find many indications that relativistic processes are responsible for much of the observed radiation reaching us from quasi-stellar objects. First we find that those spectral lines that can be observed correspond to gases at temperatures of the order of $10^{6\circ}$K. These lines are very broad and indicate the existence of bulk velocities of the order of 10^3 km/sec. There often is rapid brightening and a subsequent drop of flux, indicating that there are violent events of short duration taking place on a regular basis. Second, we add the facts that a strong radio continuum flux is observed and that the objects have a blue appearance in the visible. All this indicates violent outbursts on a supernova—or possibly far more massive—scale, in which relativistic particles are formed.

Some of the energy goes into producing bulk motion and into ionizing gases.

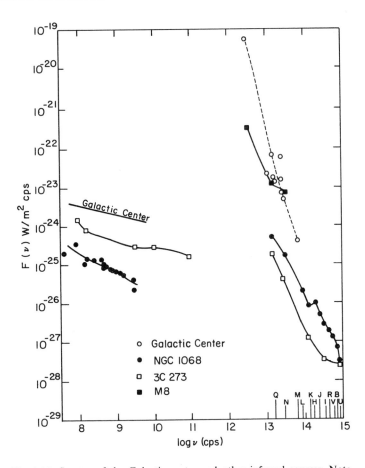

Fig. 6.16 Spectra of the Galactic center and other infrared sources. Note that these spectra run from the radio domain to the ultraviolet. This is an updated (Ha72) version of a plot first presented by Burbidge and Stein (Bu70). (With the permission of the University of Chicago Press.)

Other parts of the energy are to be found in relativistic particles that can give rise to radio and possibly infrared, visible, X-ray, and γ-radiation by means of synchrotron emission or the inverse Compton effect.

For the quasar component 3C273B, for example, we have a spectrum shown in Fig. 6.16. The bulk of the energy given off by the object lies in the far infrared at frequencies around 10^{13} Hz. There are a variety of theories proposing different origins for this flux. The most convincing ones are synchrotron radiation or the inverse Compton effect. The main difficulty with synchrotron emission lies in the fluctuations that are observed. A synchrotron radiating electron has a lifetime (see equation 6–155)

$$\tau \sim \frac{\mathscr{E}}{P(\mathscr{E})} = \frac{3m_0^4 c^5}{2\mathscr{E}e^4 B^2} \tag{6-176}$$

that for a 10^9 ev electron amounts to something of the order of 10^5 y for $B \sim 10^{-3}$ gauss, a field that would be consistent with the spectral curves of Fig. 6.16. But the observed visible fluctuations in radiation intensity are much shorter lived than a year. It is not clear whether the infrared emission also fluctuates; however, if it does, it would seem that the particles that emit radiation must be able to emit the bulk of their energy very rapidly, and the synchrotron emission or inverse Compton effect might do this well.

The inverse Compton effect has also been used to explain the presence of the Galactic gamma ray and X-ray flux and the isotropic X-ray flux that appears to be reaching us from outside the Galaxy. The idea here is that relativistic electrons can be produced readily enough by acceleration in magnetic fields, but there is no well-known way to produce highly energetic photons except by first producing energetic particles that then give up their energy in some exchange process. The inverse Compton effect is the most likely mechanism for this kind of energy conversion. The Galactic component may be due to relativistic electrons colliding with the visible flux emitted at the source of the electrons, say, in a supernova explosion. The extra-galactic flux might be due to the interaction of extra-galactic electrons with the 3°K cosmic blackbody radiation.

Interestingly a maximum brightness temperature of 10^{12}°K can be set on optically thick synchrotron emitting sources. At this brightness the inverse Compton scattering by the radiation emitted within the source quickly reduces the energy of the relativistic electrons and thus reduces the brightness temperature.

6:22 THE CHERENKOV EFFECT

We now come to a process that is primarily important for studying cosmic ray particles—the *Cherenkov effect*. This effect is not so important in the interaction of cosmic rays with other matter in the universe, as it is in the interaction of incoming particles with the earth's atmosphere. The Cherenkov effect causes these particles to decelerate radiatively and the light emitted can be used as a sensitive means for detecting the particles.

To see how the effect works, we consider a highly relativistic particle entering the earth's atmosphere. Because the particle is arriving from a region where the density has been very low and is entering a region of relatively high density, it finds that it has to make some adjustments. The presence of the electrically charged

6:22

particle produces an impulse on atoms it approaches in the upper atmosphere and will cause the atoms to radiate. The impulse comes about because the particle is moving faster than the speed of propagation of radiation in this dense medium. The electric field due to the particle therefore appears to the atoms to be switched on very abruptly; a rapidly time-varying field arises at the position of the perturbed atom. This is just the condition required to cause the atom to radiate. The relativistic particle will continue to affect atoms along its path in this way, until it has slowed down to the local speed of propagation of light. At that time the electric field changes produced in the vicinity of atoms take on a less abrupt character and the radiative effects are diminished.

There are many parallels between Cherenkov radiation and hydrodynamic shocks. Just like a supersonic object that produces sonic booms and keeps on losing energy until it slows down to the local speed of sound, the cosmic ray particle also keeps losing energy, through the Cherenkov effect, until it slows down to the local speed of light in the medium through which it is traveling.

The radiation produced in this manner is emitted into a small forward angle of full width $\Delta\theta$ (see equation 5–50 and Fig. 6.17):

$$\Delta\theta \sim 2\sqrt{1 - \frac{v^2}{c^2}}$$

just as in synchrotron emission or in any other relativistic radiation effect. The time of arrival of the radiation is also dictated by considerations similar to those found in synchrotron radiation. If the layer through which the radiation passes, before it has become sufficiently slowed down, is of thickness d, the time elapsing between the arrival of the first and the last photons of a wave train at the observer is, in analogy to equation 6–147,

$$\Delta t_c \sim \left(\frac{d}{v} - \frac{d}{c}\right) \sim \frac{d}{v}\left(1 - \frac{v}{c}\right) \sim \frac{d}{2c}\left(1 - \frac{v^2}{c^2}\right) \tag{6–177}$$

Fig. 6.17 Diagram to illustrate Cherenkov radiation.

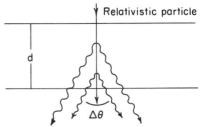

Relativistic particle

so that the corresponding frequency is of the order of

$$\omega_c = \frac{1}{\Delta t_c} \sim \frac{2c}{d}\left(1 - \frac{v^2}{c^2}\right)^{-1} \sim \frac{2c}{d}\left(\frac{\mathscr{E}}{m_0 c^2}\right)^2 \qquad (6\text{--}178)$$

If the distance traversed in the upper atmosphere is of the order of $d \sim 10^6$ cm $= 10$ km, and a proton of energy 3×10^{14} ev is considered, $\mathscr{E}/mc^2 \sim 3 \times 10^5$ and $\omega_c \sim 6 \times 10^{15}$ or $v_c \sim 10^{15}$ cps.

In many cases, an energetic primary produces a shower of secondary particles through collisions with atoms of the upper atmosphere. These secondaries also can give rise to Cherenkov radiation. The Cherenkov radiation spectrum does not normally peak at a frequency near ω_c. It depends more on atomic radiation properties of atmospheric gases and is relatively insensitive to the energy of the primary particle.

One interesting feature of Cherenkov detection is that it not only identifies the existence of cosmic ray particles, but also gives the direction of arrival with reasonable accuracy; the uncertainly $\Delta\theta$ in the direction from which the particles will appear to arrive is quite small.

Gamma rays arriving at the earth can also be detected by Cherenkov radiation if they are sufficiently energetic. The detection process is indirect in this instance, and depends on the formation of very energetic secondary charged particles in the upper atmosphere.

ADDITIONAL PROBLEM 6-17. A rotating mass has energy $I\omega^2/2$, where I is the moment of inertia and ω is the angular frequency. Suppose that the rate of change of energy is proportional to the $n + 1$ power of ω:

$$\dot{\mathscr{E}} = K\omega^{n+1}$$

Show that

$$n = \frac{\ddot{\omega}\omega}{\dot{\omega}^2} \qquad (6\text{--}179)$$

For the Crab nebula pulsars, we observe that $n \sim 2.5$. Does this more nearly match the result expected for magnetic dipole or for gravitational radiation?

ANSWERS TO SELECTED PROBLEMS

$$6\text{-}1. \qquad R_L = \frac{p_c c}{qB} = \frac{m_p v_p c}{qB}$$
$$= 10^8 \text{ cm},$$

$$1 \text{ AU} = 1.5 \times 10^{13} \text{ cm}$$

$$m_p = 1.6 \times 10^{-24} \text{ gm}$$

$$v_p = 10^6 \text{ cm/sec}$$

$$q = 4.8 \times 10^{-10} \text{ esu}$$

$$B = 10^{-6} \text{ gauss.}$$

6–2. For circular motion at the Larmor radius R_L:

$$\mathscr{E} = \frac{p_c v_c}{2} = \frac{q B v_c R_L}{2c} \tag{6-13}$$

$$\dot{\mathscr{E}} = -q\dot{B}\pi R_L^2 \frac{\omega}{2\pi} \tag{6-16}$$

Since the circular path takes the particle $\omega/2\pi$ full turns around the field.

$$\therefore \; \frac{d\mathscr{E}}{\mathscr{E}} = \frac{dB}{B}.$$

Hence the particles have a tenfold increase in energy.

6–3. $\Delta V = 2(7 \text{ km/sec}) \sim 14 \text{ km sec}^{-1}$ is the speed imparted to cosmic rays, measured in their rest frame. $\mathscr{E}_i = \gamma(v_i) m_0 c^2$, the initial energy ($\mathscr{E}_i = 10^{10}$ ev), where

$$\gamma(v_i) = \frac{1}{\sqrt{1 - v_i^2/c^2}}$$

and v_i is the initial velocity. v_f is the final velocity.

For protons say, $m_0 c^2 \sim 10^9$ ev, so that $\gamma(v_i) \sim 10$. If the energy is doubled $\gamma(v_e) \sim 20$, and

$$\left(\frac{v_i}{c}\right)^2 \cong (1 - 0.01) \Rightarrow \frac{v_i}{c} = 1 - 0.005$$

$$\left(\frac{v_f}{c}\right)^2 \cong (1 - 0.0025) \Rightarrow \frac{v_f}{c} = 1 - 0.00125$$

where v_f is the final velocity.

Hence $v_f - v_i = (.0037)c = 1.1 \times 10^3 \text{ km sec}^{-1}$.

By the law of composition of velocities, if v' is the velocity after one bounce:

$$v' \simeq (v_i + \Delta V)\left(1 - \frac{v_i \Delta V}{c^2}\right) \sim v_i - v_i^2 \frac{\Delta V}{c^2} + \Delta V - 0\left[\frac{(\Delta V)^2}{c^2}\right]$$

$$v' - v_i = \Delta V\left(1 - \frac{v_i^2}{c^2}\right) = \frac{\Delta V}{\gamma^2(v_i)}.$$

Now γ changes from 10 to 20. Using an average $\gamma^2 \sim 280$ always, $\Delta V/\gamma^2 \sim 14/280 \sim 0.05$ km sec^{-1}.

Hence the number of collisions is 2×10^4.

At a distance between bounces of 10^{17} cm, and a speed $\sim 3 \times 10^{10}$ cm sec^{-1}, the number of years to double the energy is $[10^{17}/(3 \times 10^7 \times 3 \times 10^{10})] 2 \times 10^4 \sim 2 \times 10^3$ y for protons. During this time the clouds approach and just about touch. Electrons have a higher γ value and are not energized nearly as much during this time.

6–4. $\mathscr{E} = m_0 c^2 \gamma(\omega r), \qquad \gamma(\omega r) = \dfrac{1}{\sqrt{1 - \omega^2 r^2/c^2}}.$

At $\omega r = c$, particles and field can no longer co-rotate.

6–7. Problem 4–3, gives the rms deviation of N steps of length L as $\sqrt{N} L$. In each step, the Faraday rotation angle is given by $\theta(L) = \frac{1}{2}(\omega/c) L \Delta n$. Substitution of (6–66) and the gyrofrequency from (6–62) gives the result.

6–9. If we imagine fictitious charges, by analogy with electric charges

$$H = \frac{Q}{(r - a/2)^2} - \frac{Q}{(r + a/2)^2} = \frac{2Q\mathbf{a}}{r^3} = \frac{2\mathbf{M}}{r^3}.$$

Hence $\mathbf{d}$ and $\mathbf{M}$ are analagous and by substitution we are led to the result (6–89).

6–10. $\mathbf{H} = \dfrac{\mathbf{A} \wedge \mathbf{n}}{c} \qquad \text{and} \qquad \mathbf{A} = \dfrac{1}{6c^2 R_0} \ddot{\mathbf{D}}$

$$\mathbf{H} = \frac{1}{6c^2 R_0} \frac{\ddot{\mathbf{D}} \times \mathbf{n}}{c}$$

$$\mathbf{S} = \frac{c}{4\pi} H^2 \mathbf{n} \propto \frac{(\dddot{\mathbf{D}})^2}{c^5}.$$

6–13. To show this, use the total radiated power (6–86):

$$\frac{2}{3c^3} \ddot{d}^2 = \frac{2e^2}{3c^3} \left(\frac{\dot{p}}{m_0} \right)^2. \qquad \text{(6–13) and (5–40) then give}$$

$$\dot{p} = \frac{eB}{m_0 c} p_c = \left(\frac{eB}{m_0 c} \right) \left(\frac{\mathscr{E}}{c} \right). \qquad \text{Substitute, to get the result (6–155).}$$

6–14. $B = 10^{-4}$ gauss

$$v_c = \frac{\omega_c}{2\pi} = \frac{eB}{2\pi m_0 c} = 300 \text{ Hz}$$

$$\omega_m = \gamma^2 \omega_c = 1200 \text{ MHz for } \gamma = 2 \times 10^3$$

Peak at $\omega = \omega_m/2 = 600$ MHz.

6–15. $I(v) \Delta v \propto \int_0^{\mathscr{E}} P(\mathscr{E}) \, n(\mathscr{E}) \, d\mathscr{E}$ (by 6–156) where

$P(\mathscr{E}) \propto \mathscr{E}^2 B^2$ by (6–155).

$\therefore I(v) \Delta v \propto B^2 \int_0^{\mathscr{E}} \mathscr{E}^2 \mathscr{E}^{-\gamma} \, d\mathscr{E} \propto K B^2 \mathscr{E}^{3-\gamma}.$

But $\mathscr{E} \propto \left[\dfrac{\omega_m}{B} \right]^{1/2}$ by (6–151)

$\therefore I(v) \Delta v \propto B^2 B^{(\gamma-3)/2} \omega_m^{(3-\gamma)/2}$

$I(v) \propto B^{(\gamma+1)/2} \, v \left(\dfrac{1 - \gamma}{2} \right).$

6–16. Squaring (6–162) and (6–163) and adding gives

$- 2h^2 vv' \cos \theta + h^2 (v^2 + v'^2) = m_0^2 v^2 \gamma^2 c^2 = m_0^2 c^4 (\gamma^2 - 1).$

Squaring (6–160) gives $m_0^2 c^4 (\gamma^2 - 1) = h^2 (v^2 + v'^2 - 2vv') + 2hm_0 c^2 (v - v').$

Equating these two expressions we obtain $hvv'(1 - \cos \theta) = (v - v') \, m_0 c^2$ which is equivalent to (6–164).

6–17. $\mathscr{E} = K\omega^{n+1} = I\omega\dot{\omega}$

$\therefore \dot{\omega} = \dfrac{K}{I} \omega^n, \qquad \ddot{\omega} = \dfrac{K}{I} n\omega^{n-1}\dot{\omega} \qquad \text{and} \qquad \ddot{\omega}\omega = n\dot{\omega}^2.$

Equations 6–88 and 6–96 show that $n + 1 = 4$ for the magnetic dipole, and $n + 1 = 6$ for the gravitational quadrupole radiation. The current data therefore are in closer agreement with a magnetic dipole mechanism. Since pulsars may emit predominantly gravitational radiation when they are first formed, observations leading to a value $n = 5$ could be obtained right after the formation of such an object. This would be interesting because the observations could be made in the radio domain, but would give evidence of gravitational radiation that might be too difficult to detect directly.

7

Quantum Processes in Astrophysics

7:1 ABSORPTION AND EMISSION OF RADIATION BY ATOMIC SYSTEMS

In Chapter 6, we considered a series of processes by means of which radiation could be absorbed or emitted by particles. But we restricted ourselves to situations in which the Maxwell field equations of classical electrodynamics could be applied. These equations break down on the scale of atomic systems. The electron bound to a positively charged nucleus does not lose energy because of its accelerated motion, although the classical theory of radiation predicts that it should. Instead, the ground state of hydrogen, for example, is stable for an indefinitely long period of time. Moreover, when energy is actually radiated away from one of the excited states, and the atomic ground energy state is finally reached, we always find that only discrete amounts of energy have been given off in each transition. Again, this is at variance with classical predictions.

Since the interpretation of astronomical observations depends on an understanding of transitions that occur between atomic levels, we will consider just how they take place, and what can be learned from them.

It is important in this connection that almost everything we know about stars or galaxies is learned through spectroscopic observations. Our ideas of the chemistry of the sun and of the chemical composition of other stars is based entirely on the interpretation of line strengths of different atoms, ions, or molecules. Our understanding of the temperature distribution in the solar corona is based on the strength of transitions observed for several highly ionized atoms, notably iron. Our picture of the distribution of magnetic fields across the surface of the sun is based entirely on the interpretation of the splitting of atomic lines by magnetic fields in the solar surface. What we know about the motion of gases and their temperature at different heights above the solar surface, again, is largely based on spectroscopic information. In this case, small shifts in line positions, and the shape and width of the lines yields much of the information we need. Some idea of the densities of atoms, ions, or electrons at different levels of the solar atmosphere can also be obtained from line width and shape.

Of course, for the sun, some of this information can be obtained by other means, because it is near, and can be clearly resolved. We therefore can determine velocities of gas clouds at the limb by direct observations and, eventually, we may expect to obtain radial velocity measurements from radar observations. Densities presumably can also be probed by radar and much data can be obtained by measuring the dispersion of cosmic radio waves passing near the sun. Thus direct visual observations and radio measurements based on classical theory do go a long way toward clarifying our understanding of motions and densities in the solar atmosphere. However, when it comes to more distant objects like emission line stars, or QSO's, where detailed resolution of the object does not appear possible, much new information can only be gained by an understanding of atomic processes observed using spectroscopic techniques. Of course, the dominance of quantum processes is not complete. For example, the knowledge we have about relativistic particles emitting synchroton radiation in quasars is based on classical theory only. Much of the thermal radio emission from interstellar clouds of plasma can also be understood classically (see section 6:16). However, almost everything else we know about these objects has some connection with the quantum theory of radiation.

In the next few sections we will describe how knowledge of quantum processes permits us to learn a great deal about the physical characteristics of astronomical objects. We will try to understand these processes in terms only of the elementary conditions that lie at the base of quantum theory. In general, this will only yield rough values of the parameters we need to know, but we will nevertheless be able to obtain a valid understanding of the role played by quantum processes in astronomy.

7:2 QUANTIZATION OF ATOMIC SYSTEMS

The classical theories of physics no longer apply on a scale comparable to the size of atoms, and many of our preconceptions have to be changed. However, a number of important features are shared by quantum and classical theory. Thus, in a closed system we find that:

   (a) Mass energy is always conserved.
   (b) Momentum and angular momentum are always conserved.
   (c) Electric charge is always conserved.

7:2

On the atomic scale, these conservation principles are phrased somewhat differently than in a classical situation. However, when these differences are important, we can still be sure that:

(d) As the size of the atomic system grows, the features predicted by quantum theory approach those calculated on the basis of classical physics. This is called the *correspondence principle*.

In contrast to these similarities between classical and quantum behavior, there are three major differences:

(a') *Action*, a quantity that has units of (energy × time) or (momentum × distance), is quantized. The unit of action is $\hbar$. By this we mean that in a bound atomic system action can only change by integral amounts of Planck's constant h divided by 2π; $h/2\pi \equiv \hbar$. This statement has many consequences, some of which will be described in this chapter.

(b') Even if they existed—and they do not—states of an atomic system whose characteristic action differs by an amount less than $\hbar$, cannot be distinguished. This is *Heisenberg's uncertainty principle*.

(c') Two particles having half-integral spin cannot have identical properties in the sense of having identical momentum, position, and spin direction. This is *Pauli's exclusion principle* (see equation 4:11).

The three statements (a'), (b'), and (c') are not axioms of quantum mechanics. Rather, they can be considered as useful rules that emerge from a more complicated theory of quantum mechanics that also makes quantitative predictions about the behavior of electrons, atoms, and nuclei.

The concept of action is not as familiar as the idea of *angular momentum*, which has the same units and is subject to quantization in the same way. We might therefore take a brief look at how angular momentum changes occur in atoms.

In any bound atomic system, a change of angular momentum along any given direction in which we choose to make a measurement, will always have a value $\hbar$. The direction of this angular momentum is important.* We shall therefore talk

---

* For, while the angular momentum along the measured direction can only change in steps whose size is $\hbar$, we have no such definite prescription for the changes in the transverse angular momentum that can simultaneously take place. The uncertainty principle precludes a simultaneous definitive measurement of the longitudinal and transverse angular momentum components. All that we can say is that there exist a number of *selection rules* that specify allowed changes of the *total angular momentum* of the system as will be shown in section 7:7. The rules state that the magnitude of the angular momentum squared J^2 changes by integral amounts of a basic step size $\hbar^2$. These integral amounts depend on the initial value of J characterizing the system, and on the multipole considerations mentioned in section 6:13.

7:2

about *measured angular momentum component* with the understanding that we have a definite direction in mind whenever we make a measurement.

This angular momentum quantization can be understood in more basic terms. All the fundamental particles involved in building up atoms have definite measured *spin* values. For electrons, protons, and neutrons these values are $\pm \hbar/2$. A change from one spin orientation of an atomic electron, over to another orientation, therefore, amounts to a change of one unit of $\hbar$ in the measured angular momentum component. Such a change is readily brought about by the absorption or emission of a photon, because photons have spin angular momentum components $\pm \hbar$ along the direction of their motion. Because of the quantization of photon spin components, any change at all in angular momentum of an atomic system must have a component $\hbar$. For, all the different states of an atom can be reached from any other state through a succession of photon absorption or emission processes, or through a set of spin-flip transitions for the electrons or within the nucleus.

In any case, however, quantization is intrinsic to atoms, even without this argument concerning photons. We can therefore be sure that if an atomic system has a state of zero angular momentum, then all other states must have integral values of the angular momentum component. Similarly, if the lowest angular momentum state has a value $\hbar/2$, then all other states must have half integral values of angular momentum component (see also section 7:7).

This is one way in which the statement (a′), above, provides insight into the structure of quantized systems. In addition, the principle (b′) gives us some general quantitative information. Let us consider the simplest atom, hydrogen, in terms of this principle. The energy of the lowest state can then be estimated directly. For, the smallest possible size of an electrostatically bound atom must be related to the uncertainty in momentum through

$$p^2 r^2 \sim \langle \Delta p^2 \rangle \langle \Delta r^2 \rangle \sim \hbar^2 \qquad (7\text{--}1)$$

Here we have taken the mean squared value of the radial momentum and the radial position as being equal to the uncertainty in these parameters. Through the virial theorem, applied to a system bound by inverse square law forces, we can write the energy of a state either as half the electrostatic potential energy for the interacting proton and electron, or as the negative of the system's kinetic energy (3–83). Thus the lowest energy state is

$$\mathscr{E}_1 = -\frac{Z\,e^2}{2\ r} = -\frac{p^2}{2\mu} \sim -\frac{\hbar^2}{2\mu r^2} \qquad (7\text{--}2)$$

where μ is the reduced mass of the electron. Here we have made use of equation 7–1 at the extreme right side of (7–2). By eliminating r from this equation, we can immediately write

$$\mathscr{E}_1 = -\frac{Z^2 \mu e^4}{2\hbar^2} \tag{7-3}$$

$$r = \frac{\hbar^2}{Z\mu e^2} \tag{7-4}$$

This root-mean-square radius r is called the *Bohr radius* of the atom and $\mathscr{E}_1$ is the atom's *ground state energy*. For hydrogen $Z = 1$ and $\mathscr{E}_1 = -13.6$ ev, $r \sim 5.29 \times 10^{-9}$ cm. We have proceeded here on the assumption that the electrostatic potential confines the electron to a limited volume around the proton, and have derived a solution consistent with the uncertainty principle. Nothing has been assumed about possible orbits that the electron might describe about the proton; and, in fact, the very act of setting the mean squared value of position and momentum equal to the mean squared value of their uncertainties, implied that the electron is to be found in the whole volume, not just in a well-defined orbit having a narrow range of r or p values.

PROBLEM 7-1. If we wanted to distinguish successive states having differing radial positions and momenta, the product pr for these states would have to differ by $\hbar$; otherwise, they would not be distinguishable in Heisenberg's sense. Setting $p_n r_n = n\hbar$, show that

$$\mathscr{E}_n = -Z^2 \frac{\mu e^4}{2n^2} \hbar^2 \tag{7-5}$$

If the nth radial state of the atom has a phase space volume proportional to $4\pi p_n^2 \, \Delta p_n$ and $4\pi r_n^2 \, \Delta r_n$, show that the number of possible states with principal quantum number n is proportional to n^2. We will find a great deal of use for this result!

In order to find the actual number of states corresponding to the quantum number n, we still have to invoke the Pauli exclusion principle, statement (c′) above. We know that the state $n = 1$ corresponds to only one cell in phase space. Accordingly there can only be two states, one in which the nuclear spin and electron spin are parallel and the other in which they are antiparallel. Using the result of Problem 7–1, we then see that the nth radial state comprises $2n^2$ different substates, all having the same energy to the approximation considered here.

We see from this that just the most basic concepts (a′), (b′), and (c′) suffice to tell a great deal about the structure of hydrogen and hydrogenlike atoms such as singly ionized helium, five times ionized carbon, or any other bare nucleus surrounded by only one electron.

We should, however, not pretend that all problems of atomic structure can be handled so simply. We have neglected all relativistic effects and all interactions of the spins of particles that constitute the atom. Equation 7–2, for example, makes use of Newtonian mechanics and electrostatic interactions alone. In dealing with such features, or with the interactions between particles and various types of fields, it is important to make use of the full mathematical structure provided by quantum mechanics. At the basis of any such structure, however, are the elementary principles (a) to (d) and (a') to (c'), and we shall make much use of them in the next few sections.

PROBLEM 7–2. We can show that the principles (a') to (c') also permit a determination of the size of the atomic nucleus. To see this consider the *nucleons*, that is, protons and neutrons to be bound to each other by a short-range attractive potential

$$\mathbb{V} = -\mathbb{V}_0 \quad \text{if} \quad r < r_0, \qquad \mathbb{V} = 0 \quad \text{if} \quad r \geqq r_0 \qquad (7–6)$$

Using equation 7–1 show that

$$r \sim \frac{\hbar}{[2M(\mathbb{V}_0 - \mathscr{E}_b)]^{1/2}} \qquad (7–7)$$

Here M is the nucleon mass and $-\mathscr{E}_b$ is the binding energy per nucleon, whose value is roughly 6 Mev. If $\mathbb{V}_0 \sim 2\mathscr{E}_b$ show that a typical nuclear radius is of order 10^{-13} cm. This gives a characteristic interaction cross section of $\sim 10^{-26}$ cm^2 for nucleons. We will find this to be of interest in Chapter 8 where nuclear processes inside stars are described.

We note that this nuclear radius is quite insensitive to the dependence of the potential on distance. The depth of the potential well and the binding energy determine the size of the nucleus.

7:3 ATOMIC HYDROGEN AND HYDROGENLIKE SPECTRA

The considerations of the previous section permit us to discuss some of the main features of atomic hydrogen spectra observed in astronomy. The energy of the observed spectral lines simply represents the difference in energy of the atomic levels between which a transition occurs when a photon is absorbed or emitted.

To start with one of the simplest concepts, we notice that in (7–3) and (7–5) the energy depends on the reduced hydrogenic mass. It therefore has a somewhat different value for normal hydrogen, which has only a proton in its nucleus, and for deuterium which has a nucleus composed of a neutron and a proton. The extra neutron in deuterium makes the nucleus about twice as massive as in normal hydrogen so that the deuterium reduced mass μ_D has a value

$$\mu_D = \frac{m_e m_D}{m_e + m_D} \sim \frac{2m_e m_P}{m_e + 2m_P} \sim m_e \left(1 - \frac{m_e}{2m_P} \right) \tag{7-8}$$

while

$$\mu_P \sim m_e \left(1 - \frac{m_e}{m_P} \right) \tag{7-9}$$

Subscripts e, D, and P, here represent electrons, deuterons, and protons. Corresponding to normal hydrogenic spectral lines, we would therefore expect to see deuterium lines whose energy was greater by about one part in $2m_P/m_e \sim 3700$. In the visible part of the spectrum this corresponds to a line shift of the order of 1.5 Å toward shorter wavelengths. Although such a spectral shift could be easily determined, it is interesting that until recently, despite much searching, no deuterium was ever detected in any astronomical object, anywhere.* In contrast, the terrestrial abundance of deuterium is readily measured and is roughly 2×10^{-4} by fraction of atoms. If such abundances existed elsewhere in the universe, deuterium should have been detected long ago, since many searches have been made. Its absence represents a real puzzle in cosmogony. Somehow, during the birth of the solar system, deuterium, which is normally destroyed soon after a star is born (Chapter 8), may have been formed. Alternately, deuterium might be primordial and any *deuterium* that has ever been cycled into a star has been destroyed. If all interstellar matter in our galaxy had undergone such cycling at least once, planets formed early in the Galaxy's life might be the only places where deuterium is now to be found!

The reduced mass is also helpful in differentiating between hydrogenic transitions and the spectral lines of ionized helium. Singly ionized helium He II has one electron surrounding a nucleus with charge $Z = 2$. According to equation 7–5 the energy of any given state should therefore be just four times as great as the corresponding hydrogenic energy. This integral relation would sometimes lead to an exact identity of line energies for transitions involving *principal quantum numbers* n that were twice as great for helium as for hydrogen. The difference in reduced mass, however, shifts these lines sufficiently, so that ambiguities often

---

* Recently deuterated methane CH_3D has been detected on Jupiter. Thus far, however, the observed infrared spectrum has not yielded a quantitative estimate of the deuterium abundance (Be 72). Even more recently (Je 73, Wi 73) the molecule DCN has been found in rather high abundance in the Orion Nebula. Its abundance may only be 170 times less than that of HCN. (Footnote added in proof.)

can be avoided in astronomical observations. When the Doppler line shift for a moving source is not known, identification on the basis of one or two lines may, however, not be possible, and a search may have to be made for lines of other well-known atoms or for helium lines that are not common to the hydrogen spectrum, that is, transitions involving one level with an odd principal quantum number, and one level with an even value of n.

Although we have presented this similarity of spectra as though it were a matter of difficulty, it is in fact often a great help. After years of theoretical work that explains many fine details, we understand the hydrogen spectrum well. Whenever it becomes possible to relate properties of complicated atoms to specific similar properties of hydrogen, a whole body of theoretical knowledge therefore becomes available at once, and this often leads to a better understanding of the more complex system.

Until a few years ago interest in hydrogenic spectra centered largely around transitions in which at least one of the states had a low principal quantum number, say, $n \lesssim 5$. Not much thought was given to very high lying states, and transitions to be found between such states were always thought to give rise only to very weak spectral lines. It therefore came as a surprise that transitions involving states with $n = 90, 104, 159, 166$, and many others in this same range were observable in radio astronomy (Hö 65). Not only these, but also the corresponding ionized helium states could be identified and again distinguished on the basis of reduced mass differences. These lines have permitted the observation of ionized hydrogen regions over great distances in the Galaxy. Ionized regions that are not detectable in the photographic range now are readily accessible because radio waves are transmitted through the dust clouds that extinguish visible light from all but the nearest portions of the Galaxy. Often these regions had been previously known, because dense ionized plasmas emit readily measured thermal continuum radiation (section 6:16). However, the discovery of the line radiation allowed us to deduce the radial velocity of the region and its distance within the galaxy calculated on the basis of differential rotation models (3:12).

For completeness we should still present some of the terminology often used in discussions of the hydrogen spectrum. It is useful to know that transitions involving lower states $n = 1, 2, 3$, and 4, respectively, are members of the *Lyman, Balmer, Paschen*, and *Bracket* spectral series (See Fig. 7.1). The line with the longest wavelength in each of these spectral series is termed α; the second line is called β, and so on. Thus the transition $n = 4 \rightarrow n = 2$ gives rise to the Balmer-β line in emission. Members of the Balmer spectrum sometimes are written out as Hα, Hβ, and so on. Lyman spectral lines are written as Lα, Lβ, ... or Ly-α, Ly-β ....

In Problem 7–1, we had shown that there are $2n^2$ quantum states to be associated with the nth energy level of the hydrogen atom. We must still see how these states are distinguishable from each other.

7:3

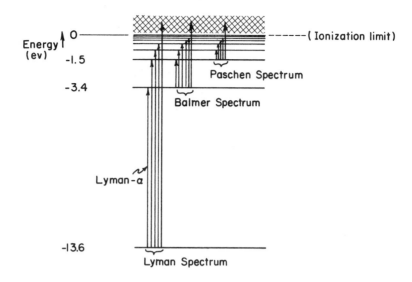

Fig. 7.1   Energy level diagram of atomic hydrogen.

The lowest or ground state actually consists of two distinct components, corresponding to the two differing orientations of the electron's spin relative to the nuclear spin direction. These two configurations have slightly differing energies and hence a transition from the higher to the lower state can occur spontaneously. In radio astronomy this transition has played a leading role. It occurs at 1420 MHz, a frequency corresponding to an energy difference of $\sim 6 \times 10^{-6}$ ev or less than one part in two million of the binding energy of the atom in its ground state ~ 13.6 ev. The distribution of hydrogen in our galaxy was first mapped by means of 1420 MHz observations. This was possible because, as already stated, radio waves are not absorbed by the dust that extinguishes visible light. It is interesting that we now have rather good maps showing the distribution of gas in the Galaxy, but no comparable map showing the distribution of stars. This comes about only because stars do not emit sufficiently great amounts of radiation in the radio part of the spectrum or in the far infrared.

The energy separation between the lowest level, in which the spins are opposed (total spin angular momentum quantum number $F = 0$) and the state with parallel spin orientation ($F = 1$), is called the *hyperfine splitting* of the ground state (Fig. 7.2). This splitting is present at all levels n, and assures a total multiplicity of $2n^2$ states at any given level.

When we take a look at the first excited state of the hydrogen atom $n = 2$, we encounter two types of sublevels whose energies happen to be close. First,

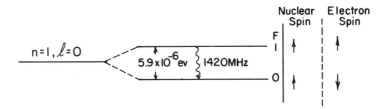

Fig. 7.2   Energy level diagram for the *hyperfine splitting* of the ground state of the hydrogen atom. The state in which the electron and proton spins are aligned has a slightly higher energy.

just as in the ground state, there are again two hyperfine states in which the electron has zero orbital angular momentum about the nucleus. A transition from this state to the ground state through the emission of a photon is forbidden, because the angular momentum of the atomic system would have to remain unchanged in such a transition; but that is not possible because the photon involved in that transition always carries off angular momentum. In tenuous ionized regions of interstellar space, the lifetime of atoms in such a state of $n = 2$ can therefore be very long. Eventually, the atom can revert to the ground state through the emission of two photons, rather than just one; but the lifetime against this two-photon decay is of the order of 0.1 sec (Sp 51b), in contrast to the usual 10^{-8} sec required for normal allowed transitions. Metastable helium atoms similarly have a two-photon decay time measured as $\sim 2 \times 10^{-2}$ sec (Va 70).

We might still ask whether the angular momentum criterion in such transitions could be satisfied if the electronic transition was accompanied by a spin flip from a parallel to an anti-parallel configuration. The coupling between the electron spin and the electromagnetic radiation is, however, low and does not suffice to make that transition as probable as the two-photon decay.

The second set of levels, within the state $n = 2$, all have an *orbital angular momentum quantum number* $l = 1$, with a corresponding (total angular momentum)^2 of $l(l + 1) \hbar^2 = 2\hbar^2$. The (angular momentum)^2 does not have the value l^2, because, aside from a well-defined angular momentum component about one (arbitrarily) chosen axis, there always remains an uncertain angular momentum about two orthogonal axes, which adds an amount $l\hbar^2$ to the (angular momentum)^2 (see section 7:7). Corresponding to $l = 1$, there are three sublevels, each split into two further, hyperfine states. One of these sublevels has an angular momentum component $\hbar$ along some given direction, the second has a component 0, and the third has a component $-\hbar$ along that direction. These three components are labeled $m = 1$, 0, and -1. The label m is called the *magnetic quantum number* because the states have differing energies when a magnetic field is applied to the atom. In the absence of a magnetic field, there is a splitting of the order of 10^{-5} ev

7:3

Fig. 7.3 Energy level diagram showing the fine structure of the $n = 2$ level of hydrogen. The labeling in the left-hand column has the following significance. The letters S and P denote the total orbital angular momenta 0 and 1, respectively. The right lower index gives the total angular momentum resulting from a vectorial addition of the electron and orbital angular momenta. The left upper index is the *multiplicity* $(2S + 1)$ of the term, where S now is the total electron spin. This two-fold meaning of S sometimes leads to confusion. As an example, the $^2P_{3/2}$ state has $l = 1$; the orbital and electron spins are parallel, giving a total spin 3/2; and, since the spin for a single electron has magnitude 1/2, the left superscript is 2.

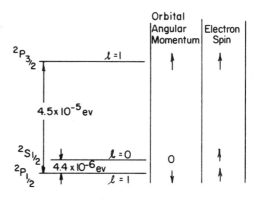

between some of these states. This is called the *fine structure* of the atom and is shown in Fig. 7.3.*

The excited state $n = 3$ again has a hyperfine split sublevel of zero angular momentum, $l = 0$. There are three such pairs of states with $l = 1$, and five pairs with $l = 2$, corresponding to magnetic quantum numbers $m = 2, 1, 0, -1$, and -2. Under normal conditions, these levels are *degenerate*, meaning that they have precisely the same energy. In an applied magnetic field, however, the energy of the states is shifted somewhat and the energy separation between states becomes proportional to the field strength H for low values of H. This splitting is called *Zeeman splitting*.

Zeeman splitting can be understood in the following way. The orbital angular momentum of the electron implies a loop current that has an associated magnetic dipole field. Depending on whether this dipole respectively is aligned along, perpendicular to, or opposed to the field, we have an atomic state with decreased, unaltered, or increased energy.

Quantitatively, the orbital angular momentum of the electron about the nucleus gives rise to a magnetic dipole moment with components along the field direction of

$$\mu_B m_i = \frac{e\hbar}{2mc} m_i, \qquad i = 0, \pm 1, \pm 2, \ldots \qquad (7\text{–}10)$$

where μ_B is called the *Bohr magneton*. The energy of a state in a magnetic field,

---

* For atoms other than hydrogen, the labeling of states does not proceed in this particular way because the spins and orbital angular momenta of the electrons interact through their magnetic moments. However, the enumeration of the different quantum states still proceeds in terms of their distinguishing characteristics — that is, in terms of the Heisenberg and Pauli principles.

is then

$$\mathscr{E} = \mathscr{E}_0 + \mu_B H m_i = \mathscr{E}_0 + \hbar \omega_L m_i, \qquad \omega_L = \frac{eH}{2mc} \qquad (7–11)$$

The state with the smallest energy has its angular momentum antiparallel to the field direction, that is, the configuration in which the quantum number m_i has its lowest value. ω_L is the *Larmor frequency*. ω_L should be compared to the gyrofrequency (6–13) that is twice as large: $\omega_c = 2\omega_L$.

We note that the classical energy of a magnetic dipole in a magnetic field would be $\mathbf{M} \cdot \mathbf{H}$. But when this expression is introduced into equation 6–18, and we seek to find the energy of a magnetic dipole aligned with the field, we obtain

$$\mathscr{E} = \mathbf{M} \cdot \mathbf{H} = \frac{e(\mathbf{v} \wedge \mathbf{r}) \cdot \mathbf{H}}{2c} = \frac{eLH}{2mc} = \frac{\omega_c H}{2} \qquad (6–18a)$$

Here we have made use of equation 6–13 to see the classical energy dependence on ω_c. We can now see why ω_L is only half the gyrofrequency. On the other hand, if we make use of the Larmor frequency in (7–11) we preserve an analogy to photons in that, for $m_i = 1$, the magnetic energy becomes $\mathscr{E} - \mathscr{E}_0 = h\nu_L = \hbar\omega_L$.

Figure 7.4 shows the splitting in the energy levels corresponding to quantum numbers $l = 2$ and $l = 1$ and gives the spectral lines that arise from a transition between such states.

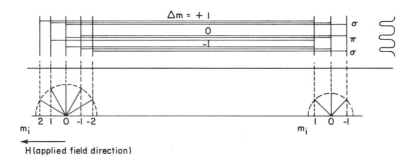

Fig. 7.4 Transitions between energy states shifted through the application of an external magnetic field. The figure shows both the orientation of the angular momentum components relative to the direction of the applied field, and the shifts in the transition energy. We note that the angular momentum components, along the field direction, always have values that are integral multiples of $\hbar$. The total length of the angular momentum vector is $[l(l + 1)]^{1/2}$, that is, nonintegral. σ and π denote the states of polarization of the emitted radiation (see text). A large horizontal distance between levels indicates a large transition energy between two states.

The Zeeman splitting provides us with useful information about the magnetic fields on the solar surface and in distant stars. In some strongly magnetic stars of spectral type A, fields higher than 30,000 gauss have been discovered. The general dipole field for the sun has a value of the order of a gauss, but the local field strength varies greatly. In sunspots, fields of 3×10^3 gauss are not unusual.

The determination of magnetic field strengths from the spectra alone would normally be very difficult, because the lines are very broad, and often overlap since the splitting is small. Fortunately, however, the lines marked σ, in Fig. 7.4 have a different polarization from the line marked π. The *magnetographic method* makes use of this polarization to separate out the different components through use of analyzers sensitive only to light of a given polarization. By carefully measuring the line centers of the variously polarized components, we can then obtain the energy splitting between states, even when the lines are strongly broadened through disturbing effects. For solar work, spectral lines of iron or chromium are often used. The energy splitting gives H directly through equation 7–11.

Interstellar magnetic fields have been measured in a similar way by means of radio observations. The principle of this technique is identical to that used in solar work.

Since a considerable number of theories depend on the existence of an interstellar magnetic field, the only direct way of measuring it is presented here in some detail.

In neutral hydrogen regions of interstellar space, the 21 cm line of atomic hydrogen can be used to determine the presence of magnetic fields. In such a field, the energy levels are split and three different transition lines are expected, corresponding to $\Delta m = 0, \pm 1$. Viewed along the direction of the magnetic field, only two lines appear, respectively, at frequencies $v = \omega/2\pi$ (see equation 7–11)

$$v = v_0 \pm \frac{eH}{4\pi mc}, \qquad v_0 = 1420 \, \text{MHz} \tag{7–12}$$

These two components are circularly polarized in opposite senses (Fig. 7.5). They are called the σ-components (Fig. 7.4) and appear linearly polarized when viewed normal to the field direction; the direction of polarization is at right angles to the direction of the field.

There is also an unshifted component, the π-component, which appears at frequency v_0 when viewed normal to the direction of the field; it is linearly polarized with the direction of polarization parallel to the field. Viewed along the field lines, this unshifted component does not appear at all.

The observation of Zeeman splitting is made difficult by the rapid motion of the interstellar gas atoms. This produces a Doppler broadening of the lines (section 7:6). A 1 km sec^{-1} random motion gives a frequency shift of order

7:3

Fig. 7.5 Zeeman shift of the positions of the two circularly polarized components of the 1420 MHz hydrogen line viewed along the direction of the magnetic field.

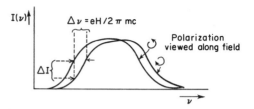

$\delta v/v \sim 3 \times 10^{-6}$. At the 21-cm line frequency of 1420 MHz, this corresponds roughly to 4×10^3 Hz. In contrast, the frequency split Δv due to the magnetic field is $2.8 \times 10^6 H$ Hz between the two σ-components. This means that a field of order 10^{-5} gauss, only gives a split of $\Delta v \sim 30$ Hz.

Normally such a splitting would be all but impossible to observe in the presence of the overriding Doppler broadening. As already indicated, however, the saving feature in the situation is the difference in polarization. By working at the edge of the line, where the slope is steep, the difference in intensity ΔI of the two polarized components can be accentuated (Fig. 7.5). This technique has established the existence of fields as high as $\sim 5 \times 10^{-5}$ gauss, at least in some dense clouds in Orion (Ve 69). In other regions the field is much lower than 10^{-5} gauss (Ve 70).

7:4 SPECTRA OF IONIZED HYDROGEN

(a) Positive Ions

Hydrogen can become ionized through the absorption of a photon whose energy is 13.6 ev or higher. Once the minimum energy required to loosen the electron from the proton is reached, the excess energy can always be absorbed in the form of translational kinetic energy of the electron and proton. This feature is important in determining the appearance of very hot stars. We never observe ultraviolet photons beyond the ionization limit—the *Lyman limit*. Whatever photons of this kind are emitted are immediately absorbed in the gas surrounding the star; and if there is not enough gas there, then the absorption will certainly take place in interstellar space, between the star and the earth.

We might think that the recombination of electrons with protons would then always regenerate the ultraviolet photons. However, this occurs only part of the time. Frequently the recombination leaves the atom in one of its excited states, with a subsequent cascade through lower excited states down to the ground state. In this process, a number of less energetic photons are created. If this type of energy degradation does not occur at the first recombination, it normally will take place on a later occasion. There will be ample opportunity, since the mean free path for ionizing radiation is very short, so that probability of ionizing an atom is extremely large. At energies somewhat higher than the ionization limit, the absorption cross section is of order 10^{-15} cm^2. Hence, even if the interstellar density is only 10^{-1} atoms cm^{-3}, the mean free path for absorption is only of order 10^{16} cm, in contrast to typical interstellar distances of order 10^{18} cm or more. In a random walk, an ionizing photon would then have 10^4 opportunities for ionization and recombination in crossing 10^{18} cm. The probability that an ionizing photon would penetrate the full 10^{18} cm without ever ionizing a single hydrogen atom is of order e^{-100}.

(b) Negative Ions

A hydrogen atom can become ionized not only through the loss of an electron, but also by gaining one to become the negative ion H$^-$, the hydride ion.

$$H + e \rightarrow H^- + h\nu \tag{7–13}$$

The structure of this ion is somewhat similar to that of the neutral helium atom in having two electrons bound to a nucleus. The second electron is only weakly bound, since the first electron is quite effective in screening out the nuclear charge. The binding energy is 0.75 ev and there is only one bound state. All transitions from or to this state therefore involve the continuum, that is, a neutral hydrogen atom and a free electron (Fig. 7.6).

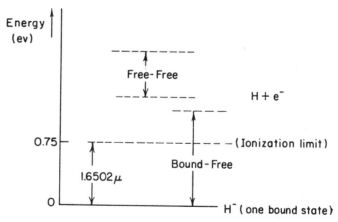

Fig. 7.6 Energy level diagram of the H$^-$ ion. There is only one bound state with binding energy 0.75 ev. All radiative transitions must be either between the bound state and a state with a free electron, or else between two free states. (A wavelength of 1 μ is 10^{-4} cm.)

Because the 0.75 ev binding energy is so low, visible starlight, from cool stars like the sun, can be absorbed. This absorption continues out to 1.65μ wavelength, where the absorption due to bound-free transitions no longer can occur. However, at longer wavelengths, absorption can take place, since the H^- ion also can have free-free transitions—absorption in which energy is taken up by the translational energy of an unbound electron in the presence of a hydrogen atom. This is no small effect. The H^- ion plays an important role in the transport of energy through the solar atmosphere (Ch 58)!

We should still explain how H^- even happens to exist in the atmosphere of cool stars. This comes about because metal atoms such as sodium, calcium, magnesium, which have low ionization potentials, can be easily ionized even by the light of cool stars. Some of the electrons generated in this way attach themselves to hydrogen atoms, thus forming the H^- ions.

Of course, the absorption due to H^- is always accompanied by subsequent reemission. It is interesting, therefore, that most of the light we receive from the sun is due to a continuum transition in which atoms and electrons recombine to form the hydride, that is, the H^- ion!

Many other elements of course also have ions that play an important role in astrophysics. Their physical properties are often similar to those exemplified by hydrogen.

Doubly ionized helium has physical properties very much the same as singly ionized hydrogen, except that the larger nuclear charge makes a difference in the transition energies involved. Singly ionized helium mimics many of the spectroscopic properties of the hydrogen atom, as already mentioned (7:3). But in a highly ionized medium, in which free electrons are abundant, such helium ions also play a role similar to that of the protons, because both types of particles are singly charged.

Molecules can of course be ionized too and molecular ions have characteristic spectra of their own.

7:5 HYDROGEN MOLECULES

In general, molecules can have three types of quantized states. First, there is the possibility that atoms in the molecule vibrate relative to one another. In that case the *vibrational energies* are quantized. Second, there is the possibility of quantized rotation. This means that the angular momentum is quantized. Third, just as in atoms, there exist different quantized electronic states.

The binding energy between atoms is relatively weak. By this we mean that the energy required to separate two atoms that have formed a molecule is normally

Fig. 7.7 Vibrational levels in the hydrogen molecule H_2 ground electronic state. The lengths and positions of the lines along the abscissa indicate the range of separations d between nuclei in the molecule. The equilibrium separation is denoted by d_e. There are only 14 vibrational states below the energy continuum. Each of these states can be split into a number of substates corresponding to different amounts of angular momentum.

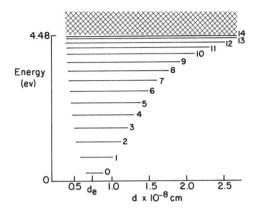

smaller than the energy required to ionize an atom. Only some large alkali atoms have ionizing energies lower than the highest molecular binding energies. Correspondingly also, the radiative transitions between excited vibrational states tend to occur at lower energies than those found for the transitions involving the lower energy levels of most atoms. Characteristically, transitions involving electronic transitions, that is, transitions between electronic excited states, in atoms or molecules, occur in the visible and ultraviolet part of the spectrum. Vibrational transitions occur in the near infrared part of the spectrum, roughly at wavelengths between 1 and $20\,\mu$, and rotational transitions occur in the far infrared $\lambda \gtrsim 20\mu$ and microwave spectrum.

Of course this is just a rule of thumb and is not strictly obeyed. We already know that hydrogen atoms can have electronic transitions that reach way out to the very lowest energies associated with radio wavelengths. Those were discussed in section 7:4. It is true, however, that pure vibrational transitions do not often occur in the visible spectra of astrophysically important substances. And pure rotational spectra are not expected to occur at wavelengths shorter than a few microns.

For many purposes, the vibrations of two atoms relative to one another can be treated as harmonic oscillations. These vibrations are quantized. Figure 7.7 shows the energy levels for hydrogen H_2 in the lowest electronic state of the molecule. If the vibrations become too violent, the molecule dissociates into two separate atoms. The dissociation energy is 4.48 ev. This corresponds to an energy just higher than the 14th excited vibrational state.

The ground state of the molecule is not at zero energy. Rather, as already encountered in the case of photons (section 4:13), there is a characteristic shift from zero equal to roughly half the energy difference between ground state and first excited state. This displacement is characteristic of all vibrational effects.

7:5

Hydrogen molecules may be a major constituent of interstellar space and may be the predominant constituent of dark clouds in the galactic plane. Unfortunately we know none of this for sure!

First, there is the difficulty of detecting the gas: Hydrogen is a symmetric dipole molecule, and as we already saw in section 6:13, symmetric configurations can at best radiate if they have a quadrupole moment. Hydrogen does have such a moment, but the transition probability for this type of transition is many orders of magnitude less than for the more usual dipole radiation of unsymmetric molecules. For this reason molecular hydrogen is hard to observe. This is true both of the vibration and of the rotation spectrum.

The vibrational spectrum of H_2 has now been observed in the atmosphere of Jupiter, where the optical depth is very great. There remains, however, the important question about whether the dark regions of interstellar space contain molecules of hydrogen. Some H_2 has been detected through absorption of light in an allowed electronic transition (Ca 70) but we do not know how abundant H_2 actually is in the Galaxy.

Normally, we would expect to find large amounts of this gas, since it is stable at low temperatures. The problem, however, is that hydrogen is readily dissociated, even by ultraviolet photons that are not energetic enough to ionize hydrogen atoms, and that therefore pass through unionized atomic gas without much hindrance. Such radiation may, however, still be absorbed by interstellar grains. Inside a dark cloud where sufficient grain absorption could shield against ultraviolet radiation, hydrogen molecules may be abundant.

This makes H_2 hard to detect, since direct visual access is ruled out and no radio transitions of neutral molecular hydrogen are known. The vibrational and rotational spectra, again, are hard to observe because of the weak quadrupole emission and absorption. Molecular hydrogen has a rotational spectrum that is largely concentrated in the far infrared. The transition from the first excited state down to the lowest (see Fig. 7.8) corresponds to a wavelength of 84 microns, or 0.084 mm. Observations from above the atmosphere would be needed to detect this transition, since the atmosphere strongly absorbs in this spectral domain (He 67).*

PROBLEM 7–3.   If a system of particles has a *moment of inertia*

$$I = \sum_j m_j r_j^2 \tag{7-14}$$

about its axis of rotation, show that the energy and angular momentum are related by

$$\mathscr{E} = \frac{\omega L}{2} = \frac{L^2}{2I} \tag{7-15}$$

7:5

Fig. 7.8 Rotational energy level diagram of H_2 for the ground vibrational state. Alternate states have nuclear spins aligned: parallel, ortho (0), or antiparallel, para (P).

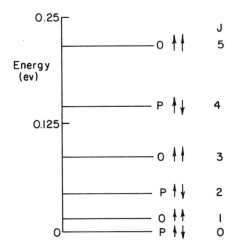

provided we are talking about a classical, rigid, nonrelativistic rotator. ω is the angular frequency of rotation. From this, show that the energy carried away by each quantum of radiation would have to be

$$\delta\mathscr{E} = \hbar\omega \qquad (7\text{--}16)$$

due to quantization $\hbar$ of angular momentum. In quantum theory where the total angular momentum is given (see section 7:7), by $\hbar[J(J + 1)]^{1/2}$, show that the energy of each state is

$$\mathscr{E} = \hbar^2 \frac{J(J + 1)}{2I} \qquad (7\text{--}17)$$

and the energy of the quantum released in the transition $J \rightarrow J - 1$ is

$$\delta\mathscr{E} = \hbar^2 \frac{J}{I} \qquad (7\text{--}18)$$

For massive objects rotating rapidly, show that this is equivalent to the classical formula.

PROBLEM 7–4. If an interstellar molecule has a rotational energy kT in thermal equilibrium with surrounding gas (section 4:18), say $T \sim 100°K$, what is the range of frequencies it will radiate if typical values of atomic weights are 10^{-23} g and typical molecular radii are 2 Å. Make use of some of the expressions obtained in the previous problem to show that radiation may be expected in the far infrared or submillimeter region.

PROBLEM 7-5. Set up an expression for the probability that a molecule will be found in a rotational state J, in a molecule with moment of inertia I about a given spin axis, when it is in thermal equilibrium with a gas at temperature T. From this, convince yourself that a molecule in a cool interstellar cloud ($T \sim 100°K$) cannot be excited into very high rotational states.

PROBLEM 7-6. If interstellar grains are in thermal equilibrium with the surrounding gas ($T \sim 100°K$), and have typical radii, 10^{-5} cm with typical masses 10^{-15} g, at what frequency could they be expected to radiate away angular momentum? This process is, in fact, not impossible because small inhomogeneous grains can be expected to have appreciable electric dipole moments. Observations would, however, have to be made from above the atmosphere, and terrestrial broadcast bands as well as interstellar plasma absorption might interfere.

PROBLEM 7-7. It has been suggested that massive cosmic objects might exist having such high angular momenta, that contraction to high density becomes impossible. In such a situation the object might slowly cool down without ever becoming a star because its central temperature does not get high enough to sustain nuclear reactions.

Such an object might, however, be capable of losing angular momentum through systematic emission of circularly polarized radiation.

An object like this might also emit gravitational radiation. Since the graviton carries away twice the angular momentum of photons, see how the formulas of Problem 7–3 would change if applied to gravitons. Try the same thing for neutrinos that carry away only half the angular momentum of photons. Under normal conditions, in any of these cases, the probability for the emission of a quantum is quite small as the discussion of transition probabilities in the next sections will show. If electromagnetic radiation were in fact emitted, it would not be transmitted by the interstellar medium, because the expected rotational frequency ω for massive objects would be small. Why not? What would happen to the energy? (Re 71)

7:6 THE INFORMATION CONTAINED IN SPECTRAL LINES

An excited atomic system, left to itself, will spontaneously jump into a state of lower energy. The mean time required before such a transition occurs varies from one specific situation to another, and depends on such factos as the symmetries of the states, the dimensions of the system, and so on (see section 7:7). If the system

7:6

has a total probability P of leaving the excited state in any unit time interval, then its total life in the excited state is $\delta t = 1/P$. The energy of the state therefore cannot be determined with arbitrarily great accuracy. The limited time in the excited state implies that the energy can only be determined to an accuracy $\delta\mathscr{E}$ given by the uncertainty principle

$$\delta\mathscr{E} = \frac{\hbar}{\delta t} \qquad (7\text{-}19)$$

Improbable transitions therefore have a narrow natural line width. The accuracy to which the transition energy between two states i and k can be determined depends on the lifetime both of the upper and the lower states; the total frequency width of the line δv, then, is the sum of the widths of the two levels. This total width is usually denoted by

$$\delta\omega = \frac{(\delta\mathscr{E}_i + \delta\mathscr{E}_k)}{\hbar} = \gamma = 2\pi\delta v \qquad (7\text{-}20)$$

γ is called the *natural line width* for the transition.

We will show in section 7:8 that the spontaneously decaying atom emits radiation with a spectral distribution or *line shape*

$$I(\omega) = I_0 \frac{\gamma}{2\pi} \frac{1}{(\omega - \omega_0)^2 + \gamma^2/4} \qquad (7\text{-}21)$$

The intensity drops to half the maximum value at frequencies $\omega_0 \pm \gamma/2$.

In astronomical situations the natural line width is seldom directly observed; but deviations from this width can give us a great deal of information, and it is useful to list the various line broadening effects:

(a) Doppler Broadening

This effect is of random motions of the atoms or molecules whose emission is observed. For small velocities the frequency shift of the radiation is roughly proportional to the line of sight velocity component v_r (see section 5–44):

$$\Delta\omega = \omega_0 \frac{v_r}{c}, \qquad v_r \ll c \qquad (7\text{-}22)$$

There are two kinds of motion that can contribute to the Doppler broadening—the thermal velocity of the emitting atoms within a cloud, and the turbulent velocities peculiar to the clouds that are superposed along a line of sight. Sometimes these two effects can be resolved.

We can, for example, observe the absorption of stellar radiation by interstellar sodium atoms. Sodium absorbs very strongly in the yellow part of the visible spectrum, in a pair of lines known as the sodium D lines at 5890 and 5896 Å. If these lines are examined at very high resolution, so that shifts of the order of one part in 10^6 are detected, then measurements of velocities as low as $\sim 3 \times 10^4$ cm sec^{-1} can be distinguished. What we then observe is a series of discrete broadened absorption lines due to individual clouds absorbing light along the line of sight, and a definite characterizing width for individual lines representing a given cloud. Whether this individual line width is entirely due to thermal motions, or also partially due to turbulent motions on a smaller scale within each cloud, cannot immediately be recognized (Ho 69).

PROBLEM 7-8. If atoms of mass m move with velocities determined by Maxwell-Boltzmann statistics, the probability of observing a given line of sight velocity v_r is proportional to exp $(- v_r^2 \, m/2kT)$ (see equation 4–56). Show that the line shape for thermal Doppler broadening therefore should be of the form

$$I(\omega) \, d\omega = I_0 \exp - \left[\frac{mc^2 (\Delta\omega)^2}{2\omega_0^2 kT} \right] d\omega \qquad (7\text{--}23)$$

See also Problem 4–27 and show how the line width can immediately be related to the temperature, through equation 7–22.

In general the width at half maximum, for Doppler broadening

$$\delta = \omega_0 \left[\frac{2kT (\ln 2)}{mc^2} \right]^{1/2} \qquad (7\text{--}24)$$

is much greater than the natural line width γ. However, it drops off exponentially and, therefore, much faster than the natural line width. The observed wings of very strong lines, for example *Lyman-α* in interstellar space, therefore generally are due to natural line width and to the other causes listed below (Je 69).

(b) Collisional Broadening

In the relatively dense atmospheres of stars, atoms or ions often suffer a collision while they are in an excited state. Because any given collision may induce a transition to a lower state, the collective effect of such collisions is to increase the total transition probability. Thus, if the spontaneous transition rate were γ and the number of collision induced transitions in unit time is Γ

$$\frac{\delta\mathscr{E}}{\hbar} = \gamma + \Gamma \qquad (7\text{--}25)$$

The emitted line has the spectral intensity distribution of the natural line shape, except that γ is replaced by $\gamma + \Gamma$.

(c) Other Types of Broadening

There are other effects due to interactions with neighboring atoms that can cause shifting and splitting of states through the influence of electric fields (Stark effect), through resonance coupling between atomic systems, and so forth. These processes all lead to line broadening, but at low densities their effects are small.

PROBLEM 7-9. To obtain a better feel for the relative importance of the Doppler and collision line widths for visual spectra obtained from stellar atmospheres, show that: If the atmosphere has a density n, thermal velocities $(3kT/M)^{1/2}$, and collision cross section σ,

$$\frac{\Gamma}{\delta} \sim \frac{n\sigma c}{\omega_0} \tag{7-26}$$

where factors of order unity are neglected. For visible light $\omega_0 \sim 10^{15}$ and collision cross sections have typical dimensions of atoms, 10^{-16} cm^2. $n \ll 10^{21}$ cm^{-3} in normal stellar photospheres and, hence, $\Gamma \ll \delta$.

7:7 SELECTION RULES

In section 7:2 we had indicated that the interaction of an atomic system with radiation obeys certain conservation principles. Compliance with these principles requires that certain transitions be forbidden, while others are allowed. The rules that tell us about the permitted transitions are called *selection rules*. We had already seen roughly how these rules come about. Here we will examine the question in somewhat geater depth.

When any two atomic systems combine to form a bigger system, the addition of angular momentum takes place so that, along any arbitrarily chosen direction z, the final angular momentum J_{zf} is the sum of the two initial angular momenta along that direction.

$$J_{zf} = J_{z1} + J_{z2} \tag{7-27}$$

where subscripts 1 and 2 refer to the two individual initial systems.

Because the precise measurement of the z-component precludes a simultaneous precise measurement of the transverse components, equations of the form (7-27) do not exist for the x and y directions. The equation is true whether we interpret the symbols J_{zi} as the z-component of the angular momentum or only as the quantum number for this angular momentum component that, when multiplied by $\hbar$, represents the actual angular momentum component. In what follows below, we will interpret the symbol J_{zi} as a quantum number. The values that these quantum numbers can take on are zero, half integer, or integer.

A second statement can be made about the addition of the squares of the angular momentum values. The (angular momentum)^2 is also a precisely measurable quantity that can, however, be measured simultaneously with J_{zi}. This second statement is somewhat more complicated. Nevertheless, reduced to its essentials it says that the allowed values of (angular momentum)^2 always take on numerical values

$$\text{(angular momentum)}_i^2 = J_i(J_i + 1)\hbar^2, \qquad i = 1, 2, f, \qquad (7\text{–}28)$$

where the relationship between J_i and J_{zi} requires that J_{zi} takes on values

$$J_{zi} = J_i, J_i - 1, J_i - 2, \ldots, 1 - J_i, -J_i \qquad (7\text{–}29)$$

We had said that the z direction can be arbitrarily chosen. Let us choose it to represent the direction along which a photon approaches the atomic system. In the scheme used here, we can assign subscript 1 to the photon, 2 to the initial atomic state, and f to the atomic state after photon absorption. The choice of the z direction then shows that

$$J_{zf} = J_{z2} \pm 1 \qquad (7\text{–}30)$$

since $J_{z1} = 1$. This tells us that an atomic system with half integer values of J_{zi} must have J_i half integer, and that no transitions to integer values of J_{zi} or J_i are possible through photon absorption or emission. Similarly a system with integer angular momentum quantum numbers always maintains these properties under photon absorption.

Let us still try to understand why equation 7–28 takes on the particular form it does. We know that the maximum value that J_{zi} can take on is J_i. For this particular state, equation 7–28 states that there is an additional amount of angular momentum $J_i\hbar^2$ to be associated with the transverse angular momentum components. These components can therefore never be zero, unless J_{zi} itself is zero also and, in that case, they must be zero. The transverse angular momentum contribution comes about because the uncertainty principle does not permit a simultaneous precise measurement of two or more angular momentum components.

We note that the quantum numbers J_{zf} and J_{z2}, alternately, may be taken to correspond to the magnetic quantum numbers we previously labeled m. Equation 7–30 then gives a selection rule which states $\Delta m = \pm 1$ when the direction of photon emission is along the magnetic field lines. This is why only two Zeeman shifted lines are observed along that direction. Along a direction of emission perpendicular to the field, equation 7–30 still is true, but the association of J_{z2} with m is then no longer valid. A division of photons into groups having $J_{z1} = \pm 1$, that is, in terms of left- or right-handed polarized light, would then mix the contributions from different magnetic energy levels m. This would happen because the photons from the various levels m, viewed along that direction, are plane polarized; as already stated in section 6:12, plane polarized light can be considered as a superposition of left- and right-handed polarized components. Hence, viewed along the direction perpendicular to the field, Δm may have values 0, as well as ± 1, even though (7–30) still is obeyed.

Equation 7–30 leads to one other selection rule, which is very important: It is impossible for an atomic system to undergo transitions between angular momentum states whose values are zero through absorption or emission of one photon. It is easy to see that this must be true. If $J_2 = 0$, then $J_{z2} = 0$ also, and similarly if $J_f = 0$, $J_{zf} = 0$. But by (7–30) both these z-components cannot be zero simultaneously. Hence the selection rule as stated must be true. This rule is absolutely inviolable, no matter whether electronic or vibrational transitions are involved. It is always true!

$$J = 0 \nrightarrow J = 0 \tag{7–31}$$

In quantum mechanics, selection rules like these are linked to the symmetry of the atomic system. When the system has sufficiently complex symmetries, the selection rules also become correspondingly complex. We have only shown one or two of the simplest selection relations here, but it is worth remembering that even the more complex appearing rules actually are basic symmetry statements, which become relatively simple when viewed in terms of the appropriate symmetry. The angular momentum selection rules discussed here are based on the rotational symmetry of atomic systems.

Equation 7–31 holds true only for transitions involving a single photon. In the very improbable two photon transitions, it is possible for such a transition to occur, and the angular momentum carried off by the individual photons is then oppositely directed. Such transitions are possible in tenuous nebulae in interstellar space where atoms in an excited state of zero angular momentum can exist undisturbed for long periods of time (Va 70, Sp 51b). In laboratory systems, where pressures are higher, such excited states normally become de-excited through atomic collisions.

An interesting feature of angular momentum quantization is that the existence of quantized states for all matter implies a lack of interactions with radiation having nonquantized angular momentum. Whatever fields, electric, magnetic, weak or strong nuclear, gravitational, or others that may exist in the universe, should therefore have associated radiation that is quantized in terms of half integer spin angular momentum $\hbar$, or multiples thereof. For example, when gravitational waves are discovered and examined, we are confident we will find them to have quantized spin angular momentum. The current prediction is that their spin should be $2\hbar$ twice that of photons (Gu 54).

7:8 ABSORPTION AND EMISSION LINE PROFILE

In estimating the amount of natural interstellar hydrogen along the light path between an ultraviolet emitting star and the earth, we need to know both the shape and the total strength of the Lyman-α absorption line. These two pieces of information will permit us to determine the amount of hydrogen in terms of the observed absorption line width. Intrinsically the calculation of line strength and shape is a quantum mechanical problem. Classical theory, however, permits us to calculate the line shape on the basis of a harmonic oscillator model. The model also yields the right order of magnitude for the line strength. However, we must take care not to take this classical model too seriously, because by itself it does not lead to quantized energy states for atomic systems.

We will first derive an expression for the emission line profile using semi-classical methods.

We start with equation 6–88 for the total energy, $I = 2e^2\ddot{r}^2/3c^3$, radiated by a charged oscillator per second. We can see that this intensity corresponds to a force

$$\mathbf{F} = \frac{2}{3}\frac{e}{c^3}\dddot{\mathbf{d}} \tag{7-32}$$

because the average work done by that force, in unit time, is then

$$\langle \mathbf{F} \cdot \dot{\mathbf{r}} \rangle = \left\langle \frac{2e}{3c^3}\dddot{\mathbf{d}} \cdot \dot{\mathbf{r}} \right\rangle = \left\langle \frac{2}{3c^3}\frac{d}{dt}(\dot{\mathbf{d}} \cdot \ddot{\mathbf{d}}) - \frac{2}{3c^3}\ddot{\mathbf{d}}^2 \right\rangle = I \tag{7-33}$$

Here the term containing $\dot{\mathbf{d}} \cdot \dot{\mathbf{d}}$ vanishes because $\dot{d}$ and $\dddot{d}$ are exactly out of phase in simple harmonic motion. The *damping force F* is small compared to the harmonic force; the oscillation in other words lasts over many cycles. We can therefore write the equation of motion as

$$m\ddot{\mathbf{r}} = -m\omega_0^2\mathbf{r} + \frac{2}{3}\frac{e^2}{c^3}\dddot{\mathbf{r}} \tag{7-34}$$

This equation is very much in the spirit of (6–107), except that we have a damping (instead of a harmonic driving) force here. Since the damping is weak, the motion is almost harmonic and we can approximate

$$\dddot{\mathbf{r}} = -\omega_0^2\dot{\mathbf{r}} \tag{7-35}$$

We then rewrite (7–34) as

$$\ddot{\mathbf{r}} = -\omega_0^2 \mathbf{r} - \gamma \dot{\mathbf{r}} \quad \text{with} \quad \gamma = \frac{2}{3} \frac{e^2 \omega_0^2}{mc^3} \ll \omega_0 \tag{7–36}$$

that has the approximate solution:

$$\mathbf{r} = \mathbf{r}_0 e^{-\gamma t/2} e^{-i\omega_0 t} \tag{7–37}$$

since $\gamma \ll \omega_0$.

The oscillating dipole thus sets up an oscillatory field of the form

$$\mathbf{E}(t) = \mathbf{E}_0 e^{-\gamma t/2} e^{-i\omega_0 t} \quad \text{(RP)} \tag{7–38}$$

where only the real part enters into physical consideration. This is not monochromatic anymore because it changes with time; and only time invariant oscillating fields can be strictly monochromatic. A time dependent change in intensity affects the frequency spectrum. The total field is now written in terms of an integral over the entire range of frequency components:

$$\mathbf{E}(t) = \int_{-\infty}^{\infty} E(\omega) e^{-i\omega t} d\omega \tag{7–39}$$

By a theorem from *Fourier theory*, an integral of this form can be inverted to give

$$E(\omega) = \frac{1}{2\pi} \int_{-\infty}^{\infty} E(t) e^{i\omega t} dt \tag{7–40}$$

If we now introduce the field (7–38) into this equation and note that $E(t)$ is defined only for time $t \geq 0$, we can readily integrate to obtain

$$E(\omega) = \frac{1}{2\pi} \frac{E_0}{i(\omega - \omega_0) - \gamma/2} \quad \text{(RP)} \tag{7–41}$$

We then can obtain the spectral line intensity (see Fig. 7.9):

$$I(\omega) = |E(\omega)|^2 = I_0 \frac{\gamma}{2\pi} \frac{1}{(\omega - \omega_0)^2 + \gamma^2/4}$$

$$I(\nu) = I_0 \left(\frac{\Gamma}{2\pi} \right) \left[(\nu - \nu_0)^2 + \frac{\Gamma^2}{4} \right]^{-1}, \quad \gamma \equiv 2\pi\Gamma \tag{7–42}$$

where I_0 is the total intensity integrated over all frequency space and $\omega = 2\pi\nu$, $\omega_0 = 2\pi\nu_0$:

$$I_0 = \int_{-\infty}^{\infty} I(\omega) d\omega = \int_{-\infty}^{\infty} I(\nu) d\nu \tag{7–43}$$

7:8

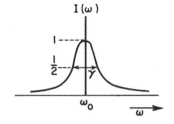

Fig. 7.9   Natural width of a spectral line. The curve is normalized to peak intensity $2I_0/\pi\gamma$ (see equation 7—42).

This type of line shape is sometimes called the *Lorentz profile*, and Γ and γ are called the *natural line width*. This is the full frequency width at half maximum. We have not yet shown that the absorption and emission profiles are the same. That question will, however, be taken up in section 7:9.

7:9 QUANTUM MECHANICAL TRANSITION PROBABILITIES

Much astrophysical information is to be obtained from the intensity of absorption or emission lines. The strengths of the lines can define the density of given atoms, ions, or molecules in a source—or along the line of sight to a source—and the ratio of the strengths of lines can be used to determine the excitation temperature of gases through application of the Saha equation (section 4:16).

However, in order to obtain this information in useful form, we must first relate the intensity of the spectral absorption or emission lines to the number density of atoms or ions in different population levels; and we can only do that if we know the transition probability between states of the system.

Very roughly the transition probability depends on three factors: (a) on the symmetry properties of the atomic system, (b) on its size in relation to the wavelength to be absorbed or emitted, and (c) on the statistics of the radiation field. The first of these factors includes the selection rules discussed in section 7:7, the statements about charge-to-mass ratio made in section 6:13 and other similar restrictions. The second factor represents the relative probability for dipole, quadrupole, and higher multipole radiation that is size dependent as seen, for example, in (6–93). The third factor, to be discussed now, depends only on the radiation field and is quite general for any transition regardless of the atomic or nuclear system involved.

If we want to compute the probability that an atomic system will undergo a transition from some state i to another state j, this probability will be proportional to the number of ways in which a change in the photon field can occur. For example, the probability for emission of a photon with radial frequency ω

is (see equation 4–65a) proportional to

$$\frac{\omega^2 \, d\omega \, d\Omega}{(2\pi c)^3} = \frac{\nu^2 \, d\nu}{c^3} \, d\Omega \qquad (7\text{–}44)$$

for photons polarized in one particular sense. Here $d\Omega$ is the increment of solid angle. This factor, considered in isolation, makes transitions in the optical domain, where $\omega \sim 3 \times 10^{15}$, much more probable than transitions, say, in the radio region at $\omega \sim 3 \times 10^9$.

Equation 7–38 holds only for *spontaneous photon emission*. In general, the emission probability is proportional to $n_\omega + 1$, where n_ω is the number density of photons per phase cell that already have the momentum and polarization characterizing the photon to be emitted. This preferential emission of photons along directions of photons already present to the vicinity of the atom is called *stimulated* or *induced emission*. It is the exact opposite of ordinary *photon absorption*. The number of absorptions is again proportional to $n_\omega + 1$, if n_ω is taken to be the number density per phase cell of photons left after the atom has reached the upper state. We therefore see that the *transition probability* $P(\omega, \theta, \phi)$ per unit solid angle and frequency range quite generally obeys the relation

$$P(\omega, \theta, \phi) \, d\Omega \, d\omega = [n(\omega, \theta, \phi) + 1] \frac{\omega^2}{(2\pi c)^3} \, d\Omega \, d\omega \qquad (7\text{–}45)$$

Here $n(\omega, \theta, \phi)$ is the probability per unit frequency range that a photon state is occupied when the atomic system is in its upper energy state and ω is the mean transition frequency. Let us now return to the factors (a) and (b) mentioned earlier. These factors have to be evaluated quantum mechanically. In general, the result of such calculations is a *matrix* with elements U_{ij} giving the *transition amplitude* between any two states i and j of the atomic system. The actual transition probability between these two states is proportional to $|U_{ij}|^2$.

The prescription for obtaining the transition probability per unit solid angle is to multiply the product $|U_{ij}|^2 \, P(\omega, \theta, \phi)$ by the numerical factor $2\pi/\hbar$. Thus

$$\text{Transition probability per unit time} = \frac{2\pi}{\hbar^2} |U_{ab}|^2 \, [P(\omega, \theta, \phi)] \, d\Omega$$

$$= \frac{2\pi}{\hbar^2} |U_{ab}|^2 \, [n(\omega, \theta, \phi) + 1] \frac{\omega_{ab}^2}{(2\pi c)^3} \, d\Omega \qquad (7\text{–}46)$$

Since the energy of the states is quite narrowly defined $n(\omega, \theta, \phi)$ will normally not change appreciably over the bandwidth of the line. The matrix elements already include the integration over the frequency bandwidth that appeared explicitly in equation 7–45. The transition probability (7–46) thus includes an integration over the entire frequency range ω and, specifically, includes consideration of strongly absorbed or emitted photons at the line center as well as the less readily absorbed and emitted photons in the line wings. More precisely stated, it includes consideration of the line shape (7–42).

We still need to relate the quantum mechanical transition probability to equation 6–86 that expressed the intensity I absorbed by an oscillating dipole in terms of the second time derivative of the dipole moment d.

$$I = \frac{2}{3c^3} \ddot{d}^2 = \frac{2e^2\ddot{r}^2}{3c^3} \tag{6–86}$$

Since $\mathbf{r}$ has the time dependence (7–37), equation 6–86 is readily rewritten as

$$I = \frac{2e^2\omega^4}{3c^3} \langle r^2 \rangle = \frac{32\pi^4}{3c^3} e^2 v^4 \langle r^2 \rangle \tag{7–47}$$

where the brackets $\langle \, \rangle$ indicate a time average. The intensity I is related to the spontaneous transition probability of quantum mechanics; and we must therefore set $n = 0$ in equation 7–46, if a comparison with (7–47) is to be made. In the dipole approximation the matrix element U_{ab} will make a contribution

$$|U_{ab}|^2 = 2\pi\hbar\omega_{ab}e^2|r_{ab}|^2 \sin^2\theta \, d\Omega \tag{7–48}$$

where an integration has been carried out over the possible directions of polarization. The physical meaning of $e^2|r_{ab}|^2$ will be discussed below. The total intensity now is given by the product of the transition probability (7–46) and the photon energy $\hbar\omega_{ab}$:

$$I \, d\Omega = \hbar\omega_{ab} \cdot \frac{2\pi}{\hbar^2} \frac{\omega_{ab}^2}{(2\pi c)^3} \cdot 2\pi\hbar\omega_{ab}e^2|r_{ab}|^2 \sin^2\theta \, d\Omega \tag{7–49}$$

where θ represents the angle between the vector $\mathbf{r}$ and the direction of propagation of the emitted radiation. Integrating over all angles of emission, we obtain the total intensity of spontaneously emitted radiation

$$I = \frac{4}{3} \frac{e^2}{c^3} \omega_{ab}^4 |r_{ab}|^2 = \frac{64}{3} \pi^4 \frac{e^2}{c^3} v_{ab}^4 |r_{ab}|^2 \tag{7–50}$$

We see that the formula obtained quantum mechanically is almost the same as the classical expression. We only have to replace the time average $\langle r^2 \rangle$ by $2|r_{ab}|^2$ if we wish to obtain identical forms. This connection is consistent with the correspondence principle. We note, however, that $e^2|r_{ab}|^2$ is not an exact quantum mechanical analogue to the mean square dipole moment. For each individual state of the atomic system, a or b, the dipole moment would be given by expressions involving the diagonal matrix elements er_{aa} or er_{bb}, respectively.

The quantities er_{ab}, instead, denote a property that is influenced both by the initial and the final state of the system. They have no exact classical analogue and we therefore need not be surprised that a factor 2 appears in equation 7–51. In fact, we had no reason to expect complete identity of classical and quantum mechanical forms. After all, the quantum theory of radiation is supposed to go a step beyond the classical results in order to provide an understanding of abrupt transitions. Its results must therefore differ from those of classical theory in some essential form.

Thus far we have only a formal solution that does not yet allow us to estimate the strength of an emission or absorption line. We can however still make use of equation 7–36 for that purpose. We note that γ^{-1} is a time constant, so that γ taken by itself is equivalent to a transition probability. By setting the value for γ equal to the transition probability (7–46), we therefore are able to estimate U_{ab} and also an absorption cross section for radiation. We write

$$\gamma = \frac{2}{3}\frac{e^2\omega_{ab}^2}{mc^3} = \frac{2\pi}{\hbar^2}\frac{\omega_{ab}^2}{(2\pi c)^3}\int|U_{ab}|^2\left[n(\omega,\theta,\phi)+1\right]d\Omega \qquad (7\text{–}51)$$

For spontaneous emission, $n(\omega,\theta,\phi)$ can be set equal to zero. The value for γ that is used here has of course been derived on the basis of a dipole radiator model, and the integral on the right-hand side of equation 7–51 will therefore contain the same factor 2/3 that already came up in the evaluation of the classical expression 6–85.

$$\int|U_{ab}|^2\,d\Omega = \frac{2}{3}|U_{ab}|^2\,4\pi \qquad (7\text{–}52)$$

Hence

$$|U_{ab}|^2 = \frac{\pi e^2\hbar^2}{m} \qquad (7\text{–}53)$$

This information suffices for us to find the atomic system's absorption cross section for radiation.

Let this cross section be

$$\sigma = \int\sigma(\omega)\,d\omega \qquad (7\text{–}54)$$

and let the atomic system be surrounded by an isotropic photon gas of density $n'(\omega,\theta,\phi)$. The total number of photons absorbed can therefore be expressed as

$$n'(\omega,\theta,\phi)\,\sigma c\,d\Omega = \frac{2\pi}{\hbar^2}|U_{ab}|^2\left[n(\omega,\theta,\phi)+1\right]\frac{\omega_{ab}^2}{(2\pi c)^3}\,d\Omega \qquad (7\text{–}55)$$

Here the left side represents the number of photons per unit frequency range of a continuous spectrum intercepted by the cross-sectional area in unit time, and the right-hand side gives the probability for absorption of a photon, as expressed in (7–46). We can cancel the photon densities in equation 7–55 if we follow the procedure of letting $n(\omega, \theta, \phi)$ stand for the fractional number of photon states occupied when the atomic system is in its upper state. Because we have taken n' to represent the number density of photons present before absorption, that is, when the atomic system still is in its lower state, it is clear that

$$n'(\omega, \theta, \phi) = [n(\omega, \theta, \phi) + 1]\frac{\omega_{ab}^2}{(2\pi c)^3} \tag{7–56}$$

The factor $\omega_{ab}^2/(2\pi c)^3$ appears because n is a number density per phase cell while n' is a density per unit volume of three-dimensional space.

From (7–53) and (7–55) it then follows that

$$\sigma = \frac{2\pi^2 e^2}{mc} = 2\pi^2 r_e c \tag{7–57}$$

where

$$r_e \equiv \frac{e^2}{mc^2} \tag{6–168}$$

The cross section (7–57) has precisely the value we would obtain by classical means if we modified equation 6–107 to include a radiative reaction force (see equation 7–32)

$$F_{rad} = \frac{2}{3}\frac{e^2}{c^3}\,\dddot{r} \tag{7–58}$$

representing the force on the moving charge due to the fact that it is radiating.

A series of remarks is in order now:

(1) The cross section obtained here holds only for atomic systems for which an oscillating charged dipole represents a satisfactory description. This must be strongly emphasized! Each type of atom or molecule has its own structure and therefore will interact with photons in its own way. However, an essential feature shared by many atomic systems is that electrons are bound to a nucleus or core. In a stable quantum state the electron then resists the efforts of an applied electromagnetic field to move it from its equilibrium position or, more accurately, from its equilibrium orbital distribution within the atomic system.

In this respect, the electron behaves as though it were harmonically bound to the more massive core. This justifies the use of the classical dipole approximation as a guide to the quantum treatment. However, it does so only for atoms or molecules having a dipole moment and for wavelengths long compared to the atomic dimensions. The limitations that held for classical radiators therefore hold equally well in the quantum limit. That was already pointed out in section 6:13, but perhaps it is worth stating again.

(2) No atom behaves precisely like a classical harmonic oscillator. Its cross section, therefore, is not precisely that given in (7–57). We can define an *oscillator strength f* that represents the actual absorption strength of a given line in units of $2\pi^2 e^2 (mc)^{-1}$. A value $f = 1$ represents an absorption equal to that of the classical dipole.

(3) As already noted in equation 7–54, the cross section of the atomic system varies with frequency. The frequency distribution is of the form (7–42), as indicated earlier.

PROBLEM 7–10. Show that if the absorption cross section is

$$\sigma_{ab}(\omega) = \frac{2\pi e^2}{mc} f_{ab} \frac{\gamma/2}{(\omega - \omega_{ab})^2 + (\gamma/2)^2}$$

$$\sigma_{ab}(v) = \frac{2\pi e^2}{mc} f_{ab} \frac{\Gamma}{2} \left[(v - v_{ab})^2 + \left(\frac{\Gamma}{2} \right)^2 \right]^{-1}, \qquad \Gamma = \frac{\gamma}{2\pi} \tag{7–59}$$

the total cross section obtained in (7–57), multiplied by an oscillator strength f_{ab}, is obtained on integrating over all frequencies.

---

(4) The identity of absorption and emission cross section is already implied in the form that equation 7–46 takes. This is true both of the magnitude of the absorption and also of its spectral distribution. When no radiation field at all appears to be present, that is, $n(\omega, \theta, \phi) = 0$, we still have the vacuum field or zero level photon population present; and this field induces the spontaneous emission of radiation, discussed in more detail in section 7:10.

(5) The numerical values associated with various kinds of transition also are of interest. The absorption cross section (7–57), corresponding to unit oscillator strength, has a value $\sigma \sim 0.17$ cm^2 sec^{-1}. This cross section, when multiplied by the radiative flux *in unit frequency interval*, gives the total amount of radiation absorbed by an atom. Sometimes it is useful to know the maximum absorption at the line center. The peak absorption cross section is then a useful quantity to know.

PROBLEM 7–11. Show that the maximum absorption cross section has the value $\sigma(\omega_{ab}) = (3\lambda^2/2\pi) f_{ab}$, so that the apparent size of the atom at resonance is roughly a factor of two lower than the wavelength squared, for $f_{ab} \sim 1$. λ is the wavelength of the radiation.

PROBLEM 7-12. What fraction of the radiation in an emission or absorption line lies within the bandwidth γ defined by the natural line width?

PROBLEM 7-13. Show that the spontaneous transition probability is roughly $\gamma \sim (5\lambda^2)^{-1}$ in cgs units. It therefore has a value of 10^8 sec^{-1} for visible light.

In a different spirit we can use (7–50) to write the transition probability w as

$$w \sim \frac{e^2}{c^3} \frac{\omega_{ab}^3}{\hbar} |r_{ab}|^2 \sim \frac{e^2}{c^3\hbar} \left(\frac{me^4}{\hbar^3} \right)^2 \left(\frac{\hbar^2}{me^2} \right)^2 \omega_{ab} \sim \left(\frac{e^2}{c\hbar} \right)^3 \omega_{ab} \sim \frac{1}{(137)^3} \omega_{ab} \qquad (7–60)$$

where we have made use of equation 7–3 for a hydrogenlike atom to roughly estimate the radiated frequency ω_{ab} and have set the Bohr radius (7–4) equal to $|r_{ab}|$. The *fine structure constant*

$$\alpha = \frac{e^2}{\hbar c} \sim \frac{1}{137} \qquad (7–61)$$

taken to the third power then appears in the last element of equation 7–60. The transition probability for visible radiation is of order 10^8 sec^{-1} and, correspondingly, we can see from (7–60) that it should be of order 10^{11} sec^{-1} for X-rays, $\sim 10^{14}$ sec^{-1} for γ-radiation, and $\lesssim 10^4$ sec^{-1} for radio waves. Interestingly (7–60) is independent of the mass of the emitting particle. It does not have to be an electron, but can be an ion. Of course the lifetime of the state is just the reciprocal of the transition probability.

The magnitude for oscillator strengths can vary greatly. For the hydrogen Lyman series we have values Lα(0.42), Lβ(0.08), Lγ(0.03), Lδ(0.01), and so on. Occasionally f values are slightly larger than 1.0, or at the other extreme values of 10^{-10} or even less can also occur. The oscillator strengths must therefore be evaluated individually for any given atom or molecule and will depend strongly on the structural properties of the atomic system.

The f values for different transitions in an atom or molecule are not independent. In particular, a given atom cannot have an arbitrarily large number of strong absorption or emission lines: If we sum f values for all possible transitions, between all possible states in an atom or ion, we should obtain a number equal to the total number of electrons in the atom. If the atom has strongly bound inner electrons, then the sum should equal the number of the more weakly bound valence electrons. This is the *Thomas-Kuhn sum rule* (A163*). For hydrogen the sum of all f values should equal unity.

PROBLEM 7–14. For many years astronomers believed that atoms could be repelled by sunlight, to a sufficient extent to account for the long comet tails we observe.

For a small molecule, having an oscillator strength $f = 1$ and a mass $m = 5 \times 10^{-23}$ g, calculate the ratio of solar radiative repulsion to gravitational attraction. For comet tails the observed repulsive acceleration corresponds to an effective ratio of the order 10^2 to 10^3. Assume that all of the sunlight is roughly evenly distributed between 4×10^{-5} and 7×10^{-5} cm wavelengths. Does it appear likely that radiation produces this repulsion?

Magnetohydrodynamic forces are currently thought primarily responsible for the acceleration of tail constituents; but it is still possible that radiation pressure plays some role in the formation of the tails. None of the current theories are without difficulties.

---

(6) Atomic systems that do not have a dipole moment can at best undergo transitions through quadrupole or magnetic dipole radiation. The transition probabilities for such processes are of order $(r/\lambda)^2$ smaller, where r is a typical dimension of the atomic system. This is consistent with what we found in section 6:13, since $r/\lambda \sim r\omega/c \sim v/c$. From (7–60) we also see that $(r\omega/c)^2 \sim (1/137)^2$ for atoms. In rough agreement with these estimates, we find that actual transition probabilities for magnetic dipole and for quadrupole transitions are of order 10^3 and $1 \sec^{-1}$ (He50).*

7:10 STIMULATED EMISSION, COHERENT PROCESSES, AND BLACKBODY RADIATION

Stimulated emission is not a mechanism in the same sense as, say, electric dipole, quadrupole, or synchrotron emission, which we discussed in Chapter 6. It is a process that will work through the aid of any one of these mechanisms. Stimulated emission can occur whenever an electromagnetic wave of frequency $v = \mathscr{E}/h$ impinges on a particle in an excited state at energy $\mathscr{E}$ above some other state.

The electromagnetic wave or photon at frequency v can then stimulate or induce the emission of an additional photon that has exactly the same polarization, direction of propagation, and frequency as the stimulating photon. The characteristics of this newly formed photon are indistinguishable from those of the stimulating photon in the sense already discussed in sections 4:11 to 4:13, and we say that the radiation is *coherent*.

Let us first show the role that *stimulated* (sometimes called *induced*) *emission* plays in the process of blackbody emission. Again, blackbody radiation is not

7:10

a mechanism. It is a process that depends on the existence of radiative mechanisms, and we can obtain a blackbody spectrum through a wide variety of mechanisms. That is why the appearance of a characteristic blackbody spectrum in an astrophysical source can only tell us something about the surface temperature of the source—nothing about the physical processes that actually give rise to the observed radiation.

Blackbody radiation always is the result of a succession of emission and absorption processes. There are two basic requirements: First, the temperature of absorbing particles must be constant in the vicinity of the emitting surface so that the photons emanating from the surface are in thermal equilibrium with particles at one well-defined temperature; second, for this equilibrium to become established, we require that the assembly of absorbing particles at constant temperature be large enough so that a succession of absorption and re-emission steps occur before energy escapes from the surface of the assembly.

The number density $n(\mathscr{E})$ of particles in an excited energy state $\mathscr{E}$ is given by the Boltzmann distribution (4–47) in terms of n_0, the number density in a lower state:

$$n(\varepsilon) = n_0 e^{-\mathscr{E}/kT} \tag{7–62}$$

To determine the equilibrium conditions between photons and particles we can proceed in a way that is different from the approach taken in Chapter 4. We can look for the conditions for which the number of photons absorbed by the assembly of particles is just equal to the number emitted per unit volume, because those are the conditions in equilibrium. To analyze this situation we will consider only the transitions occurring between two given energy states of the particle. The presence of other states will not alter the conclusions.

A particle in the lower of the two states can absorb a photon and transit to the upper energy state. In the upper state, the particle can either spontaneously emit a photon or else it can be induced to emit a photon through stimulation by radiation of appropriate spectral frequency. In equilibrium, the sum of the induced and spontaneous downward transitions must equal the number of upward transitions. Let the probability of emitting a photon of frequency v be $A(v)$ in unit time interval. Let the probability of absorbing a photon be $n(v, T)\,cB(v)$ where $n(v, T)$ is the photon density at temperature T and frequency v, and $B(v)$ is a transition probability for unit time; then it is easy to see that the probability for stimulated emission for a given excited particle equals the probability for absorption by some other particle in the lower state. As illustrated in Fig. 7.10, this is a consequence of time reversal symmetry that holds for all electromagnetic processes.

We are now ready to write the equation for equilibrium between absorption and emission of photons in the frequency range dv around frequency v.

$$n(v, T)\, B(v)\, n_0 c\, dv = [A(v) + n(v, T)\, cB(v)]\, n(\mathscr{E})\, dv \tag{7–63}$$

Fig. 7.10 In part (*a*) a photon of frequency v stimulates the emission of a similar photon, while the particle energy drops by an amount $\mathscr{E} = hv$. In (*b*), the time-reversed process takes place. The particle transits to the higher energy state $\mathscr{E}$ by absorbing a photon. As discussed in the text, spontaneous emission of radiation (*c*) can be considered caused by radiation oscillators in the ground state and corresponds to the time-reversed process of absorption of energy from a singly excited radiation oscillator (*d*).

Combining this with equation 7–62 gives

$$n(v, T)\, dv = \frac{A(v)/B(v)\, c}{e^{hv/kT} - 1}\, dv \qquad (7\text{–}64)$$

If we considered spontaneous transitions as events produced by the ground state of the radiation field, which was discussed in section 4:13, then we can set $A(v)$ equal to the number density of radiation oscillators multiplied by the transition probability per unit time $B(v)$. Basically this is equivalent to stating that all emission processes are induced, that the probability for emission is proportional to the sum of all populated radiation oscillator states, that the ground state of each oscillator ($n = 1$) is always populated and that some radiation oscillators containing what we have called photons are in higher states n. From the phase space enumeration of the number density of radiation oscillators at frequency v (equation 4–65a) we then see that

$$A(v) = \frac{8\pi v^2}{c^2}\, B(v) \qquad (7\text{–}65)$$

To relate these coefficients to earlier work, we note that (7–50) gives

$$\frac{I}{hv} = \frac{64}{3h}\pi^4 \frac{e^2 v^3}{c^3}\left| r_{ab} \right|^2 = \int A(v)\, dv = A_{ab} \qquad (7\text{–}66)$$

7:10

PROBLEM 7-15.   According to the correspondence principle, the transition probability should be related to the classical radiation intensity in the limit of large atomic systems. In the ionized regions of interstellar space, transitions often occur between highly excited states of atomic hydrogen (Ka59, Hö65). Show that the correspondence argument leads to

$$\frac{d\mathscr{E}}{dt} = h\nu A_{n,n-1} = \frac{\omega^4 e^2 a_n^2}{3c^3} \tag{7-67}$$

where a_n is the Bohr radius in the nth state. Show that this gives

$$A_{n,n-1} = \frac{64\pi^6 m_e e^{10}}{3c^3 h^6 n^5} = \frac{5.22 \times 10^9}{n^5} \tag{7-68}$$

PROBLEM 7-16.   Show that $B(\nu)$ differs from $\sigma_{ab}(\nu)$, (7-59) only by a factor of 2. Derive a relation between A_{ab} and f_{ab}.

We see now, that

$$n(\nu, T)\, d\nu = \frac{8\pi\nu^2}{c^3}\, \frac{d\nu}{e^{h\nu/kT} - 1} \tag{7-69}$$

This corresponds to equation 4-71 for blackbody radiation, and shows that the blackbody process depends heavily on the concept of stimulated emission.

The process we described is stable and self-regulating. If $n(\nu, T)$ is lower than the value given in equation 7-69, the spontaneous emission exceeds the sum of absorption and stimulated emission processes; and in this way the population of photons becomes increased until it reaches the value given by (7-69). Conversely, if $n(\nu)$ is high, absorption will lower it back to the equilibrium value.

It is worth noting that the *Einstein Coefficients* $A(\nu)$ and $B(\nu)$ are sometimes defined slightly differently—for example, in terms of emitted or absorbed energy. This difference is not important as long as we always adhere to a consistent scheme.

7:11   STIMULATED EMISSION AND COSMIC MASERS

Let us ask what would happen if the relationship between n_0 and $n(\mathscr{E})$ was not given by the Boltzman relation (7-62). For small deviations from the thermal equilibrium the photon bath would tend to cause n_0 and $n(\mathscr{E})$ to come back to equilibrium. But if the number of particles in the upper state starts to exceed the number in the lower state, then an entirely different process comes into play.

Clearly, this situation can never come into existence under conditions of thermal equilibrium, because $\exp(-\mathscr{E}/kT)$ is always less than unity for positive values of $\mathscr{E}$ and T. A *population inversion*, $n(\mathscr{E}) > n_0$, can therefore only be brought about by an artificial process. Sometimes one describes a population inversion as a state of negative temperature, for then the exponential term in equation 7–62 can exceed unity. However, this is primarily a descriptive term and does not define any physical process.

To see what happens if population inversion is brought about, we note that the probability for stimulated emission always exceeds the probability of absorption in this case. In any given transition a radiation oscillator is therefore more likely to rise to a higher energy state than to a lower one. As the radiation propagates through the assembly of particles, it therefore becomes amplified. Moreover, since the emitted photons have the same characteristics as the stimulating photons, the amplified radiation is coherent. In the laboratory the process described here corresponds to a *maser*. When the process occurs on a cosmic scale we therefore talk about maser processes.

Cosmic maser action will be maintained provided that the pumping of energy into the assembly of particles, that is, the rate of excitation of the upper levels can keep up with the downward spontaneous and induced transitions, and maintain the density of particles in the upper energy state $n(\mathscr{E})$ greater than that in the lower state n_0.

The pumping process can take several forms. We might have a very energetic photon excite particles into an energy state $\mathscr{E}'$ from which the transition probability to the ground state is low and the transition probability for going to the state with energy $\mathscr{E}$ is high. This type of maser is called a *three-level maser*. Another means for producing a population inversion can come about chemically. Suppose that a molecule is formed in the interaction of two atoms and that it is formed in a high energy state. If the formation rate is high enough, then a population inversion can be maintained and maser action can set in.

There appear to exist regions of interstellar space in which OH radicals and/or water vapor molecules H_2O have certain energy states pumped up to population inversion. The pumping action is not understood, and it is curious that different regions of space show a variety of differing OH levels inverted. We therefore seem to have a number of different pumping mechanisms that evidently come into play under differing conditions.

Cosmic masers emit extremely intense coherent radiation. Since all induced photons travel along the same direction, they appear to come from an improbably compact region (Fig. 7.11). The radiation reaching our telescopes arrives contained in a well-defined, extremely small, solid angle.

The smallness of the observed solid angle is misleading. It may not represent the actual size of the cloud, but might represent only the dimension of the region

7:11

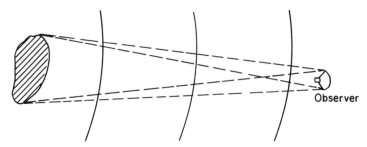

Fig. 7.11   The observer may not see the entire cloud of particles (dashed line) in coherent radiation. He may see only that portion in which the coherent wave originated. That can be a much smaller region, possibly a volume as small as λ^3, where λ is the wavelength of the maser radiation.

in which the coherent radiation originated. This volume can be seen, from phase space arguments, to equal h^3/p^3, or $c^3/v^3 = \lambda^3$, where λ is the wavelength of light. Some masers have by now been resolved by long baseline radio interferometers, and this indicates that masing may be initiated over larger regions in space.

Two types of masers are recognized. There are OH masers whose positions coincide with those of Mira variable stars. Such a maser's luminosity is roughly $10^{-4} \, L_\odot$ and it shows variability, over periods of months, synchronized with the star's pulsations. We also recognize OH and H_2O masers associated with dust clouds in or near H_{II} regions. The H_2O masers can have luminosities up to $\sim L_\odot$, and show variability over periods of weeks. These variations are far too slow to be due to scintillation—refraction of radiation by moving clouds in interplanetary or interstellar space.

Both types of masers may be pumped by strong infrared radiation fluxes associated with the masing regions.

One characteristic of the interstellar masers is that the radiation is highly polarized. As already explained, stimulated emission always involves formation of photons with the same sense of polarization as the stimulating photon. This means that all the photons derived from a given progenitor will have identical polarization and the radiation is 100 % polarized.

To realize how quickly the intensity of a beam increases as it traverses a cloud in which the population is inverted, we note that for a gain g per interaction, and for an optical depth n in the cloud, the outgoing beam will have an intensity of

$$N = g^n$$

Suppose that the gain is 1.1—that is, the probability for stimulated emission is 10 % higher than for absorption. After 100 mean free paths, the total number of photons in the beam will be 10^4. After 200 such successive absorption-emission processes, the number would have reached 10^8, and so on. An emitting region, therefore, need not have an opacity in excess of several hundred in order to emit extremely bright maser radiation. This radiation is all concentrated into a very narrow spectral bandwidth and in this narrow spectral range it therefore has the equivalent brightness of a fantastically hot body, with T up to 6×10^{13}°K (Ra71).*

7:12 STELLAR OPACITY

In the interior of a star, where matter is highly ionized, the interaction of photons with matter often determines the rate at which energy is transported through the star.

To be sure, the photon interaction tells us nothing about neutrino energy transport, which may be very important in some stellar models. The neutrinos, however, simply escape, leaving the central regions of the star at the speed of light. In contrast to the photons, neutrinos do not affect stellar structure except in the limited sense that the nuclear reactions in the center of a star heat the stellar material by a reduced amount whenever neutrino production occurs. We then speak about *neutrino energy losses*.

In the absence of such losses, three mechanisms act to transport energy from the center of the star, where nuclear reactions occur, out to the periphery where radiation is emitted into space. The transport can occur through convection of stellar matter, through conduction of heat by a degenerate electron gas, or through transfer of radiation. These processes will be discussed in Chapter 8. Each one plays an important role in different types of stars. Here, we will only be concerned about the factors that determine the rate at which photons transport energy.

A γ-photon liberated in a nuclear reaction almost immediately yields its energy to the stellar material through ionization of neutral or partially ionized atoms, or through collisions with electrons. So strong is the interaction of radiation with matter, that energy initially liberated at the center of a star normally requires tens of thousands of years before finally escaping at the stellar surface. The star is highly opaque to radiation, and it is interesting to see how the opacity originates, since many features of stellar structure and evolution depend on this physical property.

We will be interested in four distinct types of interaction of radiation with matter: (a) Thomson or Compton scattering of radiation by free electrons (6:14, 6:20); (b) *free-free absorption* or emission (section 6:16); (c) *bound-free interactions* in which an electron undergoes a transition between a bound and a free state; (d) *bound-bound* transitions, that is, the excitation or de-excitation of atoms or ions by photons.

In order to compute the mean opacity of stellar matter, we will have to proceed in three steps. First, we will need to know the interaction cross section of radiation with matter for each of the four processes. This gives us the opacity due to the individual interactions through a simple proportionality relationship. However, the total opacity of stellar matter is not just the sum of the individual opacities, and a suitably chosen mean opacity must be computed, properly weighting the individual contributions made by processes (a) to (d) and also taking induced emission into account. Stimulated emission decreases the opacity because the energy transport rate is increased.

The contributions of the various processes to the opacity depend very strongly on the temperature. In the cool surface layers of a star, where atoms are only partially ionized, the opacity may be dominated by bound-bound and bound-free transitions. At high temperatures where ionization may be nearly complete, the opacity due to free-free interactions becomes dominant. At the highest temperatures where induced emission reduces the opacity due to factors (b) through (d), electron scattering plays a dominant role.

We will let *extinction* denote the amount of radiation eliminated from a beam of light through absorption or scattering. We can then define the extinction $\not{E}$ of a slab of matter of unit thickness, through which radiation passes at normal incidence, as

$$\not{E} = \kappa \rho \qquad (7\text{--}70)$$

where the *opacity* of the substance is denoted by the symbol κ and ρ represents the density. The opacity for radiation at a particular spectral frequency v is denoted by $\kappa(v)$. Summing over the opacity contributions of processes (a) to (d), at any given frequency, we can therefore write a total opacity $\kappa^*(v)$ at frequency v:

$$\kappa^*(v) = \kappa_e + \left[\kappa_{ff}(v) + \kappa_{bf}(v) + \kappa_{bb}(v) \right] \left[1 - e^{-hv/kT} \right] \qquad (7\text{--}71)$$

where the subscripts respectively mean electron scattering, free-free, bound-free and bound-bound. κ_e is frequency independent and therefore becomes predominant as the temperature increases. κ^* represents the true opacity with induced emission taken into account.

The proper averaging of $\kappa^*(v)$ over the entire range of frequencies depends on our purpose. In the case of stellar energy transport, which will be discussed in Chapter 8, we effectively need to know the mean free path of radiation as it travels through the star. Since this is inversely proportional to the opacity, we are effectively averaging $1/\kappa^*(v)$ over the entire spectral range. This average, however, must still take into consideration that the radiation spectrum is not flat, and that the energy transport rate will therefore also depend on the radiation spectrum defined by the local temperature at any given point of the star. We will consider this later, in Chapter 8. For the present we will only show how

$\kappa^*(v)$ depends on processes that occur on an atomic scale, and how the individual opacities depend on atomic interaction cross sections for radiation.

(a) Scattering by Free Electrons

At temperatures low enough so that the photon energy does not correspond to the electron rest mass energy, that is, $T \ll mc^2 k^{-1} \sim 10^{10}{}^\circ\text{K}$, relativistic effects can be neglected and the scattering cross section is simply the Thomson cross section

$$\sigma_e = \frac{8\pi}{3}\left(\frac{e^2}{mc^2}\right)^2 = 6.65 \times 10^{-25}\,\text{cm}^2 \qquad (6\text{--}102)$$

This is frequency independent. At the centers of highly dense stellar cores, the temperature may however become large enough so that the Klein-Nishina cross section for Compton scattering (6–167) gives a more accurate representation, and a frequency dependence then does exist. At the density of the sun, $\rho \sim 1\,\text{g cm}^{-3}$, the number of electrons per cubic centimeter is of order 10^{24}, so that the mean free path of radiation between electron scattering events is only of the order of 1 cm. If n_e is the number of electrons in unit volume, the opacity for scattering is given by

$$\kappa_e \rho = \sigma_e n_e \qquad (7\text{--}72)$$

(b) Free-free Interactions

This process was discussed for tenuous plasmas in section 6:16, but the same theory describes the denser plasmas inside stars. We note that the classical expression (6–137) must unwittingly contain the induced emission factor $[1 - \exp(-hv/kT)]$ that at long wavelengths approaches hv/kT. If we, therefore, divide (6–137) by this factor and also by the number densities, we obtain an absorption coefficient per ion, for unit density of ions and electrons. It will have the form

$$\alpha_{ff} = \frac{8}{(6\pi)^{1/2}} \frac{Z^2 e^6}{cm^2 hv^3 v} \ln[\ldots] \qquad (7\text{--}73)$$

where we have set $v = (3kT/m)^{1/2}$ and have assumed that the argument of the logarithmic function has the same character as in (6–137). The actual, quantum mechanically correct, result is (To47)

$$\alpha_{ff} = \frac{4\pi e^6}{3\sqrt{3}\,chm^2} \frac{Z^2}{v} \frac{g_{ff}}{v^3} \qquad (7\text{--}74)$$

where Z is the effective charge of the ion considered and g_{ff} is called the Gaunt factor. It contains the logarithm in (6–137) and is of order unity for most cases of interest.

(c) Bound-free Absorption

Quantum mechanically, one can also compute an absorption coefficient for bound-free transitions, when only one electron per atom is active in absorbing radiation. This has the form (Cl68)*

$$\alpha_{bf} = \frac{64\pi^4 m e^{10} Z^4}{3\sqrt{3}\, ch^6 n^5} \frac{g_{bf}}{v^3} \tag{7–75}$$

where n is the principal quantum number and g_{bf}, the Gaunt factor, again is of order unity and only mildly depends on n and v. This equation of course only holds when the photon energy exceeds the ionization energy χ_n in the nth state,

$$h\nu > \chi_n \sim \frac{2\pi^2 m e^4 Z^2}{n^2 h^2} \tag{7–76}$$

(d) Bound-bound Transitions

These have cross sections already discussed in the last section. They depend strongly on the actual structure of the individual atom, and do not give rise to a continuum absorption cross section as do the factors (a) to (c). As discussed in the next section, these cross sections play an important role in determining the radiative transfer rate through a stellar atmosphere; but they do not play a significant role in the stellar interior, where processes (a), (b), and (c) dominate radiative transfer rates (Chapter 8).

The opacity of low density ionized matter also is a measure of the radiant power emitted from unit volume at given temperature T. If the chemical composition of the plasma has the "cosmic abundance" character (Fig. 1.7, Table 1.2), Figure 7.12 shows the radiated power; it assumes that self-absorption by the plasma can be neglected. At high densities, where that assumption no longer is valid, the plasma would simply radiate with a brightness characteristic of any blackbody at temperature T.

We note that for these low densities forbidden line emission dominates. *Forbidden transitions* are those for which dipole (and sometimes higher multipole) radiative transitions are not allowed by the *selection rules*. When a plasma is

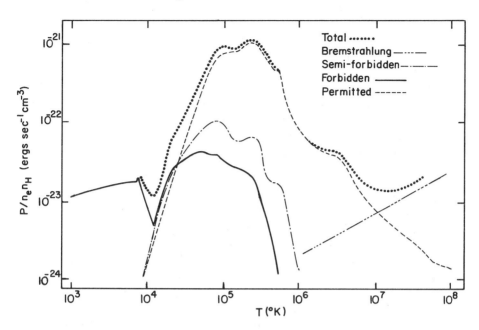

Fig. 7.12 Radiated power for unit volume of a low density ionized gas in collisional equilibrium. The power indicated is for electron and hydrogen concentrations of $n_e = n_H = 1$ cm^{-3}. n_H represents the total number density for hydrogen atoms and protons. To obtain the total power radiated from unit volume, the ordinate would have to be multiplied by $n_e n_H$. The plasma considered has a chemical composition typical of cosmic sources (see Table 1.2, Fig. 1.7). (After D.P. Cox and E. Daltabuit, Co71a. With the permission of the University of Chicago Press.)

at sufficiently low densities so that collisions are rare, a forbidden radiative transition with a correspondingly low transition probability may imply a metastable lifetime of seconds or years for the excited metastable atom. Since collisions are rare in the low density medium, these transitions may take place nevertheless. As the gas density increases collisions start dominating the deexcitation of the atom and the forbidden lines disappear.

If we know the lifetimes of a number of different metastable atoms in a hot interstellar nebula, we can often conclude a great deal about temperature and density conditions, by studying the forbidden lines.

In stars and dense stellar atmospheres, collisions between atoms and ions are frequent and no forbidden lines are expected; but even the earth's upper atmosphere is tenuous enough so that forbidden oxygen lines appear in auroral spectra.

7:13 CHEMICAL COMPOSITION OF STELLAR ATMOSPHERES—THE RADIATIVE TRANSFER PROBLEM

In order to determine the abundance of various chemical elements in the atmospheres of stars, we must be able to correctly interpret the observed spectra of these stars. This interpretation is a complicated process. First, it depends on the correct choice of a model of the stellar atmosphere. By this we mean that we have to choose an effective temperature T_e, a value for the star's surface gravity and a parameter ξ_t representative of the turbulent velocity in the atmosphere.

In interpreting individual Frauenhofer (absorption) lines in terms of the number density n_i of a given atom or ion, i, in a given state we find that the theory always yields expressions proportional to $n_i f g$, where f is the oscillator strength of the transition and g represents the statistical weight of the lower energy level—the level from which the transition takes place. For hydrogen, helium, and other one- and two-electron ions, the f values can be quantum mechanically computed. For such complex spectra as those of iron, however, f values must be obtained through laboratory experiments.

Another important parameter required for an abundance determination when the absorption line is very strong is the *damping constant* γ, which represents broadening of the line due to the intrinsically finite lifetime of the states and due to the shortening of this life through collisions with electrons and atoms.

We can define an *equivalent width* W_λ of a Frauenhofer line, which represents the total energy absorbed in this line, divided by the energy per unit wavelength emitted by the star in its continuum spectrum around wavelength λ. Figure 7.13 shows this relationship.

For very weak lines the amount of radiation absorbed and, hence, the equivlent width, depends linearly on the abundance n_i and on the product gfn_i. As W_λ approaches the Doppler width due to thermal and turbulent motion, the absorption line becomes saturated, and the *curve of growth* (Fig. 7.14), which represents the growth of W_λ with increasing material traversed, flattens out. For still stronger lines absorption in the wings of the lines becomes possible. Here the parameter γ determines the amount of radiation that is absorbed.

In determining the abundance of various chemical elements in stars, we have to keep in mind that the population of various atomic or ionic energy states depends quite critically on the atmospheric temperature and also to some extent on the surface gravity that determines the pressure. The Boltzmann and Saha equations are applied in these computations, on the assumption that the atmosphere is in thermodynamic equilibrium. Often, the f values for a given transition are not well enough understood; but in some such situations we can at least

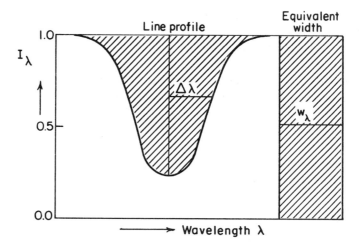

Fig. 7.13 Profile and equivalent width W_λ of a Fraunhofer line. The intensity of the continuum has been made equal to 1. The area under the line profile is equal to that of a completely "black" strip in the spectrum of width W_λ, usually measured in milliangstroms (after A. Unsöld, Un69).

obtain an idea of the relative abundances of an element in a given star compared to its abundance in the sun.

To relate quantitatively the total flux from a star or nebula to its chemical and physical properties, we proceed in the following way: The *intensity* $I(v)$ of radiation at spectral frequency v changes as it crosses a layer of matter of thickness dx. There is a loss of intensity through absorption and a corresponding gain through emission. For normal incidence on the layer, the total intensity change is

$$\frac{dI(v)}{dx} = -\kappa(v)\,\rho I(v) + j(v)\,\rho \qquad (7\text{-}77)$$

where the first term represents extinction (see equation 7–70) and $j(v)$ is the emission per unit mass. $j(v)$ may strongly depend on the radiation intensity itself as in the case of strong scattering or induced emission. When $j(v)$ is entirely due to induced emission, the opacity $\kappa^*(v)$ in (7–71) is the difference between $\kappa(v)$ and $j(v)/I(v)$.

Equation 7–77 can be rewritten as

$$\frac{1}{\rho\kappa(v)}\frac{dI(v)}{dx} = -I(v) + J(v) \qquad (7\text{-}78)$$

7:13

where $J(v) \equiv j(v)/\kappa(v)$ is called the *source function* and (7–78) is called the *equation of transfer* (Ch50)*.

In section 8:7 we will discuss the transfer of radiation from the center of a star to its periphery. It will then be necessary to consider not only normal incidence on a layer, but also incidence at other azimuthal angles θ. For arbitrary angles of incidence (θ, ϕ) we can express the energy density of radiation as

$$\int \rho(v)\, dv = \frac{1}{c} \int\int I(v, \theta, \phi)\, d\Omega\, dv, \qquad (7\text{--}79)$$

where $I(v, \theta, \phi)$ is the intensity in the direction (θ, ϕ).

The radiative flux depends on the intensity in a related way:

$$F = \int F(v)\, dv = \int\int I(v, \theta, \phi) \cos\theta\, d\Omega\, dv \qquad (7\text{--}80)$$

If we consider $I(v, \theta, \phi)$ to be a distribution function that specifies the angular distribution of radiation at frequency v, then $\rho(v)$ and $F(v)$ involve the zeroeth and first moments of this function. The second moment leads to the radiation pressure

$$P = \int P(v)\, dv = \frac{1}{c} \int\int I(v, \theta, \phi) \cos^2\theta\, d\Omega\, dv \qquad (7\text{--}81)$$

This relation follows from the discussion of sections 4:5 and 4:7. The radiation pressure will be important in the theory of stellar structure, where hydrostatic equilibrium requires a balance between gravitational forces and pressure gradients. In some stages of stellar evolution, notably in stages leading to planetary nebulae, these gradients depend more strongly on radiant than on kinetic gas pressures. Radiant pressures also play a role in determining the atmospheric structure, particularly of giant and supergiant stars.

Let us still describe the factors that determine the shape of a spectral absorption or emission line seen in a star's atmosphere. We have already discussed factors that lead to broadening of a line. However, we still should mention that for gas at a given temperature T, the emission line intensity $I(v)$ will normally not exceed the blackbody intensity at that temperature and at frequency v. Stimulated and spontaneous emission will therefore tend to increase the brightness of an emission line in its wings, as radiation is transferred through the star's atmosphere. The center of the line may already have become saturated—reached its peak intensity—close to the surface of the star. This effect will lead to emission line broadening. Similarly absorption lines become broadened on passage through the cool outer portions of a star's atmosphere because absorption in the wings becomes increasingly probable as more matter is traversed. We talk about a

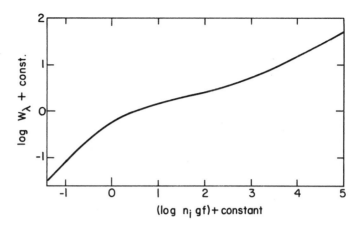

Fig. 7.14 Curve of growth showing increasing absorption with increasing amount of material traversed in a stellar atmosphere. The equivalent width W_λ is plotted against $n_i g f$ where n_i is the abundance of an element in a particular energy state, f is the oscillator strength of the Fraunhofer line, and g is the statistical weight of the absorbing state (after A. Unsöld, Un69).

curve of growth for a spectral line. By this we mean a plot of the equivalent width W_λ (Fig. 7.13) against the product $n_i f$ of the number of atoms n_i in a column of unit area through the atmosphere, and the oscillator strength of the transition f. Sometimes, as in Fig. 7.14, the curve of growth is a plot of W_λ not against $n_i f$ but against some other specified function—in this case $\log n_i g f$, where g is the statistical weight of the absorbing state.

ANSWERS TO SELECTED PROBLEMS

7–1. $E_n = -\dfrac{p_n^2}{2\mu} = -\dfrac{n^2\hbar^2}{2r_n^2\mu} = -\dfrac{Ze^2}{2r_n}, \qquad E_n = -\dfrac{Z^2 e^4 \mu}{2n^2\hbar^2}.$

Number of states $= \dfrac{\text{phase space volume}}{\text{volume of unit cell}} = \dfrac{16\pi^2 n^2 \hbar^2 \cdot \hbar}{\hbar^3} \propto n^2.$

7–2. $r \sim \dfrac{\hbar}{\sqrt{p^2}}$ where $\dfrac{p^2}{2M} = \mathbb{V}_0 - \mathscr{E}_b.$

7–3. $\mathscr{E} = \dfrac{1}{2}\sum_i m_i v_i^2$

$$I = \sum m_i r_i^2$$

$$\text{and } \mathscr{E} = \frac{\omega L}{2} = \frac{L^2}{2I}$$

$$\delta\mathscr{E} = \frac{L}{I}\delta L = \omega h.$$

Since $L = \hbar\{J(J+1)\}^{1/2}$,

$$L = \sum m_i v_i r_i$$

$$L = \omega I, \quad \omega L = \sum m_i v_i r_i \omega$$

$$\mathscr{E}_J = \frac{L^2}{2I} = \frac{\hbar^2(J+1)J}{2I} \qquad \text{and} \qquad \delta\mathscr{E} = \mathscr{E}_J - \mathscr{E}_{J-1} = \hbar^2\frac{J}{I}.$$

For rapidly rotating massive objects $\hbar J \sim L$ and $\delta\mathscr{E} = L\,\delta L\,/I$.

7-4. $\frac{1}{2}I\omega^2 = \frac{3}{2}kT, \quad T = 100°\text{K}$

$$I \sim \frac{2}{5}mr^2 \sim \frac{2}{5}(10^{-23}\text{ gm})(2 \times 10^{-8})^2$$

$$\omega \sim 5 \times 10^{12}\text{ sec}^{-1}, \qquad \nu = \frac{\omega}{2\pi} \sim 8 \times 10^{11}\text{ Hz}, \qquad \lambda \sim \frac{3}{8}\text{ mm}.$$

7-5. In (4:18) we stated that the rotational excitation probability would be proportional to a Boltzmann factor. Its form is

$$\exp\left(-\frac{\hbar^2 J(J+1)}{2IkT}\right)$$

which is small for large J and low T.

7-6. $E = \frac{1}{2}I\omega^2, \quad I = \frac{2}{5}(10^{-15})(10^{-10})\text{ g cm}^{-2}$

$$= kT \sim 1.4 \times 10^{-14}\text{ cgs}$$

$$\therefore \omega \sim 8 \times 10^5\text{ sec}^{-1}$$

$\nu \sim 10^5$ Hz, which is slightly below the AM radio band.

7-7. $\delta\mathscr{E}_{\text{grav}} = 2\hbar\omega, \quad \delta\mathscr{E}_{\text{neutrino}} = \frac{1}{2}\hbar\omega$ and results of Problem 7-3 are applied directly.

7-8. Probability $\propto \exp\left(-\frac{v_r^2 m}{2kT}\right)$

$$f(v_r) = \left(\frac{m}{2\pi kT}\right)^{3/2}\exp\left(-\frac{mv_r^2}{2kT}\right)$$

from 7–22, $v_r = \dfrac{\Delta\omega}{\omega_0} c$

$$I(\omega) = I_0 \exp\left(-\frac{mc^2 \Delta\omega^2}{2\omega_0^2 kT} \right)$$

From Problem 4–27 $\langle v_r^2 \rangle = \dfrac{kT}{m} = \dfrac{\langle \Delta\omega^2 \rangle}{\omega_0^2} c^2.$

7–10. $\sigma_{ab}(\omega) = \dfrac{2\pi e^2}{mc} f_{ab} \dfrac{\gamma/2}{(\omega - \omega_{ab})^2 + (\gamma/2)^2}$

$$\int_{-\infty}^{\infty} \frac{(\gamma/2)\, d\omega}{(\omega - \omega_{ab})^2 + (\gamma/2)^2} = \pi$$

$$\therefore\ \sigma = \int_{-\infty}^{\infty} \sigma_{ab}(\omega)\, d\omega = \frac{2\pi^2 e^2}{mc} f_{ab}$$

7–11. $\sigma_{\max} = \dfrac{2\pi e^2}{mc} f_{ab} \dfrac{1}{\gamma/2}$

where $\dfrac{\gamma}{2} = \dfrac{e^2 \omega_{ab}^2}{3c^3 m}$, $\omega_{ab} = \dfrac{2\pi c}{\lambda}$

$$\therefore\ \sigma_{\max} = \frac{3}{2} \frac{\lambda^2}{\pi} f_{ab}.$$

7–12. $I_1 = \dfrac{I_{ab}}{\pi} \displaystyle\int_{\omega_{ab} - \gamma/2}^{\omega_{ab} + \gamma/2} \dfrac{(\gamma/2)\, d\omega}{(\omega - \omega_{ab})^2 + (\gamma/2)^2} = \dfrac{I_{ab}}{2}$

$I_2 = \dfrac{I_{ab}}{\pi} \displaystyle\int_{-\infty}^{\infty} \dfrac{(\gamma/2)\, d\omega}{(\omega - \omega_{ab})^2 + (\gamma/2)^2} = I_{ab}$

$I_1/I_2 = 1/2.$

7–13. $\gamma = \dfrac{2}{3} \dfrac{e^2}{mc^3} \omega_{ab}^2 \sim \dfrac{2}{3} \dfrac{e^2}{mc^3} \dfrac{4\pi^2 c^2}{\lambda^2} \sim \dfrac{1}{5\lambda^2}.$

7–14. $F_{\text{rad}} = \displaystyle\int \sigma(\omega) \dfrac{I(\omega)\, d\omega}{c} = \dfrac{\sigma}{\Delta c} I_{TOT} = \dfrac{\sigma}{c} \dfrac{L_\odot}{4\pi r^2 \Delta}$

where Δ is the frequency bandwidth.

$$F_g = \frac{GM_\odot m}{r^2}, \qquad \sigma = \frac{2\pi^2 e^2}{mc} f \sim \frac{2\pi^2 e^2}{mc}$$

$$\therefore \frac{F_{rad}}{F_{grav}} = \frac{(\sigma/\Delta c)(L_{\odot}/4\pi r^2)}{GM_{\odot}m/r^2} = \frac{\sigma L_{\odot}}{\Delta 4\pi c G M_{\odot}m}.$$

7–15.   From (7–47) $\dfrac{d\mathscr{E}}{dt} = \dfrac{2e^2\omega^4\langle r^2\rangle}{3c^3} = \hbar\omega A_{n,n-1}$ (by 7–47 and 7–66)

$$\mathscr{E}_n = \hbar\omega = \frac{m\,e^4}{2\,\hbar^2}\left\{\frac{1}{(n-1)^2} - \frac{1}{n^2}\right\}$$

for n large $\omega \cong \dfrac{me^4}{\hbar^3}\dfrac{1}{n^3}.$

Take $a_n^2 = (n^2\hbar^2/me^2) = 2\langle r^2\rangle$ from (7–2) and (7–5), to obtain $A_{n,n-1} = 64\pi^6 m e^{10}/3c^3 n^5 h^6.$

7–16.   In equations 7–63 and 7–65 we called the number of photons of a given polarization absorbed per particle per second $n(v, T)\,cB(v)$. In defining σ, we talked about photons of either polarization, so that $\sigma(v) = 2B(v)$.

$$\frac{I}{h\nu} = \int A(v)\,dv = A_{ab} \qquad (7\text{–}66)$$

$$I(\omega) = \frac{I_{ab}}{\pi}\frac{\gamma/2}{(\omega - \omega_{ab})^2 + (\gamma/2)^2} \qquad (7\text{–}42)$$

$$\sigma(\omega) = \frac{2\pi e^2}{mc}f_{ab}\frac{\gamma/2}{(\omega - \omega_{ab})^2 + (\gamma/2)^2} \qquad (7\text{–}59)$$

$$\therefore A(\omega) = \frac{1}{\pi}\frac{A_{ab}(\gamma/2)}{(\omega - \omega_{ab})^2 + (\gamma/2)^2}, \quad A(v) = \frac{8\pi v^2}{c^2}B(v)$$

$$\therefore A(\omega) = \frac{2\omega^2}{\pi c^2}B(\omega)$$

$$= \frac{\omega^2}{\pi c^2}\sigma(\omega)$$

$$\therefore A_{ab} = \frac{8\pi^3 e^2 v^2}{mc^3}f_{ab}.$$

8

Stars

8:1 OBSERVATIONS

We do not know how stars are formed, nor just how they die. But we think we understand the structure of the most commonly found stars and the mechanisms by means of which they generate the energy we see as starlight.

How sure can we be of this understanding? How correct is the theory of stellar structure? Such questions are difficult to answer. Many of the most important stellar processes go on deep in a star's interior, while the observations available to us mainly concern themselves with surface characteristics. Conditions in the star's central regions must therefore be inferred, and our evidence is indirect.

Generally the merit of a theory is judged by the number of unrelated observations it can explain, and when the observations are indirect, a larger than usual body of data is desirable. For the theory of stellar structure and evolution, we have several different classes of observations:

(a) We have measurements on the masses of a variety of stellar types. However, the number of precision measurements is small. Each such measurement involves the detailed analysis of a stellar binary system (see section 3:5).

(b) We can determine the luminosity of a star quite accurately provided the star is near the sun where its distance is easily measured and interstellar extinction is negligible. Even then however the bolometric magnitude is not well established because far infrared fluxes and the far ultraviolet magnitudes measured from rockets, are known only for a small handful of stars. The few available observations suggest frequent sizeable deviations from blackbody behavior.

(c) The surface temperature of stars can be obtained in three different ways.

(i) We can observe a color temperature (section 4:13);

(ii) we can determine the effective temperature if the star's angular or linear diameter is known (section 4:13); or

(iii) we can observe the strengths of spectral lines representing transitions between various excited atomic states. Since the relative population of excited states is governed by the (temperature dependent) Boltzmann factor, the temperature can be computed directly, provided the relevant transition probabilities are known. These probabilities can be calculated or, preferably, observed in the laboratory.

These three independent techniques yield a satisfactory estimate of stellar surface temperature, but the temperature in the interior of stars remains unobserved.

(d) We can determine the angular diameter of a star by interferometric means (section 4 :12). It can also be obtained indirectly if both the apparent magnitude and the temperature of the star are known. But we can determine the linear diameter of the star only if we know both the angular diameter and the star's distance. In some cases eclipse data from binaries can also yield stellar diameters.

The luminosity, diameter and surface temperature of a star are interrelated and involve no more than two independent parameters, provided the star's spectrum is simple enough to allow the assignment of a single representative temperature.

(e) The chemical makeup of stars is spectroscopically determined. The abundance of the different elements obtained in this way refers only to the surface layers of the stars. Using current observational techniques, we cannot directly verify conjectures about the composition in the stellar interior.

We find that the abundance of elements on the surfaces of stars varies from one type of star to the next. Normally, hydrogen is by far the most abundant constituent. By mass its concentration lies close to 70%. Helium, the next most abundant element, has a concentration of 25 to 30% by weight. Neon, oxygen, nitrogen, and argon follow in order of decreasing abundance; together they account for about 2% of the total mass. Carbon, magnesium, silicon, sulfur, and iron are next; each of these has an approximate abundance of the order of one part per thousand, by weight.

One task of a theory of stellar evolution must be the correct prediction of the abundance of chemical elements in different types of stars. The interior of stars is the most plausible place for heavier elements to be formed from the relatively pure hydrogen—or more probably from a hydrogen-helium mixture—that seems to have been the main constituent of the Galaxy at the time the earliest stars formed in globular clusters.

The relative abundances of the various isotopes of different elements are repeatedly found in similar ratios in stars, in the interstellar medium, in meteorite fragments, and in the earth's crust. The similarity of these ratios cannot be accidental, and the detailed explanation of the hundreds of known abundance ratios provides a severe task for the theory of stellar evolution.

8:1

(f) For a limited number of stars we have measurements on the surface magnetic fields. Peculiar stars of spectral type A are the best studied members of this group. Their high fields, which may be of order 10 kilogauss in strength, are relatively easy to measure. White dwarfs too are expected to have strong fields. $B \sim 10^5$ to 10^8 gauss may not be unusual, and a detailed study of such objects is likely to be valuable (Ke70). Eventually theories of stellar evolution will have to take magnetic fields into account. Current theories have not yet incorporated magnetic effects to any great extent.

(g) The surface rotational velocities of stars can be statistically determined from a study of different spectral classes. While projection effects preclude an analysis of the rotational velocity of individual stars, the statistical study of large numbers of stars has indicated that the young O and B stars have extremely high rotational velocities, that these velocities slowly drop until late stars of spectral type A are reached, and that redder stars of types G to M have very low rotational velocity (see Fig. 1.9 and Table A.4). We are only just starting to consider effects of rotation on stellar evolution.

(h) Similarly, a statistical study of stellar velocities in our galaxy (see Table A.6) tells us that the origins of stars of differing spectral types must be dissimilar. Presumably these stars were formed at different epochs in the Galaxy's life. A final theory of stellar evolution will have to consider such age differences for stars, and will have to take into account that the chemical composition of the interstellar gas from which stars are formed must change as the Galaxy ages.

(i) For a number of different stars, notably K-giants, O-stars, and nuclei of planetary nebulae, we now have data on mass loss. For the first two of these spectral types, the information comes from spectral measurements of gas outflow. In the case of planetary nebulae, we actually see the accumulation of ejected gas. Such evidence is important both for studying unstable states of stars, and for understanding the chemical changes that the interstellar gas may suffer when material, which has undergone nuclear reactions in stars, is returned to interstellar space. Nova and supernova ejecta also yield important data in this respect. Unfortunately we do not as yet have enough information to judge whether violent but infrequent explosive events dominate or merely contribute to the cycling of matter between stars and interstellar space.

(j) For the sun, we also have data on internal rotation rates (Di67b) obtained from solar oblateness studies; and we know of a lack of neutrino emission (Da68). For other stars such data are completely lacking. Similarly, we have information on solar cosmic ray and X-ray emission; X-ray data do exist in indirect form for a small number of other stars. In each of these cases, however, it is unclear how the circumstellar regions from which such radiation reaches us are affected by conditions prevailing inside the stars.

(k) Finally, a very important body of statistical information is contained in the Herzsprung-Russell and color magnitude diagrams (see Figs. A.3, 1.4, 1.5, and 1.6).

The distribution of stars within quite narrow confines on a H-R diagram sets a relatively detailed standard that must be met by any theory of stellar structure and evolution. The theory must prohibit the appearance of stars with characteristics corresponding to the empty regions of the diagram. Moreover, the relative density of stars in the populated portions of the H-R diagram must also be explained by such a theory. Finally, the theoretical evolution of a model star must always take it from one well-populated part of the diagram to another, without straying into an empty region or spending much time in a sparsely populated domain. For, if a correct model star strayed in this way, we would surely observe similar stray stars in nature.

The detailed structure of the H-R diagram then provides us with a truly severe observational criterion that must be met by theories of stellar evolution. An acceptable theory must explain the significance of the main sequence, the existence of the red giant and horizontal branches, the meaning of the variable turn-off point that has the main sequence joining the red giant branch at differing locations in different groups of stars, and a variety of other features.

We find then that we know too little about stellar masses and diameters to have these parameters play a hypercritical role in the theory of stellar structure, but the Hertzsprung-Russell diagrams that are available for a large number of different stellar groups and populations, and the tables of chemical abundances compiled for many astronomical objects, provide a wealth of observational detail against which to gauge the merit of our theories.

The purpose of the present chapter is to outline the main ideas involved in our current theories of stellar structure, and to show the extent to which these theories fit observations.

8:2 SOURCES OF STELLAR ENERGY

We have shown how one determines the overall characteristics of stars—their radii, masses, and luminosities. Knowing these, we can now ask ourselves what might cause stars to shine so brightly? Clearly they shine because they are hot. But why are they hot? What is the source of energy that can heat a star and replenish the energy that it loses so readily in the form of starlight?

Before these problems can be discussed, we may want to answer a somewhat different question: How much energy does a typical star radiate away in the course of its life? Here we may want to know the average luminosity of the star and its age at death—whatever form that death might take. It is not easy to decide how old a given star is, because stellar ages are far greater than the few hundred years during which reliable astronomical observations have been carried out; but two pieces of information are useful.

First, stars like the sun, that occupy positions on the lower main sequence of the Hertzsprung-Russell diagram, are found not to have noticeably changed either in brightness or in color since photographic techniques became well established around the turn of the century.

Second, the sun must be older than the earth which, as judged from the abundance of the radioactive uranium isotope U^{238} and its decay products, is more than four aeons old. The sun is thought to be of the order of 4.5 æ old. From paleontological evidence, we surmise that the temperature of the sun cannot have varied a great deal in the past 3×10^9 y during which life has existed on earth. Fossil remains that we find today indicate that liquid water must have been present on earth during this entire interval. Had the sun been somewhat cooler or hotter during this epoch, the oceans might have frozen or evaporated away, and the observed early forms of life would have died out.

We can therefore assume that, to rough approximation, the sun has radiated at its present rate for about five aeons. Since its luminosity is $L_\odot = 4 \times 10^{33}$ erg sec^{-1}, the total radiated energy emitted by the sun thus far is $\sim 6 \times 10^{50}$ erg. Because the solar mass is $M_\odot = 2 \times 10^{33}$ g, this amounts to an an energy-to-mass ratio of 3×10^{17} erg g^{-1}.

We ask ourselves whether this much energy could be made available by chemical reactions, or else through slow gravitational contraction which, as seen from equation 4–136, yields radiant energy in quantities that are of the order of the increase in absolute value of the potential energy gained by the contracting body.

Neither of these sources turns out to be adequate. The energy yield of chemical reactions normally does not exceed 100 kilocalories, or about 4×10^{12} erg per gram. If the sun had to depend on chemical sources, its total age would be no greater than $\sim 5 \times 10^4$ years. This is a factor of 10^5 too short!

If we assume, for purposes of a rough estimate, that the sun has the same density throughout, its total potential energy would be

$$\mathbb{V} = -\int_0^{R_\odot} \left(\frac{4\pi}{3}\right) \rho r^3 \frac{G}{r} (4\pi\rho r^2) \, dr = \frac{3}{5} \frac{M_\odot^2 G}{R_\odot} \qquad (8\text{--}1)$$

which amounts to $\sim 2 \times 10^{48}$ erg. This corresponds to 10^{15} erg per gram, still two or three orders of magnitude short of the required energy. Even a hundredfold density increase at the center of the star could not change this result significantly.

We cannot, on these grounds alone, rule out a very dense central core with $\rho \sim 10^{15}$ g cm^{-3} and radius $R \sim 10^5$ cm: Approximately the right amount of gravitational energy would then be available. But while this source of energy seems to be important, for very compact stars, it appears to play no significant role for normal stars.

The only remaining source of energy involves nuclear reactions. The high abundances of hydrogen and helium found in the universe strongly suggest that hydrogen is transmuted into helium in the center of stars. The energy released per gram of hydrogen is very large in this reaction.

We note that the mass difference between four hydrogen atoms and one atom of helium is

$$4m_H - m_{H_e} = 0.029m_H \qquad (8-2)$$

The transmutation of hydrogen into helium therefore includes a mass loss of the order of 7×10^{-3} g for each gram of converted hydrogen. Since the energy given off in the annihilation of mass m is mc^2, this amounts to an energy liberation of 6×10^{18} erg/g—ample compared to the amount required, even if only a fraction of a star's hydrogen content is converted into helium (Be39).

If we now ask about the lifespan of stars on the main sequence and about the rate at which stars are born, we can proceed in the following way: Let us first assume that we know the lifespan τ_i for a given type, i, of main sequence star. Let the number density in the Galaxy be ϕ_i for this kind of star. We can then define a birthrate function—usually called the *Salpeter birthrate function* ψ_i

$$\psi_i = \frac{\phi_i}{\tau_i} \qquad (8-3)$$

giving the rate of star formation in unit volume of the Galaxy. For disk population stars the formation rate will of course be high only in and near the Milky Way disk, while the birthrate will be negligible in the halo.

We can also obtain a very rough estimate of the age of a star when it moves off the main sequence. Suppose that a certain fraction of the stellar mass $f(M)$ needs to be exhausted of hydrogen before the star moves onto the red giant branch. If the initial composition of the stellar material contains a fraction (by mass, not by number of atoms) X, in the form of hydrogen available for nuclear conversion, the energy $\mathscr{E}$ liberated by the star while it still resides on the main sequence is

$$\mathscr{E} = 6.4 \times 10^{18} f(M) XM \text{ erg} \qquad (8-4)$$

The numerical factor gives the energy in ergs liberated by one gram of hydrogen converted into helium. The time taken to expend this energy is just the energy $\mathscr{E}$ divided by the star's luminosity L. Now Fig. 8.1 shows that the mass-luminosity relation for main sequence stars is roughly $L = L_\odot (M/M_\odot)^a$, where $3 \lesssim a \lesssim 4$.

Fig. 8.1 Mass-luminosity diagram for main sequence stars. (After M. Schwarzschild, Sc58b. From Martin Schwarzschild, *Structure and Evolution of the Stars*, copyright © 1958 by Princeton University Press, p. 16.)

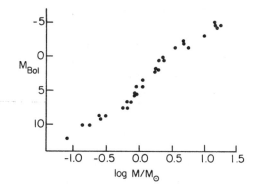

Taking $a \sim 3.5$, the star's life τ on the main sequence becomes

$$\tau = 6.4 \times 10^{18} X f(M) \left(\frac{M_\odot}{M} \right)^{5/2} \frac{M_\odot}{L_\odot} \qquad (8\text{–}5)$$

$f(M)$ is of order 15 % for stars with solar composition for which $X \sim 0.7$. Inserting these two numerical values, making use of the mass-luminosity relation once more, and converting the time scale to years we find

$$\tau \sim \left(\frac{L_\odot}{L} \right)^{5/7} \times 10^{10} \text{ y} \qquad (8\text{–}6)$$

The sun should therefore have a total lifespan of 10^{10} y, while O stars, which are some ten thousand times brighter, should survive only a few million years.

8:3 REQUIREMENTS IMPOSED ON STELLAR MODELS

Granted that sufficient energy is available from hydrogen "burning" (8–2) and perhaps from other nuclear reactions, we still need to investigate whether the hypothesis of nuclear energy conversion also fits all the other characteristics observed in stars. These are:

(a) Conditions inside stars must be compatible with adequate nuclear reaction rates. These rates must be such that the energy generation rate is of the order of the observed luminosity of stars. The energy released at the star's surface must further be predominantly in the form of visible, ultraviolet or infrared radiation since most of a normal star's radiation is actually emitted at these wavelengths. If a predominant fraction of the generated energy were channeled elsewhere, say into neutrino emission, then we would still be faced with the problem of accounting for the visual starlight.

(b) We will find that nuclear reaction rates depend on the temperature, density, and chemical composition of the matter in stars, and it will be important to discover whether the values of these parameters, needed to maintain a given luminosity, are compatible with stable stellar structure.

Pressure equilibrium, for example, must be maintained throughout the star. But this pressure is determined by two factors. First, the pressure in any region is determined by the local temperature, density, and chemical composition. The relationship between these quantities is summarized in an equation of state such as the ideal gas law or some similar expression. Second, the local pressure must be just able to support the weight of material lying overhead—matter at larger radial distance from the center of the star. This is called the condition of *hydrostatic equilibrium.*

If the temperature and density are too high, the local pressure becomes too large and the star expands. If the pressure is too low, the star will contract. We will see that any appreciable deviation from pressure equilibrium leads to a readjustment that takes no more than about an hour. A star that lives for many *aeons* must therefore be very close to pressure equilibrium throughout, unless the star pulsates.

(c) The energy generated at the center of a star must be able to reach the surface within a time small compared to the age of the universe or the evolutionary age of the star; otherwise, the whole life of the star would have to be described by transient conditions, and the stable characteristics of main sequence stars could not be explained.

(d) The temperature at any given distance from the center of the star must not only lead to the correct pressure (condition b), it must also be compatible with adequate energy transfer rates to assure that the luminosity just equals the rate of energy generation (condition c).

(e) At the center of the star the *luminosity* must be zero. This means that there is no finite outflow of energy, no mysterious source pouring out energy from an infinitesimal volume about the center of the star.

At the same time, there can be no more than an infinitesimal mass enclosed in an infinitesimal volume about the center of the star. These two requirements impose boundary conditions on the differential equations implied by requirements (a) to (d).

(f) At the surface of the star the pressure and temperature can usually be taken to be very small compared to values found in the central regions. This follows from the equation of state and from condition (b) that required pressure balance throughout the star. It is a statement of the fact that stars have high internal pressures and that the boundary between star and surrounding empty space is relatively sharp. Nevertheless, some caution has to be observed in applying this last condition; and differences will arise between early spectral-type stars—where energy is transported in the surface layers primarily by radiation—and late spectral types whose surfaces are convective.

8:4 MATHEMATICAL FORMULATION OF THE THEORY

The requirements described above can be summarized in a number of differential equations. In giving this formulation we will find it convenient to follow a procedure slightly different from that of section 8:3.

(a) The change of pressure dP on moving a distance dr outward from the center of a star is

$$dP = -\frac{\rho GM(r)}{r^2} dr \qquad (8-7)$$

where ρ is the local density and $M(r)$ is the mass enclosed by a surface of radius r. This increment of pressure is produced by the gravitational attraction between $M(r)$ and the mass $\rho\, dr$ enclosed in the incremental volume of height dr and unit base area. G is the gravitational constant.

(b) The change of mass $dM(r)$ on moving a distance dr outward from the center is

$$dM(r) = 4\pi r^2 \rho\, dr \qquad (8-8)$$

(c) The change of luminosity $L(r)$ within an increment dr at distance r from the center of the star is

$$dL(r) = 4\pi r^2 \rho \varepsilon\, dr \qquad (8-9)$$

where ε is the energy generation rate per unit mass.

(d) In general, this generation rate is a function of the local density ρ, temperature T, and the mass concentrations X_i of elements i. Hydrogen and helium mass concentrations are usually labeled X and Y, respectively:

$$\varepsilon = \varepsilon(\rho, T, X, Y, X_i) \qquad \text{where} \qquad i = 1, \ldots, n \qquad (8-10)$$

when n elements other than hydrogen and helium are present in significant amounts.

(e) The local temperature is related to the pressure, density, and chemical composition. We will find it convenient to write this in the form

$$P = P(\rho, T, X, Y, X_i) \qquad i = 1, \ldots, n \qquad (8-11)$$

8:4

because it will facilitate comparison of pressures derived from expressions 8–7 and 8–11. The right side of this equation is an expression of the equation of state that often is well approximated by the ideal gas law (4–37).

(f) Next, the temperature gradient must be related to the parameters that assure a stable luminosity profile throughout the star. Two possibilities arise here:

(i) If the star has a low opacity κ,

$$\kappa = \kappa(\rho, T, X, Y, X_i), \qquad i = 1, ..., n \quad\cdot \tag{8–12}$$

so that light can travel long distances within the star before being absorbed or scattered, then no large temperature gradients can arise. In this case the transfer of energy is achieved by radiation alone. The photons are emitted, scattered, absorbed, re-emitted many times; and their energy and number density changes as they diffuse through the star in a complex random walk that eventually takes them from the center to the star's surface. There they start on their long journey through space.

(ii) If the opacity is high, this random walk may be excessively slow. The center of the star then becomes too hot and the stellar material starts to convect. A convective pattern of heat transfer sets in and, as we will see below, the temperature gradient is given by the so-called *adiabatic lapse rate* that depends on the ratio of heat capacities of the material, $\gamma = c_p/c_v$ (see section 4 :18).

Corresponding to these two alternatives, we can derive temperature gradients of the form

$$\frac{dT}{dr} = F_1[\kappa, L(r), T, r] \qquad \text{for radiative transfer} \tag{8–13}$$

or

$$\frac{dT}{dr} = F_2(T, P, r, \gamma) \qquad \text{for convective transfer} \tag{8–14}$$

(g) The two boundary conditions implied by (e) and (f) in section 8:3 are

(i) at $r = 0$, $M(r) = 0$, and $L(r) = 0$ \hfill (8–15)

(ii) at $r = R$, $T \ll T_{central}$, and $P \ll P_{central}$ \hfill (8–16)

where R is the star's radius. For purposes of computing hydrostatic pressures, the relations (8–16) are tantamount to writing

$$T(R) \approx 0, \qquad P(R) \approx 0 \tag{8–17}$$

Equations 8–7 to 8–17 constitute the foundations of the theory of stellar structure. We will examine them in greater detail throughout much of this chapter.

8:4

One point, however, is particularly interesting and should be mentioned now. The equations state nothing about the physical source of the generated energy. The overall structure and appearance of the star can therefore give no clue about whether nuclear reactions indeed are responsible for stellar luminosities, or which particular reactions predominate at any given evolutionary stage. We have to derive this information by indirect means—mainly by looking at the debris ejected by stars that become unstable, or by spectrally analyzing stars whose surfaces expose matter previously evolved at the center.

8:5 RELAXATION TIMES

Suppose we could artificially change the temperature or pressure within a star. After this perturbation stopped, the star would again relax to its initial temperature and pressure equilibrium.

The time required to relax is called the *relaxation time*. We will find that the relaxation time in response to a pressure change is very much faster than the time required to re-establish temperature equilibrium.

(a) We first wish to estimate the time required to reach pressure equilibrium. Let the perturbed pressure $P_p(r)$ differ from the equilibrium pressure $P(r)$ by a fractional amount f

$$P_p(r) - P(r) = fP(r) \tag{8–18}$$

This pressure acts on a mass $M - M(r)$ lying at radial distance greater than r, with a force $F = 4\pi r^2 f P(r)$. As a result, that material moves with an acceleration

$$\ddot{r} = \frac{4\pi r^2 f P(r)}{M - M(r)} \tag{8–19}$$

We suppose that a displacement Δr amounting to a fraction g of the total radius R is required to relieve the pressure difference

$$\Delta r = gR \tag{8–20}$$

Then the time required to obtain this displacement with the acceleration given in equation 8–19 is of the order of

$$\tau_P \sim \left(\frac{2\Delta r}{\ddot{r}} \right)^{1/2} = \left[\frac{gR[M - M(r)]}{2\pi r^2 f P(r)} \right]^{1/2} \tag{8–21}$$

Let us compute the approximate value of τ_P. We can estimate $P(r)$ and $M(r)$ by assuming a uniform density throughout the star, and considering a star with

8:5

one solar mass $M = M_\odot = 2 \times 10^{33}$ g contained in one solar radius $R = R_\odot = 7 \times 10^{10}$ cm. The density then is $\rho \sim 1$, and from (8–7)

$$P(r) = - \int_R^r \frac{4\pi}{3} \rho^2 r G \, dr = \frac{2\pi}{3} \rho^2 G (R^2 - r^2) \qquad (8\text{--}22)$$

Let us choose $r \sim R/2$, then

$$P\left(\frac{R}{2}\right) \sim 10^{15} \text{ dyn cm}^{-2}$$

$$M(R) - M\left(\frac{R}{2}\right) \sim 2 \times 10^{33} \text{ g}$$

and

$$\tau_P \sim 5 \times 10^3 \sqrt{\frac{g}{f}} \text{ sec}$$

For small perturbations, g/f can be expected to be of order unity and the relaxation time is of the order of an hour.

We note that the speed of propagation of pressure information is roughly $(P/\rho)^{1/2}$ (see equation 4–31). This speed is of order 3×10^7 cm/sec. Pressure information can therefore be conveyed over distances $R_\odot$ in $\sim 2 \times 10^3$ sec, a time comparable to the pressure adjustment time.

PROBLEM 8–1. (a) Show that the temperature $T(R/2)$, under the conditions assumed above, is $\sim 10^7 \, °\text{K}$.

(b) One difference between a planet and a star (Sa70a) is that for planets Coulomb forces on electrons and ions are more important than gravitational forces. The opposite is true of a star. Let $\mathscr{E}_c$ be a typical Coulomb interaction energy. Show that a planet's mass M_p is of order

$$M_p \lesssim \frac{1}{\rho^{1/2}} \left[\frac{\mathscr{E}_c}{GAm_H} \right]^{3/2} \sim \frac{R\mathscr{E}_c}{GAm_H} \qquad (8\text{--}23)$$

where A is an average atomic mass, measured in atomic mass units, and m_H is the mass of the hydrogen atom.

(c) If $\mathscr{E}_c$ is of the order of binding energies of solid material, about 10^{-11} erg, show that Jupiter, which consists largely of hydrogen, lies near the upper limit of the mass range for planets.

(b) Next we wish to estimate the time taken to transport heat from one point within the star to another. If the transport process is radiative, the time can be computed as a random walk process; we only need to know the mean free path of radiation, and that is given (see 7:12) by the opacity of the material κ. When a beam of n photons passes through a layer of thickness dl, a fraction dn will be absorbed or scattered by the material. The loss of photons from the beam can then be expressed as

$$dn = - n\kappa\rho \, dl \tag{8–24}$$

where ρ is the density of the material. We note that we have not gone into detail about the scattering process. Some processes strongly favor light scattering into a forward direction. Such scatterers make a medium much less opaque than isotropic scattering centers. We will assume here that the scattering is isotropic. Alternatively, we could count a photon as being lost from the beam only after a large number of collisions has increased its angle with the original direction of propagation to some large value, say $90°$, so that all memory of the original direction is lost. We made a similar assumption about electron scattering in (6:16). We wish to calculate the mean free path of the photons under such conditions. Integrating equation 8–24 we obtain

$$n = n_0 e^{-\kappa\rho l} \tag{8–25}$$

The mean distance $\langle l \rangle$ traveled by a photon before it is absorbed or strongly scattered is then

$$\langle l \rangle = - \frac{\displaystyle\int_0^{n_0} l \, dn}{n_0} = \int_0^{\infty} l\kappa\rho e^{-\kappa\rho l} \, dl = \frac{1}{\kappa\rho} \tag{8–26}$$

For a star like the sun, $\kappa\rho$ is of order unity, and the mean free path is of the order of a centimeter. To traverse a distance of the order of the solar radius $R \sim 10^{11}$ cm, we would require 10^{22} steps, which would cover a total distance $\sim 10^{22}$ cm. The total time taken is $\sim R^2\kappa\rho/c$ sec when we do not count the time required between absorption and reemission. The time constant therefore is at least of the order of 10^{11} sec, several thousand years.

(c) Energy can also be transported by convection, provided high enough thermal gradients can be set up. A buoyancy force then accelerates a hot blob of matter upward and returns cooler matter down toward the center of the star. For nondegenerate matter we can take $\Delta\rho$, the density difference between hot and cold material, to be

$$\Delta\rho \sim \frac{\rho}{T} \Delta T \tag{8–27}$$

the upward force on unit volume of the hotter material is

$$F(r, \rho, \Delta T) = \frac{M(r)G}{r^2} \frac{\rho}{T} \Delta T \tag{8–28}$$

which leads to a convective motion accelerated at a rate

$$\ddot{r} = \frac{M(r)G}{r^2} \frac{\Delta T}{T} \tag{8-29}$$

If the blob travels a distance of the order of one-tenth of a solar radius, the time required is

$$t = \left[\frac{2R_\odot}{10\ddot{r}} \right]^{1/2} \sim 3 \times 10^6 \text{ sec} \sim 1 \text{ month} \tag{8-30}$$

for $M_r \sim 10^{33}$ g, $T \sim 10^7$ °K, and $\Delta T \sim 1$°K

This is a rather fast transport rate. It sets in whenever radiative heat transfer is too slow to maintain thermal equilibrium within the stars. We will return to this stability problem in section 8:9, where we will also justify the choice of $\Delta T \sim 1$°K.

(d) If the center of a star is degenerate, the electrons can readily transport heat at a rate much faster than is possible by either radiative or convective means. This comes about since the electrons cannot give up their energy to other particles: All the lower electron energy states already are filled and there is no space for another electron that is about to lose energy. The mean free path for electrons therefore becomes extremely long, and heat transport proceeds swiftly. In the limiting case an electron could traverse the entire degenerate region and not lose energy until it reaches the nondegenerate surroundings. If the span in question amounts to a distance of the order of $R_\odot/10^2$, the traversal times at $T \sim 10^7$ °K would be of the order of one second. This represents the thermal relaxation time for the degenerate core of a star.

8:6 EQUATION OF STATE

The equation of state needed to define the pressure in terms of temperature, density, and composition depends on whether (i) conditions at the center of a star are nondegenerate or degenerate, and (ii) the temperature is high enough to involve relativistic behavior.

(a) Nondegenerate Plasma

At the high temperatures found within stars all but the heaviest elements are completely ionized. Under these conditions the electrons and ions are far apart compared to their own radii since electrons and bare nuclei have radii of order

10^{-13} cm. The ideal gas law can therefore be expected to hold:

$$P = nkT \qquad (4\text{--}38)$$

where n is the number of particles in unit volume.

In Table 8.1 we enumerate the contribution to the number density by the various particles.

Table 8.1 Number Densities

	Number of Ions	Number of Electrons
Hydrogen	$\dfrac{X\rho}{m_H}$	$\dfrac{X\rho}{m_H}$
Helium	$\dfrac{Y\rho}{4m_H}$	$\dfrac{Y\rho}{2m_H}$
Others	$\dfrac{Z\rho}{\langle A \rangle m_H}$	$\dfrac{Z\rho}{2m_H}$

The symbols X, Y, Z, represent the concentration, by mass, of hydrogen, helium, and heavier elements, respectively. $\langle A \rangle$ is the mean atomic mass of the heavier elements. In the last column of the table the number of electrons contributed by the heavier elements is obtained on the hypothesis that the number of electrons per atom is $\langle A \rangle/2$. This is a fairly good approximation for the less massive elements. The number of ions contributed by the heavier elements amounts to a negligibly small fraction of the total population—only about one part per thousand. The total number density of particles to be inserted in the ideal gas law relation, therefore, is roughly

$$n = \frac{\rho}{m_H}\left[2X + \frac{3}{4}Y + \frac{1}{2}Z \right] \qquad (8\text{--}31)$$

and the equation of state reads

$$P = \frac{\rho kT}{m_H}\left[2X + \frac{3}{4}Y + \frac{1}{2}Z \right] \qquad (8\text{--}32)$$

At first we might think that P represents the total pressure; but it does not. It is only the pressure contribution due to particles. A further pressure, due to the presence of electromagnetic radiation, must be added to the particle pressure to give the total pressure. This is true both for the nondegenerate and the degenerate situation.

We had already found in section 4:7 that the radiation pressure has a value numerically equal to one-third of the energy density. Inside the star that density is aT^4; the refractive index is practically unity, and the relationship of Problem 4–21 reduces to equation 4–72. Hence

$$P_{Rad} = \frac{aT^4}{3} \tag{8–33}$$

The equation of state for nondegenerate matter now reads

$$P_{Total} = \frac{\rho kT}{m_H} \left(2X + \frac{3}{4}Y + \frac{1}{2}Z \right) + \frac{aT^4}{3} \tag{8–34}$$

(b) Degenerate Plasma

The maximum number of electrons that can occupy unit volume is (see section 4:11)

$$n_e = \frac{8\pi}{3} \frac{p_0^3}{h^3} \tag{8–35}$$

where p_0 is the momentum corresponding to the Fermi energy. The number density of electrons also can be written as

$$n_e = \left(X + \frac{1}{2}Y + \frac{1}{2}Z \right) \frac{\rho}{m_H} \equiv \frac{1}{2}(1 + X) \frac{\rho}{m_H} \tag{8–36}$$

since

$$X + Y + Z = 1 \tag{8–37}$$

The pressure contribution due to isotropically moving electrons is then given through equations 4–27, 4–28, and 4–30, as

$$P_e = \int_0^{p_0} \int_0^{2\pi} \int_0^{\pi/2} 2n_e(p)\, p \cos\theta\, v \cos\theta \sin\theta\, d\theta\, d\phi\, dp \tag{8–38}$$

$$= \frac{1}{3} \int_0^{p_0} \frac{8\pi p^2}{h^3} pv\, dp \tag{8–39}$$

where $n_e(p)$ is the number density of electrons having momenta in the range p to $p + dp$.

(i) In the nonrelativistic case $v = p/m_e$ and the electron pressure is

$$P_e = \frac{8\pi}{15} \frac{p_0^5}{m_e h^3} \tag{8-40}$$

Substituting for p_0 from (8–35) and (8–36), equation 8–40 becomes

$$P_e = \frac{h^2}{20m_e m_H} \left(\frac{3}{\pi m_H}\right)^{2/3} \left(\frac{(1 + X)}{2}\rho\right)^{5/3} \tag{8-41}$$

(ii) In the relativistic case $v \sim c$, and equation 8–38 integrates to

$$P_e = \frac{2\pi c}{3h^3} p_0^4 \tag{8-42}$$

Using the same device to eliminate p_0, we obtain

$$P_e = \frac{hc}{8m_H} \left(\frac{3}{\pi m_H}\right)^{1/3} \left(\frac{1 + X}{2}\rho\right)^{4/3} \tag{8-43}$$

To obtain the total pressure we need to add the pressures P_i contributed by individual ions. These normally are non degenerate, as was pointed out in section (4:15).

$$P_i = \left(X + \frac{1}{4}Y\right)\frac{\rho kT}{m_H} \tag{8-44}$$

Finally we have to add the radiation pressure from equation 8–33 to obtain

$$P_{Total} = P_e + P_i + P_{Rad} \tag{8-45}$$

8:7 LUMINOSITY

We have estimated the time required for a star to recover from a thermal perturbation when a range of different conditions prevails within the star. However, we still have to ask ourselves "When does each of these different conditions predominate? Under what circumstances is radiative heat transfer dominant? When is convection a major contributor, and what are the conditions that favor degeneracy?" These are the problems we must look at next. When we obtain an answer we will also be able to quantitatively express the rates of energy transfer that add to give the total luminosity of a star.

8:7

Fig. 8.2   Illustration to show relation between luminosity and temperature gradient.

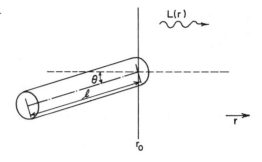

The total flux at radial distance $r = r_0$ from the star's center is the difference between the outward and inward directed energy flow. Let the temperature at r_0 be T_0. Radiation passes through the surface r_0—which can be assumed to be plane since the radiation mean free path is very small compared to r_0—at all values of azimuthal angle θ (Fig 8–2). The unattenuated flux originating at a distance l along direction θ, which would pass through the surface in unit time, would be

$$aT^4(l, \theta) \cdot c \cos \theta \cdot \frac{2\pi \sin \theta \, d\theta}{4\pi} \qquad (8\text{--}46)$$

where

$$T(l, \theta) = T_0 - \frac{dT}{dr} l \cos \theta \qquad (8\text{--}47)$$

$c \cos \theta$ represents the cylindrical volume from which radiation crosses unit area of the surface in unit time, and $2\pi \sin \theta \, d\theta / 4\pi$ gives the fraction of the total solid angle at (l, θ) containing the radiation that will pass through the appropriate unit area, at r_0.

In actuality, radiation from (l, θ) does not reach r_0 unattenuated. A photon originating at l only has a probability $\pi(l)$ of reaching r_0:

$$\pi(l) = \kappa\rho e^{-\kappa\rho l} \qquad (8\text{--}48)$$

This follows from (8–24) and also gives the proper normalization

$$\int_0^\infty \pi(l) \, dl = 1 \qquad (8\text{--}49)$$

The radiant flow through unit area is now formally given as

$$F(r_0) = \int_0^\infty \int_0^\pi a \left(T_0 - \frac{dT}{dr} l \cos \theta \right)^4 \cdot c \cos \theta \cdot \frac{2\pi \sin \theta \, d\theta}{4\pi} \pi(l) \, dl \qquad (8\text{--}50)$$

8:7

However, to obtain the actual radiative flux we must decide what value of κ to use in equation 8–48. Equation 8–24 did not take into account stimulated emission which, as explained in section 7:12, is important. On the other hand, if we use $\kappa_*(v)$ from (7–71) we have to properly average over all frequencies to arrive at a suitable mean opacity. This will be done in section 8:8 below.

PROBLEM 8–2. The luminosity $L(r)$ at any radial distance within the star is

$$L(r) = 4\pi r^2 F(r) \tag{8–51}$$

Show by integration that to first order

$$L(r) = -\frac{16\pi ac}{3\kappa\rho} r^2 T^3 \frac{dT}{dr} \tag{8–52}$$

8:8 OPACITY INSIDE A STAR

In section 7:12 we had already discussed the four sources of opacity: electron scattering, free-free transitions, free-bound transitions, and bound-bound transitions. However, we have not yet indicated how to compute the mean opacity obtained from these four contributing factors. It is that opacity that has to be used in expression 8–52.

Two factors enter into consideration of the mean opacity to be used in this expression. First, we have to average over all radiation frequencies; but clearly those frequencies at which the radiation density gradient is greatest should receive greater weight in the averaging process if the opacity is to give an accurate assessment of the radiative transfer rate. Secondly, those frequency ranges in which the opacity is smallest potentially make the greatest contribution to energy transport. We therefore will be more interested in averaging $1/\kappa(v)$ rather than $\kappa(v)$.

Let us write (8–52) in its more fundamental form, involving energy density $\rho(v)$ of radiation at frequency v and temperature T (see equation 4–71).

$$L(r, v) = \frac{-4\pi r^2}{3\rho\kappa^*(v)} c \frac{d\rho(v)}{dr} \tag{8–53}$$

Here we have defined a contribution $L(r, v)$, at frequency v, to the total luminosity $L(r)$ at r; and we have set the total energy density U equal to the blackbody energy density. $\kappa^*(v)$ is the opacity at frequency v that takes account of stimulated emission:

$$L(r) = \int_0^\infty L(r, v)\, dv \quad \text{and} \quad U = \int_0^\infty \rho(v)\, dv = aT^4 \quad (8\text{--}54)$$

We can neglect bound-bound transitions that play a negligible role in the stellar interior. Equation 7–71 therefore simplifies to

$$\kappa^*(v) = [\kappa_{bf}(v) + \kappa_{ff}(v)](1 - e^{-hv/k}) + \kappa_e \quad (8\text{--}55)$$

and we can define a mean opacity

$$\frac{1}{\kappa} = \frac{\displaystyle\int_0^\infty \frac{1}{\kappa^*(v)} \frac{d\rho(v)}{dT} \frac{dT}{dr}\, dv}{\displaystyle\int_0^\infty \frac{d\rho(v)}{dT} \frac{dT}{dr}\, dv} \quad (8\text{--}56)$$

called the *Rosseland mean opacity*, in which (4–71) can be substituted for $d\rho(v)/dT$. The Rosseland mean opacity, as can be seen from (8–53), does indeed favor the frequencies important to the transfer process, by using the energy density gradient $d\rho(v)/dr$ as a weighting function for $1/\kappa^*(v)$, which is a measure of the mean free path at frequency v. The opacity at any frequency is the sum of contributions from bound-free (bf) and free-free (ff) transitions, and from electron scattering (e).

$\kappa_{bf}(v)$ and $\kappa_{ff}(v)$ themselves are sums over the opacity contributions of the individual states of excitation n, of the various atoms and ions A present at radial distance r, in the star

$$\kappa_{bf}(v)\rho = \sum_{A,n} \alpha_{bf} \left(\frac{X_A \rho}{A m_H}\right) N_{A,n} \quad (8\text{--}57)$$

$$\kappa_{ff}(v)\rho = \sum_A \int \alpha_{ff} \frac{X_A \rho}{A m_H} n_e(v)\, dv \quad (8\text{--}58)$$

Here $X_A\rho/Am_H$ is the number density of atoms of kind A, X_A is the abundance by mass of atoms or ions with mass number A, m_H is the mass of a hydrogen atom, and $N_{A,n}$ is the fraction of these atoms or ions in the nth excited state. $n_e(v)$ is the number density of electrons in a velocity range dv around v. α_{bf} and α_{ff} are the atomic absorption coefficients defined in (7–74) and (7–75). As shown in (7–72),

$$\kappa_e \rho = \sigma_e n_e \quad (8\text{--}59)$$

where the right side is the product of the electron number density and the Thomson (or—at high energies—the Compton) scattering cross section.

8:8

To evaluate $N_{A,n}$ we make use of the *Saha equation* (4–105) that, for high ionization, leads to

$$N_{A,n} = n^2 \left[n_e \frac{h^3}{2(2\pi m_e kT)^{3/2}} e^{\chi_n/kT} \right] \qquad (8\text{–}60)$$

where we have considered that most of the ions are in the $r + 1$st ionization state. We can understand this equation in the following way:

χ_n is the energy needed to ionize the atomic species A from the nth excited state; m_e is the electron mass. Using a Bohr atom model this energy is (7–5)

$$\chi_n \sim \frac{2\pi^2 e^4 m_e}{h^2} \frac{Z'^2}{n^2} \qquad (8\text{–}61)$$

where Z' is the effective charge of the ion considered. Equation 8–61 assumes that all the excited atoms of a given species A will be in the same state of ionization at radial distance r from the star's center. In our present notation, this means that in equation 4–105 $n_r/n_{r+1} = N_{A,n}$. We note that $N_{A,n}$ is proportional to n^2. This is because the statistical weight g_r—the number of sublevels—of the nth excited state is $2n^2$ (see Problem 7–2). From section 4:16 we also have $g_e = 2$. Similarly, the ion can also exhibit two states of polarization $g_{r+1} = 2$. But for any given final state there are only two possible combinations of polarization, $g_{r+1}g_e = 2$.

Making use of equation 7–75 for α_{bf}, with χ_n from (8–61) substituted into this expression, we can now obtain

$$\kappa_{bf}(\nu) = \frac{2}{3} \sqrt{\frac{2\pi}{3}} \frac{Z'^2 e^6 h^2 \rho(1 + X) Z}{c A m_H^2 m_e^{1.5}(kT)^{3.5}} \left[\frac{1}{n} \frac{\chi_n}{kT} e^{\chi_n/kT} \left(\frac{kT}{h\nu} \right)^3 g_{bf} \right] (8\text{–}62)$$

Here Z is the metal abundance, by fraction of the total mass. We have summed (8–57) only for these constituents, since hydrogen and helium do not contribute significantly to the bound-free transitions. The summation over states has been neglected, since the lowest state n usually contributes most. We have used an electron density from Table 8.1:

$$n_e = \tfrac{1}{2}(X + 1) \frac{\rho}{m_H} \qquad (8\text{–}63)$$

Equation 8–62 can be considerably simplified, if approximate values of the opacity suffice. For example, we can restrict our attention to those levels for which $\chi_n/kT \sim 1$, $h\nu/kT \sim 1$, since this makes use of the frequencies and ionization potentials that will contribute most to the opacity.

Constituents that would be ionized at lower temperatures, $\chi_n \ll kT$, already are almost fully ionized and have too few bound electrons to be effective, while

8:8

those with higher χ_n values absorb too few of the photons present. Similarly the photons of frequency $v \sim kT/h$ are weighted most favorably by the Rosseland mean.

For most elements we can also choose a typical value $Z'^2/A \sim 6$.

With these approximations we obtain *Kramer's Law of Opacity* for bound-free absorption:

$$\kappa_{bf} = 4.34 \times 10^{25} Z(1 + X) \frac{\rho}{T^{3.5}} \frac{\langle g_{bf} \rangle}{f} \qquad (8\text{--}64)$$

where $\langle g_{bf} \rangle$ is the mean Gaunt factor, which is always of order unity, and f contains correction factors—all of order unity also—that arise because of the approximations we have made. For free-free transitions, we can similarly obtain expressions (Sc58b)*:

$$\kappa_{ff} = \frac{2}{3} \sqrt{\frac{2\pi}{3}} \frac{e^2 h^6 (X + Y)(1 + X) \rho}{cm_H^2 m_e^{1.5} (kT)^{3.5}} \frac{1}{196.5} g_{ff}$$

$$= 3.68 \times 10^{22} \langle g_{ff} \rangle (X + Y)(1 + X) \frac{\rho}{T^{3.5}} \qquad (8\text{--}65)$$

where $\langle g_{ff} \rangle$ is the mean Gaunt factor (7–74). We note that if we had taken $\kappa(v)$ in equation 6–137 and substituted into (8–56) for $\kappa^*(v)$, we would have obtained an opacity expression proportional to $e^6 n^2 c^{-1} (m_e kT)^{-1.5}$ and a weighted mean function proportional to v^{-2} that would be proportional to $h^2 (kT)^{-2}$. This is just the dependence found in (8–65). For electron scattering, (8–59) combined with the number of free electrons (8–63), we obtain

$$\kappa_e = \frac{4\pi}{3} \frac{e^4}{c^4 m_H m_e^2} (1 + X) \sim 0.19(1 + X) \qquad (8\text{--}66)$$

Electron scattering is the main contributor to the opacity at low densities and high temperatures, where the interaction between electrons and ions is weakened.

Figure 8.3 shows the relative importance of scattering and absorption processes for different densities and temperatures. At high densities, where electrons become degenerate, heat is transferred most rapidly through conduction by these electrons.

Figure 8.4 shows the opacity as a function of temperature in stars whose composition is similar to that of the sun.

Thus far we have discussed radiative transfer only in the interior of a star. However, the equations of radiative transfer also play a dominant role in the transport of energy through stellar atmospheres (section 7:13).

8:8

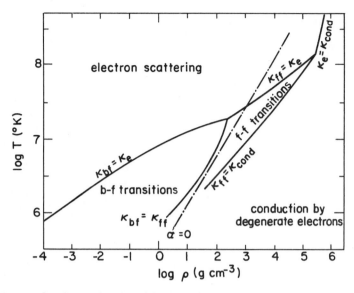

Fig. 8.3 Opacity as a function of density and temperature in a star of population I composition. The diagram is divided into four regions characterized by different mechanism of energy transport. The sources of opacity that dominate these mechanisms are electron scattering, bound-free transitions, free-free transitions, and the effective opacity that would describe the energy transport by degenerate electrons. The dashed line shows where the degeneracy *parameter* α (see equation 4-93) equals zero (after Hayashi, Hoshi, and Sugimoto, Ha62c).

PROBLEM 8-3. Using equations 8–7, 8–52, and the ideal gas law, show that the luminosity of stars should be roughly proportional to M^3.

We find, in reality, that main sequence stars more nearly obey the *mass-luminosity relation* (Fig 8.1):

$$L \propto M^a \qquad 3 \lesssim a \lesssim 4 \qquad (8–67)$$

Presumably this relation holds in main sequence stars because radiative transfer dominates there, while convective transfer (section 8:9 below) is more important in the giants, and degenerate electron transfer dominates in compact stars and in compact stellar cores.

However, radiative transfer is always present, even when these other processes dominate. The total energy transfer rate is then the sum of all the different transfer rates.

Fig. 8.4 Opacity for stars whose composition is similar to that of the sun. Each curve represents a different density value ρ, measured in g cm^{-3}. (After Ezer and Cameron, Ez65. With permission of the editors of *Icarus. International Journal of Solar System Studies*, Academic Press, New York.)

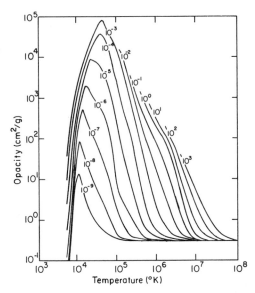

8:9 CONVECTIVE TRANSFER

Let us establish the conditions under which the temperature gradient becomes so large that the medium starts to convect, and the spherically symmetrical temperature distribution about the stellar center becomes unstable.

Consider an element of matter at some density ρ'_1 and pressure P'_1 surrounded by a region with exactly the same characteristics (ρ_1, P_1) (see Fig 8.5):

$$\rho'_1 = \rho_1 \qquad P'_1 = P_1 \tag{8-68}$$

The element is then moved to a new position, subscript 2, where its final pressure P'_2 equals the ambient pressure P_2:

$$P'_2 = P_2 \tag{8-69}$$

Using (8–30) we found that a convective motion of this kind is fast compared to the time required for radiative heat transfer, and we can therefore consider the process to be adiabatic. Equation 4–123 then implies that

$$\rho'_2 = \rho'_1 \left(\frac{P_2}{P_1} \right)^{1/\gamma} \tag{8-70}$$

Fig. 8.5 Convective outward motion of a low density "bubble." When thermal gradients become too high, such convective motion sets in and becomes the dominant vehicle for heat transport.

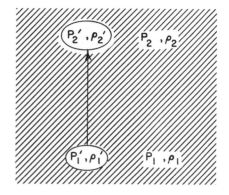

For the highly ionized plasma in a star, the ratio of heat capacities is $\gamma = c_p/c_v = 5/3$.

If the initial displacement of the element was upward, and we find that $\rho'_2 > \rho_2$, then the element will be forced down toward its initial position, 1; the medium is then stable. However if $\rho'_2 < \rho_2$, then the initial displacement leads to further motion along the same direction—upward—and the medium is unstable. Convection can then set in.

The condition for stability therefore is

$$\rho_2 < \rho'_2 = \rho'_1 \left(\frac{P'_2}{P'_1} \right)^{c_v/c_p} = \rho_1 \left(\frac{P_2}{P_1} \right)^{c_v/c_p} \tag{8-71}$$

where we have made use of expressions 8–68, 8–69, and 8–70. This can be rewritten as

$$\frac{d\rho}{\rho} < \left(\frac{P + dP}{P} \right)^{c_v/c_p} - 1 = \frac{c_v}{c_p} \frac{dP}{P} \tag{8-72}$$

In terms of radial gradients this becomes

$$\frac{1}{\rho} \frac{d\rho}{dr} < \frac{c_v}{c_p P} \frac{dP}{dr} \tag{8-73}$$

which, with the ideal gas equation 4–37, leads to the stability condition

$$\frac{dT}{dr} > \frac{T}{P} \left(1 - \frac{c_v}{c_p} \right) \frac{dP}{dr} \tag{8-74}$$

Both dP/dr and dT/dr have negative values. The right side of (8–74) is called the adiabatic temperature gradient, and we conclude that stability will prevail when

8:9

the absolute value of the temperature gradient is less than that of the adiabatic gradient. When the absolute value of dT/dr becomes larger than the absolute value of the adiabatic gradient, instability sets in and heat is transferred by convection.

To compute the heat transfer rate we have to know four quantities: the velocity v of the buoyant element, its heat capacity C, its density, and the temperature differential ΔT between the element and the surroundings to which it finally imparts this temperature. The heat transport rate per unit area is then

$$H = C\rho v\, \Delta T \qquad (8\text{-}75)$$

Here C is the heat capacity under the assumed adiabatic conditions. Using the acceleration given in equation 8–29 and assuming transport over a distance one-tenth of a solar radius, the mean velocity v is of order $\left[\ddot{r}\,\dfrac{R_\odot}{10}\right]^{1/2}$

$$H \sim C\rho \left[\frac{GM(r)}{Tr^2}\frac{R_\odot}{10}\right]^{1/2} (\Delta T)^{3/2}$$

$$\sim C\rho \left[\frac{GM(r)}{Tr^2}\right]^{1/2} \left(\frac{d\,\Delta T}{dr}\right)^{3/2} \left(\frac{R_\odot}{10}\right)^2 \qquad (8\text{-}76)$$

The distance $R_\odot/10$ is chosen somewhat arbitrarily since we do not know how to estimate the cell size. Convective theories are currently seeking to solve that problem. We take $d\,\Delta T/dr$ to be the difference between the actual and the adiabatic gradient. The equations we have obtained hold equally well for the upward convection of hot matter and downward convection of cool material. For a given gradient $d\,\Delta T/dr$ we can now obtain the order of magnitude of H if the heat capacity is known. We have not yet discussed the equation of state, although we have assumed an ideal gas law above. For a completely ionized plasma the heat capacity is roughly known, even though the process described here proceeds neither at constant pressure nor at constant volume. We will, however, not be far wrong in taking $2RT$ per gram, where R is the gas constant (see equation 4–34).

We now wish to see at what gradients the convective flux exceeds radiative transfer. This can be done by checking the value of $d\,\Delta T/dr$ at which the total convective flux equals the luminosity. With $r \sim R_\odot/2$, $M(r) \sim M/2$, $C \sim 2 \times 10^8$ ergs/g, $\rho \sim 1$ g cm^{-3}, $T \sim 10^7$ °K, and $L \sim 10^{34}$ erg sec^{-1}, we have

$$L = \pi R_\odot^2 H \sim \pi R_\odot^2 C\rho \left(\frac{GM(r)}{Tr^2}\right)^{1/2} \left(\frac{R_\odot}{10}\right)^2 \left(\frac{d\,\Delta T}{dr}\right)^{3/2} , \quad \frac{d\,\Delta T}{dr} \sim 10^{-10}$$

$$(8\text{-}77)$$

8:9

The average temperature gradient for a star is of order $T_c/R \sim 10^7/10^{11} \sim 10^{-4}$, where T_c is the central temperature. The required excess gradient is only of the order of one millionth of the total gradient. Over a distance $\Delta r \sim R_0/10$, the excess temperature drop corresponds to $\sim 1°K$, the figure we had previously used in establishing the time constant for convective transport in equation 8–29 and 8–30.

We have now dealt with all the differential equations discussed in section 8:4; but we still have to derive the energy generation rate through nuclear reactions that take place at the center of a star. This is done in the next section.

8:10 NUCLEAR REACTION RATES

The nuclear reactions that take place in stars are largely reactions in which two particles approach to within a short distance, become bound to each other, and at the same time release energy. These exergonic processes are the ultimate source of energy for the star.

Let us look into the various factors that determine the reaction rate. We will assume that two kinds of particles are involved and will label them with subscripts 1 and 2, respectively. The reaction rate then is proportional to:

(i) the number density n_1 of nuclei of the first kind,

(ii) the number density n_2 of nuclei of the second kind,

(iii) the frequency of collisions, which depends on the relative velocity v with which particles approach each other, and

(iv) on the velocity dependent interaction cross section $\sigma(v)$ that normally is proportional to $1/v^2$. However, in order for a reaction to occur, the Coulomb barrier, which bars positively charged particles from approaching a nucleus, must be penetrated. This makes the reaction rate proportional to

(v) the probability $P_p(v)$ for penetrating the Coulomb barrier that has an exponential form

$$P_p(v) \propto \exp - \left(\frac{4\pi^2 Z_1 Z_2 e^2}{hv} \right) \qquad (8\text{--}78)$$

Here $Z_1 e$ and $Z_2 e$ are the nuclear charges.

Once the nuclear barrier has been penetrated, there is a probability P_N for nuclear interaction. This is insensitive to particle energy or velocity, but does depend on the specific nuclei involved. We therefore introduce a factor proportional to

(vi) P_N the probability for nuclear interaction. For the interaction of two protons, this interaction is known from theory. For all other reactions laboratory data has to be used to evaluate the probability. The rate of the process further is proportional to

(vii) the distribution of velocities among particles. This can be assumed to be Maxwellian since the nuclei normally are not degenerate. Equation 4–59 gives

$$D(T, v) \propto \frac{v^2}{T^{3/2}} \exp - \left(\frac{1}{2} \frac{m_H A' v^2}{kT} \right) \qquad (8\text{–}79)$$

where $A' = A_1 A_2/(A_1 + A_2)$ is the reduced atomic mass, measured in atomic mass units.

We can now write down the overall reaction rate in unit volume

$$r = \int_0^\infty n_1 n_2 v \sigma(v) P_p(v) P_N D(T, v) \, dv \qquad (8\text{–}80)$$

This integral is readily evaluated because of the narrow range of velocities in which the product of P_p and D is high. Outside this velocity range the integrand is too small to make a significant contribution to the integral. We proceed in the following way. The integral in equation 8–80 has the form

$$\int_0^\infty v \exp - \left(\frac{a}{v} + bv^2 \right) dv \qquad (8\text{–}81)$$

The integrand has a sharp maximum at the minimum value of the exponent. We take the derivative of the exponent with respect to v and equate to zero. This gives the value v_m

$$v_m = \left(\frac{a}{2b} \right)^{1/3} = \left(\frac{4\pi^2 Z_1 Z_2 e^2 kT}{h m_H A'} \right)^{1/3} \qquad (8\text{–}82)$$

To evaluate the integral, however, we still need to estimate the effective velocity range over which the integrand is significant. For order of magnitude purposes it will suffice to take a range between points where the value of the integrand has dropped a factor of e. This happens at v values for which

$$\left(\frac{a}{v} + bv^2 \right) - \left(\frac{a}{v_m} + bv_m^2 \right) = 1 \qquad (8\text{–}83)$$

Since the deviation from v will be small, we set

$$v = v_m + \Delta \qquad (8\text{–}84)$$

and substitute in equation 8–83. Terms linear in Δ drop out, but the quadratic

8:10

terms yield

$$\left(\frac{a}{v_m^3} + b\right) \Delta^2 = 3b \, \Delta^2 = 1 \tag{8-85}$$

$$\Delta = \pm \sqrt{\frac{1}{3b}} = \pm \sqrt{\frac{2kT}{3A'm_H}} \tag{8-86}$$

The integral (8–80) can now be readily evaluated. First, however, we would like to lump all the proportionality constants into a single constant B and relate velocity to temperature everywhere.

We note that

$$\Delta \propto T^{1/2} \tag{8-87}$$

and that the integrand is proportional to

$$v_m \cdot \frac{1}{v_m^2} \cdot \frac{v_m^2}{T^{3/2}} = \frac{v_m}{T^{3/2}} = T^{-7/6} \tag{8-88}$$

where we have made use of the relation (8–82). This means that $r \propto T^{-7/6}\Delta \propto T^{-2/3}$

We can set

$$n_1 = \frac{\rho_1}{m_1} = \frac{\rho}{m_1} X_1 \quad \text{and} \quad n_2 = \frac{\rho}{m_2} X_2 \tag{8-89}$$

where X_1 and X_2 are the concentrations and m_1 and m_2, the masses of nuclei of species 1 and 2. Absorption of factors m_1 and m_2 into the proportionality constant B then yields the reaction rate

$$r = B\rho^2 X_1 X_2 T^{-2/3} \exp{} - 3\left(\frac{2\pi^4 e^4 m_H Z_1^2 Z_2^2 A'}{h^2 kT}\right)^{1/3} \tag{8-90}$$

Thus far we have developed an estimate of reaction rates without much thought about the individual reactions involved, the required temperatures and densities, and the resulting energy liberation rate. We now return to these points.

We first ask ourselves how much energy would be needed for two particles to interact. It is clear that a nuclear reaction can only take place if the particles approach to within a distance of the order of a nuclear diameter $D \sim 10^{-13}$ cm. However, since both nuclei are positively charged, they tend to repel each other and the work required to overcome the repulsion is

$$E = \frac{Z_1 Z_2 e^2}{D} \sim 2 \times 10^{-6} Z_1 Z_2 \text{ erg} \sim Z_1 Z_2 \text{ Mev} \tag{8-91}$$

This might lead one to think that temperatures of the order of 10^{10} °K would be required for nuclear reactions to proceed. This is far higher than the 10^7 °K temperature we had estimated in Problem 8–1.

Two factors allow the actual reaction temperature to be so low. First, a small fraction of the nuclei with thermal distribution $D(T, v)$ has energies far above the mean. Second, two particles have a small but significant probability of approaching each other by tunneling through the Coulomb potential barrier rather than going over it. This probability is quantum mechanically determined and is included in the function $P_p(v)$.

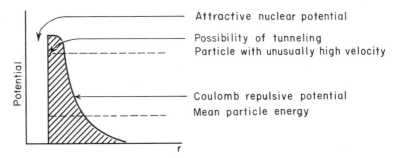

Fig. 8.6   Energies involved in nuclear reactions.

These two factors suffice to allow nuclear reactions to proceed at mean energies some 10^3 times lower than those employed to produce nuclear interactions in the laboratory. The main difference is that in the laboratory speed is essential, while the star is in no hurry. A reaction that will have a probability of transmuting a given particle after some 10 billion years is sufficiently fast to produce a luminosity found for many stars such as the sun. In the laboratory, however, we want higher reaction rates so that results be obtained within a few minutes or, at most, hours. This prolongation of the available time by a factor of $\sim 10^{14}$ is the prime difference that permits low temperature generation of energy and transmutation of the elements at the center of stars at cosmically significant rates.

8:11 PARTICLES AND BASIC PARTICLE INTERACTIONS

A number of basic particles are involved in most nuclear reactions that take place in stars. We list their properties in Table 8.2.

The spin of a particle tells us the type of statistics it obeys. Integral spin implies obedience to the *Bose-Einstein* statistics, while a half integral spin labels a particle as a *Fermion*.

8:11

Table 8.2 Some Particles That Take Part in Many Stellar Nuclear Reactions.

Particle	Symbol	Rest Mass		Charge esu	Spin	Mean Life (sec)	Class
		g	Mev				
Photon	γ	0	0	0	1		Photon
Neutrino	ν	0	0	0	$\frac{1}{2}$		Lepton
Anti- neutrino	$\bar{\nu}$	0	0	0	$\frac{1}{2}$		Antilepton
Electron	e	9×10^{-28}	0.511	-5×10^{-10}	$\frac{1}{2}$		Lepton
Positron	e^{+}	9×10^{-28}	0.511	$+5 \times 10^{-10}$	$\frac{1}{2}$		Antilepton
Proton	$\mathscr{P}$	1.6×10^{-24}	938.256	$+5 \times 10^{-10}$	$\frac{1}{2}$		Baryon
Neutron	$\mathscr{N}$	1.6×10^{-24}	939.550	0	$\frac{1}{2}$	1.1×10^{3}	Baryon

A number of basic conservation laws govern all nuclear reactions:

(a) Mass-energy must be conserved (section 5:6).

(b) The total electric charge of the interacting particles is conserved.

(c) The number of particles and antiparticles must be conserved. A particle cannot be formed from an antiparticle or vice versa. But a particle-antiparticle pair may be formed or destroyed without violating this rule. In particular:

(d) The difference between the number of *leptons* and *antileptons* must be conserved (*conservation of leptons*); and

(e) The difference between the number of *baryons* and *antibaryons* must be conserved (*conservation of baryons*).

With these rules in mind we enumerate some of the most common nuclear reactions found in stars:

(i) Beta Decay

A *neutron* as a free particle, or as a nucleon inside an atomic nucleus can decay giving rise to a proton, an electron, and an *antineutrino*:

$$\mathscr{N} \to \mathscr{P} + e^{-} + \nu \tag{8-92}$$

This reaction often is *exergonic* and can proceed spontaneously. When (8–92) proceeds in reverse, we speak of inverse beta decay.

$$\mathscr{P} + e^{-} \to \mathscr{N} + \nu \tag{8-92a}$$

8:11

(ii) Positron Decay

Here a proton gives rise to a neutron, positron, and neutrino. This process is *endergonic*—requires a threshold input energy—since the mass of the neutron and positron is considerably greater than the proton mass.

$$\mathscr{P} \rightarrow \mathscr{N} + e^+ + \nu \qquad (8\text{--}93)$$

In principle all these reactions could go either from left to right or right to left; but normally the number of available neutrinos or antineutrinos is so low that only the reaction from left to right need be considered.

(iii) $(\mathscr{P}, \gamma)$ Process

In this reaction a proton reacts with a nucleus with charge Z and mass A, to give rise to a more massive particle with charge $(Z + 1)$. Energy is liberated in the form of a photon, γ:

$$Z^A + \mathscr{P} \rightarrow (Z + 1)^{A+1} + \gamma \qquad (8\text{--}94)$$

A typical reaction of this kind involves the carbon *isotope* C^{12} and nitrogen isotope N^{13}

$$C^{12} + H^1 \rightarrow N^{13} + \gamma \qquad (8\text{--}95)$$

(iv) (α, γ) and (γ, α) Processes

In processes of this kind an *α-particle*—helium nucleus—is added to the nucleus or ejected from it. The excess energy liberated in adding an alpha particle is carried off by a photon. The energy required to tear an alpha particle out of the nucleus also can be supplied by a photon. These two processes are particularly important in nuclei containing an even number both of protons and neutrons— the *even-even nuclei*. These nuclei are particularly stable and play a leading role in the processes that lead to the formation of heavy elements.

(v) $(\mathscr{N}, \gamma)$ and $(\gamma, \mathscr{N})$ Processes

Such processes involve the addition or subtraction of a neutron from the nucleus. A photon is emitted or absorbed in the reaction and assures energy balance.

8:11

8:12 ENERGY GENERATING PROCESSES IN STARS

A variety of different energy generating processes can take place in stars. We will enumerate them in the succession in which they are believed to occur during a star's life.

(a) When the star first forms from the interstellar medium it contracts, radiating away gravitational energy. The amount of energy available from this process was already computed in (8–1). During this stage no nuclear reactions take place.

(b) When the temperature at the center of the star becomes about a million degrees, the first nuclear reactions set in. From the discussion of section 8:10 it is clear that these reactions will not set in sharply as some given temperature value is exceeded. The temperature is not a threshold in this sense, even though threshold energies are involved. Instead, we can think of a critical temperature at which reactions will proceed at a certain rate. We will choose to define the critical temperature T_c as that temperature at which the mean reaction time becomes as short as five billion years. Because of the rapid increase in reaction rates with temperature, the reactions will become completely exhausted in a very short time (on the scale of a billion years) if T_c is exceeded. The first reactions to occur are those that destroy many of the light elements initially in the interstellar medium and convert them into helium isotopes. We list the reactions and the energy released in each reaction (Sa55). Note that this energy is carried away by photons or neutrinos, but we have not specifically shown this here:

$$
\begin{array}{lll}
D^2 + H^1 \rightarrow He^3 & 5.5 \text{ Mev} & \\
Li^6 + H^1 \rightarrow He^3 + He^4 & 4.0 & \\
Li^7 + H^1_1 \rightarrow 2He^4 & 17.3 & \\
Be^9 + 2H_1 \rightarrow He^3 + 2He^4 & 6.2 & \left.\begin{array}{l}\\ \end{array}\right\} \text{two-step} \\
B^{10} + 2H^1 \rightarrow 3He^4 + e^+ & 19.3 & \left.\begin{array}{l}\\ \end{array}\right\} \text{reactions} \\
B^{11} + H^1 \rightarrow 3He^4 & 8.7 &
\end{array}
\tag{8–96}
$$

These reactions have lifetimes of the order of 5×10^4 y at respective temperatures $\sim 10^6$, 3×10^6, 4×10^6, 5×10^6, 8×10^6, and 8×10^6 °K. At temperatures ranging from about half a million degrees to five million degrees, these reactions would take place rapidly as the star contracts along the Hayashi track (Fig 8.11)— a fully convective stage in the star's pre-main sequence contraction. Because these temperatures are so low, the elements burn up everywhere including the surface layers of the stars, where they can be destroyed because convection takes place.

With few exceptions these light elements are only found in small concentrations

8:12

in the surface layers of stars. However, large concentrations of, say, Lithium can be found in some stars; these are called lithium stars and constitute a puzzle! The theory of stellar evolution seeks to explain such anomalies by providing coherent ideas on the origin of the chemical elements.

None of the reactions listed in (8–96) contribute a large fraction of the total energy emitted by stars during their lifetime. However, they are of interest in connection with the theory of the formation of elements; and the low abundance of these chemical elements in nature provides one test of the accuracy of our notions.

(c) When the temperature at the center of the star reaches about 10 million degrees, hydrogen starts burning (Be39). The reactions and mean reaction times for any given particle are given below for $T = 3 \times 10^7 \, °K$. The amount of energy liberated in each step is also given.

$$
\begin{array}{llll}
H^1 + H^1 \rightarrow D^2 + e^+ + \nu, & 1.44 \text{ Mev}, & 14 \times 10^9 \text{ y} & \\
D^2 + H^1 \rightarrow He^3 + \gamma, & 5.49 \text{ Mev}, & 6 \text{ sec} & (8\text{--}97) \\
He^3 + He^3 \rightarrow He^4 + 2H^1, & 12.85 \text{ Mev}, & 10^6 \text{ y} &
\end{array}
$$

The first and second reactions have to take place twice for each one of the third reactions. Not all of the energy liberated contributes to the star's luminosity. Of the energy liberated in the first step 0.26 Mev is carried away by the neutrino and is lost. The total contribution to the luminosity is therefore 26.2 Mev for each helium atom formed.

This set of reactions is the main branch of the *proton-proton reaction*. Other branches are shown in Fig. 8.12. Hydrogen burning can also take place in a somewhat different way, making use of the catalytic action of the carbon isotope C^{12}. This set of reactions comprises the *carbon cycle* or a more elaborate scheme sometimes called the *CNO bi-cycle*, since carbon, nitrogen, and oxygen are all involved; the CN portion is energetically the more significant (C168)*. The reaction times are given for 15×10^6 and $20 \times 10^6 \, °K$.

	Mev	$15 \times 10^6 \, °K$	$20 \times 10^6 \, °K$	
$\rightarrow C^{12} + H^1 \rightarrow N^{13} + \gamma$	1.94	$\sim 10^6 \text{y}$	$\sim 5 \times 10^3 \text{y}$	
$N^{13} \rightarrow C^{13} + e^+ + \nu$	2.22	15 min		
$C^{13} + H^1 \rightarrow N^{14} + \gamma$	7.55	$2 \times 10^5 \text{y}$	$2 \times 10^3 \text{y}$	
$\rightarrow N^{14} + H^1 \rightarrow O^{15} + \gamma$	7.29	$2 \times 10^8 \text{y}$	10^6y	(8--98)
$O^{15} \rightarrow N^{15} + e^+ + \nu$	2.76	3 min		
$N^{15} + H^1 \rightarrow C^{12} + He^4$	4.97	10^4y	30y	
$N^{15} + H^1 \rightarrow O^{16} + \gamma$	12.1	4×10^{-4} of		
		$N^{15}(\mathscr{P}, \alpha) C^{12}$ rate		
$O^{16} + H^1 \rightarrow F^{17} + \gamma$	0.60	$2 \times 10^{10} \text{y}$	$5 \times 10^7 \text{y}$	
$F^{17} \rightarrow O^{17} + e^+ + \nu$	2.76	1.5 min		
$O^{17} + H^1 \rightarrow N^{14} + He^4$	1.19	$2 \times 10^{10} \text{y}$	10^6y	

8:12

Fig. 8.7 Nuclear energy generation rate as a function of temperature (with $\rho X^2 = 100$. $X_{CN} = 5 \times 10^{-3}X$ for the p-p reaction and carbon cycle, but $\rho^2 Y^3 = 10^8$ for the triple-α process) (Sc58b, from Martin Schwarzschild, *Structure and Evolution of the Stars*, copyright © 1958 by Princeton University Press, p. 82.)

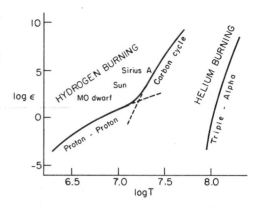

Many of the reaction times still are somewhat uncertain. The second part of the cycle occurs about 4×10^{-4} times as often as the first, since the $N^{15}(\mathscr{P}, \alpha) C^{12}$ reaction is about 2.5×10^3 times more probable than the $N^{15}(\mathscr{P}, \gamma) O^{16}$ reaction.

In the decay of the N^{13} particle 0.71 Mev is carried off by the neutrino; and in the O^{15} decay 1.00 Mev is lost on the average. The total energy made available to the star per helium atom formed is therefore only 25.0 Mev, slightly less than the energy available from the proton-proton reaction. The reaction rates given here are for total concentrations of the carbon and nitrogen isotopes amounting to $X_{CN} \sim 0.005$. The relative predominance of the proton-proton reaction and the carbon cycle as a function of temperature is given in Fig. 8.7. The C^{13} formed in the CN cycle can act as a source of neutrons as can other particles with mass number $4n + 1$. We will see that in reactions (8–102) and (8–103) below. For example, Ne^{21} can be produced as follows:

$$
\begin{array}{llll}
Ne^{20} + H^1 \to Na^{21} + \gamma & 2.45 \text{ Mev} & 10^9 \text{y at } 3 \times 10^7 \, ^\circ K & \\
Na^{21} \to Ne^{21} + e^+ + \nu & 2.5 & 23 \text{ sec} & (8\text{–}99)
\end{array}
$$

The hydrogen-burning reactions we have discussed contribute the energy given off by the star during its long stay on the main sequence. Once the hydrogen at the center of the star is largely depleted, helium burning can set in as described in the next paragraph. In general, hydrogen burning will continue in a shell surrounding this depleted core.

(d) When the hydrogen burning phase of a star is completed, no further nuclear energy generating processes may be available for some time and the star slowly contracts (Fig 8.8). Its central temperature rises continually as a result, until at about $10^8 \, ^\circ K$, helium burning sets in (Sa52). In this process three alpha particles are transmuted into a carbon nucleus. Two steps are involved:

$$
\begin{array}{lll}
He^4 + He^4 \to Be^8 + \gamma & -95 \text{ kev} & \\
Be^8 + He^4 \to C^{12} + \gamma & +7.4 \text{ Mev} & (8\text{–}100)
\end{array}
$$

8:12

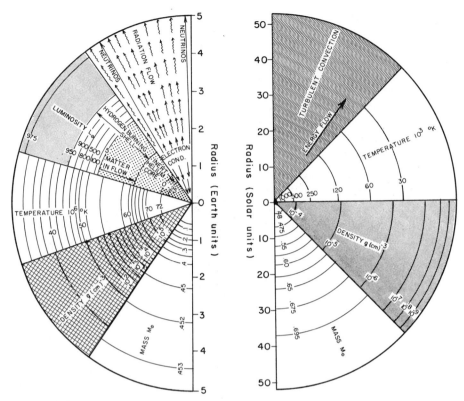

Fig. 8.8 Shell structure of a red giant star in whose central regions hydrogen has become depleted (after Iben, Ib70). The section on the left is a blown up version of the little disk in the center of the drawing at the right. (By permission of Scientific American Inc.)

The first reaction is *endergonic*. Energy has to be supplied to make it proceed. The Be^8 nucleus is unstable, and decays back into two alpha particles. An equilibrium is set up between alpha particles and Be^8 particles in which the concentration of Be^8 is quite small, of the order of 10^{-10} times the concentration of alpha particles. This particular abundance is determined by the lifetime of the metastable Be^8, the density and energy (temperature) of the helium and by the magnitude of the (negative) binding energy, -95 kev.

(e) The star's core does not stay in the helium burning phase for a very long time, because the available amount of energy is small ($\sim 10\%$), compared to the energy generated in the hydrogen-burning phase. At higher internal temperatures a succession of (α, γ) processes may set in to form O^{16}, Ne^{20}, and Mg^{24}. This type of process is called the α-*process*. After depletion in the core, helium burning may continue in a shell surrounding the depleted core. This shell is surrounded by a hydrogen-burning shell.

(f) At higher temperatures yet, 10^9°K, reactions may take place among the C^{12}, O^{16}, and Ne^{20} nuclei. At this stage there would be no supply of free helium, but these particles can be made available through a (γ, α) reaction. The densities at this stage are of the order of $\rho \gtrsim 10^6$ g/cc. A typical reaction is

$$2Ne^{20} \rightarrow O^{16} + Mg^{24}, \qquad 4.56 \text{ Mev} \tag{8-101}$$

Mg^{24} can capture alphas to form Si^{28}, S^{32}, A^{36}, and Ca^{40}.

That this process actually takes place may be partly confirmed by the relatively large natural abundance of these isotopes compared to isotopes of the same substances, or neighboring elements in the periodic table (Fig. 8.10).

This chain eventually terminates in the iron group of nuclei that have the largest stability in that the mass per nuclide is at a minimum for these elements. During the time that an equilibrium concentration is being reached between these even-even nuclei, the expected temperature and density are

$$T \sim 4 \times 10^9 \text{ °K}, \qquad \text{and} \qquad \rho \sim 10^8 \text{ g/cc}$$

This process is called the *equilibrium* or *e-process*. The α and e processes probably occur rapidly—perhaps explosively.

(g) In a second generation star—one that has formed from interstellar gases containing appreciable amounts of the heavier elements—we may find that Ne^{21} is produced. At high temperatures, in the helium core, we can then have the exergonic reaction

$$Ne^{21} + He^4 \rightarrow Mg^{24} + \mathcal{N}, \qquad 2.58 \text{ Mev} \tag{8-102}$$

Similarly from the carbon cycle there will be some C^{13} available and we may have the reaction

$$C^{13} + He^4 \rightarrow O^{16} + \mathcal{N}, \qquad 2.20 \text{ Mev} \tag{8-103}$$

taking place.

These neutrons are preferentially captured by the heavy nuclei, particularly those in the iron group, and these can then be built up into heavier elements yet. There are of the order of a hundred C^{13} and Ne^{21} nuclei available for each iron group element and, hence, an abundance of neutrons is at hand. Elements as heavy as Bi^{209} can be built up in this way. The chain only ends at Po^{210}, which is unstable and undergoes α-decay. In addition, light nuclei such as Ne^{22} also can be built up and, with the exception of the even proton-even neutron nuclei, most particles with $24 \leq A \leq 50$ are believed to have been built up through neutron capture. This neutron process is *slow*; it is therefore called the *s-process*.

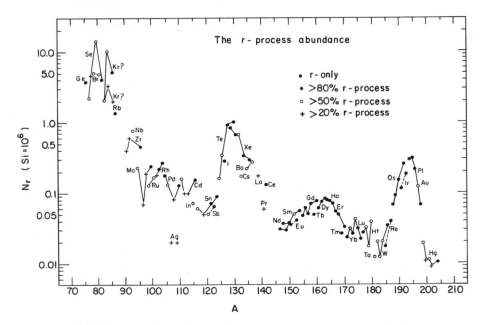

Fig. 8.9 r-process abundances estimated for the solar system by subtracting the calculated s-process contribution from the total observed abundances of nuclear masses. Isotopes of a given element are joined by lines, dashed lines for even Z and solid lines for odd Z. Three peaks, and a broad rare earth bump, are the most prominent features of this plot. The question marks indicate a high uncertainty about the relative roles of the r- and s-processes in producing krypton. (After Seeger, Fowler, and Clayton, Se65. With the permission of the University of Chicago Press.)

Neutron capture at this stage typically requires several years to several thousand years. This time scale is slow compared to beta decay rates, and only those elements can be built up that involve the addition of neutrons to relatively stable nuclei.

Evidence for a stage with abundant neutrons comes from peaks in the abundance curve at mass numbers $A \sim 90$, 138, and 208. These nuclei have closed shells of neutrons with $N = 50$, 82, and 126.

The stage of stellar evolution in which the s-process becomes effective probably is represented by the segment EF in the H-R diagram of Fig. 1.6. In this repeated red giant phase, a series of helium shell flashes may produce a convective behavior in which the helium-rich shell above the helium-burning layer is mixed with the carbon-rich core, and in which eventually even the outermost hydrogen-rich layer can be convected into the core (Sc70). In this situation C^{12} and H^1 (see equation 8–98) can yield C^{13} which produces the required neutrons through interaction with He^4.

(h) In addition to the slow neutron process, it is likely that neutrons can also

8:12

be added to heavy nuclei in a *rapid process* (*r-process*) that takes place at least in some stars. Some such process is required, in any case, if a theory of the buildup of chemical elements is to explain the existence of elements beyond Po^{210} that α-decays with a half-life of only 138d:

If a star runs out of all energy sources, a rapid implosion can take place on a free fall time scale that as we saw earlier, corresponds to times of the order of 1000 sec. Extremely high temperatures then set in and iron group nuclei can be broken up into alpha particles and neutrons; for, in Fe^{56} there are 4 excess neutrons for 13 alphas. All this takes place at temperatures of order 10^{10} °K and with neutron fluxes of order 10^{32} $cm^{-2}sec^{-1}$. The *r*-process can build up elements to about $A \sim 260$ where further neutron exposure induces fission that cycles material back down into the lower mass ranges.

In the breakup of iron group elements into helium, the ratio of specific heats γ becomes less than 4/3 so that an implosion occurs (section 4:20). This is accompanied by γ-photon production, pair formation, and electron-positron pair annihilation which at these high pressures can give to large neutrino fluxes. The neutrino flux lifts off the outer layers and the rapid neutron process then takes place while the star is again expanding—explosively. It is believed that this is the process involved at least in some types of supernova explosion. A neutron star forms from the central imploding core.

Detailed computations based on neutron capture cross sections and nuclear decay times, both measured in the laboratory, show that many features of the abundance curve in the region between $A = 80$ and $A = 200$ can be explained if the *r*-process occurs (Fig. 8.9). This leads us to believe that the sequence of events described above is at least roughly correct. Two related comments might still be made.

(i) Proton-rich isotopes are relatively rare although they can be produced in ($\mathscr{P}$, γ) processes (sometimes called *p*-processes) or in a (γ, $\mathscr{N}$) reaction. Such nuclei could be produced if hydrogen from the outer layers of a star could come into contact with hot material from the core in convective processes. Generally, however, the *r*-process can account semi-quantitavely for the abundance ratios of many of the heavier elements.

(ii) The uranium isotopes U^{235} and U^{238} might be expected to arise in roughly equal abundances in the *r*-process. However, their present ratio as found in the earth is of the order of 0.0072. This would be expected on the basis of their respective half lives of 0.71 and 4.5 æ if we assume a common time of formation some six aeons ago. This gives us a scheme for dating the origin of terrestrial material. We still face many uncertainties, some of which are illustrated by Problem 10–14 in section 10:13.

A brief stage in which carbon, oxygen, and silicon are successively burned at

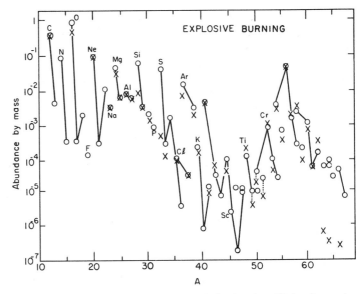

Fig. 8.10 Production of elements through explosive burning. Circles show solar system abundances by mass, crosses are computed values. Explosive burning of carbon at 2×10^9 °K and density $\rho \sim 2 \times 10^5$ g cm^{-3} contributes to atomic masses up to $A \sim 30$. Oxygen burning at 3.6×10^9 °K, at densities of $\sim 5 \times 10^5$ g cm^{-3}, contributes up to atomic mass values $A \sim 50$. Silicon burning at temperatures in the range 4.7 to 5.5×10^9 °K and density 2×10^7 g cm^{-3} accounts for the more massive nuclear abundances shown. This is a composite plot based on the work of Arnett and Clayton (Ar70). It is possible that similar abundances could be obtained by Wagoner's explosive nucleosynthesis (see section 8:16) (Wa67). Note that solid lines join different isotopes of the same element.

higher and higher temperatures, during the explosion of a star in the mass range of 20 to 40 $M_\odot$, has been suggested (Ar70) as responsible for the observed abundances of elements in the range $20 \leqq A \leqq 64$. Initially carbon in the helium depleted core of a star would undergo fusion reactions of the kind:

$$
\begin{aligned}
C^{12} + C^{12} \rightarrow Na^{23} + \mathscr{P} \qquad & 2.238 \text{ Mev} \\
Mg^{23} + \mathscr{N} \qquad & -2.623 \text{ Mev} \qquad (8\text{–}104) \\
Ne^{20} + \alpha \qquad & 4.616 \text{ Mev}
\end{aligned}
$$

These reactions would take place at a temperature of 2×10^9 °K. The initial density would be of order 10^5 g cm^{-3}. The reactions are assumed to last for about a tenth of a second, after which the explosion has cooled the stellar matter enough to stop the processes. At higher temperatures, 3×10^9 °K, oxygen also burns and thereafter silicon Si^{28} disintegrates. In this latter process, the silicon splits into seven α's, which are absorbed by other Si^{28} nuclei to form increasingly massive nuclei in the range up to Fe^{56}. These processes will take place if ignition of the nuclear fuel takes place at a temperature of $\sim 5 \times 10^9 \,^\circ K$ and a peak density of $2 \times 10^7 \, g \, cm^{-3}$ in the helium depleted core. The exact abundance ratios found at the end of the explosive process depend in part on the neutron excess, that is, the fractional number of neutrons in excess of protons in the nuclei initially present. A neutron excess

$$\eta = \frac{(n_{\mathcal{N}} - n_{\mathcal{P}})}{(n_{\mathcal{N}} + n_{\mathcal{P}})} \tag{8–105}$$

of ~ 0.002 gives remarkably good agreement as shown, for example, by Fig. 8.10. It is not clear what stellar configuration should be assumed during this explosion. Are the different substances arranged in concentric shells, or somehow irregularly distributed over the core? These processes are not sufficiently well understood.

8:13 THE HERTZSPRUNG-RUSSELL DIAGRAM AND STELLAR EVOLUTION

We believe that stars form from the interstellar medium. Initially a cool cloud of interstellar gas contracts, giving off thermal radiation in the far infrared. As the contraction proceeds the cloud temperature rises. According to Hayashi the surface temperature then remains constant for a long period as the star becomes smaller and smaller. This means that the brightness of the star decreases during this stage, but the color remains constant. The *Hayashi track* of the star on a Hertzsprung-Russell diagram is therefore represented by a nearly vertical line. Iben has indicated that this stage is followed by a nearly horizontal leftward motion toward the main sequence (Fig. 8.11). As the light elements are burned, the track may undergo some short-lived changes, but finally the star settles down for a protracted stay on the main sequence.

Actually, as shown in Fig. 1.6, there is a very slight motion through the Hertzsprung-Russell diagram, even during this hydrogen burning phase. Over a period amounting to several aeons, a star moves over from the initial *zero-age main sequence* to point B. The star becomes brighter and larger. We believe that the sun had a zero-age luminosity of $2.78 \times 10^{33} \, erg \, sec^{-1}$ and a radius of $6.59 \times 10^{10} \, cm$, compared to its current values of $3.90 \times 10^{33} \, erg \, sec^{-1}$ and $6.94 \times 10^{10} \, cm$, 4.5 æ later (St65).

When hydrogen burning is completed at the center of the star, it may still continue in a thin shell surrounding a central core of helium. The core contracts until its temperature becomes sufficiently high to produce helium burning.

Fig. 8.11   Contraction of stars toward the main sequence. The path of the stars across this Hertzsprung-Russell diagram proceeds toward the left. The left end of the curves roughly coincides with the main sequence. The star with mass $15M_\odot$ completes the transit shown here in $\sim 6 \times 10^4$ y; the $0.5\,M_\odot$ star in 1.5×10^8y. The steep portion on the right is called the *Hayashi track* (Ha66. After Iben, Ib65. With the permission of the University of Chicago Press.)

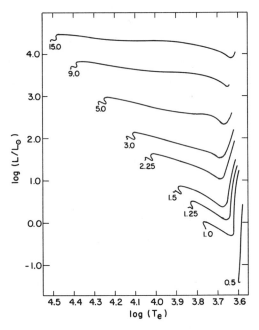

During this stage the star expands and becomes both brighter and redder. The track *BC* in Fig. 1.6 shows how the star moves upward and to the right in the color-magnitude diagram. This branch of the diagram is the *red giant branch*.

Up to now the theory is well understood. What happens next is not as well known. Somehow, the red giant stars shed some of their mass and eventually end up to the left of the main sequence and below it, in the form of *white dwarfs*!

We believe (Sc70)* that when point *C* in Fig. 1.6 is reached, helium burning through the triple alpha process sets in. This probably occurs in a flash. The star is forced to expand, but it keeps burning helium at a slower rate in the convecting central core, while hydrogen burns in an outer shell. The loop *DE* probably represents this stage, which would last some 10^8y in a star of mass $\sim 1.2M_\odot$. As indicated in section 1:5, a series of less well understood stages then follows in rapid succession. Probably there is a stage of carbon burning and, as already indicated, this may be rapid. We associate these stages in which helium burning goes on in a shell surrounded by an outer hydrogen-burning shell, with the *horizontal branch* stars. These must be short-lived because few of them are observed. The *cepheid variables* may be associated with this stage of nuclear evolution. These stars change periods at a rate that changes substantially over a time of the order of a million years.

8:13

The theory of stellar evolution allows us to draw one interesting conclusion about the age of *globular cluster stars*. Using plausible stellar models, we can compute what the hydrogen-burning time scale should be for stars having different masses. A cluster contains stars with a variety of masses; but at any given time only stars having a particular color and magnitude will be about to leave the main sequence to enter the red giant branch. Now since stellar luminosities and masses are related (see Problem 8–3), we can use the *turn-off point* where stars leave the main sequence as an indicator of the mass of these stars that are just completing the hydrogen-burning phase. The age of these stars can then be calculated and this defines the time when the cluster of stars must initially have been formed—assuming of course that all the stars were born roughly at the same time.

Ages of globular clusters derived in this way amount to about 10 aeons. The computed ages are not identical for all different clusters, indicating that they were formed at different epochs and perhaps during different phases of the life of the Galaxy.

The last stage in the evolution of stars with low mass seems to be the *white dwarf stage* in which the stellar interior becomes degenerate and no further contraction takes place. The star then gradually cools down over a long period of time, but no further nuclear reactions take place.

Massive stars do not go through the white dwarf stage. Their ultimate fate is not known. Perhaps they explode leaving only a small remnant that could become a white dwarf. Perhaps they shed mass nonexplosively. Perhaps they keep contracting until they become degenerate neutron stars. Such stars are described in more detail in section 8:16.

8:14 EVIDENCE FOR STELLAR EVOLUTION, AS SEEN FROM THE SURFACE COMPOSITION OF SOME STARS

As already discussed in section 1:6, the spectra of a wide variety of stars show atmospheric compositions very similar to that of the sun. This is shown in Table 8.3 where the small differences are within the expected errors of the observations and of the data reduction. The table presents the logarithm of the ratios of the number densities of heavier elements relative to hydrogen where, for comparison, $\log n_H$ is set equal to 12.00 in all the stars.

The similarity in abundances for stars of as widely differing ages as a B0 star, which probably formed only a few million years ago, and a red giant or the planetary nebulae, which should be among the oldest objects in the galaxy and hence seven or more aeons old, indicate that the interstellar medium may not have changed much during the time between formation of these various stars.

Table 8.3 Abundances log n for "Normal" Stars Relative to log $n = 12$ for Hydrogen. Compiled from various sources by A. Unsöld (Un69)*.

Atomic number	Element	Abundance: log n					
		Sun		τ Scorpii B0 V	ζ Persei Bl Ib	Planetary Nebulae	Log $(n/n_\odot)$ ε Virginis G8 III
		Goldberg, Müller, Aller (1960, 1967)	Various Recent Sources				
1	H	12.00	12.00	12.00	12.00	12.00	0.00
2	He		11.2	11.12	11.31	11.25	
3	Li	0.90	0.97 $\leqq$ 0.38				
4	Be	2.34					
5	B	3.6					
6	C	8.51	8.51 8.55	8.21	8.26	8.7	− 0.12
7	N	8.06	7.93	8.47	8.31	8.5	
8	O	8.83	8.77	8.81	9.03	9.0	
9	F					5.5	
10	Ne			8.98	8.61	8.6	
11	Na	6.30	6.18				+ 0.30
12	Mg	7.36	7.48	7.7	7.77		+ 0.04
13	Al	6.20	6.40	6.4	6.78		+ 0.14
14	Si	7.24	7.55	7.66	7.97		+ 0.13
15	P	5.34	5.43				
16	S	7.30	7.21	7.3	7.48	8.0	+ 0.09
17	Cl					6.5	
18	A			8.8		6.9	
19	K	4.70	5.05				+ 0.10
20	Ca	6.04	6.33				+ 0.10
21	Sc	2.85					− 0.07
22	Ti	4.81					− 0.07
23	V	4.17					− 0.04
24	Cr	5.01					0.00
25	Mn	4.85					+ 0.07
26	Fe	6.80		7.4			+ 0.01
27	Co	4.70					− 0.03
28	Ni	5.77					+ 0.03
29	Cu	4.45					+ 0.06
30	Zn	3.52					+ 0.05
31	Ga	2.72					
32	Ge	2.49					
37	Rb	2.48	2.63				
38	Sr	3.02	2.82				+ 0.02
39	Y	3.20					− 0.17

Table 8.3 (*continued*)

Atomic number	Ele-ment	Sun		τ Scorpii B0 V	ζ Persei Bl Ib	Planetary Nebulae	Log $(n/n_\odot)$ ε Virginis G8 III
		Abundance: log n					
		Goldberg, Müller, Aller (1960, 1967)	Various Recent Sources				
40	Zr	2.65					− 0.15
41	Nb	2.30					
42	Mo	2.30					
44	Ru	1.82					
45	Rh	1.37					
46	Rd	1.57					
47	Ag	0.75					
48	Cd	1.54					
49	In	1.45					
50	Sn	1.54					
51	Sb	1.94					
56	Ba	2.10	1.90				− 0.09
57	La	2.03					− 0.08
58	Ce	1.78					− 0.08
59	Pr	1.45					+ 0.37
60	Nd	1.93					+ 0.06
62	Sm	1.62					+ 0.01
63	Eu	0.96					
64	Gd	1.13					
66	Dy	1.00					
70	Yb	1.53					
72	Hf						
82	Pb	1.63	1.93				+ 0.18

This lack of evidence for chemical evolution is somewhat puzzling although the constancy of the chemical makeup of the medium could be attributed to a lack of mixing of the surface layers of the stars with matter from the stellar interior. This would need to be coupled with a return only of surface material to the interstellar medium whenever a supernova or other unstable star ejects matter. Of course (see Table 1.1 or section 1:6) some very metal-deficient stars are known to exist, and these exceptional objects do suggest chemical evolution. Clearly the evidence is ambiguous.

In a small fraction of the observed stars, some material from the stellar center does seem able to reach the surface in appreciable quantities. This may happen regularly in some *close binary stars* where surface material systematically flows from one component to its companion, exposing lower layers in which nuclear evolution is considerably advanced. There may also be other processes that permit the convection of matter from the interior to the surface of individual stars (Un69)*.

At any rate overabundances of helium, carbon, or of the metal elements is exhibited by some components of close binaries. For these, the spectroscopically determined abundance ratios are consistent with the processes already mentioned: the burning of hydrogen into helium through the proton-proton reaction or through the CNO cycle, the burning of helium into carbon C^{12}, and the burning of carbon into heavier elements. Evidence for the e-, s-, and r-processes also appears to be accumulating.

The *helium stars* (Table 8.4) show a systematic increase of the helium abundance over stars shown in Table 8.3. In fact, the nominally assumed abundance $n_{He} = 11.61$ represents complete burning of every set of four hydrogen nuclei to produce a helium nucleus, if an initial hydrogen abundance $\log n_H = 12.00$ is assumed, and if the initial helium abundance is taken comparable to the values in Table 8.3: $\log n_{He} = 11.2$.

A comparison of the two tables also shows that, although in the star HD 160641 the CNO composition remained almost unchanged, the star HD 30353 shows a reduction of carbon and oxygen relative to nitrogen. This would perhaps indicate that the proton-proton reaction dominates the hydrogen burning in the first-named star, while the second star has undergone the CNO cycle, which has left most of the carbon and oxygen in the form of nitrogen. The equilibrium abundance of nitrogen in this cycle would in any case, be expected to be high because of the slow conversion of N^{14} into O^{15} shown in (8–98). The relatively high oxygen abundance shown by the star's spectrum indicates that other processes might also be at work. This is also true in HD 168476 and HD 124448, where helium burning may have replenished some of the carbon converted in the CNO bi-cycle.

The *carbon stars*, red giants that exhibit anomalously large surface abundance of carbon, either in atomic form or as a component of small radicles, give evidence that the *triple-alpha process* plays an important role. Various types of carbon stars show differing histories. Some have heavy element abundances reduced relative to hydrogen by an order of magnitude. These also exhibit elements, from barium Ba on, which are overabundant relative to iron by an order of magnitude. This overabundance must have been produced by neutron addition.

Interesting information can also be obtained from the ratio of C^{12} to C^{13} as determined from the spectra of diatomic radicles. Slightly shifted from the spectra associated with the C^{12} isotope, one finds a weaker band structure from radicles containing C^{13}. On earth and in the sun the C^{12}/C^{13} ratio is ~ 100. But for the CNO bi-cycle the lifetimes given in section 8:11 indicate an equilibrium ratio of $\sim 5:1$, and in most carbon stars this is the ratio actually observed; other

Table 8.4 Abundances of Elements in Helium Stars Relative to $\log n_{He} = 11.61$ (after A. Unsöld Un69).

Atomic Number	Element	Abundances: $\log n$				
		HD 160641	HD 168476	HD 124448	BD + 10° 2179	HD 30353
1	H		< 7.1	< 7.8	8.49	7.6
2	He	11.61	11.61	11.61	11.61	11.6
6	C	8.66	9.16	9.01	9.51	6.2
7	N	8.77	8.35	8.38	8.67	9.2
8	O	8.91	< 8.3	< 8.4	< 8.2	7.5
10	Ne	9.42	9.05			8.5
12	Mg	7.61	7.53	7.75	7.2	
13	Al		6.19	6.61	5.8	
14	Si	7.61	7.12	7.21	7.42	7.6
15	P		6.06			
16	S		6.75	7.19		7.8
18	A			6.9		
20	Ca		6.00	6.40	5.91	
21	Sc		4.3			
22	Ti		5.98	6.3		
23	V		4.65			
24	Cr		5.20	4.8		
25	Mn		4.57	4.84		
26	Fe		7.42	7.58		
28	Ni		5.4	5.2		

carbon stars show solar abundances. Where the ratio is low, it indicates that somehow the carbon formed through helium burning is later subject to the CNO bi-cycle. But because the carbon:nitrogen:oxygen abundance ratios found in individual carbon stars frequently are inconsistent with the CNO bi-cycle (Th72) it may be necessary to invoke other explanations.

Another class of stars, the barium stars, exhibit an unusually strong doublet due to barium at wavelengths of 4554 and 4934 Å. In general, for such stars elements with atomic number $Z > 35$ are overabundant by about an order of magnitude compared to the solar abundance. In these stars iron with $Z = 26$ is normal, but strontium with $Z = 38$ already is overabundant. These observations are consistent with neutron irradiation.

Similar features hold for the S stars where, again, the great strength of ZrO, LaO, YO bands, and the lines of atomic zirconium, barium, strontium, lanthanum, and yttrium show overabundances in the high Z range. That these elements originate from neutron processes also is consistent with the presence of technetium lines. Tc^{99} has a half-life of 2×10^5y, similar to the life computed for stellar evolution along the red giant branch. This isotope can be produced in the s-process. We can therefore conjecture that the isotope has been formed on the surface of the star within the last few 10^5y, or more recently; or else it was formed in the star's interior during this time and brought out to the surface. Had it, however, been present when the star first formed several aeons earlier, no measurable traces would have remained by this time. This gives direct evidence for current nucleosynthesis in stars, and because we only know of the neutron processes for forming such heavy elements, it supports our belief that the s-process takes place in the stellar interior.

The rare earth element promethium, whose longest-lived isotope has a half-life of only 18 y, has been tentatively identified in the atmosphere of HR465, a star that exhibits spectral lines of many other rare earth elements. This would show that nuclear reactions must be going on to replenish the decaying promethium (A170). Since convection times (8–30) are of the order of a month, this promethium need not have been generated very close to the surface, although that too might be possible. These observations still need confirmation.

8:15 THE POSSIBILITY OF DIRECT OBSERVATIONS OF NUCLEAR PROCESSES IN STARS

Thus far we have described the currently expected sequence of nuclear events that may be going on in the interior of stars, and have identified these events with phases in the life of a star represented by different portions of the Hertzsprung-Russell diagram. Since the sequence of events, the variety of possible reactions, and the number of assumptions required are so large, direct verification of the postulated nuclear reactions would be highly desirable.

The most promising observations that can be made in this respect are measurements on neutrinos emitted in the nuclear reactions. As already indicated *neutrinos* carry off somewhere around 2 to 6% of the hydrogen-to-helium conversion energy, depending on whether the proton-proton reaction or the CN cycle predominates. The neutrinos can escape from a star, virtually without hindrance, because in a 1 cm² column, one stellar radius deep, the neutrino would typically encounter $\rho R/m_h \sim 10^{35}$ nuclei, where $\rho \sim 1$ g cm^{-3}, $R \sim 10^{11}$ cm and $m_H \sim 10^{-24}$ g. Since

8:15

the neutrino interaction cross section with nuclei normally is of order 10^{-45} cm^2, only one neutrino in 10^{10} would be intercepted on its way out of the star.

An attempt to directly observe neutrinos from the sun has been carried out by Raymond Davis, Jr. and co-workers at the Brookhaven National Laboratory (Da68). Their experiment was based on the large absorption cross section for neutrinos exhibited by the chlorine isotope Cl^{37}, in the reaction

$$Cl^{37} + \nu \rightarrow Ar^{37} + e^{-} \tag{8–106}$$

This reaction requires a minimum neutrino energy of 0.81 Mev. The argon isotope Ar^{37} is radioactive and makes itself evident through a 34-day half-life, 2.8 kev *Auger* (*X-ray*) transition that can be recorded. The reaction cross section for this process is only large for high energy neutrinos, however, and so not all of the neutrinos emitted by the sun could be counted in this way. The experimenters figured they would only be able to observe the neutrinos from the decay of the boron isotope B^8, which is formed in very small quantities according to the predictions of nuclear theory. The neutrino given off in that decay can have energies as high as 14 Mev. The scheme, which first gives rise to boron, is shown in Fig. 8.12, together with the probability for the occurrence of each reaction.

In order to detect this process, the group made use of chlorine in the form of 520 metric tons of C$_2$Cl$_4$. The argon produced in this liquid was purged out of the tank with helium gas and trapped at low temperatures to separate it from the helium. The trapped argon was then released into a counting chamber. The cross section for neutrino capture from the boron decay was $\sigma \sim 1.35 \times 10^{-42}$ cm^2,

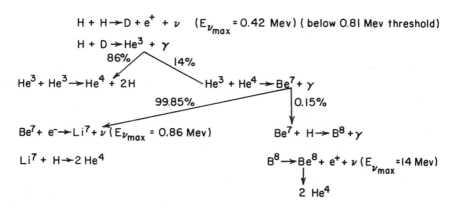

Fig. 8.12 Nuclear reactions leading to the production and decay of B^8 in the sun, showing relative probabilities for the different reactions and the energy of neutrinos produced. The branching ratios are temperature dependent. The ratios shown are expected at temperatures of $\sim 1.5 \times 10^{7\circ}$K at concentrations $X = 0.726$, $Y = 0.26$, $Z = 0.014$ (Ba72).

8:15

and the expected number of captures amounted to somewhere between 2 and 7 per day. The actual observed upper limit to the counts attributable to Ar^{37} decay was somewhat lower than that, limiting the flux at the earth to less than $2 \times 10^6 \, cm^{-2} \, sec^{-1}$ neutrinos from this reaction. For the 0.86 Mev neutrino, $\sigma \sim 2.9 \times 10^{-46} \, cm^2$ and the contribution from the Be^7 electron capture was small compared to B^8.

This experiment apparently also yielded an upper limit of about 10% for the contribution by the CNO cycle to solar energy production. If it played a bigger role, the neutrinos produced by O^{15} decay in this cycle would have been detected.

After this experiment had been completed, theorists re-examined their predictions on the basis of improved measurements of the sun's composition and nuclear reaction probabilities, but the expected flux was not substantially changed. Meanwhile experiments continued with a somewhat larger amount (610 tons) of C_2Cl_4 and a production rate of order $0.3 \pm 0.2 \, Ar^{37}$ atoms was obtained after subtraction of $0.2 \, Ar^{37}$ atoms formed per day by cosmic rays and fast neutrons (Da71). The predicted value was a production of 2 atoms of Ar^{37} daily by solar neutrinos. This discrepancy is being pursued; a resolution of this problem could lead to a clearer understanding of nuclear evolution in stars!

8:16 THE POSSIBILITY OF ELEMENT SYNTHESIS IN OBJECTS EXPLODING FROM AN INITIAL TEMPERATURE OF $10^{10} \, {}^\circ K$

One observation that has not been satisfactorily answered by the theory of nuclear evolution in stars is the detection of some heavy elements in the atmospheres of the earlierst stars that can be found in the Galaxy. The indications are that these stars, which are found as members of *globular clusters*, contain matter that has already undergone some nuclear processing. This would imply that the earliest stars formed within the galaxy were not formed from hydrogen or hydrogen-helium mixtures alone, but must have been formed from matter that already had an admixture of heavier elements. From a cosmological viewpoint this may have important implications. If the earliest stars we know contain heavier elements, then where did those come from? Is it necessary or even possible that these elements were formed during some primordial exploding state of the Universe, or can we explain their presence in other ways?

Before becoming too concerned, we should consider the possibility that the observed heavy elements are no more than a surface contamination. We ask whether there is a possibility that stars might have picked up a sufficient amount of interstellar material, simply on passage through dense interstellar clouds. The present composition of the interstellar material certainly would have a high enough abundance of heavy elements to satisfy observations. However, when the calculations are made (and these take a form similar to Problem 3–8), we find that the *accretion rate*, or capture rate, for interstellar matter, is too small to give rise to the observed metal abundances.

Computations by Wagoner (Wa67) and by others indicate that the observed abundances of the heavy elements in globular cluster stars can be explained by the following hypothesis: At some time before the globular cluster stars were formed, galactic hydrogen collected into one or more massive objects whose central temperatures rose to 10^{10}°K before exploding to spew the material back into the Galaxy. Under the right conditions of expansion following such an explosion, the correct abundance ratios are found. The thought here is that such an explosion might occur on a galaxy-wide scale and perhaps represent a phenomenon of the *quasar* type—very compact and luminous. The material cycled through such a stage would then be available for subsequent star formation in globular clusters. These massive objects could comprise an entire galactic mass, or be smaller—that is, they could be the size of a globular cluster or even a massive star. Either way this idea could explain the presence of heavier elements in the oldest stars within our Galaxy.

This theory might also explain the apparently high amount of helium present in the earliest stars, the halo stars in the Galaxy (Wa71)*, although backers of an evolutionary model of the universe feel that this helium may have been formed during early stages of cosmic evolution when matter was at a high density and temperature.

If the massive object represents the whole universe, in an intially compact state, we can argue that the presently observed 3°K blackbody background radiation, if present primordially, determines the temperature of the earliest evolutionary stages. The rate at which the radius of the universe and hence the temperature and density changed can be calculated. We can then use nuclear reaction rates to determine the composition of the material when the density finally became so low that no further nuclear changes took place.

That composition, of course, depends on the constituents initially present in the universe. In particular, the final neutron-proton ratio will be strongly affected by the number of neutrinos and antineutrinos initially present. If the *electron neutrinos* are very abundant the reaction

$$v + \mathcal{N} \rightleftarrows \mathcal{P} + e^- \tag{8–107}$$

will go mainly toward the right and produce a large abundance of protons. When the neutrinos are less dense, two other reactions also play a role

$$e^+ + \mathcal{N} \rightleftarrows \mathcal{P} + v \tag{8–108}$$

$$\mathcal{N} \rightleftarrows \mathcal{P} + e^- + v \tag{8–109}$$

8:16

and produce roughly equal neutron and proton densites. For large antineutrino abundances, these reactions run from right to left. The neutrinos considered here of course are electron neutrinos—not muon neutrinos. *Muon neutrinos* influence the development only very indirectly through their contribution to the cosmic expansion rate that is influenced by the total density of matter (section 10:9).

The ratio of protons to neutrons determines the final abundance of elements present. In most cosmological models the evolution stops at mass 7 amu, because nuclei with mass 8 are unstable. Here the triple-α process does not take place because densities are too low. The possible processes are shown in Fig. 8.13.

Detailed calculations for differing initial densities of the various elementary particles predict different abundances of deuterium, He^4, He^3, and Li^7, which might be measurable at the present time by observations of the intergalactic medium. Such observations are very difficult. No intergalactic gases have ever been identified to date. As astronomical methods improve, however, identification of these gases should become possible.

8:17 COMPACT STARS

Thus far we have dealt with stars whose densities are roughly comparable to the sun's, except late in life when central portions of these stars become very compact.

We now turn to stars that are orders of magnitude more compact: the *white dwarf* and *neutron stars*. The structure of these stars can be considered from the same general viewpoint that allowed us to understand processes in the interior of ordinary stars. Before proceeding in this direction, we should, however, review one particular argument that we had brought out to demonstrate the importance of nuclear reactions in stellar interiors. In (8–1) we had shown that the potential energy per unit mass of stellar substance is $\sim 3MG/5R$, while the available nuclear energy is of the order of $10^{-2}c^2$, if matter-antimatter annihilation is ruled out. It follows that very compact stars may be able to liberate amounts of gravitational energy in excess of the normal nuclear energies available. This will happen when

$$R \lesssim \frac{MG}{10^{-2}c^2} \sim 10^7 \, \text{cm} \qquad (8-110)$$

for stars of one solar mass. This is still larger than R_S, the *Schwarzschild radius*, at which the conventional expression for potential energy per unit mass would become equal to c^2:

$$R_S = \frac{2MG}{c^2} \qquad (8-111)$$

8:17

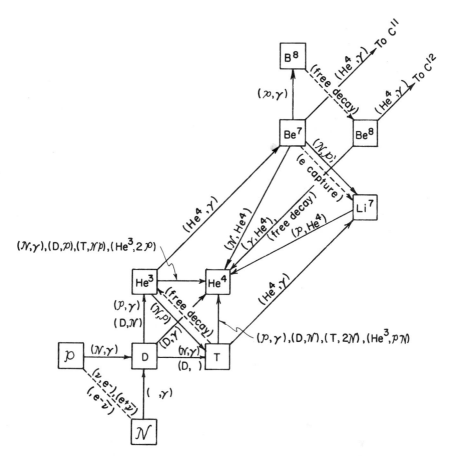

Fig. 8.13 Reactions of importance in nucleosynthesis during early stages of an evolving universe. The exoergic directions are indicated, although rates are often rapid in both directions. The other incoming and outgoing particles, those not shown in squares, are indicated in parentheses. Dashed arrows indicate the beta reactions. Sometimes there are competing reactions leading from one nucleus to another (after R. Wagoner, Wa67).

Since white dwarfs typically have masses of the order of 10^{33}g, the corresponding Schwarzschild radius would be

$$R_S \sim 10^5 \text{ cm} = 1 \text{ km}$$

which is small compared to the white dwarf radii computed below, in section 8:18, but comparable to the neutron star radii ~ 10 km, discussed in section 8:19.

8:17

8:18 WHITE DWARF STARS

We had suggested that the compact stars are so dense that matter becomes degenerate in the interior. At the surface of a white dwarf, the density is lower and no degeneracy will be found in the outer layers of such a star. However, the actual thickness of the nondegenerate layer is small, and we will be able to progress by treating the star as completely degenerate throughout.

To proceed with this computation, we first note that most of the pressure in the stellar interior must be provided by the degenerate electrons. The partial pressure of the nuclei is very small. This comes about because under degenerate conditions, the lowest momentum states of the electron gas are always filled. The more compact the star, the higher the Fermi energy of the electrons, and the higher the electron gas pressure. Since only the electrons are degenerate, as discussed in section 4:15, the nucleon pressure is relatively low. We will neglect it here and take the total pressure to equal the electron pressure, $P = P_e$.

In section 8:6 we had given the electron pressure for a nonrelativistic and for a completely relativisitic degenerate electron gas, respectively, as:

$$P = \frac{h^2}{20m_e m_H} \left(\frac{3}{\pi m_H} \right)^{2/3} \left(\frac{(1 + X)}{2} \rho \right)^{5/3} \quad \text{nonrelativistic} \quad (8\text{–}41)$$

and

$$P = \frac{hc}{8m_H} \left(\frac{3}{\pi m_H} \right)^{1/3} \left(\frac{(1 + X)}{2} \rho \right)^{4/3} \quad \text{relativistic} \quad (8\text{–}43)$$

In general, there will exist an important transition region where the gas is neither highly relativistic nor completely unrelativistic. In that region the pressure can be shown (Problem 8–4) to take the form:

$$P = \frac{8\pi m_e^4 c^5}{3h^3} f(X), \qquad \rho = \mu_e \frac{8\pi m_H m_e^3 c^3}{3h^3} x^3 \qquad (8\text{–}112)$$

with

$$\mu_e = \frac{2}{1 + X} \qquad (8\text{–}113)$$

The function $f(x)$ is

$$f(x) = \frac{1}{8} \left[x(2x^2 - 3)(x^2 + 1)^{1/2} + 3 \sinh^{-1} x \right] \qquad (8\text{–}114)$$

8:18

where

$$x = \frac{p_0}{m_e c} \qquad (8\text{--}115)$$

PROBLEM 8-4.   For a degenerate relativistic gas, all momentum states (4–65) are filled, and equation 5–30 holds. Using equations similar to (4–27) through (4–30) as a guide, show that

$$P = \frac{1}{3} \int_0^{p_0} pv(p)\, n_e(p)\, dp \qquad (8\text{--}116)$$

$$= \frac{8\pi}{3 m_e h^3} \int_0^{p_0} \frac{p^4\, dp}{[1 + (p/m_e c)^2]^{1/2}} \qquad (8\text{--}117)$$

Setting $\sinh u = p/m_e c$, show that

$$P = \frac{8\pi m_e^4 c^5}{3h^3} \int_0^{u_0} \sinh^4 u\, du \qquad (8\text{--}118)$$

We see that (8–118) has the same coefficient as (8–112). We can also show by integration that the integral in (8–118) equals the expression 8–114 for $f(x)$.

Small values of $x, x \ll 1$, correspond to the lower density portions where the gas is nonrelativistic, while high x-values correspond to the frequently relativistic central portions of the star.

PROBLEM 8-5.   Evaluate $f(x)$ in the limits $x \ll 1$ and $x \gg 1$ and show that equations 8–41 and 8–43 are obtained.

Equation 8–112 is computed on the basis of statistical mechanics and involves no assumptions concerning stars. It is simply an *equation of state* for a partially relativistic degenerate gas, no matter where it may be found. We should note that the pressure is temperature independent in this equation; it only depends on x, which is a measure of the momentum at the Fermi energy of the electron gas; and that is only density dependent. The mathematical problem of computing conditions at the center of the star can therefore be separated into two portions, a hydrostatic and a thermodynamic one.

The hydrostatic equilibrium conditions are the same as those obtained earlier:

$$\frac{dP}{dr} = -\rho\, \frac{GM(r)}{r^2}, \qquad \frac{dM(r)}{dr} = 4\pi r^2 \rho \qquad (8\text{--}7, 8\text{--}8)$$

To integrate these equations, we assume that we know what the chemical composition of the white dwarf is, because that composition determines the value of μ_e in equation 8–113. We next choose an arbitrary central density ρ_c for the star, and then integrate the hydrostatic equations outward from the star's center until we reach a radius r where the pressure has dropped to zero. In this model computation, that radius represents the surface of the star. The value of $M(r)$ at this radial distance corresponds to the total mass of the star and the value of r represents the actual stellar radius.

Clearly this procedure can be repeated for a range of different central densities, and we therefore obtain a whole family of stellar models with differing central densities. Similarly, we can obtain a new family of models having different chemical composition; but no recomputation is required here, because a change of chemical composition is mathematically equivalent to a simple change of variables. We will show this:

If the initially computed values are denoted by primes, and new variables— corresponding to a new chemical composition—are denoted by unprimed symbols, we find that the required interrelations are

$$P = P'$$

$$\rho = \frac{\mu_e}{\mu_e'}\rho'$$

$$M(r) = \left(\frac{\mu_e'}{\mu_e}\right)^2 M(r') \qquad (8\text{–}119)$$

$$r = \left(\frac{\mu_e'}{\mu_e}\right)r'$$

We can readily see that substitution of these expressions into equations 8–112, 8–7, and 8–8 leaves the form of these equations unchanged; a change in the chemical composition is therefore equivalent to a change in central density. Consequently we are dealing with a one-parameter family of models, because everything about a given star can be described entirely in terms of an equivalent central density. We present the results of the described computations in the form of Table 8.5.

The argument presented thus far neglects a number of correction factors that act to lower the last few mass values in the table by $\sim 20\%$ (Sc58b). However, we will not be concerned here with factors of that order of accuracy. Instead, we will concentrate on the overall properties of these stars that are:

(1) The larger the mass of the white dwarf, the smaller is its radius.

(2) For masses comparable to the sun, the white dwarf's radius is a factor of $\sim 10^2$ smaller than $R_\odot$.

8:18

Table 8.5 Central Densities, Total Mass, and Radius of Different White Dwarf Models, Taking $\mu_e = 2$ (Negligible Hydrogen Concentration).[a]

$\log \rho_c$	$M/M_\odot$	$\log R/R_\odot$
5.39	0.22	-1.70
6.03	0.40	-1.81
6.29	0.50	-1.86
6.56	0.61	-1.91
6.85	0.74	-1.96
7.20	0.88	-2.03
7.72	1.08	-2.15
8.21	1.22	-2.26
8.83	1.33	-2.41
9.29	1.38	-2.53
∞	1.44	$-\infty$

[a]See text for comments. (After M. Schwarzschild Sc58b.) From *Structure and Evolution of the Stars* (copyright © 1958 by Princeton University Press) p. 232.

(3) There exists an upper limit to the mass—the *Chandrasekhar limit*—above which no stable white dwarf configuration exists, because even infinite central pressure cannot keep the star from further collapse. The actual limit, taking into account the above-cited 20% correction, should be $\sim 1.2 M_\odot$.

The reasons for the limit are apparent if we consider that there are quite different relations between central pressure and density in the relativistic and nonrelativistic limits. From (8–41) and (8–43):

$$\text{nonrelativistically:} \quad P \propto \rho^{5/3}, \quad \frac{dP}{dr} \propto \rho^{2/3} \left(\frac{d\rho}{dr} \right) \qquad (8\text{–}120)$$

$$\text{relativistically:} \quad P \propto \rho^{4/3}, \quad \frac{dP}{dr} \propto \rho^{1/3} \left(\frac{d\rho}{dr} \right) \qquad (8\text{–}121)$$

At the same time, the gravitational pressure gradient is

$$\frac{dP}{dr} \propto - \frac{\rho(r)}{r^2} \int_0^r \rho(r') 4\pi r'^2 \, dr' \qquad (8\text{–}122)$$

In the crudest approximation, we can set the density equal to the stellar mass divided by the cube of the radius R so that

$$\text{nonrelativistically} \quad \frac{dP}{dr} \propto \frac{M^{5/3}}{R^6}$$

$$\text{relativistically} \quad \frac{dP}{dr} \propto \frac{M^{4/3}}{R^5} \qquad (8\text{–}123)$$

$$\text{gravitational} \quad \frac{dP}{dr} \propto \frac{M^2}{R^5}$$

We note that the dependence of the relativistic pressure gradient on radius has the same power as the gravitational force. Both increase as R^{-5} as the star contracts. This means that once a relativistic white dwarf core is forced to contract by hydrostatic pressure, the counterforce produced through contraction increases at the same rate as the gravitational attraction, and that tends to compress the star even further. There is, therefore, no way in which the star can come into equilibrium. On the other hand, a nonrelativistic gas at the center of a white dwarf can always adjust itself through contraction, until the gravitational forces compressing the star are countered.

We then have the following situation. For small stellar masses, the central pressure is determined more nearly by the nonrelativistic approximation, and the star can find a stable equilibrium position. For more massive objects, the central density becomes so high during contraction that the relativistic regime is reached, and further contraction no longer leads to a situation of equilibrium. The Chandrasekhar limit is therefore symptomatic of the transition from a predominantly nonrelativistic to a predominantly relativistic central core (Ch39)*.

We still must think about the appearance of white dwarf stars in the Hertzsprung-Russell diagram. We recall the definition of the effective temperature of a star

$$L = \sigma T_e^4 (4\pi R^2) \qquad (4\text{–}76)$$

and rewrite this in terms of the solar luminosity and surface temperature:

$$\log \frac{L}{L_\odot} = 4 \log \frac{T_e}{T_{e\odot}} + 2 \log \frac{R}{R_\odot} \qquad (8\text{–}124)$$

If we then use the white dwarf radii and masses from Table 8.5, we can obtain plots of L against T_e as a function of different mass values. Choosing five different representative masses, we obtain the curves shown in Fig. 8.14.

The agreement with observations is satisfactory, and we can feel reasonably confident that the discussion pursued here is at least roughly correct. This is important! For, white dwarfs have a local number density of $\sim 2.5 \times 10^{-2}$ white dwarfs pc^{-3}—10% to 20% of the mass in our neighborhood of the Galaxy

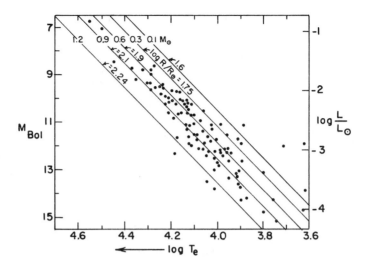

Fig. 8.14 White dwarf Hertzsprung-Russell diagram. Lines of constant radius are shown. Also shown are the masses based on completely degenerate core models containing elements having $\mu_e = 2$ (after Weidemann We68).

(We68). They are objects we should come to understand well, if we are to learn how stars die!

8:19 NEUTRON STARS AND BLACK HOLES

For some stars the most advanced stage of evolution appears to contain a core that consists of densely packed neutrons. We can imagine the evolution toward this state in the following way (Sa67)*.

The equations in (8–112) are applicable as long as we are dealing with a dense star, but as the star evolves, the value of μ_e changes. As hydrogen becomes depleted, we saw μ_e assumes a value of 2. This holds, for example, for a star in which the major constituent is C^{12}. But as the chemical elements evolve toward the more neutron-rich species, equations 8–36 and therefore 8–113 no longer hold, and for a star rich in Fe^{56} we find $\mu_e = 2.15$. Now the Chandrasekhar limiting mass is porportional to μ_e^{-2}, as seen from (8–119), so that we can draw a number of curves of mass against central density—as in Fig. 8–15. In these curves, we assume the lowest possible temperature and we show plots for stars of differing chemical composition. Corresponding to each chemical composition is a different Fermi-energy E_F for the electrons at the center of the star, as a direct consequence of the star's changed central density.

8:19

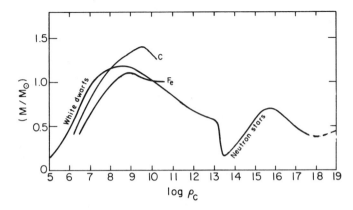

Fig. 8.15   Mass of a cold star, as a function of central density. The full curve is for a representative initial chemical composition and assumes relativistic hydrostatic equilibrium. The curves labeled C and Fe assume pure carbon and iron composition (after Ru71a, Sa67). Negative slopes on the curves indicate regions where there are no hydrostatically stable configurations. The density is given in units of g cm^{-3}. The dashed portion of the curve near central densities of the order 10^{18} g cm^{-3} is very uncertain, because the physical state of matter at these densities is uncertain. (Reprinted with permission of the publishers of The American Mathematical Society, from *Lectures in Applied Mathematics*, copyright © 1967, Volume 10, Part 3, "Stellar Structure," p 33.)

Table 8.6   Maximum Central Density and Electron Fermi Energy for Compact Stars with Different Chemical Composition (after E. E. Salpeter, Sa67).

	C^{12}	S^{32}	Fe^{56}	Sn^{120}
$\log \rho (\text{g cm}^{-3})$	10.6	8.2	9.1	11.5
$E_F (\text{Mev})$	13	1.7	3.7	24

Reprinted with permission of the publishers of The American Mathematical Society, from *Lectures in Applied Mathematics*, Copyright © 1967, Volume 10, Part 3, "Stellar Structure," p. 34.

As the central density increases, for a given composition, the electron Fermi energy always increases up to the point where inverse beta decay takes place and drives the electrons into the nuclei. This is what produces the increasingly neutron-rich elements cited in Table 8.6. The symbolic reaction is

$$\mathcal{P} + e \rightarrow \mathcal{N} + \nu \qquad (8\text{--}125)$$

The reverse reaction cannot take place if the Fermi energy is high enough, because all the electron states into which the radioactive nucleus might decay are already occupied. This gives otherwise unstable nuclei an environmentally induced stability.

The value of μ_e, which is an effective nuclear mass per free electron, also increases during contraction. When the Fermi energy reaches 24 Mev, the density is $\rho \sim 10^{11.5} \mathrm{~g~cm}^{-3}$ and $\mu_e \sim 3.1$. At this stage free neutrons become energetically favorable so that a further increase in density leads to an increased partial density of neutrons, a practically constant density of ions, and a constant electron Fermi energy of 24 Mev.

As the density increases, E_F increases to the point where reaction (8–125) proceeds rapidly and the electrons are driven into the nuclei causing the collapse of the central core, because the electron pressure no longer increases at a sufficient rate during the contraction.

In Fig. 8.15 the curves for stars containing C^{12} and Fe^{56} show a maximum mass at ρ_c values where inverse beta decay first sets in and μ_e increases. At the extreme lower right, the curve for free neutrons is shown. It has a maximum just beyond $\rho_c \sim 10^{15} \mathrm{~g~cm}^{-3}$.

The reason for this maximum is relatively easy to understand if we compute the mass expected on the basis of a nonrelativistic neutron gas and an extreme-relativistic gas, respectively (Sa67)*.

The virial theorem gives the ratio of pressure P to density ρ in terms of stellar mass as

$$3 \left\langle \frac{P}{\rho} \right\rangle \sim \frac{M}{R} \propto M^{2/3} \langle n^{1/3} \rangle \tag{8–126}$$

where mean values are denoted by brackets, and n is the number density of neutrons. As can be seen from (8–35), (8–40), and (8–42)

$$P \propto n^{5/3} \qquad \text{nonrelativistic} \tag{8–40a}$$

$$P \propto n^{4/3} \qquad \text{extreme relativistic} \tag{8–42a}$$

Similarly, the ratio of mass density to number density is

$$\langle \rho \rangle / \langle n \rangle \sim m_N \qquad \text{nonrelativistic} \tag{8–127}$$

$$\langle \rho \rangle / \langle n \rangle \sim \frac{E_F}{c^2} \qquad \text{extreme relativistic} \tag{8–128}$$

because in the extreme case, the rest-mass energy can be neglected. But since $E_F \propto n^{1/3}$, one then has

$$\langle \rho \rangle \propto \langle n \rangle^{4/3} \qquad \text{extreme relativistic} \tag{8–129}$$

and from equation 8–126 we then find

$$M^{2/3} \propto 3 \left\langle \frac{P}{\rho} \right\rangle \langle n^{1/3} \rangle^{-1} \propto \begin{cases} \langle n^{1/3} \rangle & \text{nonrelativistic} \\ \\ \langle n^{1/3} \rangle^{-1} & \text{extreme relativistic} \end{cases} \tag{8–130}$$

This means then that the mass first increases as $\langle n \rangle^{1/3}$ and then decreases as $\langle n \rangle^{-1/3}$ as the density continues to increase.

The cores of massive neutron stars probably contain a variety of *mesons*, *baryons*, and *hyperons* in addition to the neutrons. At these very high densities, general relativistic effects must also be taken into consideration, because, for example, the Newtonian expression for potential energy no longer has meaning. This region is of greatest interest, because potentially it is in these last stages that a star can give off by far its greatest amount of energy by converting a large fraction of its mass into some form of radiation, perhaps gravitational. The star then turns into a *black hole*. However, before we turn to this ultimate form of stellar death, it is worthwhile mentioning a number of important considerations that we have neglected.

At the very high densities involved in neutron stars, the nuclei arrange themselves in a *crystal lattice*. The equation of state and structural properties we have assumed will therefore not be strictly correct. In parts of neutron stars a *superfluid* state may also exist. One model (Fig. 8.16) pictures a solid crust of nuclei floating on a superfluid layer. In summary, a variety of different crystalline or fluid phases may exist at different depths in the star. These will have to be understood more thoroughly before we can claim to understand the structure of neutron stars (Sa70a and Ru71b).

An additional matter of importance is the magnetic field. If a star like the sun were to collapse to a radius of a few kilometers, its magnetic field would be $\sim 10^{12}$ gauss. Is the field really that strong in neutron stars? And if it is, does this strong field permeate the entire interior or does it exist only at the surface? If superconducting effects were present in the core of the neutron star, magnetic fields might be excluded just as they are in laboratory superconductors. But how could that exclusion take place in a rapidly collapsing star?

We also know that an ordinary star that suddenly collapses would have to rotate rapidly if angular momentum is to be conserved. In the region surrounding the neutron star, the rapidly rotating magnetic field is thought responsible for accelerating charged particles to cosmic ray energies and perhaps for radiating away some of the star's rotational energy. The rotation presumably also affects the star's structure, since perfect spherical symmetry no longer holds.

If the Crab Nebula is at all representative, the neutron star, pulsar and supernova phenomena all seem to have a common origin. There may of course exist

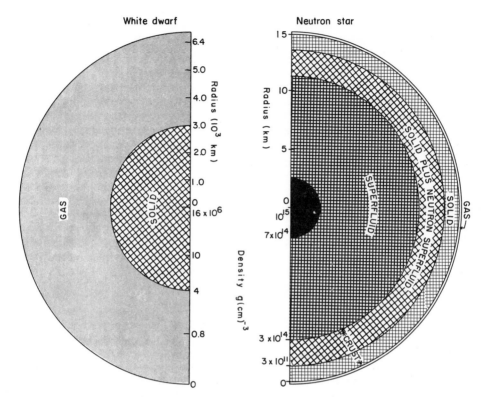

Fig. 8.16 Density and structure of white dwarf and neutron stars. The diagram on the left is a white dwarf model. The shell structure on the right represents a neutron star. Note that the white dwarf radius is 6400 km while that of the neutron star is only 15 km (after Ruderman Ru71b). At the neutron star's center we expect to find mesons and hyperons. (By permission of Scientific American Inc.)

a number of different processes that lead to supernova explosions since a number of different types of supernovae have been observed. Perhaps only some of these lead to the formation of neutron stars and pulsars. Others might be responsible for the formation of heavy elements and their return to the interstellar medium (Ar70) without necessarily also producing a compact remnant. This, in fact, raises the whole question of the interrelationship of supernovae with ordinary and/or recurrent novae, with planetary nebula stars that have shed a great deal of mass and left a hot central star behind, and finally with white dwarfs that might be the remnants of the planetary nebulae. How are all these related? Is the deciding factor between a white dwarf and a neutron star death merely a matter of whether the star's final mass is less or greater than the Chandrasekhar limiting mass (Ch64a)? We do not yet know!

Similarly, we also do not know much about black holes. If a neutron star is massive enough, does it collapse once more into a final state where, as discussed in Chapter 5, it no longer can send any radiation that will be received outside the star? What happens to the magnetic field when a black hole is formed, and to electric charge? How can we talk about conservation of baryons in the universe when all these particles disappear without a trace, giving off only photons or gravitational radiation and neutrinos as the collapse takes place? What are the correct dynamic equations in such a collapse? Are they the equations of general relativity, or not? Again a thorough discussion is needed; we are just beginning to grope our way (Op39a, Op39b, Ru71a, Pe71, and Ke63)!

8:20 VIBRATION AND ROTATION OF STARS

We know from the virial theorem that the absolute value of the potential equals twice the kinetic energy per unit mass. In a star this kinetic energy is represented by the thermal motions of the atomic particles whose speeds are of order of the speed of sound v_s. We can therefore write

$$\frac{GM}{R} \sim v_s^2 \tag{8-131}$$

where G is the gravitational constant, M is the mass, and R the radius of the star. We can now estimate a very rough order of magnitude of the stellar vibration frequency by noting that the period P_{vib} should be comparable to the time it takes to transmit information about pressure changes, across the entire dimension of the star. This time equals $2R/v_s$, and we can write (Sa69)

$$P_{vib}^{-1} = \nu_{vib} \sim \frac{v_s}{2R} \sim \sqrt{\frac{GM}{4R^3}} \sim \sqrt{G\rho} \tag{8-132}$$

where ρ is the density of the stellar material.

We also note that the maximum rotational frequency is determined by the equilibrium between centrifugal and gravitational forces, since at higher frequencies the star would be torn apart. Nonrelativistically,

$$R\nu_{rot,max}^2 = \frac{GM}{(2\pi R)^2}$$

and

$$\therefore P_{rot,min}^{-1} \sim \frac{1}{\pi}\sqrt{G\rho} \tag{8-133}$$

Table 8.7 Approximate Relation Between Stellar Density, Pulsation Period, and Minimum Rotational Period.

Star	Density ρ	P_{vib}	$P_{rot,min}$
Neutron star	10^{15} g cm^{-3}	10^{-4} sec	3×10^{-4} sec
White dwarf	10^7	1	3
RR Lyrae star	10^{-2}	$10^{4.5}$	10^5
Cepheid Variable	10^{-6}	$10^{6.5}$	10^7

We note that the vibrational periods for neutron stars ought to be about a 10th of a millisecond and that the vibration period for white dwarfs should be of the order of a second. Of course these stars can have a wide range of densities, and only representative periods are given in Table 8.7. These can vary by over an order of magnitude. When the Crab Nebula pulsar was discovered, having a period of only 33 milliseconds, it became clear that white dwarfs could not be considered to play a role in the pulsar phenomenon, because the vibration frequencies of white dwarfs would be too low. On the other hand the rotational period of a neutron star could well be several milliseconds. The discovery that the period of the Crab Nebula pulsar is increasing, so that it could have been closer to the minimum expected rotational period of a neutron star, shortly after the supernova explosion in the year 1054 AD, further supports the theory that pulsars are rapidly rotating neutron stars that are losing angular momentum and slowing down as time goes on. Since magnetic pressures would tend to disrupt rapidly rotating neutron stars, the actual minimum rotation period will be somewhat greater than shown in Table 8.7.

Table 8.7 also shows that RR Lyrae variables and Cepheid variables have periods consistent with the very simple vibration picture discussed in the present section. Such stars indeed are pulsating. This can be judged by the periodic Doppler line shifts and color temperature changes. The observed periodicity is the period of that pulsation. The nova remnant DQ Her (Nova Herculis, 1934) has a 71 sec period; and periodic behavior has been seen in some white dwarfs, although these periods are too long to represent fundamental pulsations.

ADDITIONAL PROBLEMS

The following set of problems uses greatly simplified stellar models, mainly to show that we can obtain reasonable order of magnitude estimates of stellar luminosities and lifetimes even without using sophisticated computing methods.

8:20

Reaction rates for the proton-proton reaction and the carbon cycle have the form given in equation 8–90. Schwarzschild (Sc58b) gives the energy generated per unit time and unit mass of matter for the proton-proton reaction:

$$E_{pp} = 2.5 \times 10^6 \rho X^2 \left(\frac{10^6}{T} \right)^{2/3} \exp \left[-33.8 \left(\frac{10^6}{T} \right)^{1/3} \right] \qquad (8\text{–}134)$$

and Clayton (Cl68) gives a similar expression for the CN cycle

$$E_{CN} = 8 \times 10^{27} \rho X X_{CN} \left(\frac{10^6}{T} \right)^{2/3} f \exp \left[-152.3 \left(\frac{10^6}{T} \right)^{1/3} \right] \qquad (8\text{–}135)$$

where f is an electron screening factor, $f \sim 1$.

For problems 8–6 to 8–8, consider an initial concentration $X_{CN} = 0.005X$ and $Y = 0.12$.

8–6. Consider the following model of a B1 star, that is, a massive young star. Its mass $M = 10M_\odot$ and its radius, $R = 3.6R_\odot$. Its central density is 10 g cm^{-3} and the bulk of energy generation takes place within a radial distance $0.1\,R$. Assume constant density for this region, and a temperature $2.7 \times 10^{7\circ}\text{K}$. Determine the rate of energy generation by the star and its surface temperature, assuming that the star radiates like a blackbody at its given radius. How long can the star exist in its present state without being appreciably changed by the nuclear burning, that is, how soon will it use up the hydrogen in its central core at the present rate?

8–7. Repeat this for a star like the sun in its early main sequence stages, assuming burning in the central region out to $0.2\,R$. Take a central density of about 55 g cm^{-3}, and a central temperature $10^{7\circ}\text{K}$; assume hydrogen burning.

8–8. At the present time the concentration of hydrogen, by mass, at the center of the sun is about 70%, throughout a central region out to roughly $0.11\,R$. The density is about 10^2 g cm^{-3} in the center. Take the mean temperature throughout the region to be $1.5 \times 10^{7\circ}\text{K}$ and recalculate the parameters for this star.

8–9. A red giant star whose radius is 100 solar radii, is in an evolutionary state where the inner hydrogen has all been exhausted but helium burning has not yet set in. The main energy source is hydrogen burning that takes place in a thin shell surrounding the inert helium core (Fig. 8.8). Let hydrogen burning take place at a radial distance ranging from 1.8 to 2×10^9cm. The mean density in this layer is about 50 g cm^{-3}. The temperature is $5 \times 10^{7\circ}\text{K}$. Calculate the above parameters. Take $X_{CN} = 10^{-3}X$, $X \sim 0.5$.

8–10. A white dwarf star is thought to shine by virtue of its stored thermal energy. Assuming its mass to be $0.45\,M_\odot$, its radius $0.016\,R_\odot$, and its density roughly uniform throughout (however, see Fig. 8.16). calculate how long the star could radiate at its present rate. Its luminosity is $10^{-3} L_\odot$.

8–11. For a star in which radiative transfer dominates, the energy density at each point is very nearly the blackbody radiation density (at all frequencies v) even though the radiant flux $L(r)$ is predominantly outward directed from the center of the star. Show how closely the radiation density ρ really equals the blackbody radiation density for a given opacity value $\kappa(v)$, by considering the higher order terms in equation 8–50.

8–12. In section 8:2 we argued that the sun's energy must be nuclear, because equation 8–1 did not provide enough potential energy to account for the total solar luminosity over the past aeons. Show that the argument actually was incorrect, by working out the potential energy of a structure in which roughly half the sun's mass is uniformly distributed throughout a sphere of radius $R_\odot$ and half the mass is concentrated in a core of radius 10 km. It is interesting that some years ago, astrophysicists were sure that most of the radiant energy in the universe was directly related to thermonuclear reactions. With the discovery of pulsars, these ideas changed. It is possible that gravitational collapse has contributed at least as much energy as nuclear reactions.

ANSWERS TO PROBLEMS

8–1. (a) $P = nkT$

$$P\left(\frac{R}{2}\right) \sim 10^{15} \text{ dyne cm}^{-2}, \qquad n = \frac{\rho}{m_H} = \frac{1}{1.67 \times 10^{-24}} \text{ cm}^{-3}$$

$$T\left(\frac{R}{2}\right) = \frac{10^{15} \cdot 1.67 \times 10^{-24}}{1.38 \times 10^{-16}} \sim 10^7 {}^\circ\text{K}.$$

(b) The gravitational potential energy is $\dfrac{GM_p m}{R} \le \mathscr{E}_c$

where m is the mass of an atom, $m = A m_H$.

$$M_p \lesssim \frac{\mathscr{E}_c R}{GA m_H}. \qquad \text{But } M_p \sim \rho R^3$$

$$M_p \lesssim \frac{\mathscr{E}_c M_p{}^{1/3}}{GA m_H \rho^{1/3}}, \qquad R = \left[\frac{M_p}{\rho}\right]^{1/3}$$

$$M_p^{2/3} \lesssim \frac{\mathscr{E}_c}{GA\rho^{1/3}m_H}, \qquad M_p \lesssim \left(\frac{\mathscr{E}_c}{GAm_H}\right)^{3/2} \frac{1}{\rho^{1/2}}.$$

(c) $\mathscr{E}_c \sim 10^{-11}$ erg $\qquad R_J = 7.1 \times 10^9$ cm $\left.\begin{array}{c} \\ \\ \\ \end{array}\right\}$ see Table 1.3

$$M_J = 2 \times 10^{30} \text{ g}$$

while $M_p \lesssim \dfrac{\mathscr{E}_c R}{Gm_H} = 10^{30}$ g

8–2. $F(r) = -\displaystyle\int_0^\infty \kappa\rho e^{-\kappa\rho l}\, dl \int_0^\pi a\left(T - \frac{dT}{dr}\, l\cos\theta\right)^4 \frac{c}{2}\cos\theta\, d(\cos\theta)$

Expanding, we have to first order in dT/dr,

$$\simeq \frac{4ac\kappa\rho}{3} \frac{dT}{dr} T^3 \int_0^\infty e^{-\kappa\rho l}\, l\, dl\,.$$

And using (8–26) and (8–51) we have

$$L(r) = \frac{16\pi acr^2}{3\kappa\rho} T^3 \frac{dT}{dr}\,.$$

8–3. $dP \propto \rho\, \dfrac{M(r)\, dr}{r^2} \qquad (8\text{–}7)$

$P = nkT \qquad$ so that $\qquad dP \propto \rho\, dT$

$\therefore\ \dfrac{dT}{dr} \propto \dfrac{M(r)}{r^2} \quad$ and $\quad T(r) \propto \dfrac{M(r)}{r}$

$L(R) \propto \left.\dfrac{R^2 T^3}{\rho} \dfrac{dT}{dr}\right]_{r=R} \propto \dfrac{M^4}{\rho R^3} \propto M^3.$

8–4. The number of particles incident on a hypothetical surface in unit time and solid angle is

$$n(\theta, \phi, p)\, v\cos\theta\, d\Omega$$

The pressure then is

$$P = \int_0^{p_0} \int_0^{2\pi} \int_0^{\pi/2} 2 \cdot p \cdot \cos\theta\, v\cos\theta\, n(\theta, \phi, p)\, d\Omega\, dp.$$

If the gas is isotropic, $n(\theta, \phi, p) = \dfrac{n(p)}{4\pi}, \qquad P = \dfrac{1}{3}\displaystyle\int_0^{p_0} pvn(p)\, dp$

where $n(p)$ is given by (4–65) since all states are filled. Now $p = mv\gamma(v)$, which can be solved for v:

$$v = \frac{p}{m\sqrt{1 + p^2/m^2c^2}}$$

$$\therefore P = \frac{8\pi}{3mh^3} \int_0^{p_0} \frac{p^4}{\sqrt{1 + p^2/m^2c^2}}\, dp$$

If $p/mc = \sinh u$ $dp/d\theta = mc\cosh u$ so that

$$P = \frac{8\pi m^4 c^5}{3h^3} \int_0^{u_0} \sinh^4 u\, du.$$

8–5. $f(x) = \dfrac{1}{8}\left[x(2x^2 - 3)(x^2 + 1)^{1/2} + 3\,\text{arc}\sinh^{-1} x \right], x = \dfrac{p_0}{m_e c}.$

If $x \ll 1$, $\sinh^{-1} x = x - \dfrac{x^3}{6} + \dfrac{3}{40}x^5 - \dots$

On expanding

$$8f(x) \approx x(2x^2 - 3)\left(1 + \frac{x^2}{2} - \frac{x^4}{8}\right) + 3x - \frac{x^3}{2} + \frac{9}{40}x^5 - \dots \approx \frac{8}{5}x^5$$

$$x \ll 1$$

Substitution of $f(x)$ and (8–113) into (8–112) then gives (8–41)

If $x \gg 1$, $\sinh x \sim \dfrac{e^x}{2}$, and $\sinh^{-1}x = \ln(2x)$

$$\therefore f(x)\frac{x^4}{4} \qquad x \gg 1;$$

substitution in (8–112) leads to (8–43).

8–6. We use a temperature $2.7 \times 10^{7}\,°\mathrm{K}$ and density $\rho = 10\,\mathrm{g\,cm^{-3}}$ in equations 8–134 and 8–135:
This gives

$$E_{pp} = 27 \text{ erg g}^{-1}\text{sec}^{-1} \quad E_{CN} \sim 3 \times 10^3 \text{ erg g}^{-1}\text{ sec}^{-1}$$

$\therefore$ The total energy generated is $4\pi/3\rho\, E_{CN}(0.54R_\odot)^3 \sim 7 \times 10^{36}$ erg sec. The total surface area of the star $4\pi(3.6R_\odot)^2 \sim 8 \times 10^{23}$ cm^2. Hence the flux crossing unit area is 0.9×10^{13} erg cm^{-2} sec^{-1} and using the blackbody law

$T^4\sigma$ = Flux across unit area,

$$T \sim 20{,}000°\text{K}$$

The total available energy per hydrogen atom is about 10^{-5} ergs with $n \sim 6 \times 10^{24}$ cm^{-3}. Hence the total energy available per cubic centimeter is about 6×10^{19} erg. The total time during which the star's core can supply energy, therefore, is about 7×10^7y.

8–7. With a temperature $10^7°$K, density 55 g cm^{-3} and hydrogen burning out to $0.2R_\odot$, we find that the proton-proton reaction predominates and yields 190 erg cm^{-3} sec^{-1}. This leads to a surface temperature of about 5000°K.

8–8. Using $X = 0.7$ and $\rho = 10^2$ g cm^{-3} with $T = 1.5 \times 10^7°$K, we again find that the pp reaction predominates, yielding about 2300 erg cm^{-3} sec^{-1}. The total energy generated is $\sim 4 \times 10^{33}$ erg sec^{-1}. This gives a surface temperature of 5900°K.

8–9. At $5 \times 10^7°$K the carbon cycle predominates. If we assume burning in a thin shell ranging from 1.8×10^9 to 2×10^9 cm, we obtain a carbon-cycle energy generation rate of 4×10^8 erg cm^{-3} sec^{-1} throughout a volume of 10^{28} cm^3. The luminosity therefore is 4×10^{36} erg $\sim 10^3 \, L_\odot$ and the surface temperature at $R = 7 \times 10^{12}$ cm is $T = (L/4\pi R^2\sigma)^{1/4} = 3300°$K.

8–10. A white dwarf can only radiate away the kinetic energy of its ions. Since the electrons are degenerate, they cannot lose energy, and provide the main support against hydrostatic pressures. Just before electron degenerate pressures start predominating, the kinetic energy of the ions is about one-tenth of the white dwarf's potential energy. The virial theorem would predict that half of the energy should be kinetic energy, but there will be two electrons or more for each ion present, and the electrons will have higher energies because of the onset of partial degeneracy. The total available ion energy therefore is of order $0.1 \, M^2G/R$, and the lifetime $\sim 0.1 \, M^2G/RL$. It may still be worth stating how the luminosity can be derived. The nondegenerate outer layers of the white dwarf permit radiative transfer. Using equation 8–52 with $T \sim 0.1 \, MGm_i/kR$ and $dT/dr \sim T/R$, we obtain the luminosity if the opacity is known. The opacity can be computed from the Kramers' expressions, although a look at Fig. 8.4 shows that at high densities the opacity is nearly independent of density and has a very approximate value $\sim 10^8/T$, in the temperature range of interest. The cooling time can therefore be expressed solely as a function of the star's mass, radius, and chemical composition (ion mass). If the luminosity for a star of mass $0.45 \, M_\odot$ is $\sim 10^{-3} \, L_\odot$, and its radius is 1.1×10^9 cm, $\tau \sim 0.1 \, M^2G/RL \sim 1.2 \times 10^{18}$ sec, and the cooling time is of the order of 40 aeons. This problem is discussed more rigorously in (Sc58b).

8–11. In Problem 8–2, $F(r) = \dfrac{4ac}{3\kappa\rho} T^3 \dfrac{dT}{dr}$ was derived by neglecting higher

order terms, since $\dfrac{dT}{dr} \ll 1$.

The next term is $\dfrac{4ac\kappa\rho}{5} \displaystyle\int_0^\infty e^{-\kappa\rho l} T \left(\dfrac{dT}{dr}\right)^3 l^3 \, dl$

$$= \frac{4ac}{(\kappa\rho)^3} \frac{T}{5} \left(\frac{dT}{dr}\right)^3 \int_0^\infty e^{-y} y^3 \, dy = \frac{24T}{5(\kappa\rho)^3} \left(\frac{dT}{dr}\right)^3$$

$$\rho(v) = \int_0^r 4acT \left(\frac{dT}{dr}\right)^3 \frac{18}{5(\kappa\rho)^2} \, dr$$

is the departure from blackbody energy density in a star.

Hence $\dfrac{d\rho}{dr} = 4ac\left[T^3 \dfrac{dT}{dr} + T \left(\dfrac{dT}{dr}\right)^3 \dfrac{18}{5} \dfrac{1}{(\kappa\rho)^2} \right]$

$$= \frac{d\rho(v)}{dr} + \frac{d\rho^*(v)}{dr}.$$

8–12. For a dense sphere

$$V = \frac{3}{5} \frac{GM^2}{R} \sim 4 \times 10^{52} \text{ erg} \quad \text{for} \quad R = 10^6 \text{ cm}, \quad M = 10^{33} \text{ g}.$$

9

Cosmic Gas and Dust

9:1 OBSERVATIONS

In this chapter we will try to define a common framework within which most of the processes involving astronomical gas and dust clouds can be understood. Before we can do that, however, we should know something about the temperature, density, state of ionization, and linear dimensions of the clouds and we should summarize the information we have about dust grains.

The methods we use to ascertain the rough quantitative values cited in Table 9.1 differ greatly from one type of medium to another. It may therefore be useful to state qualitatively how this information is derived.

(a) Extragalactic Medium.

Very little is known about the extragalactic medium. All the values for particle or field densities are upper limits; for all we know the space between galaxies is completely empty.

An upper limit to the neutral hydrogen density in intergalactic space is set by the lack of an observed Lyman-α absorption continuum for emission from quasistellar objects, QSO's. This argument assumes that the QSO's are extragalactic and at distances indicated by their red shifts as computed from Hubble's red-shift–distance relationship. Hubble's relation, however, may only hold for ordinary galaxies. The expected absorption would not correspond to a sharp line because neutral hydrogen lying at different distances between us and a QSO would absorb radiation at differently red-shifted wavelengths. We would therefore expect roughly uniform absorption over the whole wavelength range from the Lyman-α line of the QSO to the local wavelength of Ly-α, 1216 Å.

An upper limit to the number of electrons in intergalactic space can be derived from X-ray background observations on the hypothesis that these electrons are at high temperatures. Their free-free emission then has an appreciable X-ray component. However, if the electron temperatures were low, free-free emission would give lower values of the X-ray background and no X-ray flux would be observed. The precise values of upper limits obtained depend on the model adopted (see Chapter 10) for the expanding universe. We can place another upper limit on the density of the ionized extragalactic component by considering the actual ionization rates available from the known sources of ultraviolet radiation, primarily the QSO's. Again, the limit obtained depends on the cosmological model one prefers, but generally the range of number density values is $\lesssim 10^{-5}$ cm^{-3}.

Upper limits on the extragalactic magnetic field also depend on the assumptions we make. The fields cannot be so great that their pressure would exceed the gas pressure in our Galaxy. Otherwise interstellar hydrogen would actually suffer compression. Neither can the pressure be so great as to compress the extended radio sources that extend out from some galaxies. The characteristics of those sources are very uncertain, however, and therefore yield relatively little information. A coarse upper limit, probably far too high, is $B \lesssim 10^{-6}$ gauss.

(b) Quasistellar Objects

Many different models have been constructed to account for observed features exhibited by various QSO's. We have to account for the radio, infrared, and X-ray luminosity and for optical line emission. Moreover, it is not clear that all QSO's are basically similar in nature.

One model (Ca70b, Gr64), which places QSO's at cosmic distances, indicates that the visual line emission of highly excited ions can be accounted for by a nebula having an electron density of order 3×10^6 cm^{-3} and a radius of the order of 1 pc. The emission by relativistic particles in a core having the high magnetic field strength of order 10^5 gauss would then account for much of the remaining radiation. The actual strength of the magnetic field is always uncertain in such computations. First, some of the observed radiation might be produced through inverse Compton scattering so that no field at all would be needed. For the remainder we often assume that synchrotron radiation is important; however, the observed spectrum normally does not cover all wavelengths of interest and the partial information obtained does not suffice to uniquely determine the magnetic field. Basically there are two independent parameters as shown in (6–159): the energy exponent γ of the relativistic particles and the magnetic field strength. If γ is not known, the field strength also remains uncertain.

Faraday rotation within the quasistellar object is a potential means for determining magnetic field strengths, but we obtain no more than an upper limit if the irregularities in the field are unknown (section 6:12). It also is necessary to have an electron density value if the Faraday rotation is to determine a magnetic field strength. That density is only available if recombination line data exist, or free–free emission has been observed (sections 6:16, 6:17, and 7:12).

(c) The Galaxy

In the Galaxy and in some extragalactic objects, neutral hydrogen can be observed through the line absorption or emission of atomic hydrogen at a wavelength of 21 cm. Within the Galaxy we also have Lyman-α absorption of light emitted by O and B stars. These two types of data do not always agree, but the indications are that neutral hydrogen number densities in our part of the Galaxy are of order 0.1 to 0.7 cm^{-3}. Between the spiral arms, the density is lower (Je70) (Ke65).

The electron number density is determined from dispersion data (section 6:11) by using pulsars as sources of radio emission. Different frequency radio waves suffer different time delays (6–58) in traversing the distance between the pulsar and earth.

Except in the case of the Crab Nebula pulsar, the distance to the sources is not well known. Statistically, however, a self-consistent model is obtained if we assume that pulsars cluster around spiral arms (Da69). The electron density obtained then is 0.03 cm^{-3}, averaged over the arm and interarm domains in the local part of the Galactic disk.

Dust grain number densities and radii are computed in the following way. The size of the grains is estimated by the differential extinction of light at different wavelengths. Because red light is less strongly extinguished than blue, we argue that grain sizes should at least be smaller than the wavelength of red light. This argument extends also to ultraviolet wavelengths, with the result that grains are considered to have radii $\lesssim 10^{-5}$ cm. Similar results are obtained from slight coloration effects and polarization effects in reflection nebulae. For any given grain size, an effective extinction cross section can be computed and this, together with interstellar extinction data for stars within a few kpc from the sun determines an approximate dust grain number density for the solar neighborhood.

(d) H II Regions and Planetary Nebulae

The electron temperatures and densities are readily determined from free-free emission observations in the radio domain (section 6:17). Visual and radio recombination line data provide comparable information. This recombination rate can be computed from (7–75) in the following way. Let the recombination cross section for an electron in a gas at temperature T be $Q_n(T)$ where n is the principal

9:1

quantum number of the final state in the hydrogenlike ion. Consider an idealized situation of thermal equilibrium in which the number of ionizations equals the number of recombinations to this level. The recombination rate per electron is, in velocity range dv, proportional to the electron's velocity $v \sim (3kT/m)^{1/2}$; to the cross section $Q_n(v)$ and to the number density $n_e(v)$ of electrons and n_{r+1} of ionized atoms. The ionization rate is proportional to c, to α_{bf} (7–75), to the number density of atoms in the lower ionization state n_r, and to the number density of photons, at a frequency v high enough to produce both ionization and an electron velocity v. If χ_r is the ionizing energy,

$$v = \frac{1}{h}\left(\frac{m}{2}v^2 + \chi_r\right) \tag{9-1}$$

We can then write the equilibrium condition between ionization and recombination, very roughly, as

$$n_e(v)\, n_{r+1} v Q_n(v)\, dv = c \cdot n_r \cdot \frac{8\pi}{c^3} \frac{v^2 \alpha_{bf}(v)}{(e^{hv/kT} - 1)}\, dv \tag{9-2}$$

by making use of the blackbody spectrum (4–71) for the number density of photons $\rho(v)/hv$. Use of the Saha equation 4–105 and the absorption coefficient given by expression 7–75 then leads to the relation

$$\frac{g_{r+1}g_e}{g_r} \frac{[2\pi mkT]^{3/2}}{h^3} e^{-\chi_r/kT} Q_n(v)\, v\, dv = \frac{8\pi}{c^3} \frac{64\pi^4 m e^{10} Z^4}{3\sqrt{3}\, h^6 n^5} g_{bf} \frac{dv}{v[e^{hv/kT} - 1]} \tag{9-3}$$

Using the interrelationship (9–1) between variables v and v, and knowing that $g_r \propto n^2$ (see Problem 7–1) we obtain a relation for the recombination rate α_n for unit electron and ion density:

$$\alpha_n = \int_0^\infty v Q_n(v)\, dv$$

$$= \frac{g_r}{g_e g_{r+1}} \int_0^\infty \frac{2^9 e^{10} \pi^5 Z^4 m e^{\chi_r/kT}}{h^3 n^5 [6\pi mkT]^{3/2} [e^{(\chi_r + mv^2/2)/kT} - 1]} \frac{d\left(\dfrac{mv^2}{2}\right)}{\left(\dfrac{m}{2}v^2 + \chi_r\right)} \tag{9-4}$$

The fraction outside the integral is $\sim n^2$ (section 8:8).
For visible radiation (9–4) is considerably simplified since kT is small compared to χ_r so that the exponential dependence on χ_r can be neglected. The integral can then be expressed in approximate form (Za54):

$$\alpha_n = \frac{2.08 \times 10^{-11}}{T^{1/2}} \phi(T) \tag{9-5}$$

Table 9.1 Rough Characterization of Gas and Dust Aggregates.

	Representative Object	Density of Neutral Gas n_H (cm^{-3})	Electron Density n_e (cm^{-3})	Radius (cm)
Extragalactic Medium		$< 3 \times 10^{-11}$	$\lesssim 10^{-4}$	$\sim 10^{28}$ cm to cosmic horizon
Quasars	3C273		3×10^6	3×10^{18}
Spiral Galaxy Arm:	Galaxy	0.1 to 0.7 $\rbrace$	0.03	3×10^{20} thick;
Interarm Medium:		$\lesssim 0.05$		disk diameter 10^{23}
H II Region	Orion Nebula		10^4	5×10^{18}
Planetary Nebula	NGC6543		6×10^3	10^{17}
Supernova Remnant	Crab Nebula		40	5×10^{18}
H I Cloud	Heiles Cloud I (22^h 29^m5, $74° 58'$)	40 to 125	~ 0.3	10^{19}
Stellar Wind	O star δ Ori	0.14	10^8	at 1 AU from star
Solar Wind			2	at 1 AU from sun
Comet Head	Arend-Roland 1957 III	$n_{\text{molecules}}$ $\sim 10^4$		10^{10} cm
Comet Dust Tail	Arend-Roland 1957 III			10^{12} cm long
Comet Ionized Tail	Arend-Roland 1957 III		$n_{\text{ion}} \sim 2$ (at 10^{11} cm)	5×10^{12} long

where $\phi(T)$ has a value of 3.16 at 1580°K and 1.26 at $7.9 \times 10^{4°}$K. It therefore does not rapidly change with temperature.

Although the thermal velocities of ions already lead to broadening of spectral lines, we can still determine bulk velocities of turbulent motion when these are high enough to lead to an actual split appearance of spectral lines, or if the thermal broadening contribution can be computed from independent temperature data.

Dust densities in such clouds can be determined by measuring the continuum

Magnetic Field (gauss)	Turbulent or Bulk Velocity (cm sec^{-1})	Temperature (°K)	Number Densities and Radii of Grains		Remarks
			n_g (cm^{-3})	a_g (cm)	
$\ll 10^{-6}$?		$> 2 \times 10^{4}$°K if the medium produces the diffuse X-ray background	$\lesssim 10^{-15}$	$\sim 3 \times 10^{-5}$ assumed	n_H is dependent on cosmic QSO's and assumed cosmological model (Gu65)
$\sim 10^{5}$? in center	10^{8} to 10^{9}	17,000			velocities from multiple absorption spectra (Gr64) (Ca70b)
$< 10^{-5}$	10^{6}		10^{-13}	3×10^{-5}	Ly-α and 21 cm n_H data differ (Je70) (Ke65)
	4×10^{6}	10^{4}	10^{-9}	3×10^{-5}	
		8400	$\sim 3 \times 10^{-10}$	3×10^{-5}	
3×10^{-4}	10^{8}	$< 17,000$			(Co70) (Wo57)
$\lesssim 10^{-5}$	10^{4} to 10^{6}	4.5°K to 100°K	10^{-10}	3×10^{-5}	$n_H \sim 4 \times 10^{-6}$ cm^{-3} $n_{H_2} \sim 200$? (He69) (He68)
	1.4×10^{8}	10^{4}°K			(Mo67)
3×10^{-5}	4×10^{7}	10^{4} to 10^{5}°K	$\sim 10^{-13}$ at 1 AU	5×10^{-5}	Wind terminates at 10 to 100 AU; grains orbit the sun; they do not move with wind
	2×10^{5}				Molecules are: C_2, C_3, CN, NH, CH, OH, NH_2, OH^+, CH^+, and so on.
	10^{6}		$\sim 10^{-7}$	5×10^{-5} cm	
$\sim 3 \times 10^{-5}$	10^{7}				Mainly CO^+ ions, CO_2^+, N_2^+, OH^+, and so on.

radiation from the cloud in the visible part of the spectrum. Much of this radiation is likely to be starlight scattered by dust. Some assumption about particle size must then still be made, and ideally we should also know the chemical composition and physical structure of the grains. Likely candidates are silicates, iron containing minerals, or graphite grains; but no definite knowledge exists. As will be seen in section 9:6, infrared emission from HII regions and planetary nebulae may also provide data on dust densities.

9:1

Table 9.2 Absorption and Recombination Coefficients for Hydrogen and Helium.[a]

Atom	Term	a_{v_0}	f	α_n 10,000°K	Q_n 10,000°K
		10^{-18} cm^2		10^{-14} cm^3 s^{-1}	10^{-22} cm^2
H I	1s	6.3	0.436	15.8	32
	2s	15	0.362	2.3	4.7
	2p	14	0.196	5.3	11
	3s	26	0.293	0.8	1.6
	3p	26	0.217	2.0	4.1
	3d	18	0.100	2.0	4.1
	4s	38	0.248	0.4	0.7
	4p	40	0.214	1.0	2.0
	4d	39	0.149	1.0	2.0
	4f	15	0.057	0.6	1.2
	Total			43	88
He I	1s^2 ^{1}S	7.6	1.50	15.9	33
	1s2s ^{3}S	2.80	0.25	1.4	2.9
	1s2s ^{1}S	10.5	0.40	0.55	1.1
	Total			43	88
He II	1s	1.8	0.44	73	150

[a]a_{v_0} is the absorption cross section at the ionization limit; the oscillator strength f is defined in section 7:9; the recombination coefficients α_n and Q_n are defined by (9–4). (After Allen, Al64. With the permission of Athlone Press of the University of London, second edition © C. W. Allen 1955 and 1963.)

(e) Supernova Remnants

These remnants can stretch across many tens of parsecs in diameter and often show a circular arc structure. High expansion (Doppler) velocities can be measured by spectral observations; for the Crab Nebula an actual expansion can be observed by a comparison of present-day and 50-year-old photographs. The expansion velocity for the Crab is of the order of 10^8 cm sec^{-1}. For the Crab Nebula strong polarization along a direction perpendicular to the length of the continuum emitting wisps is seen. Assuming that this comes from synchrotron radiation, emitted by highly relativistic electrons spiraling in magnetic fields

9:1

that lie embedded with field lines running the length of the wisps, we can make an estimate of the magnetic field strength. It amounts to about 10^{-4} gauss. Besides relativistic electrons, there also exists lower temperature plasma that can be detected through its Hα emission lines, in the red part of the visual spectrum.

(f) Stellar Winds

The stellar wind, in say O and B stars can be detected by observing the Doppler shifted lines of highly excited ions. Initially these were observed in the ultraviolet part of the spectrum through rocket observations (Mo67). Assuming that solar abundance also characterizes the surface material in these stars, it is possible to interpret the observed line strengths in terms of total amounts of matter ejected from these stars. Typically, a massive O star seems to eject matter at a rate of one solar mass in 10^6 y, so that, during the star's lifetime, the ejected mass amounts to several percent of the total.

(g) Solar Wind

The solar wind density can be measured directly through the use of interplanetary probes that sample the plasma at appreciable distances from the earth where the earth's magnetosphere no longer interferes with observations. Magnetic fields can be measured by magnetometers carried on these spacecraft. Considerable wind variations between quiet and active periods on the sun are observed. The overall wind velocity can reach values of ~ 1000 km sec^{-1} following a solar flare, although during quiet times it does not vary greatly from a mean value around 400 km sec^{-1}. At quiet times the density amounts to a few particles per cubic centimeter. Following a flare the density can increase by an order of magnitude.

(h) Comets

These objects have three distinct parts, a roughly spherical head, a long straight tail, and a shorter curved tail. The head, frequently some 10^{10} cm in diameter, contains C_2, C_3, CN, NH, CH, OH, and NH_2 among other molecules, as well as such ions as OH^+, CH^+, and CO^+. Typical molecular densities range around 10^4 cm^{-3}. Recently a large atomic hydrogen envelope around the head has also been discovered through Ly-α observations from spacecraft. The velocities of the molecules in the head can be measured by Doppler observations and amount to a few km sec^{-1}. The head also includes a solid nucleus that is too small to be directly seen.

9:1

The long straight tails seen in comets sometimes stretch over a distance larger than an astronomical unit. They are the largest objects in the solar system but their densities are low and the total mass contained is minute. Only ions and no neutral molecules are seen in these tails. The number density of the ions is determined by the intensity of the molecular lines. The f values for the excitation have been computed and we know how to relate the observed brightness of the emission lines to the number of molecular ions in the line of sight and to the rate at which these ions would be placed into excited states by absorption of sunlight.

The shorter curved tails in comets reflect sunlight as judged by the *Frauenhofer* (absorption) *line spectrum* that mimics the sun's. Since the solar lines do not appear broadened in this reflection, the scattering particles must be slowly moving and cannot be electrons, which would have thermal velocities $\sim 10^8$ cm sec^{-1}. We conclude that the tails contain dust. This also is corroborated by the curved shape of the lagging tail. It is curved because the repulsion of the dust by sunlight and the requirement for constant orbital angular momentum about the sun produces an increasing lag for grains at increasing distance from the sun. The size of the dust grains can be roughly determined by the rate at which solar radiation pressure pushes them away from the head of the comet. Grains of different sizes will follow different paths since the radiative repulsion varies, and it is possible to derive rough estimates of grain sizes at different locations across the spread out width of the tail. The smallest grains will lie closest to the radius vector pointing away from the sun. The largest grains will be most distant from that axis. From the rough estimate of grain sizes, we can compute the number density of grains as judged from the total scattered sunlight.

Table 9.1 has been constructed to summarize some of the information we have on individual diffuse objects in the solar system, in the Galaxy, and beyond. Within each class listed in the first column, variations in size, density, and so on amounting to orders of magnitude are not uncommon, and we have to be careful not to assume that different members of a class always have identical properties.

9:2 STRÖMGREN SPHERES

In 1939, Strömgren (St39) considered the interaction of a very young star with the interstellar medium. To make matters simple he made two assumptions. First, he suggested that the star lights up rapidly to full strength; secondly, he considered the surrounding medium homogeneous throughout. These two assumptions permitted Strömgren to draw a simple picture of the development of ionized hydrogen regions around massive, ultraviolet emitting stars.

We note that if the star emits a number of photons dN_i capable of ionizing the surrounding gas, then the number of electrons that are stripped off the atoms,

Fig. 9.1   Schematic diagram of a Strömgren Sphere.
H II is the ionized gas. H I is the neutral region and δ
the thickness of the separating layer.

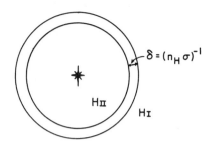

Fig. 9.1   Schematic diagram of a Strömgren Sphere.
H II is the ionized gas. H I is the neutral region and δ
the thickness of the separating layer.

over the same time interval, will also be dN_i if equilibrium is maintained. This assertion is always true in practical cases, since the cross section for ionization by energetic photons is of the order $\sigma \sim 10^{-17}$ cm^2 and typical gas densities in the vicinity of young stars might be of order $n_H \sim 10^3$ cm^{-3}. At these densities a photon can only travel a distance of order $(n_H\sigma)^{-1} \sim 10^{14}$ cm through the neutral medium before it ionizes an atom. But this is a small distance compared to the radii of ionized regions normally observed. These radii range from $\sim 10^{16}$ to $\sim 10^{20}$ cm; hence, practically no ionizing photons can escape through the gas without becoming absorbed.

Although an energetic photon can only travel a short distance in the neutral medium, its path through the ionized gas is very long. Here it occasionally is scattered; but the scattering cross section is relatively small—Thomson cross section is only $\sim 6.7 \times 10^{-25}$ cm^2 (6–102)—and we can assume that the ionizing photons proceed undisturbed through the ionized gas immediately surrounding the star, until they hit the boundary region where the neutral gas commences. The thickness of the boundary region also is the mean free ionizing path:

$$\delta = (n_H\sigma)^{-1} \qquad (9\text{–}6)$$

We will call the neutral clouds, H I regions, and the ionized clouds, H II regions.

We picture an ionizing star, as surrounded by an H II region, which is separated from an embedding H I region by a thin layer of thickness δ. This is shown in Fig. 9.1. If the gas is homogeneous, the surfaces of separation are spherical, and the sphere containing the ionized gas is called the *Strömgren sphere*.

We now ask ourselves, how quickly the sphere becomes established. To do this, we note that the number of atomic particles in a shell of radius R and thickness dR is $4\pi R^2 n_H\, dR$; so that when the star emits dN_i ionizing photons, the radius of the region grows by an amount dR, given by

$$\frac{dN_i}{dt} = 4\pi R^2 n_H \frac{dR}{dt} \qquad (9\text{–}7)$$

Here we have gone through the additional step of formally dividing both sides by dt, to obtain the time rate of development.

Equation 9–7, however, is only true during the initial stages of growth. It neglects the recombination of ions and electrons that also has to occur in the interior of the Strömgren sphere. For, if an electron and ion recombine to form an atom, a new ionizing photon will be required to separate the two particles. This photon never reaches the boundary R and it will therefore not contribute to the growth of the region. The recombination rate per unit volume is proportional to the product $n_e n_i$, since each electron has a probability of colliding that is proportional to the number of ions it encounters. In addition, the recombination rate per unit volume is also proportional to the recombination factor $\alpha \sim 4 \times 10^{-13}$ cm^3 sec^{-1}, which depends (9–5) on the temperature of the ionized gas, normally $\lesssim 10^4 °$K, and represents the sum over recombination factors leading to all states n (see column 5 of Table 9.2). The full equation, satisfied by the gas therefore, is

$$\frac{dN_i}{dt} = 4\pi R^2 n_H \frac{dR}{dt} + \frac{4\pi}{3} R^3 n_i n_e \alpha \qquad (9–8)$$

During the late developmental stages, this very simple model would predict that dR/dt eventually becomes zero as the sphere grows so large that the star emits photons only just fast enough to keep up with the total number of recombinations. This will happen at an equilibrium radius

$$R_s^3 = \frac{3}{4\pi n_i n_e \alpha} \frac{dN_i}{dt} \qquad (9–9)$$

Equations 9–7 and 9–9 are the extreme cases covered by equation 9–8. They describe the initial growth and final equilibrium value of the radius of the Strömgren sphere as long as the very simplest assumptions are retained.

A number of comments are needed:

(1) Equation 9–8 has to be used in conjunction with some model for ultraviolet emission by stars. If the star's luminosity and temperature are known, then we can readily estimate N_i from the Planck blackbody relation (4–71). In actual fact, the ultraviolet spectrum of a very hot star does not closely approximate a blackbody because absorption by the star's outer atmosphere changes the spectrum of the escaping radiation. This is called the *blanketing effect*. Despite the blanketing effect, however, the blackbody approximation gives roughly the correct magnitude for the number of ionizing photons to be inserted in equation 9–8. At present, the actual ultraviolet spectrum down to wavelengths of about 1000 Å is, in any case, only known for a small number of the brightest ultraviolet stars. The next years should bring in much new information, however, since

ultraviolet observations from above the atmosphere are now more readily carried out. The orbiting observatory *Copernicus* is yielding such data.

(2) Many times the recombination of an electron with an ion will yield a photon that is still capable of ionizing another atom. In fact, the only time that this will not be true is if the recombination first takes the atom into a high lying excited state. A recombination directly into the ground state, on the other hand, always yields a photon capable of further ionization. For this reason the second term on the right side of equation 9–8 is an upper limit on the loss of ionizing photons through recombination. Similarly the radius R_s of equation 9–9 is a lower limit for an equilibrium value. The effect turns out to be more important in very dense regions than in tenuous gases surrounding a star. For very dense regions the true R_s value may be more than 10 times greater than that given by (9–9). For values of n_H around 10^4 to 10^5 cm^{-3}, typical of the denser ionized regions normally encountered, the radius R_s is a factor of 2 to 3 higher than predicted by (9–9).

(3) A very quick consideration shows that equations 9–8 and 9–9 cannot be completely correct because they neglect the problem of pressure equilibrium. This can be seen rather easily. For, the ionized region must in any case have at least twice as many particles per unit volume as the nonionized region, that is, at least one ion and one electron for each atom. This means, according to the ideal gas law (4–37), that the pressure on the inner side of the boundary separating ionized from nonionized regions would be at least twice as great as the pressure on the outside; and that only if the temperature was the same on both sides. In practice, the temperature of the H I region is likely to be of order 70°K while the temperature of the H II region normally amounts to $\sim 7000°$K. The total pressure inside the separating boundary is therefore of order 200 times greater than the pressure outside and expansion must therefore proceed quite rapidly.

(4) If we were to draw the very simplest picture of an expansion, we would proceed by visualizing the process in terms comparable to the inflation of a balloon. If the mass of the surrounding H I region is M, the mass per unit area at the separating surface is $M/4\pi R^2$, and the pressure inside is $2n_i kT_i$. T_i is the temperature of the ionized region, and the factor 2 assumes that the number of ions n_i equals the number of electrons. Neglecting the small gas pressure on the outside of the sphere, we obtain the outward acceleration of the boundary as

$$\ddot{R} = \frac{2n_i kT_i}{(M/4\pi R^2)} \tag{9–10}$$

This can be integrated if we first multiply both sides by $\dot{R}$. Then

$$\frac{\dot{R}^2}{2} = \frac{8\pi}{3} \frac{n_i kT_i}{M} R^3 \tag{9–11}$$

9:2

which leads to a development time scale of order

$$t \sim \left(\frac{3M}{16\pi n_i k T_i R} \right)^{1/2}$$

(9–12)

If we take M roughly equal to one solar mass $\sim 2 \times 10^{33}$ g, $n_i \sim 10^4$ cm^{-3}, $T_i \sim 10^{4\circ}$K, and $R \sim 10^{17}$ cm, we find that

$$\dot{R} \sim 3 \times 10^5 \frac{\text{cm}}{\text{sec}}, \quad t \sim 3 \times 10^{11} \text{ sec}$$

(9–13)

This velocity has to be compared with the random speed of atoms in the cool medium that only is $\sim (3kT/m_H)^{1/2} \sim 1.5 \times 10^5$ cm/sec, at the low temperature of the H I region. The correct dynamics, therefore, cannot be described by equations 9–10 through 9–13 because pressure is normally propagated at roughly the speed of sound in H I regions. If the expansion on the inner edge of the region proceeds faster than the speed of sound, the outer portions of the region will not be aware that a pressure is being exerted at the inner boundary, and will therefore not move. As a result, the quantity of material actually accelerated at any given instant will not have the mass M used in the subsonic approximation (9–10) and the actual velocity $\dot{R}$ will be considerably higher. The equations of supersonic hydrodynamics must therefore be used and these will be explained in section 9:3 below.

(5) Before proceeding to the dynamical treatment of expanding H II regions, it is interesting to point out that equation 9–7 may still hold well for extremely early stages of development, because dR/dt is so high there that the *ionization front*, that is, the region separating ionized and neutral regions, proceeds into the medium at velocities that can be orders of magnitude greater than the speed of sound in the medium. There is then no possibility at all for major instanteous adjustments of density in response to pressure differences between ionized and nonionized regions. This also will be discussed in section 9:3.

(6) It is interesting that the expansion produced by gas pressure brings down the density of ionized material in the H II region and therefore decreases the recombination rate per unit volume. The factor $n_e n_i \alpha$ of the second term in equation 9–8 decreases as R^{-6}, because both n_i and n_e decrease as R^{-3} when only expansion due to excess pressure (in contrast to expansion through further ionization) is involved. This means that the second term on the right of (9–8) is always reduced by pressure induced expansion, thereby giving rise to a higher value for the expansion velocity $\dot{R}$ of the boundary.

(7) Finally, it is important that our whole concept of the development of a Strömgren sphere has been based on a picture in which the central star suddenly brightens and produces ionizing radiation. But that does not at all correspond to the development of massive stars shown in Fig. 8.11. There we had shown that

a massive O or B star takes some 6×10^4 y to contract to the main sequence and for a good fraction of that time it is bright without emitting much ionizing radiation. Davidson (Da70a) has argued that during the contraction stage, light pressure pushes gas and dust away from the star. This happens because a dust grain with radius a, accelerated by light pressure to a velocity v with respect to the gas, suffers collisions at a rate $n_H v \pi a^2$ with the atoms and, hence, suffers a drag (momentum loss) amounting to a deceleration

$$\dot{v}_d = -\frac{n_H m_H v^2 \pi a^2}{(4\pi/3) a^3 \rho} \tag{9-14}$$

where ρ is the grain's density.

The radiative acceleration, for a star having luminosity L, is

$$\dot{v}_r = \frac{L}{4\pi c R^2} \frac{\pi a^2}{(4\pi/3) a^3 \rho} \tag{9-15}$$

so that equilibrium is established at a velocity

$$v \sim \left[\frac{L}{4\pi c R^2 n_H m_H} \right]^{1/2} \tag{9-16}$$

which has a value of $\sim 1.5 \times 10^6$ for $L \sim 10^{38}$ erg sec^{-1}, $R \sim 10^{17}$ cm, and $n_H \sim 10^4$. This velocity is set up in a time

$$\tau \sim \frac{v}{\dot{v}} = \frac{(4/3) \rho a}{\sqrt{(L/4\pi c R^2) n_H m_H}} \tag{9-17}$$

If $\rho \sim 3$ g cm^{-3} and $a \sim 10^{-5}$ cm, $\tau \sim 2.5 \times 10^9$ sec. From this we see that grains reach equilibrium velocity in a matter of a century, while the contraction of the star to the main sequence takes tens of thousands of years.

The grains drag the gas out to quite large distances through this process. For, a radiative pressure $(L/cR^2 4\pi)$ acting, say, on a column of length R and hence of mass $n_H m_H R$, would produce a mean acceleration of order

$$\dot{v} \sim \frac{L}{4\pi c n_H m_H R^3} \tag{9-18}$$

and for the same conditions chosen above, but with $R \sim 3 \times 10^{17}$ cm, $\dot{v} \sim 10^{-6}$ cm sec^{-2}; in 3×10^4 y a distance of order R would be covered. Davidson therefore argues that when the star begins copious emission of ionizing radiation, most of the gas already has been pushed to large distances. The ionization of course still occurs, but it takes place at the edge of the low density cavity in which the star now finds itself. What may happen then is that the newly ionized gas flows inward to the star, rather than outward away from it. One does not yet know!

9:3 SHOCK FRONTS AND IONIZATION FRONTS

In the preceding section we gave one example of supersonic flow in the vicinity of hot stars. There are many other such examples. The *wind* or stream of gas that continuously sends stellar material out into surrounding space blows at supersonic velocities ranging from some thousands of kilometers per second for the hottest O stars, down to speeds of the order of 400 km sec^{-1} for stars like the sun.

This fast stream of gas coming from the sun interacts with the earth's *magnetosphere* to give rise to a wide variety of different effects, ranging from the colorful *aurora borealis* to the nuisance of poor radio wave propagation in the broadcast band. The solar wind also gives rise to the long ionized tails of comets.

On a larger scale, supersonic phenomena are encountered in stellar explosions of all kinds, ranging from the small outbursts that regularly occur in flare stars to the explosion of supernovae and the explosive ejection of gas from the nuclei of galaxies.

This brief list shows that supersonic velocities are quite normal in astrophysics. In fact, virtually all phenomena that take place outside stars, or solid bodies like planets and grains of dust, involve velocities greater than the speed of sound.

In this section, we will be concerned with the equations that describe the interaction of an H II region with the surrounding neutral medium; but the treatment is general and can be applied to many other supersonic phenomena in astronomy.

We will assume that a star has suddenly undergone an increase in brightness, that it rapidly ionizes the surrounding medium, and that a *shock front*, or an *ionization front*, or both, move outward into the cool H I region at supersonic velocity. We will see below just how these fronts behave.

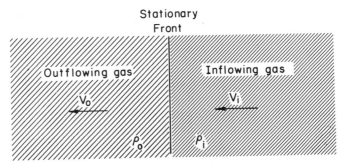

Fig. 9.2   Conditions on the two sides of a shock or ionization front.

There are two ways of considering a front—or dividing region—between the expanding ionized gas and the still unperturbed neutral hydrogen region. We can either consider the front as moving out into neutral gas at some velocity v, or else we can pretend that the neutral gas is moving into a stationary front at velocity, $-v$. After passing through the front, the gas is compressed and possibly ionized; and there will be energy changes—mainly heating.

We will adopt this second point of view, that the front between the two regions is stationary, and we will make a number of demands on the gas flowing through the front, Fig. 9.2.

(a) We require that the mass flow into the front equals the mass flowing out. This is just the continuity condition. If the inflow density and velocity are ρ_i and v_i and the outflow density and velocity are characterized by ρ_0 and v_0, this requirement reads

$$\rho_i v_i = \rho_0 v_0 \equiv I \qquad (9\text{--}19)$$

Here, I is the *mass flow* through unit area in unit time.

(b) We can consider the front to be a surface that absorbs inflowing gas and emits outflowing gas. In that case the pressure of inflowing material would be $\rho_i v_i^2$ even if the inflowing gas had no intrinsic pressure due to random motion of the atoms. $\rho_i v_i^2$ simply is the momentum transferred to the surface per unit area and unit time through absorption of the inflowing particles—and that is what we mean by pressure. Similarly the back pressure due to the seemingly emitted outflowing gas would be $\rho_0 v_0^2$. Since (9–19) has to be satisfied and since v_0 generally differs from v_i, these two pressures will not normally be equal and opposite. We still have two thermal pressures due to the random thermal motions of the gas atoms adding to the overall pressure acting on each side of the front. If the front is not to be accelerated—and we have assumed that there is a constant inflow and outflow velocity here—then momentum conservation requires that the overall pressures on the two sides of the front be equal and oppositely directed.

$$P_0 + \rho_0 v_0^2 = P_i + \rho_i v_i^2 \qquad (9\text{--}20)$$

This is the condition of steady flow.

(c) On passing through the front, the energy content of the gas changes. A number of different sources contribute to the overall energy. For the inflowing gas there is:

(i) kinetic energy due to the bulk flow, $v_i^2/2$ per unit mass flowing into the front;

(ii) internal energy per unit mass U_i (see section 4:18);

(iii) the work done on unit volume as it flows into the surface. Since we are picturing the gas as being stopped—absorbed—at the front, this is equivalent to the gas being compressed down to zero volume as it approaches the front.

The work done in unit time then involves a volume equal to the velocity v_i,

$$\text{work/time} = P_i v_i \tag{9-21}$$

The product on the right of (9-21) is called the *enthalpy* of the gas. Reduced to unit volume, the enthalpy numerically just equals the pressure P_i. Reduced to unit inflow mass, it is P_i/ρ_i.

As it crosses the front, energy Q per unit mass may be fed into the fluid, so that the actual energy gain per unit mass of inflowing material is Q.

In unit time, a mass $\rho_i v_i$ of inflowing material crosses the front. This must contain energy equal to the energy contained in the gas flowing out of the front in unit time except that the outflow energy can be greater by an amount Q. The outflow energy consists of terms similar to those described under (i), (ii), and (iii). We therefore have the energy conservation equation

$$Q + \left(\frac{v_i^2}{2} + U_i + \frac{P_i}{\rho_i} \right) - \left(\frac{v_0^2}{2} + U_0 + \frac{P_0}{\rho_0} \right) = 0 \tag{9-22}$$

where we have again used subscripts o to denote outflow.

PROBLEM 9-1. Show that the internal energy can be written as

$$U = c_v T = \frac{R}{\gamma - 1} \frac{P}{R\rho} = \frac{P}{(\gamma - 1)\rho} \tag{9-23}$$

(see section 4:18).

This leads to

$$\left[\frac{v_i^2}{2} + \left(\frac{\gamma_i}{\gamma_i - 1} \right) \frac{P_i}{\rho_i} \right] - \left[\frac{v_0^2}{2} + \left(\frac{\gamma_0}{\gamma_0 - 1} \right) \frac{P_0}{\rho_0} \right] = -Q \tag{9-24}$$

and since both gases consisting of neutral atoms and gases containing only electrons and atomic ions have γ values of 5/3, we can finally write (9-24) in the form

$$\left[\frac{v_i^2}{2} + \frac{5}{2} \frac{P_i}{\rho_i} \right] - \left[\frac{v_0^2}{2} + \frac{5}{2} \frac{P_0}{\rho_0} \right] = -Q \tag{9-25}$$

Equations 9-19, 9-20 and 9-25 describe the motion of a front into a monatomic medium.

Let us first examine the structure of an *ionization front*. We will assume that J ionizing photons are incident on the front per unit area in unit time. The mean energy of these photons is $\chi_r > \chi_0$ where χ_0 is the energy required for ionizing atoms (Ka54)*. Since J is the number of ionizing photons incident on the front that divides the ionized from the neutral gas, it follows that J atoms are flowing into the front in unit time and J ions plus J electrons are streaming out.

By referring to (9–19) we see that the mass flow across the front is related to the flux of ionizing photons through the relation

$$I = mJ \qquad (9\text{–}26)$$

where m is the mass of the neutral atoms. For a pure hydrogen cloud $m = m_{\mathrm{H}}$. We now define the ratio of densities

$$\frac{\rho_0}{\rho_i} \equiv \Psi \qquad (9\text{–}27)$$

Then, by (9–19) and (9–20)

$$P_0 = P_i - \frac{\rho_i v_i^2 (1 - \Psi)}{\Psi} \qquad (9\text{–}28)$$

From (9–19), (9–25), and (9–28) we then obtain

$$\left[5\frac{P}{\rho} + v^2 + 2Q \right] \Psi^2 - 5\left[\frac{P}{\rho} + v^2 \right] \Psi + 4v^2 = 0 \qquad (9\text{–}29)$$

where we have dropped all subscripts, but the pressure, density, and velocity all refer to the inflowing material. We note that the energy supplied at the ionizing front goes partly into ionization and partly into kinetic energy of the particles. The part that goes into kinetic energy on the average is $\chi_r - \chi_0$ per ion pair. Per unit mass of ionized material this relationship corresponds to a mean square velocity

$$u^2 = \frac{2[\chi_r - \chi_0]}{m} = 2Q \qquad (9\text{–}30)$$

Note that Q in this sense represents only the heating energy, not the energy needed to overcome atomic binding.

This binding energy has not been specifically included in the formalism presented here. Essentially, we have been able to neglect it by concentrating only on those photons having $\chi_r > \chi_0$. For a higher binding energy, fewer photons are available to ionize material.

We also note that the speed of sound in the neutral medium has an isothermal velocity

$$c = \left[\frac{\gamma P}{\rho} \right]^{1/2} = \left[\frac{5P}{3\rho} \right]^{1/2} \qquad (9\text{–}31)$$

so that equation 9–29 can now be written entirely in terms of the three velocities

v, u, and c, and in terms of the ratio Ψ

$$[3c^2 + v^2 + u^2]\Psi^2 - [3c^2 + 5v^2]\Psi + 4v^2 = 0 \tag{9-32}$$

We can treat this as a quadratic equation in Ψ that can have a pair of coincident roots, two positive roots, or a pair of complex roots depending on whether

$$(3c^2 + 5v^2)^2 \gtreqless 16v^2(3c^2 + v^2 + u^2) \tag{9-33}$$

or on whether

$$9(c^2 - v^2)^2 \gtreqless 16v^2u^2 \tag{9-34}$$

There are real roots under two conditions. The first condition is that

$$3(c^2 - v^2) \leqq -4vu \tag{9-35}$$

which means that v is greater than some critical speed v_R:

$$v \geqq v_R = \frac{1}{3}\left(2u + \sqrt{4u^2 + 9c^2}\right) \tag{9-36}$$

This requires that the flux of ionizing photons be larger than a critical value J_R; by (9-19) and (9-26)

$$J \geqq J_R = \frac{n}{3}\left(2u + \sqrt{4u^2 + 9c^2}\right) \tag{9-37}$$

where n is the initial number density of atoms in the neutral medium. R stands for "rarefied" and D for "dense."

The second condition under which real roots exist is if

$$3(c^2 - v^2) \geqq 4vu \tag{9-38}$$

which implies velocities of the front less than a critical value v_D and an ionizing flux below J_D:

$$v \leqq v_D = \frac{1}{3}\left(-2u + \sqrt{4u^2 + 9c^2}\right) \tag{9-39}$$

$$J \leqq J_D = \frac{n}{3}\left(-2u + \sqrt{4u^2 + 9c^2}\right) \tag{9-40}$$

The two critical speeds v_R and v_D correspond, for a given ionizing flux, to densities

$$\rho_R = \frac{I}{v_R} \quad \text{and} \quad \rho_D = \frac{I}{v_D} \tag{9-41}$$

If the gas ahead of the ionizing front has a value ρ_R or ρ_D, only one possible value of ρ_0 can result, that is, the density in the ionized medium behind the front has a fixed value. Since (9-32) is quadratic in $\Psi = \rho_0/\rho$, we see that for $\rho < \rho_R$ and also for $\rho > \rho_D$ there are two possible values which ρ_0 could assume, and for intermediate values of the initial density, $\rho_R < \rho < \rho_D$, there is no permissible value. This means that the ionization front cannot be in direct contact with the undisturbed H I region.

We therefore have the following development of an ionized hydrogen region around a star that suddenly flares up and emits ionizing radiation. Initially, the interface between the ionized and neutral region is very close to the star; the flux J still is very high and well above the critical value J_R. We then have what is called the *R-condition*. The rate at which the front moves into the neutral medium (or vice versa according to the formalism used here) is $v = J/n$. This is just what equation 9–7 stated. However, as the ionization front moves further from the star, the value of J decreases and the front slows down until the critical velocity v_R is reached. This velocity has the approximate value

$$v_R \sim \frac{4}{3}u \qquad (9\text{–}42)$$

because the mean energy of the photons is so high that the excess energy carried off by the ionized particles makes them move at velocities much higher than the speed of sound in the undisturbed neutral medium. The temperature in the ionized medium simply is much higher than in the neutral gas. Typical temperatures in ionized regions of interstellar space are between 5000 and 10,000°K, while H I regions probably have temperatures a factor of 10^2 lower.

When the critical velocity v_R is reached, the ionization front no longer has direct contact with the undisturbed medium. It is now moving slowly enough that a shock front, signaling the impending arrival of the ionized region, precedes the ionization front, and in so doing compresses the medium to a density greater than that of the undisturbed state.

Essentially, this just means that the ionization heats the gas that then expands into the neutral medium fast enough so that a compression wave travels into the neutral gas at a speed exceeding the local speed of sound. A shock front therefore precedes the ionization front into the neutral medium and modifies the density in this medium so that the boundary conditions (9–19), (9–20), and (9–22) once again are satisfied at the ionization front. As the ionization front moves still farther from the star, the velocity with respect to the undisturbed neutral medium drops below the lower critical value v_D. Here a very gradual expansion is going on, no shock is propagating into the neutral medium and the ionization front once again is in direct contact with the undisturbed medium. This is called the *D-condition*.

PROBLEM 9–2. Show that $v_D \sim 3c^2/4u$. $\qquad\qquad$ (9–43)

We should still note that the conditions at a normal shock front are identical to those across an ionization front except that the ionization energy is not supplied, that is, $Q = 0$. The equations derived here therefore have a wide range of applications. It is also worth noting that usually a magnetic field is present and that the energy balance and pressure conditions must then also include magnetic field contributions. *Hydromagnetic shocks* are particularly significant because under conditions where interparticle collisions are rare, the magnetic fields are the main conveyors of pressure information throughout the medium. Pressure equilibrium between gas particles is established through mutual interaction via magnetic field compression.

At the interface between H II and H I regions we sometimes see bright rims that outline the dark, dust-filled regions not yet ionized (Po56). The bright rims generally are located at the edge of the nonionized matter and appear pointed toward the direction of the ionizing star that normally is of spectral type earlier than O9. It is possible (Po58) that these rims occur when ionizing radiation arriving at the H I region satisfies the *D*-critical condition and sets up an ionization front that moves into the neutral gas without being preceded by a shock wave.

9:4 FORMATION OF MOLECULES AND GRAINS

One of the puzzles about interstellar grains is their origin. The density in interstellar space is so low that grain formation appears impossible there. To see this, consider the growth rate of a grain. Let its radius at time t be $a(t)$. Interstellar atoms and molecules impinge on the grain with velocity v. If the number density of heavy atoms, having mass m is n, the growth rate of the grain is

$$4\pi a^2 \frac{da}{dt} = \frac{\pi a^2 nmv}{\rho} \alpha_s \qquad (9\text{--}44)$$

where ρ is the density of the interstellar atoms after they have become deposited on the grain's surface, α_s is the sticking coefficient for atoms impinging on the grain, and the left side of (9–44) represents the grain's volume growth. Taking $v \sim \sqrt{3kT/m}$, with $T \sim 100°K$ and $m \sim 20 \, amu \sim 3 \times 10^{-23}$ g, $\rho \sim 3$ g cm^{-3} and $n \sim 10^{-3}$ cm^{-3} and the maximum value $\alpha_s = 1$, we obtain

$$\frac{da}{dt} \sim \frac{n\sqrt{3kTm}}{4\rho} \alpha_s \sim 10^{-22} \text{ cm sec}^{-1} \qquad (9\text{--}45)$$

To grow to a size of 10^{-5} cm, the available time would have to be 10^{17} sec $\sim$ 3×10^9 y. With more realistic values of α_s, the required length of time increases to an age greater than that of the Galaxy. Here we have neglected the deposition of hydrogen on a grain, since pure hydrogen would normally evaporate rapidly.

Of course, there exist regions in space where the number density of atoms like oxygen, nitrogen, carbon, and iron is ~ 1 cm^{-3}. The Orion region is about that dense. If there were no destructive effects there, grains could perhaps form in a time $\sim 3 \times 10^6$ y, if $\alpha_s \sim 1$. Furthermore, if the temperature in a dense cool cloud could become low enough so that hydrogen could solidify on the grains without rapid re-evaporation, the growth rate could be still higher by two or three orders of magnitude.

Thus far we have neglected destructive effects. For example, in HII regions, radiation pressure often accelerates small grains to higher velocities than large grains. Intercollisions of grains can then take place at such high velocities, $\gtrsim 1$ km sec^{-1}, that both of the colliding grains are vaporized. The vapors then have to recondense. In addition, *sputtering* by fast moving protons can knock atoms off a grain's surface after they have become attached. Such destructive effects tend to reverse the growth implied by equation 9–45, or at least will decrease the growth rate. This destruction would be stronger for substances like ice, where molecules are bound weakly, rather than for strongly bound substances like silicates or graphite. These three substances all may be constituents of interstellar grains.

Another—less catastrophic—destructive effect depends on the *vapor pressure* of the material from which the grains are formed. In thermal equilibrium at a temperature T the vapor pressure gives the rate at which molecules or atoms evaporate from a grain's surface. The equilibrium vapor pressure is that pressure of ambient vapor at which the growth rate is just equal to the evaporation rate. This partial pressure, P_{vap}, is related to the vapor density through the equation of state, so that if Dalton's law (4–38) holds, the molecule mass impact rate is

$$nmv \sim \left(\frac{m}{kT} \right)^{1/2} P_{\text{vap}} \qquad (9\text{–}46)$$

per unit area. This then must also represent the evaporation rate from the surface. We see that the pressure in the ambient space must exceed the vapor pressure, if the grain is to grow. Hydrogen has a vapor pressure of about 10^{-7} torr at 4°K (Table 9.3). This amounts to about 10^{11} molecules cm^{-3}. Of course, this is a density much higher than any expected for interstellar space. On the other hand, grains are never likely to be cooler than 4°K. This means therefore that hydrogen cannot very well remain on grains, unless it is chemically bound by the presence of other substances or else adsorbed on the basic grain material. For other substances this would not be true. Silicates and graphite, for example, have such low vapor pressures that no appreciable evaporation off grains would occur in periods of the order of the life of the Galaxy. For ice the situation is somewhat more complicated. On approaching close to an individual star, H_2O molecules could evaporate as a grain's temperature rose. Basically this is what happens when a comet approaches the sun from the outer portions of the solar system. The surface warms until water, ammonia, and other ices evaporate. Near most stars sputtering through collisions with atoms is the dominant destroying mechanism; only on very close approach to a star can evaporation be significant!

It is possible that grain formation primarily takes place in the atmospheres of very cool giant stars or Mira variables that eject gas into interstellar space. The atmospheres of these stars are dense so that n in (9–45) is high. Formation must, however, take place in a period as short as a month, because after that, the outflowing gas soon becomes too rarefied. With $n \sim 10^5$ cm^{-3} for the heavier atoms, $T \sim 2 \times 10^{3\circ}$K, $m \sim 12$ for carbon, it might be possible to form a graphite grain at a rate $da/dt \sim 3 \times 10^{-14}$, a grain with $a \sim 10^{-7}$ cm would form in a month. This would only be a nucleus that could subsequently grow, possibly because radiation pressure would keep it moving slightly faster than the gas flowing out from the star (see equation 9–15).

The fact that Mira variables and cool giants often seem to emit an excessive amount of infrared radiation (Jo67a) would indicate a close association of dust with these stars, and very possibly this is due to the formation and ejection of dust from their atmospheres.

Geisel, Kleinmann, and Low (Ge70) have found that the nova Ser 1970 became brighter in its infrared emission as it dimmed in the visible part of the spectrum. Presumably over a period of weeks, dust formed in the gas ejected by this nova. Since similar behavior was also observed in novae Aql 1970 and Del 1967, novae may generally be responsible for appreciable dust formation.

Finally, the infrared emission from planetary nebulae may indicate the presence of dust; perhaps the ejection process there still permits adequate formation to proceed at rates governed by equation 9–45.

The formation of molecules presents problems similar to dust formation. In the past few years, microwave techniques have been used to discover an increasing number of interstellar molecules, NH_3 (ammonia), CO, H_2O, HCN (hydrogen cyanide), H_2CO (formaldehyde), CN (cyanogen), HC_3N (cyanoacetylene), the hydroxyl radical OH, and many others. To be sure, being smaller aggregates, the molecules form first before the larger grains ever have a chance of forming. However, complicated molecules having several heavy atoms would perhaps not be expected to form readily once smaller stable molecules had already formed. For, these smaller molecules would then not readily accept an additional atom. The theory of molecule formation under the extremely rarefied conditions of interstellar space has, however, not yet been firmly tackled, and little is understood about the formation of molecules as complex as, say, formaldehyde and formic acid. These larger molecules normally are found associated with dense dark clouds in the vicinity of H II regions. It is not clear whether ionized or neutral regions, or perhaps only the boundaries of such regions, are most conducive to the formation of molecules.

A number of destructive effects should still be mentioned. Molecules can be destroyed through dissociation that often is a consequence of absorption of, or ionization by, ultraviolet photons. Calculations indicate that the prevalent galactic starlight may destroy molecules like CH_4, H_2O, NH_3, and H_2CO in times of the order of a hundred years, unless the molecules are shielded from the light, inside strongly absorbing dust clouds (St72). Ionization by energetic electrons or cosmic ray particles could produce similar effects. Collisions of interstellar clouds that can produce energetic particles could therefore produce destruction of molecules; however, we could also argue conversely that the high densities at the contact face of two colliding clouds could lead to more rapid formation of molecules. Such competing formation and destruction rates must be compared in a variety of specific settings to see whether the overall situation is conducive or destructive to molecule formation.

9:5 CONDENSATION IN THE PRIMEVAL SOLAR NEBULA

In section 1:7 we presented some current views on the origin of the solar system.

PROBLEM 9-3. Just after the sun first formed, it may have been surrounded by a dense cloud of gas from which the planets eventually condensed. As a first step, small grains probably formed. Suppose the mass was evenly distributed throughout the nebula, that its total mass was equal to twice that of all planets combined (see Table 1.3), that the radius was 10 AU and that the initial abundance was similar to the solar abundance (Table 1.2). Making use of Table 9.3, calculate an approximate distance from the sun at which iron would have condensed. Do the same for carbon. Would water or ice have been able to condense within the nebula? Note that the sun may have been on the Hayashi track (Fig. 8.11) at that time. Assume its luminosity was ten times greater than now.

PROBLEM 9-4. The action of light pressure would have tended to produce homogeneity in the solar nebula: If the outward directed flux amounted to $1.4 \times 10^7/r^2$ erg cm^{-2} sec^{-1} at r astronomical units from the sun, find the orbital velocities of two grains, both orbiting the sun at the distance of Jupiter, both having a grain radius $s \sim 10^{-3}$ cm, one particle having density 2 g cm^{-3}, and the other having density 4 g cm^{-3}. Assume the grains to be spherical and black, absorbing light with a cross section πs^2. Note how the orbital velocity differs as a function of s. The difference is greatest for small s. Grains with large density

Table 9.3   Relation Between Temperature and Vapor Pressure Compiled from (Ro65), (Du62), and (Le72).[a]

	10^{-11}	(1 torr = 1.33×10^3 dyn cm^{-2}) 10^{-10}	10^{-9}	10^{-8}	10^{-7}	torr
C	1695	1765	1845	1930	2030	°K
Fe	1000	1050	1105	1165	1230	

Most solid substances obey a pressure-temperature relationship of the form $\log_{10} P = A - B/T$,

at low pressures:
P (in torr):
$$\begin{cases} \text{carbon} & A = 12.73 & \text{and } B = 4.0 \times 10^{4}\,°K \\ \text{iron} & A = 9.44 & B = 2.0 \times 10^{4} \\ \text{NaCl} & A = 7.9 & B = 8.5 \times 10^{3} \end{cases}$$

For water the following data are available:
$$\text{H}_2\text{O} \begin{cases} 7 \times 10^{-9} & 3 \times 10^{-10} & 7.4 \times 10^{-15} & 1.4 \times 10^{-22} & \text{torr} \\ 143.2 & 133.2 & 123.2 & 90.2 & °K \end{cases}$$

For hydrogen:
$$\text{H}_2 \begin{cases} 3.1 \times 10^{-7} & 8.8 \times 10^{-9} & 7.5 \times 10^{-11} & 4.5 \times 10^{-13} & \text{torr} \\ 4.0 & 3.5 & 3.0 & 2.6 & °K \end{cases}$$

[a] Parts reprinted by special permission from Rosebury, *Handbook of Electron Tube and Vacuum Techniques*, 1965, Addison-Wesley, Reading Mass. Parts copyright © 1962, John Wiley and Sons. We note that the inner planets consist primarily of low vapor pressure material and that, by and large, the outer planets contain more volatile substances.

---

and/or size differences would therefore intercollide more frequently and be destroyed. Grains with nearly identical properties would tend to persist longer.

PROBLEM 9-5. After small particles and chunks were formed through condensation, a second stage of condensation seems to have taken place in which gravitational attraction played a dominant role. Prior to this time, particles presumably had acquired almost identical low eccentricity, low inclination orbits at any given distance from the sun, and the relative velocities of these grains must have been small. This would have come about because high or low velocity grains would be eliminated preferentially through more frequent destructive collisions with other bodies.

(a) At what size would a body whose density ρ is 3 g cm^{-3} have a gravitational capture cross section that is twice as large as its geometric cross section? Assume

9:5

a relative velocity V_0 for particles to be captured. The result of Problem 3–7 may be useful.

(b) Derive the growth rate of a body with $\rho = 3$ g cm^{-3} moving through a nebula whose density is $\rho_0 = 3 \times 10^{-12}$ g cm^{-3}. Let its relative velocity be $V_0 = 1$ km sec^{-1} and start at a time when its gravitational capture cross section is twice its geometrical cross section.

(c) Show that the mass growth for a spherical gravitating body, whose capture cross section is much greater than its geometric cross section, and whose density $\rho = 3M/4\pi R^3$ has a fixed value, is proportional to $M^{4/3}$ or R^4. More massive bodies therefore have a higher mass capture rate than lower mass bodies whose geometric capture cross section only allows them to capture mass at a rate proportional to R^2.

PROBLEM 9–6.   Suppose that a grain stays spherical as it grows through capture of matter. It moves through the solar nebula at $V_0 = 1$ km sec^{-1}, escapes destructive collisions by chance, and grows from a radius of $\sim 10^{-8}$ cm—one molecule—up to 10 km. If the nebular density is 3×10^{-12} g cm^{-3}, of which a 1 % nonvolatile fraction can be captured, and the particle's density is 2.5 g cm^{-3}, show that the growth time is roughly 10^8 y.

9:6 INFRARED EMISSION FROM GALACTIC SOURCES

Infrared emission has now been observed for a variety of different sources. Cool supergiants often seem to be surrounded by dust layers that obscure much of their visible radiation, and these circumstellar clouds emit strongly in the near infrared. At longer wavelengths, principally beyond 10 microns—1 micron = $1 \mu = 10^{-4}$ cm—we find that planetary nebulae seem to emit strongly, as do H_{II} regions. The Galactic center also is a very strong emitter of radiation. Much of this emission seems to peak around wavelengths of order 100μ although only coarse spectroscopic data exists for this region (see Figs. 6.16 and 9.4).

The near infrared emission from circumstellar dust is relatively easy to explain. Equation 4–78 specifies the temperature that a grain will assume at a given distance R from a star. Some uncertainty always arises because we do not know the emissivity of grains in the visual or in the infrared region. Both the equilibrium temperature, which obeys the relation

$$T = \left(\frac{\varepsilon_a}{\varepsilon_r} \frac{L_\odot}{16\pi\sigma R^2} \right)^{1/4}$$

(4–78)

and the emission spectrum therefore remain somewhat uncertain. In any case, not all of the grains will be at precisely the same distance from the parent star, and we would therefore expect a rather broad emission spectrum representing emission from grains located at different distances from the star and emitting at different effective temperatures.

The emission process associated with planetary nebulae and H II regions is somewhat different. Here there is relatively little dust. Often the obscuration within the ionized region is not noticeable, and yet strong infrared emission is observed. How can the dust be responsible?

To explain this phenomenon we have to return to the discussion of ionization equilibrium given in section 9:2. We had noted there that, for an equilibrium Strömgren sphere, the recombination rate equals the ionization rate. Each time an electron recombines with a proton to form a hydrogen atom we have two possibilities. Either the recombination leaves the atom in the first excited state $n = 2$, or else it leaves it in a higher level from which a cascade of photon emission processes eventually places the atom into state $n = 2$ or else $n = 1$. Any photon emitted through a transition from $n = 2$ to $n = 1$, or from any higher excited state down to $n = 1$, has a very high probability of being reabsorbed by another hydrogen atom in the $n = 1$ state in the H II region. The photon then wanders through the H II region in a random walk interrupted by successive absorptions and re-emissions (see Problem 4–4). The mean free path of the Ly-α photons in this walk may only be 0.03 times the radius of the H II region, so that the photon would have to crisscross the region ~ 30 times before finally reaching the boundary of the ionized cloud. As a result, each Ly-α photon has about 30 times as

Fig. 9.3 Plot of infrared emission from 45 to 750μ against 2 cm wavelength radio emission from ionized hydrogen regions. (After Harper and Low, Ha71. With the permission of the University of Chicago Press.)

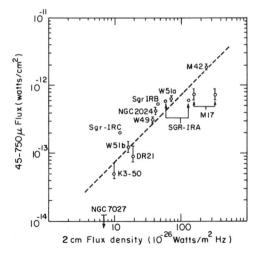

high a probability of being absorbed by a grain as does visible starlight that passes straight through the H II region and out into the surrounding cool cloud. We have acted here as though all of the Lyman spectrum photons were Ly-α radiation. This is almost true. A Ly-β photon emitted by one hydrogen atom is likely to be absorbed and to give rise to an Hα photon in an atomic transition $n = 3$ to $n = 2$, succeeded by a Ly-α transition $n = 2$ to $n = 1$.

As a rule of thumb, we can state that for each ionizing photon, the H II region eventually must produce one recombination. That recombination, followed by a succession of emission, reabsorption, and re-emission processes, eventually has to give rise to one photon of the Balmer spectrum, and one Ly-α photon. The oscillator strength for absorption of the Ly-α photon is near unity so that the effective cross section per hydrogen atom is large. Even if the fractional neutral hydrogen density is very low, Ly-α absorption usually is large enough to trap the Ly-α photons in the nebula.

PROBLEM 9-7. For an H II region in which each ionizing photon has unit optical depth on passing a distance equal to the radius, R, relate n_H, the number density of neutral atoms, to the absorption coefficient a_v listed in Table 9.2.

(a) With reference to Problem 7-11 determine the mean free path for Ly-α absorption with respect to R.

(b) The absorption bandwidth, γ, usually is small compared to the Doppler frequency shift $\Delta v \sim v \langle v^2 \rangle^{1/2}/c$ where $\langle v^2 \rangle^{1/2}$ is the root mean square velocity in the H II region. For $\langle v^2 \rangle^{1/2} \sim 30$ km sec^{-1}, what is the Ly-α absorption mean free path (see Problem 7-13).

Since each ionizing photon gives rise to a Ly-α quantum of radiation and since this radiation is likely to be absorbed by a grain before it can escape to the boundary of the H II region, we might expect that all the radiation converted into Ly-α would (Kr68) eventually be absorbed by a grain and the energy would be thermally radiated in the far infrared. Most of the ionizing photons given off by a star have an energy less than twice the Ly-α energy. Hence we conclude that more than half the ionizing energy emitted by a star would eventually find its way into infrared radiation. This seems to be at least approximately true. We can make a comparison to the free-free emission from ionized regions observed in the radio domain. This free-free emission can be directly related to the expected recombination line intensities, through equations like (9-5) or (6-141) and (6-139). We can therefore derive a proportionality relationship between the expected free-free radio emission from the H II region and the far infrared flux from grains. Harper and Low (Ha71) have checked this proportionality for a number of regions. The agreement is reasonably good although the infrared

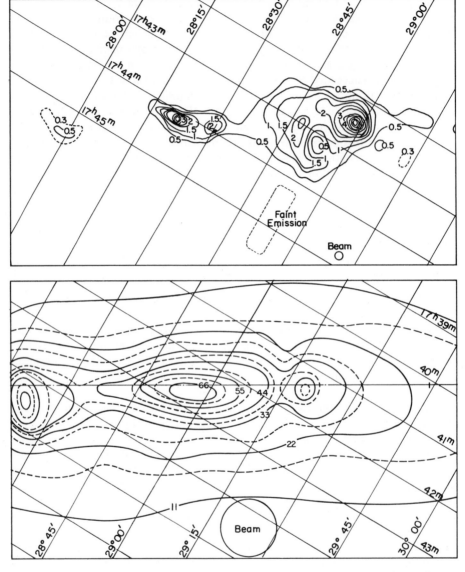

Fig. 9.4   2 cm radio and 100μ infrared emission radiation from the Galactic center (After Hoffmann, Frederick and Emery, Ho71 and Terzian, Te68). Note the similarity in positions of the two main peaks near 29°, $17^h 42.5^m$ and near 28°20′, $17^h 44^m$. The infrared luminosity is about 10^6 times the radio emission from this region. (See also the X-ray emission sources near the Galactic center Fig. A.7.) (With the permission of the University of Chicago Press, and from *Interstellar Ionized Hydrogen*, Y. Terzian, Ed., (1968), W.A. Benjamin, Inc., Reading, Massachusetts.)

9:6

emission observed by them was systematically somewhat high. A number of factors can account for this discrepancy. Figure 9.3 shows Harper and Low's results while Fig. 9.4 gives a comparison of radio and infrared emission maps in the Galactic center region. The correlation is so good that a connection between infrared and radio fluxes seems reasonably well established.

In Seyfert galaxies, where strong infrared emission from nuclei is sometimes observed, the infrared emission mechanism might well be the one we have described. However, infrared radiation can also arise through such processes as synchrotron radiation or inverse Compton scattering, and it is known that the Seyfert galaxies can be strong emitters of X-rays so that highly energetic electrons are likely to be present. We still lack the observational evidence to distinguish between these mechanisms.

9:7 STAR FORMATION

In section 1:4 we presented a number of problems associated with star formation. Three primary processes have to take place if a star is to form: (1) The gas from which it forms has to radiate away energy so that the total energy of the protostar can continuously decrease—giving rise to an increasingly compact configuration. (2) Angular momentum must somehow be reduced, from the high values associated with distended hydrogen clouds moving in differential rotation about the Galactic center, to the low values observed in stars. (3) The magnetic field observed by Faraday rotation and Zeeman splitting methods must somehow be removed from the contracting cloud of interstellar gas to give the relatively small magnetic field values actually seen at the surface of stars.

To these three requirements we should add one more that apparently is met rather easily: (4) A relatively rapid initial compression of the gas is required to trigger contraction. This compression must be accompanied by cooling, and the total energy loss must be large enough so that the cloud can no longer expand back to roughly its initial diameter. The compression, in other words, has to be strongly inelastic. Some such triggering mechanism is needed because turbulent motions, normally present in the interstellar medium, would otherwise disrupt the contraction process before it had even properly started. These turbulent motions are produced by the expansion of ionized regions and by radiation pressure that acts on the gas via the dust grains (Ha62a).

This particular compression criterion can apparently be met through the strongly compressive shock waves formed around H II regions. Although these expanding regions tend to endow the cool clouds with turbulent motion, they also can produce rapid compression, as we saw in section 9:3. This compression can lead to continued contraction when coupled with strong radiative cooling through such processes as grain emission, radiation from molecular hydrogen that may be abundant in dark clouds (Go63), or perhaps emission from H_2O vapor that has a high dipole moment and can be rotationally excited by low energy collisions.

Criteria (1) and (4) can therefore apparently be met. The shedding of angular momentum and dissociation of magnetic field, however, still pose unsolved problems. We will only be able to show one or two possible ways out of the dilemma.

Let us consider the angular momentum problem first. We might argue that we overestimated the total angular momentum initially present in the protostellar matter. The differential rotation effect might perhaps not be as large as initially estimated.

One way in which the angular momentum due to differential rotation might be low would be to gather only that material that had uniform angular momentum about the Galactic center. If, for example, we assumed all gas particles to be in circular motion about the center, then we would gather material only from a ring at some fixed radius R_c from the center. This ring would have a circumference $2\pi R_c$. The areal density—viewed normal to the Galactic plane—of gaseous matter in the solar neighborhood, is of order 10^{20} atoms cm^{-2}, or $\sigma \sim 10^{-4}$ g cm^{-2}. The length $2\pi R_c$ in our neighborhood is $\sim 2 \times 10^{23}$ cm, so that a ring of width

$$W \sim \frac{M_\odot}{2\pi R_c \sigma} \sim 10^{14} \text{ cm} \tag{9-47}$$

would be needed to form a star of one solar mass. Using the figures for differential rotation used in section 1:4, we would find then that the differential velocity would be of order 3×10^{-2} cm sec^{-1} across this width and that the angular momentum per unit mass, once contraction had taken place, would only be of order 3×10^{12} cm^2 sec^{-1}. This is a factor of 3×10^3 less than the angular momentum per unit mass of solar material and a factor of 3×10^5 less than the corresponding angular momentum figure for the whole solar system. What this means is that mass would have to be gathered only from a shorter segment some 600 pc along an arc of radius R_c, that W could be increased to $\sim 10^{16}$ cm, thus increasing the angular momentum per unit mass to roughly 3×10^{16} cm^2 sec^{-1} and that a solar system or star with reasonable angular momentum characteristics might thus be formed.

What we conclude is that star formation with reasonable angular momentum characteristics would be possible if stars were formed from segments of thin cylindrical shells of matter initially orbiting about the Galactic center in a narrow range of orbital angular momentum values. Whether such a process is possible on other grounds is an entirely different question and much harder to answer.

First, it is not likely that a gravitational contraction within such platelike segments could proceed (Eb55); gravitational contraction normally is most effective if the contracting volume is roughly spherical, for then a small compression in volume can amount to a relatively large change in potential energy.

Second, the magnetic field problems are not averted in such a configuration. If the magnetic field lines were largely normal to the radius R_c—and this might be favored by shear produced through differential rotation—then contraction along the magnetic field would be unresisted by the field, but the remaining contraction perpendicular to the field direction would still necessitate compression of field lines. Here, we are still concerned about compression from a dimension of order 100 pc down to solar system $\sim 10^{15}$ cm dimensions, and this alone would produce a field of the order of a gauss for a region 10^{15} cm in diameter. Compression into an object the size of the sun would then still require compression in two dimensions and give a final field of order 10^9 gauss.

The magnetic field would therefore have to be dissipated in some way or other; there is no immediately apparent way of avoiding the problem in the way we avoided some of the angular momentum difficulties by contraction along a region of constant radial distance from the Galactic center. We would not be able to wholly avert the magnetic field difficulties even if we took the extreme case involving the gathering of matter along a complete circular arc around the galactic center. Somehow the magnetic fields must be dissipated.

We should still mention that many means of avoiding the angular momentum difficulties have been suggested. It might, for example, be that magnetic coupling of the contracting matter to external gases produces a viscous drag that slows the rotation of contracting gases. In this case, the magnetic field would actually help to overcome one of the difficulties of star formation while, on the other hand, introducing another stumbling block of its own.

We will discuss magnetic fields in more detail in section 9:9, where we will speculate on the origins of interstellar magnetic fields. Some of the problems involved in forming magnetic fields are identical to those encountered in trying to destroy them.

Literally hundreds of ideas on how stars are formed have been advanced in the past decade. However we still are far from any real solution. Here we have only presented one or two simple concepts; the actual answer to the problem may turn out to be much more complex. On the other hand, we might also find that we have misinterpreted some observational results and that actual conditions in interstellar space are really far more conducive to star formation than we have thus far believed.

9:8 ORIENTATION OF INTERSTELLAR GRAINS

When the light from stars close to the Galactic plane is analyzed, it is found to be both reddened and polarized. This has been interpreted (see section 6:15) as

9:8

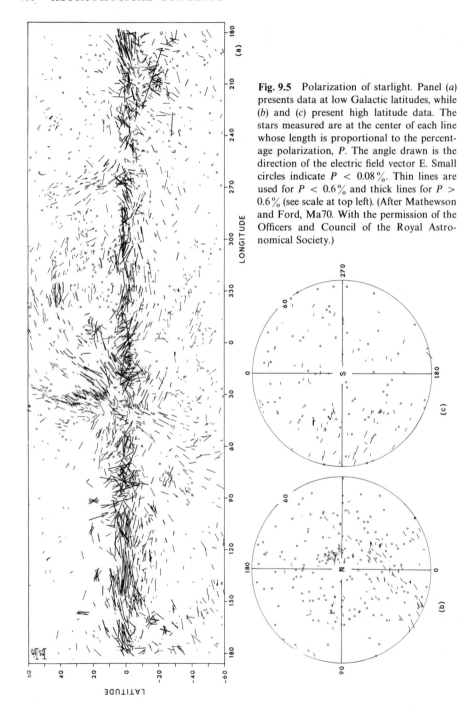

Fig. 9.5 Polarization of starlight. Panel (*a*) presents data at low Galactic latitudes, while (*b*) and (*c*) present high latitude data. The stars measured are at the center of each line whose length is proportional to the percentage polarization, *P*. The angle drawn is the direction of the electric field vector E. Small circles indicate $P < 0.08\%$. Thin lines are used for $P < 0.6\%$ and thick lines for $P > 0.6\%$ (see scale at top left). (After Mathewson and Ford, Ma70. With the permission of the Officers and Council of the Royal Astronomical Society.)

due to the alignment of elongated or flattened grains, spinning in such a way that their long axes are close to planes perpendicular to the Galactic disk. The transmitted starlight is polarized in a direction predominantly parallel to the plane (Fig. 9.5).

How can grains become aligned in this way? In Problem 7–6, we showed that grains have angular frequencies of the order of 10^5 Hz. It therefore makes no sense to talk about the orientation of a stationary particle. There is, however, a different kind of orientation involving preferred directions of a grain's angular momentum vector.

This can be illustrated by a simple process. Suppose that a grain is quite elongated, like a stick, and that it is systematically moving through the gas with directed velocity $\mathbf{v}$. The gas is so tenuous that individual gas atoms can be considered as colliding with the stick, one at a time. Let us first assume that the gas atoms have no random velocity of their own. The root mean square value of the angular momentum transferred to the stick in any given collision then is

$$\delta L = \left[\frac{1}{a} \int_0^a (mvr)^2 \, dr \right]^{1/2} \sim \frac{mva}{\sqrt{3}} \qquad (9\text{–}48)$$

where a is half the length of the stick, m is the mass of the atoms, and v their approach velocity to the grain. The root mean square (rms) angular momentum after N collisions can be obtained from a random walk calculation (as in equations 4–12, 4–13) and is $N^{1/2} \delta L$. If the density of a grain is ρ, its width is s, and its length $2a$, we can compute the number of collisions N required to appreciably change any initial angular momentum the grain may have. This is

$$N = \frac{M}{m} = 2\rho s^2 \frac{a}{m} \qquad (9\text{–}49)$$

where M is the grain mass. This equation merely states that the grain will have an appreciable rotational velocity change after it has sustained collisions with a number of atoms N, whose total mass is equal to M. A random angular momentum accompanies this systematic change. The final root mean square angular momentum, $\langle L^2 \rangle^{1/2}$, which a typical grain will acquire,

$$\langle L^2 \rangle^{1/2} = N^{1/2} \delta L \sim \left[\frac{2m\rho a^3}{3} \right]^{1/2} sv \qquad (9\text{–}50)$$

If we take m to be the mass of atomic hydrogen, 1.6×10^{-24} g, $v = 10^5$ cm sec^{-1}, $a = 10^{-5}$ cm $= 3s$, $\rho = 1$ g cm^{-3}, we obtain $L \sim 10^{-20}$ g cm^2 sec^{-1}.

PROBLEM 9–8. Show that the angular velocity of the grain is $\omega \sim 10^6$ rad sec^{-1} in this case, and that $L \sim 10^{-20}$ g cm^2 sec^{-1} also represents the thermal equilibrium value at a temperature of 100°K.

Since we know angular momentum to be quantized in units of $\hbar$ we see that typically the angular momentum quantum number will have a value

$$J \sim 10^7 \hbar \tag{9-51}$$

We note that the direction of the angular momentum acquired in this process must lie perpendicular to **v** since all impulses on the stick produce changes in angular momentum $\delta\mathbf{L}$ whose directions lie in this plane. The gas systematically streaming past the grain therefore produces a preferred orientation of the angular momentum axis.

Let us now consider, in contrast to the systematic drift velocity of the grain through the gas, that there also is a random velocity u possessed by the atoms in the gas. Then the random walk process will endow the grain with an angular momentum greater by a factor of $\sim [(u^2/v^2) + 1]^{1/2}$ than the value given by (9–48). In this case, the systematic angular momentum will only be of the order of $[(u^2/v^2) + 1]^{-1/2}$ times the random angular momentum. In many cases $u \gg v$, and only slight preferential orientation of angular momentum can take

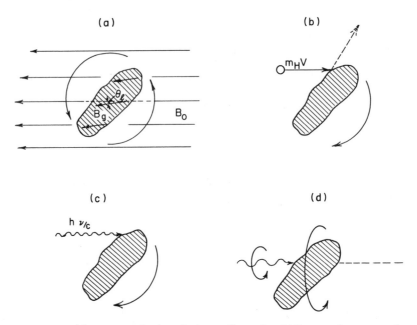

Fig. 9.6 Alignment mechanisms for interstellar grains: (*a*) Process of paramagnetic relaxation; (*b*) alignment by streaming through gas, or through a photon field (*c*). In process (*c*) the photon's linear momentum causes the grain to spin; in process (*d*) the photon's intrinsic spin angular momentum is of importance (see text).

place. This, however, may be enough, because the starlight is only partially polarized—not completely.

We still have to examine which orientations of the angular momentum vector will be preferred. To do this, we consider how an individual collision affects an elongated grain. We note that when $v \gg u$, collisions are much more probable if the length of the stick is perpendicular to the direction of **v**. In that configuration more projected area is exposed to the incident gas atoms. The sense of any collisions, then, is to produce a change in angular momentum perpendicular both to the direction of gas flow and to the length of the grain. A rotation about an axis whose direction is along **v** is therefore very unlikely, and it is the absence of angular momentum orientations having this sense that can produce a partially preferred orientation of the sticks with their long axes parallel to the direction of flow. In this process, suggested by Gold (Go52), orientations perpendicular to the direction of flow simply tend to be missing (Fig. 9.6b). This general conclusion holds true even if $v < u$.

Since the sense of polarization is such that light whose electric vector **E** oscillates perpendicular to the plane of the Galaxy is preferentially scattered or absorbed, we assume that the grains also are oriented with their long axes perpendicular to the Galactic plane. They then absorb or scatter light, somewhat like a radio antenna, parallel to the long axis, along which electrons can most readily flow in response to an oscillating E field.

What has been said here about collisions with gas atoms, holds equally well for collisions with photons, Fig. 9.6(c). Here the atomic momentum mv has to be replaced by the photon momentum $h\nu/c$ if we are to compute δL by means of equation 9–48. The photon effect can be expected to be strong close to bright stars.

For photons Harwit (Ha70) has proposed another effect that will dominate for small grains. We stated in section 7:2 that each photon must have an intrinsic angular momentum $\hbar$. When a photon is absorbed by a grain, *intrinsic angular* momentum $\hbar$ therefore is transferred, Fig. 9.6(d). This means that per photon $\delta L = \hbar$.

We now are in a position to compare these three effects. Per collision, angular momenta:

$$\delta L_g \sim \frac{mva}{\sqrt{3}}, \quad \delta L_e \sim \frac{h\nu a}{\sqrt{3}c} \quad \text{and} \quad \delta L_i \sim \hbar \qquad (9\text{--}52)$$

are transferred, respectively, by the gas atoms, extrinsic photon and intrinsic photon angular momentum processes. The last term predominates over the second, if

$$\frac{a}{\sqrt{3}} \lesssim \frac{\lambda}{2\pi} \qquad (9\text{--}53)$$

where $\lambda = c/v$ is the wavelength of the photons. The evidence from light scattering suggests that a $\sim 5 \times 10^2$ Å $= 5 \times 10^{-6}$ cm, so that the intrinsic effect of photons dominates over the extrinsic for visible light. In that case $mva/\sqrt{3} \sim 10^{-29}v$ and the effect of individual atoms is greater for all systematic velocities v in excess of 10^2 cm sec^{-1}, provided that gas collisions are as frequent as photon collisions. This, however, is never true, as we will now show.

If the interaction cross section for gas and photons is taken to be the same, the number of collisions per second is going to be equal to nv for the gas and Nc for photons. N is the number density of starlight—mainly visible and near infrared—photons. Typical values for the gas density n range from 1 to 10^3 cm^{-3} in interstellar space.

The systematic velocities v are harder to assess. In a collision between gas clouds, a grain may find itself streaming through gas with a systematic velocity v as high as 10^6 cm sec^{-1}. But this will last only for a time of order

$$t = \frac{M}{2sanvm} \tag{9–54}$$

the time in which a grain suffers collision with an equal mass of gas. For $n = 10$ cm^{-3}, $v = 10^6$ and with the previously used parameters, we have $t \sim 2 \times 10^{11}$ sec. High velocities may therefore occur $< 1\%$ of the time because the lifetime between acceleration or collisions of clouds is of the order of 10^{14} sec, and it is these accelerations that can initiate systematic motion of grains through ambient gas. The radiation effect on the other hand is always present. Within the Galaxy, $N > 0.02$ cm^{-3} so that $Nc \sim 6 \times 10^8$ sec^{-1} cm^{-2} as contrasted to $nv \sim 10^7$. The photon effects dominate if $\sqrt{3}\,Nch > nmv^2a$ or if $Nhv > mnv^2$.

Thus far we have considered only orienting effects of gas and photons. Next we must consider the randomizing effects. When v is small compared to u, the randomizing angular momentum per atomic collision, that is, the step size in the random walk process is $mua/\sqrt{3}$. The number of collisions per unit time is proportional to the grain cross section that, without defining it further, we will call σ. It also is proportional to the product nu, the gas density times velocity. After some time τ, the collisional random walk process will produce a root mean square angular momentum

$$L_g \sim \frac{mau}{\sqrt{3}}[\sigma nu\tau]^{1/2} \tag{9–55}$$

During the same time interval, the number of thermal photons given off by the grain can be shown to be $\sim 1.5 \times 10^{11}\ T^3$ cm^{-2} sec^{-1}. Each of these has the ability to endow the grain with angular momentum $\hbar$ in some random direction.

At typical interstellar grain temperatures of, say, 15°K, we then have

$$L_p \sim [5 \times 10^{14} \sigma \tau]^{1/2} \hbar \qquad (9\text{-}56)$$

Putting $a = 5 \times 10^{-6}$ cm, $u = 10^5$ cm sec^{-1}, $n = 10$ cm^{-3}, we find

$$\frac{L_g}{L_p} \sim 2 \times 10^{-2} \qquad (9\text{-}57)$$

Only for large grains, or for dense regions with $n \gtrsim 10^3$, or for a combination of these, is the gas effect equal to that of the re-emitted radiation.

We note now that the random photon effect, if photons are emitted roughly at wavelengths 100 times longer than the absorption wavelength, will make the $\mathbf{L_p}$ vector about 10 times larger than the component $\mathbf{L_i}$ due to any systematic radiation anisotropy effects. The photon-orienting effect depends on an intrinsic asymmetry in the radiation field to produce alignment. The illumination by photons coming from directions close to the Galactic plane is about 10 times greater than the illumination perpendicular to the plane—as we know the Milky Way is bright only in a narrow band. If the angle that the angular momentum axis of a typical grain makes with respect to some line lying in the Galactic plane is θ, then the average value of $\cos^2 \theta$ becomes of the order $L_i^2/L_p^2 \sim 0.01$. Observed figures indicate that this ratio probably is more nearly 0.02. However, without knowing much about the ratio a/s for actual interstellar grains, we cannot compute an expected ratio accurately, and these figures will have to suffice until we learn more about the structure of grains.

We next come to an effect first proposed by Davis and Greenstein (Da51). In this process, a grain is bombarded by ambient gas in interstellar space and is thus set spinning about an arbitrary axis. We can now postulate that the grain material is paramagnetic. Such materials, when placed into a magnetic field, set up an internal field whose direction is (Fig. 9.6a) parallel to the external field. The internal field cannot, however, change instantaneously. If the grain rotates with angular velocity ω about a direction perpendicular to the magnetic field, the internal field is forced—again at frequency ω—to change its direction relative to an axis fixed in the grain. However, since this readjustment of direction does not proceed instantaneously, a slight misalignment of the internal and external field arises as shown in Fig. 9.6(a). The interaction of the induced internal field with the externally applied field attempts to compell parallelism by opposing the rotational motion. This drag torque is proportional to the external field $\mathbf{B}$, to the internal field that also is proportional to $\mathbf{B}$, to the grain volume, V and to ω.

$$\text{Torque} = KVB^2\omega \qquad (9\text{-}58)$$

This represents the situation when the grain is spinning about an axis perpendicular to the direction of **B**. When the grain spins about an axis parallel to **B**, the induced field does not need to change its direction relative to the external field and no drag force arises.

For an arbitrary spin direction, that component of the spin whose axis is perpendicular to the field will therefore be damped out in a time

$$\tau = \frac{I}{KVB^2} \qquad (9\text{--}59)$$

where I is the moment of inertia about the spin axis. On the other hand, a component whose spin axis is parallel to **B** remains undamped. This damping process is called *paramagnetic relaxation*.

An aspherical grain whose spinning motion is slowed tends to align itself in such a way that the *axis of greatest moment of inertia* becomes parallel to the angular momentum axis. This inertia axis is perpendicular to the long axis of an elongated grain, and the net effect of paramagnetic relaxation is to align elongated grains with their long axes perpendicular to the magnetic field. For substances with which we are familiar in the laboratory, and for grain temperatures that typically might be of order $10°K$, in interstellar space at large distances from the nearest stars, the value of K is probably less than 10^{-12}. As we will see, this leads to rather high values of the relaxation time, unless the magnetic field is at least of order 10^{-5} gauss. But such field strengths are roughly a factor of 3 higher than the general field estimated to be present in the Galaxy. And the relaxation times for the more probable fields $\sim 3 \times 10^{-6}$ gauss (Fig. 9.9) are a factor of 10 too long. Jones and Spitzer (Jo67b) have therefore suggested the possibility that cooperative effects between iron atoms in grains might produce what they have termed *superparamagnetism*, to yield a sufficiently high value of K. Experimental evidence for the existence of such substances does not yet seem to exist.

The relaxation time τ must now be compared to the relaxation time produced by random gas collisions and also to the relaxation time due to re-emission of infrared photons. A comparison to the gas effect (9–54) shows that

$$\frac{\tau}{t} \sim \frac{I}{M} \frac{2sanum}{KVB^2} \sim 10 \qquad (9\text{--}60)$$

where we have taken $I \sim Ma^2/3$, $n = 10$ cm^{-3}, $u \sim 10^5$ cm sec^{-1}, $3s = a \sim 5 \times 10^{-6}$, $V \sim s^2a$, $B \sim 10^{-5}$ gauss, $K \sim 10^{-12}$. The relaxation therefore is of order $\tau/t \sim 10\%$. For more reasonable B values it is ~ 0.01, comparable to the photon alignment. Even then, we have probably overestimated K, and the effect may be still weaker.

We are therefore faced with a situation in which we could account for the alignment magnetically, if the Galactic magnetic field is strong and lies predominantly parallel to the Milky Way plane. If the motion of gas relative to grains were perpendicular to the plane, the Gold process would give the correct orientation, and we know that the photon flux is in the right direction for the photon process. But in each of these mechanisms we have to strain the properties or the expected parameters of the grains, or the gas, or the magnetic field, or a combination of these, to obtain rough quantitative agreement. It is not unlikely that all of these processes play important roles in the alignment, but since none of the processes is entirely free of difficulties, it also is possible that we have overlooked some dominant factor of greater overall importance!

9:9 ORIGIN OF COSMIC MAGNETIC FIELDS

Magnetic fields are known to exist in stars and in the interstellar medium. Stars like the sun have typical surface magnetic fields of the order of one gauss, but in some A stars the surface fields may reach magnitudes of the order of 40,000 gauss. The fields in the interstellar medium of course are much weaker, typically of the order of 10^{-6} gauss. But there are great variations. In some regions of the Galaxy no magnetic fields at all have been determined in measurements that should have detected fields of strength 10^{-6} gauss, while in other regions, quite strong fields exist. In the Crab Nebula supernova remnant, for example, the field strength is believed to be as high as 10^{-4} gauss, and fields of the order of 5×10^{-5} gauss have been observed by Verschuur in Orion (Ve70).

Where does this field come from? Is it primordial? Is its origin a cosmological question dating back to some early stages of the universe? We do not know!

If magnetic fields are not related to the introduction of matter into the universe—either primordially or on a continuing basis—then they should be generated at some subsequent time. We can then envisage two possible alternatives.

(1) Magnetic fields are formed in the interstellar or intergalactic medium and find their way into stars primarily because stars are formed from the interstellar material, or

(2) magnetic fields are formed in stars and the field lines might then be introduced into the interstellar medium as mass is ejected from the stars. This might be consistent with the high strength of the Crab Nebula field. It would also be consistent with the observation that the solar wind carries along magnetic fields. Whether some portion of this field becomes detached from the sun and strays out into the interstellar medium is not known. But many stars have much more massive winds than the sun, and the outflow of magnetic fields may be a customary accompaniment to the outflow of mass.

Fig. 9.7 Two magnetic field configurations with the same net flux. Configuration (*a*) has low field strength everywhere. Configuration (*b*) has high field strength in some places. In this figure, the field strength is taken proportional to the number of lines crossing unit length of the abscissa. This would be representative for a situation in which the field lines were embedded in sheets normal to the plane of the paper.

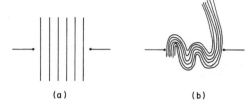

(a)   (b)

Once a magnetic field exists in very weak form, it can be strengthened by turbulent motion of the gases, and by other types of motion of the medium in which the fields are embedded. The net magnetic flux crossing any given fixed surface cannot be increased in this way, but by folding the field direction many times, local fields of greatly increased strength can be formed without an accompanying high net flux (Fig. 9.7). Turbulent motion therefore obviates the need for strong initial fields. Small, seed magnetic fields can be amplified by turbulent stretching and folding of the field lines.

Let us see how big this effect could be. If a field B_0 initially existed in some location within the Galaxy, the flow of gas at a velocity v could have stretched out field lines maximally at that velocity. Folding the field back on itself also could maximally occur at velocity v, so that the ability of a turbulent motion to amplify the field is limited by the speed of the motions.

Essentially the amplification of the field is given by the ratio of the initial volume $\mathscr{V}_0$ containing the seed fields to the final volume $\mathscr{V}_f$ that would have been obtained through stretching the region through a rectilinear motion, at velocity v

$$\therefore \frac{B_f}{B_0} = \frac{\mathscr{V}_f}{\mathscr{V}_0} \tag{9-61}$$

where B_f is the final magnetic field strength obtained through stretching and folding in a constant volume $\mathscr{V}_0$.

Within the Galaxy explosive velocities of order 10^3 km sec^{-1} are sometimes observed. We can choose this to represent the maximum turbulent velocity. The initial dimension of the Galaxy is ~ 30 kpc along a diameter and ~ 100 pc perpendicular to the disk. If the stretching motion were to go on for 10^{10} y at 10^3 km sec^{-1}, a distance of 10 Mpc would be covered, and a turbulent folding would increase the magnetic field strength respectively by a factor of 300, or 10^5 depending on whether the turbulent motion took place predominantly within the Galactic plane or perpendicular to it.

Since the field in the Galaxy is estimated to have a value of order 3×10^{-6} gauss, at the present epoch, the initial seed fields must at least have had values of the order of 3×10^{-11} gauss. There seems no way to escape this conclusion.

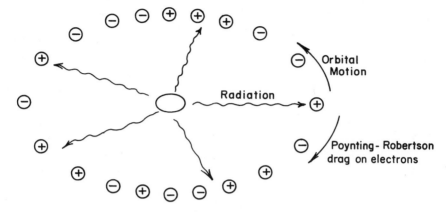

Fig. 9.8 Poynting-Robertson drag on electrons produced by a luminous body acting on an orbiting plasma. The drag on protons is much weaker, and a small net current can therefore be induced.

A primordial field of this magnitude must therefore have been present initially, or else some mechanism must have existed for producing this field. A number of processes have been suggested for setting up such a *seed field* that later could grow in strength through turbulent motion.

One relatively easily understood process could be provided by the Poynting-Robertson effect that could slow down electrons orbiting about a luminous massive object, while leaving protons almost unaffected. The current produced in this way will set up a weak magnetic field (Ca66).

This type of effect applies different forces to electrons and protons and is sometimes called a *battery effect*, because it sets up a current like a battery does. Here we will only show that the Poynting-Robertson effect, Fig. 9.8, would be able to provide sufficient energy for setting up the magnetic field. Some, or perhaps even most of the energy may, however, end up in some form other than magnetic energy. A complete analysis of such effects is very complicated and depends in detail on the interaction of the electrons with protons, on the direct differences in radiation pressure acting on electrons and protons, on the resulting tendency for positive and negative charges to slightly separate along a radial direction from the light source, and so on. A great deal of work remains before we can be sure that battery effects really are the generators of interstellar magnetic fields, but the effect discussed here should serve as an illustration of the type of mechanism that could perhaps be effective.

We note that the Poynting-Robertson drag on an electron is large because the Thompson cross section (see equation 6–102) is $(m_p/m_e)^2 \sim 3 \times 10^6$ times larger for electrons than for protons. Here m_e and m_p are the electron and proton mass. The deceleration is even stronger for electrons, being a factor $(m_p/m_e)^3$ times larger than for protons, because for electrons the drag force acts on a smaller mass. From (5–48) we see that the orbital angular momentum L, for an electron in a circular orbit, changes at a rate

$$\frac{1}{L}\frac{dL}{dt} = \frac{L_s}{4\pi R^2}\frac{\sigma_T}{m_e c^2} \qquad (9\text{–}62)$$

where L_s is the source luminosity, R is the distance from the source, and σ_T is the Thompson cross section. The work dW/dt done on an electron in unit time is equal to the force F acting on it, multiplied by the distance v through which the electron moves in unit time.

$$\frac{dW}{dt} = Fv = \left(\frac{1}{R}\frac{dL}{dt}\right)\frac{L}{Rm_e} = \frac{L_s \sigma_T L^2}{4\pi R^4 m_e^2 c^2} \qquad (9\text{–}63)$$

The largest number of electrons that could be slowed down in this way would be $N \sim 4\pi R^2/\sigma_T$, since if we had more electrons than this, some would shadow the others. Hence the maximum total work that can be done on the clouds is

$$\frac{NdW}{dt} \sim \frac{L_s}{R^2}\frac{L^2}{m_e^2 c^2} = L_s \frac{v^2}{c^2} \qquad (9\text{–}64)$$

This gives the maximum work that can go into building up a magnetic field over volume $\mathscr{V}$:

$$NW \sim \mathscr{V} \int \frac{d}{dt}\left(\frac{B^2}{8\pi}\right) dt \sim \frac{B_f^2}{8\pi}\mathscr{V} \qquad (9\text{–}65)$$

where $B^2/8\pi$ (see section 6:10) is the instantaneous magnetic field energy density. For the Galaxy, gas is contained in a disk volume $\mathscr{V} \sim 3 \times 10^{66}$ cm^3 and $B_f \sim 3 \times 10^{-6}$ gauss.

$$\therefore NW = \frac{L_s v^2}{c^2}\tau \sim 10^{54}\text{ erg} \qquad (9\text{–}66)$$

Let us first see whether the total field could have been produced had the Galaxy at one time been as bright as a QSO for, say, 3×10^6 y. Taking typical QSO internal velocities $v \sim 10^8$ cm sec^{-1}, $\tau \sim 3 \times 10^6$ y $\sim 10^{14}$ sec; we would require $L_s \sim 10^{45}$ erg sec^{-1} at peak efficiency to produce a field of $\sim 3 \times 10^{-6}$ gauss.

This does not seem too unreasonable for producing the entire flux, so that perhaps we would not even need the subsequent turbulent amplification. However, if we want to do the same thing in our Galaxy right now, we find that there are so few electrons that only $\sim 10^{-5}$ of the total flux would be scattered by electrons near the center where velocities are $\sim 10^8$ cm sec^{-1}. There $L_s \lesssim 10^{43}$

erg sec^{-1}, so that the overall rate of work done on electrons is decreased by $\sim 10^7$. In 10^{14} sec a seed field of $\sim 3 \times 10^{-10}$ gauss could be formed for the Galaxy—in 10^{10} y, a field of $\sim 2 \times 10^{-8}$ gauss. Such a seed field could be amplified by turbulence as indicated above.

Other processes that could produce similar currents by acting differentially on electrons and ions have been suggested. One such process could work on the basis of viscous drag (Br68).

Such processes could also act to destroy magnetic fields. If the orbiting plasma contains an initial magnetic field, the drag acting on the electrons may be in a direction that produces a current that opposes the magnetic field. Such processes might play a role in decreasing the magnetic fields during protostellar stages.

Dynamo effects, which also can produce magnetic fields, should just be mentioned, for completeness: however, not much is known about the role they may play in setting up cosmic fields. This is a difficult theoretical problem.

9:10 COSMIC RAY PARTICLES IN THE INTERSTELLAR MEDIUM

Cosmic ray particles, mainly high energy electrons and protons, contribute an energy density of about 10^{-12} erg cm^{-3} to the interstellar medium. This compares to a mean starlight density of $\sim 7 \times 10^{-13}$ erg cm^{-3} and a kinetic energy of gas atoms, ions, and electrons ranging from about 10^{-13} erg cm^{-3} in the low density cool clouds, to roughly 10^{-9} erg cm^{-3} in high density H II regions.

Is there any interaction between the cosmic rays, the gas, and the radiation field?

This interaction, in fact, is quite strong. In this section we will estimate that strength under different conditions. The interaction usually implies an energy loss for a cosmic ray particle.

Such losses can be divided in the following way (Gi69), (Gi64):

(a) Highly relativistic electrons having energies $\mathscr{E} \gg mc^2$, lose energy to the interstellar medium through a number of different processes that sometimes are collectively referred to as *ionization losses*. These comprise (i) the ionization of atoms and ions, (ii) the excitation of energetic atomic or ionic states, and (iii) production of Cherenkov radiation. These effects are not always separable and their interrelationship will be determined in part by the electron energy and in part by the nature of the medium. Neutral and ionized gases give rise to different loss rates. Table 9.4 gives expressions for these losses and for other cosmic ray losses discussed below.

(b) These ultrarelativistic electrons also can suffer *bremsstrahlung* losses. These occur when electrons are deflected by other electrons or by nuclei of the medium.

Table 9.4 Energy Losses of Cosmic Ray Particles in the Interstellar Medium (After Ginzburg, Gi69).[a]

(a) Ionization Losses for Electrons with $\mathscr{E} \gg mc^2$:

in a plasma

$$-\frac{d\mathscr{E}}{dt} = \frac{2\pi e^4 n}{mc}\left\{\ln\frac{m^2 c^2 \mathscr{E}}{4\pi e^2 n\hbar^2} - \frac{3}{4}\right\} = 7.62 \times 10^{-9} n \left\{\ln\left(\frac{\mathscr{E}}{mc^2}\right) - \ln n + 73.4\right\} \text{ ev sec}^{-1}$$

in a neutral gas

$$-\frac{d\mathscr{E}}{dt} = \frac{2\pi e^2 n}{mc}\left\{\ln\frac{\mathscr{E}^3}{mc^2\chi_0^2} - 0.57\right\} = 7.62 \times 10^{-9} n \left\{3\ln\left(\frac{\mathscr{E}}{mc^2}\right) + 20.2\right\} \text{ ev sec}^{-1}$$

Electron Radiation Losses:

for plasma

$$-\frac{d\mathscr{E}}{dt} = 7\times10^{-11} n\left\{\ln\left(\frac{\mathscr{E}}{mc^2}\right) + 0.36\right\}\frac{\mathscr{E}}{mc^2} \text{ ev sec}^{-1}$$

for neutral gas

$$-\frac{d\mathscr{E}}{dt} = 5.1\times10^{-10}\frac{\mathscr{E}}{mc^2} \text{ ev sec}^{-1}$$

Electron Synchrotron and Compton Losses:

$$-\left[\left(\frac{d\mathscr{E}}{dt}\right)_s + \left(\frac{d\mathscr{E}}{dt}\right)_c\right] = 1.65\times10^{-2}\left[\frac{H^2}{8\pi} + \rho_{ph}\right]\left(\frac{\mathscr{E}}{mc^2}\right)^2 \text{ ev sec}^{-1}$$

(b) for nuclei in neutral gas

$$-\frac{d\mathscr{E}}{dt} = 7.62\times10^{-9}Z^2 n\left\{4\left[\ln\left(\frac{\mathscr{E}}{mc^2}\right)\right] + 20.2\right\} \text{ ev sec}^{-1}, \quad \text{if} \quad Mc^2 \ll \mathscr{E} \ll \left(\frac{M}{m}\right)^2 Mc^2$$

$$-\frac{d\mathscr{E}}{dt} = 7.62\times10^{-9}Z^2 n\left\{3\left[\ln\left(\frac{\mathscr{E}}{mc^2}\right)\right] + \ln\frac{M}{m} + 19.5\right\} \text{ ev sec}^{-1}, \quad \text{if} \quad \mathscr{E} \gg \left(\frac{M}{m}\right) Mc^2$$

for plasma

$$-\frac{d\mathscr{E}}{dt} = 7.62\times10^{-9}Z^2 n\left\{\left(\ln\frac{W_{max}}{mc^2}\right) - (\ln n) + 74.1\right\} \text{ ev sec}^{-1} \quad \text{if} \quad Mc^2 < \mathscr{E} \ll \left(\frac{M}{m}\right) Mc^2$$

where $W_{max} = 2mc^2\left(\frac{\mathscr{E}}{mc^2}\right)^2$ if $\mathscr{E} \gg \left(\frac{M}{m}\right)Mc^2$

$= \mathscr{E}$ if $\mathscr{E} \gg \left(\frac{M}{m}\right)Mc^2$

ρ_{ph} is the energy density of photons, M is the nuclear mass, Z the nuclear charge, $\mathscr{E}$ is the particle energy, n is the density of electrons in the medium, and χ_0 is the ionization energy.

[a]Reprinted with the permission of Gordon and Breach Science Publishers, Inc. New York, London, and Paris.

The deflection amounts to an acceleration that causes the particle to radiate. Again the loss rates differ for a plasma in which hydrogen is predominantly ionized and for a neutral gas.

(c) Synchrotron and Compton losses (see sections 6:18 and 6:20) are interrelated loss rates, respectively, proportional to the energy density of the magnetic and radiation fields. That these two processes can be considered to be similar can be seen from a simplified argument. Imagine two electromagnetic waves—photons—traveling in exactly opposing directions in such a way that their magnetic field vectors are identical in amplitude and frequency and their electric fields are exactly opposite in amplitude, but again at the same frequency. The electric field and the Poynting vector **S** both cancel for these two waves at certain times, and we are left with a pure magnetic field whose energy density is equal to the total energy in the radiation field. At this point synchrotron loss should occur, and this synchrotron loss should then be equivalent to the losses from inverse Compton scattering off the two protons of equivalent energy. The combined expression for these losses is given in Table 9.4.

(d) For protons and nuclei in the cosmic ray field, we again have ionization losses given in Table 9.4. Synchrotron and Compton losses should be less than those of electrons by the ratio of masses taken to the fourth power $\sim 10^{13}$. There also are a variety of interactions between cosmic ray nuclei and the nuclei of the interstellar gaseous medium and grains.

Table 9.5 gives these interactions for several different groups of nuclear particles interacting with an interstellar gas composed of 90% hydrogen and 10% helium by number of atoms.

TABLE 9.5   Cross Sections, mean free paths Λ, and absorption paths λ.[a]

Cosmic Ray Particle	Cross Section for Collision	Λ Mean Free Path	λ Absorption Path
$\mathscr{P}$	3×10^{-26} cm^2	72 g cm^{-2}	— g cm^{-2}
α	11	20	34
Li, Be, B	25	8.7	10
C, N, O, F	31	6.9	7.8
$Z \geq 10$	52	4.2	6.1
Fe	78	2.8	2.8

[a]For cosmic ray particles in different groups of elements, interacting with an interstellar medium which consists of 90% hydrogen and 10% helium (in number density of atoms) (see text). (After Ginzburg Gi69. Reprinted with the permission of Gordon and Breach Science Publishers, Inc. New York, London, and Paris.)

9:10

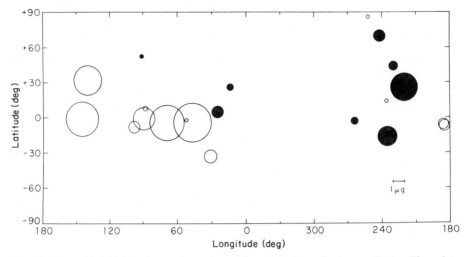

Fig. 9.9 Magnetic field direction at the sun's position, plotted in galactic coordinates. These data actually are mean line-of-sight magnetic field components for pulsars, as judged from their rotation and dispersion measures. For fields greater than 0.3 μgauss the circle diameter is proportional to the field strength. When the field has a direction toward the observer (positive rotation measure), the circles are filled. When they are away from the observer, they are unfilled. The diameter for 1 μgauss is indicated in the figure. The observations are consistent with a relatively uniform field of about 3.5 μgauss directed along the local spiral arm. Note that the directions of greatest field strength are toward longitudes $\sim 60°$ and $\sim 240°$, although there are large variations. These also are roughly the directions of the stellar and gaseous components in the local spiral arm (Bo71). (From Manchester, Ma72b. With the permission of the University of Chicago Press.)

The mean free path Λ gives the distance traveled between nuclear collisions. Essentially a proton travels until it has passed through an effective layer thickness containing 72 g of matter per cm^2. In a cool cloud with density of order 10^{-23} g cm^{-3} this would amount to a distance of order 2 Mpc. Since the cosmic ray particles describe spiral paths in the Galaxy's magnetic field, they traverse such a distance in about 6×10^6 y. Cosmic ray nuclei in the more massive groups suffer collisions more rapidly. Their absorption path length $\lambda = \Lambda/(1 - P_i)$ (where P_i is the probability that the collision will again yield a nucleus belonging to the same initial cosmic ray group) is somewhat longer, as shown in column 4 of Table 9.5.

PROBLEM 9-9. If the energy loss per collision of a cosmic ray nucleon with a nucleus of the interplanetary medium leads to an energy loss comparable to the total energy of the nucleon,

$$\left(-\frac{d\mathscr{E}}{dt} \right)_{nucl} = cn\sigma\mathscr{E} \tag{9–67}$$

Fig. 9.10 Proton and alpha particle cosmic ray flux at the earth. At any given energy the proton flux is about 100 times as intense as the electron flux. (Compiled from various sources by P. Meyer, Me69.) (Error bars have been omitted.) At higher energies the flux continues to drop, the flux J obeying the power law $dJ/dE \propto E^{-\gamma}$ with $\gamma \sim 2.6$. This appears to be true of all the nuclei of different elements. Their flux energy-curves are quite similar.

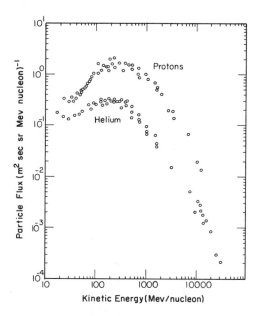

show that this loss dominates the other processes listed in Table 9.4 for cosmic ray nuclei.

PROBLEM 9-10. Using the above loss rate for protons and using the loss rates for cosmic ray electrons having the spectrum shown in Fig. 9.11, calculate roughly how fast the electron and the proton cosmic ray components lose energy and estimate how fast the interstellar medium is being heated by cosmic rays. This cosmic ray heating effect is thought to be important.

The observed flux of cosmic ray protons and alpha particles incident on the earth's atmosphere is shown in Fig. 9.10. Similar data exists for many of the elements. Roughly 90% of the nuclear component of the cosmic ray flux at the top of the atmosphere consists of protons. Alpha particles make up $\sim 9\%$ and the remaining particles are heavier nuclei. Curiously, there is a great excess of Li, Be, B, and He^3 that have a low overall abundance in other cosmic objects, presumably because they are easily destroyed at temperatures existing at the center of stars (section 8:12). We can account for the presence of these elements if they are produced through collisions of carbon, nitrogen, and oxygen cosmic ray particles with hydrogen nuclei in the interstellar medium. The lighter elements then are the *spallation products* of the more massive parent particles. To obtain the amount of these low mass elements observed and also to obtain the correct He^3/He^4 ratio, cosmic ray particles with energies in excess of 1 Gev($= 10^9$ev)

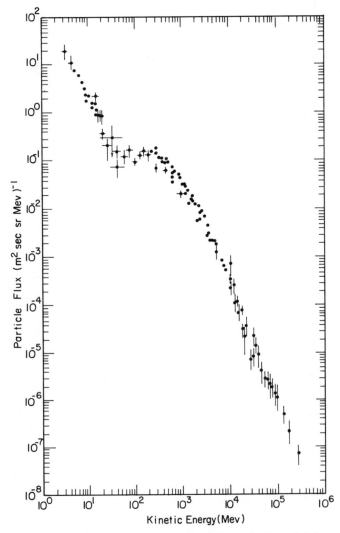

Fig. 9.11   Cosmic ray electron spectrum at the earth. Compiled from various sources by P. Meyer (Me69). Electron and positron abundances are comparable. 1 Gev = 10^9 ev.

would have had to pass through ~ 3 g cm^{-2} of matter (Re68c). This suggests an age of about 2×10^6 y if the particles have been spiraling within the Galaxy all this time. Evidently this represents the mean time taken for cosmic ray particles to diffuse out of the Galactic disk.

9:10

Fig. 9.12 Diffuse X-ray spectrum observed from above the earth's atmosphere. (Compiled from various sources by A.S. Webster and M.S. Longair, We71. With the permission of the Officers and Council of the Royal Astronomical Society.)

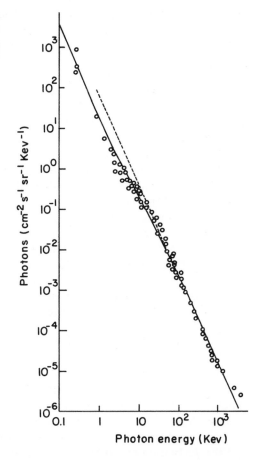

Fig. 9.12 Diffuse X-ray spectrum observed from above the earth's atmosphere. (Compiled from various sources by A.S. Webster and M.S. Longair, We71. With the permission of the Officers and Council of the Royal Astronomical Society.)

We also find that the heavy elements are represented far more abundantly in the cosmic ray flux than in meteorities or in the solar atmosphere. This suggests that these highly energetic particles originate in supernova explosions, pulsars, or white dwarfs, where high concentrations of heavy elements will have been produced during advanced stages of stellar evolution (Co71b).

The cosmic ray flux in the Galaxy appears to be quite steady. Meteorites and lunar surface samples have been analyzed for tracks left by heavy nucleons. The total flux, as well as the relative abundance of heavy nuclei, cannot have drastically changed over the past 5×10^7 y.

Electrons and the somewhat less abundant positrons have fluxes that, at any given energy, amount to about 1% of the proton flux (Figs. 9.10 and 9.11). Interestingly the spectrum of the X-ray flux arriving at the earth (Fig. 9.12) is roughly similar to that of the cosmic ray electrons. This suggests that the X-rays are formed by inverse Compton scattering off electrons by the cosmic millimeter and sub-millimeter radiation flux. However, other mechanisms might also be possible (Po71).

9:11   X-RAY GALAXIES AND QSO'S

A number of galaxies, notably the *Seyfert galaxies* NGC 1275 and NGC 4151, the massive elliptical galaxy and radio source M87, the radio source Centaurus A, and the quasistellar object 3C 273 all are powerful sources of X-ray emission. The X-ray flux from NGC 1275 amounts to 2.4×10^{44} erg sec^{-1} (Gu71), and 3C 273, if at the cosmic distance indicated by its red shift, would emit 1.5×10^{46} erg sec^{-1} (PO71). These objects emit more energy at X-ray frequencies than at all others combined and the flux from NGC 1275 is comparable to the visible flux emitted by normal spirals in the form of starlight.

3C 273 and the two Seyfert galaxies are powerful emitters of infrared energy (K170a) (K170b). It therefore is likely that relativistic electrons in some of these compact sources, inverse-Compton scatter the intense infrared radiation into the X-ray frequency range (We71). Radio emission (Fig. 9.13) would take place by virtue of synchrotron emission given off by the energetic electrons spiraling in their local magnetic field.

The source of infrared emission might be grains thermally radiating energy supplied by the bombardment of relativistic cosmic ray particles.

Whether a given component of the radiation is due to thermal emission, inverse Compton scattering, synchrotron radiation, or some other mechanism, should become apparent when we know more about the brightness fluctuations of these objects in different wavelength ranges. As discussed in section 6:21 some mechanisms of emission have intrinsically faster onset or decay times than others, and the fluctuation rate may therefore permit us to decide which spectral ranges of emission are connected by one and the same emission mechanism and what that mechanism could be.

ADDITIONAL PROBLEMS

9–11. A star composed of hydrogen alone is limited to a maximum brightness. When its luminosity-to-mass ratio exceeds $4 \times 10^4 L_\odot / M_\odot$, its surface layers are blown off. This is what may happen in the development of *planetary nebula* envelopes. (a) Show that this occurs when the radiative repulsion of electrons exceeds the star's gravitational attraction for protons so that ionized hydrogen no longer is bound. (b) Show that for a star predominantly composed

9:11

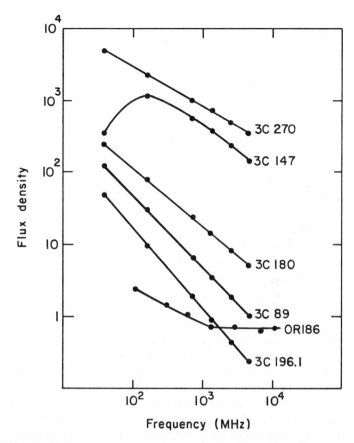

Fig. 9.13 Radio spectra of several extragalactic objects. Note that some have the regular slope discussed in section 6:19. Others show considerable curvature. (Compiled from Kellermann, Pauliny-Toth, and Williams Ke69 and Jauncey, Niell, and Condon, Ja70a). Flux density is measured in units of 10^{-26} wm^{-2} Hz^{-1}. (With the permission of the University of Chicago Press.)

of He^4, C^{12}, O^{16}, or Si^{28}, the luminosity-to-mass ratio can be as high as $8 \times 10^4 L_\odot / M_\odot$ before an outer envelope is blown off.

9–12. The *solar wind* is produced by the high temperature $\sim 2 \times 10^6{}^\circ$K of the solar corona. The wind velocities that are ~ 400 km sec^{-1} are higher than the speed of sound that, for protons, ~ 130 km sec^{-1} for ionized hydrogen at that temperature. Show that this difference can be partly accounted for because randomly projected velocities in a confined gas all become projected onto a radial direction if the gas is allowed to freely expand into a much larger volume where collisions between particles no longer are important. A similar process probably plays a role in all stellar winds. Assume equipartition of energy between protons, electrons, and magnetic fields in the corona.

9–13. In a *comet* ionized gas is propelled into a straight tail pointing away from the sun. This seems to happen in the following way. We think that molecules like CO, initially in the comet's head, suffer charge exchange with protons of the solar wind. In this process, which has a very high cross section $\sim 10^{-15}$ cm^2, an electron is transferred from the CO molecule to the proton. The newly formed CO^+ ion now is forced to travel along with the magnetic field embedded in the solar wind. The magnetic field is predominantly transverse to the wind direction. Make use of the solar wind data of Table 9.1 to compute the velocity of ions in the comet's tail, if roughly half the protons on any given magnetic line of force undergo charge exchange (Ha62b).

9–14. The *X-ray source* Sco X-1 is believed to be a Galactic object in which plasma with electron density $\sim 10^{16}$ cm^{-3} radiates from a volume of radius $\sim 10^9$ cm at a temperature of 5×10^{7}°K. Several models have been suggested. In one model, gravitational energy of matter falling onto the surface of a white dwarf produces the energy. To account for the fast accretion rate of matter and the high radiation intensity it produces, this matter would have to be falling into the white dwarf at a high rate, say from a close binary star syphoning matter onto the white dwarf's surface. Compute the infall rate of matter required and convince yourself that the energies produced in the infall are of the right order of magnitude by comparing gravitational to thermal energies (Pr68). Many other models involving white dwarfs and neutron stars have been proposed. We hope to be able to distinguish between these possibilities by the differing predictions the models make.

ANSWERS TO PROBLEMS

9–1. By (4–117) $c_v = R/(\gamma - 1)$. The ideal gas law states $T = P/R_\rho$. This gives the result (9–23).

9–2. Since $u \gg c$, we can write

$$v_D = \frac{1}{3}\left(-2u + 2u \sqrt{1 + \frac{9c^2}{4u^2}} \right) \sim \frac{2u}{3}\left(-1 + 1 + \frac{9c^2}{8u^2} \right)$$

9–3. (a)

$$\rho_{tot} \sim \frac{4 \times 10^{30}}{\frac{4\pi}{3}(1.5 \times 10^{14})^3} \sim 4 \times 10^{-13} \text{ g cm}^{-3}$$

from Table 1.2

$$\rho_{Fe}/\rho_{tot} \sim 8.9 \times 10^5/4.7 \times 10^{10} \sim 2 \times 10^{-5}; \quad \rho_{Fe} \sim 8 \times 10^{-18} \text{ g cm}^{-3}$$

$$P_{Fe} = \frac{\rho_{Fe}}{m_{Fe}} kT \sim 10^{-11}T, \text{ with } T \sim \left(\frac{L}{16\pi R^2 \sigma}\right)^{1/4} \text{ by (4–78).}$$

For $T \sim 10^3{}^\circ K$, $P_{Fe} \sim 10^{-8}$ dyn cm$^{-2} \sim P_{Fe,vap}$

Hence iron can condense at distances greater than

$$R \sim \left(\frac{L}{16\pi\sigma T^4}\right)^{1/2} \sim 10^{12} \text{ cm}$$

(b) Carbon will condense ~ 3 times closer to the sun.

(c) The temperature out to 10 AU would have been greater than $T \sim 160^\circ K$. Using the H_2O data in Table 9.3 to derive approximate coefficients A and B for H_2O, we see that the vapor pressure $P_{H_2O} \gtrsim 10^{-7}$ torr $\sim 10^{-4}$ dyn cm^{-2} throughout the nebula. On the other hand, from Table 1.2, assuming all oxygen to be in the form of H_2O,

$$\rho_{H_2O} \sim 1.5 \times 10^{-15} \text{ g cm}^{-3}, \quad P_{H_2O} \gtrsim \frac{\rho_{H_2O}}{m_{H_2O}} kT_{min} \sim 10^{-4} \text{ dyn cm}^{-2}$$

H_2O could therefore have condensed near the edge of the nebula; but since the temperature rises rapidly on approach to the sun, we cannot expect H_2O condensation nearer to the sun than the major planets. We should note, however, that the presence of other substances such as ammonia NH_3 can lower the vapor pressure, since H_2O molecules become more strongly bound.

9–4. For circular motion the balance of forces leads to:

$$r\omega^2 = \frac{3}{4\pi\rho s^3}\left(\frac{4\pi M \rho s^3 G}{3r^2} - \frac{L\pi s^2}{4\pi c r^2}\right)$$

$$\text{velocity} = \omega r = \left[\frac{1}{r}\left(MG - \frac{3 \times 4 \times 10^{34}}{16\pi\rho s c}\right)\right]^{1/2}$$

The velocity difference is

$$\Delta v = \left[\frac{1}{r}\left(MG - \frac{3 \times 4 \times 10^{34}}{32\pi sc}\right)\right]^{1/2} - \left[\frac{1}{r}\left(MG - \frac{3 \times 4 \times 10^{34}}{64\pi sc}\right)\right]^{1/2}$$

for $\rho = 2$ and 4 g cm^{-3}, and $s \sim 10^{-3}$ cm, $\Delta v \sim 10^5$ cm sec^{-1}.

9–5. (a) From Problem 3–7 we see that a body with radius s and density ρ has a total capture cross section equal to twice its geometric cross section

$$\frac{4\pi}{3}\frac{2G}{V_0^2}\rho R^2 = 1$$

(b) $\quad \dfrac{dM}{dt} = 4\pi\rho R^2 \dfrac{dR}{dt} = \rho_0 V_0 \pi R^2 = \rho_0 V_0 \pi \left[R^2 + \dfrac{2MGR}{V_0^2}\right]$

$$\int \frac{4\rho\, dR}{\rho_0 V_0 \left[1 + \dfrac{8\pi}{3}\dfrac{GR^2}{V_0^2}\right]} = \int dt = \tau$$

Initially $\dfrac{dM}{dt} = 2\rho_0 V_0 \pi R^2 = \dfrac{3}{4}\dfrac{\rho_0}{\rho}\dfrac{V_0^3}{G} \sim 10^{10}$ g sec^{-1}.

(c) For large bodies $\dfrac{dM}{dt} \propto R^4$ or $M^{4/3}$, since $M \propto R^3$.

9–6. Initially $\dfrac{dM}{dt} \sim 4\pi s^2 \dfrac{ds}{dt}\rho = 10^{-2}\rho_0 V_0 s^2$

$$s = 10^{-2}\frac{\rho_0 V_0}{4\pi}t \sim 3 \times 10^{-10}t$$

for a particle with radius s. From Problem 9–5(a) we see that gravitation takes over when $s \sim [3V_0^2/8\pi\rho G]^{1/2} \sim 10^8$ cm, so that gravitation can be neglected for a body with $s \lesssim 10^6$ cm $= 10$ km. The growth time then is

$$t \sim \frac{10^6}{3 \times 10^{-10}} \sim 3 \times 10^{15} \text{ sec} \sim 10^8 \text{ y.}$$

9–7. $n_H R a_v = 1$.

Since most atoms are in the lowest state, we use $a_v \sim 6.3 \times 10^{-18}$ cm^2 so that

$$R \sim \frac{1.6 \times 10^{17}}{n_H} \text{ cm.}$$

(a) For Ly-α, the oscillator strength f is 0.42 so that

$$\sigma \sim \frac{3\lambda^2 f}{2\pi} \sim 3 \times 10^{-11} \text{ cm}^2.$$

This would give an absorption distance

$$\frac{1}{\sigma n_H} \sim \frac{3 \times 10^{10}}{n_H} \text{ cm} \sim 2 \times 10^{-7} \text{ R}.$$

However,

(b) the Doppler shift may make an atom unable to absorb radiation at the central emission frequency of a different atom. The mean absorption cross section therefore is $c\sigma\gamma/v\langle v^2\rangle^{1/2}$ and the mean free path is

$$\sim \frac{v\langle v^2\rangle^{1/2}}{c\gamma\sigma n_H} \sim \frac{10^{13}}{n_H}.$$

9–8. For a gas at $T \sim 100°K$, $kT \sim 1.4 \times 10^{-14}$ erg, and the grain that has moment of inertia $I \sim 10^{-26}$ g cm^2 has $\omega \sim 10^6$ rad sec^{-1} to make $kT \sim I\omega^2$. The angular momentum $L \sim I\omega$ therefore has a thermal equilibrium value that also is $L \sim 10^{-20}$ g cm^2 sec^{-1}. This is no coincidence. The "random" angular momentum acquired in $N = M/m$ collisions becomes the "systematic" angular momentum to be altered by the next generation of N collisions. These collisions endow the grain with a random angular momentum of the same magnitude as its initial value, but oriented in some other arbitrary direction.

9–9. Table 9.5 shows a collision mean free path of order 20 g cm^{-2}, which for 10^{-23} g cm^{-3} gives a path of order 2×10^{24} cm and a life $\sim 6 \times 10^{13}$ sec.

$$\therefore \frac{d\mathscr{E}}{dt} \sim 1.6 \times 10^{-14}\mathscr{E} \text{ sec}^{-1}.$$

The loss rates in Table 9.4 are of order $10^{-6} Z^2$ ev sec^{-1}, so that collisional losses dominate for all nuclei with energies in excess of $\sim 10^9$ ev.

9–10. For electrons typical losses from Table 9.4(a) for the most significant part of the energy range covered in Fig. 9.11 are 10^{-6} ev sec^{-1}. At $\sim 10^8$ ev the life is $\sim 10^{14}$ sec. Taken together with the result of Problem 9–9, this indicates a cosmic ray energy loss to the interstellar medium of 10^{-12} erg cm$^{-3}/10^{14}$ sec or $\sim 10^{-26}$ erg cm^{-3} sec^{-1}.

If integrated over a gas-containing volume of 10^{66} cm^3 in the Galaxy, this would indicate an eventual radiation loss of order 10^{40} erg sec^{-1}. The total luminosity of the Galaxy is 10^3 to 10^4 times higher; but only about 10% of this luminosity may contribute to heating the interstellar medium; and in the darkest clouds, where radiation does not readily penetrate, cosmic ray heating may be a dominant factor.

9–11. The radiative repulsion of an electron at distance R from a star is

$$\frac{L\sigma_T}{4\pi R^2 c} \sim 1.8 \times 10^{-36} \frac{L}{R^2}$$

where σ_T is the Thomson cross section (6–102).
The gravitational attraction for a proton is $m_H(MG/R^2)$.
For the sun, this ratio of repulsion to attraction is

$$\frac{\sigma_T L_\odot}{4\pi c M_\odot G m_H} \sim (3 \times 10^4)^{-1}.$$

For a star with $L/M \sim 3 \times 10^4$ greater than the sun, electrons are repelled and pull the protons along. For He^4, C^{12}, and so on each electron pulls a proton plus a neutron, and twice this luminosity-to-mass ratio would be needed.

9–12. The total energy per proton in the corona is $3kT/2$. Because of the magnetic field, protons and electrons will be moving together in the solar wind expansion, and the energy of electron random motion can be transferred into expansion velocity of the protons. The total magnetic energy can also decrease, at the expense of proton velocity, since $B \propto r^{-2}$ and the energy that is proportional to $B^2 r^3$ will be proportional to r^{-1}. The three sources of energy, protons, electrons, and magnetic fields, now supply energy $9kT/2$ for each hydrogenic mass m_H streaming away from the sun. For $T = 2 \times 10^6 °K$, we have $v \sim (9kT/m_H)^{1/2} \sim 4 \times 10^7$ cm sec^{-1}.

9–13. If half the protons undergo charge exchange, then the momentum carried by the others must be shared with an equal number of CO^+ ions. These are 28 times as massive as a proton. Very little momentum is transferred by the proton that exchanges charges so that only the remaining protons supply momentum. A total slowing down of ~ 29 times should occur. For an initial velocity of 400 km sec^{-1} for protons, the final ion velocities would be ~ 14 km sec^{-1}. Actually observed velocities are higher, implying either less complete charge exchange—more protons per ion—or an additional pressure from solar wind protons piling into the more slowly moving plasma, and accelerating it away from the sun through pressure transfer by the magnetic field embedded in the proton–CO^+ plasma.

9–14. The energy of a proton freely falling onto a white dwarf star's surface is $mMG/R \sim 10^{-7}$ erg for $R \sim 10^9$ cm. This is $\sim 10^5$ ev and X-rays up to this energy can therefore be given off by the protons. The actual energy at $5 \times 10^{7} °K$ is 5×10^3 ev. The plasma radiates like an optically very thin gas as can be guessed from an extrapolation of Figs. 8.3 and 8.4 and equations 8–64 and 8–65.

10

Structure of the Universe

10:1 QUESTIONS ABOUT THE UNIVERSE

In preceding chapters we discussed the appearance of stars and stellar systems and we looked in some detail at the immediate surroundings of the sun, the one star to which we have easy access. Now we want to examine the environment in which the sun, the stars, and the stellar systems are embedded. We want to learn about the properties of the universe.

The first questions we would impulsively ask are:

(1) What is the shape of the universe?
(2) How big is it?
(3) How massive is it?
(4) How long has it existed?
(5) What is its chemical composition?

These are some of the simplest and most basic questions we would normally ask about any object and we will find it easy to prescribe conceptually simple observational tests for determining the corresponding properties of the universe. In practice, however, it is very hard to follow these prescriptions, and, for this reason most of the above questions have thus far received only partial answers.

We can approach the problem in two distinct ways. The first is the observational approach. We attempt to observe what the universe is "really" like. The other approach is synthetic. We construct hypothetical models of the universe and see how the observations fit the models. This second procedure might at first glance seem superfluous. It might seem that all we need are observations. That is not so!

Any observation has to be interpreted and can only be interpreted within the framework of theory, even if that theory consists of nothing more than the prejudices that constitute common sense. Common sense itself implies a model. It is three dimensional; time measurements can be completely divorced from distance measurements; bodies obey the Newtonian laws of motion; there are laws of conservation of energy and of momentum. We can attempt to see how far common sense can take us in cosmology. We will find that it is quite useful at times, but it can lead to great misconceptions if applied uncritically.

10:2 ISOTROPY AND HOMOGENEITY OF THE UNIVERSE

If we look out into the universe as far as the best available telescopes allow, we find that no matter which direction we look essentially the same picture presents itself. We find roughly the same kind and number of galaxies at large distances in all directions. There may be statistical variations occasionally, but they appear to be random. The general coloration of galaxies also is the same whichever direction we look. The only systematic color differences we detect are those associated with distance, but the universal red shift of the spectra of distant galaxies does not appear to be dependent on direction.

Strictly speaking, all this is true only when we look at fields of view outside the plane of our own Galaxy. The Milky Way absorbs light so strongly that we always have to make allowances for its presence.

Independence of direction is called *isotropy*. As far as we can tell the universe is isotropic. There are no indications of any preferred directions, except for the flow of time (see section 10:15 below).

Next we take into account all those effects associated with distance. We ask ourselves whether conditions at large distances from us appear to be different from local conditions. Is the universal red shift the only effect we see, or are there other distance dependent factors? If the red shift indeed is the only effect, then we can postulate an expanding model to explain all observations. The red shift is taken to be a Doppler shift caused by the recession of distant galaxies. We imply that if it were not for this velocity produced shift, distant parts of the universe would appear identical with our local environment. In such a model no structural differences exist in different parts of the universe and the universe might be taken to be *homogeneous*.

At this point of the discussion we should stop to reconsider. Our argument is not strictly correct! We have forgotten to take into account the fact that the universe is very large and that we obtain all our information by means of light signals that sometimes take billions of years to reach us. A distant galaxy we view today appears not as it would to a local observer stationed near that galaxy, but rather as it would have looked to such an observer many aeons ago when the galaxy was younger. Therefore, would we not expect that distant galaxies would look younger and younger, the farther away we looked?

10:2

Not necessarily! And this is the place where theoretical models begin to become important. We have to consider the possible existence of two entirely different models, an *evolving* model and a *steady state* model. These models will have differing histories.

In most evolving models, matter starts out densely confined. At some more or less narrowly defined stage, galaxies are formed. They recede from each other giving rise to a cosmic expansion and to the observed red shift. In this model more distant galaxies should appear younger and younger. Thus far we have not yet stated how the appearance of galaxies changes with age or how to tell an old galaxy from a young one. No one really knows. But since energy is continually being emitted in the form of star light, we would expect in time some changes to take place in the appearance of galaxies. We conclude that the distant galaxies should at least appear "different" from those nearby. If some such difference, no matter what kind, could be observationally established, we would have strong evidence in support of an evolving universe.

Let us now consider the steady state model. The picture presented here is one of a universe that has always existed and will continue to exist. As distant galaxies stream away from us, new matter is created everywhere and the newly created matter gives rise to new galaxies. Through this replenishment the density of matter can always be kept constant. Any depletion due to the cosmic expansion is exactly compensated by the creation of new matter.

In the steady state model two galaxies created at approximately the same time will find themselves receding from one another at ever increasing speed. As this separation occurs new, younger galaxies are formed in the intervening space. These younger galaxies themselves will recede from each other making room for succeeding generations of galaxies. Any chosen volume of space will, in this way, contain galaxies of varying ages. There will be relatively many young galaxies and decreasingly many old galaxies, because the old galaxies have had time to drift far apart and therefore the number density of old galaxies must be low.

In a steady state universe, the assortment of galaxies in a given volume will be roughly the same at epochs separated by many aeons. There will always be a mixture of young and old galaxies occupying any given volume and the ratio of these galaxies will remain the same. For this reason it does not matter whether we view a distant region today, or several aeons from now. It will always look roughly the same even though the individual galaxies occupying that region will no longer be the same.

The important consequence of this idea is that the view of a distant region observed by us today should, in the steady state model, be roughly the same as the view that a local observer within that region would experience. Consequently the delay introduced by the travel time of light should not affect the age distribution of galaxies observed in a given region of space. That age distribution should be identical with the age distribution an observer sees in the local region around him. The steady state theory therefore predicts that distant galaxies generally look just like nearby galaxies; the evolutionary theories, on the contrary, state that there should be differences in the appearance of galaxies at differing distances. Such differences have not been established thus far, but this may only be because telescopes cannot yet yield accurate observations of regions so distant that differences would show up.

In principle, we can observationally distinguish between steady state and evolutionary models by looking for age differences in distant galaxies. In practice, however, we do not yet have the means for reliably accomplishing this comparison.

Another test for distinguishing steady state and evolutionary models can also be described. In a steady state universe the mean intergalactic distance never changes from one epoch to another, or from one location to another. In evolutionary models the opposite is true. We usually postulate that most galaxies were formed at a particular epoch in an evolutionary universe. The intergalactic distances should therefore have been closer in the past than they are now. By observing the separation between distant galaxies we would be measuring their separation many aeons ago, at a time when these galaxies were still close together. The number density of galaxies at a large distance should, therefore, be different from the nearby density. Number density counts have been attempted both by visual and radio-astronomical techniques. The latest radio-astronomical results are controversial. They may, or may not, indicate a small change of number density with distance. This is still debated (Sh68) and may not be significant. A well-established difference in number density would rule out the possibility that the universe can be described by a steady state model.

It is clear that observations on the apparent homogeneity of the universe can be very important; both the numbers of galaxies and the types of galaxies observed in different regions can yield information about the past history and further evolution of the cosmos. But an understanding of the observations can only be achieved by careful study of different cosmological models. The common sense approach that might deny the possibility of continuous creation is no longer good enough. The fact that we have never seen matter created, and that it therefore is not part of our common sense point of view, is of little importance. It would only be important if the required rate of creation of matter were higher than actually observed.

10:3 COSMOLOGICAL PRINCIPLE

Some postulates about the universe have to be made before any theory can be developed. These postulates or axioms must then be shown to be internally consistent, but we hope they will also be borne out by observational evidence related to predictions made by the postulates.

10:3

One of these postulates is the cosmological principle (Bo52) that appears in various forms. The main hypothesis here is that our position in space and time is not unusual. Hence our local physics, and our locally made observations of the universe should not markedly differ from those made by other observers located in different regions of the universe.

The *Perfect Cosmological Principle* states that for any observer located at an arbitrary position, at an arbitrary time in the history of the universe, the cosmos will present exactly the same aspect as that observed by an observer at some other location at the same or even at some completely different epoch. This principle is very far reaching and, in particular, it leads to the development of the steady state universe.

Many cosmologists will not accept this perfect cosmological principle, but favor a more restricted *cosmological principle* stating that at any epoch (suitably defined) observers at different positions in the universe will observe identical cosmic features—except for small local variations.

In a sense all these principles are extensions of the Copernican hypothesis that we should in no way consider ourselves favored observers.

Although the cosmological principle applies only in a statistical sense, since obviously one galaxy looks different from its neighbors, it will nevertheless be very useful when used in conjunction with a number of simple abstract concepts.

The first of these is that of a substratum. The *substratum* in any cosmic model is a matrix of geometrical points all of which move in the idealized way required by the model. Real galaxies will have random velocities measured with respect to the substratum. On the other hand we would expect the mean motion of distant galaxies to be zero as seen by an observer who is at rest with respect to the substratum. We might also expect that the $3°K$ component of the microwave background radiation would appear isotropic to such an observer. A state of rest relative to the substratum can therefore be determined in a number of practical ways. Such a state plays a fundamental role in cosmology and it therefore is useful to call a particle that is at rest with respect to the substratum a *fundamental particle* and an observer who is similarly stationary a *fundamental observer*.

A fundamental observer may have a watch on him. Such a watch, stationary with respect to the substratum, measures *proper time* for the observer. The time measured by locally moving clocks will be different from this time. The proper time of a fundamental observer can be considered to define a *world time* scale that could be used by all fundamental observers for intercomparing measurements. For example, in describing the evolution of a cosmic model we normally think of a *world map* that describes the appearance of the universe at one particular world time. In contrast, we can also think of a *world picture* that is just the aspect the universe presents to a particular fundamental observer at any given time. To see the difference between these concepts, we note that all galaxies are at rest in a world map, but the map may be expanding. On the other hand, in a world picture, distant galaxies would appear to recede from the observer—at least at the present epoch.

10:4 CREATION OF MATTER

The most surprising part of the steady state theory of cosmology is its suggestion that matter is continually being created. It is not being created from something else, the way we might form water from oxygen and hydrogen in the laboratory.

This matter is created from nothing!

Or is it? Some theories postulate a new field from which this matter would be created, but thus far this has been done mainly to keep the conservation laws of physics intact. The new field—called the C-field—in that sense is an artifact.

An observer in a steady state universe would expect to see matter created locally and we might wonder whether the rate of creation might be observed directly. We can readily compute what that rate should be. Consider a spherical volume with radius r. This radius is expanding at some rate directly proportional to r. Call this rate Hr.

$$\frac{dr}{dt} = Hr \tag{10–1}$$

The rate at which the volume expands is

$$\frac{d(4\pi r^3/3)}{dt} = 4\pi r^2 \left(\frac{dr}{dt} \right) = 4\pi r^3 H \tag{10–2}$$

If the density of the sphere is to be maintained constant at some value ρ_0 during this expansion, the increased volume must be filled with matter at density ρ_0 so that the rate of matter creation is $4\pi r^3 H\rho_0$ in a sphere of radius r. Dividing by the volume of the sphere, we find the rate of matter creation in unit volume to be $3H\rho_0$. The quantity H is the *Hubble constant*. It is a measure of the expansion of the universe and must be observationally determined. The best present estimate is roughly

$$H = 75\,\mathrm{km\,sec^{-1}\,Mpc^{-1}} = 2.5 \times 10^{-18}\,\mathrm{sec^{-1}} \tag{10–3}$$

This means that an object at a distance of one Mpc (*megaparsec* $= 10^6$ parsec) has a typical recession velocity of 75 km/sec. An object N times this distance has N times that velocity.

An estimate of the value of ρ_0 can be obtained by taking the number density of galaxies (see sections 2:10 and 2:11) and multiplying by a typical galactic mass. The density so obtained is $\sim 10^{-30}\,\mathrm{g\,cm^{-3}}$. It is a lower limit to the density since there might exist large amounts of nonluminous unobserved matter in the universe. We can now assess the creation rate as

$$3H\rho_0 \sim 10^{-47} \, \text{g cm}^{-3} \, \text{sec}^{-1} \tag{10–4}$$

If matter is created in the form of hydrogen, this would imply a creation rate of the order of one atom in each one liter volume, every five aeons. At the moment there is no way of measuring such small creation rates.

10:5 HOMOGENEOUS ISOTROPIC MODELS OF THE UNIVERSE

Observations made to date do not indicate that there are any preferred directions or unusually dense regions in the universe. The data are compatible with a homogeneous, isotropic model of the universe, that is, a universe in which there are no select locations or directions.

An observer placed at any location in the universe would see distant galaxies red shifted, in apparent recession, no matter what direction he chose to observe.

In order to construct a model of such a universe, we assume that the red shift indeed indicates a genuine expansion. This assumption has become entrenched in cosmology, primarily through default. When the red shift of distant galaxies was first discovered, a number of explanations were advanced. One by one the alternate hypotheses have been dropped as being incompatible with observations, or unlikely on other grounds. The velocity-induced red shift has been the only hypothesis that could not be discarded. It has survived and may well be the genuine source of the red shift. Still, from time to time the search for alternate explanations continues and probably will continue until the recession hypothesis can be established on a firmer basis.

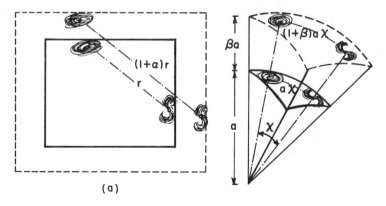

Fig. 10.1 (*a*) Expansion in a flat universe.   (*b*) Expansion in a curved universe.

It is important to have some means of visualizing a model of the universe in which an observer at each point sees all other observers in distant galaxies receding from him. A simple model in two dimensions consists of a rubber sheet (Fig. 10.1). Let spots be painted on this sheet in some random manner. If the sheet is now stretched in length L and width W, by fixed amounts of αL and αW, respectively, then all distances are increased by a fractional amount α. If the spots on the sheet represent galaxies, then a galaxy that initially was at some distance $r = (x^2 + y^2)^{1/2}$ from a given galaxy, will later be at distance $(1 + \alpha) r = \{[(1 + \alpha) x]^2 + [(1 + \alpha) y]^2\}^{1/2}$ where x and y are the components of the separation along the L and W directions, respectively.

A flat rubber sheet is not the only two-dimensional model for an expanding homogeneous isotropic universe. Take a rubber balloon and paint spots on it to represent galaxies. At a given instant let the radius of the balloon be a. Let the angle subtended by two galaxies be χ at the center of the balloon. The distance between galaxies measured along the surface is the arc length $a\chi$. If the balloon is expanded the angle χ remains constant, but the radius increases to some new value say $a' = (1 + \beta) a$, where β is the fractional increase in the radius. The distance between galaxies now is $(1 + \beta) a\chi$, and we note that the fractional increase is independent of χ. This means that if the universe is homogeneous and isotropic at a given instant, an isotropic expansion will keep it that way.

If the time rate of change of β is $\dot\beta$, the recession velocity between the two galaxies is $a\dot\beta\chi$, which increases linearly with angle χ. Dividing the velocity of separation by the distance, we obtain the ratio $a\dot\beta\chi/a\chi = \dot\beta$. We talk about a linear distance-velocity relation because, for increasing separation, the recession velocity increases in proportion to the separating distance.

In section 2:10 and Fig. 2.4, we saw that distant galaxies and clusters of galaxies obey such a relation—at least approximately.

A sphere of radius a is described by the equation

$$x_1^2 + x_2^2 + x_3^2 = a^2 \tag{10-5}$$

where x_1, x_2, x_3 are three mutually orthogonal Cartesian coordinates. An element of length dl on the sphere is given by

$$dl^2 = dx_1^2 + dx_2^2 + dx_3^2 \tag{10-6}$$

Eliminating the coordinate x_3 by means of equation 10–5, we find

$$dl^2 = dx_1^2 + dx_2^2 + \frac{(x_1 \, dx_1 + x_2 \, dx_2)^2}{a^2 - x_1^2 - x_2^2} \tag{10-7}$$

In terms of spherical polar coordinates, we can write dl^2 as

$$dl^2 = a^2 (d\theta^2 + \sin^2\theta \, d\phi^2) \tag{10-8}$$

10:5

We can repeat this procedure for a four-dimensional sphere in an exactly analogous way. Here we do not deal with a two-dimensional surface or a three-dimensional space. Rather, we work with a space showing isotropy and homogeneity in three dimensions; and analogously to the three-dimensional approach of equations 10–5 to 10–8, we want to investigate the properties of a three-dimensional *hypersurface* on a four-dimensional *hypersphere*. Problem 10–1 will show that the relation corresponding to equation 10–8 then has the form

$$dl^2 = a^2 [d\chi^2 + \sin^2\chi(\sin^2\theta \, d\phi^2 + d\theta^2)] \qquad (10\text{–}9)$$

PROBLEM 10–1. Show how relation (10–9) is obtained by starting with an equation for a hypersphere

$$x_1^2 + x_2^2 + x_3^2 + x_4^2 = a^2 \qquad (10\text{–}10)$$

Continue by showing that in terms of three-dimensional polar coordinates, we have

$$dl^2 = dr^2 + r^2 \, d\theta^2 + r^2 \sin^2\theta \, d\phi^2 + \frac{(r \, dr)^2}{a^2 - r^2} \qquad (10\text{–}11)$$

Then substitute the new variable

$$r = a \sin \chi \qquad (10\text{–}12)$$

Consider a sphere of radius R in a conventional three-dimensional space. On a two-dimensional spherical surface the distance along the sphere is given by $R\theta$. The length of a circle in these coordinates is $2\pi R \sin \theta$. At increasing distance from the origin, the size of the circle increases to a maximum value $2\pi R$ at a distance $\pi R/2$. After that it decreases and shrinks to a geometric point at the *antipodal position*—at distance πR.

PROBLEM 10–2. Show that on a four-dimensional hypersphere
  (i) the ratio of the circumference of a circle to its radius is less than 2π;
  (ii) the surface area of a sphere is

$$S = 4\pi a^2 \sin^2\chi \qquad (10\text{–}13)$$

  (iii) As the angle χ increases the sphere grows and the surface of the sphere increases reaching a maximum value $4\pi a^2$ at distance $\pi a/2$ before shrinking to a point at distance πa. Show, moreover, that the element (10–9) defines the total volume

$$V = \int_0^{2\pi} \int_0^{\pi} \int_0^{\pi} a^3 \sin^2 \chi \sin \theta \, d\chi \, d\theta \, d\phi \qquad (10\text{–}14)$$

so that

$$V = 2\pi^2 a^3 \tag{10-15}$$

We can choose a parameter λ

$$\lambda = \frac{1}{a^2} \tag{10-16}$$

which defines the *curvature* properties of a space. When the *radius of curvature* is infinite, $\lambda = 0$ and the space has zero curvature. Such a space is called *flat* or *Euclidean*. When $\lambda > 0$, we talk of a space of positive curvature. We can also define spaces of negative curvature for which $\lambda < 0$ if, as in (10–17) below, we replace the right side of (10–10) by $-a^2$. Note that the two two-dimensional universes described above have differing curvature constants. The sheet model is Euclidean. The balloon model has positive curvature.

PROBLEM 10–3.   In a space of negative curvature—a *hyperbolic space*, sometimes called a *pseudospherical space*:

$$x_1^2 + x_2^2 + x_3^2 + x_4^2 = -a^2 \tag{10-17}$$

where a is real.
  Show that
  (i)

$$dl^2 = r^2 (\sin^2\theta \, d\phi^2 + d\theta^2) + (1 + r^2/a^2)^{-1} dr^2 \tag{10-18}$$

where r can have values from 0 to ∞.
  (ii) Defining $r = a \sinh \chi$ (where χ goes from 0 to ∞).

$$dl^2 = a^2 \{ d\chi^2 + \sinh^2\chi \, (\sin^2\theta \, d\phi^2 + d\theta^2) \} \tag{10-19}$$

Show that the ratio of the circumference of a circle to its radius is greater than 2π.
  (iii) Show that the surface of a sphere is

$$S = 4\pi a^2 \sinh^2\chi \tag{10-20}$$

which increases without limit.
  (iv) The volume of the space is

$$V = \int_0^{2\pi} \int_0^{\pi} \int_0^{\infty} a^3 \sinh^2\chi \sin\theta \, d\chi \, d\theta \, d\phi \tag{10-21}$$

which is infinite.

10:5

To summarize, we note that a space of positive curvature has a finite volume and is closed. Increasing χ beyond a value π returns us to a region already defined by χ values between 0 and π. The space of negative curvature is *open*. The volume of a closed space is finite and given by equation 10–15. The volume of an open space is infinite.

A steady state universe can only exist in a flat space. This has to be so because in a curved expanding space the radius of curvature would be constantly changing and that means that the number of galaxies observed at different distances would continually be changing and would, in fact, be a measure of the evolutionary state (age) of the universe. This does not mean that creation of matter might not be possible in a curved universe; it does mean, however, that such a universe would be observably evolving.

We often see the curvature of cosmological models described by a constant k that can assume values $+1, 0,$ or -1. k is the *Riemann curvature constant* and the values respectively describe universes of positive, zero, and negative curvature. k denotes the algebraic sign of the parameter λ of equation 10–16.

In our balloon model of a universe, a galaxy close to an observer subtends a large angular diameter. At larger distances galaxies of the same size subtend smaller and smaller angular diameters reaching a minimum value when their distance is $\pi a/2$, a is the radius of curvature of the balloon. After that the angular diameter once again increases until it reaches a maximum value of 2π when seen at the antipodal point of the balloon, that is, at a distance πa. An observer can then look in any direction he pleases and see that particular galaxy at one and the same distance from him. Figure 10.2 illustrates these effects.

These effects are found on three-dimensional hypersurfaces in exact analogy. Observations have been made to detect whether there exists a minimum angular diameter for radio-astronomical sources. Thus far these observations have not been successful because some QSO's have intrinsically very small radii and available techniques are not yet well enough developed to measure these small angles accurately.

10:6 MEASURING THE GEOMETRIC PROPERTIES OF THE UNIVERSE

It is possible, at least in principle, to determine the size and curvature of the universe on the basis of astronomical observations (Ro55)*, (Ro68)*. The simplest quantitative relationship between directly observed quantities and the more abstract geometrical properties of the universe have been derived for models in which complete homogeneity and isotropy pervade. These relationships will be described here; their greatest asset is an independence of the specific dynamics—for example, general relativity—used to describe the evolutionary properties

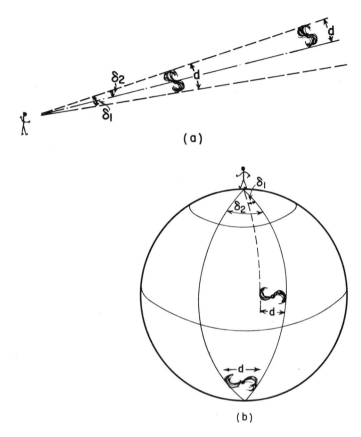

Fig. 10.2 (*a*) Distance–angular-diameter relation in a flat space. (*b*) Distance–angular-diameter relation on the surface of a three-sphere.

of the cosmological models. In effect, we can obtain a geometric description of the universe as it appears at the present world time, without being able to—or, for that matter, needing to—make any statements about how the universe evolved before reaching this state or how it will evolve in the future. At our present stage of understanding, this restricted approach is useful. We do not yet know the size or sense of curvature of the universe; and it is welcome that we could at least obtain that much information, without the additional complication of first needing to know the dynamic laws describing cosmic evolution.

One can show on the basis of group theoretical arguments (Ro33) (Wa34) that the most general metric describing homogeneous isotropic spaces is

$$ds^2 = c^2\,dt^2 - dl^2 \tag{10-22}$$

10:6

the Robertson-Walter metric, for which

$$dl^2 = a^2(t)\{d\chi^2 + \sigma^2(\chi)[d\theta^2 + \sin^2\theta \, d\phi^2]\} \tag{10-23}$$

where dl^2 is the metric of a homogeneous, isotropic three-dimensional space. The function $\sigma(\chi)$ has the form sin χ, χ or sinh χ, depending on whether the *Riemannian curvature* of the three-space is $k = 1, 0,$ or -1.

In this notation:

(a) the world line of a stationary galaxy is a curve, with χ, θ, and ϕ constant, along which ds measures the world time interval dt;

(b) the world line of any light signal is a *null-geodesic*, meaning that it is characterized by $ds = 0$.

(c) If we choose a specific world time—$t = $ constant, $dt = 0$—we can measure spatial distances within the universe with the aid of the metric $-ds^2$. The curvature of the universe k/a^2 will then be fully determined if we can ascertain the value of k and of $a(t)$. To this end, consider a world diagram representing an observer O located at $(\chi, \theta, \phi) = (0, 0, 0)$ and a galaxy at $(\chi_0, \theta_0, \phi_0) = $ constant. As $a(t)$ changes with time, a constant interval dl, in the auxiliary three-space (10–23) will lead to a changing value of $ds^2 - c^2 \, dt^2$. In particular, for a light beam traveling from t_1 to t_0 (Fig. 10.3) we can set $ds = 0$ and then the integration of equation 10.22 along a fixed line of sight (θ, ϕ) yields

$$\int_{t_1}^{t_0} \frac{c \, dt}{a(t)} = \chi \tag{10-24}$$

This is the relation between *distance parameter* χ and emission time t_1 as measured by observer O at time t_0. We will keep referring to a *distance parameter* here in-

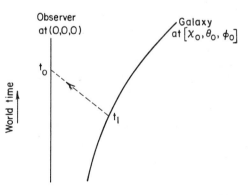

Fig. 10.3 Relation between galaxy and observer. (After H.P. Robertson, Ro55. Courtesy of the Publications of the Astronomical Society of the Pacific.)

stead of a *distance*. The reason is that it is not quite clear just what we would like to call "distance." Perhaps the *cosmic distance* (see equation 10–12)

$$r = a(t_0)\,\sigma(\chi) \tag{10-25}$$

is a useful measure. This represents the distance in the auxiliary three-space at the world time of observation, t_0.

There are other alternatives. We will see that the apparent luminosity of distant objects makes the quantity $a(t_0)\,\sigma(\chi)(1 + z)$ a useful measure of distance. Here z is a measure of the red shift defined by equation 10–26 below and the *luminosity* of a galaxy is derived further on in equation 10–32. This discussion is brought up at the present stage only because it is annoying not to have an exact analogue to all the concepts we normally like to attribute to distance; but in the more general mathematical spaces that are useful in cosmology we cannot expect to have all these properties embodied in a single parameter.

If the light emitted in the interval t_1 to $t_1 + dt_1$ is received between times t_0 and $t_0 + dt_0$, the increment dt_0 can be evaluated either by differentiating equation 10–24 or equivalently by setting $ds = 0$ and replacing $d\chi$ by the constant distance parameter χ in equation 10–22. If the frequency of the emitted signal is v_1 and that of the received signal v_0 then

$$v_0\,dt_0 = v_1\,dt_1$$

because the total number of oscillations in the wave is conserved during propagation. In terms of wavelength $\lambda = c/v$, we have

$$z \equiv \frac{\lambda_0 - \lambda_1}{\lambda_1} = \frac{dt_0}{dt_1} - 1 = \frac{a(t_0)}{a(t_1)} - 1 \tag{10-26}$$

Equation 10–26 defines the measured *red shift parameter z* for radiation emitted at time t_1.

PROBLEM 10-4.   Equation 10–26 is not yet in a useful form because we do not know how $a(t)$ varies with time. However, if we make the assumption that $a(t)$ varies regularly, so that a Taylor expansion may be used to determine $a(t_1)$ in terms of derivatives of $a(t_0)$, show that

$$z = \frac{\dot{a}_0}{a_0}(t_0 - t_1) + \left[\left(\frac{\dot{a}_0}{a_0}\right)^2 - \frac{1}{2}\frac{\ddot{a}_0}{a_0}\right](t_0 - t_1)^2 + \dots \tag{10-27}$$

where $\dot{a}_0$ and $\ddot{a}_0$ are the first and second time derivatives of $a(t)$ evaluated at t_0, the time of observation.

10:6

We can next ask about the angular diameter δ subtended by a galaxy as observed by O. If the intrinsic, locally measured diameter of the galaxy is D, we can define a parameter distance $d\chi$ across it such that $D = a_1 \, d\chi$. But at world time t_1, when the center of the galaxy has a fixed position (χ, θ, ϕ), equation 10–23 shows that the total parameter length of a circle, drawn transverse to the line of sight and traversing the galaxy's major diameter is $2\pi a_1 \sigma(\chi)$. This length corresponds to the full range of values that θ and ϕ can take. It follows that the linear diameter of the galaxy amounts to a segment $D/2\pi\sigma(\chi) \, a_1$ of a full circle, and that the angular diameter therefore is

$$\delta = \frac{D}{a_1 \sigma(\chi)} \tag{10–28}$$

To convert this into values of $a(t)$ measured at the present epoch, we can still make use of equation 10–26 to obtain

$$\delta = \frac{(z+1) \, D}{a_0 \sigma(\chi)} \tag{10–29}$$

The second relation of interest to observational cosmology is the dependence on χ of the number of galaxies $N(\chi)$ whose parametric distance is less than a given value χ, or equal to it. If n is the density of galaxies in the auxiliary three-space defined by the metric dl^2, (10–23), then n is independent of t if evolutionary effects are neglected: in a homogeneous model it is also independent of χ:

$$N(\chi) = 4\pi n \int_0^\chi \sigma^2(\chi) \, d\chi \tag{10–30}$$

where the more general function σ now represents the functions sin and sinh that held for the special case $k = +1$ and $k = -1$ in relations (10–13) and (10–20). Fig. 10.5 on page 450 illustrates these ideas.

PROBLEM 10–5. Show that (10–30) can be expanded into the series relation

$$N(\chi) = \frac{4\pi n}{3} \chi^3 \left(1 - \frac{k}{5}\chi^2 + \ldots \right) \tag{10–31}$$

Both relations (10–29) and (10–31) depend on a knowledge of z, if observations are to be interpreted. However z often is hard to measure at great distances because galaxies there are quite faint. We might therefore prefer to deal with the total observed flux, a readily determined quantity, rather than with the red shift parameters. To do this, we need to know more about the apparent luminosity of distant galaxies as seen by an observer O at our epoch. To determine that we have to make the further assumption that photons are conserved and the energy is related to frequency by the Planck expression $\mathscr{E} = hv$ with h a universal constant independent of world time.

If L_1 is the bolometric luminosity of the galaxy at the time of emission, then the apparent bolometric luminosity $\mathscr{L}$ observed by O is

$$\mathscr{L} = \frac{L_1}{4\pi a_0^2 \sigma^2(\chi)} \cdot \frac{1}{(1+z)^2} \tag{10-32}$$

Here the first term represents the geometrical dilution of radiation, since $4\pi a_0^2 \sigma^2$ is the surface area of an auxiliary three-space drawn about the emitting galaxy at the distance of the observer (equations 10–13 and 10–20). The second term represents the reddening. The term $(1 + z)$ appears squared. One such factor is just due to the decrease in spectral frequency and, hence, the decrease in energy per photon as viewed by the observer. The second factor enters because all conceivable frequencies are diminished, including the rate at which photons emitted by the galaxy arrive at the observer. In unit time interval the observer sees $(1 + z)$ fewer photons than were emitted at the galaxy in unit time. This corresponds to a general slowing down of clocks and again means that less energy arrives at O in unit time.

We now proceed to write the angular diameter relation (10–29) purely in terms of observable quantities. If the luminosity $\mathscr{L}$ in equation 10–32 is divided by the square of the angular diameter δ (10–29), we obtain

$$\log \mathscr{L} = 2 \log \delta - 4 \log (1 + z) + \log \frac{L_1}{4\pi D^2} \tag{10-33}$$

Observations normally are carried out in a fixed spectral frequency range Δv in which the radiation detector is sensitive. Let $\Delta\mathscr{L}$ be the rate of reception of photons at O in this spectral range. ΔL_1 is the comparable emission rate for photons in the emission frequency range Δv_1 that, when red shifted, becomes superposed on the received spectral band Δv. Then

$$\log \Delta\mathscr{L} = 2 \log \delta - 3 \log (1 + z) + \log \frac{\Delta L_1}{4\pi D^2} \tag{10-34}$$

We now return to equation 10–32 and convert it into a relation between bolometric magnitudes. We now write

$$m = M_1 + 5 \log \left[\sigma(\chi)(1 + z)\frac{a_0}{10} \right] \tag{10-35}$$

10:6

where a_0 is now measured in parsecs and the division by 10 is necessitated by the fact that absolute magnitudes always refer to the apparent magnitude of an object at a distance of 10 pc. In terms of present values, we replace the magnitude M_1 by its expanded form

$$M_1 = M_0 - \dot{M}_0(t_0 - t_1) + \ldots \qquad (10\text{--}36)$$

PROBLEM 10-6. Show that if we further expand in powers of z and χ, relation (10–35) can be written as

$$m = M_0 - 45.06 + 5\log\left(\frac{a_0 z}{\dot{a}_0}\right) + 1.086\left(1 + \frac{a_0\ddot{a}_0}{\dot{a}_0^2} - 2\mu\right)z + \ldots \qquad (10\text{--}37)$$

where

$$\mu = 0.46\,\frac{\dot{M}_0 a_0}{\dot{a}_0} \qquad (10\text{--}38)$$

is a measure of the change of the magnitude of the galaxy.

The quantity

$$q_0 \equiv -\frac{a_0\ddot{a}_0}{\dot{a}_0^2} \qquad (10\text{--}39)$$

often appears in cosmology. In equation 10–37 the logarithmic term gives rise to the linear red-shift–distance relation. The Hubble constant

$$H \equiv \frac{\dot{a}_0}{a_0} \qquad (10\text{--}40)$$

that is not constant if $\ddot{a}_0 \neq 0$, allows us to write (equations 10–25, 10–27)

$$cz = a_0 H\chi \qquad (10\text{--}41)$$

in the first approximation, with the consequence that cz can be thought of as the linear velocity of the galaxy and $a_0\chi$ as its distance. If such a linear velocity relation had held throughout the past, the universe would have had to start from a point origin a time $1/H$ ago.

q_0 is sometimes called the *deceleration parameter*. Its observational value is very uncertain. The data do not permit a differentiation between cosmic models

Fig. 10.4 The red-shift–magnitude relation (after Hoyle and Sandage Ho56). Curve *A* corresponds to $q_0 \sim 2.5$. Curve *B* is $k = 0$, $q_0 = \frac{1}{2}$. Curve *C* represents the steady state model (see also Figure. 2.4). (Courtesy of the Publications of the Astronomical Society of the Pacific.)

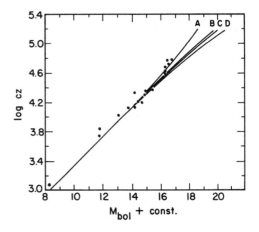

having distinct q_0 values. We can tabulate the q_0 values in terms of the curvature k of space.

PROBLEM 10-7. From the definition of q_0, and the fact that in steady state cosmologies $a(t)$ is proportional to e^{tH}, show that for a steady state cosmos, $q_0 = -1$.

Table 10.1.

	k	q_0
For exploding models with zero cosmological constant and pressure (see section 10:9 below) $\left\{\begin{array}{c} \\ \\ \\ \end{array}\right.$	$+1$ 0 -1	$> 1/2$ $= 1/2$ $0 \leqq q_0 < 1/2$
Steady state model	0	$q = -1$

The observational data are plotted in Fig. 10.4 (see also Fig. 2.4). No clear distinction between models is possible.

With the information about the apparent magnitude of galaxies, contained in (10–37), we can return to expression 10–31 in an attempt to determine the

10:6

number of galaxies that would be observed at magnitudes out to a given apparent magnitude m. On replacement of distances by magnitudes in expression 10–31 one can obtain (Ro55)

$$\frac{1}{N}\frac{dN}{dm} = 1.382 \left\{ 1 - (1 - \mu) z + \left[\frac{3}{2} + \frac{kc^2}{5\dot{a}_0^2} + \frac{a_0 \ddot{a}_0}{2\dot{a}_0^2}(1 + \mu) \right. \right.$$

$$\left. \left. - \frac{7}{2}\mu + \mu^2 - K \right] z^2 + \dots \right\} = 0.4 \frac{d(\log N)}{d(\log S)}$$

$$(10\text{–}42)$$

where the expansion is in terms of the red shift z and the meaning of log S is discussed below. The linear term in z depends only on the rate of change of the absolute magnitude of galaxies μ, while the quadratic term depends in addition on the term

$$K \equiv 0.46 \frac{\ddot{M}_0}{H^2} \qquad (10\text{–}43)$$

In radio astronomy we often talk (section 2:11) about the logarithm of the local flux S obtained from a galaxy, and equation 10–42 then leads to values of the slope $d \log N / d \log S$ characteristic of individual models on a log S-log N plot (Fig. 2.7). Since log S is $(2.5)^{-1}$ times a "radio magnitude" for a radio source, we see that the right side of (10–42) could be written as $0.4 \, d(\log N)/d(\log S)$ with K and μ now interpreted as evolutionary parameters in the radio frequency domain.

The relations (10–34), (10–37), and (10–42) all suffer from a common drawback in that separate information such as that inherent in expressions 10–38 or 10–43 must be obtained before meaningful cosmological conclusions can be reached. This problem would not be severe were it not for the fact that in most of the cosmological models considered today, curvature effects of the universe, such as those illustrated in Fig 10.5, apparently occur, if at all, only at such large distances that galaxies, and the stars inside them, presumably evolve very significantly during the time their signals take to reach us. How well we can then define the time derivative $\dot{M}_0$ or $\ddot{M}_0$ is not at all clear. We now know that galaxies often suffer catastrophic structural changes as evidenced, say, by the explosion of material from the nucleus of the galaxy M82 or the extremely powerful radio "jet" of the giant spherical galaxy M87. If quasistellar objects are at cosmological distances, even greater variations in brightness can be expected, and the flux may be emitted in vastly differing spectral ranges at different epochs during a QSO's evolution.

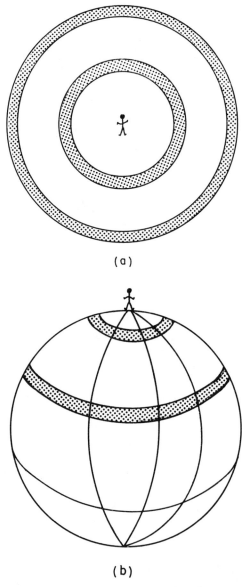

(a)

(b)

Fig. 10.5 (*a*) Distance–number-count relation in a flat space. (*b*) Distance–number-count on a spherical surface. Because the circle or surface drawn about the observer at any given distance is always smaller than the corresponding circle or surface in a flat space, the number of galaxies counted at any distance in a spherical universe will also be less than the number counted in a flat or Euclidean universe.

10:6

We must bear all this in mind in considering the observational results obtained thus far. In each case strong assumptions about the rate of galactic evolution underlie the correction factors that have to be inserted into the equations derived.

The only exception to this requirement is the steady state model where we can simply assume that locally observed conditions at the observing epoch are and have been characteristic of all portions of the universe for all time. Even here, however, we still have the additional difficulty that observations are carried out in a different spectral range from that at which radiation is emitted by distant objects. For observations in the visual part of the spectrum, we would therefore need to know a great deal about the ultraviolet behavior of sources at least in the observer's vicinity. Such information is just becoming available from observations that now can be made through use of orbiting astronomical observatories. It therefore appears that much of the cosmological interpretation of observational data may have to await a more complete understanding of the ways in which local extragalactic objects emit radiation in wavelength ranges not carefully studied thus far.

10:7 TOPOLOGY OF THE UNIVERSE

Thus far we have assumed that the universe is *simply connected*—that it has the simplest *topological* structure. In the two-dimensional models, we have talked about spherical surfaces, or planes, or hyperbolical surfaces of negative curvature.

There exist more complicated surfaces some of which can be easily constructed. If we take a rectangular sheet and label the four edges *a*, *b*, *c*, *d*, as shown in Fig. 10.6(*a*) we can, first, obtain a cylindrical surface by joining edges *a* and *b*.

But there really are several ways of joining *a* and *b*. We can give the sheet a twist, as indicated by the arrows, and obtain a Möbius strip as in Fig. 10.6(*b*).

If edges *a* and *b* are joined and edges *c* and *d* are joined also, we can obtain a torus or a Klein bottle depending on whether the sheet is given no twist or one twist— Fig. 10.6(*c*).

The Möbius strip and Klein bottle are *re-entrant*. Starting on the exterior side of a surface, we are able to reappear at the starting point, but on the interior of the surface, without ever crossing an edge or perforating the surface. There is, however, a change in directions. An arrow pointing in a particular direction appears reversed when it returns to its starting point (on the opposite surface of the strip).

These models have by no means been investigated thoroughly. There are many peculiarities. In some re-entrant models, a right-handed glove reappears as a left-handed glove on reaching its starting point after a traversal of the universe.

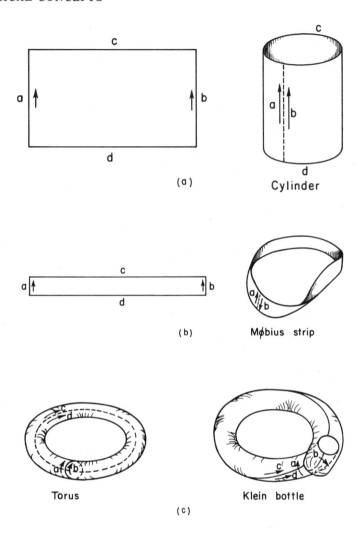

Fig. 10.6

In other models an observer may return to his own past. In still others he might return to his starting point with his arrow of time reversed with respect to his surroundings. Some spaces of negative curvature are not open when complex topological forms are allowed. Clearly much remains to be done in the topological study of the universe (He62).

10:7

10:8 DYNAMICS ON A COSMIC SCALE

In Chapter 3 the mass of galaxies was estimated using the Newtonian laws of physics. The masses of clusters of galaxies computed in this way are roughly consistent with their masses estimated on the basis of brightness. On this scale of events Newtonian dynamics therefore is probably not in error by as much as an order of magnitude. However, there are some features of Newtonian theory that lead to difficulties for phenomena on a large scale:

(a) The propagation time of gravitational signals becomes large so that forces no longer can be assumed to be transmitted instantaneously. The laws of motion should be modified to take this time lag into account.

(b) The laws of special relativity, which are well verified in the laboratory, should hold locally throughout the universe. Again, this feature is not incorporated in Newtonian dynamics and should be.

These and other difficulties are eliminated when general relativistic field equations are used to replace Newton's laws of motion. That does not mean however that general relativity itself might not have drawbacks in treating phenomena on a very large scale. General relativity has been tested out on a scale no larger than the solar system $[o(10^{13} \text{ cm})]$ and it is not clear that the same laws hold on the scale of the universe $o(10^{28})$ cm. Few laws of physics span such large ranges.

General relativity may also run into difficulties in very compact exploding objects—like the universe itself when it was only $\sim 10^{-23}$ sec old—or compact imploding objects like black holes. Bahcall and Frautschi (Ba71b) have pointed out that velocity differences of the order of the speed of light would exist, in such dense exploding or imploding states, across distances of 10^{-13} cm, the size of an elementary particle. This shows that quantum effects should be incorporated into a theory that deals with highly compact states.

10:9 SOME SIMPLE MODELS OF THE UNIVERSE

If we understood the proper dynamics to be used in dealing with phenomena on the scale of the entire universe, we could describe the evolution of different models and give their histories. Despite the fact that the problem of dynamics has not yet been finally solved, we can make the best use of provisional information in order to at least enumerate a number of theoretical models of evolving universes. We can then fit the observed characteristics of the universe to such models and an effort can be made to reject those that do not fit observations.

10:9

(a) Steady State Universe

First described by Bondi and Gold (Bo48) and by Hoyle (Ho48), this universe is flat and provides the same aspect at all times and in all places. The expansion rate is uniform in space and time. Old galaxies and young ones are statistically distributed in some fixed ratio at all distances from an observer.

(b) Static Universe of Einstein

Before the expansion of the universe had been discovered, Einstein (Ei17) proposed a cosmic model based on his general relativistic field equations. This model was static (not expanding). General relativistic dynamics allowed Einstein to calculate the density of such a universe, in terms of its radius, since a static situation will arise for one density value only, if the pressure in the universe is taken to be negligibly small. In fact, most relativistic models assume the pressure to be negligible for dynamic purposes. Observations allow this assumption, and the calculations are greatly simplified if pressure can be neglected. The Einstein universe is spherical ($k = 1$) and has constant radius of curvature a (Fig. 10.7, page 458).

In 1930, Lemaître and Eddington discovered that the Einstein Universe is unstable (Ed30). A small deviation from the perfect conditions postulated by Einstein would result in either a continuing expansion or else an accelerating collapse. They set up a model invoking this instability. It is particularly interesting because galaxies might be able to form at the unstable stage.

(c) De Sitter Model

Shortly after Einstein proposed the static model in 1917, de Sitter (deS17) pointed out that the general relativistic field equations permitted the description of a second model, one that existed in a flat space, $k = 0$. This was an expanding model. At first it had no more than academic interest; but in the late 1920s, Hubble's discovery (Hu29) of the cosmic expansion revived interest in the model. Its main drawback is that the density of such a universe must be zero. However cosmic densities are low anyway and this has not been considered an overriding difficulty.

(d) Eddington Model

This model of a universe starts out in an Einstein state. It then becomes perturbed by processes involving the formation of the galaxies from an initially uniform distribution of gas, and goes over into a uniform expansion. One difficulty with

10:9

it is the uncertainty whether the formation of galaxies should not lead to an instability that would give rise to contraction rather than expansion. The model is interesting because it focused attention on the fact that cosmology is not just a matter of geometry. A model must also be able to account for the physical state of matter found in the universe. Galaxies are likely to have condensed out of an initially uniform gas: If this gas was in rapid expansion, how was it possible to counteract the expansion in order to force the gas to contract into galaxies? We do not know the answer; but Eddington and Lemaître tried to make a plausible guess.

(e) Lemaître model

Lemaître (Le50) has suggested another model. The universe starts out in a highly contracted state and initially expands at a rapid rate. The expansion is slowed down and brought to a halt in a state that is nearly identical with the Einstein state. Galaxies form at this stage and give rise to a new expanding phase that continues indefinitely (Fig. 10.7, page 458).

An interesting feature of Lemaître's model is the high initial density of the universe. We can compute the temperature and pressure that must have existed at that time, and thus establish the nuclear reactions that should have taken place. We obtain information about the chemical composition of matter at early stages of the universe before galaxies had a chance to form. We may expect this composition to be identical with the chemical composition of the surface material in some of the oldest stars observed in the galaxy.

This points out the importance of the chemical characteristics of the universe. A cosmological model must be able to simulate not only the overall density and pressure of the universe, or the existence of galaxies, it also must be able to predict in some detail the chemical composition of matter from which the earliest stars were formed. The chemical composition of later stars need not be predicted in this way, because nuclear reactions take place in the interior of stars so that early generations of stars can produce heavier chemical elements that later are distributed in interstellar space through ejection or explosion. Subsequent generations of stars can incorporate this newly formed matter and their atmospheres will contain these chemical elements.

(f) Friedmann Models

The relativistic cosmological models described so far have had one feature in common. They all involve a nonzero *cosmological constant* Λ in the relativistic field equations to be discussed below (10–44, 10–45). This constant corresponds to a tension in the cosmic substrate so that work has to be done on the universe in order to expand it; alternately work can be derived during an expansion, depending only on the value of Λ. Friedmann (Fr22) sets this constant equal to zero, essentially denying its existence.

Whether such a constant should be used has been a hotly debated point for many years. Thus far its use or neglect has been a matter of taste. There is some hope, however, that the actual value of Λ can be determined in part from a study of the dynamics of galaxies within a cluster. The presence of a Λ term produces changes in the virial equation, so that the kinetic energy of galaxies would no longer need to be exactly half of $-\mathbb{V}$, the potential energy of the cluster of galaxies (Ja70b). Just exactly how readily such effects could be observed with present techniques is still controversial.

The Friedmann models can have Riemann curvature $k = -1$ or $+1$. Some of the models start in an extremely dense state and continue to expand. Others start out in a dense state, expand, eventually start contracting, and collapse back into the initial dense state. This cycle might repeat itself and such models then oscillate. The state of the nuclear matter after repeated cycles is not known; some attempts have been made to see whether the observed chemical composition of the universe is consistent with such cycling. In principle an oscillating universe could have existed in the indefinite past and might continue to exist into the indefinite future—much as a steady state universe does. Although the oscillating model need not postulate the creation of matter, it must be able to assure the proper chemical composition following a collapse phase.

Models (b) to (f) all have similar mathematical forms: Robertson (Ro33) and Walker (Wa34) found that in an isotropic homogeneous space the field equations of Einstein reduce to two simple differential equations in the radius of curvature a:

$$\kappa \rho = -\Lambda + 3\left(\frac{k + \dot{a}^2}{c^2 a^2}\right) \tag{10-44}$$

$$\frac{\kappa P}{c^2} = \Lambda - \left(\frac{2a\ddot{a} + \dot{a}^2 + k}{c^2 a^2}\right) \tag{10-45}$$

Here κ is $8\pi G/c^2 = 1.86 \times 10^{-27}$ cm g^{-1}, where G is the (Newtonian) gravitational constant; κ is sometimes called the *Einstein gravitational constant*. Λ is the cosmological constant, and ρ and P are the density and pressure of the universe. Dots represent differentiation with respect to world time.

PROBLEM 10-8. For *Einstein's universe* a = constant and $k = 1$.
   (i) Show that

$$\Lambda = \frac{1}{c^2 a^2} + \frac{\kappa P}{c^2} \tag{10-46}$$

and that the density of the universe is fixed at

$$\rho = \frac{2}{c^2 a^2 \kappa} - \frac{P}{c^2} \tag{10–47}$$

We know that $P/c^2 \ll \rho$, at present. If we lived in an Einstein universe, Λ would have to have a value $\sim c^{-2} a^{-2}$, and $\rho \sim 2(c^2 a^2 \kappa)^{-1}$.

(ii) Show that if $k = 0$ and $P = 0$ a static universe would require $\Lambda = 0$ and $\rho = 0$, leaving a undefined.

PROBLEM 10–9. De Sitter's universe is flat and empty, $k = \rho = P = 0$. Show that the scale factor a of the expanding universe obeys

$$a = a_0 e^{(\Lambda c^2/3)^{1/2} t} \tag{10–48}$$

and the age of the universe judged from the Hubble constant H is

$$\frac{1}{H} = \sqrt{\frac{3}{\Lambda c^2}} \tag{10–49}$$

PROBLEM 10–10. From the field equations with cosmological constant $\Lambda = 0$ and with zero pressure, obtain the relations

$$\frac{\ddot{a}}{a} = -\frac{4\pi G \rho}{3} \quad \text{and} \quad H^2 q_0 = \frac{4\pi G \rho}{3} \tag{10–50}$$

and, from (10–40), thus obtain also

$$(2q_0 - 1) = \frac{k}{H^2 a^2} \tag{10–51}$$

PROBLEM 10–11. Prove the instability of the Einstein universe for $P = 0$. Note that an infinitesimal expansion makes $\rho < 2(\kappa c^2 a^2)^{-1} - Pc^2$, so that even with $\dot{a} = 0$, we have $\ddot{a} > 0$ and the expansion must continue. The proof for an initial contraction is similar.

PROBLEM 10–12. Prove for a Friedmann universe ($\Lambda = 0$) that

(a) with $k = 1$ and an initially dense universe for which $P/c^2 = \rho/3$, the solution has the parametric form

$$a = b_0 \sin x, \quad t = b_0 (1 - \cos x) \tag{10–52}$$

where x is a parameter.

Fig. 10.7 Some cosmological models. The scale of $a(t)$ and of t is not the same for different curves. The only important consideration in this figure is the shape of each curve, rather than its exact dimensions.

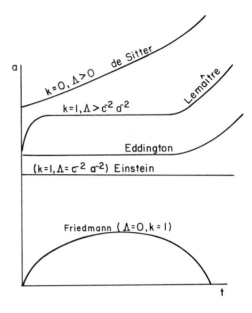

(b) For late stages of the universe, with $k = 1$, $P = 0$

$$a = a_0(1 - \cos x), \qquad t = a_0(x - \sin x) \qquad (10\text{–}53)$$

Note that x grows monotonically with t, so that $(1 - \cos x)$ eventually must approach zero. The universe first grows, but later on collapses.

(c) For the dense stage, of a *hyperbolic universe* $(k = -1)$

$$a = b_0 \sinh x, \qquad t = b_0(\cosh x - 1) \qquad (10\text{–}54)$$

(d) For the late stage of a hyperbolic universe

$$a = a_0(\cosh x - 1), \qquad t = a_0(\sinh x - x) \qquad (10\text{–}55)$$

10:10 OLBERS'S PARADOX

Suppose that a Euclidean space were uniformly filled with stars. Light emitted by stars in a shell at distance r to $r + dr$ from an observer would be proportional to $4\pi r^2\, dr$. Of this light, a fraction proportional to $1/r^2$ would be incident on the observer's telescope, since light intensity drops as the inverse square of the distance. From each spherical shell of thickness dr an observer would therefore receive an amount of light proportional to dr alone. On integrating out to infinite distance, we find that the light received by the observer should have infinite brightness. This infinity arises only because we have not taken into account the self-shadowing of stars. A foreground star will prevent an observer from seeing a star in a more distant shell, provided both stars lie along the same line of sight. When shadowing is taken into account we find that the sky should only be as bright as the surface of a typical star, not infinitely bright. Of course, that still is much brighter than the daytime sky; and the night sky is fainter still.

To someone who strongly believed in Euclidean space and in the infinite size and age of the universe this would appear paradoxical. Olbers who first advanced the argument in 1826 saw that such a cosmological view could not be held.

If we try to circumvent the argument by introducing curved space, no advance can be made. In such a space the area of a sphere drawn about an observer is of the form of equation 10–13 or 10–20—the surface area $S = 4\pi a^2 \sigma^2(\chi)$ is a function of distance χ alone. The number of stars in a spherical shell is proportional to $S(\chi)\,d\chi$. But the amount of light reaching the observer from that shell is also reduced by a factor $S(\chi)$, and these two factors cancel to give the same distance independence obtained for a flat space.

We could next argue that interstellar dust might absorb the light. But in an infinitely old universe, dust would come into radiative equilibrium with stars and would emit as much light as was absorbed. The dust would then either emit as brightly as the stars, or else it would evaporate into a gas that either transmitted light or else again emitted as brightly as the stars.

As long as galaxies themselves are distributed more or less randomly, this argument for a bright night sky remains valid, and only the overall space density of stars in the universe need be taken into account. That stars aggregate in galaxies, rather than being homogeneously distributed, does not affect the validity of the argument.

Unless we wish to suggest that no laws of physics hold for phenomena on such a scale—and then of course we must resign ourselves never to bother with cosmology—we are forced into one of three conclusions (Bo52), (Ha65):

(a) The density or luminosity of stars at large distances diminishes.
(b) The constants of physics vary with time.
(c) There are large systematic motions of stars that give rise to spectral shifts.

Argument (a) would hold, for example, if the universe were very young—stars would only have been radiating a short time.

Argument (b) forms the basis of some cosmologies that postulate that such quantities as the gravitational constant might vary from one epoch to the next. Since these constants affect the rate at which stars emit light, it might be that stars only started shining brightly in recent times. The universe would then not be filled with as much radiation as Olbers calculated.

10:10

Argument (c) states that an expanding universe need not be bright since the radiation from distant galaxies is less intense by the time it reaches the observer. As photons reach the observer from points closer to the *cosmic horizon*, where the red shift of galaxies approximates infinity, their energy and arrival rate approaches zero. In fact, no energy will reach the observer from beyond a certain distance. What that distance is depends on the individual cosmological model and is discussed in the next section.

Arguments (a) and (c) are most commonly accepted as solutions to Olbers's paradox. This paradox is useful because it places quite stringent conditions on cosmological models. A model that is to be taken seriously must assure that the night sky remains dark.

10:11 HORIZON OF A UNIVERSE

When a man on a cruising ocean liner wants to determine the distance of the horizon, he need only drop a buoy overboard and determine the buoy's range at the last instant before it disappears over the horizon. If the man is quick enough, he may then be able to shin high up on the ship's mast and briefly see the buoy again before it finally disappears over the horizon a second time. Two points are worth noting.

First, the distance of the horizon depends on the position of the observer. If a preferred horizon distance is to be defined, it should be selected in terms of some fundamental observer placed at some specific height above the ocean surface.

Second, no matter how high above the surface the observer climbs, there is an absolute horizon beyond which he can never see. He cannot see further than halfway to the antipodal point. His absolute horizon divides the surface of the earth into two hemispheres.

An observer placed at a given location in a universe will also be able to define a horizon beyond which he cannot see. In fact, there are a number of ways in which a horizon can be specified. The distance to the horizon may depend on the speed at which the observer is moving, so that a horizon is best defined in terms of a fundamental observer who is at rest with respect to the mean motion of galaxies in his local environment.

W. Rindler (Ri56)* has given a classification of horizons in different cosmological models. He defines three kinds of horizons: an *event horizon*, a *particle horizon*, and finally an *absolute horizon*:

(a) In some cosmological models, galaxies at large distances recede from an observer at ever-increasing speeds. The steady state universe is an example of such a model. In such a universe there will exist a world time t_1 (Fig. 10.8a) at which the distance of a particular galaxy P from the observer A increases at pre-

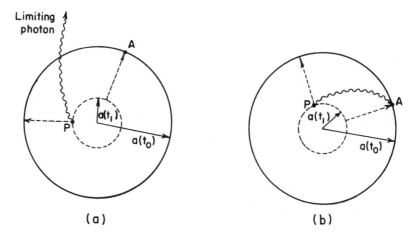

Fig. 10.8 (*a*) *Event horizon.* Light emitted at *P* will reach *A* at world time ∞. Events which occur beyond *P* will lie beyond *A*'s event horizon, as will events at *P*, after t_1. Events at *P* before t_1 reach *A* in finite world time. The limiting photon asymptotically moves parallel to *A* for increasing values of $t_0 - t_1$.

(*b*) *Particle horizon.* Effective locus of light in a four-dimensional expanding spherical universe. A particle formed at *P* at world time t_1 does not appear on *A*'s horizon until time t_0.

cisely the speed of light.† Before t_1 the galaxy can emit radiation that eventually reaches the observer; but after t_1 radiation emitted by the galaxy cannot reach the observer because the intervening distance is increasing at a rate greater than the speed of light. Events prior to t_1 can therefore be transmitted to the observer whereas events subsequent to t_1 must forever remain hidden from him. Events occurring just before t_1 will appear highly red shifted and dilated in time. The time dilatation dictates that events occuring right at time t_1 will only become accessible to the observer an infinite time after t_1 and of course events occuring after t_1 remain altogether inaccessible to him. It is interesting to note that particles that have at some time been visible to an observer remain so forever although they appear increasingly faint and red shifted. We can now define an *event horizon* as a hypersurface in space and time that divides all events into two classes: those that have been, are, or will be observable to a given observer, and those forever inaccessible to him.

(b) In other cosmological models a different type of horizon is important.

---

† There is no conflict between such a rapid increase in distance and the special relativistic statement that speeds greater than the velocity of light cannot be attained. Special relativity is valid only as a *local theory* holding the vicinity of any given point in the universe. By vicinity we mean a domain small enough so that the cosmic expansion can be neglected. Over larger domains where the expansion is significant, a less confined theory is required. One approach is provided by the general theory of relativity.

Take an explosive model in which matter initially is highly compact. At zero time the universe suddenly explodes. Two particles, P and A, that initially were quite far apart may recede from each other at nearly the speed of light. Because of the large separation, light initially emitted by particle P may not reach A for a long time. In particular, light emitted by P at time t_1 may not reach an observer at A until time t_0. Before t_0 the observer is quite unaware of the existence of particle P. After t_0 he can receive messages emitted at P. Essentially particle P enters the observer's horizon at time t_0 (Fig. 10.8b).

We can now define a *particle horizon* for any fundamental observer and for given world time t_0. It is a surface that divides all fundamental particles into two classes: those that have already been observable and those that have not.

It is clear that there can exist models in which both particle and event horizons exist. The Lemaître model is of this kind. Because there is an initial explosion from a compact state, a particle horizon will exist; because there is a subsequent acceleration of galaxies, following the period of galaxy formation in the Einstein state, an event horizon will also come into play.

(c) We may wonder about the distance of the horizon from a moving observer. Clearly, if the observer accelerates himself toward a fast receding galaxy, his event horizon can be extended. A number of results can be proven (Ri56):

(i) In a model without event horizon, a fundamental observer can sooner or later observe any event.

(ii) In a model having an event horizon, but no particle horizon, an observer can be present at any one specified event, provided he is willing to travel and provided he starts out early enough.

Statement (i) depends on the inability of particles to recede at a speed greater than light when no event horizon exists. Statement (ii) hinges on the fact that any given particle must have been within an observer's event horizon at some time in the distant past.

(iii) In a model with both event and particle horizons an observer originally attached to a fundamental particle finds that there exists a class of events absolutely inaccessible to him, no matter how he travels through space. This class of events defines an absolute horizon as shown by the following argument.

Suppose a fundamental observer were placed at some position A in the universe. There can then exist a critical particle P that initially recedes at exactly the speed of light and that enters A's particle horizon at time $t = \infty$. Let the initial distance (10–25) between P and A be D. Next, consider a fundamental observer at a point B situated along the line of sight AP but at a distance D beyond P. Again, P will enter B's particle horizon at $t = \infty$. By moving at the speed of light toward P, observer A would find P stationary and he would reach P at $t = \infty$. B would be receding at the speed of light relative to P and would, therefore, enter A's particle horizon at time $t = \infty$; but all particles beyond B would for-

Fig. 10.9 The absolute horizon for an observer at A (see text).

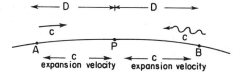

ever remain inaccessible to A. Position B defines an absolute horizon for a fundamental observer at an initial position A (Fig. 10.9).

10:12 COSMOLOGICAL MODELS WITH MATTER AND ANTIMATTER

Antimatter is hard to detect at a distance: Antihydrogen gives off a spectrum identical to hydrogen and hence a distant galaxy would look the same whether it was composed of matter or antimatter.

For these reasons we have no way of knowing whether our universe consists only of normal matter—protons and electrons as observed here on earth—or whether there might not be large amounts of antimatter in the universe.

Either way, there are difficulties. If there is no antimatter, or very little, then how are we to account for this cosmic asymmetry? What was it that decided the universe should consist predominantly of protons and electrons, when the probability of forming antiprotons and protons should seemingly be the same. We see no obvious explanation!

We might at first think that protons and antiprotons could have been formed randomly and with equal probability. In that case, just as in a random walk process (Chapter 4) we would expect an excess of either particles or antiparticles. If annihilation on a large scale had subsequently led to the destruction of matter through the reaction

$$\mathscr{P} + \overline{\mathscr{P}} \rightarrow \text{pions} \rightarrow \text{muons} + \text{neutrinos} \longrightarrow \text{electrons} + \text{positrons}$$
$$\text{gammas} + \text{neutrinos} \quad (10\text{--}56)$$

we might now be left with only those particles that had been produced in preeminence, apparently the protons. This idea founders because, in a random walk process the fluctuation—the deviation from equality after N^2 steps—is N. Since the universe now contains $N \sim 10^{78}$ protons, the initial number of protons and antiprotons would have had to be 10^{156}. The destruction of such a vast number of particles would however have produced an overwhelming amount of radiation and there certainly is no evidence for that. The fluctuation hypothesis must therefore be discarded.

10:12

It is still possible to argue that matter and antimatter might somehow become separated into galaxy or star-sized objects that had low probability of interaction. In that case, we would have the problem of explaining how this separation comes about. Matter and antimatter do not seem to repel each other gravitationally (Sc58a, see section 3:7) and some other explanation must be found.

If it really is abundant we might search for cosmic antimatter in two ways. First, cosmic ray particles produced in regions of antimatter might reach the earth from time to time. The fraction of antiparticles in the cosmic ray flux would then be an indicator of cosmic abundance. Unfortunately, the highest energy cosmic ray particles have not been analyzed in this respect. Such experiments are not yet possible. The lower energy cosmic rays do not show antiprotons in abundance, but these particles may have a more local origin and the presence of antigalaxies can therefore not be excluded in this way.

The other way of identifying antiprotons would be through their annihilation at boundaries where matter and antimatter meet. This annihilation would eventually lead to $\lesssim 100$ Mev gamma rays. Thus far we have no good observational data on that radiation. The question therefore still is open and it is not impossible that the universe could consist half of matter, half of antimatter.

A theory combining the ideas of H. Alfvén and O. Klein (K171)* suggests such a state. The primordial substance consisting half of matter, half of antimatter is called *ambiplasma* in this theory. This model of the universe is taken to be composed of isolated "metagalaxies" so that different portions can be separated by cosmic horizons and do not communicate. The theory then postulates that our local portion started out quite large, collapsed gravitationally until the ambiplasma started reacting violently as the density increased and the probability for proton-antiproton annihilation became large. A substantial part of the matter was thus destroyed and this made sufficient energy available for the subsequent expansion of the universe.

Some interesting conclusions can be drawn if we assume only that the contraction did not proceed up to the point where the Schwarzschild radius R_s (8–111) would have been reached. This means that the radius R at the time of greatest contraction (see Figure 10.10) obeys the inequality:

$$\frac{2MG}{R} \lesssim c^2 \qquad (10\text{--}57)$$

Here M is the mass of the observed portion of the universe.

An argument suggests that R was not much greater than R_s: If we take $M \sim 10^{54}$ g, our best guess from present day observations, then we find $R \sim 10^{26}$ cm at the time of greatest concentration. If much of the matter is to have been destroyed suddenly about the time the radius was R, we must have had an anni-

Fig. 10.10 Radius of the *Klein-Alfvén Universe* as a function of time.

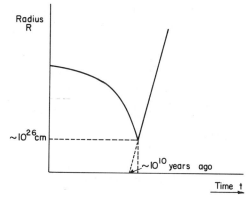

hilation probability of order unity as each particle moved through distance R:

$$n\sigma R \sim 1 \qquad (10-58)$$

Here σ is the cross section for annihilation and for collisions that randomize the motion and permit eventual annihilation. n is the number density at that epoch:

$$n \sim \frac{M}{R^3 m_H} \qquad (10-59)$$

m_H is the proton mass. Combining the last two equations gives

$$\frac{M\sigma}{R^2 m_H} \sim 1 \qquad (10-60)$$

If this is combined with the inquality for the radius, we obtain, after cancelling terms, that

$$\frac{\sigma}{4M m_H} \gtrsim \frac{G^2}{c^4} \qquad (10-61)$$

This of course is a verifiable statement, as the authors of the theory point out. Inserting the known values for σ, m_H, G, and c we find that for $\sigma \sim 10^{-25}$ cm^2

$$M \lesssim 10^{55} \text{ g} \qquad (10-62)$$

which appears to agree with observations. The authors thus feel that a relation between cosmic and atomic quantities has been established and that of course is very interesting.

It is intersting to note that if M were somewhat larger than postulated above. or

$$M > \frac{c^4 \sigma}{4G^2 m_H} \qquad (10-63)$$

10:12

the collapse would go through the Schwarzschild singularity and we would not be here today. We only escape this fate because M is small enough—although barely so!

Even if our portion of the universe had collapsed there might have been others left that remained intact because their masses had been low enough. Hence not everything would be lost in such a universe that has many separate, almost independent portions. A universe in which there were many black holes would conceptually not be too different. It also would have portions between which communication could not be established.

10:13 GALAXY FORMATION

The question of galaxy formation is closely related to the problems of cosmic evolution. The study of galaxies and their past history is therefore not only of intrinsic interest, but also is a means of gaining insight into the type of universe we inhabit.

Two almost unrelated problems actually are involved. First, we have to understand how a sufficiently large amount of matter can be aggregated in a small enough volume to give rise to galaxies. It is of course possible that the galaxies were or are spontaneously formed and that matter, at least for some time, flows out of the volume occupied by the nucleus. This would be acceptable in some forms of the steady state theory. However, usually we have considered galaxies as objects formed from the general cosmic medium, and we have sought inherent instabilities capable of giving rise to galaxies in such a medium. The stability question here is somewhat more involved than in the star formation problem (section 4:20), first, because there are difficulties in defining a suitable potential energy per unit mass of cosmic material and, second, because the rapid expansion of the universe seems to endow the medium with such stability against contraction that sufficiently rapid growth of condensations is hard to understand. We will discuss this problem in some detail below; but first we note that there is a second way in which galaxy formation entails deeper insight into the evolution of the universe. For, on the assumption that enough matter can be brought together by some means or other, we can next ask how this matter behaves once it has aggregated. How does the halo containing old stars form in the galaxy, and how does the galactic disk evolve? If that were understood (see Fig. 10.11a for one possibility) we would better understand the formation of chemical elements and so be able to define a proper age for the Galaxy as shown in section 10:14 below. We would be able to understand at what epoch and perhaps at which cosmic density the initial instabilities set in. Our understanding of dynamic processes on a cosmic scale would therefore be sharpened appreciably.

10:13

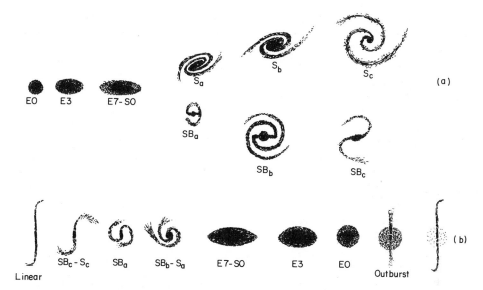

Fig. 10.11 Possible evolutionary trends in galaxies: (*a*) An evolutionary sequence due to Hubble. (*b*) A pattern in which a symmetric outburst from a nucleus develops into a barred spiral that winds itself into a regular spiral and then diffuses into an elliptical and eventually spherical aggregate before another outburst takes place. It is also possible that spiral galaxies always remain spiral and that ellipticals always remain elliptical. We do not yet know.

Let us look at some of the alternatives more closely.

(a) Possibilities That Galaxies Are Formed Through Explosions

In the late 1950s Ambartsumian in the Soviet Union suggested that galaxies might be formed, not through the condensation of extragalactic material, but rather through the outflow of material from selected regions. He envisaged galaxies as being formed in pairs through an explosive process. In support of this hypothesis we can cite a number of observational facts.

(i) Galaxies never appear to occur singly. They are only found in pairs or in larger aggregates.

(ii) Some pairs or multiple galaxies are joined by bridges of luminous matter. In a few cases the velocities of the galaxies along the radial direction alone are of the order of several thousand kilometers per second so that it is not likely that these galaxies are gravitationally bound. They seem therefore to have originated recently—perhaps as complete or nearly completely formed galaxies.

(iii) In general, the masses of galaxies that are members of a physically well-isolated group or cluster seem to be smaller than the mass that would be required to bind the galaxies gravitationally. The virial expression (3–38) for the relationship between potential and kinetic energy could therefore only be obeyed if there were a great deal of unseen matter between the galaxies binding them together. The total mass needed would be about an order of magnitude greater than the visible mass if Newtonian dynamics still holds on that scale.

(iv) That some large-scale explosive activity takes place in galaxies is shown by studies of such galaxies as M82 (Ly63) and M87. The first of these is found to be ejecting large quantities of hydrogen from its nucleus into interstellar space. The second object is shooting out a jet of gas containing relativistic particles that can be observed through the synchrotron radiation they emit.

(v) Quasistellar objects too are sometimes associated with jets (Ha63). The high velocity absorption lines observed in these objects suggest that matter is being exploded outward at speeds of the order of 10^4 to 10^5 km sec^{-1}. These quasars might be representing the formation of galaxies.

If this type of explosive origin is the way in which galaxies are actually formed, see Figure 10.11(b) for example, then much of the difficulty encountered in the condensation theories might be obviated. However, there does not seem to exist enough data at present to permit quantitative estimates of the consequences of such catastrophic formation. We have no good theoretical guides against which to test the hypothesis of explosive origin.

(b) Formation of Galaxies in a Steady State Universe

The formation of galaxies in a steady state universe presents problems if an explosive origin is ruled out. For, in that case, the rate of formation of condensations is completely specified by the value of Hubble's constant H. This comes about because in a steady state universe every feature must reproduce itself—somehow—within a time (see section 10:4)

$$\tau_s = (3H)^{-1} \tag{10–64}$$

To demonstrate the difficulty, we take the most favorable simple case in which a galaxy of mass M is already present and the formation of a new galaxy merely requires that extragalactic matter fall into the galaxy at a rate sufficiently great so that at the end of time τ_s the total gravitationally bound mass be $2M$. Once the mass of a galaxy is doubled, we could envisage a subsequent split into two—the formation of an additional galaxy. Clearly this is more favorable than the spontaneous formation of galaxies in a part of space where no initial attracting "seed" galaxy is present. But the accretion of enough matter before such a fission is a prerequisite.

10:13

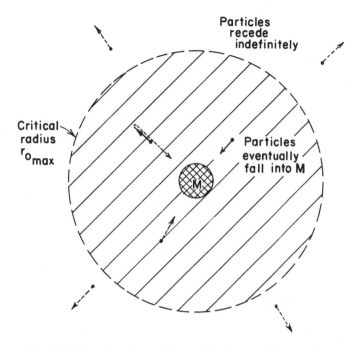

Fig. 10.12 Accretion of matter into a galaxy in a steady state universe.

The most favorable situation will be that in which the external gas has no velocity other than its recession velocity. If there were additional thermal velocities, these also would act to resist gravitational contraction. We therefore assume that the cosmic gas is created at zero temperature. First, we set up a model in which Newtonian dynamics is still obeyed. This can be done because the velocities and masses considered are small. The potential will be central. Close to the mass M a particle will be predominantly under the influence of M. At large distances it will be predominantly under the influence of the cosmic repulsion (Fig. 10.12). Conservation of energy then requires the velocity relationship

$$\left(\frac{dr}{dt}\right)^2 - \left(\frac{dr_0}{dt}\right)^2 = H^2(r^2 - r_0^2) + 2GM(r^{-1} - r_0^{-1}) \qquad (10\text{--}65)$$

where G is the gravitational constant, r is the instantaneous particle distance from M, and r_0 is an initial distance. We imagine the situation in which, before some time t_0, all particles have velocities Hr. At $t = t_0$ the gravitational field

10:13

of M is switched on. For $t > t_0$ we then have

$$\left(\frac{dr}{dt}\right)^2 = H^2r^2 + 2GM(r^{-1} - r_0^{-1}) \tag{10-66}$$

Second, if r_0 is small, the attractive force of the galaxy is able to decelerate and eventually reverse the initial outflow of matter. Particles, initially at large distances will forever continue to flow away. But there will be some terminal particle initially at some distance $r_{0\,max}$, whose velocity will just be reduced to zero at $t = \infty$. The maximum amount of matter that can possible be gathered into the galaxy in time τ_s then is $4\pi r_{0\,max}^3 \rho/3$, if matter created within the sphere of radius r_0 is neglected. We will return to this point later after computing $r_{0\,max}$.

We do this by noting that particles, initially at distances less than $r_{0\,max}$, will have negative velocities $V \equiv (dr/dt) < 0$ at $t = \infty$, while those initially beyond $r_{0\,max}$ will have $V > 0$. We therefore seek a solution to equation 10–66 that has $V^2 = 0$ and $d(V^2)/dr = 0$ simultaneously.

PROBLEM 10–13.   Show that this happens for

$$r_{0\,max} = \left(\frac{8GM}{27H^2}\right)^{1/3} \tag{10-67}$$

Different versions of the steady state theory propose overall densities of $3H^2/8\pi G$, and $3H^2/4\pi G$, and some versions propose no definite density at all. If we take the larger value $3H^2/4\pi G$, we find that the amount of matter initially contained in the sphere $(4\pi/3)\,r_{0\,max}^3$ is $8M/27$. This represents the maximum amount of matter that can fall into the condensation of mass M, in time τ_s, no matter what the value of M might be.

If we now add to this all the additional mass that might have been created within $r_{0\,max}$ during time τ_s, we could conceivably double the accreted amount of matter, but we would still have a total mass well below the required amount M. One can show more rigorously that galaxy formation would require a density of at least $\rho_{min} = 15H^2/4\pi G$ for a steady state universe in which galaxies are formed through gravitational forces alone (Ha61). For current estimates of Hubble's constant, this corresponds to a density of about 10^{-28} g cm^{-3} which is more than two orders of magnitude greater than the density estimated to be resident in condensed matter (galaxies). Although not absolutely ruled out by current observations, this density is nevertheless very high. If galaxies form in a steady state universe, through the collapse of intergalactic gas, they probably form nongravitationally.

10:13

(c) Formation of Galaxies in an Evolutionary Universe

The difficulties inherent in steady state universes are also present in evolutionary models. The cosmic expansion makes condensation of matter very difficult if only gravitational self-attraction is considered. A difficulty in the case of evolving universes is that no large-scale condensations initially exist to provide condensation nuclei around which galaxies could form. Such nuclei are possible only in steady state universes since those have always existed. In an evolving universe, however, we start out with a hypothetically uniform medium in rapid expansion and this medium may not be sufficiently unstable to form galaxies.

If we deal with the set of Friedmann universes, then the process of galaxy formation appears to present great difficulties. In a classical paper published in 1946, Lifshitz (Li46) analyzed the stability of this family of cosmic models and showed that instabilities could simply not grow rapidly enough to give us anything like the concentration of matter observed in galaxies. He based his conclusion on an analysis of the most general type of perturbations—seed disturbances—that might grow into larger condensations. His assumption, however, was that these disturbances grow independently of each other. More recently the nonlinear superposition of a number of such perturbations has been considered. There, the possibilities seem more favorable, and perhaps galaxy formation would be possible (Ko69).

Otherwise, evolving universes appear capable of forming galaxies through gravitational condensation only in Eddington-Lemaître models where a sufficiently long sojourn in the Einstein state might permit galaxies to form. In that case, however, formation would only have been possible in the past. The apparent formation of galaxies at the present time, as suggested by the observations (i) to (v) in subsection (a) above could not be explained by this means.

Moreover, many cosmologists feel uncertain about whether the cosmic constant Λ could actually be nonzero in the Einstein field equations. It is this constant, of course, that allows the Lemaître universe to go through a quiescent stage before proceeding to a second phase of rapid expansion.

From all this, it is clear that the gathering of enough matter to form a galaxy is difficult in any cosmological model. So stringent are the conditions that perhaps quite severe limitations will be found that all cosmological models must obey if galaxies are to exist at all. One aim of theories of galaxy formation is to ascertain these limitations and thus to further sharpen our understanding of the universe we inhabit.

(d) Formation of our own Galaxy

In our own galaxy we do have some additional clues about how the initial birth took place. The information is primarily given by the orbital parameters of the very oldest stars in the Galaxy. These stars are deficient in metals, and this gives them an excessively large ultraviolet magnitude relative to more recently formed metal-rich stars. The difference between U and B magnitudes, $\delta(U - B)$, increases with increasing age.

We can now look for the orbital characteristics of stars in the solar neighborhood. This was done by Eggen, Sandage, and Lynden-Bell (Eg62). As indicated in Fig. 10.13, they found that the oldest stars in the Galaxy have highly eccentric, low angular momentum orbits with high velocities perpendicular to the Galactic plane. All this suggest an initial, almost radial collapse toward the center—or

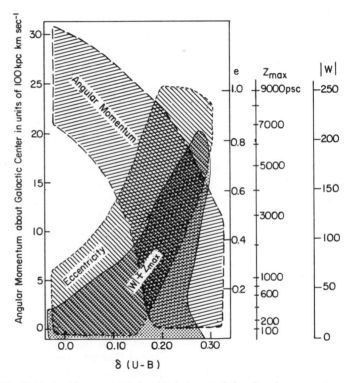

Fig. 10.13 Age of stars and their orbital characteristics: Angular momentum about the Galactic center, eccentricity e of the orbits about the center, the velocity $|W|$ perpendicular to the Galactic plane, and height z_{max} risen above the plane. Large values of $\delta(U - B)$ are associated with old stars in the Galaxy. (After Eggen, Sandage, and Lynden Bell, Eg62. See text. With the permission of the University of Chicago Press.)

10:13

possibly an initial, almost radial ejection from the center. As pointed out by Oort (Oo65), the angular momentum per unit mass, measured about the Galactic center, is a factor of eight lower for highly metal-deficient RR Lyrae stars than for disk or spiral arm stars in the neighborhood of the sun. How the disk population could have been formed from matter initially related to the halo population therefore is a real puzzle. Perhaps the origins of the halo and disk stars are quite distinct. Von Hoerner (voH55) found that globular clusters have orbital characteristics similar to the halo stars.

An approximate age for the Galaxy can be obtained on the assumption that *carbonaceous chondrites*, a class of meteorites that often impact on the earth, represent material from which the solar system was formed. We analyze for the content of thorium and uranium isotopes, Th^{232}, U^{235}, and U^{238}. These elements are formed in the r-process (section 8:12) in a ratio 1.6:1.6:1. Their α-decay *half-lives* are 1.4×10^{10}, 7.1×10^{8}, and 4.5×10^{9} y, respectively. The present $Th^{232}:U^{238}$ ratio is somewhat uncertain, but the best guess seems to be 3.3:1. The $U^{238}:U^{235}$ ratio is 1:0.007.

PROBLEM 10–14. If 60% of the uranium and thorium was formed in the birth of the Galaxy and 40% was formed continuously between the birth of the Galaxy and the solar system birth $\sim 5 \times 10^{9}$ y ago, show that the Galaxy would only be $\sim 7 \times 10^{9}$ y old. If all the metal formation had taken place continuously, at an even rate, the Galaxy would be 20 aeons old. This age still is controversial (Di69).

10:14 DO THE CONSTANTS OF NATURE CHANGE WITH TIME?

How do we know that the speed of light has been the same throughout the history of the universe? Or has it, in fact, always been the same? Does Planck's constant or the gravitational constant, or the charge of the electron change very slowly but yet significantly on a cosmic scale? In fact, can one even answer such questions?

The first person to worry about this problem and to come up with some quantitative indicators was P. A. M. Dirac (Di38). He noted that the constants of nature could be arranged in groups that were dimensionless numbers of order 10^{39} or $(10^{39})^2 = 10^{78}$. In general, such numbers can be constructed by taking the ratio of a cosmic and a microscopic quantity. We do not, of course, expect to get ratios of precisely 10^{39} in each case, but the exponents cluster remarkably closely around the numbers 39 and 78.

Dirac argued that if this was no coincidence, then it indicated that microscopic and macroscopic—atomic or subatomic and cosmic—quantities were interrelated. Since the universe is expanding, then because of the changes in the size of the cosmos, there should be corresponding changes on an atomic scale. In fact, we can see how large this change would be by constructing some dimensionless quantities from the ratio of the radius of the universe to atomic or nuclear lengths:

(a) The radius of the universe is believed to be of order 10^{28} cm.

(b) The radius of the Bohr orbit of an atom is

$$\frac{\hbar^2}{me^2} = 5 \times 10^{-9} \text{ cm} \tag{7-4}$$

(c) The Compton wavelength of the electron is

$$\lambda_c = \frac{\lambda_c}{2\pi} = \frac{\hbar}{mc} = 4 \times 10^{-11} \text{ cm} \tag{6-166}$$

(d) The classical electron radius is $\dfrac{e^2}{mc^2} = 3 \times 10^{-13}$ cm $\qquad$ (6-168)

(e) Nuclear dimensions also are of the order of 10^{-13} cm $\qquad$ (7-7)

By taking the ratios of (a) to (b) through (e), we get numbers ranging from 10^{36} to 10^{41}.

Dirac argued that since these numbers are dimensionless, they should somehow not change with time. They should stay constant as the universe evolves because a pure dimensionless number has no time or length dependence built into it. Just how the interaction between cosmic and microscopic quantities takes place to keep these dimensionless quantities constant in an evolving universe is not known. Generally, gravitational fields have been considered the only suitable candidates ever since Einstein tried to incorporate Mach's principle into his general relativistic theory of gravitation. In fact, Dirac's hypothesis of constant dimensionless numbers is another version of Mach's principle and its quest to unite the very large-scale behavior of matter with what happens on a local scale. That gravitation may in fact play a role is hinted at by a dimensionless length that can be constructed through use of atomic and gravitational physical constants. We note that none of the ratios (b) through (e) included the gravitational constant G. We can however construct the ratio

(f) $$\frac{\hbar^2}{m^2 m_p G} = 10^{31} \text{ cm} \tag{10-68}$$

Here m_p is the proton mass and this length represents the radius that a hydrogen atom would have if electromagnetic forces were absent and only gravitational forces held the atom together. The ratio of the gravitational to the electromagnetic

Bohr orbit then is just

$$\frac{e^2}{mm_pG} = 10^{39} \qquad (10\text{--}69)$$

This brings up the point that electromagnetic forces and gravitational forces also differ in strength by the same 39 orders of magnitude; for, the ratio of the lengths (b) and (f) happens to be the ratio of electromagnetic force to gravitational force of attraction between proton and electron: $F_E/F_g = e^2/mGm_p$.

If this ratio is to be constant, then we would expect the mass and charge of the electrons to change if cosmic evolution affected the value of the gravitational constant.

What about the ratio, say, of cosmic to atomic or nuclear mass? This is a particularly interesting quantity because the mass of the universe divided by the mass of the proton gives just the number, N, of atoms in the universe. We obtain

$$N = \frac{M}{m_p} = 10^{78} \qquad (10\text{--}70)$$

The puzzling thing about this ratio is that the flow of particles across the cosmic horizon would clearly destroy its constancy on a time scale of the order of an inverse Hubble constant $T = H^{-1} = 4 \times 10^{17}$ sec. Over a period of 10^{10} y, N would change appreciably. Would not that perhaps destroy Dirac's argument that these large dimensionless constants must not change?

There are two ways to answer that. First, we might decide that outflow beyond the cosmic horizon does indeed weaken Dirac's argument. On the other hand, we can also state that Dirac's ideas fit perfectly into a steady state theory of the universe. There the constant replenishment of matter keeps N constant and, in fact, keeps all cosmic parameters constant so that microscopic quantities need not change at all.

This is an attractive feature of the steady state theory because, as we shall see, however crude the observational indications we do possess, they all suggest that the physical constants of nature have not changed with time.

Once we have talked about the dimensionless ratios of forces, masses, and lengths, we have effectively constructed most, if not all, of the independent quantities that are available from the ratios of the parameters: time, length, and mass. After all, we normally express all physical entities in terms of these three basic parameters.

Nevertheless, a couple of other dimensionless numbers will still be presented, partly because of their general interest and partly because all the ratios thus far constructed depend on the electromagnetic and gravitational properties of matter; we have not worried much about strong and weak nuclear interactions.

We therefore might still look at dimensionless constants constructed from the ratios of different times scales:

(a') The cosmic time scale, as already mentioned, is $T = H^{-1} = 4 \times 10^{17}$ sec.

(b')
$$\frac{e^2}{mc^3} \sim 10^{-23} \text{ sec} \tag{10-71}$$

(c')
$$\frac{\hbar}{mc^2} \sim 10^{-21} \text{ sec} \tag{10-72}$$

These last two numbers are not entirely independent. They are related through the fine structure constant $\sim 1/137$. These short times are characteristic of nuclear interactions where time scales of the order of 10^{-22} sec are common. Again, we note that the ratio of the cosmic to microscopic time scales is of order 10^{38} to 10^{41}.

The fact that nuclear dimensions and time scales are not too different from some of the purely electromagnetic ones comes about because of the comparable strength of *strong* and *electromagnetic interactions*. The difference is only about three orders of magnitude, while here we talk about some 39 orders.

The weak interactions have not been considered at all here and, perhaps, they would not even fit into a coherent picture of interactions between cosmos and atom. Perhaps we must not expect too much from a simple idea!

Finally, we can still mention the dimensionless constant that can be formed using only gravitational and cosmic parameters

$$\rho_0 \frac{G}{H^2} \approx 1 \tag{10-73}$$

where ρ_0 is the density of the universe. This is an observational result. Although it is true that the value of $\rho_0 \sim 10^{-28}$ g cm^{-3}, which would be needed to make the relation exact, is high compared to the estimated density of galactic mass $\sim 10^{-30}$ g cm^{-3}, we must remember that galactic mass may provide only a fraction of the total mass content of the universe. Such a density also represents the value that would be needed to make the universe spherical as, for example, in an Einstein or related model of the universe.

So much for an enumeration of the coincidences that make the dimensionless numbers of nature 10^{39} or small powers (including the zeroeth power) of this number. Many scientists regard them as no more than coincidences. Others feel they are suggestive of underlying basic relationships that should be explained. We will not enter this argument, but rather go on to the observational searches for evolutionary changes in the constants of time that have been inspired by Dirac's ideas.

How fast would we expect the constants to change? We do not know of course, but the inverse Hubble constant H^{-1} might provide a suitable unit of time in terms of which changes should be measured.

If the gravitational action of the universe produces all of the changes on a microscopic scale, it might be interesting to see whether the gravitational constant had appreciably changed during recent aeons. Edward Teller (Te48) first analyzed this question. He looked for climatological changes on earth as a function of time. These would be related to the solar luminosity changes and changes in the earth's orbit resulting from time variations in the constant G. This argument is quite complex, but indicates changes—if any—of $(dG/dt)/G \lesssim 10^{-10}$ per year.

More recent radar observations of the constancy of the orbits of Venus and Mercury (Sh71) indicate similar results and involve no assumptions beyond Newtonian gravitational theory. These results will probably be refined by an order of magnitude in the next few years.

An interesting study on the possible variability of Planck's constant with world time was done by Wilkinson (Wi58). He was interested in the integrated effect of changes in Planck's constant over a period of the order of the age of the earth. The age of the earth and of meteorites can be determined separately from a number of different radioactive decay schemes, some of which involve alpha-particle emission and others beta-decay. These two processes have quite different physical bases, and we would not expect the ages given by beta- and alpha-decay schemes to be the same if the constants of nature varied appreciably.

The evidence cited by Wilkinson comes from a study of *paleochroic haloes*. These haloes are spherical shells observed in rocks that have small inclusions of radioactive material. As the material decays, any alpha particles will give rise to a thin visible shell at the end of the particle's path through the rock where most of the energy is dissipated. Corresponding to individual alpha-particle velocities v, we then obtain individual shells. These shells are easily identified with given α-decay schemes. Two interesting statements can then be made.

(i) The physics of charged particle transit through material, a process that is purely electromagnetic, is invariant over a period of order 2×10^9 y or perhaps slightly more. Otherwise the shells would be diffuse, not thin. This is interesting because the α-decay scheme discussed by Wilkinson involves both electromagnetic and nuclear forces.

(ii) Some alpha-emitting nuclei also have the possibility of emitting through beta-decay. The ratio of these two decay rates is called the *branching ratio*. Wilkinson was able to make the statement that if the branching ratio had increased or decreased by an amount of order 10 over the past 2×10^9 y, we would have found that certain alpha particle that caused haloes should have been absent and others again much stronger. Since no such anomalies were found, any changes taking place over the past few aeons in the many fundamental physical constants involved probably were small.

The search for changes in the fundamental physical constants has only begun. This study may eventually lead to a better understanding of the universe.

10:15 THE FLOW OF TIME

We tend to think that time always increases. But with respect to what? And what do we mean by "increase"?

The simplest answer would be to say that time is that which is measured by a clock. We know how clocks work, and that is that! Of course (section 3:10) there are different types of clocks and we might wish to compare them to see whether they all are running at the same rate or whether there might be, say, a systematic slowing down of one type of clock relative to the others.

Here we have assumed that all possible clocks will always run in one direction only. In that case, however, we would never be able to decide whether time is running "forward" or "backward" because these two directions would be indistinguishable.

For gravitational and electromagnetic processes, we do not know how to define a direction of time's flow. The physics that describes the orbiting of the earth around the sun holds equally whether the earth moves in a direct or retrograde orbit about the sun. Under a time reversal, too, the earth and all the other planets would return along the same orbits that had led them to their current positions in the solar system. But such orbits would be no different from a set of future orbits that could have been predicted from a simple reversal of all velocities involved.

Similarly we could use the orbital motion of an electron in a magnetic field to define time. Here it is interesting that the orbit in which the electron travels is identical to one that a positron would travel if it were going along the same path backward in time.

Both these examples show a basic symmetry that seems to pervade all natural physical processes: that if we reverse the flow of time T, reverse the sign of the electric charge of matter C, and reverse the sign of all positions and motions, (this is also called an inversion) P, then the observed results are indistinguishable from an original process in which none of these reversals or reflections took place. The operation P is called the *parity operation*; C is called *charge conjugation*; and T is the *time-reversal operation*. A fundamental theorem of physics requires that under the combined operations CPT all physical processes remain invariant.

Because of these symmetries, it is apparently impossible for us to know whether we are living in a world in which time is running forward and the universe is expanding, or whether time is running backward and the universe is contracting. These cosmic motions are independent of electric charge so that a charge conjugation would not be noticeable either. We would just assume we were made up of matter, but actually it might be what we currently call antimatter.

How, then, do we actually determine the direction in which time is flowing?

For a long time it was suggested that the second law of thermodynamics defines the direction of time in a unique way. The law states that as time increases, any isolated system tends toward increasing disorder. Light initially concentrated near the surface of a star flows out to fill all space. The reverse never happens. Light-filling space simply does not converge and flow into a single compact object. Such ordered motions, although strictly permitted by a simple time reversal argument, do not happen often. They are possible but highly improbable. The second law of thermodynamics basically states that as time increases, greater randomness comes about because there are many states of a system in which the system is disordered and only few in which it has a high degree of order. If any given state is as likely to occur as any other state, then the chances are that the evolved system will be found in one of the many disordered states rather than in one of the very few ordered configurations.

We could argue, however, that the second law of thermodynamics really is a consequence of the cosmic expansion: An undisturbed system shows no systematic evolution with time. The cosmic expansion provides just the disturbance needed. Because the universe expands, there is always more empty space being created and starlight can flow into this volume to fill it. Distant galaxies are red shifted, the sky looks black, and thermodynamic imbalance is actively maintained. In a static universe, equilibrium could be attained and the direction of time lost.

Let us look at the reverse of all this: If the universe were to collapse, distant galaxies could be approaching us at great speeds and we would observe them highly blue shifted. The night sky would be bright and perhaps light would flow from the night sky into stars rather than the other way around. Under these conditions would the second law still hold? Or would we find that physical processes tended toward greater order as the universe collapsed?

If tendency toward disorder depends on the cosmic expansion, as suggested by T. Gold (Go62), then the flow of time is well correlated with the flow toward order; however, we still have no way of telling whether time is running forward or backward because a reversal of time might be correlated with cosmic collapse, decrease of randomness, blue shifting of galaxies, and so on.

What we think we see in the universe then depends on how we define the flow of time. If we first fall down the steps and then hurt ourselves, the universe is expanding, time flows "forward," and physical systems tend toward randomness.

10:15

If we first hurt ourselves and then tumble upstairs, time is actually running "backward," the universe is contracting and physical systems are becoming more and more ordered. It may just be a matter of definition.

Since we do not like to leave such a fundamental question in such an unsatisfactory state of arbitrariness, we wish that there might be more straightforward ways of determining the direction of the flow of time. A possibility of this sort has come into sight in the past few years.

In 1956 Yang and Lee (Le56) pointed out that parity might be violated in weak interactions. This was swiftly verified in a variety of experiments. It was then realized that invariance under a combination of operations *CP* seemed to hold universally true. This would mean that reversing the charge on an object would produce a mirror image of its initial physical behavior. In electromagnetic cases this certainly is true. A positron moving in a negative direction traces out the same path as an electron moving in a positive direction at the same velocity through the same electric and magnetic fields. The same rule seemed true also for other types of interactions, including the "weak" interactions of which beta-decay is an example. *CP* invariance, together with the above-mentioned *CPT* invariance, indicated that the laws of physics still would remain invariant in time.

In the mid-1960s, however, a group at Princeton University (Ch64b) discovered that violations of *CP* symmetry occasionally occurred in the decay of neutral *K*-mesons. This would indicate that time reversal might also be violated in these reactions since otherwise *CPT* invariance would not hold and the laws of physics then would be in very fundamental difficulties. If time reversal symmetry really were violated, a preferred direction of time could at least be defined, and the discussion of the flow of time might be made easier. A series of experiments searching for time reversal asymmetries has therefore been attempted. Thus far no violations of symmetry have been found.

To show the type of behavior we are searching for, an example may be useful.

If we look at the decay products of the lambda particle (which always has zero charge)

$$\Lambda^0 \rightarrow \mathscr{P} + \pi^- \tag{10–74}$$

into a proton and a negatively charged pion, we are dealing with a weak interaction process. Initially the spin of the lambda particle can be considered to be upward, as shown in Fig. 10.14(*a*). The proton moves away perpendicular to that spin direction, and its own spin is unknown. It may have components *a*, *b*, *c* along three mutually perpendicular directions as shown. Under time reversal all the spin directions change and so does the direction of the velocity. Magnitudes of the components, however, do not change. Rotating the whole process about the *c*-axis by 180° gives picture (*c*), in which the Λ° particle has the same initial spin direction, the proton again comes off in the same direction, but the component

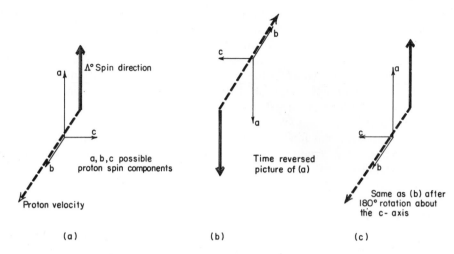

Fig. 10.14 Decay of a Λ^0 particle with spin in an "upward" direction into a proton and pion. The proton velocity vector is shown by a dashed line. Its spin components are indicated by solid lines and the lambda spin direction by iterated arrows. The decay seen in a time reflection transformation is shown in (b) and after a further 180° rotation in (c). If the process is to appear time-reversal invariant, the proton spin component c must have zero value (after Overseth Ov69). (By permission of Scientific American Inc)

c of the spin is now reversed. Since the 180° degree rotation that has been introduced is only a matter of viewpoint, we would expect that pictures (a) and (b) would be identical if time reversal symmetry held. For this to be true, however, the c component of the proton spin would have to vanish. Time-reversal symmetry in Λ^0-decay would therefore only be preserved if the c component of the proton spin always had a value zero. Experiments in which the proton is scattered off carbon targets allow one to determine the direction of the proton spin. Such experiments thus far have shown the c component to actually be zero, except for a predictable deviation due to the strong interaction of the pion and proton at the moment of decay. This calculable correction can be subtracted from the experimentally observed c values to find that the c component would actually have zero value, as required by time-reversal invariance.

At the same time, one does find violations of parity for the Λ^0 decay, because under such an inversion all linear momenta change directions; but spins, which can be thought of as the vector product of two directed quantities, do not change sign. If parity were to remain unviolated, the proton velocity and the spin component b should always have the same relationship and that could happen only if b were to have the value zero. In actual fact its value is not zero. Sometimes it lies along the velocity vector and sometimes in the opposite direction but one of these directions is preferred. Parity is therefore violated in this process, but we do not yet know whether the combination *CP* is violated also. A complication is that *T* symmetry is often difficult to check where *CP* violation is known to occur and vice versa. There appears, however, to be a possibility that *T* symmetry violations will be uncovered in the next few years and a preferred direction of time might then become established. What is clear is that the time-symmetry problem is one of observational physics and need not necessarily remain in the domain of philosophy.

It also appears that fundamental processes on a nuclear scale seem closely related to important structural features of the universe. We wish we understood the interrelation!

ANSWERS TO PROBLEMS

10–1. $x_4^2 = a^2 - r^2$, $r^2 = x_1^2 + x_2^2 + x_3^2$

$$dx_4^2 = \frac{(r - dr)^2}{a^2 - r^2}, \quad dx_1^2 + dx_2^2 + dx_3^2 = dr^2 + r^2\, d\theta^2 + r^2 \sin^2\theta\, d\phi^2.$$

This leads to (10–11) which gives (10–9) on substituting (10–12).

10–2. The radius of the circle is $a\chi = $ constant.
From (10–9) we see that an element at distance $a\chi$ has length $dl^2 = a^2 \sin^2 \chi (\sin^2 \theta\, d\phi^2 + d\theta^2)$:

(i) If we chose $\theta = \pi/2$, as the plane of the circle, the length of the circle is $\oint dl = \int_0^{2\pi} a \sin \chi\, d\phi = 2\pi a \sin \chi$. The ratio of circumference to length therefore is $\chi^{-1} 2\pi \sin \chi \le 2\pi$

(ii) The area of an element on the sphere is

$$d\sigma = (a \sin \chi \sin \theta\, d\phi)(a \sin \chi\, d\theta).$$

The area of the whole sphere therefore is

$$\iint a \sin \chi \sin \theta\, d\phi a \sin \chi\, d\theta = 4\pi a^2 \sin^2\chi.$$

(iii) The element of three-dimensional volume suggested by (10–9) is

$$dV = (a\, d\chi)(a \sin \chi\, d\theta)(a \sin \chi \sin \theta\, d\phi) = a^3 \sin^2\chi \sin \theta\, d\theta\, d\phi\, d\chi$$

from which (10–14) and (10–15) follow.

10:15

10–3. (i) $x_4^2 = -a^2 - r^2,$ $\quad dx_4^2 = -\dfrac{r^2\, dr^2}{(a^2 + r^2)}$

$$dl^2 = dr^2 + r^2(\sin^2\theta\, d\phi^2 + d\theta^2) - \frac{r^2\, dr^2}{a^2 + r^2}.$$

This is equivalent to (10–18). With $r \equiv a \sin h\chi$

(ii) $\quad dl^2 = a^2 \sin h^2\chi\,(\sin^2\theta\, d\phi^2 + d\theta^2) + \dfrac{a^2\,(d\sin h\,\chi)^2}{1 + \sinh^2\chi}$

but $1 + \sinh^2\chi \equiv \cosh^2\chi$ and $d\sinh\chi = \cosh\chi\, d\chi$, so that we obtain (10–19). For $a\chi = $ constant, $\theta = \pi/2$, $dl = a \sinh\chi\, d\phi$, and a circle has circumference $2\pi a \sinh\chi \geq 2\pi a\chi$.

(iii) The element of surface is then seen to be

$$d\sigma = (a \sinh\chi \sin\theta\, d\phi)\,(a \sinh\chi\, d\theta)$$

for a surface $a\chi = $ constant. Integration over all values of θ and ϕ gives (10–20).

(iv) Similarly the volume element $dV = (a \sinh\chi\, d\theta)\,(a \sinh\chi \sin\theta\, d\phi)$ $(a\, d\chi)$ which leads to (10–21).

10–4. $1 + z = a(t_0)/a(t_1)$. A Taylor expansion in $(t_0 - t_1)$ gives

$$1 + z = a_0[a_0 + \dot{a}_0(t_1 - t_0) + \ddot{a}_0(t_1 - t_0)^2/2]^{-1}$$

which, on expanding in $(t_1 - t_0)$ leads to (10–27).

10–5. $\sinh^2\chi = \left(\chi + \dfrac{\chi^3}{6} + \dots\right)^2 = \chi^2 + \dfrac{\chi^4}{3} + \dots = \chi^2 - \dfrac{k\chi^4}{3} + \dots, k = -1.$

$\quad\quad\ \sin^2\chi = \left(\chi - \dfrac{\chi^3}{6} + \dots\right)^2 = \chi^2 - \dfrac{\chi^4}{3} + \dots = \chi^2 - \dfrac{k\chi^4}{3} + \dots, k = 1.$

$\quad\quad\ N(\chi) = \dfrac{4\pi}{3} n\chi^3\left(1 - \dfrac{k}{5}\chi^2 + \dots\right).$

10–6. On inverting (10–27), we have

$$\Delta \equiv (t_0 - t_1) = z\frac{a_0}{\dot{a}_0}\left[1 - z\left(1 - \frac{\ddot{a}_0 a_0}{2\dot{a}_0^2}\right)\right].$$

The integral (10–24), then becomes

$$\chi \sim \int_{t_0 - \Delta}^{t_0} \frac{c\, dt}{a_0 - \dot{a}_0(t_0 - t)} \sim \frac{c}{a_0}\left[\frac{za_0}{\dot{a}_0} - \frac{z^2}{2}\frac{a_0}{\dot{a}_0}\left(1 - \frac{\ddot{a}_0 a_0}{\dot{a}_0^2}\right)\right].$$

With $\sigma(\chi) \sim \chi$ we now have

$$5\log\left(\frac{\sigma(\chi)\,a_0(1+z)}{10\text{pc}}\right) = 5\log\left\{\left[\frac{a_0 z}{\dot{a}_0}\right]\left[\frac{c}{10\text{pc}}\right]\left[1+\frac{z}{2}\left(1+\frac{a_0\ddot{a}_0}{\dot{a}_0^2}\right)\right]\right\}$$

To first order $\dot{M}_0(t_0-t_1) = -\dot{M}_0\dfrac{za_0}{\dot{a}_0}$ and $\log_{10}(1+A) = \dfrac{A}{2.303}$.

$A \ll 1$ so that $m = M_0 - 45.06 + 5\log\dfrac{a_0 z}{\dot{a}_0} + \dfrac{2.5}{2.303}\left(1+\dfrac{a_0\ddot{a}_0}{\dot{a}_0^2}-2\mu\right)z.$

10–7. $q_0 = -\dfrac{a_0\ddot{a}_0}{\dot{a}_0^2}.$

$\dot{a}_0 = a_0 H, \qquad \ddot{a}_0 = a_0 H^2 \qquad \therefore q_0 = -1.$

10–8. (i) $\Lambda = (c^2 a^2)^{-1} + \dfrac{\kappa P}{c^2}$ since $\dot{a} = \ddot{a} = 0$ in (10–45).

(10–47) then follows from (10–44) and (10–45).

(ii) The result follows from substitution of the given values in (10–45) and (10–44).

10–9.   (10–44) and (10–45) yield

$$\Lambda = \frac{3\dot{a}^2}{c^2 a^2} = \frac{2a\ddot{a}+\dot{a}^2}{c^2 a^2}$$

which has the solution (10–48). The age is $a/\dot{a} = H^{-1} = \sqrt{3/\Lambda c^2}$.

10–10. (10–44) and (10–45) give $-2a\ddot{a} = \dot{a}^2 + k = \dfrac{\kappa\rho c^2 a^2}{3}$ so that

$$-\frac{\ddot{a}}{a} = \frac{4\pi\rho G}{3} = -\frac{\dot{a}^2}{a^2}\left(\frac{\ddot{a}a}{\dot{a}^2}\right) = H^2 q_0 \text{ and } 2q_0 - 1 = \frac{-2a\ddot{a}}{a^2} - 1 = \frac{k}{\dot{a}^2} = \frac{k}{H^2 a^2}.$$

10–11.   Initially, by (10–44), (10–45) $6\ddot{a} = ac^2[2\Lambda - \kappa\rho]$. If ρ decreases because of a disturbance, $\ddot{a} > 0$, the universe expands, which decreases ρ further, and so on (Ed30).

10–12. (a) $\dfrac{\kappa P}{c^2} = -\left[\dfrac{2a\ddot{a}+\dot{a}^2+1}{c^2 a^2}\right] = \dfrac{\kappa\rho}{3} = \dfrac{1+\dot{a}^2}{c^2 a^2}$

so that initially $\ddot{a}a = -(1+\dot{a}^2)$.
We try the solution (10–52) that gives

$$\dot{a} = \cot x, \qquad \ddot{a} = \frac{-1}{b_0 \sin^3 x}$$

which satisfy the differential equation above.

(b) The trial solution (10–53) satisfies the equation

$$2\ddot{a}a = -(\dot{a}^2 + 1)$$

which follows from (10–45) at this stage of evolution. We see this since

$$\dot{a} = \sin x \,[1 - \cos x]^{-1} \qquad \ddot{a} = -[a_0(1 - \cos x)^2]^{-1}$$

(c) The equation to be satisfied now is $\ddot{a}a = 1 - \dot{a}^2$ that, following the above procedures, is satisfied by (10–54).

(d) Similarly (10–55) satisfies $2\ddot{a}a = -\dot{a}^2 - 1$ for the late stages of a hyperbolic universe.

10–13. The two requirements imposed on (10–66) are

$$V^2 = 0 = H^2 r^2 + 2GM\left[\frac{1}{r} - \frac{1}{r_{0\,max}}\right]$$

and

$$\frac{dV^2}{dr} = 0 = 2H^2 r - \frac{2GM}{r^2}; \qquad r^3 = \frac{GM}{H^2}.$$

Substituted into the first relation this gives $r = \dfrac{3r_{0\,max}}{2}$ and hence $r_{0\,max}^3 = \dfrac{8}{27}\dfrac{GM}{H^2}$.

10–14. For continuous formation of Th^{232} and U^{238} at rates dn_T/dt and dn_U/dt, respectively, since a time t in the past, the present abundance ration R should be

$$R = \frac{\displaystyle\int_0^t \frac{dn_T}{dt} 2^{-t/\tau_T}\, dt}{\displaystyle\int_0^t \frac{dn_u}{dt} 2^{-t/\tau_U}\, dt}$$

where τ_T and τ_U are the half-lives, in aeons.

$$\therefore R = \frac{14}{4.5} 1.6 \frac{[1 - e^{-0.69t/\tau_T}]}{[1 - e^{-0.69t/\tau_U}]} = 3.3$$

$\therefore t \sim 20$ æ, but this does not agree with the $U^{238}:U^{235}$ ratio.

If 60% of the material had formed at time t and 40% between t and 5 æ ago, the ratio would be

$$R = 1.6\left[0.6\frac{e^{-0.69t/\tau_T}}{e^{-0.69t/\tau_U}} + 0.4\frac{\tau_T}{\tau_U}\left[\frac{e^{-3.45/\tau_T} - e^{-0.69t/\tau_T}}{e^{-3.45/\tau_U} - e^{-0.69t/\tau_U}}\right]\right] = 3.3$$

which gives $t \sim 7$ æ.

11

Life in the Universe

11:1 INTRODUCTION

Since prehistoric times, men have always wondered about where they came from and where life originated. As it became apparent that the earth was just one planet orbiting the sun, that the sun was just one star among some $\sim 10^{11}$ in our galaxy, and that the Galaxy itself was only one such object among $\sim 10^{11}$ similar systems populating the universe, it became clear that life on other planets, near some other star, in some other galaxy was possible. The cosmological principle (section 10:3) also makes this idea philosophically attractive.

It would suggest that life is some very general state of matter that must be prevalent throughout the universe. The probability of finding some form of life, however primitive, on other planets either within the solar system or around nearby stars seems very high from this point of view. Nevertheless, it has not been possible to make unequivocal predictions about where life should exist, mainly because we do not yet understand the thermodynamics of living organisms and what different forms life may take.

11:2 THERMODYNAMICS OF BIOLOGICAL SYSTEMS

In thermodynamics one distinguishes between three types of systems. *Isolated systems* exchange neither energy nor matter with their surroundings. *Closed systems* exchange energy but not matter, and *open systems* exchange both matter and energy with their surroundings. Biological systems always are open; but in carrying out some of their functions, they may act like closed systems.

The processes that go on in living systems are also characterized by a certain

11:2

type of time dependence. Some physical processes could take place equally well whether time runs forward or backward. If we viewed a film of a clock's pendulum, we would not be sure whether the film was running forward or backward. Only if the film also showed the ratchet mechanism that advances the hands of the clock, would we be able to tell that the film was running in the right direction. The pendulum motion is reversible but the action of a ratchet is an *irreversible process*. Life processes are invariable irreversible.

In an irreversible process, *entropy* always increases. Entropy is a measure of disorder. If a cool interstellar grain absorbs visible starlight and re-emits the radiation thermally, it does so by giving off a large number of low energy photons. In equilibrium the total energy of emitted photons equals the energy of the absorbed starlight; but the entropy of the emitted radiation is larger. The increased entropy is a measure of the disorder associated with a large number of low energy photons moving in unpredictable, arbitrary directions. The initial state of a single photon carrying a large amount of energy is more orderly and, hence, characterized by a lower entropy.

Biological systems thrive on disorder. They convert order in their surroundings into disorder. In so doing, however, they also increase their own internal degree of order. The entropy in the surroundings increases, the internal entropy can decrease, but the total entropy of system plus surroundings always increases. The second law of thermodynamics, which states that the overall entropy change— of the entire universe—in any process is always positive, is therefore not violated.

It may seem strange that biological systems can increase their internal order in this way; but actually we encountered a similar process in the alignment of interstellar grains (section 9:8). We saw there that anisotropic starlight arriving at a grain primarily from directions lying within the plane of the Galaxy, tended to orient the grain so it was spinning with its angular momentum axis lying in the plane. An oriented set of grains shows greater order than randomly oriented dust; and the decrease in the grains' entropy is produced through the absorption of low entropy, anisotropic starlight and emission of high entropy isotropic infrared radiation.

These interstellar dust grains are in a state of *stationary nonequilibrium*. Such a state is characterized by transport of energy between a source at high temperature (the stars) and a sink at low temperature (the universe). There is no systematic change of the system in time, although statistical fluctuations in the orientation, angular momentum, and other properties of the grains do take place.

We hope that the study of stationary nonequilibrium processes will lead to a better understanding of the behavior of biological systems (Pr61). For, when a plant absorbs sunlight—photons whose energy typically is ~ 2 ev—and thermally re-emits an equal amount of energy in the form of 0.1 ev photons, it is acting like a stationary nonequilibrium system and, in fact, a wide range of biological processes seem to fit into this pattern.* Pendulum clocks also are stationary systems. Low entropy energy in the form of a wound-up spring is irreversibly turned into high entropy heat. As Schrödinger pointed out (Sc44), living organisms and clocks have a thermodynamic resemblance.

Nonequilibrium characterizes virtually every astrophysical situation since energy is always flowing out of highly compact sources into vast empty spaces. Any biological system stationed near one of these sources could make good use of this energy flow. It would therefore seem that the conditions necessary for the existence of life in one form or another would be commonplace. Maybe life does abound; but perhaps it is of a form we do not yet recognize.

Fred Hoyle (Ho57) has speculated that interstellar dust clouds might be alive. From a thermodynamic viewpoint the situation would be ideal. We know that dust clouds absorb roughly a 10th of the starlight emitted in a spiral galaxy like ours. The grain temperatures are so low that a maximum increase in entropy can be produced. What is uncertain, however, is whether the grains are not too cold to make good use of the available energy. At the 10 to 20°K temperatures that might be typical of interstellar grains, the mobility of atoms within the grains is so low that the normal characteristics we associate with life might be ruled out (Pi66).

A thermodynamically similar scheme has been suggested by Freeman Dyson (Dy60) who has proposed that intelligent civilizations would build thin shells around stars to trap starlight, extract useful energy, and then radiate away heat in the infrared. Perhaps some infrared sources are such objects (Sa66b).

Our experience on earth is that life will proliferate until stopped by a lack of energy sources, a lack of raw materials, or an excess of toxins. It would perhaps be surprising if no form of life had adapted itself sufficiently to make use of the huge outpouring of energy that goes on in the universe and apparently is just going to waste.

A search for unknown forms of life might concentrate on striking examples of nonequilibrium. A Martian astronomer, for example, would find only two pieces of evidence for life on earth. The first is a radio wave flux that would correspond to a nonequilibrium temperature of some millions of degrees. This is produced by radio, television, and radar transmitters. The second is an excess of methane, CH_4, which is very short-lived in the presence of atmospheric oxygen. It is converted into CO_2 and H_2O. Its nonequilibrium concentration, which could be spectroscopically detected from Mars, is rapidly replenished by methane bacteria that live in marshes and in the bowels of cows and other ruminants (Sa70b).

---

*The photochemistry of green plants of course is an enormously varied process that also centers about the buildup of large molecules.

11:2

11:3 ORGANIC MOLECULES IN NATURE AND IN THE LABORATORY

Granted that we do not specifically know how to search for exotic forms of life could we not find indications of extraterrestrial life in the form familiar on earth? All our living matter contains organic molecules of some complexity—proteins and nucleic acids, for example—and we might expect to find either traces of such molecules or at least of their decay products.

The last few years have shown two quite distinct locations in which such complex molecules are found. There may be many more. First, observations of interstellar molecules by means of their microwave spectra have revealed the existence of such organic molecules as hydrogen cyanide, methyl alcohol, formaldehyde, and formic acid (section 9:4). Water vapor and ammonia also exist in interstellar space, and it is quite likely that much larger and more complex molecules will be found in the next few years.

Second, an analysis of a meteorite—a *carbonaceous chondrite*—that fell near Murchison in Australia on September 28, 1969 showed the presence of many hydrocarbons and of 17 amino acids, including six that are found in living matter (Kv70). One such amino acid was alanine. It has the form

$$
\text{Alanine:} \qquad
\begin{array}{c}
\overset{\displaystyle O}{\overset{\|}{}} \\
CH_3-CH-C-OH \\
| \\
NH_2
\end{array}
\qquad (11\text{--}1)
$$

All *organic acids* are marked by the group of atoms

$$
\begin{array}{c}
\overset{\displaystyle O}{\overset{\|}{}} \\
-C-OH
\end{array}
$$

and *amino acids* contain the additional characterizing *amino group* NH_2.

Contamination by terrestrial amino acids seems to be ruled out by three features of these observations.

(a) Alanine can occur in two different forms. One in which the CH_3, NH_2, COOH, and H surrounding the central carbon atom are arranged in a configuration that causes polarized light to be rotated in a left-handed screw sense. The other in which polarized light would be rotated in the opposite sense. These are respectively labeled L- and D-alanine. The symbol L stands for levo—left—and D for dextro—right.

All amino acids can be derived from alanine. If derived from L-alanine such an acid is called an L-amino acid, and if derived from D-alanine, a D-amino acid. All amino acids found in proteins are L-amino acids. Although not all of them rotate light in a left-handed screw sense, they all can be structurally derived from L-alanine.

The Murchison meteorite showed D- and L-forms in essentially equal abundances. These amino acids are therefore very unlikely to have been biogenic contaminants.

On earth, amino acids are overwhelmingly in the left-handed form. Why that should be so is a mystery, because chemically speaking the right- and left-handed forms are equally probable. They are simply mirror images of each other. Perhaps evolutionary considerations have played a role. It might be that primitive life existed in a racemic mixture—having both L and D forms—and that the L-form won out in a competition for the raw materials essential to life.

It may in fact be impossible for racemic life to exist in an effective way. The search for nutrients would be inefficient. A bolt in search of a nut is more readily satisfied if all nuts and bolts have a right-handed thread; trying to match nuts and bolts from a racemic mixture would be extremely vexing.

(b) A second distinctive feature of the Murchison material was that the ratio of carbon isotopes C^{13} to C^{12} was about twice as high as normally found in terrestrial material. This too indicated that contamination could be ruled out.

(c) Finally some of the amino acids found in the material consisted of nonprotein amino acids. This could not have been a contaminant.

Biogenic molecules and molecules needed for the existence of life therefore seem prevalent elsewhere. They occur naturally not just on earth.

We still have to ask how these molecules arise. Is their fabrication simply achieved under normal astrophysical conditions? The answer to this seems to be "Yes."

A series of experiments that had their foundations in the work of Miller (Mi57a, Mi59) has shown that amino acids and other molecules found in living organisms can be produced artificially if mixtures of gases such as ammonia NH_3, methane CH_4, and water vapor H_2O are irradiated with ultraviolet radiation, subjected to electrical discharges or shocks, to X-ray, γ-ray, electron- or alpha-particle bombardment. Thus far these molecules always have been produced in racemic mixtures. Since all the gases that are used in the experiment are atmospheric constituents on a number of planets and are expected universally on cosmic abundance grounds, it seems that solar ultraviolet, X-ray, and cosmic ray bombardment, occasional irradiation by nearer supernova explosions, and other natural sources of irradiation should have been able to produce molecules of biological interest within planetary atmospheres.

This should be true not only in the solar system, but in other similar systems.

Perhaps the planets around Barnard's star have had sufficiently similar histories so that life would be expected there too.

Although such molecules are readily formed by energetic bombardment, they are also readily destroyed by it. A biogenic molecule formed in the atmosphere might therefore be destroyed unless it were rapidly removed to a safer place. On earth, rain could have washed such molecules out of the atmosphere and into the oceans where they would be shielded from destructive irradiation by a protective layer of water.

We note that the conditions for forming life—or highly ordered biogenic molecules—are those that seemed thermodynamically favorable (section 11:2). There is a source of low entropy energy in solar ultraviolet or cosmic ray irradiation and a possiblity of converting this energy into a higher entropy form through collisions with atmospheric molecules or through radiation at long wavelengths.

11:4 ORIGINS OF LIFE ON EARTH

Before we can make a rough guess about the origins of life on earth, we should know something about the earth's atmosphere during the aeons immediately following the birth of the solar system.

Initially, the atmosphere seems to have contained no molecular oxygen. The most prevalent atmospheric molecules probably were those strongly reduced by the presence of abundant amounts of hydrogen: methane, ammonia, water, and ethane.

Over one or two aeons, the atmosphere became less rich in hydrogen. Free oxygen appeared, perhaps as hydrogen was freed from water vapor through ultraviolet dissociation in the upper atmosphere. Such hydrogen atoms may have escaped the atmosphere altogether although how that happened does not seem well established (Va71).

The forms of life that would be formed under the earliest reducing conditions would be anaerobic. Presumably the very first organism to be formed found itself in a rich environment of large organic molecules (Op61a, b) that had been built up by ultraviolet irradiation and other bombarding mechanisms discussed in (section 11:3). Such an organism could feast and procreate at will, until the supply of organic molecules dwindled. Such organisms that obtain energy by breaking down pre-existing molecules are called *heterotrophs*. Clearly they would be at a disadvantage compared to *autotrophs*, organisms that in addition could also make use of energy in other forms; autotrophs that make use of sunlight are called *photoautotrophs*. The autotrophs probably soon took over. Initially they must have been anaerobes, but as hydrogen kept escaping at the top of the atmosphere and oxygen became more prevalent, the anaerobes came to be at a dis-

11:4

advantage compared to aerobes that form the basis on which all the higher organisms of today have evolved. When the oxygen concentration in the atmosphere became roughly one percent of its present abundance, respiration should have become a more efficient process than fermentation and the aerobes may have originated at that time.

Living organisms naturally suffer mutations in their genetic makeup—that is, in the code that defines the makeup of the progeny. The mutation rate can be artificially increased through X-ray and other destructive bombardment. The aerobes probably arose from anaerobes through such mutative processes. Being able to make use of atmospheric oxygen, they soon became the dominant form of life. The anaerobes nowadays can proliferate only under conditions where atmospheric oxygen is somehow excluded.

The balance between stability and mutability seems to be particularly important in forms of life that succeed. Without mutability an organism cannot adapt to changes in its environment; but without some stability, higher forms could not evolve either. In Darwin's theory of survival of the fittest, these fittest are likely to be produced through occasional mutations of rather stable forms. The death of individuals seems essential, in order that life forms may evolve. Yet, for life to evolve optimally, each fit individual should attempt to survive—resist death. Presumably there is an optimal eugenic life span. This will vary from species to species. Some male spiders are devoured immediately after mating. For men, a longer life span must be desirable since they are needed to help rear the young.

We think that an eventual grouping of small organisms into larger ones led to multicellular forms and eventually to the higher forms of life encountered today. Interestingly, irreversible thermodynamics should play a role not only in the metabolism of life, as in section 11:2. It is likely that the growth of more highly organized forms of life through mutations and consequent changes in metabolic forms and metabolic rates can also be described using the methods of irreversible thermodynamics.

The next few years sould show major advances in the thermodynamic approach to theoretical problems of life, in laboratory studies of the formation of ever more complex biological forms and in our searches for extraterrestrial evidence for life-supporting molecules.

11:5 COMMUNICATION AND SPACE TRAVEL

If life exists elsewhere in the universe, perhaps it also shows intelligence. If it is intelligent, perhaps it has organized into a civilization. How should we exchange information with it and how would others be likely to get in touch with us? (Sh66) (Dr62)

11:5

This is a problem in communications. How does one most effectively send messages over large distances? How does the enormous time lag between sending and reception of electromagnetic signals affect the problem of communication? These questions are actively being studied, but no single optimum way has yet been discovered. Much depends on what we would like to do best.

If you like to travel, perhaps a rocket journey at relativistic speeds would suit you. But then you must decide how to stay alive during the long trip. Suggestions have been made for deep-freezing spacemen who would undertake the journey. Thus far, nothing much bigger than a frog has been successfully frozen and revived, and it is not clear whether the technique could be developed for large mammals. Unmanned spaceflight or flights in which several generations would pass before landing also are possibilities.

Alternately, we might be willing to restrict ourselves to communicating through transmission of radio or visible signals, or perhaps infrared or X-ray messages. Is there any one electromagnetic frequency that is optimal? If such a frequency is found, we still must ask ourselves whether its characteristics are optimal only because of our particular technological resources, or whether there is some more fundamental reason for choosing this particular means of communicating. It is clear that if we choose the wrong frequency for transmitting our signals, no one is likely to receive them. We also are likely to miss messages sent by other civilizations, if we do not know to which frequency we should tune our receiver. We cannot tune in on all frequencies because that might require an insurmountable financial effort. We have to second-guess the correct frequency in order to make the initial communication contact more probable.

What about tachyons? In section 5:12 we mentioned these particles that would travel faster than light. Clearly, if tachyons existed they would have many desirable properties. They could travel at millions of times the speed of light and would therefore make meaningful two-way conversations a real possibility. Moreover, tachyons require only low transmission energy (5–56) and might therefore be very economical. Finally, tachyons would apparently free us from the limitations imposed by cosmic horizons (section 10:11). The one disadvantage might be that tachyons seem not to readily interact with normal matter—otherwise they should probably have been detected by now. Construction of suitable transmitters and receivers might therefore be difficult.

There are an apparently endless set of questions to be answered before an effective means for communicating with other civilizations can exist. We need only note that if tachyons could be readily produced and received, other civilizations would probably use them to the exclusion of all other means of communicating. Yet we do not even know whether tachyons can exist.

To show some of the questions that need be considered it may be worth thinking about the following two problems.

11:5

PROBLEM 11-1. A spaceship slowly accelerates on its voyage from earth to a distant galaxy. As it accelerates to ever higher speeds, it suffers collisions with interstellar gas and dust, with photons criss-crossing space, with magnetic fields, and with cosmic ray particles. Estimate the effects of these and other possible particles and fields of interstellar and intergalactic space on the momentum of the spaceship, electric charges deposited on the ship and the effect of these charges, the abrasion and ablation effects on the hull of the ship, heating effects, and so on. What are the most serious limitations? Almost everything discussed in Chapters 6 and 9 bears on this problem.

PROBLEM 11-2. The rate at which messages can be transmitted and received is normally proportional to the area of the transmitter A and to the solid angle Ω subtended by the receiver at the point of transmission. Let us now assume that the transmitted particles or waves have a momentum range Δp, and that the number of message bits that can be transmitted per unit time—the *bit rate*—equals the number of phase cells contained in the beam sent out during that period (4–65).

(a) For an electromagnetic wave show that the bit rate is

$$\text{Photon bit rate} = A\Omega \frac{v^2}{c^2} \Delta v \qquad (11-2)$$

where v is the frequency, Δv is the bandwidth of the transmitted beam, and the antenna only transmits photons of one polarization.

(b) For a tachyon system, show that if (4–65) is applicable,

$$\text{Tachyon bit rate is} \qquad \left| \frac{A\Omega}{h^3} m^3 c^4 \frac{\Delta N}{N^3} \right| \qquad N \gg 1 \qquad (11-3)$$

where we have assumed that transmission occurs for tachyon velocities ranging from $V = Nc$ to $V = (N + \Delta N)c$, where N is a large number. If the tachyon mass is of the order of the electron mass and the radiation frequency is that of visible light, show that the tachyon bit rate for $N \lesssim 10^7, \Delta N \sim 0.5N$ is several orders of magnitude greater than the electromagnetic bit rate. Show that for $N \sim 10^8$, however, the bit rate and energy expenditure would be comparable to visible light. Equation 5–56 is useful in tackling this problem.

PROBLEM 11–2 ANSWER

This problem is highly speculative, particularly in view of some of the difficulties cited in Chapter 5:

We assume that phase space arguments determine the distinguishability of tachyons and that the number of distinguishable tachyons transmitted per unit time determines the bit rate. For a receiver with area A and solid receiving angle Ω, the volume from which tachyons are received per unit time is ANc, where N is the tachyon speed measured in units c, the speed of light. The momentum space volume occupied by these tachyons is $\Omega p^2 dp$, per mode of polarization. The number of distinguishable tachyons incident on the detector in unit time (here referred to as bit rate) therefore would be

$$\left| \frac{ANc\,\Omega p^2 dp}{h^3} \right|$$

We make use of the relativistic expression

$$\mathscr{E}^2 = p^2c^2 + m^2c^4 = m^2c^4(1 - N^2)^{-1}$$

relating energy $\mathscr{E}$ and rest mass m to momentum and velocity. This leads to the (imaginary) momentum value

$$p = \frac{N}{\sqrt{1 - N^2}}\,mc$$

and the bit rate obtained for a velocity range dNc reads

$$\left| \frac{A\Omega}{h^3}\,m^3c^4\,\frac{dN}{N^3} \right| \qquad \text{for} \qquad N \gg 1$$

The corresponding expression for electromagnetic radiation is

$$\frac{A\Omega v^2 dv}{c^2}$$

where dv is the frequency of the radiation. If we take the frequency to be that of visible light, and take m to be an electron mass, the tachyon bit rate is seen to be many magnitudes greater than the electromagnetic bit rate, as long as N remains less than about 10^7 and $dN/N \sim dv/v$. At that speed, the energy per tachyon would be about $10^{-7}mc^2$ corresponding to about 0.1 ev while the visual radiation would require a transmission energy about an order of magnitude higher.

If $N \sim 10^8$, the bit rate and energy expenditure per message is comparable to that for visible light, but communication across the universe can be achieved in times of the order of 100 years.

A question about the stability of tachyons has recently (Be71) been raised. If they exist but are unstable they would not be suitable information carriers. Clearly our ideas on tachyons still are speculative.

Epilogue

At crucial points in this book we have been stopped by unsolved problems. Some of the most important questions that remain unanswered are:

(1) Do the laws of physics as we know them apply on the scale of the universe?

(2) Is there a connection between the structure of the universe and the structure of elementary particles?

(3) Does the universe have a beginning and an end in time, and what exactly is time?

(4) How are galaxies born and how do they die?

(5) How are stars formed and how do they die?

(6) What is the origin of cosmic magnetic fields?

(7) Is there a basic preference for matter over antimatter in the universe?

(8) What is the origin of life and do other intelligent civilizations exist?

(9) Are we even asking the right kind of questions?

Appendix A

Astronomical Terminology

A:1 INTRODUCTION

When we discover a new type of astronomical entity on a photographic plate of the sky or in a radio-astronomical record, we refer to it as a new *object*. It need not be a star. It might be a galaxy, a planet, or perhaps a cloud of interstellar matter. The word "object" is convenient because it allows us to discuss the entity before its true character is established. Astronomy seeks to provide an accurate description of all natural objects beyond the earth's atmosphere.

From time to time the brightness of an object may change, or its color might become altered, or else it might go through some other kind of transition. We then talk about the occurrence of an *event*. Astrophysics attempts to explain the sequence of events that mark the evolution of astronomical objects.

A great variety of different objects populates the universe. Three of these concern us most immediately in everyday life: The sun that lights our atmosphere during the day and establishes the moderate temperatures needed for the existence of life, the earth that forms our habitat, and the moon that occasionally lights the night sky. Fainter, but far more numerous, are the stars that we can only see after the sun has set.

The objects we detect can be divided into two groups. Many of them are faint, and we would not be able to see them if they were not very close to the sun; others are bright, but at much larger distances. The first group of objects, taken together with the sun, comprise the *solar system*. They form a gravitationally bound group orbiting a common center of mass. Within the solar system the sun itself is of greatest astronomical interest in many ways. It is the one star that we can study in great detail and at close range. Ultimately it may reveal precisely what nuclear processes take place in its center and just how a star derives its energy. Complementing such observations, the study of planets, comets, and meteorites may ultimately reveal the history of the solar system and the origins of life. Both of these are fascinating problems!

A:2 THE SUN

The sun is a star. Stars are luminous bodies whose masses range from about 10^{32} to 10^{35} g. Their *luminosity* in the visual part of the spectrum normally lies in the range between 10^{-4} and 10^4 times the sun's energy outflow. The *surface temperatures* of these stars may range from no more than $\sim 1000°K$ to about $50,000°K$. Just how we can determine the relative brightness of stars will be seen later in this Appendix. The determination of temperatures is discussed in Chapter 4.

The sun, viewed as a star, has the following features:

(a) Its radius is 6.96×10^{10} cm. Although occasional prominences jut out from the solar surface, its basic shape is spherical. The equatorial radius is only a fractional amount of 5×10^{-5} larger than the polar radius: $[(r_{eq} - r_{pol})/r] \approx 5 \times 10^{-5}$ (Di67b).

(b) The sun emits a total flux of 3.9×10^{33} erg sec^{-1}. Nearly half of this radiation is visible, but an appreciable fraction of the power is emitted in the near ultraviolet and near infrared parts of the spectrum. Solar X-ray and radio emission make only very slight contributions to the total luminosity.

(c) The sun's mass is 1.99×10^{33} g.

(d) We recognize three principal layers that make up the sun's atmosphere. They are the photosphere, chromosphere, and corona.

(i). The *photosphere* is the surface layer from which the sun's visible light emanates. It has a temperature of about 6000°K.

(ii). The *chromosphere* is a layer some ten to fifteen thousand kilometers thick. It separates the relatively cool photosphere from the far hotter corona.

(iii). The *corona* extends from 1.03 $R_\odot$, or about 20,000 km above the photosphere, out to at least several solar radii. The outer boundary has not been defined and there is some reason for believing that the corona merges continuously into the interplanetary gas that streams outward from the sun at speeds of several hundred kilometers per second. This streaming ionized gas, mainly protons and electrons, is called the *solar wind*. The temperature of the corona is $\sim 1.5 \times 10^6°K$.

(e) Sunspots and sunspot groups, cool regions on the solar surface, move with the sun as it rotates, and allow us to determine a 27-day rotation period. This period is only an apparent rotation rate as viewed from the earth which itself orbits about the sun. The actual rotation period with respect to the fixed stars is only about twenty-five and a half days at a latitude of 15° and varies slightly with latitude; the solar surface does not rotate as a solid shell. The sun exhibits an 11-year *solar cycle* during which time the number of sunspots increases to a maximum and then declines to a minimum. At minimum the number of spots on the sun may be as low as zero. At maximum the number of individual sunspots or members of a sunspot group may amount to 150. There are special ways of counting to arrive at this *sunspot number* and a continuous record is kept through the collaborative effort of a number of observatories.

The 11-year cycle actually amounts to only half of a longer 22-year cycle that takes into account the polarity and arrangement of magnetic fields in sunspot pairs.

(f) A variety of different events can take place on the sun. Each type has a name of its own. One of the most interesting is a *flare*, a brief burst of light near a sunspot group. Associated with the visible flare is the emission of solar cosmic ray particles, X-rays, ultraviolet radiation, and radio waves. Flares also are associated with the emission of clouds of electrons and protons that constitute a large component added to the normal solar wind. After a day or two, required for the sun-to-earth transit at a speed of $\sim 10^3$ km sec^{-1}, these particles can impinge on the earth's *magnetosphere* (magnetic field and ionosphere), giving rise to *magnetic storms* and *aurorae*. These disturbances tend to corrugate the ionosphere and make it difficult to reflect radio waves smoothly. Since radio communication depends on smooth, continuous ionospheric reflection, reliable radio communication is sometimes disrupted for as long as a day during such *magnetic storms*.

A:3 THE SOLAR SYSTEM

A variety of different objects orbit the sun. Together they make up the *solar system*. The earth is representative of planetary objects. *Planets* are large bodies orbiting the sun. They are seen primarily by reflected sunlight. The majority emit hardly any radiation by themselves. In order of increasing distance from the sun, the planets are Mercury, Venus, Earth, Mars, Jupiter, Saturn, Uranus, Neptune, and Pluto. All the planets orbit the sun in one direction; this direction is called *direct*. Bodies moving in the opposite direction are said to have *retrograde orbits*. Table 1.3 gives some of the more important data about planets. It shows that the different planets are characterized by a wide range of size, surface temperature and chemistry, magnetic field strength, and so on. One of the aims of astrophysics is to understand such differences, perhaps in terms of the history of the solar system.

A:3

Besides the nine planets we have listed, there are many more minor planets orbiting the sun. They are sometimes also called *planetoids* or *asteroids*. Most of them travel along paths lying between the orbits of Mars and Jupiter, a region known as the *asteroidal belt*. The largest asteroid is Ceres. Its radius is 350 km. Its mass is about one ten-thousandth that of the earth.

Many of the smaller known asteroids have diameters of the order of a kilometer. These objects number in the thousands and there must be many more orbiting masses that are too small to have been observed. Among these are bodies that might only be a few meters in diameter or smaller. From time to time, some of these approach the earth and survive the journey through the atmosphere. Such an object that actually impacts on the earth's surface is called a *meteorite*. Meteorites are studied with great interest because they are a direct means of learning about the physical and chemical history of at least a small class of extraterrestrial solar system objects.

Even smaller than the meteorites are grains of dust that also circle the sun along orbits similar to those of planets. From time to time a grain of dust may enter the atmosphere. Much of it may burn through heat generated by its penetration into the atmosphere, and the particle becomes luminous through combustion and can be observed as a *meteor*, historically called a *shooting star*.

In contrast to meteoritic material, meteoric matter does not generally reach the earth's surface in recognizable form. However, some fragments do appear to survive and are believed to contribute to a shower of fine dust that continually rains down on the earth. Most of this dust has a micrometeoritic origin. *Micrometeorites* are micron- (10^{-4} cm) or submicron-sized grains of interplanetary origin that drift down through the atmosphere and impinge on the earth's surface. They have a large surface-to-mass ratio and are easily slowed down in the upper atmosphere without becoming excessively hot. Once they have lost speed they slowly drift down through the air. Some of these grains may be formed in the burnup of larger meteors; others may come in unchanged from interplanetary space. Collections of these grains can be made from the arctic snows or deep ocean sediments, far from sources of industrial smoke.

The identification of this extraterrestrial dust is not simple. It is difficult to distinguish cosmic dust from dust generated in, say, volcanic explosions. Nevertheless, some researchers have claimed that the amount of dust deposited on the earth each day is of the order of several hundred or several thousand tons. These figures may be compared with dust collected by means of special satellite-borne devices; but the inter-comparison has been difficult because the measuring techniques employed in artificial satellites differ so widely from those used on earth.

A cloud of micrometeoritic dust exists in the space between the planets and possibly also as a tenuous dust belt about the earth. The dust reflects sunlight and gives rise to a glow known as the *zodiacal light*. The zodiacal light can be seen, on very clear days, as a tongue-shaped glow jutting up over the western horizon after sunset or the eastern horizon before sunrise. The glow is concentrated about the *ecliptic*, the plane in which the earth orbits the sun.

We recognize that these planetary and interplanetary objects continually interact. There are indications that planets and their satellites often collide with huge meteorites, objects that are as large as the asteroids. The surface of both Mars and the moon are pockmarked by what are believed to be impact craters. The earth too shows vestiges of such bombardment; but our atmosphere erodes away and destroys crater outlines in a time of the order of several million years, whereas on the moon erosion times are of the order of billions of years.

We should notice that in talking about planets, asteroids, meteorites, meteors, and micrometeoritic dust grains we are enumerating different-sized members of an otherwise homogeneous group. The major known difference between these objects is their size. Other differences can be directly related to size. For example, it is clear that planets may have atmospheres while micrometeorites do not. But this difference arises because only massive objects can retain a surrounding blanket of gas. The gravitational attraction of small grains just is not strong enough to retain gases at temperatures encountered in interplanetary space. The different names, given to these different-sized objects, have arisen because they were initially discovered by a variety of differing techniques; and although we have known the planets, meteorites, meteors, and other interplanetary objects for a long time, we have just recently come to understand their origin and inter-relation.

A set of objects similar to the planets are the *satellites* or *moons*. A satellite orbits about its parent planet and these two objects together orbit about the sun. In physical makeup and size, satellites are not markedly different from planets. The planet Mercury is only four times as massive as our moon. Ganymede, one of Jupiter's satellites, Titan, one of Saturn's satellites, and Triton, one of Neptune's satellites, all are nearly twice as massive as the moon. Titan even has an atmosphere. Many other satellites are less massive; they look very much like asteroids. An extreme of the moon phenomenon is provided by the rings of Saturn that consist of clouds of fine dust—micrometeoritic grains, all orbiting the planet like minute interacting moons.

Evidently there are great physical similarities between satellites and planetary objects of comparable size. The main difference lies in the orbital motion of the two classes of objects. It is interesting that an asteroid could be captured by Jupiter to become one of its satellites. The reverse process might also be possible.

The somewhat vague distinction between planets and interplanetary objects is not unique. Differences between stars and planets also are somewhat vague. We talk about *binaries* in which two stars orbit about a common center of gravity.

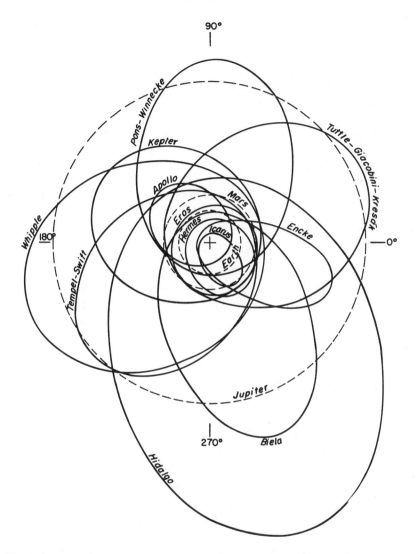

Fig. A.1 Comparison of planetary, asteroidal, and short-period cometary orbits. Although the Earth, Mars, and Jupiter have nearly circular orbits, the orbits of the asteroids, Icarus, Hermes, Eros, Apollo, Kepler, and Hidalgo are appreciably eccentric, as are those of comets Encke, Pons-Winnecke, Tempel-Swift, Whipple, Tuttle-Giacobini-Kresak, and Biela. Comets are named after their discoverers. Many comets and asteroids have aphelion distances near Jupiter's orbit, and Jupiter has a controlling influence on the shape of the orbits and may have "captured" comets from parabolic orbits into short-period orbits.

Often one of these objects is much smaller than the other, sometimes no more than one thousandth as massive. Jupiter also has about one thousandth the mass of the sun. Perhaps, as some astronomers have suggested, Jupiter should more appropriately be called a star. Clearly, size alone does not provide an appropriate distinction between stars and planets. Possibly a more pertinent criterion for recognizing a star can be formulated in terms of internal structure, and attempts to clearly state such criteria are under way (Sa70a).

We should still mention one final class of objects belonging to the solar system: the comets. Their orbits are neither strictly planetary, nor are they at all similar to those of satellites. Some comets have elliptic orbits about the sun. Their periods may range from a few years to many hundreds of years. Other comets have nearly parabolic orbits and must be reaching the sun from the far reaches of the solar system. Comets are objects which, on approaching the sun from large distances, disintegrate through solar heating: gases that initially were in a frozen state are evaporated off and dust grains originally held in place by these volatile substances become released. The dust and gas, respectively, are seen in reflected and re-emitted sunlight. They make the comet appear diffuse (Fig. A.2). Comet tails are produced when the gas and dust are repelled from the sun through electron and proton bombardment and by the pressure of sunlight. The dust from a comet tail produces a meteor shower when the earth passes through the remnants of the tail (Wa56).

A:4 STELLAR SYSTEMS AND GALAXIES

Before we turn to a description of individual stars, we should first consider the groupings in which stars occur.

Stars often are assembled in a number of characteristic configurations, and we classify these systems primarily according to their size and appearance. Many stars are single. Others are accompanied by no more than one companion; such pairs are called *binaries*. Depending on their separation and orientation, binary stars can be classified as *visual, spectroscopic* or *eclipsing*. The limit of visual resolution of a binary is given by available optical techniques. Refinements are continually being made, but interferometric techniques now allow us to resolve stars about 0.01 arc seconds apart (Ha67). For smaller separations, we cannot use interferometric techniques. The two stars in such a close pair constitute a spectroscopic binary and have to be resolved indirectly by means of their differing spectra. We sometimes encounter a special but very important type of spectroscopic binary in which the stars orbit about each other roughly in a plane that contains the observer's line of sight. One star may then be seen eclipsing the other and a change in brightness is observed. An eclipse of this kind becomes

Fig. A.2 (*a*) The Andromeda galaxy, NGC 224, Messier 31, a spiral with two smaller companion galaxies, one of which, NGC 205, an elliptical galaxy is shown enlarged (*d*). The barred spiral galaxy (*b*) is NGC 1300. Its spiral classification is SBb. These three pictures were photographed at the Mount Wilson Observatory. The globular cluster (*c*) is Messier 3 (M3), also known as NGC 5272. The comet (*e*) is comet Brooks, and the photograph was taken October 21, 1911. Only the region of the comet around the head is shown. The elongated streak at about 45° to the comet's tail is a background star. This shows that the comet's direction of motion is at 45° to the tail. The tail does not trail behind the head. Photographs (*c*) and (*e*) were taken at the Lick Observatory.

probable only when the two companions are very close together, no more than a few radii apart. We call such systems *eclipsing binaries*. Binaries are important because they provide the only means of determining accurate mass values for stars (other than the sun). How these masses are determined is shown in the discussion of orbital motions (section 3:5).

Close binaries also are important because if one of the two stars begins to expand, as it moves onto the red giant branch of the Hertzsprung-Russell diagram (section A:5g and Fig. 1.5), the surface material may become more strongly attracted to the companion star. An exchange of mass can take place, and it becomes possible for us to see some of the portions of the uncovered star that previously were in its interior. This allows us to check for systematic production of the heavy elements in the star and thus also to test the theory of chemical evolution and energy production in stars (section 8:13).

Binary stars are not the only known close configuration. There exist many ternaries consisting of three stars; and higher multiple systems are not uncommon. Perhaps one in every five "stars" shows evidence of being a binary and about one in every 20 "stars" shows evidence of being a higher order system. For stars more massive than the sun, these fractions are considerably higher; single stars occur only in one third of the observed cases. The proportion of observed binaries also differs for stars with differing spectral features.

Sometimes stars form an aggregate of half a dozen or a dozen members. This is called a stellar *group*. There also exist stellar *associations* that are groups of some 30 or so stars mutually receding from one another. Such stars appear to have had a common point of origin in the past. They are thought to have been formed together and to have become separated shortly after formation. By observing the size of the association and the rate at which it is expanding, we can determine how long ago the expansion started and how old the stars must be.

There are two principal groupings called clusters: *galactic clusters* and *globular clusters*. The galactic clusters usually comprise 50 to several hundred stars loosely and amorphously distributed but moving with a common velocity through the surrounding field of stars. In contrast, globular clusters (Fig. A.2) are much larger, containing several hundred thousand stars and having a very striking spherical (globular) appearance. Stars in a cluster appear to have had a common origin. We think they were formed during a relatively short time interval, long ago, and have had a common history.

Clusters are not just populated by single stars. Binaries and higher multiples and groups of stars often form small subsystems in clusters. Normally stars and clusters are members of *galaxies*. These are more or less well defined, characteristically shaped systems containing between 10^8 and 10^{12} stars (Fig. A.2). Some galaxies appear elongated and are called elliptical or E galaxies. Highly elongated ellipticals are classed as E7 galaxies. If no elongation can be detected and the galaxy has a circular appearance, it is called a globular galaxy and is classified as E0. Other numerals, between 0 and 7, indicate increasing apparent elongation. The observed elongation need not correspond directly to the actual elongation of the galaxy because the observer on earth can only see a given galaxy in a fixed projection.

Elliptical galaxies show no particular structure except that they are brightest in the center and appear less dense at the periphery. In contrast, *spiral galaxies* (S galaxies) and *barred spiral galaxies* (SB) show a strong spiral structure. These galaxies are denoted by a symbol O, a, b, or c following the spiral designation to indicate increasing openness of the spiral arms. In this notation, a compact spiral is designated SO and a barred spiral with far-flung spiral arms and quite open structure is designated SBc (see Fig. A.2). (See also Fig. 10.11(a), page 467.)

In comparing galaxies on a photographic plate, one expects nearby galaxies to appear larger than more distant objects; this means that the angular diameter of a regular galaxy can be taken as a rough indicator of its distance. When the spectra of different galaxies are correlated with distance, we find that a few nearby galaxies have blue shifted spectra (Bu71b); but all the more distant galaxies have spectra that are systematically shifted toward the red part of the spectrum. Galaxies at larger apparent distances, as judged from their diameters or brightness, are increasingly red shifted! This correlation is so well established that we now take an observation of a remote galaxy's red shift as a standard indicator of its distance.

Sometimes galaxies do not fall into the well-defined E, S, or SB patterns. They appear more as randomly shaped patterns. Such galaxies are classified as irregular galaxies, designated by the symbol Ir. Peculiar galaxies of one kind or another are denoted by a letter p following the type designation, for example, E5p.

Of course, galaxies do not contain stars alone. Between the stars there exists interstellar gas and dust. In some spiral galaxies the total mass of dust and gas is comparable to the total stellar mass observed in galaxies. The exact mass ratios are not known because one is not sure whether all the gas in existence has yet been detected.

Dust clouds can be detected through their extinction, which obscures the view of more distant stars. Moreover, dust near hot ionized regions absorbs radiation and emits it at long infrared wavelengths. This emission apparently is so effective that some galaxies radiate more strongly in the infrared, than in all other spectral ranges combined (Kl70a,b).

Gas also may be detected in absorption or through emission of radiation. Through spectroscopic studies in the radio, infrared, visible, and ultraviolet parts of the spectrum, many ions, atoms, and molecules have been identified, and the temperature, density, and radial velocity of such gases has been determined.

Galaxies are not the largest aggregates in the universe. There exist many pairs

A:4

and groups of galaxies. Fig. 1.10 shows one such group. Our *Galaxy*, the *Milky Way* to which the sun belongs, is a member of the *Local Group* that contains somewhat more than a dozen galaxies. The Andromeda nebula and the Galaxy are the largest members. The other members of the group have a combined mass about one tenth the Galactic mass (Table 1.4).

Larger *clusters of galaxies* containing up to several thousand galaxies also exist. Grouping on a larger scale than the scale of clusters has thus far not been definitively established and may not exist.

The scheme of classification of galaxies leaves a number of borderline cases in doubt. Small E0 galaxies are not appreciably different from the largest globular clusters. Double galaxies sometimes cannot be distinguished from irregular ones; and the distinction between a group or a cluster of galaxies may also be a matter of taste. But the classification is useful nevertheless; it gives handy names to frequently found objects without making any attempt to provide rigorous distinctions.

When we transcend the scale on which clusters of galaxies are found, we seem to be dealing with the universe as a whole. No further subdivisions are apparent. The universe can best be described as consisting of randomly grouped clusters of galaxies.

Crossing the vast spaces between the galaxies are quanta of electromagnetic radiation and highly energetic cosmic ray particles that travel at almost the speed of light. These are the carriers of information that permit us to detect the existence of the distant objects.

There is one overwhelming feature that characterizes the galaxies. Everywhere, at large distances, the spectra of galaxies and clusters of galaxies appear shifted toward the red, long wavelength, end of the spectrum. The farther we look, the greater is the red shift. Most astrophysicists attribute the *red shift* to a high recession velocity. The galaxies appear to be flying apart. The universe expands!

A:5 BRIGHTNESS OF STARS

(a) The Magnitude Scale

One of the first things we notice after a casual look at the sky is that some stars appear brighter than others. We can visually sort them into different brightness groups. In doing this, it becomes apparent that the eye can clearly distinguish the brightness of two objects only if one of them is approximately 2.5 times as bright as the other. The factor of 2.5 can therefore serve as a rough indicator of apparent brightness, or *apparent visual magnitude*, m_v of stars.

Stars of first magnitude, $m_v = 1$, are brighter by a factor of ~ 2.5 than stars of second magnitude, $m_v = 2$, and so on. The magnitude scale extends into the region of negative values; but the sun, moon, Mercury, Venus, Mars, Jupiter and the three stars, Sirius, Canopus, and Rigel Kent are the only objects bright enough to have apparent visual magnitudes less than zero.

Normally it would be cumbersome to use a factor of 2.5 in computing relative brightnesses of stars of different magnitudes. Since this factor has arisen not because of some feature peculiar to the stars that we study, but is quite arbitrarily dependent on a property of the eye, we are tempted to discard it altogether in favor of a purely decimal system; but a brightness ratio of 10 is not useful for visual purposes. As a result, a compromise that accommodates some of the advantages of each of these systems is in use. We define a magnitude in such a way that stars whose brightness differs by precisely five magnitudes, have a brightness ratio of exactly 100. Since $100^{1/5} = 2.512$, we still have reasonable agreement with what the eye sees, and for computational work we can use standard logarithmic tables drawn up to the base 10.

(b) Color

The observed brightness of a star depends on whether it is seen by eye, recorded on a photographic plate, or detected by means of a radio telescope. For different astronomical objects the ratio of radiation emitted in the visible and radio regions of the spectrum varies widely. The spectrum of an object can be roughly described by observing it with a variety of different detectors in several different spectral regions. The apparent magnitudes obtained in these measurements can then be intercompared. Several standard filters and instruments have been developed for this purpose so that we may intercompare data from observatories all over the world. The resulting brightness indicators are listed below:

m_v denotes *visual brightness.*

m_{pg} denotes *photographic brightness.* The photographic plate is more sensitive to blue light than the eye; nowadays this brightness is usually labeled B, for blue. If a photographic plate is to be used to obtain the equivalent of a visual brightness, a special filter has to be used to pass yellow light and reject some of the blue light.

V or m_{pv} denotes photovisual brightness obtained with a photographic plate and the above mentioned yellow transmitting filter. This brightness is generally denoted by V for visual. m_{pg} and m_{pv} are older notations.

U denotes the *ultraviolet brightness* obtained with a particular ultraviolet transmitting filter (Table A.1).

I denotes an *infrared brightness* obtained in the photographic part of the infrared. At longer wavelengths photographic plates no longer are sensitive, but a number of infrared spectral magnitudes have been defined so that results obtained with lead sulfide, indium antimonide, and other infrared detectors might be intercompared by different observers. These magnitudes are labeled J, K, L, M, N, and Q.

Table A.1 lists the wavelengths at which these magnitudes are determined.

Table A.1 Effective Wavelength for Standard Brightness Measurements.

Symbol	Effective Wavelength	Symbol	Effective Wavelength
U	0.3540μ	K	2.2μ
B	0.4330	L	3.4
V	0.5750	M	5.0
R	0.6340	N	10.2
I	0.8040	Q	20
J	1.25		

1μ (micron) $= 10^{-6}$m $= 10^{-4}$cm $= 10^4$ Å (angstrom units)

m_{bol} denotes the total apparent brightness of an object, integrated over all wavelengths. This *bolometric magnitude* is the brightness that would be measured by a bolometer—a detector equally sensitive to radiation energy at all wavelengths.

(c) Color Index

The difference in brightness as measured with differing filters gives an indication of a star's color. The ratio of blue to yellow light received from a star is given by the difference in magnitude—logarithm of the brightness—of the star measured with blue and visual filters. This quantity is known as the *color index*:

$$C = B - V$$

Differences such as $U - B$ also are referred to as color indexes.

The intercomparison of colors involved in producing a reliable color index can only be achieved if we can standardize photographic plates and filters used in the measurements. And even then errors can creep into the intercomparison. For this reason some standard stars have been selected to define a point where the color index is zero. These stars are denoted by the spectral-type symbol A0, (cf A:6).

A:5

(d) Bolometric Correction

Normally the bolometric brightness of a star can only be obtained by indirect means. We may be able to measure the apparent visual brightness; but to estimate the total radiative emission of the star, over its entire spectrum, we must make some assumptions about the surface temperature and emissivity. This estimate is obtained in terms of a factor known as the *bolometric correction*, BC, defined as the difference between the bolometric and visual magnitudes of a star. The bolometric correction always is positive

$$\text{BC} = m_v - m_{\text{bol}}$$

Estimation of the bolometric correction is not simple. The color index $B - V$ can be used to obtain a rough assessment of the surface temperature. We can then estimate the total radiation output, on the assumption that we understand how electromagnetic radiation is transferred through a star's atmosphere. Generally this assumption is not warranted. In the last decade, stellar brightness values, obtained in the ultraviolet and infrared regions of the spectrum, have not generally agreed with predictions based on earlier theoretical models. Of course, these new data are being incorporated into the theory of stellar atmospheres and more reliable bolometric correction values are becoming available. In the meantime, provisory tables of bolometric corrections for stars with different spectral indices are in use.

(e) Absolute Magnitude

For many purposes we need to know the absolute magnitude rather than the apparent brightness of a star. It is therefore important to convert apparent magnitudes into absolute values. We define the *absolute magnitude* of a star as the apparent magnitude we would measure if the star were placed a distance of 10 pc from an observer. (1 pc $= 3 \times 10^{18}$ cm. See section 2:2, page 54.)

Suppose the distance of a star is r pc. Its brightness diminishes as the square of the distance between star and observer. The apparent magnitude of the star will therefore be greater, by an additive term $\log_{2.5} r^2/r_0^2$, than its absolute magnitude.

$$m = M + \log_{2.5} \frac{r^2}{r_0^2} = M + 5 \log \frac{r}{r_0}$$

where the logarithm is taken to the base 10 when no subscript appears. Since $r_0 = 10$ pc, we have the further relation,

$$M = m + 5 - 5 \log r \tag{A-1}$$

A:5

Thus far no attention has been paid to the extinction of light by interstellar dust. Clearly the apparent magnitude is diminished through extinction and a positive factor A has to be subtracted from the right side of equation A–1 to restore M to its proper value

$$M = m + 5 - 5 \log r - A \tag{A–2}$$

Obtaining the star's distance, r, often is less difficult than assessing the interstellar extinction A. We discuss this difficulty in section A:6a below.

The detector and filter used in obtaining the apparent magnitude m, in equation A–2, determines the value of the absolute magnitude M. We can therefore use subscripts v, pg, pv, and bol for absolute magnitudes in exactly the same way as for apparent magnitudes.

(f) Luminosity

Once we have obtained the bolometric absolute magnitude of a star, we can obtain its total radiative emission, or luminosity, L, directly in terms of the solar luminosity:

$$\log\left(\frac{L}{L_\odot}\right) = \frac{1}{2.5}\left[M_{\text{bol}_\odot} - M_{\text{bol}}\right] \tag{A–3}$$

The luminosity of the sun, $L_\odot$, is 3.8×10^{33} erg sec^{-1} and the solar bolometric magnitude, $M_{\text{bol}_\odot}$, is 4.6. The luminosity of stars varies widely. A supernova explosion can be as bright as all the stars in a galaxy for a brief interval of a few days. The brightest stable stars are a hundred thousand times more luminous than the sun. At the other extreme, a white dwarf may be a factor of a thousand times fainter than the sun; and even fainter stars may well exist.

(g) The Hertzsprung-Russell Diagram

One of the most useful diagrams in all astronomy is the Hertzsprung-Russell, H-R, diagram. It presents a comparison of brightness and temperature plotted for any chosen group of stars. We will see in Chapter 2 that the diagram is valuable in estimating the dimensions of galaxies and intergalactic distances; more important, such H-R diagrams, obtained for different stellar age groups, provide the main empirical foundation for the theory of stellar evolution.

H-R diagrams can take many different forms. The color index is an indicator of a star's surface temperature, as shown in Chapter 4. Hence the abscissa sometimes is used to show a star's color index, and we then speak of a *color-magnitude*, instead of an H-R, *diagram*. The ordinate can show either M_v, or M_{bol} or lumi-

A:5

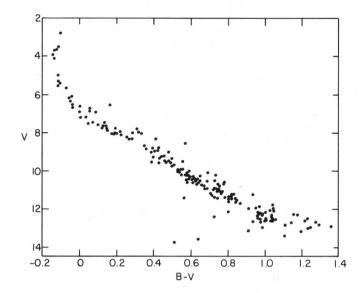

Fig. A.3 Color-magnitude diagram of the Pleiades cluster stars, after correction for interstellar extinction effects. The Pleiades cluster contains some of the most recently formed stars in the Galaxy. (After Mitchell and Johnson, Mi57b. With the permission of the University of Chicago Press.)

nosity. When only an intercomparison of stars, all of which are known to be equally distant, is needed, it suffices to plot the apparent magnitude. Figure A.3 shows such a plot of the Pleiades cluster stars. Figure 1.6 plots the characteristics of M3. M3 is an old globular cluster in our Galaxy, while the Pleiades are among the most recently born Galactic stars. This difference is reflected in the difference of the two diagrams.

These two figures, as well as Fig. 1.4, show that stars are found to fall only in a few select areas of the H-R and color-magnitude diagrams. The largest number of stars cluster about a fairly straight line called the *main sequence*. This is particularly clear for the Pleiades cluster. The main sequence runs from the upper left to the lower right end of the diagram, or from bright blue down to faint red stars. To the right and above the main sequence (Fig. 1.5) there lie bright red stars along a track called the *red giant branch*. There also is a *horizontal branch* that joins the far end of the red giant branch to the main sequence. These two branches show up particularly in Fig. 1.6. In the horizontal branch, we find some stars that periodically vary with brightness. Finally, some faint *white dwarf stars* lie below and to the left of the main sequence. The rest of the diagram usually is empty.

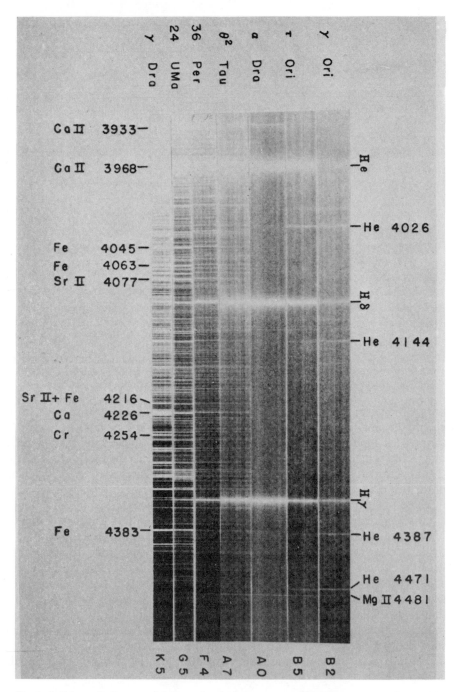

Fig. A.4 Schematic diagram of spectra of typical stars representing different spectral types. The four top spectrograms were taken by A. Slettebak (after Keenan, Ke63b. With the permission of the University of Chicago Press.)

A:6 CLASSIFICATION OF STARS

(a) Classification System

The classification of stars is a complex task, primarily because we find many special cases hard to fit into a clean pattern. Currently a "two-dimensional" scheme is widely accepted. One of these "dimensions" is a star's spectrum; the other is its brightness. Each star is therefore assigned a two parameter classification code. Although the object of this section is to describe this code, we should note that the ultimate basis of the classification scheme is an extensive collection of spectra such as those shown in Fig. A.4. Each spectrum is representative of a particular type of star.

Stars are classified primarily according to their spectra, which are related to their color. Although the primary recognition marks are spectral, the sequence of the classification is largely in terms of decreasing stellar surface temperature— that is, an increase in the star's radiation at longer wavelengths. The bluest common stars are labeled O, and increasingly red stars are classed according to the sequence (Table A.2)

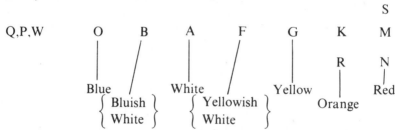

Over 99% of all stars belong to the basic series B, A, F, G, K, and M. Stars with designation O, R, N, and S are comparatively rare and so also are spectral types: Q denoting *novae*—stars that suddenly brighten by many orders of magnitude becoming far brighter than any nonvariable star.

P denotes *planetary nebulae*, hot stars with surrounding envelopes of intensely ionized gases, and W refers to Wolf-Rayet stars, which exhibit broad emission bands of ionized carbon, nitrogen, and helium. A few of these stars have been found to be members of close binaries.

The classes R and N denote stars containing unusually strong molecular bands of diatomic carbon, C_2, and cyanogen, CN. The S stars are characterized by bands of titanium oxide, TiO, and zirconium oxide, ZrO.

Stars classed as W, O, B are sometimes said to be *early types*, while stars of class G, K, M, R, N, S are designated *late types*.

A:6

The transition from one spectral class to another proceeds in 10 smaller steps. Each spectral class is subdivided into 10 subclasses denoted by arabic numerals after the letter. A5 lies intermediate between spectral types A0 and A9; and F0 is just slightly redder than A9.

Table A.2 Spectral Classification of Stars.[a]

Type	Main Characteristics	Subtypes	Spectral Criteria	Typical Stars
Q	Nova: sudden brightness increase by 10 to 12 magnitudes			T Pyx Q Cyg
P	Planetary nebula: hot star with intensely ionized gas envelope			NGC 6720 NGC 6853
W	Wolf-Rayet stars		Broad emission of OIII to OVI, NIII to NV, CII to CIV, and HeI and HeII.	
O	Hottest stars, continuum strong in UV (O5 to O9)		OII λ 4650 dominates HeII λ 4686 dominates } emission Lines narrower } lines Absorption lines dominate; only HeII, CII in emission SiIV λ 4089 at maximum OII λ 4649, HeII λ 4686 strong	BD + 35°4013 BD + 35°4001 BD + 36°3987 ζ Pup, λ Cep 29 CMa τ CMa
B	Neutral helium dominates	B0 B1 B2 B3 B5 B8 B9	CIII/4650 at maximum HeI λ 4472 > OII λ 4649 HeI lines are maximum HeII lines are disappearing Si λ 4128 > He λ 4121 λ 4472 = Mg λ 4481 HeI λ 4026 just visible	ε Ori β CMa, β Cen δ Ori, α Lup π^4Ori, α Pav 19 Tau, ϕ Vel β Per, δ Gru λ Aql, λ Cen
A	Hydrogen lines decreasing from maximum at A0	A0 A2 A3 A5	Balmer lines at maximum CaII K = 0.4 Hδ K = 0.8 Hδ K > Hδ	α CMa S CMa, ι Cen α PsA, τ^3 Eri β Tri, α Pic
F	Metallic lines becoming noticeable	F0 F2 F5 F8	K = H + Hδ G band becoming noticeable G band becoming continuous Balmer lines slightly stronger than in sun	δ Gem, α Car π Sgr α CMi, ρ Pup β Vir, α For

[a]Compiled mainly from Keenan (Ke63b) (based on Cannon and Pickering Ca24) and also from Allen (Al55). This table, which is based on the Henry Draper classification scheme, is a rough guide to the spectral features of stars. The classification of stars, however, remains an ongoing process and changes occur. (With the permission of Athlone Press of the University of London, second edition © C. W. Allen 1955 and 1963, and with the permission of the University of Chicago Press.)

A:6

Table A.2 Spectral Classification of Stars (continued):–

Type	Main Characteristics	Subtypes	Spectral Criteria	Typical Stars
G	Solar-type	G0	Ca λ 4227 = Hδ	α Aur, β Hya
	spectra	G5	Fe λ 4325 > Hγ on small-scale plates	κ Gem, α Ret
K	Metallic lines	K0	H and K at maximum strength	α Boo, α Phe
	dominate	K2	Continuum becoming weak in blue	β Cnc, v Lib
		K5	G band no longer continuous	α Tau
M	TiO bands		TiO bands noticeable	α Ori, α Hya
			Bands conspicuous	ρ Per, γ Cru
			Spectrum fluted by the strong bands	W Cyg, RX Aqr
			Mira variables, Hγ, Hδ	χ Cyg, o Cet
R,N	CN, CO, C$_2$ bands		CN, CO, C$_2$ bands appear instead of TiO. R stars show pronounced H and K lines.	
S	ZrO bands		ZrO bands	R Gem

The classification scheme also allows us to denote a star's luminosity class by placing a Roman numeral after the spectral type. Each of these luminosity classes has a name:

> I — Supergiant
> II — Bright Giant
> III — Normal Giant
> IV — Subgiant
> V — Main Sequence
> — Subdwarf
> — White dwarf

The sun has spectral type G1 V indicating that it is a yellow main sequence star.

Sometimes we find classes I, II, and III collected under the heading "giant" while stars of group V are called "dwarfs." Letters "g" or "d" are placed in front of the spectral class symbol to denote these types. Similarly placed letters "sd" and "w" denote subdwarfs and white dwarf stars. Another classification feature concerns supergiants, which are often separated into two luminosity classes Ia and Ib depending on whether they are bright or faint.

A letter "e" following a spectral classification symbol denotes the presence of emission lines in the star's spectrum. There is one exception to this rule. The combination Oe5 denotes O stars in the range from O5 to O9; it has no further connection with emission.

A letter "p" following the spectral symbol denotes that the star has some form of peculiarity.

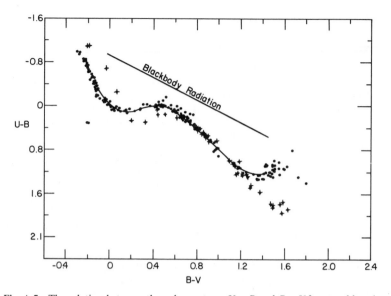

Fig. A.5 The relation between the color systems $U - B$ and $B - V$ for unreddened main-sequence stars (dots) and little reddened supergiants and yellow giants (crossed dots). The line along which blackbody radiators would fall is also shown. (After Johnson and Morgan, Jo53. With the permission of the University of Chicago Press.)

The color designation (stellar spectral type) given here is nearly linear in the color index $B - V$. It is not however linear in $U - V$ nor do the $U - V$ values decrease monotonically with increasingly late spectral type. Small differences in color indexes exist for giants and main sequence stars of the same spectral type. This unfortunate difficulty has arisen for historical reasons and should be corrected in future revisions of the spectral classification scheme. We might still see how well stellar colors approach those of a blackbody. The closeness of fit is shown in Fig. A.5, called a color-color diagram.

Four factors are responsible for the rather large deviations from a blackbody. (i) For stars around spectral type A, where the fit to the blackbody spectrum is poorest, absorption by hydrogen atoms in their first excited states produces absorption. We talk about the *Balmer jump* in connection with the sharp rise in absorption at wavelengths corresponding to the Balmer continuum produced by these excited atoms in the outer atmosphere of a star. (ii) Cool stars have H^- ions in the outer atmospheres. These ions absorb radiation selectively, making the star appear bluer. (iii) The relatively high abundance of metals in population I stars produces a number of absorption lines that change the color of a star, moving it toward the lower right of Fig. 1.5. (iv) Finally, no star looks completely

A:6

black, because its outer layers are not equally opaque at all wavelengths. Light at different wavelengths therefore reaches us from different depths within the star, and these levels are at differing temperatures. The resulting spectrum of starlight therefore corresponds to a mixture of temperatures, rather than to blackbody radiation at one well-defined temperature.

Determination of the spectral type of a star by means of its color index alone would be very difficult, since proper account would have to be taken of the changes in color produced by interstellar dust. Small dust grains tend to absorb and scatter blue light more strongly than red. Light from a distant star therefore appears much redder than when emitted. To discover the true color index of the star a correction has to be introduced for interstellar reddening. However, in order to make this correction, we have to know how much interstellar dust lies along the line of sight to a given star, and to what extent a given quantity of dust changes the color balance. None of this information is normally available. Instead, we have to make use of a circular line of reasoning. We know that nearby stars of any given spectral type exhibit characteristic spectral lines either in absorption or in emission. Since these stars are near, there is little intervening interstellar dust, and their spectra can be taken to be unreddened. We can therefore draw up tables listing the spectral line features of each color class. A distant star can then be classed in terms of its spectral lines rather than its color index and the color index can be used to verify the class assignment. If the color is redder than expected, we have an indication of reddening by interstellar dust. Whether dust is actually present can then be checked—in many instances—by seeing whether other stars in the immediate neighborhood of the given object all are reddened by about the same amount. If they are, we have completed the analysis. The results give the correct spectral identification of stars in the chosen region and, in addition, we are given the extent to which interstellar dust changes the color index. A similar analysis can also be applied to determine the extent to which the overall brightness of the star is diminished through extinction by interstellar dust. This analysis allows us to determine the amount of obscuration in all the spectral ranges for which observations exist.

As already stated, the color and spectrum of a star depends on its surface temperature. Table A.3 gives the effective temperature for representative stars. As discussed in Chapter 4, the effective temperature is measured in terms of the radiant power emitted by the star over unit surface area. Since our information about the ultraviolet and infrared emission of stars is still quite incomplete, we can expect (Da70b) that the values given in this table will change over the next few years. The uncertainties for O stars are particularly great and an effective temperature is not listed for them. The provisory nature of the data is emphasized in the table heading.

By analyzing the spectra of stars we can obtain their speed of rotation from the

A:6

Table A.3   Provisory Effective Stellar Temperatures.[a]

Types	T_e (°K)					
B0	27,000					
B2	20,000					
B5	16,000					
B8	12,500					
A0	10,400					
A3	8500					
F0	7200					
	Main-Sequence Subgiants		Giants		Supergiants	
	V	IV	III	II	Ib	Ia
F5	6700	6600	6500	6350	6200	....
G0	6000	5720	5500	5350	5050	....
G5	5520	5150	4800	4650	4500	....
K0	5120	4750	4400	4350	4100	....
K5	4350	....	3700	3600	3500	....
M0	3750	....	3500	3400	3300	....
M2	3350	....	3100	2050	....	....

[a]Adapted from Keenan Ke63b. See also text. (With the permission of the University of Chicago Press.)

broadening of stellar spectral lines. If the axis of rotation of a star is inclined at an angle i, relative to the line of sight, we then obtain a measure of $v_e \sin i$, where v_e is the equatorial velocity of the star. Only those stars whose spin axes are perpendicular to the line of sight will exhibit the full Doppler broadening due to the rotation of the star; but by analyzing the distribution of line widths, we can statistically determine both the rotational velocity and the distribution function of the angle i (Hu65). Table A.4 gives some typical values of v_e for different types of stars. Figure 1.9 shows the angular momentum for unit stellar mass for these stars. (See page 33.)

A:6

Table A.4 Stellar Rotation for Stars of Luminosity Class III and V (After Allen A164).

Spectral Type	Mean v_e (km sec^{-1})	
	III	V
O5		190
B0	95	200
B5	120	210
A0	140	190
A5	170	160
F0	130	95
F5	60	25
G0	20	< 12
K,M	< 12	< 12

(b) Variable stars

Two main types of variable stars can be listed. *Extrinsic variables* such as (i) close binary stars whose combined brightness varies because one star can eclipse the other, and (ii) stars in nebulosities that are eclipsed by clouds or that illuminate passing clouds from time to time. These are called *T-Tauri variables.* They are named after the star in which variable features of this kind were first detected.

The second class of variable stars contains *intrinsic variables*—stars whose luminosity actually changes with time. The brightness variations may be repetitive as for periodic variables, erratic as for irregular variables, or the behavior may be semiregular. The distinction is not always clearcut. A brief summary of some characteristics of the pulsating variables is given in Table A.3. These stars are important in the construction of a reliable cosmic distance scale.

Other types of intrinsic variables include exploding stars such as novae, recurrent novae, supernovae, dwarf novae, and shell stars.

The brightness of a nova rises 10 to 12 magnitudes in a few hours. The return to the star's previous low brightness may take no more than a few months, or it may take a century. Both extremes have been observed. The absolute photographic brightness at maximum is about −7.

Table A.5 Properties of Pulsating Variables.

Type	Range of Period, P	Spectral Type	Mean brightness M_v and variation ΔM_v	Remarks
RR Lyrae (Cluster Variables)	< 1d	A4 to F4	$M_v = 0.6$ $\Delta M \sim 1.0$	Found in the halo of the Galaxy
Classical Cepheids	1–50	F to K	$M_v = -2.6$ to -5.3 $M_v, \Delta M_v$ depend on P $\Delta M_v \sim 0.4$ to ~ 1.4	Found in the disk of the Galaxy
W Virginis stars (Type II Cepheids)	> 10	F,G	$M_v =$ one or two mag. less luminous than Class. Ceph. of similar period $\Delta M_v = 1.2$	Halo population
Mira Stars (Long Period Variables)	100–1000	Red giant	$M_v \sim$ from -2.2 to 0, $\Delta M_v =$ from 3 to 5 for increasing period	Intermediate between disk and halo
Semiregular Variables	40–150	Red giant	$M_v = 0$ to -1 $\Delta M_v \sim 1.6$	Disk population

Recurrent novae brighten by about 7.5 magnitudes at periods of several decades. Their peak brightness is about the same as that of ordinary novae. The brightness decline usually takes 10 to 100 days but sometimes is outside this range.

Supernovae are about 10 magnitudes brighter than novae. The brightness may become as great as that of a whole galaxy. Two types have been recognized. Type I has $M_v = -16$ at maximum. Type II has $M_v = -14$ at maximum and exhibits the spectrum of an ordinary nova.

On exploding, a supernova can thrust about one solar mass of matter into interstellar space at speeds of order 1000 km sec^{-1}. Often these gaseous shells persist as *supernova remnants* for several thousand years. On photographic plates they appear as filamentary arcs surrounding the point of initial explosion.

Dwarf novae brighten by about 4 magnitudes to a maximum absolute brightness of $M_v = +4$ to $+6$. Their spectral type normally is A. Their outbursts are repeated every few weeks.

Shell stars are B stars having bright spectral lines. The stars seem to shed shells. A rise in brightness of one magnitude can occur.

Flare stars brighten by ~ 1 magnitude in a brief time. They then relapse. These stars are yellow or red dwarfs of low luminosity. The flares may well be similar to those seen on the sun, except that they occur on a larger scale.

R Coronae Borealis stars are stars that suddenly dim by as much as eight magnitudes and then gradually return to their initial brightness. At maximum the spectrum is of class R, rich in carbon.

The variable stars are not very common, but they are interesting for two reasons. First, some of the variable stars have a well-established brightness pattern that allows one to use them as distance indicators (see chapter 2). Second, the intrinsic variables show symptoms of unstable conditions inside a star or on its surface. In that sense the variable stars may provide an important clue to the structure of stars and perhaps to the energy balance, or imbalance, at different stages of stellar evolution.

T-Tauris and novae, which apparently eject material that forms dust, are found to be strong emitters of infrared radiation.

A:7 THE DISTRIBUTION OF STARS IN SPACE AND VELOCITY

We judge the radial velocities of stars by their spectral line shifts; the transverse velocity can be obtained, for nearby stars, from the *proper motion*—the angular velocity across the sky—and from the star's distance, if known. We find that stars of differing spectral type have quite different motions. Stars in the Galactic plane have low relative velocities, while stars that comprise the *Galactic halo* have large velocities, relative to the sun. These latter objects are said to belong to population II, while those orbiting close to the plane are called population I stars. In practice there is no clear-cut discontinuity between these populations (Ku54). This is rather well illustrated by the continuous variation in velocities given in Table A.6. A star's velocity is correlated with the mean height, above the Galactic plane.

The question of real interest is whether the stars were formed and have always traveled in their present orbits. In that case we would expect the trend of velocities to define the sequence in which stars were formed from the interstellar gas. Alternatively, these different velocities may have been acquired in time, as a result of distant encounters among stars (section 3:14). In that case the stars all may have been formed at low velocities, in the galactic plane. We do not know, but we hope that studies of stellar dynamics may clear up this important question!

By noting the distribution of stars in the solar neighborhood, we at least obtain some idea about how many stars of a given kind have been formed in the Galaxy. If we can compute the life span of a star, as outlined in Chapter 8, then we can also judge the rate at which stars are born. For the short-lived stars, such birth rates represent current formation rates; and we can look for observational evidence to corroborate estimates of longevity, once the spatial number density of a given type of star has been established (Fig. A.6). Such studies still are in relatively preliminary stages, because we are not quite sure what the

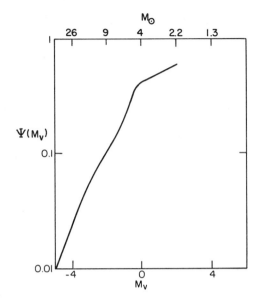

Fig. A.6 Present formation rate, ψ, of bright stars per square parsec of area (projected onto the Galactic plane) during 10^{10}y. The mass of stars of different brightness is shown. (After M. Schmidt, Sc63. With the permission of the University of Chicago Press.)

appearance of a star should be at birth, particularly if it still is surrounded by some of the dust from which it has been formed (section 1:4) (Da67).

A:8 PULSARS, RADIO STARS, AND X-RAY SOURCES

(a) Pulsars

With the exception of the Crab Nebula pulsar, pulsars have thus far been identified only in the radio wavelength region. They emit sharp pulses at roughly one-second intervals. The remarkable feature is the regularity of the pulse. For many of these objects the rate is constant to one part in 10^8.

Within each pulse, there are subpulses that march, relative to the overall envelope, with regularities both in their phase and in their changing sense of polarization. We hope that detailed analysis of these marching pulse trains will give us greater insight into the emission mechanism.

The coherence and pulse rates already tell us that the source is small compared to normal stars. We think we are dealing with *neutron stars*, stars whose cores consist of closely packed, degenerate neutrons. In such a star the mass of the sun is packed into a volume about 10 km in diameter.

The most widely accepted model for the pulsars has a neutron star rotating with a period equal to the interval between the main pulses. The method of

A:8

Table A.6 Stellar Velocities Relative to the Sun, and Mean Height Above Galactic Plane.[a]

Objects	Velocity[b], v km sec^{-1}	Density, ρ $10^{-3} M_\odot \mathrm{pc}^{-3}$	Height, h pc
Interstellar clouds			
large clouds	8		
small clouds	25		
Early main sequence stars:			
O5–B5	10 ⎫		50
B8–B9	12 ⎭ 0.9		60
A0–A9	15	1	115
F0–F9	20	3	190
Late main sequence stars:			
F5–G0	23		
G0–K6	25	12 ⎫	350
K8–M5	32	30 ⎭	
Red giant stars:			
K0–K9	21	0.1	270
M0–M9	23	0.01	
High velocity stars:			
RR Lyrae variables	120	10^{-5}	
Subdwarfs	150	1.5	
Globular clusters	120–180	10^{-3}	

[a]Stellar velocities collected by Spitzer and Schwarzschild from other sources (Sp51a). Densities, ρ, and Heights, h, after Allen (A164). (With the permission of the University of Chicago Press, and the Athlone Press of the University of London, second edition © C.W. Allen 1955 and 1963.)
[b]Root mean square value for component of velocity projected onto the Galactic plane.

generating the pulses, however, has not yet been settled. In all the theories we have, the radiation is emitted in a direction tangential to the charged particles moving with the rotating star and, hence, there is a loss of angular momentum and a corresponding slowdown of the star's rotation and of the pulse rate. Careful measurements indeed show this slowdown in a number of pulsars (Go68).

From time to time a discontinuous change in the period can also occur; and this kind of change is not yet understood. There are other remarkable features that remain unexplained: Giant pulses, thousands of times brighter than a normal pulse, but appearing only about once in ten thousand pulses; null pulses, where there is no intensity at all, appearing at times; sudden changes in the pulse structure, with an equally sudden flipping back to the original pulsing mode. All these provide puzzles we must seek to fathom.

There are two pulsars associated with known gaseous *supernova remnants*. One is in the constellation Vela. The other is a star in the Crab Nebula, remnant of a supernova seen in 1054 AD. It was identified, more than 25 years before the pulsar's discovery, as the stellar remnant of the supernova. This pulsar is the only one for which not only radio waves have been detected but also pulsed visual and X-radiation.

It is interesting that the Crab Nebula pulsar has the shortest known pulse period, 0.033 sec, and that the pulsar in the direction of the constellation Vela also has a short period, 0.089 sec. Since pulsation rates slow down, these two fast pulsing objects presumably are very young. Using the present slowdown rate, we can make a rough linear extrapolation of the Crab pulsar's period, backward in time, and see that this is indeed the object that exploded in 1054 AD. Curiously, other supernova remnants have been thoroughly searched for pulsars— but to no avail. Only the Crab and Vela remants appear to have associated pulsars.

A particularly interesting feature of pulsars is that they may also be responsible for cosmic ray particles. The charged particles, which give rise to the observed pulsed electromagnetic radiation, are believed to be highly relativistic; and it is

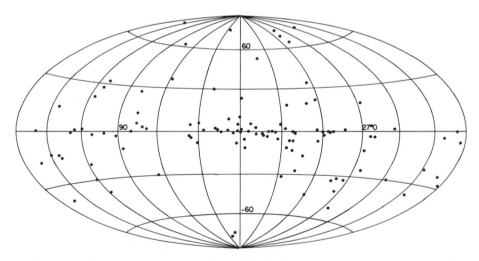

Fig. A.7 The X-ray sky—1971 (Gi72). Map of the X-ray sky as known in 1971 plotted in galactic coordinates. Note the concentration toward the galactic center and galactic plane. The source at the center has an extent of 2° and coincides both with the extended radio and infrared sources shown in Fig. 9.4. (With the permission of the University of Chicago Press.)

entirely possible that some of the particles generated in pulsars have energies as high as the highest energy *cosmic rays* (section 5:10) observed. In that case, it may well be that cosmic rays are a more local phenomenon than had long been believed. If this hypothesis is true, we might expect an anisotropy in the cosmic ray flux at the highest energies. It is also true that we would expect a different chemical composition in cosmic rays than found in ordinary stellar material, because the material of a neutron star has probably undergone severe nuclear changes during the evolution of the star.

One interesting feature of the pulsars is the arrival time of pulses at different radio frequencies. Although equally spaced, these pulses arrive at slightly different times because the refractive index of the interstellar medium, and of any outer layers of the pulsar, are slightly different for differing radio frequencies. This allows us to compute the number of electrons along the line of sight to the object. We can then make a rough estimate of the distance of the object and its radio brightness (section 6:11).

The pulsars—some 60 of them were known in 1971—are concentrated toward the plane of the Galaxy. Based on arguments of distance and concentration, we are sure that these objects are stars within our own Galaxy. Other galaxies, however, doubtlessly contain pulsars too.

(b) Radio Stars

The first radio star to be discovered was the sun (Re44). Its radio emission is very weak and we detect it clearly only because we are near. For more than a decade after the sun's detection, all discovered radio sources were extragalactic radio galaxies or quasars, or involved nebulosities such as supernovae or ionized hydrogen regions in the Galaxy. However, in recent years, several new classes of radio stars have been observed. Besides pulsars, novae and X-ray emitting stars have also been detected at radio frequencies. Red supergiants, red dwarf flare stars, and a blue dwarf companion to a red supergiant have also been studied. These objects are very faint and only the most sophisticated techniques available to us are suitable for their study (Hj71).

(c) X-Ray Stars

A number of extragalactic sources are known to emit X-rays: M87, a spherical galaxy that emits radio waves and strikingly exhibits jets of relativistic particles apparently shot out from its center; 3C 273, the brightest quasar; and a couple of other radio galaxies and Seyfert galaxies. In addition, X-rays have been detected from the direction of several clusters of galaxies. Generally, however, most X-ray sources observed to date are galactic. Figure A.7 shows a clustering of the sources about the galactic plane. These sources are associated with stars and fall into several groups.

(i) The Crab pulsar emits extremely regular pulses with a 0.33-sec period. We have detected no other pulsar at X-ray frequencies.

(ii) Centaurus X-3 is a source that pulses semiregularly. For periods of a day and a half it pulses with a period that slowly increases from 4.84 to 4.87 sec. Then it suddenly drops in intensity in a period of an hour, only to start all over again a half day later. Sources of this kind appear to be close binary stars.

(iii) There are a variety of X-ray sources that seem to have some regularity to their brightness changes that occur on the scale of seconds. But we have not yet been able to define this regularity and it may be spurious.

(iv) Novalike sources that flare up for a month or so and then die away. About two of these have been discovered each year. Some X-ray stars may be associated with white dwarf stars, or with planetary nebulae, or neutron stars. We do not yet know. Some astrophysicists have suggested that they might be associated with *black holes*, stars in an ultimate state of collapse (section 8:19).

A:9 QUASARS AND QUASISTELLAR OBJECTS (QSO's)

These objects (Ha63, Gr64) are listed separately because their nature is not yet properly understood.

Quasars have many features in common with some types of radio galaxies; in particular the visible spectra bear a strong resemblance to the nuclei of *Seyfert galaxies*, which are spiral galaxies with compact nuclei that emit strongly in the infrared and exhibit highly broadened emission lines from ionized gases. In both the quasars and Seyfert nuclei, we find highly ionized gases with spectra indicating temperatures of the order of 10^5 to $10^{6}°K$ and number densities $\sim 10^6$ cm^{-3}. The conditions resemble those found in the solar corona. In the quasars and Seyfert nuclei the spectra of these gases show velocity differences of the order of 1000 or 2000 km/sec, indicating either (a) that gases are being shot out of these objects at high velocity, (b) that they are falling in at high speed, (c) that there is a fast rotation, or (d) that there is a great deal of turbulent motion present. Most likely, a combination of two or three factors is involved.

The quasars show brightness variations on a time scale of months and are, therefore, believed to be less than a light-month $\sim 10^{17}$ cm in diameter. This argument, however, is weak, because it assumes that the brightness changes are coherent, while actually we may be dealing with independent outbursts in different portions of a much larger object in which independent outbursts occur on the time scale of weeks.

A:9

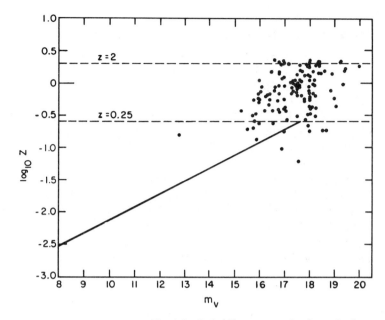

Fig. A.8   Red-shift–apparent visual magnitude relation for all QSO's with $z = \Delta\lambda/\lambda$ known in 1969. Note the difference between this plot and Fig. 2.4 (after Burbidge and Burbidge, Bu69).

Very strikingly the quasars also show spectra that are highly red shifted, indicating either (a) great cosmic distance, (b) great recession velocity of an object recently ejected from our galaxy, or (c) a gravitationally red shifted, nearby object. The second alternative seems attractive to a number of astronomers who point out that there is no correlation between brightness and red shift, Fig. A.8, quite in contrast to the finding for regular galaxies. This would also be true of alternative (c).

If quasars are at the cosmic distances implied by their red shifts, they must be fantastically luminous. Some may emit more than 10^{46} erg sec^{-1}—a hundred times more than our galaxy! Since the sizes of these objects are so small, the surface brightness must be some 10 orders of magnitude greater than that of normal galaxies!

The terms "quasar" and "QSO" are often used interchangeably. However, some astronomers reserve "quasar" for those QSO's that emit strongly at radio frequencies, and use "QSO" to denote both radio quiet and powerful radio emitters—that is, the whole class of these compact objects.

A:10 PHOTONS AND COSMIC RAY PARTICLES

The earth, the solar system, and the galaxy all are bathed in streams of photons and highly relativistic particles. Inside the galaxy there are more photons than outside, since starlight and infrared emission give a stronger local illumination; but there is a microwave component with the spectrum of a blackbody at 3°K that seems to be equally strong inside the galaxy and in the ambient universe (Pe65).

   Cosmic ray particles, highly energetic electrons and nucleons, constitute a denser energy bath in the earth's vicinity than starlight and microwave photons combined; but we do not know how the particles are distributed in extragalactic space. Table A.7 shows the energy densities of these components. X-rays and γ-rays, highly energetic photons, have far smaller energy densities than visible and microwave radiation.

Table A.7 Energy and Number Densities of Photons and Cosmic Rays.

	Cosmic Ray Particles	Visible Light	Microwaves
Energy density in Galaxy (ergs cm^{-3})	10^{-12}	$\sim 2 \times 10^{-13}$	$\sim 5 \times 10^{-13}$
Extragalactic energy density (ergs cm^{-3})	2	$\sim 2 \times 10^{-14}$	$\sim 5 \times 10^{-13}$
Number density in Galaxy (cm^{-3})	$\sim 10^{-9}$	$\sim 10^{-1}$	$\sim 10^{3}$
Extragalactic number density (cm^{-3})	2	$\sim 10^{-2}$	$\sim 10^{3}$

Appendix B

Astrophysical Constants*

B: 1. PHYSICAL CONSTANTS

Velocity of light	$c = 2.998 \times 10^{10}$ cm sec^{-1}
Planck constant	$h = 6.626 \times 10^{-27}$ erg sec
Gravitational constant	$G = 6.67 \times 10^{-8}$ dyn cm^2 g^{-2}
Electron charge	$e = 4.803 \times 10^{-10}$ esu
Mass of electron	$m_e = 9.1096 \times 10^{-28}$ g
Mass of proton	$m_p = 1.6724 \times 10^{-24}$ g
Mass of hydrogen atom	$m_H = 1.6733 \times 10^{-24}$ g
Mass of neutron	$m_N = 1.6747 \times 10^{-24}$ g
Atomic mass unit	$amu = (1/12)\, m_{C^{12}} = 1.661 \times 10^{-24}$ g
Avogadro's number	6.0222×10^{23}
Boltzmann constant	$k = 1.380 \times 10^{-16}$ erg deg^{-1}
Electron volt	$ev = 1.602 \times 10^{-12}$ erg
Stefan-Boltzmann constant	$\sigma = 5.67 \times 10^{-5}$ erg cm^{-2} deg^{-4} sec^{-1}
Rydberg constant	$R_\infty = 2.17992 \times 10^{-11}$ erg

B: 2. ASTRONOMICAL CONSTANTS

Year $= 3.156 \times 10^7$ sec

Astronomical unit $\qquad$ AU $= 1.496 \times 10^{13}$ cm
(mean sun earth distance)

---

*(Al64), (Ta70).

Parsec, pc $= 3.086 \times 10^{18}$ cm $= 2.06 \times 10^5$ AU $= 3.261$ light years

Solar mass	$M_\odot = 1.99 \times 10^{33}$ g
Solar radius	$R_\odot = 6.96 \times 10^{10}$ cm
Solar luminosity	$L_\odot = 3.9 \times 10^{33}$ erg sec^{-1}

Star with $M_{bol} = 0$ radiates 3.02×10^{28} watt

Aeon $= 1$ æ $= 10^9$ y

List of References

Articles and books that are cited in the text are referred to by the first two letters of the author's name and the last two digits of the year in which the publication appears. An article by Johnson published in 1928 is designated (Jo28). Where several individuals have co-authored a work, only the first author's name is used. A publication that appeared prior to 1901 carries a numeral designation 00, for example Newton's *Principia* carries the designation (Ne00). An asterisk following a given designation, for example (He67)*, implies that the reference is important for the entire section in which it is cited.

The following abbreviations have been used:

Ann. New York Acad. Sci.: Annals of the New York Academy of Sciences.
Ann. Rev. Astron. and Astrophys.: Annual Reviews of Astronomy and Astrophysics.
A.J.: Astronomical Journal (USA).
Ap.J.: Astrophysical Journal (USA). If the letter L precedes the page number, the article appeared in the affiliated journal Astrophysical Journal Letters. The Astrophysical Journal also publishes a supplement, denoted by "Suppl."
Astrophys. Let.: Astrophysical Letters.
B.A.N.: Bulletin of the Astronomical Society of the Netherlands.
I.A.U.: International Astronomical Union. The organization issues symposium proceedings and a variety of other publications.
J. Roy. Astron. Soc. Can.: Journal of the Royal Astronomical Society of Canada.
J. Phys. USSR: Journal of Physics of the Soviet Union.
MNRAS: Monthly Notices of the Royal Astronomical Society (London).
PASP: Publications of the Astronomical Society of the Pacific.
Phil. Mag.: Philosophical Magazine (London).
Phys. Rev.: Physical Reviews (USA).
Phys. Rev. Let.: Physical Review Letters (USA).
Proc. Nat. Acad. Sci.: Proceedings of the National Academy of Sciences (USA).
Proc. Roy. Soc.: Proceedings of the Royal Society (London).

PASJ: Publications of the Astronomical Society of Japan.

Quart. J. Roy. Astron. Soc.: Quarterly Journal of the Royal Astronomical Society (London).

Rev. Mod. Phys.: Reviews of Modern Physics (USA).

Soviet A.J.: Astronomical Journal (Soviet Union). This journal also appears in English translation.

Z. Astrophys.: Zeitschrift für Astrophysik (Germany).

(Aa60) S. Aarseth, "The Rotation of a Barred Galaxy under Gravitational Forces," *MNRAS, 121*, 525 (1960).

(Aa61) S. Aarseth, "The Rotation of Barred Galaxies under Gravitational Forces II," *MNRAS, 122*, 535 (1961).

(Al55) C.W. Allen, *Astrophysical Quantities*, Athlone Press, London (1955).

(Al63) L.H. Aller, "Astrophysics" in *The Atmospheres of the Sun and Stars,* 2nd ed., Ronald Press, New York (1963).

(Al64) C.W. Allen, *Astrophysical Quantities*, 2nd ed., Athlone Press, London (1964).

(Al68) T. Alväger and M.N. Kreisler, "Quest for Faster than Light Particles," *Phys. Rev., 171*, 1357 (1968).

(Al70) Margo F. Aller and C.R. Cowley, "The Possible Identification of Prometheum in HR46," *Ap.J., 162*, L145 (1970).

(Ar65) H. Arp, "A Very Small, Condensed Galaxy," *Ap.J., 142*, 402 (1965).

(Ar70) W.D. Arnett and D.D. Clayton, "Explosive Nucleosynthesis in Stars," *Nature, 227*, 780 (1970).

(Ar71) H. Arp, "Observational Paradoxes in Extragalactic Astronomy," *Science, 174*, 1189 (1971).

(As71) M.E. Ash, I.I. Shapiro, and W.B. Smith, "The System of Planetary Masses," *Science, 174*, 551 (1971).

(Ba71a) E.S. Barghoorn, "The Oldest Fossils," *Scientific American* May (1971), p. 30.

(Ba71b) J.N. Bahcall and S. Frautschi, "The Hadron Barrier in Cosmology and Gravitational Collapse," *Ap.J., 170* L81 (1971).

(Ba72) J.N. Bahcall and R.L. Sears, "Solar Neutrinos," *Ann. Rev. Astron. and Astrophys.* (1972).

(Be39) H. Bethe, "Energy Production in Stars," *Phys. Rev., 55*, 434 (1939).

(Be71) A. Bers, R. Fox, C.G. Kuper, and S.G. Lipson, "The Impossiblity of Free Tachyons" in *Relativity and Gravitation*, C.G. Kuper and A. Peres (Eds.) Gordon and Breach, New York (1971).

(Be72) R. Beer, C.B. Farmer, R.H. Norton, J.V. Martonchik, and T.G. Barnes, "Jupiter: Observations of Deuterated Methane in the Atmosphere," *Science, 175*, 1361 (1972).

(B161) A. Blaauw, "On the Origin of the O- and B- type Stars with High Velocities, (The 'Run-away' Stars), and Some Related Problems," *B.A.N., 15,* 265 (1961).

(Bo48) H. Bondi and T. Gold, "The Steady State Theory of the Expanding Universe," *MNRAS, 108,* 252 (1948).

(Bo52) H. Bondi, *Cosmology,* Cambridge University Press, Cambridge, England (1952).

(Bo64a) S. Bowyer, E. Byram, T. Chubb, and H. Friedman, "X-ray Sources in the Galaxy," *Nature, 201,* 1307 (1964).

(Bo64b) S. Bowyer, E. Byram, T. Chubb, and H. Friedman, "Lunar Occultation of X-ray Emission from the Crab Nebula," *Science, 146,* 912 (1964).

(Bo71) B.J. Bok, "Observational Evidence for Galactic Spiral Structure" in *Highlights in Astronomy,* DeJager, Ed. I.A.U. (1971).

(Br68) P.F. Browne, "The Generation of Magnetic Fields by Viscous Forces in a Non-Uniformly Rotating Plasma," *Astrophys. Let., 2,* 217 (1968).

(Br72) A.H. Bridle, M.M. Davis, E.B. Fomalont, and J. Lequeux "Counts of Intense Extragalactic Radio Sources at 1400 MHz," *Nature Physical Science, 235,* 123 (1972).

(Bu60) E.M. Burbidge, G.R. Burbidge, and K.H. Prendergast, "Motions in Barred Spiral Galaxies II. The Rotation of NGC 7479," *Ap.J., 132,* 654 (1960).

(Bu69) G.R. Burbidge and E.M. Burbidge, "Quasistellar Objects—A Progress Report," *Nature, 224,* 21 (1969).

(Bu70) G.R. Burbidge and W.A. Stein, "Cosmic Sources of Infrared Radiation," *Ap.J., 160,* 576 (1970).

(Bu71a) E.M. Burbidge and W.L.W. Sargent, "Velocity Dispersion and Discrepant Redshifts in Groups of Galaxies," from Pontifical Academy of Science, *Scripta Varia No. 35: Nuclei of Galaxies* (1971).

(Bu71b) E.M. Burbidge and P.M. Hodge, "Is NGC 4569 a Member of the Virgo Cluster?" *Ap.J., 166,* 1 (1971).

(By67) E.T. Byram, T.A. Chubb, and H. Friedman, "Cosmic X-ray Sources, Galactic and Extragalactic," *Science, 152,* 66 (1967).

(Ca24) A.J. Cannon and E.C. Pickering, "The Henry Draper Catalogue," *Ann. Astron. Obs. Harvard College, 91* to *99* (1924).

(Ca66) D. Cattani and C. Sacchi, "A Theory on the Creation of Stellar Magnetic Fields," *Nuovo Cimento, 46B,* 8046 (1966).

(Ca68) A.G.W. Cameron "A New Table of Abundances of the Elements in the Solar System," in *Origin and Distribution of the Elements,* L.H. Arens, Ed., Pergamon Press, Elmsford, New York (1968), p. 125.

(Ca70a) G.R. Carruthers, "Rocket Observations of Interstellar Molecular Hydrogen," *Ap.J., 161,* L81 (1970).

(Ca70b) A. Cavaliere, P. Morrison, F. Pacini, "A Model for the Radiation from the Compact Strong Sources," *Ap.J., 162,* L133 (1970).

(Ch39) S. Chandrasekhar, *An Introduction to the Study of Stellar Structure,* University of Chicago Press, Chicago (1939), Dover, New York (1957).

(Ch43) S. Chandrasekhar, *Principles of Stellar Dynamics,* University of Chicago Press, Chicago (1943), Dover, New York (1960).

(Ch50) S. Chandrasekhar, *Radiative Transfer,* University of Chicago Press, Chicago (1950), Dover, New York (1960).

(Ch58) S. Chandrasekhar, "On the Continuous Absorption Coefficient of the Negative Hydrogen Ion," *Ap.J., 128,* 114 (1958).

(Ch64a) H.Y. Chiu, "Supernovae, Neutrinos and Neutron Stars," *Annals of Physics, 26,* 364 (1964).

(Ch64b) J.H. Christenson, J.W. Cronin, V.L. Fitch, and R. Turlay, "Evidence for the 2π Decay of the K_2^0 Meson," *Phys. Rev. Let., 13,* 138 (1964).

(Ch70) W.Y. Chau, "Gravitational Radiation and the Oblique Rotator Model," *Nature, 228,* 655 (1970).

(Cl68) D.D. Clayton, *Principles of Stellar Evolution and Nucleosynthesis,* McGraw-Hill, New York (1968).

(Co57) T.G. Cowling, *Magnetohydrodynamics,* Interscience, New York (1957).

(Co68) C.C. Counselman III and I.I. Shapiro, "Scientific Uses of Pulsars," *Science, 162,* 352 (1968).

(Co69) E.J. Conklin, "Velocity of the Earth with Respect to the Cosmic Background Radiation," *Nature, 222,* 971 (1969).

(Co70) R. Cowsik, Y. Pal, and T.N. Rengarajan, "A Search for a Consistent Model for the Electromagnetic Spectrum of the Crab Nebula" *Astrophys. and Space Sci., 6,* 390 (1970).

(Co71a) D.P. Cox and E. Daltabuit, "Radiative Cooling of a Low Density Plasma," *Ap.J., 167,* 113 (1971).

(Co71b) R. Cowsik and P.B. Price, "Origins of Cosmic Rays," *Physics Today,* October (1971), p. 30.

(Co72) J.J. Condon, "Decimetric Spectra of Extragalactic Radio Sources," Doctoral Dissertation in Astronomy, Cornell University (1972).

(Da51) L. Davis and J.L. Greenstein, "The Polarization of Starlight by Aligned Dust Grains," *Ap.J., 114,* 206 (1951).

(Da67) K. Davidson and M. Harwit, "Infrared and Radio Appearance of Cocoon Stars," *Ap.J., 148,* 443 (1967).

(Da68) R. Davis, Jr., D.S. Harmer, K.C. Hoffman, "Search for Neutrinos from the Sun," *Phys. Rev. Let., 20,* 1205 (1968).

(Da69) K. Davidson and Y. Terzian, "Dispersion Measures of Pulsars," *A.J., 74,* 849 (1969).

(Da70a) K. Davidson, "The Development of a Cocoon Star," *Astrophysics and Space Science, 6,* 422 (1970).

(Da70b) J. Davis, D.C. Morton, L.R. Allen, and R. Hanbury Brown, "The

Angular Diameter and Effective Temperature of zeta Puppis," *MNRAS, 150,* 45 (1970).

(Da71) R. Davis Jr., L.C. Rogers, and V. Radeka, "Report on the Brookhaven Solar Neutrino Experiment," *Bull. Amer. Phys. Soc. Ser. 11, 16,* 631 (1971).

(De68) S.F. Dermott, "On the Origin of Commensurabilities in the Solar System-II-The Orbital Period Relation," *MNRAS, 141,* 363 (1968).

(deS17) W. deSitter, "On Einstein's Theory of Gravitation and its Astronomical Consequences," *MNRAS, 78,* 3 (1917).

(Di31) P.A.M. Dirac, "Quantized Singularities in the Electro-magnetic Field," *Proc. Roy. Soc., A, 133,* 60 (1931).

(Di38) P.A.M. Dirac, "A New Basis for Cosmology," *Proc. Roy. Soc., A, 165,* 199 (1938).

(Di67a) R.H. Dicke, "Gravitation and Cosmic Physics," *Am. J. of Physics, 35,* 559 (1967).

(Di67b) R.H. Dicke and H.M. Goldenberg, "Solar Oblateness and General Relativity," *Phys. Rev. Let., 18,* 313 (1967).

(Di69) R.H. Dicke, "The Age of the Galaxy from the Decay of Uranium," *Ap.J., 155,* 123 (1969).

(Dr62) F.D. Drake, "Intelligent Life in Space," MacMillan, New York (1962).

(Du62) S. Dushman and J.M. Lafferty, "Scientific Foundations of Vacuum Technique," John Wiley, New York (1962).

(Dy60) F.J. Dyson, "Search for Artificial Stellar Sources of Infrared Radiation," *Science, 131,* 1667 (1960).

(Eb55) R. Ebert, "Über die Verdichtung von H$_I$ Gebieten," *Z. Astrophys., 37,* 217 (1955).

(Ed30) A.S. Eddington, "On the Instability of Einstein's Spherical World," *MNRAS, 90,* 668 (1930).

(Eg62) O. Eggen, D. Lynden-Bell, and A. Sandage, "Evidence From the Motions of Old Stars that the Galaxy Collapsed," *Ap.J., 136,* 748 (1962).

(Ei05a) A. Einstein, "On the Electrodynamics of Moving Bodies," *Ann. d. Phys., 17,* 891 (1905), translated and reprinted in *The Principle of Relativity,* A. Sommerfeld, Ed., Dover, New York.

(Ei05b) A. Einstein, "Does the Inertia of a Body Depend Upon Its Energy Content?" *Ann. d. Phys., 18,* 639 (1905) translated and reprinted in *The Principle of Relativity,* A. Sommerfeld, Ed., Dover, New York.

(Ei07) A. Einstein, "On the Influence of Gravitation on the Propagation of Light," *Jahrbuch für Radioakt. und Elektronik, 4,* 1907 (translated in *The Principle of Relativity,* A. Sommerfeld, Ed., Dover, New York.

(Ei16) A. Einstein, "The Principle of General Relativity" *Ann. d. Phys., 49,* 769 (1916), translated and reprinted in *The Principle of Relativity*, A. Sommerfeld, Ed., Dover, New York.

(Ei17) A. Einstein, *Kosmologische Betrachtungen zur Allgemeinen Relativitäts-theorie,"* S.B. Preuss. Akad. Wiss, (1917). p. 142.

(Ez65) D. Ezer and A.G.W. Cameron, "The Early Evolution of the Sun," *Icarus, 1,* 422 (1963).

(Fa71) J. Faulkner, "Ultrashort-Period Binaries, Gravitational Radiation, and Mass Transfer I. The Standard Model, with Applications to WZ Sagittae and Z Camelopardalis," *Ap.J., 170,* L99 (1971).

(Fo62) W.A. Fowler, J.L. Greenstein, and F. Hoyle, "Nucleosynthesis During the Early History of the Solar System," *Geophys. J., 6,* 148 (1962).

(Fr22) A. Friedmann, "Über die Krümmung des Raumes," *Z. Phys.,* 10, 377 (1922).

(Fr46) J. Frenkel, *Kinetic Theory of Liquids,* Oxford University Press, Oxford, England (1946).

(Fr69) G. Fritz, R.C. Henry, J.F. Meekins, T.A. Chubb, and H. Friedman, "X-ray Pulsars in the Crab Nebula," *Science, 164,* 709 (1969).

(Ga00) Galileo Galilei, *Dialogues Concerning Two New Sciences,* translated by H. Crew and A. de Salvio, Northwestern University Press, (1946).

(Ge70) S.L. Geisel, D.E. Kleinmann, and F.J. Low, "Infrared Emission from Nebulae," *Ap.J., 161,* L101 (1970).

(Gi62) R. Giacconi, H. Gursky, F. Paolini, and B. Rossi, "Evidence for X-rays From Sources Outside the Solar System," *Phys. Rev. Let., 9,* 439 (1962).

(Gi64) V.L. Ginzburg and S.I. Syrovatskii, "The Origin of Cosmic Rays," Pergamon Press, Elmsford, New York (1964).

(Gi69) V.L. Ginzburg, "Elementary Processes for Cosmic Ray Astrophysics," Gordon and Breach, New York (1969).

(Gi72) R. Giacconi, H. Gursky, E. Kellogg, S. Murray, E. Schreier, and H. Tananbaum. "The Uhuru Catalog of X-Ray Sources," *Ap.J.* 178, 281 (1972).

(Go52) T. Gold, "The Alignment of Galactic Dust," *MNRAS, 112,* 215 (1952).

(Go62) T. Gold, "The Arrow of Time," *Am. J. of Phys., 30,* 403 (1962).

(Go63) R.J. Gould and E.E. Salpeter, "The Interstellar Abundance of the Hydrogen Molecule I. Basic Processes," *Ap.J., 138,* 393 (1963).

(Go68) T. Gold, "Rotating Neutron Stars as the Origin of the Pulsating Radio Sources," *Nature, 218,* 731 (1968).

(Go69) T. Gold, "Rotating Neutron Stars and the Nature of Pulsars," *Nature, 221,* 25 (1969).

(Gr64) J.L. Greenstein and M. Schmidt, "The Quasi-Stellar Radio Sources 3C48 and 3C273," *Ap.J., 140,* 1 (1964).

(Gr66) K. Greisen, "End to the Cosmic Ray Spectrum," *Phys. Rev. Let., 16,* 748 (1966).

(Gr68) J.M. Greenberg, "Interstellar Grains" in *Nebulae and Interstellar Matter,* B.M. Middlehurst and L.H. Aller, Eds., University of Chicago Press, Chicago (1968).

(Gu54) S.N. Gupta, "Gravitation and Electromagnetism," *Phys. Rev., 96,* 1683 (1954).

(Gu65) J.E. Gunn and B.A. Peterson, "On the Density of Neutral Hydrogen in Intergalactic Space," *Ap.J., 142,* 1633 (1965).

(Gu66) H. Gursky, R. Giacconi, P. Gorenstein, J.R. Waters, M. Oda, H. Bradt, G. Garmire, and B.V. Sreekantan, "A Measurement of the Location of the X-ray Source Sco X-1," *Ap.J., 146,* 310 (1966).

(Gu69) J.W. Gunn and J.P. Ostriker, "Acceleration of High-Energy Cosmic Rays by Pulsars," *Phys. Rev. Let., 22,* 778 (1969).

(Gu71) H. Gursky, E.M. Kellogg, C. Leong, H. Tananbaum, and R. Giacconi, "Detection of X-rays from the Seyfert Galaxies, NGC 1275 and NGC 4151 by the Uhuru Satellite," *Ap.J., 165* L43 (1971).

(Ha54) R. Hanbury Brown, and R.Q. Twiss, "A New Type of Interferometer for Use in Radio Astronomy," *Phil. Mag., 45,* 663 (1954).

(Ha61) M. Harwit, "Can Gravitational Forces Alone Account for Galaxy Formation in a Steady State Universe?" *MNRAS, 122,* 47; *123,* 257 (1961).

(Ha62a) M. Harwit, "Dust, Radiation Pressure and Star Formation," *Ap.J., 136,* 832 (1962).

(Ha62b) M. Harwit and F. Hoyle, "Plasma Dynamics in Comets II," *Ap.J., 135,* 875 (1962).

(Ha62c) C. Hayashi, R. Hoshi, and D. Sugimoto, "Evolution of the Stars," *Progress of Theor. Physics,* Suppl. 22 (1962).

(Ha63) C. Hazard, M. B. Mackey, and A. J. Shimmins, "Investigation of the Radio Source 3C 273 by the Method of Lunar Occultations," *Nature, 197,* 1037 (1963).

(Ha65) E.R. Harrison, "Olbers' Paradox and the Background Radiation Density in an Isotropic Homogeneous Universe," *MNRAS, 131,* 1 (1965).

(Ha66) C. Hayashi, "Evolution of Protostars," *Ann. Rev. Astron. and Astrophys., 4,* 171 (1966).

(Ha67) R. Hanbury Brown, J. Davis, L.R. Allen, J.M. Rome, "The Stellar Interferometer at Narrabri Observatory-II. The Angular Diameters of 15 stars," *MNRAS, 137,* 393 (1967).

(Ha70) M. Harwit, "Is Magnetic Alignment of Interstellar Dust Really Necessary?," *Nature, 226,* 61 (1970).

(Ha71) D.A. Harper and F.J. Low, "Far Infrared Emission from H_{II} Regions," *Ap.J., 165,* L9 (1971).

(Ha72) M. Harwit, B.T. Soifer, J.R. Houck, and J.L. Pipher, "Why Many Infrared Astronomical Sources Emit at $100\,\mu$m," *Nature Physical Science, 236,* 103 (1972).

(He50) G. Herzberg, *Molecular Spectra and Molecular Structure I. Spectra of Diatomic Molecules* 2nd ed., Van Nostrand, Princeton, New Jersey, (1950).

(He62) O. Heckmann and E. Schücking, "Relativistic Cosmology," in *Gravitation*, L. Witten, Ed., John Wiley, New York. (1962).

(He67) G. Herzberg, "The Spectra of Hydrogen and Their Role in the Development of Our Understanding of the Structure of Matter and of the Universe," *Trans. Roy. Soc.*, Canada *V*, Ser *IV*, 3 (1967).

(He68a) C.E. Heiles, "Normal OH Emission and Interstellar Dust Clouds," *Ap.J., 151*, 919 (1968).

(He68b) A. Hewish, S.J. Bell, J.D.H. Pilkington, P.F. Scott, and R.A. Collins, "Observation of a Rapidly Pulsating Radio Source," *Nature, 217*, 709 (1968).

(He69) C.E. Heiles, "Temperatures and OH Optical Depths in Dust Clouds," *Ap.J., 157*, 123 (1969).

(Hi71) J.M. Hill, "A Measurement of the Gravitational Deflection of Radio Waves by the Sun," *MNRAS, 153*, 7p (1971).

(Hj71) R.M. Hjellming and C.M. Wade, "Radio Stars," *Science, 173*, 1087 (1971).

(Ho48) F. Hoyle, "A New Model for the Expanding Universe," *MNRAS, 108*, 372 (1948).

(Ho56) F. Hoyle and A. Sandage, "Second Order Term in the Redshift-Magnitude Relation," *PASP, 68*, 306 (1956).

(Ho57) F. Hoyle, "The Black Cloud," Signet, New York (1957).

(Hö65) B. Höglund, and P.G. Mezger, "Hydrogen Emission Line $n_{110} \rightarrow n_{109}$: Detection at 5009 Megahertz in Galactic HII Regions," *Science, 150*, 339 (1965).

(Ho69) L.M. Hobbs, "Regional Studies of Interstellar Sodium Lines," *Ap.J., 158*, 461 (1969).

(Ho71) W.F. Hoffmann, C.L. Frederick, and R.J. Emery, "100 Micron Map of the Galactic-Center Region," *Ap.J., 164*, L23 (1971).

(Hu29) E. Hubble, "Distance and Radial Velocity Among Extragalactic Nebulae," *Proc. Nat. Acad. Sci., 15*, 168 (1929).

(Hu65) S. Huang, "Rotational Behavior of the Main-Sequence Stars and its Plausible Consequences Concerning Formation of Planetary Systems I and II," *Ap.J., 141*, 985 (1965) and *150*, 229 (1967).

(Ib65) I. Iben, Jr., "Stellar Evolution-I. The Approach to the Main Sequence," *Ap.J., 141*, 993 (1965).

(Ib70) I. Iben, "Globular-Cluster Stars" *Scientific American*, July 1970, p. 27.

(Já50) L. Jánossy, *Cosmic Rays*, 2nd ed., The Clarendon Press, Oxford, England (1950).

(Ja70a) D.A. Jauncey, A.E. Niell, and J.J. Condon, "Improved Spectra of Some Ohio Radio Sources with Unusual Spectra," *Ap.J., 162*, L31 (1970).

(Ja70b) J.C. Jackson, "The Dynamics of Clusters of Galaxies in Universes with Non-Zero Cosmological Constant, and the Virial Theorem Mass Discrepancy," *MNRAS, 148*, 249 (1970).

(Je69) E.B. Jenkins, D.C. Morton, and T.A. Matilsky, "Interstellar Lα Absorption in β^1, δ, and π Scorpii," *Ap.J., 158*, 473 (1969).

(Je70) E.B. Jenkins, "Observations of Interstellar Lyman-α Absorption," *IAU Symposium* #36: *Ultraviolet Stellar Spectra and Related Ground Based Observations*, L. Houziaux and H.E. Butler, Eds., D. Reidel Publ. Co., Holland (1970).

(Je73) K. B. Jefferts, A. A. Penzias, and R. W. Wilson "Deuterium in the Orion Nebula," *Ap.J., 179*, L57 (1973).

(Jo28) J.B. Johnson, "Thermal Agitation of Electricity in Conductors," *Phys. Rev., 32*, 97 (1928).

(Jo53) H.L. Johnson and W.W. Morgan, "Fundamental Stellar Photometry for Standards of Spectral Type on the Revised System of the Yerkes Spectral Atlas," *Ap.J., 117*, 313 (1953).

(Jo56) H.L. Johnson and A.R. Sandage, "Three-Color Photometry in the Globular Cluster M3," *Ap.J., 124*, 379 (1956).

(Jo67a) H.L. Johnson, "Infrared Stars," *Science, 157*, 635 (1967).

(Jo67b) R.V. Jones and L. Spitzer, Jr., "Magnetic Alignment of Interstellar Grains," *Ap.J., 147*, 943 (1967).

(Ka54) F.D. Kahn, "The Acceleration of Interstellar Clouds," *B.A.N. 12*, 187 (1954).

(Ka59) N.S. Kardashev, "On the Possibility of Detection of Allowed Lines of Atomic Hydrogen in the Radio-Frequency Spectrum," *Astronomicheskii Zhurnal, 36*, 838 (1959), *Soviet A.J., 3*, 813 (1959).

(Ka68) F.D. Kahn, "Problems of Gas Dynamics in Planetary Nebulae," *I.A.U.*, Symposium on Planetary Nebulae, D. Osterbrock, and C.R. O'Dell, Eds., Springer-Verlag, New York (1968).

(Ke63a) R.P. Kerr, "Gravitational Field of a Spinning Mass as an Example of Algebraically Special Metrics," *Phys. Rev. Let., 11*, 237 (1963).

(Ke63b) P.C. Keenan, "Classification of Stellar Spectra" in *Stars and Stellar Systems*, Vol. 3 K.A. Strand, Ed. (1963).

(Ke65) F.J. Kerr and G. Westerhout, "Distribution of Interstellar Hydrogen" in *Stars and Stellar Systems V: Galactic Structure*, A. Blaauw and M. Schmidt, Eds., University of Chicago Press, Chicago (1965).

(Ke68) K.I. Kellerman, I.I.K. Pauliny-Toth, and M.M. Davis, "The Dependence of Radio Source Counts and the Spectral Index Distribution on Frequency," *Astrophys. Let., 2*, 105 (1968).

(Ke69) K.I. Kellermann, I.I.K. Pauliny-Thoth, and P.J.S. Williams, "The Spectra of Radio Sources in The Revised 3C Catalogue," *Ap.J., 157*, 1 (1969).

(Ke70) J.C. Kemp, J.B. Swedlund, J.D. Landstreet, and J.R.P. Angel, "Discovery of Circularly Polarized Light from a White Dwarf," *Ap.J., 161*, L77 (1970).

(Ke71) K.I. Kellermann, D.L. Jauncey, M.H. Cohen, B.B. Shaffer, B.G. Clark, J. Broderick, B. Rönnäng, O.E.H. Rydbeck, L. Matveyenko, I. Moiseyev, V.V. Vitkevitch, B.F.C. Cooper, and R. Batchelor, "High Resolution Observations of Compact Radio Sources at 6 and 18 Centimeters," *Ap.J., 169*, 1 (1971).

(Kl70a) D.E. Kleinmann and F.J. Low, "Observations of Infrared Galaxies," *Ap.J., 159*, L165 (1970).

(Kl70b) D.E. Kleinmann and F.J. Low, "Infrared Observations of Galaxies and of the Extended Nucleus of M82," *Ap.J., 161*, L203 (1970).

(Kl71) O. Klein, "Arguments Concerning Relativity and Cosmology," *Science, 171*, 339 (1971).

(Ko69) M. Kondo, "On the Formation of Condensations in an Expanding Universe," *PASJ, 21*, 54 (1969).

(Kr68) K.S. Krishna Swamy and C.R. O'Dell, "Thermal Emission by Particles in NGC 7027," *Ap.J., 151*, L61 (1968).

(Ku54) B.W. Kukarkin, *Erforschung der Struktur und Entwicklung der Sternsysteme auf der Grundlage des Studiums veränderlicher Sterne*, Akademie Verlag, Berlin (1954).

(Kv70) K. Kvenvolden, J. Lawless, K. Pering, E. Peterson, J. Flores, C. Ponnamperuma, I.R. Kaplan, and C. Moore, "Evidence for Extraterrestrial Amino-acids and Hydrocarbons in the Murchison Meteorite," *Nature, 228*, 923 (1970).

(La51) L. Landau and E. Lifshitz, *The Classical Theory of Fields*, Addison-Wesley, New York (1951).

(La72) R.B. Larson, "Infall of Matter in Galaxies." *Nature, 236*, 21 (1972).

(Le50) G. Lemaître, *The Primeval Atom*, Van Nostrand, Princeton, New Jersey (1950).

(Le56) T.D. Lee and C.N. Yang, "Question of Parity Conservation in Weak Interactions," *Phys. Rev. 104*, 254 (1956).

(Le72) T.J. Lee, "Astrophysics and Vacuum Technology," *Journal of Vacuum Science and Technology*, Jan/Feb (1972).

(Li46) E.M. Lifshitz, "On the Gravitational Stability of the Expanding Universe," *J. Phys. USSR, 10*, 116 (1946).

(Li67) C.C. Lin, "The Dynamics of Disk-Shaped Galaxies." *Ann. Rev. Astron. Astrophys., 5*, 453 (1967).

(Lo04) H.A. Lorentz, "Electromagnetic Phenomena in a System Moving with Any Velocity Less Than Light," *Proceedings of the Acad. Sci.*, Amsterdam *6*, 1904, reprinted in *The Principle of Relativity*, A. Sommerfeld, Ed., Dover, New York.

(Ly63) C. R. Lynds and A. Sandage, "Evidence for an Explosion in the Center of the Galaxy M82," *Ap.J., 137*, 1005 (1963).

(Ma70) D.S. Mathewson and V.L. Ford, "Polarization Observations of 1800 Stars," *Memoirs of the Royal Astronomical Society*, 74, 139 (1970).

(Ma72a) P. M. Mathews and M. Lakshmanan, "On the apparent Visual Forms of Relativistically Moving Objects," *Nuovo Cimento, 12 B*, 168 (1972).

(Ma72b) R.N. Manchester, "Pulsar Rotation and Dispersion Measures and the Galactic Magnetic Field," *Ap.J., 172*, 43 (1972).

(McNa65) D. McNally, "On the Distribution of Angular Momentum Among Main Sequence Stars," *Observatory, 85*, 166 (1965).

(Me68) P.G. Mezger, "A New Class of Compact, High Density H$_{II}$ Regions" in *Interstellar Ionized Hydrogen*, Y. Terzian, Ed., Benjamin, Menlo Park, Calif. (1968).

(Me69) P. Meyer, "Cosmic Rays in the Galaxy," in *Ann. Rev. Astron. Astrophys., 7*, 1 (1969).

(Mi08) H. Minkowski, "Space and Time" an address delivered to the German Natural Scientists and Physicians (1908), translated and reprinted in *The Principle of Relativity*, A. Sommerfeld, Ed., Dover, New York.

(Mi57a) S.L. Miller, "The Formation of Organic Compounds on the Primitive Earth," *Ann. New York Acad. Sci., 69*, 260 (1957).

(Mi57b) R.I. Mitchell and H.L. Johnson, "The Color-Magnitude Diagram of the Pleiades Cluster," *Ap.J., 125*, 418 (1957).

(Mi59) S.L. Miller and H.C. Urey, "Organic Compound Synthesis on the Primitive Earth," *Science, 130*, 245 (1959).

(Mo67) D.C. Morton, "Mass Loss from Three OB Supergiants in Orion," *Ap.J., 150*, 535 (1967).

(Mo68) P.M. Morse and K.U. Ingard, *Theoretical Acoustics*, McGraw-Hill, New York (1968).

(Ne00) Isaac Newton, *Mathematical Principles of Natural Philosophy* and *Systems of the World*, revised translation by Florian Cajori, University of California Press, Berkeley (1962).

(No71) P.D. Noerdlinger and V. Petrosian, "The Effect of Cosmological Expansion on Self-Gravitating Ensembles of Particles, *Ap.J., 168*, 1 (1971).

(Ny28) H. Nyquist, "Thermal Agitation of Electric Charge in Conductors," *Phys. Rev., 32*, 110 (1928).

(Od65) M. Oda, G.W. Garmire, M. Wada, R. Giacconi, H. Gursky, and J.R. Waters, "Angular Sizes of the X-ray Sources in Scorpio and Sagittarius," *Nature, 205*, 554 (1965).

(Oo27a) J.H. Oort, "Observational Evidence Confirming Lindblad's Hypothesis of a Rotation of the Galactic System." *B.A.N., 3*, 275 (1927).

(Oo27b) J.H. Oort, "Investigations Concerning the Rotational Motion of the Galactic System, Together with New Determinations of Secular Parallaxes, Precession and Motion of the Equinox," *B.A.N., 4*, 79 (1927).

(Oo65) J.H. Oort, "Stellar Dynamics" in *Galactic Structure*, A. Blaauw and M. Schmidt, University of Chicago Press, Chicago (1965), p. 455.

(Op39a) J.R. Openheimer and G.M. Volkoff, "On Massive Neutron Cores," *Phys. Rev., 55*, 374 (1939).

(Op39b) J.R. Openheimer and H. Snyder, "On Continued Gravitational Contraction," *Phys. Rev., 56*, 455 (1939).

(Op61a) A.I. Oparin, *Life, Its Nature, Origin and Development*, Academic Press, New York (1961).

(Op61b) A.I. Oparin and V.G. Fessenkov, *Life in the Universe,* Foreign Languages Publishing House, Moscow; also Twayne and Co., New York (1961).

(Ov69) O.E. Overseth, "Experiments in Time Reversal," *Scientific American,* October (1969) p. 89.

(Pa68) F. Pacini, "Rotating Neutron Stars, Pulsars and Supernova Remnants," *Nature, 219,* 145 (1968).

(Pe65) A. A. Penzias and R. W. Wilson "A Measurement of Excess Antenna Temperature at 4080 Mc/s," *Ap.J., 142,* 420 (1965).

(Pe69) J.V. Peach, "Brightest Members of Clusters of Galaxies," *Nature, 223,* 1141 (1969).

(Pe71) R. Penrose and R.M. Floyd, "Extraction of Energy from a Black Hole," *Nature, 229,* 177 (1971).

(Pi66) G.C. Pimentel, K.C. Atwood, H. Gaffron, H.K. Hartline, T.H. Jukes, E.C. Pollard, and C. Sagan, "Exotic Biochemistries in Exobiology" in *Biology and the Exploration of Mars,* C.S. Pittendrigh, W. Vishniac, and J.P. Pearman, Eds., Nat. Acad. Sci., NRC (Washington) (1966).

(Po56) S. Pottasch, "A Study of Bright Rims in Diffuse Nebulae," *B.A.N., 13,* 77 (1956).

(Po58) S. Pottasch, "Dynamics of Bright Rims in Diffuse Nebulae," *B.A.N., 14,* 29 (1958)

(Po68) G.G. Pooley and M. Ryle, "The Extension of the Number-Flux Density Relation for Radio Sources to Very Small Flux Densities," *MNRAS, 139,* 515 (1968).

(Po71) K.A. Pounds, "Recent Developments in X-ray Astronomy," *Nature, 229,* 303 (1971).

(Pr61) I. Prigogine, "Thermodynamics of Irreversible Processes," John Wiley, New York (1961).

(Pr68) K.H. Prendergast and G.R. Burbidge, "On the Nature of Some Galactic X-ray Sources," *Ap.J., 151,* L83 (1968).

(Ra71) D.M. Rank, C.H. Townes, and W.J. Welch, "Interstellar Molecules and Dense Clouds," *Science, 174,* 1083 (1971).

(Re44) G. Reber, "Cosmic Static," *Ap.J., 100,* 279 (1944).

(Re68a) V.C. Reddish, "The Evolution of Galaxies," *Quart. J. Roy. Astron. Soc., 9,* 409 (1968).

(Re68b) M.J. Rees, "Proton Synchrotron Emission from Compact Radio Sources," *Astrophys. Let., 2,* 1 (1968).

(Re68c) M.J. Rees and W.L.W. Sargent, "Composition and Origin of Cosmic Rays," *Nature, 219,* 1005 (1968).

(Re70) M.J. Rees and J. Silk, "The Origin of Galaxies," *Scientific American, 222,* (June 1970), p. 26.

(Re71) M.J. Rees, "New Interpretation of Extragalactic Radio Sources," *Nature, 229,* 312 (1971).

(Ri56) W. Rindler, "Visual Horizons in World Models," *MNRAS, 116*, 662 (1956).

(Ro33) H.P. Robertson, "Relativistic Cosmology," *Rev. Mod. Phys., 5*, 62 (1933).

(Ro55) H.P. Robertson, "The Theoretical Aspects of the Nebular Red Shift," *PASP, 67*, 82 (1955).

(Ro64a) Bruno Rossi, *Cosmic Rays,* McGraw Hill, New York (1964).

(Ro64b) P.G. Roll, R. Krotkov, and R.H. Dicke, "The Equivalence of Inertial and Passive Gravitational Mass," *Annals of Physics* (USA), *26*, 442 (1964).

(Ro65) F. Rosebury, "Handbook of Electron Tube and Vacuum Techniques," Addison-Wesley, Reading, Mass. (1965).

(Ro68) H.P. Robertson and T.W. Noonan, *Relativity and Cosmology,* Sanders, (1968).

(Ru71a) R. Ruffini and J.A. Wheeler, "Introducing the Black Hole," *Physics Today, 24*, 30, January 1971.

(Ru71b) M. Ruderman, "Solid Stars," *Scientific American,* February (1971), p. 29.

(Ry68) M. Ryle, "The Counts of Radio Sources" in *Ann. Rev. Astron. and Astrophys., 6*, 249 (1968).

(Sa52) E.E. Salpeter, "Nuclear Reactions in Stars Without Hydrogen," *Ap.J., 115* 326 (1952).

(Sa55) E.E. Salpeter, "Nuclear Reactions in Stars II. Protons on Light Nuclei," *Phys. Rev., 97*, 1237 (1955).

(Sa57) A. Sandage, "Observational Approach to Evolution-II.A Computed Luminosity Function for K0-K2 Stars from $M_v = +5$ to $M_v = -4.5$," *Ap.J., 125*, 435 (1957).

(Sa58) A. Sandage, "Current Problems in the Extragalactic Distance Scale," *Ap.J., 127*, 513 (1958).

(Sa66a) A.R. Sandage, P. Osmer, R. Giacconi, P. Gorenstein, H. Gursky, J. Waters, H. Bradt, G. Garmire, B.V. Sreekantan, M. Oda, K. Osawa, and J. Jugaku, "On the Optical Identification of Sco X-1" *Ap.J., 146*, 316 (1966).

(Sa66b) C. Sagan and R.G. Walker, "The Infrared Detectability of Dyson Civilizations," *Ap.J., 144*, 1216 (1966).

(Sa67) E.E. Salpeter, "Stellar Structure Leading up to White Dwarfs and Neutron Stars" in "Relativity Theory and Stellar Structure," Chapter 3 in *Lectures in Applied Mathematics*, Vol. 10., American Mathematical Society (1967).

(Sa68a) D.H. Sadler, "Astronomical Measures of Time," *Quarterly Journal, Roy. Astr. Soc.,* London, *9*, 281 (1968).

(Sa68b) E.E. Salpeter, "Evolution of the Central Stars of Planetary Nebulae: Theory," *IAU Symposium No. 34*, North-Holland, Amsterdam (1968) p. 409.

(Sa69a) E.E. Salpeter and J.N. Bahcall, "On the Masses of Quasi-Stellar Objects," *Ap.J., 158*, L 15 (1969).

(Sa69b) E.E. Salpeter, "Neutrinos and Stellar Evolution," from *Yeshiva University Annual Science Conference Proceedings,* Vol. II (1969).

(Sa70a) E.E. Salpeter, "Solid State Astrophysics," in *Methods and Problems of Theoretical Physics* J.E. Bowcock, Ed., North-Holland, Amsterdam (1970).

(Sa70b) C. Sagan, "Life," in *Encyclopedia Britannica* (1970).

(Sc44) E. Schrödinger, *What is Life,* Cambridge University Press, Cambridge England (1944).

(Sc58a) L.I. Schiff, "Sign of the Gravitational Mass of a Positron," *Phys. Rev. Let., 1,* 254 (1958).

(Sc58b) M. Schwarzschild, "Structure and Evolution of the Stars," Princeton University Press, Princeton, New Jersy (1958).

(Sc63) M. Schmidt, "The Rate of Star Formation II. The Rate of Formation of Stars of Different Mass," *Ap.J., 137,* 758 (1963).

(Sc70) M. Schwarzchild, "Stellar Evolution in Globular Clusters," *Quarterly J. Royal Astron. Soc., 11,* 12 (1970).

(Se65) P.A. Seeger, W.A. Fowler, and D.D. Clayton, "Nucleosynthesis of Heavy Elements by Neutron Capture," *Ap.J.* Suppl. No. 97, *11,* 121 (1965).

(Sh60) I.S. Shklovskii, "Cosmic Radiowaves," Harvard University Press, Cambridge, Mass. (1960).

(Sh66) I.S. Shklovskii and C. Sagan, "Intelligent Life in the Universe," Delta, New York (1966).

(Sh68) A.J. Shimmins, J.G. Bolton, and J.V. Wall, "Counts of Radio Sources at 2,700 MHz," *Nature, 217,* 818 (1968).

(Sh71) I.I. Shapiro, W.B. Smith, M.B. Ash, R.P. Ingalls, G.H. Pettengill, "Gravitational Constant: Experimental Bound on Its Time Variation," *Phys. Rev. Let., 26,* 27 (1971).

(Sp51a) L. Spitzer, Jr. and M. Schwarzschild, "The Possible Influence of Interstellar Clouds on Stellar Velocities," *Ap.J., 114,* 394 (1951).

(Sp51b) L. Spitzer, Jr. and J.L. Greenstein, "Continuous Emission from Planetarv Nebulae," *Ap.J., 114,* 407 (1951).

(Sp62) L. Spitzer Jr., *The Physics of Fully Ionized Gases,* Interscience, New York (1962).

(Sp71) H. Spinrad, W.L. Sargent, J.B. Oke, G. Neugebauer, R. Landau, I.R. King, J.E. Gunn, G. Garmire, and N.H. Dieter, "Maffei 1: A New Massive Member of the Local Group?" *Ap.J., 163,* L25 (1971).

(St39) B. Strömgren, "The Physical State of Interstellar Hydrogen," *Ap.J., 89,* 526 (1939).

(St65) B. Strömgren, "Stellar Models for Main-sequence Stars and Subdwarfs," in *Stellar Structure,* L.H. Aller and D.B. McLaughlin, Eds., University of Chicago Press, Chicago (1965).

(St68) F.W. Stecker, "Effect of Photomeson Production by the Universal Radiation Field on High-Energy Cosmic Rays," *Phys. Rev. Let., 23,* 1016 (1968).

(St69) T.P. Stecher, "Interstellar Extinction in the Ultraviolet II," *Ap.J., 157,* L125 (1969).

(St72) L.J. Stief, B. Donn, B. Glicker, E.P. Gentieu, and J.E. Mentall, "Photochemistry and Lifetimes of Interstellar Molecules," *Ap.J., 171*, 21 (1972).

(Ta70) B.N. Taylor, D.N. Langenberg, and W.H. Parker, "The Fundamental Physical Constants," *Scientific American*, October (1970) p. 62.

(Te48) E. Teller, "On the Change of Physical Constants," *Phys. Rev., 73*, 801 (1948).

(Te59) J. Terrell, "Invisibility of the Lorentz Contraction," *Phys. Rev., 116*, 1041 (1959).

(Te68) Y. Terzian, "Radio Continuum Observations of HII Regions" in *Interstellar Ionized Hydrogen* Y. Terzian, Ed., Benjamin Press, Menlo Park, Calif. (1968) p.283.

(Te72) Y. Terzian, *A Tabulation of Pulsar Observations*, Earth and Terrestrial Sciences, Gordon and Breach, New York (May 1972).

(Th69) D.J. Thouless, "Causality and Tachyons," *Nature, 244*, 506 (1969).

(Th72) R.J. Thompson, "Carbon Stars and The CNO Bi-Cycle," *Ap.J., 172*, 391 (1972).

(To47) C.H. Townes, "Interpretation of Radio Radiation from the Milky Way," *Ap.J., 105*, 235 (1947).

(Un69) A. Unsöld, "Stellar Abundances and the Origin of Elements," *Science, 163*, 1015 (1969).

(vdBe68) S.v.d. Bergh, "Galaxies of the Local Group," *J. of the Royal Astron. Soc. Canad., 62*, 145,219 (1968).

(vdBe72) S. v.d. Bergh, "Search for Faint Companions to M31," *Ap.J., 171*, L31 (1972).

(vdHu57) H.C. van de Hulst, "Light Scattering by Small Particles," John Wiley, New York (1957).

(v.d.Ka69) P.v.d. Kamp, "Alternate Dynamical Analysis of Barnard's Star," *Ap.J., 74*, 757 (1969).

(Va70) R.S. Van Dyck, Jr., C.E. Johnson, and H.A. Shugant, "Radiative Lifetime of Metastable $2\,^1S_0$ State of Helium," *Phys. Rev. Let., 25*, 1403 (1970).

(Va71) L.V. Vallen, "The History and Stability of Atmospheric Oxygen," *Science, 171*, 439 (1971).

(Ve69) G.L. Verschuur, "Further Measurements of Magnetic Fields in Interstellar Clouds of Neutral Hydrogen," *Nature, 223*, 141 (1969).

(Ve70) G.L. Verschuur, "Further Measurements of the Zeeman Effect at 21 Centimeters and Their Limitations," *Ap.J., 161*, 867 (1970).

(vHo55) S. von Hoerner, "Über die Bahnform der Kugelförmigen Sternhaufen," *Z. Astrophys., 35*, 255 (1955).

(vHo57) S. von Hoerner, "The Internal Structure of Globular Clusters," *Ap.J., 125*, 451 (1957).

(Wa34) A.G. Walker, "Distance in an Expanding Universe," *MNRAS, 94*, 159 (1934).

(Wa56) F.G. Watson, "Between the Planets," rev. ed., Harvard University Press, Cambridge, Mass. (1956).

(Wa67) R.V. Wagoner, "Cosmological Element Production," *Science, 155*, 1369 (1967).

(Wa71) R.V. Wagoner, "Production of Helium in Massive Objects," in *Highlights of Astronomy*, C. de Jager, Ed., I.A.U. (1971).

(We62) S. Weinberg, "The Neutrino Problem in Cosmology," *Nuovo Cimento, 25*, 15 (1962).

(We68) V. Weidemann, "White Dwarfs" in *Ann. Rev. Astron. and Astrophys., 6*, 351 (1968).

(We70) J. Weber, "Anisotropy and Polarization in the Gravitational-Radiation Experiments," *Phys. Rev. Let., 25*, 180 (1970).

(We71) A.S. Webster and M.S. Longair, "The Diffusion of Relativistic Electrons from Infrared Sources and Their X-ray Emission," *MNRAS, 151*, 261 (1971).

(Wh64) F.L. Whipple, "The History of the Solar System," *Proc. Nat. Acad. Sci., 52*, 565 (1964).

(Wi57) O.C. Wilson and M.K. Bappu, "H and K Emission in Late-Type Stars: Dependence of Line Width on Luminosity and Related Topics," *Ap.J., 125*, 661 (1957).

(Wi58) D.H. Wilkinson, "Do the 'Constants of Nature' Change with Time?" *Phil. Mag., Ser 8:3*, 582 (1958).

(Wi73) R. W. Wilson, A. A. Penzias, K. B. Jefferts, and P. M. Solomon, "Interstellar Deuterium: The Hyperfine Structure of DCN," *Ap.J., 179*, L107 (1973).

(Wo57) L. Woltjer, "The Crab Nebula," *B.A.N., 14*, 39 (1957).

(Za54) H. Zanstra, "A simple Approximate Formula for the Recombination Coefficient of Hydrogen," *Observatory, 74*, 66 (1954).

Index

IF NOT LITERATURE

Miami University, through an arrangement with the Ohio State University Press initiated in 1975, publishes works of original scholarship, fiction, and poetry. The responsibility for receiving and reviewing manuscripts is invested in an Editorial Board comprising Miami University Faculty.

If Not Literature

LETTERS OF ELINOR MEAD HOWELLS

✤ ———————————————————————— ✤

Edited by Ginette de B. Merrill and George Arms

Published for Miami University by the

OHIO STATE UNIVERSITY PRESS

Columbus

Library of Congress Cataloging-in-Publication Data

Howells, Elinor Mead, 1837–1910.
If not literature.
Includes index.
1. Howells, Elinor Mead, 1837–1910—Correspondence.
2. Howells, William Dean, 1837–1920—Correspondence.
3. Authors, American—19th century—Correspondence.
4. Wives—United States—Correspondence. I. Merrill,
Ginette de B. II. Arms, George. III. Title.
PS2033.H6 1987 818′.409 [B] 87–10816
ISBN 0–8142–0440–6

Contents

Illustrations

vii

Palazzo Giustiniani and Ca'Foscari. The Houghton Library.
Mary N. Mead (?), Le Havre. Courtesy of William White Howells.
Mary and von Dadelsen. Sketch by EMH, 13 May 1864. The Houghton Library.

Page 62
Plan of Palazzo Giustiniani apartment. The Houghton Library.

Page 63
Sketch of WDH during attempted burglary. The Houghton Library.

Following page 170
W. D. Howells in the 1870s. Howells Memorial, Kittery Point, Maine.
Winifred Howells, 1871. Herrick Memorial Library, Alfred University.
John Mead Howells. Courtesy of William White Howells.
Mrs. Reverend Guild. Sketch by EMH, 1872. Howells Memorial, Kittery Point, Maine.
The library at 37 Concord Avenue. The Houghton Library.
Mildred Howells. Herrick Memorial Library, Alfred University.
Winnie, Pilla, and Johnny. Sketch by EMH, 1 March 1874. The Houghton Library.
The W. D. Howells family, c. 1875. The Houghton Library.
Rutherford B. Hayes. WDH, *Sketch of . . . Rutherford B. Hayes,* 1876.
Lucy Webb Hayes. Howells Memorial, Kittery Point, Maine.
Winifred Howells, 1877. Taken in Quebec. Howells Memorial, Kittery Point, Maine.
Winifred as "Columbia." Herrick Memorial Library, Alfred University.
Anne Howells Fréchette. Herrick Memorial Library, Alfred University.

Following page 255
Winifred, John, and Mildred. Taken in Belmont, October 1878. Herrick Memorial Library, Alfred University.
Sketch of the Belmont house. McKim, Mead & Bigelow, 1877. Amherst College Library.
Redtop, Belmont. Exterior view with EMH, WDH, McKim, and Mead, October 1878. Courtesy of The New-York Historical Society, New York City.

Acknowledgments

The editors are greatly indebted to Professor William White Howells for his gracious cooperation in providing new materials from his collection, and for many other kindnesses. We are thankful also for permission to publish letters, diaries, photographs and sketches of Elinor Mead Howells and other members of the Mead and Howells families, as given by William White Howells for the heirs of W. D. Howells. Since this permission extends only to this book, republication of any such materials requires the same permission.

This work could not have been done without the full cooperation of institutions outstanding for their collections of Howells family papers. We are particularly grateful for the use of materials from the Howells Papers printed here by permission of the Houghton Library, Harvard University, and also for access to and permission to publish letters and illustrations from the William Dean Howells Memorial, Kittery Point, Maine.* Letters, quotations, photographs, and sketches in this publication, from the Howells/Fréchette Collection, are used with the kind permission of Herrick Memorial Library, Alfred University. We are grateful as well to the Massachusetts Historical Society in Boston for allowing us to print materials from their Howells, Mead, Noyes, and Dock Family Papers. Letters from the Howells family to

---

*The Memorial is owned and administered by Harvard University, and questions concerning access to it may be addressed to the Curator of Rare Books, the Houghton Library, Harvard University.

Mark Twain and his wife are printed courtesy of The Mark Twain Papers, The Bancroft Library, University of California, Berkeley.

The following have also generously given access to and permitted the use of letters, diaries, excerpts, photographs, artwork, and other documents: The American Antiquarian Society, Worcester, Massachusetts; the Trustees of Amherst College; Avon County Library, Bristol, England; Charles L. Burnett and Frances L. Burnett; Special Collections, Colby College Library, Waterville, Maine; James R. L. Gallagher; Katherine L. Mead Hasbrouck; the Rutherford B. Hayes Presidential Center, Fremont, Ohio; The Huntington Library, San Marino, California; Alexander R. James; the Museum of Fine Arts, Boston; The New-York Historical Society, New York City; the Ohio Historical Society, Columbus; Princeton University Library; the Schlesinger Library, Radcliffe College; the Watkinson Library, Trinity College, Hartford, Connecticut; and the Collection of American Literature, The Beinecke Rare Book and Manuscript Library, Yale University.

We warmly thank Polly Howells Werthman, Brooklyn, New York, for allowing the use of a complete letter and several excerpts and illustrations first printed in Mildred Howells' *Life in Letters of William Dean Howells*; David J. Nordloh, General Editor of the Howells Edition Editorial Board, for permitting the use of some letters as well as miscellaneous quotations from *Selected Letters of W. D. Howells* and from books printed in the Howells Edition; and Robert H. Hirst, General Editor of the Mark Twain Papers, for letting us use letters and quotations first printed in the *Mark Twain-Howells Letters*.

We owe a special debt of gratitude to many librarians without whose help this book could not have been completed. Particular thanks go to the staff of the Houghton Library: Rodney G. Dennis and Roger E. Stoddard, who went out of their way to make materials from the recently acquired Howells Memorial Collection at Kittery Point, Maine, available at the earliest possible time, and for many other courtesies, and to Susan Halpert, Marte Shaw, and Thomas Noonan for their patience and gracious assistance with questions and requests; also to Norma Higgins, Herrick Memorial Library, Alfred University, for her enthusiastic cooperation and her hospitality. We are grateful also to John Lancaster and Daria D'Arienzo, Amherst College Library; Sandra Burrows and Franceen Gaudet, National Library of Canada; Thomas A. Smith and the late Watt P. Marchman, Rutherford B. Hayes Presidential Center; the staffs of the Huntington Library and of the Massachusetts Historical Society; Connie C. Thorson, Dorothy A. Wonsmos, and others of the University of New Mexico Library; Elizabeth Shenton, the Schlesinger Library, Radcliffe College; and Jeffrey H. Kaimowitz and Margaret F. Sax, of the Watkinson Library.

The editors are pleased to thank many individuals who have contributed encouragement and advice and helped in many ways. For reading the manuscript in part or in whole and advice on it: Sara Blackburn, New York City; Edwin H. and Norma W. Cady, Duke University; John W. Crowley, Syracuse University; David L. Frazier, Miami University; the late William M. Gibson, formerly of the University of Wisconsin–Madison; Mrs. G. K. Prosser, Cambridge, Massachusetts; Randolph Runyon and Peter W. Williams, Miami University; Charlotte Dihoff, Ohio State University Press; and Nancy Woodington.

Individually we wish to acknowledge our debts to the following also for miscellaneous courtesies and advice: George Arms, to James F. Barbour, Ernest W. Baughman, Robert E. Fleming, Sam B. Girgus, Catherine L. Martin, David C. McPherson, Ivan P. Melada, Mary Bess Whidden, and Peter L. White, all of the University of New Mexico; and to William C. Dowling, Cincinnati, Ohio; Hamlin Hill, Texas A&M University; Louis G. Locke, James Madison University; Christoph K. Lohmann, Indiana University; George Monteiro, Brown University; Frederick W. Williams, Pompano Beach, Florida; and especially to my wife Elizabeth T. Arms. Ginette de B. Merrill, to Karl Beckson, Brooklyn College City University, New York; Mr. and Mrs. W. E. Benua, Columbus, Ohio; Shepard Bright Erhart, Franklin, Maine; Alan Emmet, Westford, Massachusetts; the late R. H. Ives Gammell, Boston; Mrs. John Groden, Lexington, Massachusetts; Mrs. Hugh Hencken, Boston; the late John N. M. Howells and Mrs. John N. M. Howells, Portsmouth, New Hampshire; Mrs. William White Howells, Kittery Point, Maine; Dickie Sayre Miller, Columbus, Ohio; Leland M. Roth, University of Oregon; Diana Royce, the Stowe-Day Foundation, Hartford, Connecticut; Richard Guy Wilson, the University of Virginia; and my husband, Edward W. Merrill.

Short Titles and Abbreviations

Annals	*Annals of Brattleboro, 1681–1895,* by Mary R. Cabot. Brattleboro, Vt.: E. L. Hildreth & Co. Press, 1921.
Cady, I	Edwin H. Cady, *The Road to Realism: The Early Years, 1837–1885, of William Dean Howells.* Syracuse, N.Y.: Syracuse University Press, 1956.
Cady, II	Edwin H. Cady, *The Realist at War: The Mature Years, 1885–1920, of William Dean Howells.* Syracuse, N.Y.: Syracuse University Press, 1958.
Doyle	James Doyle, *Annie Howells and Achille Fréchette.* Toronto: University of Toronto Press, 1979.
Edel, *Untried Years* Edel, *Conquest of London* Edel, *Middle Years* Edel, *Treacherous Years* Edel, *The Master*	For the five unnumbered volumes of the Leon Edel biography with the main title of *Henry James.* Philadelphia: J. B. Lippincott Co., 1953–1972.
Gibson-Arms	William M. Gibson and George Arms, *A Bibliography of William Dean Howells.* New York: New York Public Library, 1948.

Hayes, *Diary*	*Diary and Letters of Rutherford B. Hayes,* ed. Charles R. Williams. 5 vols. Columbus, Ohio: Ohio State Archaeological and Historical Society, 1922–1926.
Hay-Howells Letters	*John Hay–Howells Letters: The Correspondence of John Milton Hay and William Dean Howells, 1861–1905,* ed. George Monteiro and Brenda Murphy. Boston: Twayne Publishers, 1980.
HE	Howells Edition volumes, e.g., *Years of My Youth,* ed. David J. Nordloh. Bloomington, Ind.: Indiana University Press, 1975.
Henry James Letters	*Henry James Letters,* ed. Leon Edel. 4 vols. Cambridge: Harvard University Press, 1974–1984.
LinL	*Life in Letters of Willian Dean Howells,* ed. Mildred Howells, 2 vols. New York: Doubleday Doran & Co., 1928.
Lynn	Kenneth S. Lynn, *William Dean Howells: An American Life.* New York: Harcourt Brace Jovanovich, 1971.
Meserve	Walter J. Meserve, ed., *The Complete Plays of W. D. Howells.* New York: New York University Press, 1960.
Noyes Descendants	*Genealogical Record of Some of the Noyes Descendants of James, Nicholas and Peter Noyes.* Boston, 1904.
SL	*Selected Letters of W. D. Howells,* ed. George Arms et al. 6 vols. Boston: Twayne Publishers, 1979–1983.
Twain–Howells Letters	*Mark Twain–Howells Letters: The Correspondence of Samuel L. Clemens and William D. Howells, 1872–1910,* ed. Henry Nash Smith and William M. Gibson. 2 vols. Cambridge: Harvard University Press, 1960.
Woodress	James L. Woodress, Jr., *Howells & Italy.* Durham, N.C.: Duke University Press, 1952.

W. D. Howells and his wife Elinor are frequently but not always referred to by their initials, EMH for Elinor Mead Howells and WDH for William Dean Howells.

FAMILY GENEALOGIES

Elinor Gertrude Mead

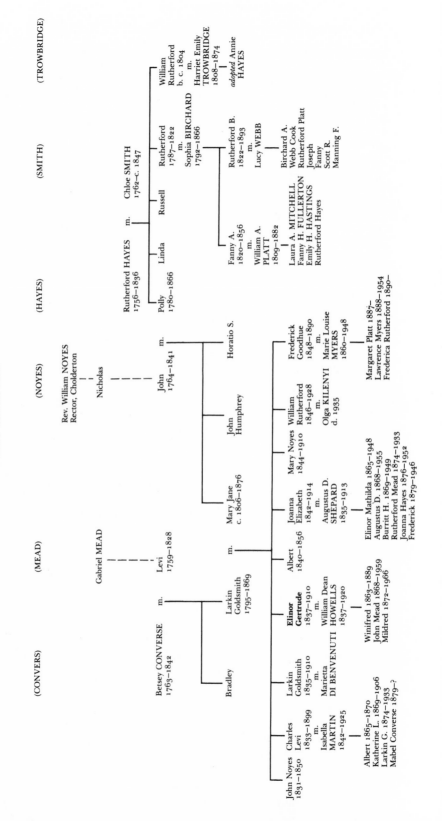

FAMILY GENEALOGIES

W. D. Howells

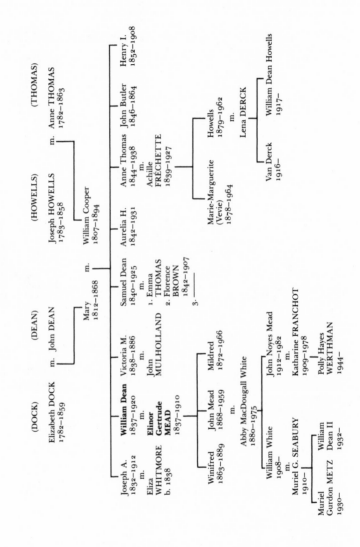

Notes on Editorial Practice

The number of letters by Elinor Mead Howells known to the editors is about 200, of which about 130 appear in this book in full, except for a few in substantial part. Those which are omitted did not have sufficient content to justify printing in the judgment of the editors and their advisers. We have also included about twenty-five passages from her Venetian diary (1863–1869), ten letters to her, and ten by immediate members of the family, mostly her husband.

A great number of Elinor's extant letters are to members of Will Howells' family, saved by them and given to Mildred Howells at the time she edited her father's letters. These are now at the Houghton Library, Harvard University. Some were retained by Anne H. Fréchette and eventually given to Herrick Memorial Library, Alfred University, along with a large and impressive archive of nearly 7,000 family letters, plus photographs and other memorabilia. Generally the Mead family saved few letters (for example, Elinor wrote weekly letters to her mother, only one of which is extant). Still, there are such happy exceptions as the letters about the building of Redtop that she wrote to her brother, who left them to the Amherst College Library, and a few family letters gathered by Elinor and given to the Massachusetts Historical Society. Most letters by Elinor to her husband and children have also disappeared, aside from those to her son John written in 1904–1905 from San Remo, which she suggested should be kept as a record of the winter spent there. Why other family letters by Elinor were not saved is not known. In spite of these and other losses, the many letters that remain provide rewarding materials.

In transcribing and printing the letters, we have made it our primary concern to provide texts as close as possible to those of the original manuscripts. Though the following details of treatment do not cover every textual situation in the book, they explain our general practice.

1. *Headings* appear in large italics and name the recipient when, as most of the time, the letter is by Elinor Mead Howells, e.g., *"To Mary J. Mead."* On those few occasions when the letter is by someone else, the headings appear with the writer's name followed by the recipient, e.g., *"Henry James, Jr., to EMH"* or *"WDH to EMH."*

2. *Addresses and dates* appear in their original locations, usually at the upper right of the letter, except that dates for diary entries appear at the upper left (as they usually do in the original). When addresses do not include the city, that has been enclosed in brackets, and when addresses are a part of printed stationery, they have been italicized. When dates are incomplete, missing elements have also been given in brackets when known from other evidence. The left-hand margins of the address and date lines are evenly blocked, with the longer or longest line run flush to the right margin.

3. *Complimentary closes and signatures* appear at the lower right of the letter, even though their original positions may be at the lower center or somewhere between. Here the lines are run flush to the right margin, with uneven indentations on the left usually resulting.

4. *Cancellations, insertions, and overwriting* are not recorded unless, as on a few occasions, they are significant, when annotations comment upon them. Judgment of significance is admittedly subjective.

5. *Misspellings* are generally preserved. When they seem likely to confuse readers, bracketed letters are introduced, as in "two napoleon[s]." A word inadvertently repeated, for example, "did did," is silently omitted. Inadvertent omissions of quotations or parentheses either at the beginning or end of the pairings are supplied. In general, brackets for misspellings and abbreviations appear in the text as infrequently as possible. The abbreviation "se" appears to stand for "some."

6. *Brackets* also are used to mark illegible words and manuscript imperfections, the nature of which is explained in notes; sometimes we have supplied conjectural readings, but when no fairly certain

guess is possible we have simply bracketed "word," and added the approximate number of words when more than one is involved.

7. *Common abbreviations* are observed, including ampersands.

8. *Superscript letters* such as *st* or *rd* following numbers or the *r* and *rs* with *Mr.* and *Mrs.* are retained. They are underscored or not as they appear, sometimes both ways within the same letter.

9. *Verbal contractions* appear as they are written ("was'nt"), for though not authorized by nineteenth-century stylebooks, irregularities are frequent enough to suggest general acceptance. Spacing ("could n't") is preserved when it definitely appears.

10. *Periods* are silently introduced into the text when there is no punctuation (except at the end of paragraphs) or when a comma or very short dash appears to serve as a period. The test here has been whether a capital letter follows or not (granting the ambiguity when the letter is a normally capitalized "I" or a proper noun). On the other hand, periods are replaced by commas when the habit of a writer indicates the use of periods or what look like periods for a comma.

11. *Commas* are silently introduced if their lack would imperil reading. Though we usually do not tamper with comma-spliced sentences, when there are three or more sequential comma splices, a semicolon is sometimes substituted for one of the commas.

12. *Dashes,* other than the very short period/dash previously mentioned, are preserved between sentences, with periods only when they appear.

13. *Postscripts* appear at the end, though they may originally have appeared at the beginning of a letter or in margins.

14. *Asterisks* are repeated (whether the letter repeats them or not) with the notation at the bottom of the page. For a second notation on the same page various signs, for example an original raised circle, have been converted to double asterisks.

15. *Paragraph indentations* appear when they exist and when the preceding line is not filled out to the right margin. When indentations do not appear after salutations, they are introduced.

16. *Library locations* of manuscripts or other sources appear at the end of the letter.

17. On the few occasions when we print letters that have already appeared in other volumes, we have derived our texts from the originals when they are available. As a result, readings sometimes differ from the previously printed texts, though these occasions are few and minor, at most a word or mark of punctuation.

Introduction

For so many women married to nineteenth-century writers in America, the acquaintance of most readers begins with the patriarchal stereotype—literary wives ritually adored for no pronounced reason by their husbands, reverently neglected by earlier biographers, and frequently treated with animosity by chroniclers until and even through the mid-twentieth century. Thus at first thought, even with Edward Wagenknecht's revealing biography of her in 1956, Frances Appleton Longfellow is the wife who had the misfortune to set fire to her dress and die before the day was over, and Olivia Langdon Clemens, in spite of all her defenders, still lingers in memory as the watchdog of her husband's cultural heresies. Increasingly, however, the stereotypes have emerged as real people, whether they were the wives and relatives of American statesmen, like Abigail Adams, Eleanor Roosevelt, and Bess Truman, or of authors, like Sophia Hawthorne, Clover Adams, and Alice James. By good luck Elinor Mead Howells did not make such a sad beginning as some of these others because her daughter Mildred, a talented artist and poet, first presented her in the *Life in Letters of William Dean Howells,* where by and large Elinor's vigorous character is shown. Still, a recent biographical study casually reverts to chauvinism—"one might say that both Howells and Mark Twain had bad luck with their wives."

For the present editors, as for others who have come to know her, it is the vibrancy of Elinor that makes the major impression, despite the limitations of the society and time into which she was born. Almost unanimous testimony points in this direction. When her cousin Rutherford B. Hayes, then a lawyer in his late thirties, visited the Meads in the summer of 1858, he singled out Elinor as "witty, chatty, and capital company. . . . not ill-natured in her satire. I like it."[1] After her visit to the Clemenses in 1875, Olivia spoke of her as "*exceedingly bright—very intellectual—sensible and nice.*"[2] More vividly, Mark Twain, in an 1880 letter to her, characterized her as "a sort of Leyden jar, or Rumkoff coil, or Voltaic battery." At times indeed she may have carried too high a charge for him, as when on another occasion he characterized her as a nonstop talker—dialogue died when she appeared, and monologue (*her* monologue) took over. And perhaps with a similar touch of envy her husband said of the elevator installed in the Beacon Street house in the 1880s that it "shoots Elinor up to her bedroom like a rocket, leaving a brilliant trail of conversation behind her."[3]

While essentially a private person, she was alert to public matters in both politics and literature. If the president she knew best scarcely qualifies as a moving force, another cousin, Edwin D. Mead, edited the *New England Magazine,* which during the crucial decade of the 1890s kept the social consciousness of New England alive to its responsibilities. The authors of the day she knew not only through her husband but at first hand, and she did not hesitate to voice complaints as well as praise. From Lowell ("awfully schoolmasterish") to Garland ("a rough one"), she saw flaws, though she still admired them, as they admired her; and while Henry James was her abiding favorite, she also recognized his fussiness. Within the family circle she encouraged Will's sister Annie in her literary and journalistic career and rejoiced in her daughter Mildred's accomplishments, even while tartly remarking "with all her native graciousness" (as her husband phrased it) that an editor of one of the Harper's magazines *had* to take a poem by her because of her father's connection with the publisher.[4]

On both sides of her family, the Noyeses and the Meads, Elinor came from old New England stock, her ancestors having emigrated from England in the mid-1630s. John Noyes, Elinor's maternal grandfather, a Vermont businessman and a member of Congress, had become the wealthiest man in Windham county. Her maternal grandmother, Polly Hayes, was a tall, gaunt, striking redhead who was warmhearted, impulsive, quixotic, passionately interested in many causes, and an indefatigable writer of letters and journals. Chloe

Smith, mother of Polly and great-grandmother of Elinor, had married Rutherford Hayes, a former blacksmith turned farmer and innkeeper in West Brattleboro, Vermont. Their grandson would become President. Chloe, the "indomitable,"[5] deeply religious, never idle, loving, dutiful, renowned for her artistic embroidery and knitting, left a diary cherished by her descendants. John Humphrey Noyes, son of John and Polly and Elinor's uncle, made his mark first in Putney, Vermont, and later in Oneida, New York, as the charismatic leader of the Community of Perfectionists and advocate of Complex Marriages, the "eccentric kinsman,"[6] whose antics would briefly cause family turmoil; while a perfectionist belief in a state of attainable sinlessness appears in Christian thought with some frequency, the view of monogamy as a sinful manifestation of selfishness was bound to prove exacerbating to neighbors.

On the paternal side, Elinor's grandfather Levi Mead had fought in the Revolution after witnessing at age sixteen the Battle of Lexington on the common near his father's farm. In 1801 he moved with his family to Chesterfield, New Hampshire, where Elinor Gertrude Mead was born on 1 May 1837. She was two when her parents settled across the Connecticut River in Brattleboro, a decade or so before the heyday of the Water-Cure which was to transform the little town into a national spa and a cultural center. "Squire" Mead, her father, was an esteemed country lawyer, later a state senator, and a leading citizen who founded the first savings bank in Vermont, serving as its earliest president and then as treasurer; he also helped establish the new high school and the town library. Active in the town and highly respected, Mary J. Mead was a fine mother of an interesting group of children, many of them with artistic talents, not surprisingly in view of the sitting-room table's being furnished with "a piece of gamboge, a cake of Prussian blue, and one of crimson lake, so that any member of the family could pause in passing through the room to do a little painting.":[7] Elinor's oldest brother, John, who succumbed to erysipelas after his third year at Harvard, was a musician and caricaturist; her brother Larkin would become a distinguished sculptor; and her younger brother Willie was later the Mead of the architectural firm of McKim, Mead & White. But the other two sons who grew into adulthood became businessmen, and unlike Elinor the younger two daughters, Joanna and Mary, do not appear to have had artistic leanings. A close-knit family who had "the gift of making others happy,"[8] they were cheerful and optimistic, friendly and outgoing, fun-loving and mutually supportive; they also worked hard and had New England consciences. Mary J. Mead briefly accepted the new religion of her brother, but soon withdrew, though her husband, denounced on one occasion by John Humphrey Noyes as a "Unitarian lawyer,"[9] successfully defended him when he was arrested on the charge of adultery in 1847 (Elinor was ten); but Polly Hayes Noyes, in spite of criticizing some of her son's tenets, remained a member of his community and recorded her spiritual concerns, as well as her reflections on the condition of women, in letters and diaries. Thus the Meads, including Mrs. Mead, appear to have remained largely untouched by John Humphrey Noyes's religious heresies.[10] Perhaps John Fiske, the popular philosopher and historian and a neighbor of the Howellses for a long time, sums up the sense of the family when he remarked, "Never yet saw a Mead that I didn't like," mentioning the brightness, sweetness, and simplicity of the whole family.[11]

Elinor's mother had gone to the well-known school of Sarah Pierce in Litchfield, Connecticut, for a year, and then with her two sisters and her brother John to the Amherst Academy. Elinor and her brothers and sisters received most of their schooling in Brattleboro, probably one reason for their father's interest in the establishment of a high school. Her own letter of 1848 does not express much enthusiasm, though her father's letter to her in the same year notes that she is "very much pleased" with the school. Two bits of evidence suggest her superiority as a student: not only did she tutor another student in mathematics, but years later, in 1861–1862, Clara Nourse, a former teacher in Brattleboro, invited her to Cincinnati to teach at the school she had recently founded, where incidentally Miss Nourse guessed Elinor's engagement before she had announced it to her family.

Though in his 1848 letter to Elinor her father (who played the violin) urged her to practice her music lessons with the expectation that she would later teach music, evidently she had turned to sketching and painting well before 1855, when letters from Brooklyn by her brother Larkin show a keen interest in her accomplishments, though as her slightly older brother he has his moments of patronizing: she will never be anybody "with her laying abed until nine in the morning." "Paint from nature as she sees it. Look sharp."[12] But all the same, Larkin reports on a couple of occasions that George Fuller, already an established painter, had expressed great hopes for her as an artist.

Elinor's artistic accomplishments, along with her sense of fun, had a part in attracting Will Howells to her, for a fairly well authenticated reminiscence recalls that her sketching a caricature of the sculptor J. Q. A. Ward on a fan when they were at a dance had first endeared her to him. On her part, she found in Will a lively newspaper columnist, an author of a book of poems and a campaign life of Lincoln, and perhaps most impressive of all a contributor of poems to the *Atlantic Monthly*. They had met in Columbus when she visited her cousin Laura Platt in the winter of 1860–1861. While no letters between them during their courtship have survived, a few letters by each and the reminiscences of themselves and others reveal a good deal of their story.[13] For example, the letter by Elinor to Lucy Webb Hayes on her engagement and Will's diary entry earlier in the year, both given in this book, provide valuable insights. Instead of Will's returning to America for the marriage, as they had first planned, Elinor sailed to Europe with her brother, was married in Paris on Christmas Eve 1862, and settled with Will in the first of their two Venetian apartments, Casa Falier. Their marriage was fulfilling and their life together brimming with fun. On 17 December 1863, Elinor gave birth to her first child, Winifred. Two more children, a boy and a girl, would follow a few years after the Howellses had left Venice for America in 1865.

From Elinor's arrival in Venice, with the encouragement of her husband, she appears to have become more actively an artist than she had been before. She drew and painted at the Academy, on street corners, inside palaces, and on excursions the couple took. One of her illustrations for a poem by her husband, "St. Christopher," appeared in *Harper's Monthly* in December 1863, and she did several illustrations for Will's Venetian poem, *No Love Lost*, which because of the vagaries of a publisher did not appear until 1869, when three of her illustrations (unsigned) were rendered as engravings.

But Will may also have pushed her more into becoming his office secretary than she wished. And is she entirely jesting or partly criticizing when she writes this complaint in the third month of her marriage? "It is with timid strokes that I begin a letter now-a-days, for everyone of them has to be submitted to the critical inspection of Mr. Howells before it leaves the house, and it's as often they go into il fuoco [the fire] afterwards as to the Post Office." She continues, "But 'It's never too late to mend', says the title of Pokey's (I knew that would escape me. It's my pet name for Mr. Howells) book. . . ." Perhaps it is all a joke—perhaps just enough of a joke to let her husband's family know of her independence. Finding Elinor too intellectual, they had not regarded the marriage with enormous delight. But a series of skillful letters from Elinor reconciled them to it.

Elinor's interest in art continued well beyond the Venetian years and indeed throughout her life. In 1869 she also took drawing lessons in Boston, at a time when her husband was thinking of writing up Winny's first year in Venice for a children's book, though the project was dropped. In 1874 she did a charming watercolor of the children for Will's birthday. While in 1876 Elinor wryly admitted to a correspondent that her art now consisted of decorating a bureau with flowers, copied at that, in the same letter she wrote perceptively about the Frank Duveneck "rage" in Boston. And she would continue to embellish her letters with sketches. Still, in the building of the two houses, that on Concord Avenue in 1872–1873 and at Belmont in 1877–1878, she turned noticeably to architecture. With the first she acted as her own architect and interior designer. Will was particularly proud of the new library, with its chestnut bookcases, frescoed ceiling, and mantelpiece, the "glory" of the room. As he reported to his father after they had moved in on 7 July, it was "splendidly carved, and set with picture-tiles and mirrors; on either jamb of the mantel" was Will's monogram "carved, and painted by Elinor, who modified and improved the carpenter's design of the whole affair."[14] With Redtop, the Belmont house, the architects were McKim, Mead & Bigelow, though Elinor's thorough participation in the building becomes evident from the many letters that she wrote her brother, a flurry of notes and telegrams (twice a day at times) arguing about plans or altering them, offering suggestions and criticism, or simply urging faster progress.

While other letters show that Elinor's awareness of the graphic arts remained as sharp as ever, it is still somewhat disappointing that she did not continue more regularly to produce the illustrations and paintings of her earlier years. Will appears to have had the highest respect and enthusiasm for her art, and as she wrote in her diary when "St. Christopher" was accepted two months before Winifred's birth, "Mr. Howells [not Pokey on this occasion] has been the means of bringing me before the public as an artist."[15] Still, he may unconsciously have placed her work in a secondary position to his own writing, in which she was fully engaged and which did provide the family income. Given the demands upon her time and energy as mother, as housekeeper (planning meals probably did not rank high, since from the Venetian to the New York years dinners were frequently sent in or eaten in restaurants), and as hostess in partly social, partly professional entertaining, and given the accepted Victorian view of a wife's place, the outcome is not surprising. A comment to her sister-in-law Annie at the time of her engagement implies Elinor's acceptance of the convention: "I believe in marrying, I believe you would be happy. But you can be happy without [marrying] as a distinguished authoress. I dont believe you'd write if you got married." Yet Annie did continue to write after her marriage to Achille Fréchette, and while she never attained wide recognition, her ability to make occasional appearances in such magazines as the *Century, Harper's Bazar,* and *McClure's* at least partly counters Elinor's foreboding.

Also ambivalent was Elinor's attitude toward going out and entertaining. In Brattleboro as a girl and young woman she had taken a full part in the social life of the place, which was especially active in the summer. Though at the beginning of the Venetian years they had little company, increasingly they entertained visitors from the United States and made trips, most notably to Florence, where Elinor's brother Larkin was, in April 1863, and to Naples and Rome in October 1864. For all their happiness in each other and in the picturesqueness of their surroundings, they were sometimes lonely and homesick and greatly missed the informal and easygoing ways of Columbus and Brattleboro. In Venice, a city in mourning under the Austrian occupation, there was no social life such as Elinor and Will had known at home, and to be seen with the Austrians meant siding against the Italians. Thus forced into isolation and worried over the fate of their country while the Civil War raged in America, the young consul and his wife were hungry for news and good talk, which indeed they enjoyed more and more from such visitors as Henry Ward Beecher, Moncure D. Conway, an old friend of both from Ohio days, and Charles Hale.

Throughout the earlier years in Cambridge, where they lived from 1866 to 1878, Elinor also fully engaged herself in the cultural activities, the more so with the move in 1870 from Sacramento to Berkeley Street, the house where they entertained the Bret Hartes at their first big Cambridge reception, perhaps partly anticipated by a remark of Will's to Henry James earlier in the year, when he spoke of Elinor's "having got very much better and having some faint aspirations" in the direction of "Cambridge society."[16] Still, after only four years in their Concord Avenue house, Elinor and Will began to resent the pressures of social involvements. They wished to put an end to this "most foolish existence,"[17] and saw in country living an advantage for both themselves and their growing children. Though at first Redtop in Belmont brought more visitors than they had had in Cambridge because of its architectural distinction, they later found in it their ideal for a time. Soon, however, Belmont seemed too isolated, and they became restless on their hilltop. As early as 1881, when Will gave up the editorship of the *Atlantic,* they began to think of going to Europe for a while.

After their relatively sequestered life in a pastoral setting, the European trip proved a joy to both, their intense activity mirrored in Elinor's letters from London. There they renewed American friendships under the sponsorship of such authors as Lowell, then minister to England, and their close Cambridge friend Henry James. Additional zest resulted from making new English acquaintances, among them the Gosses and Tademas. After a brief respite in Switzerland, they went to Florence, where lionization reached perhaps too much of a climax. "I suppose we had two hundred calls in Florence," Elinor wrote to Annie from Siena, where they fled for a few quiet weeks, and "I had the worst of it—for I had to make calls *every* afternoon—& then perhaps go out to dinner afterwards." To make matters worse, their daughter Winny overdid at parties and suffered a relapse into the illness that had been haunting her for the past two years, but after Siena they went to Venice, where the pace was less frantic and where Winny, born there, made what seemed to be substantial progress. Her father felt that she had recovered "in her native air as if by magic" and was "in a romantic rapture with the place."[18] Throughout the trip they had the pleasure of watching their other two children, John (at fourteen) and Mildred (ten), seeing Europe for the first time and responding to it. Also, for both of them the return to Europe brought pleasure, with Will no longer an apprentice but an established writer to whose career Elinor had markedly contributed.

That contribution would continue throughout her lifetime. Already in Venice she had illustrated his poems and also some of the essays that became *Venetian Life* and *Italian Journeys,* though her sketches for these books were never published. Elinor was far more aware of the history and techniques of painting than her husband, and her influence on these two works is evident from observations in her Venetian journal. Even more significantly, this influence appears to have extended in other directions, so that a kind of literary partnership, probably unique in the period, began, lasting until her death. How widely recognized this situation was at the time is hard to ascertain, but in one of her scrapbooks Elinor pasted a newspaper clipping which relates the experience of a woman who occupied the room next to the Howellses' at the Mountain House in Princeton, Massachusetts, when Will was completing *A Chance Acquaintance* during the summer of 1872. The unwitting eavesdropper told the reporter that she had "heard him read chapter after chapter aloud to his wife, she frequently interrupting him while reading with 'You must alter this passage, it won't do at all as it is,' or, 'I like that, it is very good,' etc." Elinor's daughter Mildred no doubt echoes a family tradition when she remarks of the couple's Venetian years that "she told him all about everyone and everything in Brattleboro" and that "It was this intensive view of New England that made Howells able to understand it so clearly when he went there to live, and it was his wife's vivid powers of observation and her gift for criticism that made her such a great help to him in his work. She had a wonderfully true sense of proporion both in art and literature, and though she could never argue them out, her intuitive criticisms of books and pictures were almost unerring." John had already written in a similar vein after he read an early version of a passage in *Years of My Youth* (1916). As his father reported to Mildred, "He felt . . . that it was due to mamma's memory that I should recognize the great part she had in helping me to be the writer I have become."[19]

In *Years of My Youth,* with *My Mark Twain* (1910) one of the two major books he wrote without this quasi-collaboration, he told about her centrality in his writing, a passage much revised because he still found it so difficult to write about her publicly: "We were married the next year [after their meeting], and she became with her unerring artistic taste and conscience my constant impulse toward reality and sincerity in my work. She was the first to blame and the first to praise, as she was the first to read what I wrote."[20] His own letters are full of her presence, both during her lifetime and after her death. As he wrote Henry James in 1911: "With my children I always talk gayly of their mother, not purposely, but because her life and mine were mostly a life of pleasure in the droll and amusing things. I recall our thousand and one experiences in strange character, and first of all our own characters; and the pang is no longer a sense of helpless loss, but of wonder that I did not make more of her keenly humorous criticism of all that we knew in common."[21] And he wrote to Elinor's and his

longtime friend Laura Platt Mitchell after finishing *Years of My Youth*:
"It has been a sore labor to me, and it seems to me I have not done
it in the right way, the way Elinor would have made me do it."[22]

Since the Howellses had given up Belmont with their return from
Europe in 1882, they lived in Boston, first on Louisburg Square,
partly in the hope that an active metropolitan life was what Winny
needed for her well-being. But Elinor was active too, as a series of
letters to her friend Lilian Aldrich make clear. The buying of a house
on Beacon Street in the early summer of 1884 shows that the new
arrangement suited them. Here they entertained their new British
friends the Gosses at the end of 1884, an occasion that overstrained
Elinor. To Mrs. Aldrich she wrote at the beginning of 1885, after
catching a cold at a dinner party: "I shall have to give up society
altogether I find. *My* constitution cant stand this sort of thing." Society
also took its toll on Winny, who would have made her debut that year;
they left Beacon Street for the next year, then tried it again, finally
sold it in 1887, and after a frustrating summer with Winny at Lake
George, took her first to a sanatorium in upstate New York, then on
the advice of the doctors left her there as they went to nearby Buffalo.

With the celebration of the Howellses' silver wedding anniversary
in Buffalo on 24 December 1887 (Winny, who was absent, sent a re-
membrance from her sanatorium in Dansville), it is appropriate to
review Elinor's relationships with her children up to this point and
beyond and have a brief look at their lives. Since 1863 Elinor had
devoted large parts of her Venetian diary to Winny's baby days. Beginning in 1871 and throughout the next ten years she also had kept
a scrapbook entitled "Some Sayings & Doings of Winnie & Johnny
Howells," with Mildred's name inserted as "Peelah" sometime after
her birth.[23] The jottings make up a remarkably frank account of her
children and of their relation with their mother and father. "They call
me 'The Orgress' in the most innocent way to my face." On another
occasion John's remark to Winny, "By accident *Pap*'a got rather a cross
lady," is recorded. The children liked Will's easygoing ways with them
rather better than Elinor's disciplinarian stance, though all three (even
Winny, somewhat overobedient at times) were bright and saucy with
both parents, who by and large enjoyed the situation. Thus Elinor
makes an entry that Pilla (which soon became the standard spelling
of her nickname), reproved for dawdling while dressing, remarked to
her mother, "*you* have never had the experience of a full brain," and
that her first irony occurred at the age of four when she found her

father writing a note in the nursery and observed, "I did n't *know* this was your *library!*" Among the anecdotes appears praise for all the children: Winny is a "good scholar" with "patient perseverance," Johnny is a "darling good boy" with talent in drawing, and Pilla is "brightest of all three."

Back in early 1864, upon hearing of Winny's birth, Elinor's mother had written a letter of wise advice urging her daughter not to make too much of the baby or to "show her off."[24] Like most parents with their first child they probably did, to judge from their letters and Elinor's diary on her progress. Yet up until the move to Belmont, when Winny was fifteen, she appears to have been a happy child, perhaps with a little more hypersensitiveness and obstinacy than might be regarded as normal, and certainly with an ambition to be a poet that began unusually early and that was fostered by her parents. Although early in the Belmont stay she was active in devising games and other activities for the younger children, including those of their neighbors, the Fairchilds, there are some signs of retreat into passivity, even though in her Cambridge school she was first in every subject, much to Elinor's delight. Then in November 1880, soon after a juvenile magazine had published a poem that had been praised by Longfellow (which filled Elinor with pride), Winny began to complain of headaches and fatigue and vertigo, the onset of a long and baffling disease ending only with her untimely death in 1889. For Elinor especially these nine years were fraught with feelings ranging from (false) hope to despair as Winny's illness took turns for the better or the worse, with declines outweighing advances in spite of changing treatments and physicians. Unable to make any long-range plans, the parents began more and more to realize what was happening without quite admitting it. Initial concern, worry, and sympathy gave way over the years to bafflement and dismay. Impatience, even exasperation, and a battle of wills developed between mother and daughter as Winny complained of constant pain and refused to eat or leave the house (wrapped up in shawls, she dreaded the cold even in summer). The last desperate attempt to follow doctors' orders and prevent her from starving to death—a nightmare and a total failure—left Elinor nervously exhausted herself and on the verge of a breakdown.

From the beginning, no one knew precisely what was the trouble, as the family doctor passed her on to a nervous specialist, as he was dismissed and then recalled, as a variety of cures was sought, as the parents themselves thought that they could cure her by being firm during the disastrous summer (1888) at Little Nahant which was relieved only by her momentary joy in playing the music written by a family friend for one of her poems. Here came a successful interlude for once, and perhaps she could bring off being a child and an adult at the same time, which she may have suspected her father wanted her to be. This impossible goal could have been the principal cause of her illness; and though a common explanation of Winny's malady is to say she had anorexia, at best her troubles suggest it was only one of a half-dozen symptoms, such as the vertigo and hypochondria also mentioned here.[25] In spite of Elinor's reluctance, Will, who from time to time had already been in touch with him, finally arranged to take her to S. Weir Mitchell in Philadelphia, the most famous nervous specialist at that time in the country and a fellow author as well. Probably right in diagnosing hysteria, by modern standards Mitchell, though more benign than most of his contemporaries, still treated it barbarously: he prescribed absolute rest, rubbings, and force feedings with great doses of milk and iron to fatten her and to "roast" her "out of her mania for heat."[26] He would fight her hypochondriacal delusions and be even more unyielding than her parents.

Mitchell thought that she showed improvement, and although he found his new patient's case a difficult one, he was confident that she was gaining strength. Between mid-November, when her treatment began, and 6 January, she had gone from 58 to 71 pounds, and by the end of the month, while Mitchell complained that she was still rebellious and would not admit she was any better, he was now sure that she could be cured and reported her as strong enough to "walk any distance." On 10 February Will found her "looking extremely well," though just as hypochondriachal,[27] and a week later, now having gained twenty more pounds, she was sent to the country, whether in Mitchell's program a sign of approval or a kind of exile hardly clear. The parents were totally unprepared for the blow that came when they heard of Winny's death, of cardiac arrest, on 2 March 1889 at the age of twenty-five, a day after her father's fifty-second birthday.

Perhaps because of the anxiety followed by mourning, no letters by Elinor between early 1888 and mid-1892 are known (though the lack may also result from none having been saved), but the letters by her husband often refer to their shared grief. Agonizing over Mitchell's report on Winifred (now lost) after her death, they imagined different scenarios, for they wondered if "perhaps the poor child's pain was all along as great as she fancied." Twenty days after her death, Will wrote his father, "You know there can be but one thing in my mind. Elinor and I talk it over continually. . . ." On another occasion he remarked: "I must tell you how bravely Elinor has borne it, with always a first thought for me in her own anguish. This is the more generous in her because she yielded to me all the details of this last attempt to restore Winny to health." In May they sublet the Brooks Place in Belmont near the Cambridge-Belmont line. From their hilltop they could see the other Belmont hill on which stood Redtop, where Winny's health had first failed. And the house overlooked Fresh Pond and Cambridge, where they had brought her up as a child: "her poor mother and I have gone all over the past again, and tried to retrieve the irrevocable," Will wrote his father on 26 May.[28]

Unless one counts a brief memo to her husband in early December 1890 about sending copies of the memorial volume of Winny to the Warners and Clemenses, the first extant letter by Elinor after Winny's death is one in 1892 to her son. Born 14 August 1868, John Mead Howells, after graduating from Harvard in 1891, spent a winter in the architectural office of his uncle, went to Paris, and after an unusually brief period of study passed the examinations for admission to the Ecole des Beaux-Arts. During his five years of study in Paris he saw his family fairly often either in Europe or on visits home. Much to his pleasure, Elinor wrote him regularly; though these letters have disappeared, the ones Elinor wrote to him from San Remo in 1904–1905 justify his description in a Christmas letter of 1894 from Paris: "Mama's letters are really the salt of the earth & the only thing that connect the two continents for me."[29] While still a student at the Beaux-Arts, with I. N. P. Stokes, his former Harvard classmate and now his partner, John won the competition for the University Settlement Building. Many other buildings followed, among them St. Paul's Chapel at Columbia University (1906), the Chicago Tribune Tower (1922), and the New York Daily News Building (1930). In 1907 he married Abby MacDougall White, and their first child was born on 27 November 1908. While we have no letter by Elinor on this occasion, as we do for the engagement, one by her husband reports her tremendous pleasure in her first grandson.

Elinor's youngest child, Mildred, born 26 September 1872, was eventually more active as an artist than either John or Winny. She had begun notably with an outpouring of sketches during the Italian sojourn of 1882–1883 which later appeared as *A Little Girl among the Old Masters* (1884) with commentary by her father, the reception of which is recorded in another scrapbook of Elinor's. Though the parents recognized that publication was "something hazardous, as concerned Pilla,"[30] they did not anticipate their own involvement in the reviews of the book. The several unfavorable notices of the book produced Elinor's indignant underlinings and exclamation points, along with a stern letter to one newspaper's editor by Will. But there were tributes as well, and a book at eleven was a start. Not quite seventeen at the time of Winny's death, Mildred had gone to art school during the family's stay in Buffalo, then had continued her study in New York. But in December 1890 she made her debut in Boston, though, as Will and Elinor presumably stipulated, after her two winters in Boston, where she could be "as full of gayety as she desires," she would have to start her work upon the family's return to New York.[31] In addition to her drawing, Mildred had also begun to publish with a farce, "A Class Day Sketch," appearing in *Harper's Bazar* on 13 June 1891. The 1891 farce was followed by others and by enough published poems to have made up a book (considered but not realized). In the mid-1890s Mildred exhibited watercolors at the New York Water Color Club and at the Salon du Champ de Mars in Paris. She also did some illustrating for *Scribner's*. In 1900 she worked on a school primer with a teacher friend, on a schoolbook of her father's writings, and, in 1901, on a whist calendar. Though engaged to be married in the fall of 1902, she broke off the engagement the following February, an ending that disappointed both Elinor and Will.

In this synopsis of Elinor Howells' character and personality, her illnesses have been mentioned from time to time. For the most part, biographers have created a somewhat misleading impression of Elinor's physical and emotional health, largely without sufficient reference to the time in which she lived and without recognizing that changes frequently appeared for both better and worse. Before ending this introduction, then, it is fitting to review at some length what is known, mostly from the letters of her husband to his family and friends. For certainly Elinor was not the helpless invalid that all too many accounts present her as being during the greater part of her adult life, especially after Winifred's death. Between the bumbling doctors and primitive nerve specialists of her time she did as well as most of her peers and better than many.

The two favored female maladies during Elinor's lifetime were hysteria, which some historians speculate resulted from a revolt by women against the male expectation that they play a subservient role, and neurasthenia, somewhat preferred by male physicians as a relatively minor maladjustment in which the patient exhibited deficient nervous energy while remaining cooperative with those who had charge. Elinor can hardly be regarded as exhibiting clinical symptoms of either syndrome. To the extent that she was vibrant, outspoken, and enthusiastic, unsympathetic observers might have called her an hysteric. In so far as she was at times despondent about her health and wished to find a cure, neurasthenia might better describe her condition; and since neurasthenia was originally presented as distinctively American, especially among the more intelligent and well-to-do, Elinor would have additional qualifications, even though it originated as a malady of which men were the main victims.[32] Given the impossibility of making a diagnosis at this remove, it may be most satisfactory to present what evidence we have and allow readers to draw their own conclusions. At the same time, the story of Elinor's health brings in other aspects of her life, some of only indirect bearing on her health but of relevance to her personality.

From the Venetian journal and accompanying pocket diary on her Venetian years, she appears in the best of health, energetic, eager for all the opportunities of their life in Italy. Her pregnancy brought no problems—at least none that seemed worth her comment—and though labor before Winny's birth lasted fifteen hours, the doctor was called in only a few minutes before the delivery. Eight days later, on the first anniversary of her marriage, she observes that she is "On my back in bed," but adds that she is "gradually getting along."[33] Late in February the Howellses, leaving the baby with the maid and nurse, toured the ducal cities.

During the last month in Venice, Will mentioned that Elinor "has grown very delicate in the Venetian climate,"[34] the first indication that she may not have been in good health. At any rate she made the trip home without incident, visited the Howells family in Jefferson (where Winny came down with whooping cough), probably found them a bit depressing, and was glad to get away to Brattleboro. But her stay there (necessitated by her mother's absence when a sister's baby was born) turned out tedious too, for she wished to join her husband in New York, as she at last did early in January. Shortly thereafter, when the opportunity of switching from the *Nation* to the *Atlantic* presented itself, the couple moved to Boston. Except for one remark in one of the daily letters that Will wrote her from New York when he heard that she had lost weight ("Ducky, are you trying to get fat? Do tranquillize and obese a little!"[35]), there are no suggestions of illness, though she may have suffered a miscarriage at some point in these months. We definitely know that she had a serious miscarriage during Annie's visit to Cambridge in late February, after a nonstop round of teas, shopping, and visits, an event that emerges with such roundabout allusions in the family correspondence as to make one aware of other sensitive events that may be missed. To recuperate from the illness that followed, Elinor went to Brattleboro in the spring, evidently enjoying herself so much that in the following summer she went again, this time with Will and Winny. Though there were sick people there for the Water Cure, they did not seem to dim the gaiety of the place; and the pattern of summer vacationing in the mountains or at the seaside became a regular one in the Howellses' lives, usually marked by the revival of Elinor after having been confined by winter in Cambridge, Boston, or later New York.

After John's birth in 1868, she seemed better for a while. "Elinor does not gain strength rapidly, but she is better than she has been since her miscarriage"—one of the few open references to the one in 1867.[36] But her recovery was not as rapid as had been expected, and as Will confided to Norton a year later, "Her ill-health is our one great draw-back, for it is hard for her to bear and for me to see"; and he pictured her as "quite broken up." Still, by the end of the letter he cheerily observed, "Mrs. Howells is much better than when I commenced writing."[37] As part of her recovery she had again gone to Brattleboro—"she is gaining in health"[38]—with Will and the children boarding in Boston for a few weeks; she came back the better for her Vermont vacation (was homeopathy or hydrotherapy involved?), but the death of Will's mother soon after John's birth, and the death of her own father in July the following year, probably caused both of them, especially Will, great emotional distress.

Yet there were good years too, which began with the excitement and pleasure of moving into the Berkeley Street house in September 1870. Though exhausted after the great party for the Hartes, Elinor "shot up to ninety pounds." Her letters of the winter of 1871–1872 are efferverscent with news and gossip; and she was perhaps "spinning things out," she allowed in one particularly ebullient letter, as she listed innumerable engagements, weddings, lectures, outings, and a Boston visit by Mark Twain, whom Will liked very much and saw a good deal. Though they were unable to buy the Berkeley Street house, as they had expected they might, they bought land on nearby Concord Avenue in August. Elinor designed the house sometime between then and the following month, when construction began. Her third child, Mildred, arrived on 26 September. Life continued mostly happily and healthily in the following years, except that Elinor's recovery from the birth of her third child was long; her first trip into Boston for "*eight months,*"[39] as Will underlined it, did not take place until 15 March, and in August 1875 there seems to have been some question about whether her strength would allow her to visit in Quebec. They eventually made the trip in late September, after a summer among the Shakers at Shirley Village.

With their going to Belmont in 1878 one sees clearly for the first time a pattern of restlessness and nomadism in the couple, for up to now the moves seem reasonably motivated. (This one may well have been too.) After they left Belmont, the increasing frequency of changes of dwelling—though up to the death of Winny they take place for her benefit—assumes an almost compulsive character, as the Howellses move repeatedly within New York or make wild-goose chases for a summer home at the same time. Even discounting the summer moves at a period when the custom was to move from the seaside to the mountains in midseason, and recognizing that Americans had picked up the habit of moving since the first frontier opened, Will and Elinor appear to have held a lead, as various friends observed. Elinor's delight in moving is evident, though Will's may be only slightly less, as his constant trips and changes of residence after her death suggest.

At any rate, when in November 1877 their friend Charles Fairchild, a wealthy paper manufacturer in Boston, offered to build them a house on his land on Belmont Hill, they accepted enthusiastically. In spite of the many unwanted visitors they had the first year, they also entertained the C. D. Warners and the Clemenses. But shortly before the Oliver Wendell Holmes birthday celebration in late 1879 sponsored by the *Atlantic Monthly*, Elinor suffered an unspecified back injury that did not allow her to attend and which troubled her for several months. "Elinor is slowly convalescing but is often in a very nervous condition; and she suffers a great deal of obscure, tedious pain at certain times. But she is promised thorough recovery, and she has chiefly to wait," Howells wrote his father in mid-January.[40] She did recover enough to visit her cousins the Hayeses at the White House in May. But she was in bed for the greater part of two weeks in April 1881, and that July both she and Winny, at the beginning of the latter's breakdown, were at the sanatorium-gymnasium of Dr. Dio Lewis, "for their Nerves." (At about this time Winny had exclaimed in tears, "Oh, papa, what a strange youth I'm having."[41]) Both Winny and Elinor, who had been exhausted from taking care of Winny, continued in less than middling health until Will had his own breakdown in November. What caused it and what its nature was is not part of our story, though Elinor's description of its physical symptoms constitutes the most detailed list that we have and almost literally bears out his remark to Mrs. Fields that he has been in bed "with fever, and a thousand other things."[42] What is of significance is that Elinor's letters show her ability to rise to the occasion, ignoring her illness and taking charge of his sickness and slow recovery. Winny also picked up. The members of the family were engaging in a kind of rotation of neuroses, as has been observed in clinical experience.

The pattern also continued after Will recovered, for in April 1882 he wrote a friend that Elinor had undergone her "annual spring collapse"[44] (a rest she deserved after his long and severe illness), though Winny remained well. In July the Howellses' European trip began, and, as we have seen, it was comparatively unmarred by the usual health problems for any of them, except for Winny at times. But the fall after their return, when they were living on Louisburg Square, Elinor is suffering again from her back, though the same sentence reports that Winny is "well enough to go to parties—which is something."[45] During the fluctuations of Winny's long illness, Elinor was under constant pressure and fatiguing worry, yet there is comparatively little said about her own health. Before and after the Gosses' visit to Beacon Street in December 1884 Elinor had to take to her bed; yet it is in the Beacon Street house that she makes the glorious departure from parties as she ascends in her elevator. When the end came for Winny, Elinor went into mourning and avoided most social functions. Though the Clemenses invited her to come to Hartford in the fall of 1889, she felt unable to accept until two years later; she went nowhere during the winter of 1889–1890; around September 1889 she was going through menopause ("quite poorly, from her time of life and its complications"), but by November 1890 she is reported as being "very much better," though "she still has to lie down a great part of each day"; and in the spring of 1891 she could not attend the wedding of her niece and namesake Elinor Shepard—"she dreads excitement."[46]

All these reports in letters (mostly in the weekly health bulletins that Will sent his father), whether they refer to ill health or good, require cautious reading: they may overemphasize either sickness or recovery, but the nineteenth century made a fine art out of preoccupation with health. When at the time of the invitation to visit in Hartford, Clemens told her that he had found her very much her old self when he saw her, he obviously wanted to be encouraging—but clearly felt that there was truth in his remark. She attended at least one tea associated with Mildred's debut late in 1890. Certainly she went into society much less, if not the "never" she often wrote, and Mildred increasingly served as her father's companion. After they moved to New York at the end of 1891, Elinor and Will were usually at home in the afternoon for family and friends, and more and more for their children's friends. When they lived in their apartment on Fifty-Ninth Street, they took long walks and drives around Central Park.

In 1893 Elinor tried a promising new cure, recommended by the family doctor: "If it does succeed, life will be another thing for her," Will wrote his father.[47] Whether "mind-cure" as he called it, or "hypnotism" as Elinor would have it, her new treatment worked wonders for a while. According to Mark Twain, who sketched her in notes for his wife written after his visit to the Howellses on 28 January 1894: "Mrs. H. came skipping in, presently, the very person, to a dot, that she used to be, so many years ago." That Elinor had little use for the personality of her therapist is clear from Twain's report of her remarks: "People may *call* it what they like, but it is just *hypnotism*, & that's *all* it is—hypnotism pure and simple. MIND-cure!—the *idea*! Why, this woman that cured me hasn't *got* any mind. She's a good creature, but she's dull & dumb and illiterate. . . ."[48] How long this treatment was effective is hard to say, perhaps until the trip to Europe in 1897 that had as its primary purpose her husband's taking the cure at Carlsbad. There may have been occasional relapses, as in the summer of 1896, when they left their cottage in Far Rockaway for a week in the White Mountains because Elinor "was broken down by the moving and is extremely debilitated." Or is Will using her health as an excuse for his own impatience with Long Island, where he didn't like the "horsey, rather tiresome rich people" at the Rockaway Hunt Club, while at the Fabyan House he was hearing "more meaty talk here in a day than a week anywhere else"?[49] At any event, the Far Rockaway place proved a disappointment, and it was later sold. But after the cure, the long European trip, mostly through Germany, was a huge success, as Elinor's letters to Annie and Will's own letters make clear.

Still, they both came back to New York, for which they had mixed feelings of loathing and delight, tired of their travels. "I did not feel it then," Elinor concluded in one of her letters. "I felt 10 years younger—from the excitement. But I feel it now."

In the next several years, what can be gathered from letters shows no more than a continuation of fluctuating physical and emotional health. In letters of May and June 1901 Elinor's husband reported that she was "much more broken down than usual, this spring" and that "She lies down nearly all the time, and is very feeble, though the doctor says there is nothing specific the matter."[50] York Harbor, where the family went that summer, did not have its expected effect, if one may judge from a letter that Mildred wrote in late August, which observes that her mother had been "miserable" though was now "slowly gaining." A little later, in an October letter, Will made a surprising comparison of her situation and Winny's: "With a difference of temperament, her nervous prostration has been as marked as poor Winny's." Still, by spring she had become "about the strongest of us, now," Will wrote to Aurelia on 27 April 1902, and in the summer he reported her as "tremendously well" and "in greater strength and spirits than for years." Elsewhere, "She keeps us all going," and as her niece, Vevie, who was studying art in New York and visiting the Howellses frequently over several years, exclaimed (in a letter of 3 January 1903), "I never saw Aunt Elinor so well," though shortly afterwards she suffered a temporary collapse when Mildred broke off her engagement.[51]

In England a year later Elinor was "wonderfully well, and . . . full of zest" when she arrived there at the time Will was awarded an honorary degree at Oxford and "went through the three hours [of the ceremony] on a backless bench, with the ardor of her inextinguishable youth," as Howells told Laura Platt Mitchell. But she was neither well nor happy most of the time at San Remo (although she did leave a vivid record of her stay in her letters from there). Often, very optimistic, Elinor could also get discouraged when she felt her age, "comparing what she is now with what she was a year ago . . . very gloomy at times, and then as gay as ever," as Will wrote to Aurelia in the summer of 1906. In Kittery Point, where the couple had finally found their ideal summer place in 1902 and which may have marked her later turning point, she was "uncommonly well" and active the next summer. Mama is "like a girl in her early sixties," Will wrote to Pilla away in England, as he listed her numerous engagements in a week:

teas and suppers, a lecture and a circus, and an overnight guest (an unusual burst of hospitality for her at this time). When they went to Europe again in 1908 Will described her to his brother as "the belle of the ship" who "talked with everybody, and ran to and fro like a girl of 17 instead of 70."[52] Just a year after, in January 1909, they finally moved into their new studio flat at 130 West Fifty-Seventh Street, an event long anticipated by both of them and, like the Kittery Point cottage, offering that special delight that new homes produced in her.

Suddenly, however, the bright years turned toward their end. With a good deal of reluctance Will had gone to Carlsbad again for another three-week cure in the summer of 1909, after that stopping in England on the way home. When he arrived in New York in late September, he looked forward to having dinner with Elinor, who had unexpectedly left Kittery Point. Instead, he was shocked to find her in bed after having undergone major surgery in Boston to have "some swollen glands removed" in order "to rid herself of a danger that had haunted her for 20 years." Not wishing to alarm him while he was abroad, Elinor had kept the operation from him, arranging to have her sister Mary's letters mailed from Kittery. Though nervous from the shock of the operation and in great pain, she remained alert and as ever vitally interested in the events of the day, especially the political situation in England and Canada. As often, she was reading Henry James, this time his newly published "Crapy Cornelia," one of his masterpieces, she thought. At Christmas, a few weeks later, Howells wrote James of rereading *The Tragic Muse* to the family: "My wife no longer cares for many things that used to occupy her: *hohheits* of all nations, special characters in history, the genealogy of both our families. 'Well, what *do* you care for?' I asked and I found her answer touching. 'Well, James and his way of doing things—and you.'"[53]

In the ensuing weeks Elinor's mind was blurred from the morphine she was given to alleviate her pain. The doctor came three times a day, and she had three nurses around the clock. Her husband hoped, and occasionally she rallied. "This is a glorious morning for us," he could write his sisters as late as 3 April, telling about their talk of family and reading of letters. "She is cheerful, and hopeful, and entirely rational."[54]

But their tenuous hopes were not to be realized. On 7 May Elinor died. Her letters remain. Vividly written and always entertaining, sparkling with wit and insight, they reveal a gifted and charming woman. They also reveal her place in the family and the unique role she played in her husband's career, and give a sense of the difficulties she faced as the artistic wife of a leading author and the conflicts and tensions between domestic responsibilities and her own needs. Filled with shrewd comments on the literary, artistic, and political scenes, Elinor's letters also constitute a remarkable document of her times.

Notes

1. Hayes, *Diary*, 1:532. Sources appear only for quotations that do not occur elsewhere in this book.

2. *Twain–Howells Letters*, 1:71.

3. WDH to Anne H. Fréchette, 14 September 1884 (Harvard).

4. *SL*, 4:108.

5. Robert A. Parker, *A Yankee Saint: John Humphrey Noyes and the Oneida Community* (New York: G. P. Putnam's Sons, 1935), p. 4.

6. Charles L. Mead to Rutherford B. Hayes, 17 January 1880 (Hayes Presidential Center).

7. *LinL*, 1:11.

8. Hayes, *Diary*, 1:532.

9. George W. Noyes, *John Humphrey Noyes: The Putney Community* (Oneida, New York, 1931), p. 96.

10. Still, when John Humphrey Noyes was expelled from the Oneida Community in 1879, Elinor's brother Charles not only dismissed Noyes's career as a delusion in a letter to Rutherford B. Hayes but sarcastically referred to what Charles had been "led to believe" in his youth. See Mead to Hayes, 27 December 1879 (Hayes Presidential Library).

11. *Letters of John Fiske*, ed. Ethel Fisk (New York, 1940), p. 321.

12. Larkin G. Mead, Jr., to William R. Mead, 15 May 1855 (Collection of Katherine Mead Hasbrouck).

13. See Ginette de B. Merrill, "The Meeting of Elinor Gertrude Mead and Will Howells and Their Courtship," *Old Northwest* 8 (Spring 1982): 23–47.

14. *SL*, 2:31.

15. EMH, Venetian diary, 17 September 1863, typescript, p. 21 (Harvard).

16. *SL*, 1:355.

17. *SL*, 2:186.

18. *SL*, 3:64–65.

19. The scrapbook is the one dated March 1877 (Harvard): the clipping (p. 43) ascribes the story to *The Syracuse Journal* and has a penciled notation on the side "N Y Tribune/Sep 29"; *LinL*, 1:12; WDH to Mildred Howells, 1 November 1914 (Harvard).

20. *Years of My Youth* (HE), p. 194.

21. *SL*, 5:353.

22. WDH to Laura Mitchell, 6 March 1914 (Harvard typescript).

23. The scrapbook is in the collection of Polly Howells Werthman.

24. Mary J. Mead to EMH, 5 January 1864 (Massachusetts Historical Society).

25. John W. Crowley, "Winifred Howells and the Economy of Pain," *Old Northwest* 10

(Spring 1984): 41–75, is a perspective and compassionate psychological study. Our three paragraphs, though in part indebted to it, of necessity omit much that the extended essay considers.

26. WDH to William C. Howells, 25 November 1888 (Harvard).

27. WDH to William C. Howells, 27 January and 10 February 1889 (Harvard).

28. *SL*, 3:247, 248, 249, 252.

29. John Mead Howells to his family, 25 December 1894 (Harvard).

30. *SL*, 3:90.

31. *SL*, 3:270.

32. See Elaine Showalter, *The Female Malady: Women, Madness, and English Culture, 1830–1900* (New York: Pantheon, 1985), a recent and useful history which frequently refers to the situation in America.

33. EMH, pocket diary, 24 December 1863 (Harvard).

34. *SL*, 1:219.

35. WDH to EMH, 12 December 1865 (Harvard).

36. *SL*, 1:319.

37. *SL*, 1:345, 347. The letter is written over a two-week period.

38. *SL*, 1:309.

39. *SL*, 2:20.

40. WDH to William C. Howells, 11 January 1880 (Harvard).

41. *SL*, 2:288, 289.

42. *SL*, 2:301.

43. The most recent and thorough account of WDH's breakdown appears in John W. Crowley, *The Black Heart's Truth: The Early Career of W. D. Howells* (Chapel Hill: University of North Carolina Press, 1985). See especially pp. 117–118 on the neurotic pattern within the family.

44. *SL*, 3:16.

45. *SL*, 3:80.

46. WDH to William C. Howells, 29 September 1889 (Harvard); *SL*, 3:296, 312.

47. WDH to William C. Howells, 15 October 1893 (Harvard).

48. *Twain–Howells Letters*, 2:658.

49. *SL*, 4:130, 128.

50. *SL*, 4:265, 266–267.

51. Mildred Howells to Anne H. Fréchette, 29 August 1901 (Alfred); WDH to Aurelia H. Howells, 13 October 1901, 27 April, 29 June, and 27 July 1902 (Harvard); Marie-Marguerite Fréchette to Achille and Anne H. Fréchette, 3 January 1903 (Alfred).

52. WDH to Laura Mitchell, 1 July 1904 (Harvard typescript); WDH to Aurelia H. Howells, 26 August 1906 (Harvard); WDH to Mildred Howells, 27 August 1907 (Harvard); WDH to Joseph A. Howells, 19 January 1908 (Harvard).

53. *SL*, 5:294.

54. WDH to Anne H. Fréchette and Aurelia H. Howells, 3 April 1910 (Alfred).

IF NOT LITERATURE

Elinor Mead Howells in the 1870s.

"That letter which you wrote me . . . was, if not literature, something much better."
W. D. Howells to Elinor Mead Howells, 20 December 1865, Harvard

She Has Got the Stuff in Her

1842–1863

*E*linor Gertrude Mead, the fourth
child and the first daughter of Larkin G. Mead and Mary Jane Noyes, was
born on 1 May 1837 in Chesterfield, New Hampshire. She was two years
old when the family moved to Brattleboro, Vermont. In the few existing
letters written by Elinor before her marriage, she comes out as an exuberant
and independent spirit, if sometimes a reluctant scholar. But through the
letters of others in the family her engaging personality emerges, together
with her talents and shortcomings. The oldest girl in the family, she had a
special place in the affection of her father and older brothers who, to judge
from their extant letters, frequently wrote to her, confided in her, and singled
her out for invitations.

Typically her brother Larkin prized her letters and showed them to "all,
yes all [his] friends," as he wrote from his Brooklyn studio in the winter of
1855–1856. "She has got the stuff in her," he quoted his friend the artist
George Fuller, "and will, if she chooses, make a great painter." By this
time, Elinor had apparently graduated from the local high school and
seriously taken up painting, if not with quite the "perseverance" that her
brother wanted her to have. Still, she continued to paint, though at the
same time living the busy social life of a young lady in Brattleboro. Late
in November 1860 she went to Columbus, Ohio, where she visited her cousin
Laura Platt until March. There she met her future husband, Will Howells.

3

While they probably did not become formally engaged at this time, they were on such terms that Will visited the Meads at least once when he was in the East trying to secure a consulship in September 1861.

Though they may have had a misunderstanding of some sort between then and Will's sailing for Venice in November, by February 1862 the couple had reached the point where Elinor, at this time teaching in Cincinnati, considered herself engaged. While they discussed their marriage plans, Larkin made up his mind to go to Europe for further study, and Elinor decided to sail with him. After the wedding in Paris on 24 December 1862, Elinor and Will took the train to Venice, where they arrived four days later.

Most of the letters in this chapter tell about their first year in Venice, culminating in the birth of Winifred on 17 December 1863.

To Mary J. Mead

[Brattleboro (?), c. 1842–1844][1]

dear Mo[ther]

this is the first letter I ever wrote—I will try to do better I think you will be glad to hear I am very well when you was here I went to ride with mr. Lord to mr. Brooks barn[2] I was sorry I did not get home I am sure I love my father and mother very much I hope I sh[all] always try to please them I se[nd] love to all

your daughter
Elinor Mead

Manuscript: Massachusetts Historical Society

1. On verso, in Elinor's hand, "Mrs. Mary Mead." Here and elsewhere smudges obliterate parts of words.

Mary Jane Noyes (c. 1806–1876), the daughter of John Noyes and Polly Hayes, had married Larkin G. Mead (1795–1869) in 1829.

2. The son of Judge Joseph Lord of Putney, Vt., Thomas C. Lord (d. 1851) was captain of the militia and the owner of the Vermont House, a Brattleboro inn. (He more than once got into trouble for selling liquor in the village.) William Smith Brooks (1781–1865), a manufacturer of cotton goods and the former captain of a merchant vessel, had moved to Chesterfield in 1821 and to Brattleboro in 1839, at the same time as the Meads.

To Larkin G. Mead, Sr.

[Brattleboro, October 1848]

Dear father.[1]
I go to school and enjoy myself *torable*, but not *very* well.

Yours truly.
Elinor.

Manuscript: Massachusetts Historical Society

1. As state senator Elinor's father attended the session of the legislature that began on 12 October. In the larger part of a joint letter by Larkin, Jr., and his sister, the son writes about family and local events, at one point reporting, "Elinor goes to school and I believe likes it very well she is not very well pleased with being *obliged* to go to school."
Above Elinor's contribution, Mrs. Mead commented, "I do not like it that Elinor has not written more—"

Larkin G. Mead, Sr., to Elinor

Montpelier,
Wednesday morning
[late October–early November 1848]

Dear Elinor,
Your Letter which I received yesterday was very acceptable. I know that you are going to school, improving your mind & above all to hear that you are very much pleased with the High School, gives me great pleasure.[1] The Music too—stick to that—recollecting you cannot excell in any thing without constant exertion. It may seem to you insipid to practice over these lessons, day after day—but, recollect the most celebrated Players went over the same ground—the only difference is, they probably went at it with more resolution & less pouting. I still expect to see you able to teach Music & to command the highest price for your instruction, Elinor. I wish you would just come up here & spend the next Sabbath with me. I get along through the working days, very well—having enough to keep me busy—but, when I come to sit down, my

thoughts turn towards home & the little characters I have left there. I hope to be with you all next week, tho' 'tis some uncertain.[2]

Yr father
L. G. Mead.

Manuscript: Massachusetts Historical Society

1. Mead, who also helped start the Brattleboro Library in 1842, had been selected chairman of the first committee to implement the new graded high school in Brattleboro in 1841.
2. The session did not adjourn until 13 November.

To Mary Higginson

Brattleboro' Nov. 19./56.

Dear Mary,[1]

I need not apologize to you for not writing this long time, for you know of our family affliction which prevented me, so that this is almost the only letter I have commenced for months.[2]

But last week Father & Joanna & I started very suddenly for Montreal and had such a pleasant time and saw so many curious things there. Next I thought you might like to hear about it, and as I have nothing else to tell about I am going to give you quite a lengthy description. Mother was away, when we started so I had to do a great many things Monday, and Tuesday morning we started. Mrs Archbald, a cousin of Mrs Bush's, who has been at "The Wesselhoeft"[3] all this Fall & a part of the Summer and who lives in Montreal went with us. She was a young married lady, and as all the Hotels & most of the houses were full when we got there, her husband insisted upon our going to their house to stay—during the Festival.

We got in at one o'clock at night, and next morning were up and off at nine to see the Procession. Staying at the Archbald's were Jo Archbald, a brother of Mr A's, twenty one years old and so handsome that it is of no use attempting to describe him, and Charles & Lucretia Archbald of Boston cousins of the A's. Beside these was Ned Cheney a brother of Mrs A's, the *jolliest* young man I ever saw, and, as we learned afterwards to our *great* surprise a *poet,* and an extensive writer on all subjects. Mrs Bush says he often writes for the Christian Inquirer,[4] a Unitarian paper and signs his name. Do you take it? We do not, but Joanna and I are trying to find some of the back numbers about town. He is Deputy Consul at Montreal, but he wished (Mrs Bush says) to give up all his business and devote himself entirely to literature. If you had heard that about for instance *Mr Schlesinger* (how is it spelled?) you could not have been more surprised that Joanna and I were to hear it of Ned Cheney, for he did nt say *one sentimental speech* while we were there. Now you have an idea of our *beaux,* so I will describe what we saw under their escort. The first day we went up to the top of the French Cathedral—and saw the great bell weighing eleven tons. We did not hear it rung as it is used only on great occasions. Rung for a funeral for fifty dollars I believe, and it costs much more for a wedding. The buildings in Montreal are all of gray stone, and much finer, some of them, than anything I ever saw. The streets are narrow, the vehicles odd, and it seemed so strange to see friars with gowns down to their feet and broad trimmed hats walking about the streets. One has to know French to do business there. Jo used to make inquiries in the streets, the nunneries &c in French always. He has lived there six years. There was a great banquet given to the invited guests (the gentlemen) Wednesday and the old men (I mean the *married* men) went, but the young gentlemen were there and we had a banquet of our own. In the evening there were some fireworks—but they were miserable and Ned made all manner of fun of them. He said he wished the *moon* would come out so we could see them and said we ought to have brought a *lantern* &c. Next day we went in a steamboat (*"The Muskrat"*) on an excursion on the St Lawrence—down to see the great water-works and came back on a steamboat called "The Beaver". Did you ever hear such names? We had the honor of talking with the Mayor of Montreal, and had a delightful time.

In the afternoon we went to the Review on the Champ de Mars, and that was the best thing of all. There was a Regiment from the Crimea there, (one thousand men) and they had a sham battle with the Volunteers of Montreal. Think of seeing those office[r]s distinguished at the Crimea, with medals on their breasts.! There was one old officer riding on the field who had three medals, and one he got at the battle of *Waterloo*. We talked with a Crimean soldier with a medal who fought in the trenches, and gave us frightful descriptions of the war. The band played so splendidly that Father said tears came into his eyes. It was different music from what I ever heard, as there were not many brass instruments, but there seemed to be a great many *clarinets* or whatever. Some of the Volunteers were so careless as to leave their ramrods in when they fired, and several men were taken off the field wounded. They went through a ceremony called "Saluting the Colors" I believe. A British flag was carried before the line of soldiers for them to acknowledge, the band following playing beautifully. I cannot describe all the maneuvers they went through as I do not know the right terms, but it was a splendid sight.

In the evening there were some fine fireworks. Next day we went into the Gray Nunnery, St Patrick's Church &c, & at twelve o'clock with great reluctance left Montreal. The gentlemen were down at the boat and stood on the wharf bowing till we were fairly off. O dear, I hope I shall go there again sometime for I am perfectly *delighted* with the place. I don't know as you will understand from this description what a nice time we had for I could never write all the pleasant *little* things that occurred. We rode all night coming home, and I have not got over the effects of the journey yet. We met many people we knew—James Converse of Boston & his daughter[5] just my age, our Cousins, met us at Rouse's Point. We had quite a pleasant time with a Mr Reed[6] of Boston—a young man son of the Ticknor Reed &c Reed.

Yours Cousins Lissie & Anne were in here day before yesterday and told me all about Louisa's wedding.[7] It must have been very pleasant. I have never told you about Quincy Ward's[8] visit here. Sometime I will. He visited Mr Fuller in Deerfield and Anne Higginson perfectly *raves* about him. We have great talks about him & Fuller.

Sarah Allen[9] says you are very busy—only going to three parties a week. Do write if you get time. The Pells are feeling dreadfully about the loss of the Summers on the steamer Le Lyonnais.[10] I have written this miserably—on very poor paper and in a great hurry. Please give my love to Miss Nourse.¹¹ Dinner is ready so you must excuse the ending of this letter.

Yours affectionately
E. G. Mead.

Manuscript: William White Howells

1. Mary Lee Higginson (b. 1838) was the daughter of George Higginson and Mary Cabot Lee, and the sister of Henry Lee Higginson. She was to marry Samuel Parkman Blake (1835–1904) in 1868. Elinor's daughter Mildred identifies her in a letter of 26 August 1945 to her nephew: "She was a niece of the two Miss Higginsons who lived in Brattleboro, and being a backward student my mother tutored her in Mathamatics" (William White Howells).

2. Elinor's sixteen-year-old brother Albert had died of typhoid two months before (15 or 16 September).

3. Neither the Archbalds, Mrs. Bush, Ned Cheney, nor Mr. Schlesinger has been identified. Built for the patients of the Water-Cure started by the German doctor Robert Wesselhoeft (1797–1852), the Wesselhoeft accepted guests as well, and the senior Meads, their family, and friends often boarded there. Equipped with bowling alleys and billiard rooms in a beautiful setting, it attracted a wealthy and intellectual clientele from all over the country, as it was the site of concerts, balls, and a host of other social and cultural activities, especially in its heyday before the Civil War. (For a description of the Water-Cure treatment and ambience, and for the list of famous guests who attended the Brattleboro spa, see *Annals*, especially pp. 564–575).

4. The *Christian Inquirer*, later renamed *Inquirer*, was published from 1846 to 1866 in New York. Several poems by E. M. Cheney appear in the paper.

5. James Cogswell Converse (1807–1891) was Mr. Mead's first cousin. Though he was born in Vermont, he lived in Boston from the age of nineteen, where he was a partner in the dry goods firm of Blanchard, Converse & Co. Elinor probably met his oldest daughter Elizabeth (1836–1859), who was one year older than she was.

6. John Reed, Jr., from 1849 to 1854 of Ticknor, Reed & Fields (later Ticknor & Fields), was with the Provident Institution for Savings on Tremont Place in Boston.

7. Eliza W. Channing (1834–1911) and Mary Louisa (1832–1903) were the daughters of Dr. Francis John Higginson (1806–1872), who practiced in Brattleboro beginning in 1842 and was to move to Boston in 1867. Louisa married Francis Cabot in Brattleboro on 12 November 1856. Anne (Annie) Storrow Higginson (b. 1834), their first cousin, was the daughter of Stephen Higginson (1808–1870). Her sister Agnes Gordon (b. 1838) married the painter George Fuller in 1861.

8. John Quincy Adams Ward (1830–1910), the sculptor, and George Fuller (1822–1884) were friends of the Mead family and later of WDH.

9. Sarah Fessenden Allen (d. 1891) was the daughter of the Hon. Elisha H. Allen. In 1860 she married Dr. William P. Wesselhoeft.

10. Probably a Mrs. Walden Pell of Brattleboro with six daughters; she ran a dancing school for her children and the children of her friends. The French steamer "Le Lyonnais" had sunk on its way from New York to Le Havre on 1 November, when it was rammed by another vessel.

11. For Miss Nourse, Elinor's and Mary's Brattleboro High School teacher, see 7 December 1862.

WDH, Venetian Diary

January 30th [1862][1]

There is no trace of snow on the wide, straight pavement of Broad street,[2] but the sun has thawed the frost under the bricks, and the long walk shines and swims in the light. O slow feet of mine that keep me back, and make the joy of meeting greater for delay! I loiter and think over all the droll, delicious things that have been done and said, and smile in that blissful foolishness, at which I cannot learn to mock, until my heart leaps high as I stand at the gate, and then at the door, and am all thrilled with the clangor of the bell. Is um-um-um at home? O yes, will I come in? And then the kind obscurity and obliging dull calm of the parlor, where that friendly great arm-chair receives me, and where I idly twirl my hat, and wait—only a little while.

"Dance light, for my heart it lies under your feet, love,"[3]

bounding down those broad stairs, and throbbing toward me. O ravishing little muff, and jaunty hat with shining plumes, and graceful cloak that clasps the lissome shape![4] And O, sweet fair face, with blue eyes full of roguish light, and shadowed by masses of brown hair! And O, dear absurdity of loving her! and yet not saying it, but keeping the secret to muse upon, and talk over with old Price!—We go to walk, and the January sun is warm for us. Yes the bluebirds already carol through the air, and the bees hum summer tunes, and enchant a bloom upon the locust tree, and inundate the meadows with billowy June. We do not talk much— childish nothings suffice us. L.[5] has said "You are like two children," and our idle, foolish ways confirm it. What do we care? The proper people whom our walks surprise, stare at us, and in the greenhouse where we sit,[6] the cold dark-robed beauty who comes upon us is astonished into life, and utters wonder in the icy interest of her eyes. But we do not care, and it is broad noon when she shakes my hand at the door, and I lounge back to the office. . . .

Manuscript: Harvard

1. Alone in Venice and waiting to hear from Elinor, WDH recalls in this entry of his

diary his life in Columbus just before the beginning of the Civil War. It is clear from this passage that, by the end of January 1861, he was already in love with Elinor, whom he had met only recently that winter. See Ginette de B. Merrill, "The Meeting of Elinor Gertrude Mead and Will Howells and Their Courtship," *Old Northwest*, Spring 1982. Just before the printed passage begins, Howells reminds Samuel Price (d. 1870), his roommate and colleague at the *Ohio State Journal*, that he has "an engagement to walk that morning with The Angel, to whom that best of the earth-born sends his never-delivered message of love."

2. Elinor was spending the winter as a guest of her cousin Laura Platt (1842–1916), who lived at 380 E. Broad Street in Columbus.

3. Last line and title of a popular poem by John Francis Waller (1810–1894). The poem was set to music.

4. See Larkin G. Mead, Jr.'s portrait of Elinor in "the very clothes she wore in Columbus the Winter she met W. D. H." (W. W. Howells).

5. Laura Platt.

6. In the Howells Papers at the Houghton Library, there is a sketch of Elinor and Will in a greenhouse. Entitled "In 'Joe's' Greenhouse," it bears the caption, also in Elinor's hand: "Sketched from memory, in Quincy Ward's studio, Ohio State House, March, 1861, by E. G. Mead 'as was.'" This may be the greenhouse referred to here.

To Lucy Webb Hayes

Brattleboro'
October 23, 1862

Dear Lucy,[1]

That I should be writing you of my affairs on Laura's wedding-day![2] But my engagement to Mr. Howells having, by accident, leaked out I hasten to inform you of it lest you should hear it from the *crowd*, and, at the same time, I want to tell you of my plans. If Providence permits, Lark & I leave for Europe the middle of December next, sailing from Boston.[3] Mr. H. meets us in Liverpool— then the thing is consummated and we proceed, (unless Lark makes for Rome immediately) all, to Venice where Casa Falier awaits a mistress.[4] This is all a sudden get up—not the *engagement* but the arrangement.

Lark was determined to go abroad, and, though Mr. H. did n't *ask* it and I did n't *propose* it somehow it seemed a good idea for me to go with him and so save time and money all around. I hope you wo'n't disapprove of the proceeding. It is considered *entirely respectable* by my friends generally.

What says Mrs. Webb? How is Cousin Rutherford?[5] Laura wrote he was on his way home. It is possible you were at the wedding.

How I wish *I* could have been! Your letter was *splendid,* Lucy, but you've been so taken up and I've been so busy that I did n't answer it!

I am going to New York to see Joanna,[6] Monday, in her new house. I suppose I can hardly see *you* before I leave—but you'll write to me—and you'll stand by me, wo'n't you? I believe in you. We want to hear about you, all, very much. Can you find time to write a short letter?

Goodbye—but not the last, solemn "goodbye" yet.

Your affectionate Cousin
Elinor

Manuscript: Hayes Presidential Center

1. Lucy Webb Hayes (1831–1889). The wife of Rutherford Birchard Hayes (1822–1893), future President of the United States (1877–1881), and Elinor's cousin.

2. Laura Platt (1842–1916), a cousin of Elinor, married John G. Mitchell (1838–1894), a lawyer who became a brigadier general during the Civil War.

3. Elinor and Larkin G. Mead, Jr., sailed on 10 December. In a letter to James M. Comly dated 19 December 1862 (Ohio State Historical Society), WDH's father wrote: "As to Will himself, I have a little news to give you. On the tenth of this Month, a young lady sailed in the *Africa* for Liverpool where she is to meet him and marry him. I suppose you have some acquaintance with her—which I have not. It is Miss Mead, who visited at Platt's the winter before last. Her brother the sculptor, goes out to Rome, and she accompanies him. I suppose the arrangement will do—but not knowing her I cannot well judge. He was so secretive about it that winter, that I never got to see her, which I now wish I had." William August Platt (1809–1882) was Laura's father. A leading citizen of Columbus, he was president of the Columbus Gas Company.

4. Howells had rented an apartment in Casa Falier, a small palazzo on the Grand Canal. This was their first home in Venice.

5. Maria Cook Webb (Mrs. James W.) was Lucy Hayes's mother. Both Hayes, who had recently been wounded in the left arm at the battle of South Mountain, and his wife attended Laura's wedding (Hayes, *Diary,* 2:362).

6. Elinor's sister Joanna (1842–1914) had just married Augustus Dennis Shepard (1835–1913) on 25 September. Shepard was associated with the National Bank Note Company. Their new address was 79 E. 27th Street in New York.

To Lucy Webb Hayes

Brattleboro'
Dec., 7, 1862

Dear Cousin Lucy,

The snakes arrived unharmed two days ago and I have n't written before because I've been expecting a letter from you every mail. Not that a word was necessary on the subject—for I think we understand each other—but I thought you might have something to say as a farewell speech. I'm sick of saying Goodbye myself and I wish all my friends would take your way of expressing themselves.

O Lucy, they are *so* exquisite, and then the associations (as well as the snakes) that cluster around them![1] If ever I have a shanty of my own those will make the meanest room attractive. That was a remarkable walk of ours on Fifth St. was'n't it?[2] But I did n't exactly cheat you, cousin Lucy, for it was very uncertain if I should ever be married then. I do'n't remember the month we took that walk—but I was'n't engaged when I first went to Cincinnati.[3] I hope to see you sometime dear Lucy under pleasanter circumstances than those. I'm so sorry about Aunt Hattie[4]—but I warned her. I told Katy B.[5] my experience, too, but she is very happy with Miss Nourse I believe. Have you seen her? I expected to hear from Miss Nourse before I went—but I leave town tomorrow and sail Wednesday and no letter has come yet. Nor have I heard from Laura very lately, though she has written me one splendid letter from Zanesville. How happy she is in that wild free life! We hear Cousin Rutherford has gone back to his Regiment. How you must have enjoyed having him with you, wounded though he was.[6]

Mr. Rogers spent a part of two mornings with me last week.[7] He has improved in health so that he seems another person—and Mr. Andrews is gaining rapidly, too;[8] so we are concluding that doctor is not a quack, after all. They will stay here all Winter. Mr. R. came in yesterday to see if I had heard from you. I told him I heard the first part of the Summer that Mr. & Mrs. Stephenson were coming to Brattleboro'. Why did n't they come?[9]

How is Mrs. Webb? How are you? How is little Jo?[10] How is Dr. Webb & Dr. Jim and how are you all? Are you going to write me in Venice Care of the American Consul?

I'm so glad I have your likenesses now. I wish I had more to take with me. I have about a hundred visites in all—but I mean of your family.

Lark & I are anticipating the voyage. We think it will take fever & ague out of our systems. You may think of me married a day or two before Christmas. I've a letter to a Minister in Liverpool—a

dissenter. I'm not going to be married by the *Church of England.*[11]
Lark is going to Rome or Florence. I've some beautiful presents.
Aunt Emily[12] sent me her carte de visite. It is capital. I dare say
she'll send you one. Jo is home on a visit. Mr. Shepard came too
but has gone back.[13] She stays to comfort Mother

The Meads generally send love and wish they could hear from
you oftener. Kiss Mrs. Webb and the boys for me.[14]

Goodbye for a while.

Your aff. Cousin

Elinor[15]

Manuscript: Hayes Presidential Center

1. A wedding present from the Hayeses, the snakes are probably two vases with snakes twirling around them, one of which is still at the Howells Memorial.

2. Fifth Street in Columbus is perpendicular to East Broad Street and near Laura Platt's house. This probably refers to the walk Elinor and WDH took c. 30 January 1861.

3. Elinor had arrived in Columbus on 23 November 1860. She had left for Cincinnati to visit the Hayeses between 4 and 10 March 1861 (R. B. Hayes to Sardis Birchard, 27 November 1860 and 1 March 1861 [Hayes Presidential Center]).

4. Harriet Little (Solace), b. 1820, was William A. Platt's half-sister.

5. Probably Katy Burbank, a friend of Elinor's. See 17 May 1863. The allusion is not clear. She may have been teaching at Miss Nourse's school or boarding there. A former Brattleboro teacher (Elinor had been her pupil), Clara E. Nourse had recently founded a school for girls, of which she was the principal, in Cincinnati. Elinor had taught there in 1861–1862. (See Merrill, "The Meeting of Elinor Gertrude Mead and Will Howells and Their Courtship.")

6. Col. Hayes was back with the 23rd Regiment, Ohio Volunteer Infantry, at Camp Maskell, along the Kanawha River, by 30 November. See Hayes, *Diary,* 2:366, 368.

7. Probably William K. Rogers (1828–1893), a Cincinnati friend and former law partner of Hayes in 1854. He had left the firm because of poor health. When Hayes became President in 1877, Rogers was his personal secretary.

8. Probably Dr. John W. Andrews (b. 1805), a banker, president of the Board of Control, State Bank of Ohio, and a neighbor of the Platts in Columbus. He suffered from asthma. They were presumably patients at the famous Water-Cure in Brattleboro founded by Dr. Robert Wesselhoeft.

9. Probably R. H. Stephenson and his wife. Like William K. Rogers, he was among R. B. Hayes's most intimate friends and is often mentioned in his diary.

10. Joseph Thompson Hayes, born 31 December 1861, was Lucy Hayes's baby. Dr. Joseph T. Webb (1827–1880) and Dr. James Webb were her brothers. Dr. J. T. Webb was a surgeon with the 23rd Regiment. Dr. Jim had just resigned because of poor health.

11. Elinor was then a member of the Congregational Church, as Howells wrote his father on 15 March 1863 (*SL,* 1:141).

12. Harriet Emily Trowbridge Hayes (1808–1874), of New Haven, the widow of William Rutherford Hayes (c. 1805–1852), former U.S. Consul at Barbados. W. R. Hayes was Mrs. Mead's uncle. Cartes de visite, or postcards with photographs, were popular at that time.

13. A. D. Shepard, Joanna's husband.
14. Lucy Hayes's sons were: Birchard Austin (b. 1853); Webb Cook (b. 1856); Rutherford Platt (b. 1858), and Joseph, the baby (b. 1861).
15. When Larkin G. Mead, Sr., forwarded this letter on 15 December, he enclosed "a Photograph of E. & L. taken just as they were ready to leave us. We think it a good likeness of both" (Hayes Presidential Center). The photograph is at the Howells Memorial.

To Mary J. Mead

Paris, Hotel de Louvre
Dec 24, 1862

Dear Mother:

Since Mr Howells wrote the foregoing[1] we have learned that the only through train to Venice leaves at 8 oclock P.M. and as it is too late to start to night—we'll be under the painful necessity of spending to morrow (until eight oclock) here and visiting the Louvre and whiling away the time, as best we can in this dull city. To tell you the truth, I'm in love with—Paris and dont object at all to the Hotel de Louvre. The wedding was a charming little affair altogether. T'was in Mr. Dayton's library and only six of us in all there.[2] Though I was obliged to be married in the Episcopal form (with a ring &c) it was a Methodist minister who did it.

It was rather trying to go rushing about so long trying to get some one to perform the ceremony without any success, but it is all over now and we were married on the 24ᵗʰ, in *Paris* assisted by the Legation. Mr Dayton said "You are one of the legation now, Mrs Howells." The Parliament-House in London, The chirch of Notre Dame the squares, the fountains and the Louvre of Paris have quite turned my head. And a *French* dinner is beyond anything I ever ate. The English waiters look like clergymen and patronize you so as to make you feel very awkward. Lark behaves splendidly—is *wilde* over the Louvre, talks most remarkable French aided by frantic gestures and enjoys himself hugely. I shall write all the particulars from Venice. With a great deal of love to my

mother & father (how kind they've been to me!) and all the Meads',
Ann Flannigan[3] & neighbors. I am your aff. daughter,

Elinor

this is a *copy* of my letter from Paris copied by Fred I think   EMH
Jan. 1904[4]

Manuscript: Massachusetts Historical Society

1. In his letter of the same day to Mr. Mead (*LinL,* 1:62; *SL,* 1:132), Howells relates
their tribulations, and how they came to be married in Paris, after finding that "a seven
days' residence in England was necessary to matrimony."

2. In her pocket diary (at Harvard) and in a note in the Howells, Mead, Noyes, and
Dock Family Papers at the Massachusetts Historical Society in Boston, Elinor identifies
the six. They were: the Rev. Dr. [John] McClintock (1814–1870) of the American
Chapel in Paris; Mr. [William L.] Dayton (1807–1864), Minister to France; Mr.
[William S.] Pennington (b. 1823), Secretary of the Legation; Howells, her brother
Larkin, and herself. The American Legation in Paris was then located at 3, rue de
Marignan. On the same note, Elinor pasted a sample of her wedding dress and one of
that of Howells' Welsh great-grandmother "Rees or Rhys" of Pontypool, Wales, who
married John Thomas. Elinor's wedding gown appears to have been brown with black
stripes. She also wore "a brown silk bonnet with light blue velvet inside, a talma like
the dress, and brown gloves.—Also a muff & tippet."

3. The Mead family Irish maid, according to Mildred Howells.

4. This sentence in EMH's hand appears below the copy of the letter. Fred was
Frederick Goodhue Mead, her youngest brother. The original is still unlocated.
Larkin G. Mead, Sr., added a short note about friends looking for Joanna, probably
when this copy was sent to Will and Elinor in Italy.

To William C. Howells[1]

Casa Falier [Venice],
January 4, 1863.

It was our intention to go to church this morning, but a man
came in to get the Consul's permission to send glass beads to America (We never say the *United* States now)[2] and kept us at home and
I think we'll spend our time quite as well writing home. It is only
a week ago today that we arrived in Venice,[3] but we both say it
seems like a month since. Already we are quite settled in housekeeping and I have been shown all the most interesting sights of
this wonderful city, though as every house has a history and every
church is a picture-gallery, it will take all the time we are here to
explore it fully. The churches are so cold that we are not going to
attempt visiting them at present but wait for the light & warmth

CERTIFICATE OF MARRIAGE

This certifies,

That on this *twenty fourth* day of *December* in the year of our Lord, one thousand eight hundred and *Sixty two* *William Dean Howells & Elinor Gertrude Mead* . citizens of the United States of North-America, were by me (a clergyman duly authorized), united in marriage at the legation of the United States, at Paris, France, according to the laws of the United States aforesaid, in presence of the American Minister, and of the following witnesses.

ATTEST. *John M: Clintock*

WITNESSES.

MINISTER OF THE AMERICAN CHAPEL.

PARIS.

France.

LEGATION OF THE UNITED STATES

Paris, France

Wm L. Dayton .... Envoy extraordinary and Minister plenipotentiary of the United States, at the imperial Court of France, do certify, that the parties named in the foregoing certificate, were duly united in marriage, by the reverend gentleman who has signed the foregoing certificate, at the Legation aforesaid, in presence of the undersigned and witnesses.

Given under my hand and seal, this *4* day of *December* 1862

(Howells Memorial, Kittery Point, Maine)

of the Spring. Just now we are rather taken up with the inside of Casa Falier. We have three of the cosiest little rooms you ever saw all opening into each other, and, what is rather remarkable in this city, all carpeted and two of them papered. The height doesn't correspond to one's idea of a Venetian palace being only eight feet or so, but the ceilings are elaborately frescoed and the windows elegantly draped. The parlor is a perfect gem of a room with a little balcony on the Grand Canal, the bedroom is pleasant to behold but extremely inconvenient to inhabit having no closets or corners to tuck things away in, and the dining-room, where we spend most of our time, is larger, warmer and more like home than any room I have seen in Venice. You ought to see us taking our breakfast here at the early hour of ten in the morning. Our landlord furnishes us with tea-cups, plates &c and we have bought a little tea-machine with which we can make tea on the table. Then we buy eggs, bread, salad &c for breakfast and go to the Hotel Vittoria to dine, which is living very economically and at the same time very well. I hav' n't told you of the office which is n't used much now however as it is just as convenient to transact business in the dining-room and I like this warm room for writing better than that dismal place. Mr Howells keeps me writing from morning till night you see. He left his quarterly accounts, his filing of letters and any quantity of copying for me to assist him about till I begin to think I'm rather more his *secretary* than his wife.[4]

We have great fun surprising people by Mr. Howells' suddenly introducing me to his old friends as his wife. Brunetta[5] his Italian teacher came in to see him after his return from. England and of course I was presented. He stared at me a long time with his great Italian eyes and then said "Your *sister?*" "No," said Mr Howells "my *wife.*" It was a long time before he could get the idea through his head and then he commenced laughing and scolding Mr Howells all at once in the excited manner and voice of the Italians.

Though there is no sound of wheels in this city it is the noisiest city in the world it seems to me. Everybody talks so loud & there is so much singing and crying of goods—I was going to add wares & merchandise I have got so used to writing it in the invoices. I supposed one never went out in Venice except in a gondola but we've only been twice in one since we came and once in the boat, our little, private boat which Mr. Howells propels standing in a most graceful position. I'll draw him sometime when we are out together and send you the picture. For the most part we go about through narrow, dark alleys, jostling by Austrian soldiers and Italian beggars till we get to St. Mark's where everything is bright and beautiful and everybody is promenading.

New Year's day we were standing on the balcony when we saw a man in a boat making for our house and to my surprise he came under the window and holding his hat *begged.* The idea of a beggar in his own boat was something new to me. We've only seen one American since we arrived, Mr. Delaplane an attachée of the Legation at Vienna who called New Year's day,[6] but I've not been *homesick* a bit. Perhaps when my brother goes away and I'm left with such a gloomy fellow as Mr. Howells for company, I shall be, though. This is the *gossiping* letter I promised.[7] Next time I'll try a more dignified style. With a great deal of love to all your family I am

Affectionately yours
Elinor

Manuscript: Harvard

1. Howells' letter of the same day to his sister Victoria (also at Harvard) makes it clear that Elinor's was addressed to his father.
2. Elinor refers to the Civil War.
3. As Elinor noted in her pocket diary, they arrived in Venice at 6:00 P.M. on 28 December. After a night at the Hotel Vittoria, they settled at Casa Falier with Larkin, who stayed with them until 6 January. For WDH's description of Casa Falier, see *Venetian Life,* pp. 99–100 (2d ed. 1867).
4. On Howells' use of his wife as secretary, see also his tongue-in-cheek letter of 15 March 1863 to his father (*LinL,* 1:64; *SL,* 1:140). There he writes in part: "From a primary motive of laziness and further to avoid all future cause of complaint regarding my handwriting, I have adopted Elinor as an amanuensis. As I always have a great deal to say without much disposition to say it, and as Elinor never has anything to say with the greatest possible desire to talk continually, I think you will be perfectly satisfied with the result of our arrangement. You will have no fault to find with a shortness of her letters and you will find the literary merit of the joint composition improved through my furnishing the ideas and their expression."
5. Eugenio Brunetta, whom Howells described as Biondini in "A Young Venetian Friend," *Harper's Monthly,* May 1919; *Years of My Youth* (HE), pp. 239–252. A "dear dear friend of earliest days," as Howells wrote under his photograph in the Venice album in the Howells Papers at Harvard, Brunetta was then a student. For WDH's version of Brunetta's introduction to Elinor, see *Years of My Youth,* pp. 244–245.

6. John Ferris Delaplaine (1815–1885), of New York. After studying law, he went to Europe and became attached to the American legation in Vienna. He was appointed secretary of the legation, 1866–1883.

7. "I who can't gossip, must fall back upon my old fashion of sending my love, and not saying much," Howells wrote the same day to his sister Victoria, in his first extant letter home after his marriage (Harvard).

To the Howells Family

Venice, March 4, 1863.

It is with timid strokes that I begin a letter now-a-days, for every-one of them has to be submitted to the critical inspection of Mr. Howells before it leaves the house, and it's as often they go into il fuoco[1] afterwards as to the Post office. But "It's never too late to mend,"[2] says the title of Pokey's (I knew that would escape me. It's my pet name for Mr. Howells)[3] book and I [3 (?) words][4] write for the Atlantic.

The greatest event that has happened lately is the finishing of the German Guide Book[5] (which Mr. Howells has been so long at work on, and I have copied so much on) and the arrival of a gold watch from Geneva—bought with a part of the profits of the book. It is a very nice little affair and a [g]reat relief to Mr. H., [in?] the way of weight, [against?] the huge silver one he has so long carried. We feel thoroughly posted as to Venice after such a faithful study of the Guide Book, and as soon as the churches are warm at all, we shall spend a great deal of our time in them, cultivating the acquaintance of the great masters and getting a correct idea of architecture &c. I like Venice better & better every day—the more I get to understand her history—and the peculiarities of the city. It is a selfish delightful life we lead here with nothing to do but [to] enjoy the beautiful sights & sounds with which this city abounds and which this charming climate lends its own delight.

The only drawback is we never see any Americans or anyone who sympathizes in the least with us in regard to our country. A young English girl—a particular friend—told me the other day "She was not surprised if we could take up arms against our mother-country that [w]e should fight among [ourse]lves." That was going pretty far back I thought.

At last Mr. Howells has got the flower-seeds for Annie[6] and I have labeled them as well as I could but she will have to judge them by their beauty—not by their names. The gloves as you will readily suspect are for little Willie,[7] with Aunt Elinor'[s] love—This letter—as may not [be] so evident is for Willie['s] [word] Grandfather from—here[8] Elinor got hopelessly tangled up, but I think she may be unwound to the general effect that she is your affectionate daughter. There is nothing much for me to write, except that Ive never been better in health. I hope that by this time, youve got some of the many letters we've written to you [several words] street [?]

Manuscript: Harvard

1. The fire. In an earlier letter to his father, Howells had written: "I suppose Elinor has destroyed a letter she had been writing to accompany [mine]. The production was submitted to me, and I pronounced it calculated to relieve you of the impression that I had married a too severely intellectual wife" (20 February, *SL*, 1:139).

2. Probably "Disillusion," or *No Love Lost*. See *SL*, 1:137, for the history of the publication of the book. The phrase (also the title of Charles Reade's well-known novel of 1856) does not appear in the poem, but fits the plot, and particularly the end. On Howells' *Atlantic* non-publications, see Ginette de B. Merrill and George Arms, "Howells at Belmont: The Case of the Wicked Interviewer," *Harvard Library Bulletin*, April 1982, pp. 165–168.

3. Elinor called her husband "Pokey," or even "Poke" during the early days of their marriage.

4. These and later bracketed words are missing or indecipherable, mostly because the corners of the two sheets of manuscript have disintegrated.

5. *Venice. Her Art-Treasures and Historical Associations. . . . Translated from the second German Edition of Adalbert Müller*. See *SL*, 1:140. Elinor noted in her pocket diary that she had worked on it since January, and Howells since August. The translation was finished on 2 March, and Howells' watch arrived on 4 March.

6. The Howellses were all great naturalists, and they frequently sent roots and flowers to one another.

7. William Dean Howells II, Joseph A. Howells' son, b. 1857.

8. "here . . . to you." In Howells' hand. After "you" Howells drew a line across the text to the lower right corner of the third page, where he wrote some words now missing due to the condition of the manuscript. For Elinor's impulsiveness and exuberance, see also her pocket diary entry for 24 February 1863: "A stupid day because Pokey was cross. At dinner he lectured me for asking so many questions."

Pocket Diary

March 24, 1863.[1]
Rose at 10. Breakfasted on rye bread. Read in Sismondi's Republics and Pokey read Goldoni's life.[2] Afterwards P. wrote on the little poem about the Hebrew burying-ground at the Lido and I wrote to Cousin Ann.[3] Went afterwards to the Square to hear the music. Sat outside at Florian's.[4] (Just before we went Dr Guntzberg called to see how my eruption was.)[5] I wore my brown & black, striped, woolen suit brown bonnet and camel's hair scarf. Dined at 3 1/2. Brunetta came in whilst we were eating. (We had dodged Mr Tortorini[6]—Bore No. 1, in the street) After he went Pokey & I both wrote. He to Mrs. Harding, about the trunk, and to Prof. Messadaglia[7] & I to Cousin Ann. Then we sent for *due birre*.[8] (Before the music began we strolled into the vestibule of St. Mark to see the lean kine eating the fat kine, and the history of Adam & Eve, in mosaics) After beer we carefully opened the door to hear the "cognáta" reading to Giovanna[9] in "Meditations on Death." It is Passion Week. There will be three days next week—the days in which Christ was in the ground—in which no bells will be rung. In the evening we read Kugler.[10]

Manuscript: Harvard

1. In her pocket diary (manuscript at Harvard), Elinor devoted a special section to consecutive entries on the twenty-fourth of each month, starting with 24 December 1862, the day of their wedding in Paris, and continuing until 24 January 1866, when she arrived in New York. These informal, breezy, and often amusing entries constitute a kind of newsreel of her life. A few of them are included here.

2. Jean-Charles Léonard Simonde de Sismondi (1773–1842), Swiss historian and economist. Author of *Histoire des républiques italiennes du moyen âge* (Paris, 1809–1815). *Memorie di Carlo Goldoni per l'istoria della sua vita e del suo teatro* (1831). Howells was greatly influenced by the Venetian dramatist (1707–1793), whom he mentioned in "Recent Italian Comedy," *North American Review*, October 1864, and whose *Memoirs* he edited in 1877. See also "Carlo Goldoni," *Atlantic*, November 1877.

3. Although Howells mentions the cemetery in *No Love Lost* (part 6, p. 21), the poem on the Hebrew cemetery at the Lido was not published. Cousin Ann may have been Ann Mead Lane of Troy, or New York.

4. The famous café on the Piazza San Marco.

5. Dr. Guntzberg, or Günzberg, an English-speaking German, was the Howellses' physician in Venice. His photograph is at Harvard.

6. G. Antonio Tortorini, Howells' earliest friend in Venice, also had a villa near Padua. A retired pharmacist in his fifties, he had taken care of Howells in a fatherly way during an illness and had lent him some money. He is the Pastorelli of WDH's "An Old Venetian

Friend" (*Years of My Youth*, pp. 225–237). On his photo in the Venice album at Harvard, Howells wrote this caption: "my earliest friend & generous creditor."

7. Angelo Messadaglia, a professor of political economy at the University of Padua, translated some poems of Longfellow. Howells had heard him give a paper at a scientific meeting in Venice, and was to meet him again in Padua the following summer.

8. "Two beers."

9. Giovanna, their servant at Casa Falier. In her Venetian diary (10 July 1863), Elinor identifies her sister-in-law (the cognata) as Angela Gasparini.

10. [Franz Theodor] *Kugler's Handbook of Painting* (1851). As WDH wrote his father, "We spend most of our time now in reading up the history of art. . . . So our talk is a jargon, more unintelligible on my part and less so on Elinor's, of Titians and Tintorettos, of paintings and sculptures and mosaics, of schools and of manners, and our reading naturally takes that direction, too" (15 March 1863; *SL*, 1:141–142).

To Mary D. Howells

[Venice] Thursday evening [1863].

Now I know that my dear new Mother wishes to hear from *me*,[1] particularly, I am most happy to write to her, and that it is no longer "embarrassing" let me assure her.

I was greatly amused, dear Mother, at your idea that daughters-in-law always like their fathers-in-law better than they do their mothers-in-law. My Mother always said she should like her daughters' husbands, but she was sure she could never like her *son's wives*. You can judge then how I laugh when her letters come full of praises of her about-to-be daughter, the young lady to whom my oldest brother is engaged.[2] She seems to be quite comforted for the loss of two daughters by this new child and says no more about "son's wives."

So I hope *you* will find that is not a "wise saying" about daughters-in-law, and that I can give you as much reason to change your mind in regard to that, as my adopted sister has given my mother. I am *sure* we shall like each other when we meet, however little we get on by letters. If we ca'n't come home yet, the next best way of getting acquainted will be for one of "the girls"[3]—as Mr. Howells calls them—to come out here who, knowing all you most wish to hear, will write more satisfactory letters home than either of us can. I think you would not feel her loss so much as you would her brother's gain.

Your letter arrived so late tonight that we can only write a few words in reply, but I hope these few words will comfort you. I think Sam & Johnny[4] ought to write to me now, and then I shall know the whole family.

Your loving daughter
Elinor.

Manuscript: Harvard

1. The original of Mary Dean Howells' letter has not been found.
2. Isabella Sophia Martin (1842–1925) married Charles Levi Mead (1833–1899) on 12 May 1864.
3. WDH's sisters.
4. Samuel Dean (1840–1925) and John Butler Howells (1846–1864), WDH's brothers.

Pocket Diary

[early May 1863]

Trip to Florence

Left Venice April 22 in the morning. Post from Padua to Ponte Lagoscuro, stopping at Rovigo over night. Railway to Vergato. Diligence all night over the Apennines to Pistoja. Arrived in Florence at 8 A.M. Went to Hotel de New York first. Afterwards to Dr. Slayton's, 1 Via Moro. Saw Victor Emanuel review 20000 troops. Hear nightingales in Cascine. Went up to Ombrellino to see the Jacksons.[1] Saw Mrs. Mowatt Ritchie, and she gave me a boquet. Saw Lark's Echo[2] in plaster. Went to Power's studio. Also to see Fedi's[3] splendid group. Went to Michaelangelo's house. Admired Giotto's campanile beyond anything in Florence. Did n't like Michaelangelo's David. Went to Chapel of the Medici & to Michaelangelo's Chapel. Mr. Howells kept repeating

> "Philip sleeps by unfamiliar Arno,
> And the dome of Bruneleschi."[4]

Pokey had eleven teeth filled and I two, and the bill was $52.00. Lark's one idea was to have us go to Wital's to eat—but we hated it. In the Protestant Cemetery we came upon the "Two Ladies from America," Mrs. Carter and Miss Wright of Lowell, Mass.[5] We liked John of Bologna's "Rape of the Sabine Women" as well as anything we saw. On our way to Florence we stopped a few moments at Monselice to see Mr. Tortorini, and congratulated him on having become Podesta, or Mayor, as he likes to call it. We met the Tubb's[6] at the American Chapel, and afterwards saw the Minots. Mrs. Jackson took us to Miss Blagden's,[7] the friend of Mrs. Browning. There I saw pictures & writing of the Browning family and met Thomas Trollopp.[8] Mrs. Ritchie's Greek & little Ned Powers interested us. We like Mr. Hart very much.[9] He & Lark are coming up to see us this Summer.

On the 1'st of May, Friday (My 26'th birthday) we started home— having first breakfasted with Hart & Lark at Donne's Caffè. Arrived home Sat. eve. by the hardest over night coming. Stopped at Bologna. Saw the Gallery & University going.

Manuscript: Harvard

1. John Adams Jackson (1825–1879), an American sculptor who settled in Florence, and his wife. They befriended the Howellses, whom they later visited in Cambridge. See EMH, 8 January 1868. Anna Cora Mowatt-Ritchie (1819–1870), a Swedenborgian, was an American writer and actress.

2. Larkin's little statue in plaster. William W. Corcoran (1798–1888), the Washington banker, "saw it in clay, unfinished, and bought it at once for $500—the artist's price," as Howells wrote Charles Hale on 5 April (*SL*, 1:147). "Echo is a gem," Elinor said. Hiram Powers (1805–1873), an American sculptor. Born in Vermont and raised in Ohio, he had lived in Florence since 1837.

3. Pio Fedi (1815–1892), an Italian sculptor in Florence.

4. Two misquoted lines from Tennyson's poem, "The Brook" (ll. 189–190). The passage reads:

> . . . My dearest brother, Edmund, sleeps,
> Not by the well-known stream and rustic spire,
> But unfamiliar Arno, and the dome
> Of Brunelleschi, sleeps in peace; and he,
> Poor Philip, of all his lavish waste of words
> Remains the lean P. W. on his tomb. . . .

5. "Two ladies from America" had called earlier at Casa Falier, while the Howellses were out on a picnic on the island of Torcello (EMH, pocket diary, 15 April).

6. The Howellses had entertained Captain Tubbs and his wife, and Miss Minot, for lunch at Casa Falier (EMH, pocket diary, 24 January). Tubbs, who had formerly been with the East India Company, jointly owned a gondola with Howells. The Tubbses had left Venice recently (*SL*, 1:133–134).

7. Isabella Jane Blagden (c. 1816–1873), a close friend of the Brownings. Described by Browning as a "bright, delicate, electric woman," she lived by her pen. She shared Mrs. Browning's love of dogs and flowers and her admiration for Louis-Napoléon.

8. Thomas Adolphus Trollope (1810–1892), the brother of Anthony and also a writer, had lived in Florence since 1843.

9. Joel T. Hart (1810–1877), the sculptor from Kentucky. (See EMH's letter of 17 May 1863 to Lucy Hayes about his work.) Larkin and his friend Gagliardi, the Italian sculptor in marble who had lived in Brattleboro, shared his studio until a dispute over Gagliardi's salary forced them to leave (EMH, Venetian diary, 29 August 1863).

To Aurelia H. Howells

Venice, May 15, 1863.

Dear Aurelia:—

They "clean house" in Venice as well as in Ohio and Vermont I see, for four men are beating squares of pasteboard in the yard of the palace opposite,—which pasteboards have lain so long under carpets that they send forth great clouds of dust. All the floors in Venice are either of marble mosaic or a kind of "petrified pudding," as Pokey (my name for your noble brother) calls it, composed of all sorts of bits and colors of stone thrown into a brown cement and then the whole is polished until it shines like glass. The carpets instead of being tacked down are held in place by huge hooks in the wall or "mop-board" to which rings on the carpet are fastened. For all the pasteboard underneath the carpets we have very cold floors in Venice and have to go around in fur slippers in Winter. By the way every letter I have had from America lately speaks of house cleaning till I begin to think that the large army of women-housecleaners at present would soon "clean out the rebels" if they had a chance.

This has been an eventful week and today a marked day, beginning with the postman's bringing us five letters while we were at breakfast—four from America and one from Florence. Then one day this week we heard from Mr. Conway in London who said he was coming to Venice "to see Howells"[1] as he told Emerson—which Mr. Emerson remarked was "a very anti-Ruskinian sentiment." Yesterday we had the ex-vice-consul of Constantinople (since consul at Galatz—and on his way to Göteberg, Sweden to be consul there) a jolly, young fellow of twenty-three—a Yankee from Portland, Maine—(well! I think I'll say to what or where or how we had him!)—to dinner.[2] Shall I tell you what we had for dinner? Imagine a little, round table, spread with the finest linen (Mr. Valentine furnishes us elegant table-service)[3] with the prettiest decanturs for ale and water you can fancy. First comes a course of soup of green peas—Then mutton chops and young boiled onions, young potatoes and tender little beets. (for a very grand occasion we could have little fried balls of rice & egg & chopped meat, which Giovanna makes beautifully, here) Next comes French pudding which with great difficulty I taught G. to make. (As there is only an open fireplace to cook by we can have very few things of the sort. Yeast & soda are things utterly unknown to most Italians.) At the end we had cherries and the gentlemen had coffee. Mr. Thomas said it was the most like an American dinner of anything he had had in Europe. Giovanna stares at us eating two or three vegetables together and on one plate, as Europeans always give their whole attention to one variety at once.

I would tell you something about Florence only Mr. Howells wrote such a good description of it to the Boston Advertizer that there is nothing left for me to say.[4] I will, however, send you a flower from a boquet which the lovely Mrs. Ritchie picked for me herself in Ombrellino, Galileo's garden.

My dear husband just took up the sheet I have already written and sneered at it because of "the hundredth description of our dinners." Now I'll tell you the reason why I am so elaborate on the subject of our meals. One time Annie[5] rather laughed at Italian housekeeping I thought, and I wish to impress Mr. Howells' sisters with the idea that I am an accomplished housekeeper* and that we live in a highly respectable way at Casa Falier—and that we dine *at home* now-a-days. Everything goes so smoothly that Mr Howells imagines it just happens so, and doesn't understand my general supervision at all. I beg you will take no notice whatever of that scandalous little note he had the audacity to insert in my letter. This week I have begun copying in the Academy of Fine Arts.[7] Every morning I work away two hours at my copy of Titian's John the Baptist and Mr. Howells reads Pulci[8] meanwhile. A con-

---

*Never goes into the kitchen, and don't know what's for dinner till it comes on the table.[6]

stant crowd of people keeps passing by—mostly English—but we do not allow it to disturb us in the least.⁹ You ought to hear your brother sing, Aurelia—He stands up to the piano, while I play accompaniments, and lets out his voice in a most brilliant manner on Italian songs to the great admiration of our guests. He is actually learning the contralto to quite a difficult duett at present. So you see there is something inspiring to song in the air of Italy.

We got three photographs by mail today and we had piles before—but none of them give me the pleasure that those of you & Vic¹⁰ would. And Mr. Howells talks so much of dear little Henry's beautiful face that I long to see it. I should value it very much if you would send me one of him. Please write to *me* next time.

Yours
Elinor

Manuscript: Harvard

1. Moncure D. Conway (1832–1907), an old Ohio friend of Howells, when he was editor of the Cincinnati *Dial*, later (1862–1863) edited the Boston *Commonwealth*. Before he went to England to lecture on slavery, Howells had warmly invited him to visit them in Venice. "Of course you know that I am married, and to whom. . . . I used to hear a great deal about you in letters from Cincinnati [where Elinor was teaching]. You have an additional merit in my eyes, because you met Elinor there" (24 March 1863, *SL*, 1:145). For Conway's remark to Emerson, see his answer of 6 May to Howells (quoted in part, *SL*, 1:146).

2. Identified in Elinor's pocket diary as William Thomas, he was evidently William Widgery Thomas, Jr. (1839–1927), lawyer, diplomat, author, and politician. He became enamored of Sweden and mastered its language. The founder of a Swedish colony in Maine, he later served as minister resident to Sweden and Norway.

3. Edward Valentine, the British vice-consul, and their landlord and neighbor at Casa Falier. A good friend of the Howellses, he became the godfather of their first child.

4. "From Venice to Florence and Back Again," Boston *Advertiser*, 25 May 1863.

5. Howells' sister, Anne T. Howells.

6. "Never . . . table." In WDH's hand.

7. The Academy is on the other side of the Grand Canal, just opposite Casa Falier.

8. Luigi Pulci (1432–1484), the burlesque poet. On 15 June (Venetian diary), Elinor reported that her husband was reading *Il Morgante*.

9. "I have bought Elinor a sketch-book," Howells had written his father on 15 March, "and she proposes to unite sketching with boating. Yesterday she made her first sketch in public,—a fisherman. . . . The subject was quite unconscious and sat still for a long time, in spite of the eager and applausive multitude scuffling about the elbows of the artist for the best view of her creation. At last the fisherman changed his position, to get his knife, and the artist suspended her labors amid the ill-disguised disgust of the multitude" (*SL*, 1:142).

10. Howells' sister, Victoria M. Howells, and his younger brother, Henry Israel. There are many family photographs in the Howells Papers at Harvard, the Howells Memorial

Collection at Kittery Point, Maine, and in the Howells/Fréchette Collection at Alfred University.

To Lucy Webb Hayes

Venice, May 17, 1863.

Dear Cousin Lucy:

I suppose I am never expected to wait for you to answer my letters, but just keep on writing. And, besides, that hasty letter I sent you just before leaving America hardly deserved an answer; so I am going to "address you without ceremony" and tell you all about my doings here without waiting for you to inquire. But, in the first place, I never thanked you half enough for the beautiful present you sent me. It, or rather they were the prettiest things I had given me, and the associations with them will always be so jolly![1]

I have told the story to quantities of people—but it was hardly a cheat on my part, for I was not engaged till February, and then only in a kind of half-way manner by letter.[2] To be sure letters were the whole and only arrangement; but I did not like to say anything *about* the affair until *I had* told my family, which I waited to do until I was home. Miss Nourse found it out herself, and it was impossible to conceal it from Laura[3]—but they were all who knew it until two months, about, before I left home. So I hope, Cousin Lucy, you wo'n't think I was to blame not to tell you before I told my family, and you must write to me and tell me you forgive me.

The apparently "wild-goose" chase has ended most respectably in quiet, domestic life in Venice. We live in one corner of an old palace on the Grand Canal in little rooms fitted up in snug, English style by our landlord, the English vice-consul. We have carpets in every room—a rare thing here, where the floors are all made of marble or a kind of cement filled with bits of stone and then polished, which looks like "petrified-pudding" more than anything else. Our little parlor, or bower, has a balcony over the canal where we sit evenings and hear the Venetians sing as they go by in gondolas. I call it a bower because it is so very small and is all furnished in green. It has a tiny piano, a bit of a mirror, a green and white satin sofa and pretty little pictures and ornaments. Our dining-room is not much larger, but looks out onto the canal through the acanthus tree in Mr. Valentine's (our landlord) garden, the only tree in sight, and almost the only tree on the canal. It seems to me some what like living aboard a ship here—no sound but the splashing of water except the sound of voices—and the Italians *have voices*. You would enjoy the full, rich, wild singing one hears here in the evening. The young men are so crushed and killed by the Austrian government[4] that it seems as if their only relief was to yell—but it is splendid yelling I assure you. I was going to tell you a little about *my* housekeeping—if the easy way the house-affairs go on can be ascribed in the least degree to my keeping—but I fear I have not much share in the glory—it is all Giovanna's doing. G. is a faithful old creature and the best servant I ever knew.[5] We ca'n't expect to be treated with such profound respect ever again, for no one is really a servant at home—only "help" you know? We do'n't object to being so exalted for a time, though we object to the principle. Yesterday we had the consul of Gothenberg Sweden to dinner, and everything "went off " in style—thanks to Giovanna. Giovanna, as well as the table service was provided by Mr. Valentine, so we fell in with the machinery as it was working, and consequently have things in decidedly European style—every vegetable served separately and any quantity of courses. I enjoy all this as a change, but I shall gladly go back to the dishes and manners of my childhood. Europe is the country to visit and America is the country to live in. It makes you appreciate the institutions, the education, the wealth and the wonderful prosperity of our country to see the cramped state of everything here.

But Venice! I never imagined such a charming life in the way of luxurious ease and delight of the senses as one can live here. Every where you glance perfect beauty meets your eye. The richness of the architecture is beyond anything I ever saw and the finest paintings in the world (I think) abound in every part. There is no noise to disturb you—only the music by the Austrian bands in St. Mark's Square every other evening and that delights more than it disturbs—no dust in all the city—every kind of fruit, and cheap as in Cincinnati—and then it is so nice to go about in a gondola. Do'n't you think it is a kind of charmed life? I wish the war would be over and you & Cousin Rutherford could come over here sometime in the next two years—for, if nothing happens we will perhaps be here as long as that. Mr. Howells heard from Mr. Conway in London this last week saying he was coming here very soon.[6]

We see many Americans but scarcely any we knew at home. We hear from our friends very often or we should not be contented. You will write to me at once, wont you Cousin Lucy—for you know it takes so long for a letter to come? We have heard from you twice—through Laura and through Aunt Emily,[7] but not very satisfactorily. I hope Cousin Rutherford's arm does not trouble him any more. We are very anxious to hear from Laura as it is a long, long time since she wrote. How is your dear Mother—and little Jo, and Uncles Joe and Jim and all the children?[8] And Sally Perry, and Mrs. Heron.[9] How are they all? I have not heard from Miss Nourse since I came here, though I wrote her soon after I arrived. Do you ever see her now or Katy Burbank?

We went to Florence last month to see Lark and had a grand time. Lark has already made a statuette (The Echo) and sold it to Mr. Corcoran, the Washington banker. His studio is with the Kentucky sculptor, Hart, who is making a statue of Clay for his state.[10]

I send you one of our horrid photographs, just that you may realize I am married.[11] Mr. Howells sends his love with his. Your beautiful ambrotype is as much admired here as it was at home. I show it to prove the beauty of American women. Please give a great deal of love to Mrs. Webb from her child away off in Venice[12] and a great deal of love, please, to all your family from the Howells family.

I shall hope to hear from you in July.

Your aff. Cousin
Elinor.

Manuscript: Hayes Presidential Center

1. The snakes. See her letter of 7 December 1862.
2. They became engaged in February 1862, while Howells was in Venice and Elinor in Cincinnati teaching at Miss Nourse's school. Howells tells his side of the "story"

passim in his Venetian diary (at Harvard). He probably proposed on 7 January 1862, upon receiving a letter from Elinor after a long silence perhaps due to a misunderstanding between them. On that day, he wrote in his diary: "Yesterday morning I got a letter from E. G. M. [Elinor Gertrude Mead] to my utter surprise; and it was such a letter as made all the past perfect again. To-day I answered it, in the only way I could, and now wait anxiously to hear again."

3. During Elinor's last week of teaching at Miss Nourse's school, 16–22 June 1862, Laura Platt came from Columbus to stay with her at the school. Laura may have found out about the engagement then rather than during the week of 12 April, when the two girls visited Lucy Hayes in Cincinnati (Sophia Hayes to Sardis Birchard, 15 April 1862, and Laura Platt to Lucy Hayes, 13 June 1862, Hayes Library).

4. During the Austrian occupation of Venice. In her letter of 25 March 1864 to Samuel D. Howells, Elinor told of "horrid scenes" during the conscription of young Venetians.

5. Later Elinor's opinion changed markedly, and one of the reasons the Howellses left Casa Falier was to escape Giovanna, who brought in many of her relatives to live off the inexperienced young couple. For WDH's amusing story of their relationship with her, see *Venetian Life*, pp. 111–124. A sketch of her by EMH is in the Howells Papers at Harvard.

6. Conway arrived on 5 July, according to Elinor (Venetian diary, 8 July).

7. Laura Platt Mitchell and Emily Trowbridge Hayes. See 7 December 1862.

8. Lucy Hayes's mother, baby, two brothers, and the Hayeses' three older sons. See 23 October and 7 December 1862.

9. Sallie Perry is mentioned in Hayes's diary. She was perhaps a relative of the Hon. Aaron Fyfe Perry (1815–1893), a lawyer in Cincinnati, who became a delegate to the Republican National Convention in Baltimore in 1864. A member of the U.S. House of Representatives in 1871, he later made speeches in behalf of Hayes. John W. Herron and RBH had shared an office and rooms in Cincinnati. A lifelong friend of Hayes, he became the father-in-law of President William H. Taft (Watt P. Marchman, *The Rutherford B. Hayes State Memorial* [1979], p. 8).

10. Hart had received a commission for a life-size marble statue of Henry Clay.

11. Will and Elinor had their picture taken at Sorgato's in January 1863. They sent prints to family and friends.

12. Mrs. Webb had "adopted" Elinor while she was away from home teaching in Cincinnati.

Pocket Diary

May 24'th [1863]. Sunday.
The twenty-fourths seem to be the least eventful days of the month in Venice. Breakfasted on sour cherries, fried eggs & baked (young) potatoes. Mr. Howells then went to see Mrs. Bethune[1] at Hotel de la Ville, and was afterwards going to get me some molasses in Saints Fellippo e Giaccamo and some bristol-board to make Nina[2] a paper doll of, but it was a Festa[3] and all the shops were closed. It is the first day of the feast of the pentecost. Whilst Pokey was gone I was siezed with a fit of altering things, and quite made

the office over again so it looks much better. I put the beautiful intaglio table in the centre and moved out the old one. We expect Mr. Conway soon and then we intend to put a bed in there for him and a washstand. After Pokey came home he wrote some in the office on his Venetian Sketches⁴ & I sewed on a new night-gown in the dining-room. (The lavandaja askes so much for washing my fine ones that I am making a simple sack). Brunetta came & bored Pokey. We had a splendid dinner and I (who feel much better than I did before I went to Florence) ate very heartily so I felt stupid afterwards and lay down on the parlor sofa while Pokey read Goldoni's Memoirs. Then we both went to see Mrs. Bethune, and this is all of the stupidest day we have passed for a long time. I am drawing Titian's John the Baptist at the Academy.

Manuscript: Harvard

1. Mary Williams Bethune, the widow of Dr. George Washington Bethune (1805–1862), an American clergyman who had died of apoplexy in Florence. Howells wrote the epitaph for her husband's mosaic memorial by Antonio Salviati, to be placed in his Brooklyn Dutch Reformed church. See Venetian diary, 8 July 1863; also Elinor to Sam, 1 August 1863.
2. The young daughter of their servant, Giovanna.
3. A holy day.
4. "I'm writing away on some desultory sketches which I call 'Life in Venice,'" Howells wrote to Charles Hale, the editor of the Boston *Advertiser*, on 13 June 1863 (*SL*, 1:153). Some of these were published in the *Advertiser*, and together with material from the "Letters from Venice," that had started to appear in the *Advertiser* since 27 March 1863, they constituted the nucleus of *Venetian Life*. See Gibson-Arms, pp. 87–89; Woodress, pp. 52–53; and *SL*, 1:153–154.

To Victoria M. Howells

Venice, 17/VI 1863.

Dear Vic:

Though I have not yet received your promised letter I will answer it in anticipation.

I thought I should write to Mr. Howells' family much oftener than I do, and it is not because I have not the inclination, but because I find Pokey's letters are so all-sufficient that what I add seems *mere* twaddle. Now what if I had attempted to describe our visit to Florence! Just after my letter would have arrived the Advertizer with that splendid description of the journey by your gifted brother and mine would have been regarded as nothing in comparison.[1] I used to think my girlish epistles entertaining—but I find now that if Mr. Howells adds two lines to my letters they are more esteemed than all the rest.[2] Now is n't this calculated to discourage a young woman and shut her up forever? Then if I send messages instead of writing you'll understand.

Anyone must despair of putting Venice in June into words. Even Dickens' description is vague and dream-like you know?[3] It would be sacrilege for the practical Elinor to attempt it. I can only give you our indoor life. At last we have arrived at something like system—poetical system. After breakfast—a fairy-like repast of ham, baked potatoes and coffee—Mr. Howells retires into the cool, quiet office and, on the inspiration of beautiful intaglio table and a huge, bronze crab, writes most charming sketches of Venetian character and of things happening here—all of which it is to be hoped you will see in a book sometime. Meanwhile I *arrange,* (Mr. Howells knows the meaning of that word) and sew till 1 o'clock. Then we go to the Academy or to make calls, usually, and at four we dine. Mr. Howells' office hours, like a former consul's are supposed to be from 10 o'clock till 1, but, like him, he is as liable to be out as in at that time. It is said that an old sea-captain, tired of never finding this former consul in, wrote under the notice "Ten to one you *arn't*."[4] After sunset we always go out for a stroll—often to the Square to hear the music. I trust some one of Mr. Howells' sisters will accept our invitation and come out to enjoy this beautiful life. I wish all might. I have not heard any more about my sister's coming,[5] but it is quite worth while to have bad eyes if it be the means of bringing her here. I hear, through a friend, that she would not come till November at any rate.

I went to a *pic-nic* and *danced* yesterday. We are expecting great pleasure from Mr. Conway's visit week after next.

Love to *all* the Howellses

from Elinor.

Manuscript: Harvard

1. "From Venice to Florence and Back Again," Boston *Advertiser,* 25 May 1863. For Elinor's description, see her pocket diary, 24 April.

2. On 18 May, Howells wrote to his sister Aurelia: "E's style is so exhaustive that I feel there is very little to say when she has written" (SL, 1:149).

3. Elinor may be referring to Dickens' dream about Venice in Pictures from Italy, 1844.

4. This anecdote evidently amused the Howellses. Elinor jotted it down with variants in her pocket diary, and Howells in his Italian Journeys (1867), p. 54.

5. Mary N. Mead, Elinor's youngest sister, arrived in Venice on 30 November 1863. Her studies had been interrupted because of her poor eyesight.

Pocket Diary

Wednesday, June 24'th [1863]
Last time I dated it Friday—but it was really Sunday and neither Mr. Howells nor I knew it. We paid visits, I sewed, and never dreamed it was Sunday till in making the accounts in the evening Giovanna let it out. We, of course felt deeply mortified. Both of us supposed it to be Friday. On June 24th I finished the sketch of St. Christopher for Pokey's poem,[1] and next morning it was sent off to Harper's. Sewed on my bathing-dress all day.[2] In the evening went with Miss Przemysl to visit Mr. Power[3] at the Armenian Convent. Staid a long time in the garden and looking at the Palazzo Zenobia[4] and seeing the boys gymnacticise. Padre (the little fat one) pointed out a little Egyptian who looked like a Sphynx.[5] The same padre gave us beautiful boquets of flowers. It is fearfully hot weather.

Manuscript: Harvard

1. "Saint Christopher," was published in Harper's Monthly, December 1863, with Elinor's sketch on p. 1, "a sketch of the statue on the gate-way [of the Campo di Marte], representing St. Christopher bearing the Christ child, all covered with ivy" (EMH, Venetian diary, 22 June).

2. "Sea-bathing is now the order of the day, and we go three times a week to the Lido. . . . The weather is very hot, and we just exist and nothing more" (WDH, Venetian diary, 29 June). "E. had never been sea-bathing before. She looked comical enough in the gown furnished at the dressing-rooms [of the Lido beach]," Elinor wrote of herself in the Venetian diary, 22 June.

3. Mr. Power has not been identified. Eteldride or Eteldrede Przemysl, whose photographs are in the Howells Venice album at Harvard, was of Polish origin. "She knew all English fiction and spoke our language perfectly. She smoked a large black cigarino that Conway gave her. She was a brilliant liar, and one of our greatest friends," Howells wrote under her picture. They saw her frequently.

4. Armenian monks ran a school at the Palazzo Zenobia in Venice. This must be what Elinor meant, not their convent, which is not in the city. Howells describes this visit and a subsequent one there on the occasion of their Commencement exercises

(*Venetian Life*, pp. 202–206). The Howellses were much impressed by these Armenian Fathers, whose convent in San Lazzaro, an island off Venice, they visited on 19 October. "Ate rose-leaves and saw a mummy," Elinor remarked in her pocket diary. A close friendship developed between them and two of the fathers, Padre Giacomo Issaverdanz, a scholar, and Padre Alessio, also a teacher and an artist, whose photographs are in the Howells album, together with those of other Armenians.

    5. With a small pen and ink sketch of the head of the Egyptian boy.

Venetian Diary

July 8'th [1863]

I (Elinor) must write hastily what has happened the last week, for Mr. Howells is overrun with work. He is writing an epitaph for Dr. Bethune's Mosaic Memorial,[1] which Mr. Salviati is making, and which is to be placed in his Church in Brooklyn. He is writing a letter for the Commonwealth, as an assistance to Mr. Conway, who is here, and too much distracted by Venice to write himself—and he is trying to get his monthly letter to the Daily Advertizer off. The subject this time is Fourth of July in Venice, and there is a patriotic song in it. We were going to celebrate the Fourth by ourselves at the Lido—but by the time Pietro[2] got here, it looked like rain, and the idea struck Mr. Howells of getting what Americans there were in the city together and having a celebration in the house. So we went around, in the gondola, to all the hotels and left cards for the Americans whose names we found. At the Luna was the Ex-Consul of Java, Mr Diehl, and Mr. Thomason of Kentucky—at the Europa Mrs. Swett and sister (Miss Coolidge) and at the Ville Mrs. Bethune, who could n't come, and Orville Phillips & sister who wouldn't come because they are Secessionists at heart, though pretended New York Unionists. They afterwards called & met Mr. Conway here—whose sentiments they know. To make out a party we invited Miss Prezmysl & Mr. Conway. Mrs. Swett & sister could not come as Mrs. S. was to start early next morning to Switzerland to meet her son, a Capt. in our army, who has just come over for his health. We draped the parlor with our flags, put up the Washington,[3] got a boquet of red white and blue and Mrs. B. kindly lent us Church's picture[4] of "Our banner in the sky." The toasts were good. Mr. Thomason's best. It was "The countries now struggling for their freedom—Italy, Poland and America" and ever so much more which I have forgotten—but Miss Prezmysl being a Pole and all of us sympathizing with Italy, it was just right. We sang, talked, eat fruit and drank the wine of Asti—like champaigne, only better. Sunday night after we were in bed, at eleven o'clock, Mr. Conway came.⁵ (July 5'th.) We flew up—got him a supper of sardines, eggs, pickles and kald-sloe, and made him a bed on the office sofa. . . .

Manuscript: Harvard

1. See EMH, pocket diary, 24 May. WDH gives a detailed description of the mosaic monument in his "Letter from Venice," Boston *Advertiser,* 28 July 1863.
2. The gondolier engaged to take them to the Lido.
3. A portrait of Washington.
4. The painting by Frederic Edwin Church (1826–1900), "where the sky seems to recreate the American flag."
5. The much-awaited Conway had finally arrived, and for two weeks the Howellses had a splendid time "doing Venice" with him. Elinor took him to the Academy and the Ducal Palace, and the three of them engaged in animated conversations about the U.S., the war, and their literary prospects, at Casa Falier, during long walks in the Giudecca, the Public Gardens, the Zattere quay, and over "magnificent" dinners. "Mr. C. is in raptures over Venice. I hope he will come here to live," Elinor wrote in the Venetian diary (10 July). They went bathing at the Lido, and, on an expedition to Chioggia, reported by Howells in *Venetian Life* (pp. 190–193), Elinor, "the artist in our party," stopped to sketch a sleeping boy, immediately attracting the usual crowd of street urchins. Conway went away on 21 July, "leaving Pokey and me quite used up by the fortnight's excitement" (EMH, Venetian diary, 22 July). See also Woodress, pp. 30–31, on Conway's visit, and Conway, *Autobiography* (1904), 1:428–432. Of Elinor's letter, that he hand-carried to his wife, Conway wrote, "it is before me now, as sweet a letter as woman ever wrote—picturing the enchantment of Venice and crying, 'Do come, do come!'"

Pocket Diary

July 24'th (Friday) [1863]

Drank three cups of coffee for breakfast and almost immediately afterwards started for the I. R.¹ something (old Loredan palace in St. Stefano) to give one last look at the staircase before finishing my picture of Bertha rushing up stairs,² for Mr. Howells' poem which is soon to be published in New York. Finished the picture before dinner. In the evening Mrs. Günzberg³ and a young officer came to see us and staid till 11 o'clock. The officer, who is a Ger-

man, sang "When on the stormy midnight deep" "Maedel ruck
ruck ruck" and other songs with me, told ghost stories, spoke a
little English, sang the Standard Bearer, Shubert's Serenade,
Brightest Eyes &c without accompaniment and made himself generally agreeable.⁴ Mrs. G. excused the doctor's not coming, flattered me, praised my blue silk and was as anxious as ever to take
me into German society—which I always positively decline letting
her do. We gave them peaches cut up, vanilla cream and maca-
roons.

Manuscript: Harvard

1. Imperiale Regio (imperial royal). The old Venetian palace was used as the admin-
istrative municipal building.
2. One of Elinor's illustrations for Howells' narrative poem, then called "Disillusion."
It was to have been published by Frank E. Foster, an old Columbus friend, now of the
firm of Follett, Foster and Co., in New York. Elinor had been working "diligently and
successfully" on these illustrations. When the romance finally appeared as *No Love Lost*
in *Putnam's Magazine* (1868), and in book form (1869), her two charming scenes of the
Piazzetta in Venice were used for the frontispiece and the title-page of the book, as
well as the tail-piece "a Cupid in travelling traps," (EMH, Venetian diary, 6 August
1863). The two unsigned pencil drawings of the Piazzetta from Elinor's sketchbook are
at Harvard with other artwork of hers. The picture of Bertha rushing upstairs was
evidently not used.
3. The wife of their physician, who was German. Not wishing to be considered as
Austriacanti (those favoring the Austrian rule), the U.S. Consul and his wife refused to
be seen in German society. In *Venetian Life* (pp. 18–26), Howells depicted the frustrations of foreigners "planted between two hostile camps. . . . Neutrality is solitude and
friendship with neither party; society is exclusive association with the Austrians or with
the Italians." Even the famous cafés on the Piazza were patronized by only one or the
other camp.
4. Elinor identifies him as Lieut. Kühn in her Venetian diary (26 July).

To Samuel D. Howells

[Venice, 1 August 1863]¹

Dear Sam:
(The only name by which I hear you called) I am very glad to
begin a correspondence with you, for from what I hear of the
easiness of your disposition I judge you'll not be very critical, and I
think we'll "get on" very nicely together.
We are feeling almost light-hearted these days at the good news

we hear from America. One day when we were in the deepest despair we found on our table, on returning from our evening stroll, a card sent us by Mr Blumenthal the banker,[2] on the back of which was written "Private telegram from England. Lee beaten. Vicksburg taken."[3] It was almost too much for us to believe at once, but every day has added better things, since. Last night we got particulars of the New York riots, which must have been terrible, but will not hurt our cause, I am sure, as the object of the rioters seems to have been more to plunder from innocent persons than to oppose the conscription. We hold our heads up very high now when we speak of the war, and persons here have ceased to inquire of the state of our country in the patronizing tone they used before.

Are you having hot, breathless days in Ohio? If you are allow us to pity you, for Venice is delightfully cool—cooler even than the Vermont valley at this time. The narrow streets are almost damp, and shady of course—the Canal which has a tide twice a day, is perfectly fresh, and so refreshing to look at—the little shops are full of delicious fruits and then, best of all, there is the sea-bathing. We go down to the Lido in a gondola—without noise, without dust, and plunge into the soft, warm sea, which is so delightful that we generally stay in an hour, and by that time we are strong as lions and hungry as wolves—to make two choice comparisons.

Mr. Howells says he has not told you of our presents. Some time ago, in a letter to the Boston Advertiser, Mr. Howells praised the mosaics of Mr. Salviati; which so pleased that gentleman that the other day, on departing for England, he sent around a paper-weight for Mr. Howells & an elegant pin for me. Am I not fortunate in being the wife of a journalist? Several Americans were induced by this letter to visit Mr. Salviati's establishment, and they bought a good many of his things, so he siezes Mr. Howells' hands when he meets him and overwhelms him with gratitude. He says "You are a fortune to me. I cannot understand why you are so good to me &c."

I take encouragement from what your father says that one of your sisters may come out here for a visit.[4] I have already begun to plan for it, thinking how this will suit her &c. If my sister comes

it will be in November probably! Hoping to hear from you soon &
often I am

Yours sincerely
Elinor

Manuscript: Harvard

1. Dating established by Elinor's entry for 3 August, Venetian diary.
2. Mr. Blumenthal, their banker, served now and then as acting consul. He had
obliged the Howellses, who were on friendly terms with him. He had explained to WDH
the symbolism on the Hebrew gravestones in the cemetery at the Lido, and the Howellses
had recently attended a concert of chamber music at the Blumenthals' (Venetian diary,
15 June 1863; *SL,* 1:127–128).
3. Vicksburg was taken on 4 July. The Howellses heard the news by 26 July.
4. At that time, they thought that Annie Howells might come, but she did not.

Venetian Diary

Tuesday, Aug. 18'th [1863] Noon. Yesterday was an exciting day. The
heat in the morning was terrific, but Pokey (I am going to get a
new name for him as this is undignified) went to the Bank on
business. While he was gone someone called, and, when the card
came up, it read "Henry Ward Beecher"[1] and, also, in pencil were
two other gentlemen's names Prof. J. L. Raymond and Rev. J. S.
Holme.[2] I saw them cross in their gondola to the Academy, so made
haste and dressed myself, and when Pokey returned we joined
them there.[3] They were close to the Assumption, looking at it in
the very worst light possible. We introduced ourselves, and, immediately Beecher began running on Venice—the musquitoes, the
Basilica, the architecture generally were condemned. Mr. Howells
defended them bravely, and Beecher admitted that probably one
had to be educated in oriental atmosphere in order to appreciate
it, and he is only to stay here two days. We talked & talked, and,
finally, Mr. B. & I were left alone—so we went all over the gallery
together, chatting in the most familiar manner. Mr. B. loves color,
and fully appreciated the glory of Titian's coloring—but the subjects, of course, disgusted him, and there was a struggle all the
while in his mind between the horror of the conception, and the
beauty of the execution. He was delighted with the original

sketches of the old masters—particularly da Vinci's. He likes to watch the process of thought which the drawings show, and thinks the men appear grander in their sketches than in their finished pictures. Mr. B. was in a jolly mood and had great fun with his comrades over the Emperor[4] and monarchical government generally—he pretending to have been converted to a royalist in Europe. Mr. Howells had to go home but I staid about two hours, and then they brought me home in their gondola and I gave them some books on Venetian painting to read. In the evening we went to the Malibran to see Il Trovatore, which was splendidly performed. Did n't go to sleep till 3 o'clock on account of the heat. This morning Pokey went in full dress and with the American flag on his gondola to St. Mark's to say a *tedeum* for Francis Joseph of Austria. In the midst of the service, just as the host was raised a bomb went off among the crowd inside the church. Everybody started, but the service went on. This was "Patienza's voice uttering her one word of protest." This evening Mr. Beecher & friends are coming here to see the *fresco* or gondola procession from our balcony.

Aug. 19'th Evening—O dear! Such exciting times! Mr. Beecher and Mr. Holme came at 8 o'clock. We fell into pleasant conversation at once, and the ball kept moving for an hour without the least exertion. But 9 o'clock came & 10 and no fresco. The doubt arose as to whether there would be any show—which was very embarrassing as the gentlemen had staid over on purpose to see it. We had refreshments to pass the time away—still no sign of lights nor sound of music. Giovanna was dispatched to learn if there was to be any procession—but before she returned Mr. B. said they really must go as they left the next morning at 5 o'clock. Many pleasant things were said and they left—Mr. Beecher promising to leave his photo for me at Hotel Daniele.[5] Pokey & I took rather a dispairing view of the matter & went to bed quite blue. Next day (today) we hardly mentioned the subject till evening, being very busy, (Pokey writing for the Advertiser & I drawing on the Piazzetta scene)[6] when we walked around to Hotel Daniele to get the visite.[7] There was a note awaiting us—but on touching it we both exclaimed "no photograph!" But what a note! Mr. B. said his pictures were packed away so carefully that he could n't get at them easily, but would forward his photo. from Munich—the first place he unpacked it— and added that he found on getting home the procession had moved in another direction and passed *directly in front of his hotel* &c but, said he, "that did not prevent me from enjoying the pleasantest evening at your house that I have passed on the continent" or something to that effect—only better expressed. So I shall have two letters from Mr. Beecher besides his photograph![8] I gave him the Bacchus & Adriadne by Tintoretto & Mr. Holme the Assumption. Tomorrow we go to Arquà.[9]

Manuscript: Harvard

1. Henry Ward Beecher (1813–1887), Congregational clergyman, author, and popular public speaker, was the pastor of the Plymouth Church of Brooklyn. Later involved in a notorious scandal, he was accused by Theodore Tilton of improper relations with Tilton's wife.

2. John L. Raymond (1814–1878). After studying law and theology, he became trustee and president of Vassar College. A friend of Beecher's, he accompanied him in his European lecture tour on behalf of the North. The third man is probably the Rev. John Stanford Holme, D.D. (d. 1884), a Baptist clergyman and the pastor of Trinity Baptist Church, 55th Street, near Lexington Ave., in New York City.

3. From her window at Casa Falier, Elinor could easily see their gondola cross the Grand Canal to the Academy on the opposite bank.

4. Franz Josef I (1830–1916), emperor of Austria since 1848.

5. The famous hotel frequented by kings, statesmen, lovers, and writers, among them Musset and George Sand, Dickens, D'Annunzio, Wagner. Howells had stayed at the Danieli when he first arrived in Venice in December 1861.

6. One of Elinor's illustrations for WDH's narrative poem, "Disillusion."

7. Carte de visite.

8. Beecher's two notes are among the Howells Papers at Harvard. The first one, dated 19 August from Venice, reads as follows: "My dear Mrs Howells ¶ Will you excuse me for not enclosing my photograph? I found that I had packed the parcell too skillfully to be easily found. On reaching Munich I will have my things repacked, & will then fulfill my promise—The procession came off, it seems, moving in the *other* direction from the Gove[r]nor's palace & directly in front of our Hotel! ¶ But my loss of *that* did not prevent my enjoying the pleasantest evening I've ever had on the Continent. I am very truly yours ¶ H. W. Beecher"

The second note, dated 24 August, from Munich, contained the promised photograph and read in part: "My dear Mrs Howells ¶ I enclose the promised shadow. It had contrived to secrete itself among my things, & not till I unpacked everything did it show its face. If it was unwilling to go, I am sure it cannot be a good likeness of me—. . . ."

9. A village up on the Euganean Hills, near Padua, where Petrarch had lived. In her Venetian diary (22 August), Elinor gave the following account of their first trip to Arquà that summer: "I was so tired of the style of life we had been living for some time— trying to write & draw, but with no strength or spirit, hot musquito nights without sleep, calls to be made, &c." They left Venice on the 20th. After a rainy day spent in

visiting Padua, they started off the next morning for Arquà. "Went directly to the house of Petrarch, which fully realized my expectations of a poet's dwelling. So quiet and commanding such a peaceful view. I sketched the cat, the inkstand & the chair and Pokey wrote some verses in the visitors' book. We hated to leave it, but we went to the inn for dinner,—tough chicken and native wine—and then it was time to descend." Their second trip took place in September.

Pocket Diary

September 24'th [1863].
The musquitoes of Padua kept us awake all night and we started for Arquà in rather a tired state.¹ This second trip was in order that I might make sketches of the Petrarch premises, so I determined to put it through. Made six sketches including the one of the Guido at Padua. Bought fruit at Battaglia²—but our chicken was too old for our taste, so we gave it to the custodé and the Cont.³ Returned in the rain—hung around the depot till 9 o'clock—took the wrong omnibus in Venice & had to walk way from St. Mark's at midnight.

Manuscript: Harvard

1. Their second trip to Arquà took place on 23–24 September. On 31 January 1864, they sent off to *Harper's* Elinor's six sketches "The House, Font, Grottos, Cat, Inkstand and Tomb" (Venetian diary). These were to illustrate Howells' "A Pilgrimage to Petrarch's House at Arquà," that appeared instead in the *Nation*, 30 November 1865, without Elinor's illustrations; reprinted in *Italian Journeys* (1867), pp. 216–234. Joseph Pennell illustrated the 1901 edition. Elinor's sketches do not seem to have been used. They have not been located.
2. A village three miles from Arquà.
3. There are several custodians and a number of peasants (*Cont*adini?) in WDH's sketch, which appears to fuse the two trips.

To Mary D. Howells

[Venice, 30 September 1863]
Dear Mother:
I dont believe the news of the election¹ could have created much more excitement among the Howellses at Jefferson than it did in the Howells family in Venice. No literary success ever stirred Mr.

Howells up half so much I am sure, for he was all of a tremble with laughter & joy as he read the letter. I had a distant glimpse of my future father-in-law, writing in the House, the winter I was in Columbus; but now I shall imagine him lying back in one of the luxurious seats in the Senate, or rising amid a profound hush to address the honourable assembly, while a lady wearing a Roman scarf is seen smiling from the gallery. I only hope that his services to the state wo'n't prevent him from coming to see us as he intended. With much love to all—

affectionally,
Elinor.

Manuscript: Harvard

1. Elinor meant nomination. This note follows WDH's letters to his mother of 27 and 30 September. In the first, Howells enclosed a Roman scarf as a gift for her. In the second, he offered his congratulations to his father, who had just been nominated to the state senate. The election, which the elder Howells won, was held on 18 October. The result was announced in the *Ashtabula Sentinel*, the family paper, on 28 October. See also *SL*, 1:165–166.

To Anne T. Howells

Venice, Nov. 14th, 1863.

Dear Annie:

I ca'n't let this letter go off without writing a word to tell you how much I am pleased with both your and Aurelia's photographs. I only hope Vic's will be as good, and then we shall be very proud to show our three sisters to our foreign friends. Those who have seen Aurelia's speak of its resemblance to Mr. Howells. Giovanna says yours is not so much like him. I have to just reverse Holmes' line[1] and say

"The melting black eye,*   (*meaning the *lashes,* as Mr.
The bright blue—   Howells says Aurelia's eyes are
I cannot choose   *grey*)
between the two."

We are every moment expecting a telegram from my sister saying she has arrived at Milan and will be here either tonight or very soon.² I can do nothing in this "pause of time" (this expression is to be found in "Disillusion"—a poem by W. D. Howells, about being published by Foster of New York)³ but wander around and make new arrangements in the already-arranged rooms, and Mr. Howells is more nervous on the subject than I even. He whisked off to the Square in a pouring rain awhile ago, but has finally concluded to set about studying a French lesson to occupy his mind. We went around to Florian's last night to learn if the City of New York had arrived and found Galignani⁴ full of news from America which that steamer had brought. So we judged Mary must at that time be crossing Mt. Cenis and would consequently be in Milan today. I do so hope she saw your father & Vic in New York before she sailed! She was to be in New York a week before the 31'st, and my mother remained a week after the steamer sailed, so I ca'n't help thinking some of them met.

But they didn't. Finished for   Elinor.⁵

Manuscript: Harvard

1. Holmes's line reads:

> The bright black eye, the melting blue,—
> I cannot choose between the two.   ["The Dilemma"]

2. Mary did not arrive in Venice until the evening of 30 November.

3. These were exciting times for Elinor and Will. They were getting published (both the "Saint Christopher" poem and "Disillusion" were to appear in December), and they were expecting Mary Mead and their first baby. They had engaged the services of Elizabeth Scarbro, whom they called Bettina, a *contadina* from a nearby village, and they were frantically trying to rent an extra room at Casa Falier for the consul's office in order to make room for the new arrivals in their apartment. In the Venetian diary (October) they both write of their tribulations while their bedroom was unusable because of a leak, and Elinor recounts a robbery due to their servant Giovanna's negligence and probable complicity and her sister-in-law's dishonesty. For Howells' account of the burglary, see *Venetian Life*, p. 120.

4. *Galignani's Messenger*, published in Paris.

5. "But . . . Elinor." In WDH's hand.

Venetian Diary

December 2'nd [1863]. Sunday a Courier (Sig. Rizzi) was dispatched to Milan to fetch Molly, we having learned by telegram that she was waiting at an expensive hotel there (Reale) without an escort.

Monday night, at 6 o'clock, we met her at the station with a gondola & brought her to Casa Falier where a duck dinner awaited her. In the evening Mr. Howells took her up to the Piazza. She is in fine health and spirits and Mr. Howells & she take to each other. Tonight they have gone to call on Miss Przemysl. This morning the fascinating officer[1] whom Mary met in the cars joined us in the Square and walked home with us, but we did not ask him in, not knowing who he was. He will be sure to turn up again as he often comes to Venice. Mary is delighted with her room and so very amiable that everything seems to satisfy her. She is a lovely girl and it is a privilege to have her with us. She attracts much attention in the street. The "creatura" and her sister[2] stopped short on the bridge and cried out & groaned out "O bella! bella! bella!" and ever so much more genuine *flattery* this morning as we passed. They were *really struck*, I think, for they met us suddenly and fairly gasped out their admiration.[3] Mary took one lesson in Ollendorf's Italian today and Mr. Howells was surprised and delighted at her retentive memory. I think she will make rapid progress in the language.[4]

Manuscript: Harvard

1. Elinor identifies him as Giuseppe Bechtinger, a chief physician in the Austrian army at Verona (Venetian diary, 9 December). He proposed to Mary within a few days, and WDH had to decline for her. "It appears that the Austrian whom Mary met on the train, and afterwards in the square, had really fallen in love with her," Howells adds in the Venetian diary. "He wrote to her, after returning to Verona where he is I[mperiale] R[egio] Capo Medico, and asked her to correspond, or at least reply. M[ary] handed me the letter, and I replied for her declining not only correspondence, but further acquaintance. I did not think this officer in seeking her acquaintance except through me, was acting properly. This first letter was a declaration of love. To-day in response to mine, came a letter proposing marriage."

2. Two poor weird sisters who lived in the neighborhood. Howells has described them in *Venetian Life*, pp. 115–117: "*creatura* being in the vocabulary of Venetian pity the term for a fellow-being somewhat more pitiable than a *poveretta.*"

3. Mary, who was stunning-looking, received several more proposals during her stay with the Howellses in Venice. "She, and her Venetian experiences, were the inspiration for *A Fearful Responsibility"* (Mildred Howells, *LinL,* 1:75). Bechtinger, the "fascinating officer" (see n. 1, above), was evidently the original of Ernst von Ehrhardt, Captain of the Royal-Imperial Engineers, in Howells' novel.

4. Henri G. Ollendorf, *Ollendorf's New Method for . . . Italian Language* (1859). Mary was apparently more talented in foreign languages than her sister Elinor who, as Howells put it in *Venetian Life* (p. 122), "had the rare gift of learning to speak less and less Italian every day."

Pocket Diary

December 24th [1863]
On my back in bed. Winifred 8 days old today.[1] Nina has dressed her in her best things because Miss Przemysl[2] is coming to see her. I do'n't sleep much but am gradually getting along. Mary & Mr. Howells dine out at the French Place.

This is the anniversary of our wedding—this about the hour 3 o'clock P.M.

I just add in ink (Dec. 30th) that Mr. H. & Mary dined at the Vittoria, instead of the French place, and had only Inchbold at table.[3] He made himself quite agreeable though he began by introducing Mary to himself in this way "This is Mrs. Howells' sister I believe?"

Manuscript: Harvard

1. Winifred Howells was born on Thursday, 17 December. On that day, WDH wrote the following in his wife's pocket diary: "I make one record in Elinor's book, to say that this the morning of December 17, 1863, our daughter Winifred was born at one o'clock. E's. pains commenced about 10 o'clock yesterday morning, but she got through everything very well. The midwife, Angela Vianelli, attended her very skilfully, and we did not call Dr. Günzberg till about five minutes before the birth of the child. Present in the room: Dr. G., A. V. aforesaid, Giovanna and myself. Marietta (Mr. Valentine's servant) and Bettina (the wet nurse) were in the kitchen. A. V. washed and dressed the baby first; fire in Mary's room, and W. taken there. Mary [Mead] with the Kelloggs in their lodgings." See also WDH's letters of the same day to his parents (*SL*, 1:171) and to Laura Mitchell (*LinL*, 1:80–81). Miner K. Kellogg (1814–1889), of Cincinnati, was a painter of oriental scenes. With his wife, he had recently arrived in Venice.

2. "the Przemysl, who grows more delightful and brilliant the better she is known to us" (WDH to Conway, 26 January 1864, *SL*, 1:175).

3. John William Inchbold (1830–1888), English painter and engraver. Reputed especially for his landscapes, he had been praised by Ruskin. Howells probably had him in mind when he had Mr. Rose-Black introduce Lily Mayhew to himself in *A Fearful Responsibility* (chap. 8). The English artist was ill-mannered also upon meeting Miss Lydia Blood in *The Lady of the Aroostook* (HE, p. 265).

✦ II ✦

Venetian Life and Italian Journeys
1864–July 1865

$\mathcal{I}$n the spring of *1864* the Howellses
moved from Casa Falier to Palazzo Giustiniani; they needed a more spa-
cious apartment now that Winifred and Elinor's sister Mary, who had
arrived in Venice in late November 1863, were on the scene. More enter-
taining was in order too, especially with Mary having had at least four
marriage proposals, as recorded by Elinor, the first a few days after her
arrival and the last, when "Poor Taylor popped," the night before Mary
left for home in May 1865.

In addition to Mary's disappointed suitors, the apartment was often filled
with other visitors during the rest of the Howellses' stay in Venice. Along
with their Venetian friends, a steady stream of Americans came, among
them the Pattons, aunt and uncle of Elinor's; Charles Hale, the former
editor of the Boston Advertiser, *now on his way to the consulship in Egypt;*
and Samuel P. Langley, on a tour of observatories and other scientific es-
tablishments in Europe. Between times Will was busy with his writing and
Elinor with her illustrations and the new baby, to whose progress she devoted
sections of her letters and diaries.

From the beginning of Will's and later Elinor's arrival in Venice, both
of them had avoided German society, unwilling to be viewed as condoning
Austrian rule. Though their doctor's wife, Mrs. Günzberg, urged Elinor
to enter German society, she positively declined; still, Elinor and Will al-

lowed Mary to attend a ball shunned by Italians, regarding it as a pity that she should "lose such a sight." Elinor's remarks in a letter to her brother-in-law Sam on a young Italian conscript about to be pressed into Austrian army service and his desperate attempt to escape also bear upon her view of the political servitude suffered by the Venetians.

"Letters from Venice" had begun to appear in the Boston Advertiser *in March 1863 and were still appearing in 1864, but from the beginning the plan had been to produce them as a book. So far publishers had not shown interest. Still, a letter from Lowell, received in August 1864 and quoted by Elinor in her diary—"the* most *careful and picturesque study of any part of Italy"—made up for the earlier rebuffs. Because of Lowell's letter, the London publisher Trübner finally accepted* Venetian Life *in 1865, subject to an American publisher's taking half the sheets. Meanwhile the Howellses were planning a second Italian book,* Italian Journeys, *and mostly for this reason they made a trip to Rome and other Italian cities in the fall of 1864. Though they had earlier expected to return to the United States soon after this excursion, the slowness of replies from Trübner and from the Department of State about a leave of absence delayed their leaving Venice until 3 July 1865. "Blue as indigo," Elinor had exclaimed in her diary one day in February. "All we do is play with Baby, listen for the postman's ring, and read aloud in Carlyle's Frederick the Great."*

Venetian Diary

January 10'th [1864]. A month since I have written here and *so much* has happened! In the first place we have had given us a little daughter whom we call Winifred—a name I chose as fitting Howells—though Winifred is Saxon I think, and Howells is Welsh. To-day the little thing is twenty-five days old, and already she notices objects & sounds—recognizing Nina[1] when she smacks her lips, and crying to go to her—and rises high on her feet & throws herself forward. She is very purple yet and can hardly be called pretty with so much dark hair hanging in her eyes—which eyes, by the way, cause great discussions. They are blue and the pupil (if that means the whole blue part and not merely the little black) is very large certainly—but the cut for the eye seems very small &

drawn up at the outer corners like Mr. Howells'. Well, we shall soon be able to tell all about how she is going to look if she comes on at the fast rate she has so far. She is a very large child—weighing eight pounds when she was born—and seems perfectly healthy. Her father gave her a little book New Years in which he keeps a weekly record of her progress.² She was born at 1 o'clock Thursday morning Dec. 10'th,³ 1863. Mr. Howells sat up all the rest of the night writing letters home. I was quite well the first two days, but talked & eat too much and afterwards had several feverish days when I had to lie low and drink only camomile tea. Now I eat the same as the rest—i.e. Mr. H. & Mary—and am only waiting for this dreadful cold weather to be over to go out. Miss Przemysl & Mrs. Kellogg have been to see me. Mr. Valentine, also, has spent an evening here. Mr. Howells' grandmother died the 14'th of November—⁴just before her grandchild was born. Mary has had another offer (which she has refused) from Brig. Scott⁵ who is at Geneva at school. . . .⁶

Manuscript: Harvard

1. Winnie's wet nurse. This was another nickname for Bettina.
2. Howells mentions this "pigmy diary" in a letter to Conway, 26 January (*SL*, 1:175). It has not been located.
3. An error. Winifred was born on 17 December 1863.
4. Anne Thomas Howells died of cancer in Bowling Green on 20 November, according to the obituary in the 2 December 1863 *Ashtabula Sentinel* (*SL*, 1:170).
5. Brig. Scott and his brother later visited the Howellses in Venice. See 23 June 1864.
6. Elinor continues with various news and gossip. The entry ends with the following, in WDH's hand: "Winnie was pronounced, the very night of her birth, 'el ritratto di suo pare' [her father's portrait]. In fact she is ridiculously like me.—E. wants me to set down the consolation offered me in the barber shop this morning. Somebody, whom I never saw before, but who had evidently seen E. previous to her confinement, asked me if the child were born yet. I answered, with some surprise, Yes. The next question was whether it was a boy or girl, my friend being evidently anxious for a boy. 'A girl,' said I. 'Ah, ben! un' altra volta,' (O well, another time) said my friend consolingly."

To Aurelia H. Howells

[Venice, after 9 February 1864]

[Beginning lost]

for Padre Libera¹ has written her a sonnet and so many come to see her that she holds almost daily receptions. I wish you could see her at one of these receptions—for it is then she appears best—showing already her fondness for society. The picture² will give you an idea of the way she is tied up in her bed or cushion—but it is impossible to show you by a drawing her beautiful blue eyes, her straight little nose and her long dark hair. Everybody exclaims at the quantity of hair she has—though it was dark at first it daily grows lighter and we think she will end a blonde. As soon as she has learned fairly to distinguish between her papa & mamma she shall be shown the pictures of her aunts (then Vic³ will be left out if she does n't send her picture to us shortly) and be taught to know them by name, so that she will recognize them immediately when we go home. I have just finished a cloak for her so she can be carried out the first fine day. Baby-waggons *are* of course unknown in Venice where all is stone pavement, so her nurse will carry her to the Piazzetta and promenade up and down with the other nurses there, also carrying babies.

I am glad you like her name. I would have liked to have given her the name of all her aunts—but Victoria, Aurelia, Joanna, Mary, Annie Howells would have been rather afflictive, and how could I choose between them? I am afraid Laura Mitchell is disgusted at our not calling her Laura, but I had a friend named *Angelina*⁴ who had called a child for me, and I *should* have greatly offended her if I had given my child anyone's name in preference to hers—so to avoid making a disturbance we called her Winifred—a name almost unknown except in literature. I have this morning received a letter from my cousin who is camping out with her husband, Colonel Hayes in Western Virginia. It is a wild life but she likes it and rejoices that her husband has been listed for three years more. My brother⁵ stopped here on his way back from Paris and took my sister to Florence with him. She will probably be away two months as she is going to Rome before returning. Please give a great deal of love to your Mother—*my* Mother, too—from me and tell her I am going to write to her soon. My love to all the dear unseen—though not unknown Howellses.

Affectionately
Elinor

Manuscript: Harvard

1. Dating established from Venetian diary, 10 February. That Elinor writes to Aurelia

is clear from the mention of the picture in her letter to Mary D. Howells, 27 February. Padre Libera was a Venetian priest who was also an inventor. On his photo at Harvard, Howells identifies him as the original of Don Ippolito of *A Foregone Conclusion* (1875). See also Woodress, pp. 158–160.

2. Elinor's sketch of Winny on her cushion appears here.

3. Victoria M. Howells. Howells' other sisters, Annie and Aurelia, had already sent their photographs.

4. Probably Angelina Hyde, Mrs. Lucius Henry Buckingham; one of her daughters was named Eleanor.

5. Larkin G. Mead, Jr., the sculptor, and Mary Mead left for Florence on 9 February (EMH, Venetian diary, 10 February 1864. In the same entry, Elinor mentions Padre Libera's sonnet for Winny).

To Mary D. Howells

Venice, Feb. 27'th, 1864.

My dear Mother—& baby's Grandmother:

I told Willie I was going to write to you on his birthday, because I thought perhaps you and I, more than any others, could bless the day he was born. It may be that sometime the 1'st of March will be known by the literary world as the birthday of the *great poet Howells,* but you will always think of the sweet little baby that "never made any trouble" who was born then. Our little Winifred is just such another good baby—such a patient, quiet, little thing that we feel as though we were not worthy to have her. Yesterday, for the first time, she was a little sick and moaned so pitifully that I could n't help crying myself to hear her—but this morning the pain is quite gone and she can smile again. Hereafter I suppose our moods will always hang on hers, for how can we be happy when *Winnie* is not well? This is my mother's first grandchild and your first granddaughter—as poor Eliza's baby did not live.

It was entirely a surprise to my family—as well as to yours—to hear that we had a baby—and the news kept my mother awake at least one night. My sister had no idea when she left home that she was so soon to become an aunt—but she feels the deepest interest in the little Venetian and esteems it a great privilege to know her from the very beginning. I think, Mother, you would like very much the way in which the child is done up. It is much better than having a dress all pulled up under the arms and the feet left to get cold. She wears first a little long-sleeves chemise over that a woolen shirt and then a little knit waist over that. This keeps her shoulders very warm. Then she is laid onto a thick flannel blanket, which is wrapped around her up to her arms, and turned up at the feet and, after this, she has wound around her (to keep everything snug and warm) several yards of broad bandaging. For additional protection she is then laid onto a quilted cushion like this, [sketch] (having a linen cover with a ruffle all around it) and this, also, is turned up to her waist in this manner [sketch] and, afterwards turned over at the sides and tied together up the front like the picture I sent Aurelia. It is, of course, very easy to carry and tend her in this little bed, which supports her head very nicely. Mr. Howells himself can manage her perfectly well in this arrangement. I have described it very minutely to you, as I thought it would interest you to know how comfortably the little thing was passing her babyhood. She is not tied up so tightly as to prevent her from moving her legs as some Italian babies are. Even the arms are tied down to the side with the bandage, sometimes, in Italy.

My sister is gone to Florence and we are leading our usual, quiet life; Mr. Howells studying very hard on the Italian comedies;[1] I sewing or *illustrating poems.*[2] It is a very pleasant life we lead in Italy, but I shall be glad enough to get back to the *reality* of American life. I have filled my sheet with the baby—what else could one expect? We are now going out for a walk and to mail this letter. I am going to keep you well informed of all that belongs to your grandchild, and, I wish you would advise me in regard to her. Much love to all from

your affectionate daughter
Elinor

Manuscript: Harvard

1. "Recent Italian Comedy" appeared in the *North American Review* in October 1864.
2. The Howellses had just sent off to Harper "The Mulberries" with her picture (EMH, Venetian diary, 28 February). The sketch for "The Mulberries" is at Harvard. The poem was not published until 4 January 1871, when it appeared in the *Nation* without any illustration.

To Anne T. Howells

Venice, March 8'th, 1864.

Dear Annie:

I must never let a letter go off now, without giving the last interesting particulars concerning your niece, Winifred. She was quite sick with a cold when we wrote, ten days ago, but today she is better than ever and, for the first time laughed out 'loud. It is, at last, fine weather in Venice, and when we went into the Square, at the time of the music, this afternoon, we found it full of babies whom the nurses were out airing. N.B. The Italians allow their *babies* to hear the Austrian band play.[1] These nurses are always contadine with the brownest skins and broadest backs and their costume is very picturesque. Their skirts are of a bright color and very short; (you may be sure they do'n't wear hoops—it would spoil the effect entirely) their stockings are very white, and they wear a peculiar kind of slipper with a very high heel. Their hair is stuck full of great silver pins, they have invariably a white lace kerchief over their shoulders and long black ends to their apron, behind. The babies are gracefully poised on the right arm and with the left the nurse holds a parasol.

The baby's blue (*celeste* they call it here) cloak was finished long ago and today I put the swan's down on to her little, white silk hood, so that now she is all ready to be exhibited along with the other babies—her proud papa & mamma following at a proper distance. We are thinking of making our quarterly journey next week—to Mantua, Verona, Parma &c this time—, leaving the baby to Giovanna and the nurse's care. G. is quite an old woman and has children[2] of her own so she can be trusted, and the doctor is to look in very often. We shall, also, leave word where to be telegraphed to in each city. I say we *think* of leaving the baby, but maybe I shall not wish to when the time comes.

At present the head of this house is devoted to the Italian Comedies—reading them in the soberest manner possible from morning till night—with a view to an *article* no doubt. We breakfast, dine & sup on comedy, and go to the theatre in the evening to see the charming comedies of Goldoni played. In these we see reproduced the customs of Venice a hundred years ago. Saturday evening we entertained an Italian nobleman, Count Capograsso,[3] the beau of Venice. He is young, rich & handsome. Ask Aurelia if she would like to marry an Italian?

Now that Hamburg is blockaded,[4] and we must send by the French mail, my letters will be very egotistical I fear, for I shall have to tell you the news of the Veneto-Howells family in the most concise way, and there will be no room to expand on such subjects as your being in Columbus and seeing Laura Mitchell for instance.[5] I like so much your patriotic letters, Annie—they are so hopeful.[6] Do write some more. With quantities of love to all

Elinor.

Manuscript: Harvard

1. In *Venetian Life* (pp. 23–24), Howells explained how the Italians avoided the Piazza during the concerts of the Austrian military band, as a tacit protest (*dimostrazione*) against the occupation of Venice. *Contadine:* country women.

2. Beppo (or Beppi) and Nina, whose occasional misadventures are related in *Venetian Life* and in Elinor's diaries.

3. Count Capograssi. On his photograph (at Harvard), WDH wrote in part: "He was a Dalmatian noble, charmingly intelligent." He later married Miss Przemysl. See 26 November 1865.

4. As WDH explains in his letter of the same day to Annie, Hamburg and Bremen were blockaded by the Danes in the German and Danish war (*SL*, 1:189).

5. Annie Howells visited her father in Columbus during the session of the state legislature that winter.

6. Presumably letters that Annie published from time to time in the *Ashtabula Sentinel.*

To Samuel D. Howells

Venice, March 25, 1864.

Dear brother Sam:[1]

Your being in the army makes this war quite a different thing to us and we search the papers carefully to find news of your regiment—though, to be sure, we've had to think of you in the hospital[2] a great deal of the time—and I do'n't know which is worse, to be sick in a hospital or to be making forced marches. My brother who is in Florence went once with the army six months as "special artist" for Harper,[3] and even that, without his having any military duty to perform, quite used him up. But I've no doubt that after one gets "broken in" the rough life of the camp is a good thing—just as the outdoor air is going to do the baby good when she doesn't take cold from it, as she did on her first essay. We read your description of the tramp from Nashville to Chattanooga to several persons here to give them an idea of what volunteers, young men who had been delicately brought up, were willing to suffer for the cause. Here, the young Venetians are pressed into the Austrian service, to serve for ten years, and fight against their own country. The conscription is being made now in a square through which we have to pass every day, and we witness horrid scenes— mothers falling down in fits, &c. The other day a young conscript jumped from a high window of the office of the conscription into the canal below—on hearing that he was taken. As they cannot find his body it is supposed that he swam under some building and thence escaped—perhaps out of the city into the kingdom of Italy. I like the new photograph which Vic sent very much. It gives me quite a different idea of you from what that old one did and I see a good deal of resemblance to Will, which of course I like.

Hoping to hear of your continued health and success,

affectionately
Elinor

Manuscript: Harvard

1. Elinor's letter to Sam follows WDH's own.
2. Sam's hospital was in Nashville, Tennessee.
3. Larkin G. Mead "went to the front for six months as an artist for *Harper's Weekly,* receiving forty dollars a week, and while making a drawing of a southern fort for the government barely escaped with his life, being within range of a sharpshooter, who spied him and sent a ball whizzing past his ear" (*Annals,* p. 720).

To Mary D. Howells

Venice, Apr. 25ᵗʰ, 1864.

My dear Mother:
I know you will be glad to know that little Winifred has been christened. The ceremony took place in the German Lutheran Church[1] (though the service was performed in Italian so that we could understand it) and our good friend, Mr. Valentine stood

godfather to the child. Rev. Mr. Blakeman, formerly an orthodox minister in Jefferson, and a young cousin of mine,[2] with whom he is travelling as tutor, were present, as well as a Mr. & Mrs. Wiggin of Boston. Mr. W. is a unitarian minister.

The baby behaved beautifully, not uttering a cry the whole time, but smiling approvingly both on Mr. Valentine, as he held her, and the minister. Altogether the baptism was the pleasantest *ceremony* I have seen since I came to Venice, and all the circumstances were interesting. A beautiful picture of Christ breaking bread, by Titian hung near where we stood. The only difference between this and any christening was that there were candles burning at the altar, but I could forgive that, the pastor was so sincere and everything else was so simple.

Another event is, or rather *will* be, the changing of our lodging a week from today. We are so cramped for room here that we *must* move, and we are going into one of the most beautiful old palaces in Venice.—that is into one *corner* of it—Palazzo Giustinian,[3] where the baby will have a great hall to wander about in—in arms of course. We were the more determined to make this move inasmuch as we thought Aurelia was coming out this Spring—but, to my great disappointment, Vic does n't say one word about it in her last letter, which we received a few days ago. I hope the idea is not given up. My sister longs to have her come, as they could have very jolly times together. *We* may possibly return in the fall, so I wish she could come soon.

You will see by the enclosed picture what a fat little thing Winnie is; but the photo. gives no idea of her nurse's beauty.[4] I have just finished the *seventh* little dress for the baby, and now I think her wardrobe is complete for six months at least, for I have allowed for her growing immensely. Mr. Howells wishes to tell you more about Mr. Blakeman, so, as this is to be sent by Liverpool I will cut my part of the letter short. With much love to all—

your aff. daughter
Elinor.[5]

Manuscript: Harvard

1. It was in this chapel, located in "the Campo Santi Apostoli . . . a little apart from

the church of that name," that Staniford and Lydia Blood were married . . . "before the altar under Titian's beautiful picture of Christ breaking bread" (*The Lady of the Aroostook*, HE, p. 316).

2. Elinor identifies him as Willie Trowbridge of New Haven. The ceremony, which was performed by the Rev. Dr. Wittchen, took place on 20 April. The godfather gave the baby a silver cup lined with gold, and WDH gave the minister two napoleons (about eight dollars) (EMH, Venetian diary, 17 May).

3. Palazzo Giustiniani, next to Ca'Foscari, on the other side of the Grand Canal. Strategically located at the bend of the canal, it commands a superb view on both sides. The Howellses also decided to move "to escape the encroaching nepotism of Giovanna, the flower of serving-women." Howells described the palace in "Our Last Year in Venice" (*Venetian Life*, new enlarged ed. [1872], pp. 399–412). In her first diary entry from their new address, after describing five-month-old Winifred and the christening ceremony, Elinor added: "We are in a grand palace now & have plenty of fresh air. We left the other house in excellent order & Giovanna to her fate. The baby goes to see her often & she comes here. Nina's food is sent in, which is a much cheaper arrangement. The baby is a great favorite with Mr. Da Rù, Mrs. Manning (the Simpkinson's cook) & the other servants. We had a great concert here the other night. I invited Miss Przemysl, the Count, Von Dadelsen, Mr. Valentine, Baldwin & Mr. Johnson" (17 May, Venetian diary). Darù was their new landlord; Baldwin, an American painter; and Mr. Johnson, a visitor from Boston. Evan von Dadelsen, as EMH identified him (pocket diary, typescript, p. 28), was an unfortunate suitor of Mary Mead, and this may have been the night she rejected him. There is at Harvard a photograph of a sketch by Elinor of "Mary & Von Dadelsen," dated 13 May 1864. According to Mildred Howells (*LinL*, 1:75), it represents Mary "standing upon the balcony of the Palazzo Giustinian beside a somewhat drooping young man; their backs are turned towards the artist and their figures are framed in the open window against the Grand Canal. It was afterwards discovered by the unsuspecting artist to have been the unconscious portrait of a rejected proposal." This episode in Mary's life evidently inspired a similar one in *A Fearful Responsibility* (1881) where Howells shows Lily Mayhew rejecting Mr. Andersen on the balcony, and has the U.S. Consul Hoskins, a sculptor, sketch the scene without realizing its significance: "They stood there, with their backs to the others. She seemed to be listening, with averted face, while he, with his cheek leaning upon one hand and his elbow resting on the balcony rail, kept a pensive attitude after they had apparently ceased to speak" (p. 93). After thus describing Elinor's own drawing in the novel, WDH proceeds to have Mrs. Elmore comment indignantly on the callousness of the artist: "She [Lily] was refusing him there on the balcony while that disgusting Mr. Hoskins was sketching them; and he had his hand up, that way, because he was crying" (p. 95).

4. This photograph is at Harvard.

5. In WDH's somewhat shorter letter, he writes that he can add little on Blakeman except that he was in Jefferson "a good many years ago, and that he knows nearly everybody there." He also praises Sam's response to his war experience and regrets that Aurelia will probably not come to Venice. Of the picture he observes that Elinor colored it: "The baby moved and blurred herself, and had to be touched up."

To Mary D. Howells

Venice, June 1'st, 1864.

My dear Mother:

I can, perhaps, understand my husband's grief for Johnny[1]—for I have lost two dear brothers myself[2]—but I know I cannot understand *your* deeper sorrow, and I can only tell you how much I think of you these dark days and ask you to accept my heartfelt sympathy in the great affliction. I knew you would want Sam with you after your loss, and I do hope he will be let off. It ca'n't be long now before your other absent boy will be at home I trust. We are only waiting for the baby to be old enough to bear the journey, and for it to be cold enough for us to visit Rome in safety, before leaving here for good. Already I am planning for the voyage, and am about to make the baby some warm clothes to wear on that long, cold journey. The baby will cross the Atlantic when she is just a year old, as did her grandfather before her.[3] I hope she will continue to have as perfect health in America as she has had here—for she has not known a sick moment, except for one severe cold last Winter which we wrote you about. I only wonder that in these cold, dank stone houses she does not often have a cold—but perhaps her full-bath every day prevents it. Her godfather, Mr. Valentine, takes the deepest interest in her, and comes to see her very often. Tonight the little thing has been very cunning—ringing a bell quite furiously all by herself. [sketch] She has tiny, little hands which she uses very gracefully, indeed. Everybody, when they see her say "O how pretty! just like her papa." ("O che bella! tutta papa."). She really is her father in minature—only a little softer looking, perhaps.

We heard, through my father, tonight, that Mr. Hale,[4] the Editor of the Advertiser is going to take the place of Mr. Thayer[5] late consul at Alexandria, and, as he must pass through Venice on his way there, we hope to see him soon. Mr. Baldwin,[6] an artist of great merit who is staying in Venice, asked Mr. Howells to sit for him to paint, the other day—and said when the portrait was finished he would give it to us. So we shall probably bring home a fine life-size picture of Mr. Howells.

I am very sorry Aurelia cannot come out here, but can well understand the trial such an undertaking would be to her

Mother—especially at this time. I hope she will at least visit us in our American home.

With much love to all I remain

Your affectionate daughter

Elinor.

Manuscript: Harvard

1. John Butler Howells was attending the Cleveland Institute when he came down with spotted fever and died in less than a week. On 28 April, the day after John's death, William C. Howells wrote Elinor, asking her to break the news to her husband (Massachusetts Historical Society). On 20 May, Will and Elinor replied, and other letters, such as the present one, touch on this sad event.

2. John N. Mead, in 1850, and Albert Mead, of typhoid, in 1856.

3. Joseph Howells landed in America in 1808. William C. Howells, his son, was born in Wales in 1807.

4. Charles Hale (1831–1882) served as editor of the Boston *Advertiser* from 1850 to 1864. He had printed WDH's "Letters from Venice" since March 1863 (see Gibson-Arms, pp. 19–20). Howells had written him to ask for a place on his paper, according to Elinor (Venetian diary, 17 May).

5. Alexander Wheelock Thayer (1817–1897), biographer of Beethoven, musical critic, and diplomat. In November 1864, he was appointed U.S. Consul in Trieste, where he served until 1882.

6. Perhaps the artist was the painter George Baldwin (b. 1818).

Venetian Diary

Palazzo Giustinian, June 23'd (Thursday) 1864. Dr and Mrs. Patton[1] arrived from Florence Monday night along with Annie, & Carrie Trowbridge and the two Scott brothers (of St. Alban reputation). They took rooms at the Vittoria and are doing Venice now. We see a great deal of them. We went together to Florian's the first night. Tuesday they came to see the baby, and we went to the Academy with them. Molly dined with them. Wednesday we went to the Ducal Palace with them, & today to Murano[2] & to the Lido. Aunt Emily is great on bargains, as usual, and laid in a great deal of Murano glass in different forms to give to American friends, Mr. Howells beating down the price for her. Tonight Mary is at the Piazza with them, as there is music, and Mr. Howells & I (he having first been to see Count Capogrossi who has been sick, and I having held the baby while she slept an hour so Nina could go out) are sitting quietly at home. We had a splendid bath in the

Adriatic today. Annie & Brig. Scott went in too. Last night we got a letter from Chas. Hale in London saying he would be here soon. We are hoping he will know of some place for Mr. Howells in America, and then we will most cheerfully go home, for it is agonizing not to have yet got a start in life and Winifred on our hands. She is over six months old now and as good tempered and healthy as we could wish. Aunt Emily is very happy and looking very young. The Doctor is 65 & she 55 but they do'n't look it at all.³ We like Bettina exceedingly. She is very fond of the baby & does her work well. Mr. Baldwin's⁴ picture of Pokey is poor. Hawthorne is dead—the author whom I most admire.

Manuscript: Harvard

1. The punctuation of this entry has been regularized. Harriet Emily Trowbridge Hayes, the widow of W. R. Hayes (1804–1852), married Rev. William Patton, D.D. (1798–1879) on 6 April 1864. Annie was Aunt Emily's adopted daughter. Carrie was probably Caroline Hoadley Trowbridge (b. 1849), daughter of Aunt Emily's brother Thomas. Brig. Scott, who had been at school in Geneva, had proposed to Mary Mead, who had rejected him.

2. An island near Venice famous for glass factories and shops.

3. Elinor reports in her next entry that Aunt Emily had fallen and hurt her knee getting into a gondola and was laid up for a week until they left Venice.

4. Under his picture in the album at Harvard, Howells wrote: "Baldwin American painter—'good Artist.'"

Venetian Diary

Aug. 21'st [1864]. Yesterday was quite an eventful day, beginning as early as five o'clock in the morning. At that hour I was lying awake (I did not sleep well because I ate figs before going to bed, and afterwards the fleas ate me) or half awake, my eyes being shut, but my ears distinguished sounds—at first I thought it the sound of Mary's hard breathing, in the next bed—but it became so loud that I opened my eyes, and presently I knew it came from over in one corner of the room where there is a window, and, before it a clothes-tree covered with dresses.

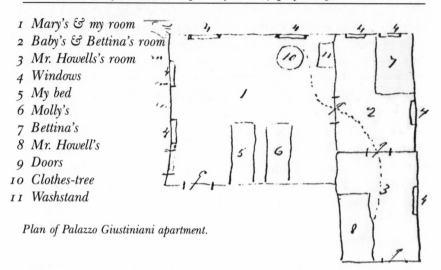

1 *Mary's & my room*
2 *Baby's & Bettina's room*
3 *Mr. Howells's room*
4 *Windows*
5 *My bed*
6 *Molly's*
7 *Bettina's*
8 *Mr. Howell's*
9 *Doors*
10 *Clothes-tree*
11 *Washstand*

Plan of Palazzo Giustiniani apartment.

I had hardly time to be frightened when I saw, as I lay on my back, the crown of a man's straw hat rise over the clothes-tree a dark straw—a regular gondolier's hat. Not seeing the face I was not terror-struck but yelled loudly "Pokey! Pokey! Come quick! Come quick. There's a man getting in the window!" Continuing to yell till he came rushing in th[r]ough Bettina's room in his shirt, only, crying "*What? WHAT?*" [sketch] and siezing the water pitcher off my wash-stand began running about the room like a madman till I could make him understand which window I meant—and by that time the thief had skedaddled.[1] At first I could n't make Mary nor Mr. Howells believe I had seen anything. Mary said it was the cat and Pokey that it was nightmare—but on looking at the wall below the window Pokey found a man could easily have gotten up on the iron grating and open blinds below and, on going down, there were mud tracks on the wall, so they were convinced and we were all very glad and thankful not to have been robbed. Undoubtedly the robber had a friend below to whom he would have thrown down our dresses & things as fast as he stole them. We shall shut our blinds on that side in the future.

In the morning Mr. Howells asked us to go to the Academy with him to look at the Carpaccios as he is writing a chapter on art for his book on Venice. We met the postman just in the Campiello,

Sketch of WDH during attempted burglary.

who handed us a letter. It proved to be the best letter we've had since we came here. It was from James Russell Lowell, saying the article on Italian Comedy was in print, begging him to write some more and praising his Letters as "the *most* careful and picturesque *study* of any part of Italy that he has ever seen" adding "It is the thing itself."[2] We met Padre Giaccomo[3] and an Armenian lady in the Academy. In the evening Miss Przemysl brought Don Antonio Pazzini to see us—& with a letter from Jo & Gus and some ices at San Stefano, the day ended. Baby says "na-na" all day. Frank Goodwin called.

Manuscript: Harvard

For Howells' humorous account, see "Our Last Year in Venice," *Venetian Life,* pp. 422–424.

2. Lowell's letter of 28 July to Howells is at Harvard (see C. E. Norton, ed., *Letters of James Russell Lowell* [1894], 1:338). WDH quoted it in part in a letter written to his father on 25 August (*SL,* 1:196–198). For his answer to Lowell, see his letter of 21 August (*LinL,* 1:84–87; *SL,* 1:193–196). Lowell's sentence read, "They are the thing itself." His letter made a profound impression on Will and Elinor, and Howells considered it to have marked "a turning point" in his life, after the frustrations with Foster and the rejections of several of his poems by the *Atlantic.*

3. Padre Giacomo Issaverdanz, the brother from the Armenian monastery on the island of San Lazzaro.

To Samuel D. Howells

Venice, Oct. 6th,/64.

Dear Brother Sam:

It was very kind of you to buy a doll for the Baby. She will be just the age to appreciate it when we go home, and she will always associate you with her first *real* doll—for she has had a queer, stumpy-looking, india-rubber cry-baby which *hardly one would* recognize as made in the human shape; and I am sure *Baby* never has known whether to stand her on the head or on the frill at the bottom of her dress. This China doll has, no doubt, the lovely red cheeks and bright black eyes which fascinate very young children, and—please tell her aunties that as she will always be dressed, they needn't be very particular about the body which they are good enough to make.

I wish you could see your little Venetian niece this morning, as she is, all ready to be produced when the great Mr. Motley[1] calls for her. We are momently expecting a call from the historian Motley, who is in the city and told Mr. Howells he should come to see us today, so we sit, dressed in our best, awaiting him and Baby, radiant in a red flannel frock and red stockings, with a new white apron on and her hair neatly curled, has to be constantly amused and watched lest she soils her apron or rumples her hair.

I am very sorry you could not remain at home this Winter to comfort your poor Mother's heart. I hope you are pleasantly situated at Columbus. When Mr. Howells told me you were in the *clerical* department I thought you were in some way assisting the *chaplains,* as I never heard the word used that way before. We think things must be looking brighter at home—but the "end is not yet" I suppose and the country may still be in distress when we come home in the Spring, which we *certainly* shall do whether there be peace or war. This week we send off all our Summer clothes to

America, by a ship which is going directly to New York from here; so you see we *ca n't* stay, only through the cold weather. The captain[2] of the ship is a very jolly man and comes to see us often—though, for fear of privateers he has English papers—and, some-day we are going out to visit his barque (which lies at the mouth of the harbor) when he will treat us to hot corn cake—something you can not, possibly, get here as there is no yeast-cooking done in Venice. The Italians eat *polenta,* which is the dough we give chickens at home, with cheese or fried fish as butter is never eaten on bread or cake here. The butter here is never salted, and you cannot salt it so it tastes well, so we have given up trying to eat it in Venice. A priest who often came to breakfast with us, used to put the piece of butter we gave him *into his coffee* and *drink* it. I am afraid we shall behave very absurdly when we get back to the land of good eating again—for everything is so bad here that we cannot help remarking when we find anything good—and buckwheat cakes, corn bread, hasty pudding, oysters, pies, cream toast and hulled corn are things we who have been in poor, half-starved Europe will appreciate.

My sister has just come in, dragging "Poor Baby"—as Mr. Howells calls her for very absurdity—in her carriage, or cart, which her papa bought her a few days ago and which she enjoys very much. I have made a miserable failure of the baby, though it may give you some idea of the carriage. [sketch] To make up for the injury *done* Baby by the picture I will send you her photograph, which, though not really good, will show you what a nice, fat baby she is. Her first tooth has, at last, appeared and she is learning to walk, so we shall not call her "baby" much longer I fancy.

I liked that which Annie wrote about the Jefferson Light Artillery very much.[3] It shows she has a gift which ought to be cultivated—and which, of all things I envy. All your family express themselves in such an easy way—I hope Winifred will inherit the gift of language which her father has so largely.

With every wish for your comfort as a soldier disabled in the service of our country,

Yours affectionately

Elinor

Now you are so situated as to be able to write more conveniently do let us hear from you oftener.

Manuscript: Harvard

1. John Lothrop Motley (1814–1877), author of *The Rise of the Dutch Republic* (1856), served as minister to Austria from 1861 to 1867. This was his second visit to Howells, who had previously helped him with his historical research. See Woodress, pp. 20–21.

2. For Elinor's account of their supper on board Captain Dyer's ship, see 16 October. According to WDH's inscription in a copy of the novel now owned by William White Howells, *The Lady of the Aroostook* was "named for love of Capt. Mark Dyer of Aroostook Co., Maine, whom I knew in Venice."

3. Annie's article had appeared in the *Ashtabula Sentinel* on 31 August.

Venetian Diary

October 16th [1864]

. . . We have been very busy since Motley was here packing off a box for America to send by Capt Dyer whose ship, the "Fanny Hamilton Gibraltar" was in Venice last week,[1] but has now started home, via Palermo where it takes on oranges & lemons. The Capt. was a very pleasant man & we saw a good deal of him—having him & Richard Clark, a little Brooklyn boy who was with him, here to tea one night, and going aboard the ship to supper one night taking Miss Przemysl along. I must write down our bill of fare as everything was New Englandish and utterly different from anything we've had in Venice.

Stuffed turkey.
Boiled potatoes.
Gravy (I had forgotten there was such a thing as gravy for fowl.)
Johnny-cake. Rather different from polenta!
Raised biscuit.
A loaf of tea-cake.

Baked custard.
Fruit &c &c
  and
CROTON WATER!!! taken aboard at New York last Summer.

The Capt. said he *had* some water which he took aboard last voyage at Messina which was better. He was *saving* his Croton for the voyage back. I went shopping with the Capt one morning & he gave me a Roman scarf. He told us a rich story about an old maid getting into his & another Capt's room at Rio Janeiro. He went to bed early having eaten lobster, woke up & saw a horrid form lying at his side—lean, toothless, hairless. Stole out & let his bedfellow Capt Lewis, who came home a little tight, go in there. Capt Lewis, supposing it to be the Capt on his side the bed hauled *her* out on to the floor—& the Capt. & some English people in the next room heard her on the floor give him *such* a lecture. Capt. Lewis thought *he* was in the wrong room & rushed up stairs into the old maid's room where a young lady was sleeping, &c &c &c. The old maid did n't appear the rest of the time they were at Rio—& Capt. Lewis thought for a year he was the only one to blame—till a lady set him right & now he has the story putting Capt Dyer in his place. . . .²

Manuscript: Harvard

1. The Howellses, who were getting ready to return to America, were sending their summer clothes back home.

2. Clemens told variants of the Captain Dyer story several times, usually attaching them to his brother Orion. To the version in *Mark Twain's Autobiography,* WDH responded, "But—but—but you really *mustn't* let Orion have got into the bed. I know he did, but—" (*Twain-Howells Letters* 2:803). Whether forty-two years earlier EMH wrote out the anecdote and called it "rich" to joke at WDH's reticence remains a question. Hamlin Hill, of Texas A&M University, also notes an occasion in 1907 when Clemens wanted to read from the "Wapping Alice" manuscript but refrained because it was "too delicately indelicate" for the company that included a teenage girl; but then a guest told still another version of the Dyer story, which was apparently acceptable. Cf. *Wapping Alice,* ed. Hamlin Hill (1981), p. 11. After listing the contents of the trunk entrusted to Capt. Dyer, Elinor cheerfully added: "We have received a jolly letter from Father saying he approves of all our plans, rejoices in us and will send what money we want. Sherman has taken Atlanta and Grant is right on Richmond. What more can we desire?"

Pocket Diary

Twenty-fourths.

February 24'th [1864].

A really exciting day! Baby a little sick in the morning but receiving an invitation from Mr. Valentine to take tea with him and some English people (Mr & Mrs Stally-something), we wrote a note to put Miss Przemysl off & accepted it. Received a rather mad note from Miss P..[1] She thinks I do not like her I guess. Baby kept getting worse—doctor happened in. Note from Mr. Val. saying Stally-s could n't come but hoped we would all the same. I wrote excusing ourselves. Postman brought lots of Adv's[2] & papers from home & five letters, which with a note from Haughton[3] (enclosing money for dinners) & those from Mr. V. & Miss P. made nine missives in one day. I wrote to Mrs. Angell[4] in Paris to thank her for the little silk stockings. Letters from: Home, Howellses, Quincy,[5] Molly (in Florence) & Mrs. Buckingham. Q. writes Mr. Piatt was in New York with a *new wife*. Baby better this evening. Ten weeks tomorrow. Never been sick before. Mrs. Kellogg called. Going to London.

Sunday, Apr. 24'th, 1864. Soon after breakfast Molly & I went to church at Palazzo Civran, as Mr. Merryweather's[6] servant came to inform us there would be service before we were up. The room was crowded with English and a stupid Englishman preached. The baby was baptized at the German Lutheran Church last week, and I liked the preacher *so* much that did he only speak in Italian on Sundays, as he did on this occasion, I should always go to hear him. After meeting Mr. Howells, Molly, the baby, Nina & I went to the Lido in a gondola. All went for the first time except Mr. H. & myself. There was troppe sole[7] for the baby and she cried a good deal. At six we went to dinner at the Vittoria (for the first time)—& then to the Square. Mary has gone to bed, early, and Pokey has been reading his 22nd Chas-Hale letter to me. (on Venetian traits).[8] A week from tomorrow we hope to change our quarters to Palazzo Giustinian.

Oct. 24th**.** A rainy day and we staid in, and Mr. Howells read aloud in Conquest of Mexico[9] while I sewed on a Winter dress for Baby. Mr. Howells is writing very hard on his Life in Venice getting it ready to offer to Trübner.[10] We intend going to Rome the 7th of November if the book is done.[11]

Manuscript: Harvard

1. "I am sick of Przemysl at present, Mr. Howells and Mary make so much of her," Elinor had written in the Venetian diary (26 January). They had been having whist parties to which she came regularly. There was "a little stiffness" with her, and a cooling-off period for a while.

2. Boston *Advertisers,* where several of Howells' "Letters from Venice" had appeared.

3. "Haughton of Boston" appears in a list of "Pleasant Americans & English met in Venice" (pocket diary, typescript, p. 47).

4. The wife of Dr. Henry C. Angell (1829–1911), a Boston ophthalmologist whom they had seen in Venice in March 1863. The Howellses stayed with them for a while when they first went to Boston in 1866.

5. J. Q. A. Ward, their sculptor friend. John J. Piatt, the Ohio poet, had married Sarah Morgan Bryan, on 18 June 1861. Howells had stayed with them in Georgetown before going to Venice. Mrs. Buckingham was perhaps Angelina Hyde Buckingham, a girlhood friend of Elinor's.

6. Perhaps an English minister.

7. Too much sun.

8. Charles Hale (editor of the Boston *Advertiser*), "Venetian Traits and Characters." See Gibson-Arms, p. 20.

9. W. H. Prescott, *History of the Conquest of Mexico* (1843).

10. "Trübner has written Mr. Howells to send his book on Venice on all ready for printing and they will consider it. Hurra! Hurra! Hurra! Lowells' letter did a great deal to help, as they about refused it the first time" (EMH, Venetian diary, 16 October). WDH also gave credit to Moncure D. Conway for this (*Literary Friends and Acquaintance* [HE], p. 88).

11. On the first page of *Italian Journeys,* Howells wrote that they left Venice on 8 November. He sent the manuscript of *Venetian Life* to Trübner & Co. in London on 13 November, from Genoa.

Venetian Diary

Rome, November 27ᵗʰ [1864], *Hotel Minerva.*
We got to Rome at 6 o'clock.[1] Mr. Taylor met us at the station and did everything for us. Now he sits smoking in Mary's room, where we all are and telling murder stories. This morning we went to see the Forum and the Coliseum, and this afternoon to the Conservatori della Mendecinti to see Virginia Gagliardi[2] whom we liked very much. One of the women took us up in to the garden from

whence we walked into and on to the Temple of Peace or Constantine's Basilica.

5 Via del Gambaro, Rome, Numero 7.
We are sitting in our new lodgings. Mr. Taloyr is spending the evening with us, of course—waiting for the black coffee, which we have ordered, to be brought up. We came here last night. Yesterday we went with Mr. Taylor to the Church of the Capuchins to see Guido's St. Michael—(*glorious!*) and to the St. M. degli Angeli to see the original pictures from which [t]he mosaics in St Peter are made (hideous Rennaissance). The form of the Church is a Greek cross—the transept is the old hall of the warm baths in the Baths of Diocletian. The columns of Egyptian marble, seven feet of which are under ground, and the largest I have ever seen unless it be those in the Piazzetta. Today we saw the immense porphry vase [sketch] taken from these baths, and for which Sixtus VII [?][3] built a room, especially, in the Vatican. At 4 o'clock we moved.[4] At 5 we dined at the Colonno on the Corso. At 8 1/2 Mr. & Mrs. Stillman[5] called for us and took us to Mrs Van Ness'[6] to a parish party. We met Mozier,[7] Freeman Gen. King our minister here, &c &c. Two young Italians played their guitars together splendidly. One of them was one of the Pope's guard—a very handsome fellow. Today we have merely been to the Vatican & rushed through St. Peter's. The Gladiator scraping himself fascinated me. I havn't said that yesterday we looked into the Capitol; saw the Marble Faun of Praxiteles—the one Hawthorne celebrates;[8] the Dying Gladiator &c. The Venus of the Capitol was not to be seen. Also, Monday, we went to Pal. Barberini where we saw the lovely Beatrice Cenci, the sharp Lucretia Cenci and Raphael's horrid Fornarina.[9] Raphael Mengs' picture of his daughter is good.

Thursday, December 1'st, 1864. We have just had a fire made in our room and are sitting lazily around. Today we have accomplished a great deal. First we breakfasted at our place in the Corso; then took a carriage and went to the Colonna palace where we met Mr. Taylor; afterwards we walked to the Corsina palace; then to the Spada, and there we took a carriage and rode out to Villa Ludovise.

Colonna:	Pretty picture of Narcissus at the Fountain, by Tin-
(showy but not	toretto. Good landscape by Salvator Rosa. Good
satisfactory)	Poussin landscape. Splendid portrait of Lucretia
	Colonna by Vandyke
Corsini.	Three famous Ecce Homos—Guido's, Guercino's &
	Carlo Dolci's—the former the best. Titians portrait
(splendid	of a woman perfectly magnificent. Also two heads
gallery!!!)	together by Giorgione, Titian's Woman taken in
	Adultery very fine coloring. Murillo's Madonna
	with Child does n't greatly please me.
At the *Spada*	we only saw the Statue of Pompey.
At *Villa L.*	we saw the splendid Mars at Rest, in the Casino.
	Also Penelope taking leave of Telemachus. Splen-
	did colossal head of Juno. Guercino's fresco of
	Fame in the New Casino is very fine.

Yesterday we went to Pal. Borghese where we saw Raphael's En-
tombment. The head of St. John in it is the most delicately painted
face I ever saw—the shadows seem to flit, and it is very lifelike.
Titian's Sacred and Profane Love is rich in color. Day before yesterday I saw Raphael's Transfiguration which is *far* beyond my
expectation. I could not get away from it for a long time, and now
I am anxious to see it again. I did n't think much of Dominichino's
St. Jerome in the same room.[10]

Manuscript: Harvard

1. Leaving Larkin in charge of the consulate (and the baby), Elinor and Will had
set out for Naples and Rome, accompanied by Mary Mead. Howells wrote up the story
of their two months' trip in *Italian Journeys*.

Mr. Taylor is no doubt the "young American who had come aboard at Leghorn," and
who "turned out afterward to be the sweetest soul in the world," as described by Howells
in *Italian Journeys* (p. 70). He had been their traveling companion until the 23rd, when
he left them in Naples to go to Rome. A list of guests in a hotel register at Capri copied
by Elinor in WDH's 1864 travel diary (at Harvard) includes, among various distinguished names, real and facetious, "G Bradshaw The Duke of New York" and a "Signor
Taylor Prof of Photography." Actually he was John Phelps Taylor (1841–1915), clergyman and educator. After graduating from Yale in 1862, he studied in Bonn, Paris,
and Venice, 1863–1865. Ordained a Congregational minister in 1868, he later became
professor of biblical theology and history and of oriental archeology at the Andover
Theological Seminary.

2. Virginia was the young daughter of Gagliardi, the marblecutter, who shared a studio with Larkin Mead in Florence. For Howells' amusing account of their visit to Virginia in the Conservatorio delle Mendicanti, a charitable institution which took in also daughters of poor people who paid a modest sum for their education, see *Italian Journeys*, pp. 152–157. Howells arranged to have a photograph taken of the little girl, "whose father was at Florence, doubly impeded from seeing her by the fact that he had fought against the Pope for the Republic of 1848, and by the other fact that he had since wrought the Pope a yet deadlier injury by turning Protestant" (p. 152).

3. There is no such Pope.

4. They moved to 5 Via del Gambero on 28 November.

5. W. J. Stillman (1828–1901), artist, journalist, and diplomat. He had applied for the Venetian consulate in 1861, but lost the appointment when Howells secured it. Laura Mack (1839–1869), his first wife, was the daughter of the Belmont teacher and abolitionist, David Mack (1803–1878). See 28 November 1882.

6. The wife of Abraham Rymer Van Nest (1823–1892), clergyman of the Dutch Reformed Church in America. He was in charge of the American Chapel in Paris (1863–1864) and of the American Chapel in Rome (1864–1865), from which he was transferred to Florence in 1866.

7. Joseph Mozier (1812–1870), an American sculptor who had lived in Rome since 1848. James Edward Freeman (1808–1884), an American painter who went to Rome after studying in New York. "They were like most expatriates," Howells commented in the 1864 travel diary. "Mr. Freeman talked well. . . . But I took most to Moshier. He is an Ohio man, and has a western jollity of soul, whatever breath of ideas he may not have." General Rufus King (1814–1876), soldier, editor, and diplomat, had been appointed U.S. minister to Rome in 1863.

8. In *The Marble Faun*. Elsewhere the Howellses also visited places and looked at pictures that Hawthorne had written about, for example Guido's *Michael* (see earlier).

9. "Guido's Beatrice Cenci . . . certainly as beautiful as it could be. It *is* wonderful; but the Fornarina of Raphael, which hangs beside it, is as uninteresting a painting as possible. I dare say it may be a great picture, but the woman it represents is of the commonplacest kind in Italy: a dull, handsome, fleshy creature, with a sort of stupid vanity and surprise at men's admiration. Nothing could be worse. The feeling of the picture is to the last degree earthly, without even the low fire of sensuality in it" (WDH, 1864 travel diary).

Anton Raphael Mengs (1728–1779), a German painter who converted to Catholicism and settled in Rome. Noted for his portraits and for his historical and religious scenes, he was compared to Raphael and Titian in his time.

10. The diary continues, with entries nearly every day through 11 December, the day before the Howellses left Rome. Their round of sightseeing included all the well-known museums and monuments, as well as a number of studios. Among the latter were those of Johann Friedrich Overbeck (1789–1869), the German painter of religious subjects, John R. Gibson (1791–1866), the English sculptor noted for statues in color, Randolph Rogers (1825–1892), the American sculptor of military monuments, and Joseph Severn (1793–1879), best remembered for his friendship with John Keats. On a fairly typical day EMH recorded going "to see the Capitoline Venus, Tarpeian Rock (*one* of them: that in the German Hospital Garden) to see the Cloaca Maxima—St Paul's (*magnificent*) MichelAngelo's Moses, Temple of Vesta, and Theatre of Marcellus taking in Rienzi's house & the Temple of Fortune on the way." That night they excused themselves from the Stillmans' reception, "we were all so tired."

To Eliza W. Howells

Rome, Dec. 7$\underline{\text{th}}$, 1864.

Dear Sister Eliza:

I am sure Annie will forgive me if I write to you instead of her, when you are to be congratulated on such an important event as having a daughter.[1] The news was quite unexpected and we felt more sympathy and pleasure from knowing of your sad disappointment before. The third generation of Howells is coming on pretty fast, and when Winifred comes home there'll be more than one apiece for the grandparents. Our little girls will be great friends I am sure and rivals for Willie's favor. I suppose that young gentlemen will patronize them quite extensively and they will admire him as little girls always do "big boys."

At the same time we heard of your having a daughter, we heard of the like good fortune having happened to our friends, the Mitchells. They have named their baby Elizabeth but call her Lily.[2] Mary is always a satisfactory name. My Mother's name is Mary, and my sister who is with me is Mary.

We are enjoying Rome exceedingly, but are very tired seeing sights when night comes. We wonder at Rome, but could never get attached to it as we have to Venice. New Rome is not much like Old Rome, and the principal interest is, of course in the ruins, (statues &c) which have been dug out from dust & rubbish twenty feet deep. I would willingly sacrifice the new city, St. Peter's and all to the ruins beneath. After Venice and Pompeii[3] the architecture of ancient Rome, even, seems somewhat dull and heavy. Venice we all conclude is the one city we would have been content to *live* in in all Italy.

Tomorrow we go to see a service in the Sistine Chapel where we hope to see the Pope. Mr. Howells must go in a dresscoat beaver & white kids, and my sister and myself in black silks and veils. Such is the etiquette of the occasion. We hope also to be delighted by hearing the pope's choir, which is considered the finest singing in the world.[4] We are making the most of our time and expect to leave here in about ten days. It is a sad thought that we shall never see Rome again, & I wish you all might see it *once*. You will excuse me if I write rather confusedly amidst the thousand and one attractions of Rome.

With much love to all and a hug for my dear little niece

Your loving Sister
Elinor

Manuscript: Harvard

1. Mary Elizabeth Howells. Eliza and Joseph A. Howells had lost a baby, a stillborn daughter, on 17 October 1863.
2. She was born on 2 October.
3. They had visited Pompeii on 18 November. Elinor found it "magnificent—better than I expected," she wrote from Naples (Venetian diary, 21 November). See WDH's "A Day in Pompeii," *Italian Journeys*, pp. 89–105.
4. In an entry of the Venetian diary for 8 December, EMH describes the Mass in the Sistine Chapel and the magnificent singing of the Pope's choir. For WDH's account of the papal mass, see *Italian Journeys*, pp. 174–177.

To William C. Howells

Venice, Christmas Day, 1864.

My dear Father:

This is day of general confusion: we having arrived only yesterday forenoon, my brother preparing to go away tomorrow, and my nurse off celebrating masses.[1] Mr. Howells is in rather an anxious state of mind about the manuscript of his book on Venice, as he hears nothing from it, though it was mailed at Genoa six weeks ago. He has written the publishers to telegraph him if it has arrived—but till he receives the telegram he is naturally somewhat uneasy—and begged me to write for him today. I think you have not been informed of our last adventure: being tipped over in a diligence in the midst of a muddy stream while returning from Rome. Our trip was one series of misfortunes from beginning to end and cost us much more than we intended it should—but, as we *saw* all we desired, we have agreed to call it satisfactory on the whole. As for the affair of the diligence: there was'n't really much danger, as the water was not high. My sister & myself escaped through a window and were carried ashore by men without getting wet in the least; but our baggage did not get out so well, one trunk being completely soaked, our books ruined and a few things lost— amongst them Mr. Howells' beaver, which he declares he saw float down the stream without a murmur. In the wet trunk was Mr. Howells' dress-suit, a new silk dress of my sister's, a quantity of Roman scarfs which we had bought for presents from friends in America, woolen dresses &c &c These, of course, are about spoiled, as the colors all ran together. Mr. Howells is going to make the most of our adventure by writing an account of it to the Advertiser.[2] As our clothes were all soiled we could make only a short stay at Florence; but we set out for home with light hearts, thinking we would soon be there, where Baby and the comforts of home would make up for all.

Imagine our disappointment on learning that the railway was not yet repaired and that a hundred miles, at least, must be made by diligence over a horrid road. After a rough journey of two days and two nights we got here—thankful that our lives were safe, and that the dear little baby had not suffered a moment while we were away. Her cheeks are as red as a rose, and, now she walks all around the room by chairs. All our photographs got wetted, and peeled off the cardboard. One, the famous "Nile" of the Vatican[3] I slip in as it will amuse you to see all those babies & the tributary streams I believe. We will write again soon in answer to Aurelia & Jo. Annie's letter to me shall be answered next after those. We are greatly interested in the new baby and want to hear all particulars of her progress in life.

Your loving daughter
Elinor

Manuscript: Harvard

1. Larkin G. Mead, Jr., had been in charge of the consulate during the Howellses' trip to Rome.
2. It appeared as "Forza Maggiore" in the *Atlantic* of February 1867 and was reprinted in *Italian Journeys*, pp. 178–195. Elinor's Venetian diary, 26 December, also provides a more extensive account of the stagecoach adventure.
3. The statue.

To Jennie Jackson

Palazzo Giustinian,
Venice, Feb. 28ᵗʰ, 1865.

My dear Mrs. Jackson:[1]

Lest I make a bad matter worse I will omit all apologies and write as though I had heard from you only last week. Dull Old Venice seems duller than ever in Carnival time, in contrast with what it used to be, and this last day is the dreariest of all. Austrian society has been rather more lively than usual this Winter, judging from the number of balls the young Countess Kinsky, who lives in this palace, has been at—about three a week.[2] A while ago a Russian Princess[3] gave a masked ball at a hotel in town, to which nearly everybody was invited, though of course, no Italians went—only the Germans and what strangers happened to be here. The wife of the Russian Consul persuaded us to let Molly go with her, as it seemed a pity she should lose such a sight. She went as La Folie, in a cap and bells—bells on her dress, "bells on her toes",—and made a (musical) sensation "wherever she went".[4] [sketch] The Countess Kinsky had a dress sent from Vienna, all complete, even to the boots—for it was the dress of a Postillion, and she had exqu[i]site little top-boots. The whole costume was done in the richest silks and was quite original and "taking". The ball was got up in a splendid manner—the Princess hiring a whole piano[5] of the hotel (which of course was an old palace) and fitting it up for the occasion; building a gallery for the music, putting in gas &c &c. Mary was introduced to crowds of princes, counts and barons, but cant remember one of their names now, and will, probably, never see them again, as we shall try to avoid them in the Piazza and elsewhere lest we get the name of "going with the Austrians".[6] As we were finishing breakfast, about ten o'clock (Venetian habits) in walked Mary—she, with many others having staid and breakfasted at the ball.

This ball, with "the American prima-donna" has formed the excitement of the Season for us. The opera is a very important thing in Venice, many of the Italians "happening in, in a quiet way" in the course of the season—though ladies of rank never go. The impressario engaged Ada Winans this year, and great things were expected [of] her. But the dampness of her lodgings gave her a dreadful cold to begin with, and the public were obliged to put up with a wretched singer for a week or so, till they would stand it no longer and said "Winans now or never"—So Winans appeared. Her great beauty and—strange to say—her *coughing between times in her singing* excited the admiration and sympathy of the volatile Italians so that she was applauded to the skies, and told us next day she was never so much "called out" before. But next time she sang no better, nor the next, and their sympathy began to flag. After five or six times they applauded her efforts—for indeed she made the most strenuous efforts and sometimes quite failed to bring out a note—no longer, and after two weeks they began to hiss. Poor Winans was frightened nearly to death and had to give up the engagement, first publishing a letter in the Gazette denouncing the Venetians as the most fickle public in the world and adding she would trouble them no longer &c &c a very foolish thing to do, of course. But she was a very pleasant person, and we stood by her as an American, naturally.[7] A few days ago I received a letter from her at Piacenza, where she has been singing with success, saying she was going to Barcelona in Spain, soon, to sing for three thousand francs a month—which, of course, is a great triumph over the Venetians she thinks. But I suppose her success in Piacenza depended on her voice as well as in Venice—she probably sang better there—and she was wrong to expect the Venetians to go on paying their money out of sympathy for her cold a whole season.

The *Baby* must have the rest of this letter. She is in a grand state of health and prosperity, having four teeth already, *almost* walking alone and saying Papa Mamma and Yes in a clear and unmistakable manner. Her hair is so long that we tie it up with blue ribbons to keep it out of her eyes. She is getting almost to[o] large to be carried, but occasionally her nurse takes her up to the Square to feed the pigeons, which sit perched on the statues and cornices on all sides, ready to come down at the first invitation, and feed out of one's hand if called. They are fed at two o'clock every day from one of the windows of the Old Procu[ra]tie, and I have actually seen them when the clock on the tower strikes two, before the window was opened, hastening to the corner of the Square where they are fed. I dont believe you like to read crossed letters, so I will stop.[8] I wish I had a photograph of Daisy. I have never been able to get her out of my mind (not that I wished to) as she sat there that morning talking to the pictures. I would not *mind* having yours, either. We all wish to be remembered to yourself and husband, and wish you would come to Venice this Spring.

Yours truly
Elinor M. Howells.

Manuscript: Howells Memorial, Kittery Point, Maine

1. The wife of John A. Jackson, the sculptor. The Howellses had visited them twice in Florence. See pocket diary, early May 1863. In her letter of 27 March to WDH (at Harvard), Mrs. Jackson mentions her daughter Daisy, as well as Elinor's "amusing account of the festivities of the Carnevale in Venice."

2. "It is said that the young countess (Theresa) Kinsky has just had a *broken engagement* with a young Heine, an editor in Vienna (nephew of the poet Heine) for which he paid her $2500.00. Perhaps this accounts for the desperate way she rushes into society—not a moment left for thought" (EMH, Venetian diary, 6 March).

3. "with a face uglier than any woman ever possessed—cross eyed and scrofulous—and no brain at all, scarcely, but rich as Croesus. (*I* make these remarks—not Mary who accepted her hospitality)" (EMH, Venetian diary, 10 February 1865).

4. "Rings on her fingers and bells on her toes, / She shall have music wherever she goes" (from the nursery rhyme "Ride a Cockhorse"). In her Venetian diary (10 February), EMH had described Mary's costume as follows: "The whole was taken from a fashion plate and was got up with scarcely any trouble in two days & consisted of a cap & bells, 'bells on her toes,' &c. Bergamo, the famous hair-dresser did her hair—coming at twelve at noon as he was so much in demand he had no other time. I had previously braided it and with *frizzing* it was perfectly uncontrolable standing out in a stunning manner behind. The front he put over rats and the ends of the back hair he merely turned up under attaching it to a braid underneath which did not show. [pen and ink sketch of profile showing her hair] The cap with two points before and two behind was of velvet and made by the first modiste in town—trimmed with gold braid and with little bells attached to the points. The over skirt of cherry satin (the exact color of the cap) was pointed & trimmed with braid & bells. The black velvet waist was trimmed in the same way, and the cherry satin shoes & the yellow kid gloves embroidered with cherry, had, also, bells attached. Bracelets & necklace of bells finished the costume. I only arranged the skirts—put on the cap and made the necklace—attaching bells to a chain. I forgot to mention the under skirt of fluted tarltan with chemisette & sleeves of the same."

5. Italian for "floor." Palazzo Dandolo was built at the end of the fourteenth century. In 1822, Giuseppe Dal Niel began to transform it into the Hotel Royal Danieli.

6. On the "Austriacanti" and the Howellses' position, see 24 July 1863.

7. Ada Winans, "the American prima-donna," has not been further identified. Howells did what he could to help out his compatriot, and recommended his scholar-friend, Mr. Barozzi, to her as a lawyer when her manager tried to break her contract (Venetian diary, 14 January).

8. Starting with "tower," the rest of EMH's letter is written crosswise over the fifth page of the manuscript.

To William C. Howells

Venice, March 19ᵗʰ, 1865.

My dear Father:[1]

If I have not written for a long time it is because I knew that not having anything to say about coming home, my letters would only bring disappointment to the dear family at Jefferson, and now, we cannot fix any time as the London publishers have not yet written, and leave of absence cannot be promised at present—but your letter seems to give the hope that we may come home to live, in the course of time, if both Government and publishers fail.[2] For my own part the Cleveland scheme appears to me the best thing that could happen, and I shall be delighted if it is carried out. The only idea in getting leave of absence is that if nothing suitable could be found for Mr. Howells to do in America—which I think highly improbable—he might return here and write something in the way of a history of Venice, or some book on Italy which he would be splendidly prepared to do, and which it seems almost a pity he should not do now he has got the language so well and understands the Italian character so perfectly. He has begun planning a book already and will write it in America from notes made here I dare say.[3] I hope his book which is in the hands of Trübner & Co. will be accepted—though they act very strange about it—not answering any of Mr. Howells' letters inquiring about it, and keeping it forever.[4] When Charles Hale was here he insisted upon sending a ballad, on a Mantuan Marquis, to the Atlantic, although Mr. Howells did not like to offer anything more to Mr. Fields. At first the ballad was "considered" but afterwards "refused because it was too long."[5] The other day Mr Howells received a letter from the editors of the Advertiser enclosing the pay ($50.00 greenbacks) for the ballad which Chas. Hale had "*sold to Bonner!*" We thought it slightly cool on Mr. Hale's part, though we know he always means well. The editor also sent money for letters he is printing descriptive of our journey to Rome—of which there will be twenty in all.[6] We are enjoying Spring in Venice—a Spring without verdure though not without sunshine and birds, for, strange to say, Venice is full of birds. Poor Baby is too fat to be carried much and will not walk alone (though she could perfectly well if she were not afraid) so that she scarcely ever gets so far as the public garden or to where there is a bit of green grass, but she remains perfectly healthy and grows like a weed. She is so very cunning now that it seems as though she never could be more so, and we regret that all her grandparents & uncles and aunts, as well, as her two cousins, cannot see her before she is any older. Her hair is the brightest gold, her cheeks are the reddest and her eyes are the bluest you ever saw—but as she gets older of course this will grow darker and paler and duller—so I wish you could see her now.

We are expecting a friend, Mr. John Taylor of Andover, who traveled with us some weeks of our journey, to make us a visit— or perhaps I should say to visit Venice, as he has never seen this lovely city. But he is going to stay here six weeks or so, and is to have a room in this palace, and we shall see a great deal of him. Mr. Howells likes him very much as he is extremely well educated and they are both pursuing the same studies in Europe. We have heard of the arrival of Capt. Dyer's ship in New York and our box on it.[7] Henry Howells has sent us money to buy him curiosities. Laura Mitchell writes she is going to live in Memphis in the Fall, and wishes we would go there to live, too. But, though I would like to be with Laura I would not like to be so far from my home, and Mr. Howells thinks it too far out of the way. Give a great deal of love to all in Jefferson and say, please, that Eliza's and Annie's letters shall be answered very soon.

Yours affectionately
Elinor

Manuscript: Harvard

1. For WDH's letter of the same day to his father, see *SL*, 1:212–213.

2. Trübner & Co. had received the manuscript of *Venetian Life* the first week of January (Venetian diary, 14 January). On 9 January Howells had written Assistant Secretary of State Frederick W. Seward to ask for a three months' leave of absence (*SL*, 1:205–207). He had just heard (on the 19th) from J. A. Garfield, his state senator, who had intervened with Seward on his behalf. According to Elinor "Mr. H.['s] Sen. also said he thought he (Pokey) would find it hard to get away from the family, once home, and proposed a plan—W. H. Smith & others are getting up a Corporation to buy out Cleveland Leader and will ask Poke to be chief editor." The Cleveland plan fell through. Of her husband's projected history of Venice (which he never completed), Elinor added: "Mr. H. has his mind full of writing a book on the Venetian Republic and would rather come back here I think—but we shall see" (Venetian diary, 22 March).

3. *Italian Journeys*.

4. Already on 25 February, Elinor had moaned in the Venetian diary: "Blue as indigo—all of us—because we dont get any letters. We ought to have heard, before this, from Trübner & Co., whether Mr. Howells' book had been accepted . . . and from the box we sent by Capt Dyer, to America. Our funds are low as well as our spirits. All we do is play with Baby, listen for the postman's ring, and read aloud in Carlyle's Frederick the Great."

5. "The Faithful of the Gonzaga" was published in the New York *Ledger* in 1865. It was the Rev. Edward Everett Hale (Charles's brother) who had sold it to Robert Bonner (1824–1899), its publisher. For Howells' letter of 19 March to E. E. Hale (1822–1909), see *SL*, 1:210–211. James T. Fields (1817–1881) was the editor of the *Atlantic Monthly*.

6. "The Road to Rome and Home Again" appeared in the *Advertiser* on 18 February, 4 March, 13 April, and 3 May 1865. Additional chapters of the book were to appear later in the *Nation* and *Atlantic*.

7. "Tonight we got news from Joanna that the Fanny Hamilton has arrived in New York—at which we all rejoice, as we began to fear she was lost. Gus [Shepard] had invited Capt. Dyer to dine with them next day, and he has said 'I do'n't care for the *dinner* but I'd like to see your wife'!!! The Capt. said he hated to have the box opened as they were full of fleas caught at Palermo" (EMH, Venetian diary, 6 March). For Captain Dyer, see 16 October 1864. Henry Craik Howells (1835–1902), of New York, also referred to as Henry C., Jr., was WDH's second cousin.

Pocket Diary

April 24'th, 1865.

This has been as eventful a day as ever a twenty-fourth was. It is poor Sheffield's worst day so far,[1] and there has been a consultation of doctors, on the subject of his spitting so much blood. We have telegraphed for his sister, Mrs. Porter[2] who is staying in Paris, to come on at once. My silk sack does not fit and a quantity of Nina's relatives have come from Dolo to visit her at this inopportune time. Her cognato actually slept with her in her bed. I took Baby away from them and had to spank the poor little thing to make her lie quietly in our bed. I think I shall send Nina away. In the evening the Walkers and Bacons[3] were here and the brothers Langley[4] came in. Mary looked at the Gazette when it came and found "Lee with all his army has capitulated"—[5] We all shouted hung the flags from the balconies—just then Mr. Howells came in to say the doctors thought Sheffield's a grave case—We all became sober—and the Austrian fresco[6] went by.

Manuscript: Harvard

1. George Sheffield, a friend of John Taylor, was the son of Joseph Earl Sheffield

(1793–1882), merchant, financier, and philanthropist. In the Venetian diary (30 April), Elinor wrote at length about Sheffield, who had arrived sick from Naples in early April to join Taylor. They both stayed at Palazzo Giustiniani. When his fever rose, she had him moved into his friend's room: "Mr. Taylor was perfectly devoted to Sheffield during his sickness, tending him day and night till he nearly got sick himself."

2. Josephine Sheffield, the wife of John Addison Porter (1822–1866), a professor of organic chemistry at Yale until 1864, when he resigned because of poor health. His father-in-law gave a million dollars to the Yale Scientific School. It was renamed after its benefactor in 1861, and Professor Porter was appointed the first Dean of the Sheffield Scientific School. Mrs. Porter and her family reached Venice on the 27th.

3. The Walkers, who knew Sheffield's family, and the Bacons were two Congregational ministers and their wives. Leonard Bacon (1802–1881) was minister at First Congregational Church, New Haven, from 1825 to 1866, before becoming professor of theology and church history, Yale Divinity School, 1866–1881. George Leon Walker (1830–1900), the son of a Brattleboro minister, graduated from Andover Theological Seminary in 1857 and married Maria Williston of Brattleboro in 1858.

4. Samuel P. Langley (1834–1906), later noted for his research on solar radiation and astronomy, and John Williams Langley (b. 1841), the chemist. They were spending the year visiting observatories, art galleries, and scientific institutions in Europe, before returning to teaching posts in America.

5. At the battle of Richmond (3 April).

6. A procession of gondolas with lights and music.

Venetian Diary

May 21'st 1865. This book, without any intention on my part will just be filled up by the time we leave Venice. We intend to cross the ocean before the September gales commence, with or without leave of absence. We would not pass another Winter abroad at any rate. Yesterday morning at 5 o'clock Molly left Venice for good, in company of old Mr. Mayo, an Englishman, to join the Oldses[1] at Milan and go with them to America. She had only a day and a half's notice—yet she went off all in good order. They will hardly hear of her coming, at home, before she is there. The night before she went away Poor Taylor popped.[2] He has behaved most singularly for a man in love—controlling his feelings to that degree that he even appeared to prefer Prof. Porter's company to hers—so that she was surprised at his offer and refused him—with the provision that he should see her again in America. He is not even to write to her while he is in Europe (he is about starting on a journey through Spain) but, after Commencement at Yale he will go to Brattleboro. Sheffield has been moved to his sister's lodgings at

Barbier's and is getting on very well. His legs are so weak, though, that he cant yet stand alone. I am going out to Tortorini's with Mr. Howells day after tomorrow.[3] We accompany Taylor to Padua. He goes on Wednesday morning to Florence. We all spend Tuesday together in Padua. Since I last wrote Lincoln has been as[sas]sinated. All the world was stirred by it. Johnson is doing well so far. Booth, the assassin, has been shot. Miss Blagden[4] and the Bracketts[5] were here last night. I gave them tea. Miss B. has given Mr. Howells letters of introduction to publishers in London.

The Mayos are the people who took Mr. Locke[6] to their house in Hampstead. Mrs. Jackson sent them to see us. Mr. Howells has lately received a very flattering letter from Dal'Ongero.[7] He thinks Mr. Howells writes better on Italy than any other foreigner.

Manuscript: Harvard

1. Mrs. T. N. Olds and her daughter (pocket diary, typescript, p. 36) have not been otherwise identified. In her letter of 27 March to Howells (see below), Mrs. Jackson mentioned a note to Elinor introducing the Mayos, who were to spend a few days in Venice with their daughters.

2. Cf. Elinor's succinct entry (pocket diary, 19 May): "After dinner, (4 1/4 o'clock we dine) on a full stomach, out on the balcony, T. popped."

3. "Out at Mr. Tortorini's at Mount Selice. Horrible Italian dinner" (EMH, pocket diary, 24 May). On Tortorini, see 24 March and early May 1863. On 6 June, Howells wrote his father: "Elinor and I went to see Tortorini at last, after three years' invitation. He is mayor of a pretty town, settled about two thousand years ago, called Monselice, and he is a person of immense importance there. His dinners were execrable, and we came away on account of them. . . . Think of rice soup, boiled beef, fried brain and strawberries with wine on them. Ugh!—" (*SL*, 1:219).

4. For Isa Blagden, see early May 1863. She is most probably the "English authoress" who "went into raptures over my things, poetry and prose, and is to send me a letter to Antony Trollope, which she hopes will get me a London publisher for my book on Venice" (WDH, 6 June; *SL*, 1:219).

5. Elinor probably meant the Brackens, close friends of Miss Blagden and the Brownings. Miss Blagden was spending a month in Venice after nursing Thomas A. Trollope's wife, Theodosia, who had just died in Florence.

6. The Rev. George L. Locke, a Boston clergyman, had dined with them at Casa Falier in February 1863 (pocket diary, typescript, pp. 46, 50). He had returned to the U.S. in October with photographs from the Howellses for Joanna Shepard and Quincy Ward, and a letter for Elinor's friend, Kitty Gannett.

7. Francesco Dall'Ongaro (1808–1873), the poet. Howells' article was "Recent Italian Comedy." On the article, Dall'Ongaro, and Mrs. Jackson, see Woodress, pp. 92–95. The letter from Dall'Ongaro mentioned by Elinor is not extant.

To Aurelia H. Howells

Palazzo Giustinian',
Venice, June 19ᵗʰ, 1865.

Dear Aurelia:

Long ago I begged Will to tell you all, that, as I was very busy preparing Baby to go home, and as the idea that we were going so soon was quite distracting I could not write any more letters; but a time has come, now, when everything is at a perfect stand-still and I cannot employ myself better than by writing. So I will try to give you some faint idea of the state of our affairs at present though—as Will says, it is of no use worrying you with them.

The fact is our hearts went to America long ago; and, last week, we sent our household goods to New York in a sailing vessel going from Trieste—so now there is nothing here but the Consular presence waiting for that dreadful leave-of-absence which never comes. We expect a letter from the State Department this week which may determine us to start at once.[1] Even if we go without the leave my brother will come here to stay and we shall not give up the hope of still receiving it in England—but all this fuss about the leave from Government is not that we really want to come back but that, with this place to fall back upon, it will be much easier for Mr Howells to ask for a place at home.

A good while since I told Bettina we were going to America in order that she might go into service if she chose and be looking for a place—but, fortunately, her husband is at this very time free from serving as a soldier and they are going to live in her native town, Dolo, he working at his trade of blacksmithing. We shall give her what few bits of furniture we have bought here—as it costs too much to carry them home, and her parting with Baby will be as much "helped off " by this housekeeping project as such a thing can ever be. To please Bettina Baby & she were photographed together last week.[2] As the picture is very good of Winnie. I will send you one that you may know how she is going to look—exactly like her father's family I am sure—but it gives no idea of how handsome Nina is.

I may as well prepare you also for a distinguished-looking individual with his hair parted in the middle and wearing only a moustache—your Venetian brother.[3]

I suppose by this time they are rejoicing at Brattleboro' over the arrival of my sister; but I have not yet learned of the safe arrival of the City of London—the ship in which she sailed—in New York—though there has been time to have heard if it had *not* arrived. We miss her a great deal—especially "Poor Baby," who, when you ask "Dove xe nagna? (where is aunty) looks very sad, stares about the room and gives her right hand the peculiar twist which, in Italian signifies, "Gone!" She makes the same motion when she has eaten all her polenta and wants some more. We find that she does not understand English at all, and that everything must be repeated in Italian before she understands it. She says papa & mamma, which are the same both languages, shakes her head for yes and no. Once she accomplished "yes" or "ess," but now she has given that up and either nods or gives a little affirmative grunt instead. She calls 'Gigi' and "Tone" (Tony) and says, "cara, cara"[4] to the cat—and calls the dog "can'," and, as she is very fond of going out of doors, has learned to say "calle" and point to the door. She has always nodded assent when we asked "Vuol andar' in calle?" (do you want to go out into the street) and she took up saying "calle" of her own accord. Day before yesterday she was a year and a half old, and yet she does not walk alone. She does not need support any longer, and has always been strong on her legs, but she seems to have a great deal of timidity or caution. She pulls herself up steps alone, and goes running about wherever there is anything to catch hold of, but she sits right down whenever you attempt to take away your hand. She is a good little creature, and has many devoted friends in Venice who will be sorry to have her go away. The other day a lady gave her a walnut shell with thimble, sissors, needle-case &c in it. But, if she is petted here I am sure she will be quite spoiled when she goes to America. My oldest brother has a son, so now Baby has three cousins: Willie, Mary and Albert.[5] I can imagine with what respect she will regard

Willie, and how she will patronize the littler ones. She will stroke little Mary's cheek and say "ca-ra, ca-ra!"

With much love to all, your affectionate sister

Elinor

Manuscript: Harvard

1. The grant of leave of absence arrived two days later.
2. This photograph is at Harvard.
3. Dated 1 May 1865 in Elinor's hand, this profile photograph of Howells is pasted on a front flyleaf of her pocket diary. It represents him with "only a moustache," and with arms crossed.
4. Dear. Gigi was probably the son of Augusta, who occasionally took care of Winny, and Tony a member of the palace household.
5. Albert (1865–1870) was the son of Belle and Charles L. Mead.

Pocket Diary

[3 July–4 August 1865]

Left Venice—for America July 3'd, 1865, Monday morning at 9 o'clock.[1] Had to pay 4 francs on 2 doz kid gloves at Desensano. Went to Bella Venezia in Milan (4'th time Mr. Howell has been there) Spent day at Clarks' 4'th of July.[2] Went to see cofin of St. Carlo. Bracelets & diamond rings hung on it. Left at 8 o'clock. Arrived at midnight at Susa. Crossed Mt Cenis and got to San Michel at 12 next day. Ate, washed and went on to Culoz. Some priests got in at Chambery and we rode 6 abreast to Aix-(les bains.) A Frenchman from Lyons who "sympathized with the South" recommended us to Hotel du Lac, at Geneva where we went and were delighted. (Thence we were recommended to Hotel du Boulanger at Berne (where we were gouged) Saw Mt. Blanc at Geneva and afterward from the cars going to Laucerne. Switzerland is sweet and cheap. They give one honey & fresh butter for nothing.[3] At Berne bought pipe with bear kicked up on it. From Berne to Fribourg Baby was petted by strangers and was cross. At Berne got her sweetmeats and wine for the journey. Met Mills & wife of Cincinnati.[4] Cross old Prussian woman & son going to Baden; she interfered with my management of Baby. Got to Manheim at 11 p.m. Went to Mayence next morning. Met Danes in the cars.

Elinor Gertrude Mead.
From an 1854 daguerreotype.

Mary Jane Mead sketched by her son
John Noyes Mead.

The Mead family, 1856 or after.

*Elinor G. Mead sketched by
her brother Larkin, 1861 or 1862.*

A true portrait of Elinor Mead
drawn by her brother
Larkin in 1861 or 2
In the very Clothes she wore
in Columbus the Winter she met Howells

In "Joe's" Greenhouse.

Sketched from memory, in Emma Ward's
Studio, Ohio State House, March, 1861,
by E. G. Mead "as was."

W. D. Howells + Elinor Gertrude Mead

*"In 'Joe's' Greenhouse."
Will and Elinor by Elinor, March 1861.*

Elinor and her brother Larkin about to sail, December 1862.

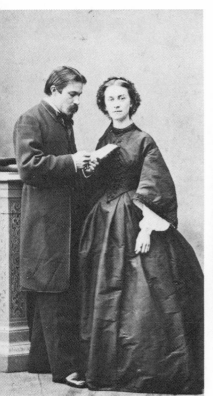

Casa Falier.

Elinor and Will, 1863.

Title page by EMH for No Love Lost.

Frontispiece by EMH for No Love Lost.

WDH sketched by Elinor, 1863.

Mary N. Mead (?), Le Havre.

Palazzo Giustiniani and Ca'Foscari.

Mary and von Dadelsen. Sketch by EMH, 13 May 1864.

Went down the Rhine with them. They were very kind to Baby. Went to Hotel du Nord in Cologne.

Coupé of railway carriage between Aix la Chapelle and Liege July 9$\underline{th}$/65.

Baby eating Nuremburg gingerbread bought at Aix. This morning we went with the Danes from Copenhagen to see the Cathedral, and afterwards in a carriage to see the Golden Chamber of relics at Sta. Ursula, taking Maria Brun along.[5] Baby was charmed with the room adorned with gold & silver busts of virgins (each of which contained a skull) and the walls tastefully covered with bones arranged so as to form an inscription about Sta. Ursula. Her teeth her arm-bones her silk hair-net, ring skull &c had particular honor. On a table, or altar, beside her skull lay that of the Scotch Prince.

Verviers

Baby ate bowl of soup. Bought St. Julien wine and apricots— which are a very good fruit in Belgium.

Liege

Bought little basket of fruit, very prettily arranged on grape-leaves: pears, plums, cherries & apricots

Cambridge, Oct. 7$\underline{th}$, 1866.

We went to Ostend & so crossed to England. Staid a fortnight in London at No. 50 Hunter st. (Cocker & Kershaw). Sailed for America the 22nd of July, and arrived home Aug. 4$\underline{th}$ (Steamer Asia[6] Capt. Moodie)

Manuscript: Harvard

1. For Howells' account of the trip back, see *SL*, 1:222–224.

2. Mary Mead had sailed to Europe with the Rev. William Clark and his family. Clark (1819–1894), a Congregational missionary, had earlier served in Turkey and in Florence, and for a time was assistant to the principal of Glenwood Seminary in West Brattleboro.

3. "Switzerland is charming, and one feels that atmosphere of freedom which is the best air in the world. I saw only proofs of industry and happiness among the Swiss, who produce a supper of cold chicken and honey of unequaled magnificence" (WDH, *SL*, 1:223).

4. Mills of Cincinnati and his wife are unidentified.

5. Maria Brun may have been one of the Danes they were traveling with.

6. The *Asia* arrived in Boston on 3 August 1865.

✤ III ✤

Beyond Our Expectations

August 1865–1867

After briefly visiting the Meads in Brattleboro, Elinor and Will travelled to Ohio for the visit Will had promised his mother to introduce his wife and their child to her and the other Howellses. But soon Will left for New York to look for work; various family obligations prevented Elinor from joining her husband until January, though Will managed several trips to Vermont, to which Elinor had soon returned from Ohio. Within a month after the couple began to live in New York, Will accepted the assistant editorship of the Atlantic, beginning on 1 March, his twenty-ninth birthday.

They had to board for a little while in Boston, and then bought a house (helped by Elinor's father, by Charles Eliot Norton, and by Norton's brother-in-law) on Sacramento Street in Cambridge. Elinor wrote to Aurelia within a week of their moving in: "We are perfectly happy. Everything is beyond our expectations." The library was sunny, their room was gorgeous, the guest-chamber very elegant, and the hall "not so bad!" In another letter after Venetian Life had appeared and received favorable reviews, she asked, "But what is fame when you have a house of your own?"

The house also meant a suburban life for them after Venice and New York. All Cambridge came to call, or so it seemed. Aurelia visited, and between January and May 1867 Annie stayed with them, endearing herself to Cambridge worthies from Longfellow and Norton to the group, all in

88

their twenties, that included Arthur G. Sedgwick, young Oliver Wendell
Holmes, and the James brothers, Henry and especially William. Unfortu-
nately the parties and other activities were too much for Elinor, who suffered
a miscarriage by the end of February and had to take to her bed. But she
recuperated in Brattleboro, where Annie also visited. The many letters be-
tween Elinor and Annie mark a lifelong friendship.

Pocket Diary

August 24ᵗʰ [1865] at Jefferson.
   On this day Baby first walked alone—from the corner of a table
to her papa. We have a lovely bed-room opening into the orchard
and everything is pleasant in the family and in the village.

Manuscript: Harvard

Pocket Diary

Sept 24ᵗʰ [1865] At Jefferson. Baby has whooping-cough badly and
I was impatient to go back to Brattleboro'.

Manuscript: Harvard

Pocket Diary

Oct 24ᵗʰ [1865] Baby nearly over her cough. Mother up at Charley's[1]
on account of weakness & nervousness after a cold. She wished to
be away from the affairs of the house. Our Irish girl being suddenly called away to see a sick sister Molly & I have had to keep
house for a few days. I've broiled beef steak, fried, boiled and
baked and washed pots and kettles. Mr. Howells wants me to come
down to New York to help look up a boarding-place. Mary is out
riding with Lieut Adams.[2]

Manuscript: Harvard

1. Charles L. Mead.
2. Lieutenant Adams has not been identified.

To Miss Przemysl

Brattleboro' Vermont,[1]
November 26ᵗʰ, 1865.

Why, Miss Przemysl, *why* have you not answered the long letter I wrote you long ago from America, nor the one Mary wrote last August? I lost the address you gave me, with the *number* of your house on it—but surely it is enough to say Care of Prof. Reynoldi, Corte San Giorgio, à San Gallo—is it not? If not they are lost, for that is the way they were directed. If you have never received them how neglected you must feel, and if you *have,* how neglected *we* feel! Mr. Howells is crazy to hear from you. You *must* write! Larkin has never spoken of your receiving any letters from us—so I fear you never got either of them. O, dear!

Mary wrote to congratulate you on your engagement.[2] I told you all we had been doing for two months after we left Venice—and now that link in the chain of your American friends' lives is lost to you forever. Don't you regret it *deeply?* They were two months of struggle and distraction, and, now, we have come out of it all into clear sailing and calm weather. Mr Howells has got an engagement on one of the best literary papers in the country—*The Nation,* just started at New York[3]—and, in January Baby & I go down there to live. My father's home, where I am staying now, is eighty miles from New York, and Mr. Howells runs up here to see me occasionally. We have a day of national thanksgiving next week, and at that time he is to be here, and Laura Mitchell with her husband, the General.[4] You know it was at Laura's house that Mr. Howells & I first met?

Manuscript: Harvard

1. This letter is probably a draft, since it is unsigned and at Harvard.
2. Mary Mead had already left Venice when the Howellses found out about Miss Przemysl's engagement to Count Capograssi just before they sailed (Venetian diary, 22 June 1865).

3. "To-day Mr. Godkin engaged me to write for the *Nation* on a salary of $40 a week," WDH had written Elinor on 17 November. On the same day, he told her that Trübner & Co. were announcing the publication of his *Venetian Life*. But he was getting more and more impatient to be reunited with his wife. "I can't write much, for I feel the *fagging* influence of the close, damp weather, and besides I'm unutterably desolate without you. When I think of the long weeks that are still to pass before we live together, I'm almost in despair, and get heartsick" (*SL*, 1:237–238).

4. John Mitchell was breveted major-general of the Ohio Volunteers for gallantry at the battle of Bentonville, N.C., on 17 March 1865 (*Years of My Youth*, p. 287). The Mitchells were to spend a week in Brattleboro with the Meads over the Thanksgiving holidays. Laura's diary (Hayes Presidential Center) records a happy visit with several confidential and lengthy talks between her and Elinor, except for one curious outburst: "Nellie thought [Charley Mead's baby] the prettiest baby in the world and I knew Lilly was prettier—so we grew warm about it and at last feeling insulted by her way of speaking to me I left her—very angry and grieved at what she had said. We were quieter all day."

Pocket Diary

Dec. 24'th 1865 Anniversary of our wedding-day. Mother is in New York with Joanna still. Coming in a few days. Father, Willie & Molly at church (Sunday) & Baby & I at home. Since I wrote the little Elinor Matilda Shepard has been born.[1] In a fortnight I am going to New York, God willing. Mary has snubbed Lieut. Adams.[2] Bob Higginson has been here and she corresponds again with Fred King. What will become of her!

Fred grows fat in New York—Joanna nurses her baby—and Willie[3] is home on a college vacation. He is a great beau in town.

Manuscript: Harvard

1. On 15 December, Elinor had noted in her pocket diary: "Received a letter from Gus Shepard saying they (he had the most to do about it I think) had named their little girl: Elinor Matilda Shepard—after Mrs. Chas. Leverich and myself—Elinor that 'with the name the child might strive to emulate the charm and graces, and in some measure be inspired with the genius possessed in such an eminent degree by her namesake.'" Joanna's friend was the wife of Charles P. Leverich (1808–1876), President of the Bank of New York from 1863 until his death. They lived in Long Island.

2. Fred King has not been identified. Bob Higginson was most probably Robert Minturn Higginson (b. 1845), son of Stephen Higginson and younger brother of Agnes G. Higginson Fuller, the wife of George Fuller, the Deerfield painter. The Higginsons and the Meads were old friends.

3. W. R. Mead was a junior at Amherst.

To William C. Howells or his family

[New York, after 20 January 1866]
[Beginning lost]
We[1] were at a birthday dinner at James Lorrimer Graham jr's
Saturday night—the Stoddards, Stedmans, Bayard Taylor & wife
& Huntington the former Paris correspondent of the Tribune
being the other guests.[2] It was a merry little party, and I laughed
so much I could scarcely eat. Tomorrow Bayard Taylor & his wife
are going to call here. Tonight we were invited to tea at Prof.
Botta's[3] (Anna C. Lynch the poetess is Madame Botta) to meet Miss
Foley the cameo-cutter, whom we knew in Rome. Bryant[4] is to be
there—but we do not care to go. It was Prof. Botta's book on Dante
that Will cut up so in the Nation—or Round Table—I forget which.
Some specimen leaves of "Venetian Life" have been forwarded to
Trübner & Co. I will enclose one.

My brother is still in Venice, but intends leaving soon.

I am so glad to hear you are really all well now. Poor Vic has
had a hard time.[5] Much love to all

Elinor

Manuscript: Harvard

1. Dating is from EMH's pocket diary, 24 January 1866. "My wife and baby, who
have been at Brattleboro' for the last three months (or centuries,) will come to-morrow,"
WDH had written his friend Comly on 8 January (Ohio Historical Society). Elinor and
Will, and Winifred, occupied a back room on the fourth story at his cousin Henry C.
Howells'.

2. "Saturday the 20ᵗʰ we were at J. L. Graham's birthday dinner. . . . It was a brilliant
affair. Plenty of wine and wit," as Elinor reported in the last of her twenty-fourths
entries of the diary. For Howells' first associations with the New York literati, see *Literary
Friends and Acquaintance*, pp. 61–79. James Lorimer Graham, Jr. (1835–1876), was an
art connoisseur and collector and a prominent figure in New York literary and artistic
circles. After traveling in Italy and living in Florence as U.S. consul general, he was
back in New York. The Grahams had visited the Howellses in Venice, and he may have
helped Larkin Mead to secure the commission for the Springfield monument (*SL*,
1:227–228). The other guests were Richard Henry Stoddard (1825–1903), the poet,
and Elizabeth Barstow Stoddard, his wife, also a poet and a novelist; Edmund C. Stedman (1833–1908), a banker, journalist and author, and later editor of Poe, and his wife,
Laura; Bayard Taylor (1825–1878), the author and traveler, who later translated
Goethe's *Faust* and became U.S. Minister to Germany shortly before his death;
William H. Huntington (1820–1885), Paris correspondent of the New York *Tribune*
(1858–1878), art collector, and philanthropist.

3. Vincenzo Botta (1818–1894), the scholar. Howells had reviewed his *Dante* (New

York, 1865) in his "Dante as Philosopher, Patriot and Poet," *Round Table*, 30 September 1865. His wife was Anna Charlotte Lynch (1815–1891), daughter of Patrick Lynch, and an author and brilliant hostess. Her home in Waverley Place was a literary salon that had been frequented by Margaret Fuller, Horace Greeley, and Edgar Allan Poe, among others. Margaret F. Foley (d. 1877) had introduced the Howellses at some of the studios they visited in Rome. In his travel diary WDH calls her "chatty and rather bright."

4. William C. Bryant (1794–1878), the poet and long-time editor of the New York *Evening Post*.

5. Victoria Howells' fiancé, Benas Anther Northway, a surgeon in the army, had died in November 1865. No letter announcing his death had yet been received in Jefferson when his body arrived there. At Victoria's request, he was buried next to her brother John.

To William C. Howells

New York, Jan. 23'd, 1866.

Dear Father:

Will went, again, to the Bible House today and saw the type, but the manager was out, and the printers could not tell him how much type there was, nor what its price was. So he will go again tomorrow, when he will find out all about it, and write, enclosing slips of the Bible stereotyped from it. There has been only one Bible stereotyped from it and it's about as good as new. Will thinks it will serve very well.[1]

We were so occupied Sunday we could not write. At 1 o'clock Will lunched with Mr Osgood,[2] one of the firm of Ticknor & Fields, who came on here charged with a message from Mr. Fields to Will. He made very handsome offers, but the matter is not yet decided. Will will not take the position of "associate editor" unless there is a prospect of his having a permanent place as chief editor by & by. Mr. Fields is going to Europe within the year, when, Mr Osgood says, it will fall into Will's hands for the time, and next year he thinks Mr Fields intends giving up the editing of it entirely. Mr. Godkin naturally advises Will not to accept the offer. It does seem retrograding to go from New York to Boston.

Since we last wrote poor Baby has had a slight attack of croup, so that we watch her very carefully now. Little Harry[3] has been very sick with it but is better now. Mr. Henry Howells sailed from

England the 13<u>th</u> it is supposed. He wrote he was intending to do so.

Please tell Aurelia[4] I should advise her to have a nice black alpacca trimmed say with green narrow ribbon. I see nothing so suitable for the time she will come on here. Empress cloth or poplin are rather too heavy. Black silk or alpacca are best for the street. Have it made any high necked, long sleeved and with no trimming around the bottom [remainder lost].

Manuscript: Harvard

1. Howells had agreed to buy some nonpareil font for his father. Cf. 28 January and 8 February (*SL*, 1:248–250).
2. James Ripley Osgood (1836–1892), then a partner of Ticknor & Fields. For Fields's offer of 11 January and Howells' answer of 14 January, see *SL*, 1:247.
3. Henry Coggeshall Howells, the two-year-old son of Henry Craik Howells, who returned from England three days later.
4. The Howellses had invited Aurelia to visit them.

To William C. Howells

New York, Feb. 4<u>th</u>/66.

My dear Father:

I dare say Will is writing you his usual Sunday letter from Boston, today, but I will write, too, partly on business. The draft was only received—though not duly acknowledged—and was put in to Mr. Coggshall's hands, in the presence of witnesses.[1] He had been, before, to see about the type—said it was "all right", and that they began, at once, to sort the English from the Welch, and that it would be expressed tomorrow, i.e. Monday. Will is to be home tomorrow night and will finish up the business himself. I got a telegram from him, yesterday, saying "Shall probably accept F's offer: Will return Monday"—so girls prepare yourselves for the disgrace of seeing your brother associate-editor of the Atlantic Monthly[2]—and you, Aurelia, to make me a long visit at my house, or rooms! (I am not going to board if I go to Boston) in Boston or Cambridge![3]

Poor Baby! How she is knocked about the world! But she manages to enjoy herself wherever she is. She is in rugged health now, and is out a long time every day taking the air. I bundle her up so that she looks like a duck as she waddles along. She wears leggins (all over her shoes as well as her legs) rubbers over these, red-flannel knickerbockers over her canton flannel drawers, a stuffed sack, her furs and a little blue silk bonnet. She looks so quaint and old in this attire that the children in the street cry out "O what a little lady!" "O look at her little hat!" [sketch] "O look at her little muff, and her rubbers, and what long hair she's got!" In the horse cars, one day when her papa was holding her a lady said "What a pretty little girl!" I am making her scarlet dress now—for *Boston*. (How do you like that, Mother?) I long to see her with the dress & stockings that match so perfectly, on, together.

Did I tell you we took dinner at Bayard Taylor's this week. There was only Will and myself there, and we had a delightful evening. Much as I dislike Taylor's novels, I like him extremely—and his German wife. They have a pretty little daughter, Lily, seven years old. Will is a great favorite amongst the literati here. Indeed where is he not a favorite?—though, perhaps I ought n't to say so. How soon will Jo⁴ be coming on? With much love to the whole Howells kin

Your aff.^{ate} daughter
Elinor

I never knew till just as Will left that Sam & Emma⁵ had written us a joint letter. Will was so engrossed he forgot to tell me. We will answer it when Will gets back. My brother Fred is up here to tea tonight.

Manuscript: Harvard

1. Henry Coggeshall was probably the son of W. T. Coggeshall (1824–1867). Cf. *SL*, 1:239, 240. From being Ohio state librarian (1856–1862), the elder Coggeshall became owner-editor of the Springfield *Republican* (1862–1865), briefly editor of the *Ohio State Journal,* and later Minister to Ecuador. Howells had the type shipped to his father, paying for it on 8 February (*SL*, 1:250).

2. WDH accepted the position on 6 February, after a brief visit to Boston to discuss conditions with Fields. For his letter of acceptance, see *SL*, 1:249. His duties were to read and proofread manuscripts, to correspond with contributors, and to write reviews.

3. On 7 February, Elinor wrote her father-in-law: "Will has accepted Mr. Fields' offer—($50 a week as associate-editor of the Atlantic) and we go to Boston in a fortnight. Shall probably live in Cambridge. Will Aurelia like to visit us then? We shall have a spare bed—, whatever we don't have" (Harvard).

4. Joseph A. Howells.
5. Emma Thomas, of Bowling Green, the first wife of Samuel D. Howells. They were divorced two years later.

To Mary D. Howells

Boston,
22 Bulfinch st.,
March 18ᵗʰ, 1866.

Dear Mother:

I dont know exactly which one of the family I owe a letter—we have been in such confusion, lately—but I am sure no one will be jealous if I write to you.

At last we are *delightfully* situated—quite to our minds—though it is'n't keeping house in Cambridge as we once fondly hoped.[1] We have the whole of a third story to ourselves consisting of four rooms: a parlor, (a sunny little nook with a bowed front, three windows, and paper all over gilt boquets. The furniture is green, the carpet green and everything in the room matches charmingly) a room opening out of it—also on the front of the house—where Will writes, and behind two beautiful bedrooms connect with folding doors. The first of these we shall give to Aurelia when she comes, and Baby will sleep there too, as she does now, in her crib. She is never known to wake up in the night, so she will be no disturbance. (We have no room for the crib with us, but, with the doors open, it is all one room.) I suppose Jo & Aurelia are in New York, today, though we have heard nothing from them as yet. I dont know as they will write before coming, but I shall look for them every day now. Tell Vic I am waiting impatiently to see the triumph of her handywork on Aurelia's clothes. No doubt Boston will secumb at once and it will take all Mr. Howells time to answer the proposals that will be rushing in from the young men, on all sides. Such is the effect of dress! Baby goes out walking with one of the servants every day on the Common, and her papa generally steals around then to get a sight of her before she comes back. We shall learn all about Mr. Eggleston[2] & Annie from Aurelia—for I feel sure there is something going to happen.

My brother Larkin is probably married by this time to an Italian, or, rather, Dalmatian girl whom he met in Venice, and they may be on their way to this country.[3] She is a great beauty and, as people remark "just the wife for a sculptor."

Have you ever yet had that photograph taken from the old ambrotype of Johnny? I think I could make something from that and the large-sized one if I had them now. Perhaps you have sent them on by Aurelia?

Next Sunday, having seen Jo & Aurelia we shall know more about you and can write more to the point. We are not going to board always nor long I hope. One great reason why I want a house is to have you come on and stay a long time with us. O dear, when I think how cramped. . . .[4]

I appreciated the delicate attention of the "onion skins" sent with the dictionaries. Many thanks. I do not use Ticknor & Fields' envelopes *generally*.[5]

Manuscript: Harvard

1. The Howellses, when they arrived in Boston on 19 February, first stayed with Dr. Henry C. Angell and his wife at 16 Beacon Street. While Elinor and Will were house-hunting with the help of Charles Eliot Norton in Cambridge, Winny had been sent to her grandparents in Brattleboro until 8 March. Unable to find a house in Cambridge, they decided to board in Boston. They lived at 22 Bulfinch Street for a while; then, as Elinor recorded in the pocket diary (typescript, p. 17): "Went to board at the Ainsworths', 10 Bulfinch St. from April 28ᵗʰ, till we moved to Cambridge, May 21ˢᵗ/66." Dr. F. S. Ainsworth was a Boston physician.

2. A letter of 11 April 1865 from Annie to Aurelia (Alfred) indicates cordiality between Annie and Benjamin Eggleston (1816–1888), a merchant and a politician in Cincinnati, also a novelist. At that time, Annie suggested that he should speak in favor of the state senate's authorizing payment of at least $7,500 for a painting of "The Battle of Lake Erie," by G. William Henry Powell (1824–1879), the Ohio painter.

3. Mead had courted Marietta di [de?] Benvenuti with neither understanding the language of the other, though they were helped by Brunetta, the young Italian friend of the Howellses. In *Years of My Youth*, pp. 245–246, WDH provides details. The marriage took place in Florence on 26 February. "Jo & Aurelia Howells were at Joanna's the day they arrived there [in New York], Tuesday, the 20'th of March," as EMH wrote in her pocket diary on 25 March.

4. One or more sheets are missing at this place in the letter.

5. These last three sentences are written across the text, at the bottom of the page.

To Aurelia H. Howells

Sacramento Street, Cambridge,[1]
May 28ᵗʰ, 1866.

Dear Aurelia:

I dont know when I've been so much entertained as I was by reading your long, descriptive letter. It was *satisfactory* in all particulars. I meant to have written you one just as long and just as particular today, but, unfortunately, in preparing dinner, I let the dumb-waiter slip, and my hand, getting in between it and the cupboard, got such a squeeze as no *mortal* ever gave it—so that I can only write with difficulty, the muscles being very stiff. I will give you all the news I can at the expense of style. I was delighted to hear how much you liked Brattleboro and the Meads. I have not had any letters from there for a week past, but expect mother & Baby tomorrow and shall hear more about your visit there. The fans were exactly right, and got here at exactly the right time, as Miss Nourse appeared the Tuesday after they arrived to take Miss Przemysl's off to Venice. Marietta's I shall send up by Mother. Lark is not coming here to exhibit his marbles, but I hope they will visit us. "Agate"[2] called the day after you left. He is lovely in every way but, poor fellow, he is in consumption. He came on to see his sister whom he placed under Mrs. Geo. S. Hillard's[3] care last December. I saw Mrs. H. the other day & told her to bring Miss Reed out here—and she said she would—or rather, it was she who proposed it. I havnt seen Frank or Moses[4] since you left. I shall make Will go to call on them. Everybody is going out of town. Thursday the Fieldses & Mrs F's mother[5] (not Mr. F's aunt) & sister are going to Brattleboro, to the Goodale or Dummer Farm for a month. Mr F. told Will he should pick out some articles and then leave him to "run the machine."

We came out here Monday—rainy, and carpets did not all fit of course. The long breadth by the door of my bed-room lacked some inchs—which I thought singular as I had ordered it to be 5½ yds long. Then the four breadths would not stretch 12 ft & four inchs. The dining-room carpet had to be bothered with. The furniture men all started & all got frightened by a sudden shower, and backed out by backing into sheds for the night. So, with the exception of Baby's crib Will's desk (we left our nice, new set to be brought by Lawrence & Wild's[6] men, along with other things we had bought of them) our mattress & trunks and the crockery there was nothing in the house at night. But we determined to stay here at any rate so we camped down on the floor with our mattress, and the box from Brat. fortunately arriving before night with a feather bed for Katy (our girl—got from an intelligence office—a great prize) in it she camped down too. Next morning everything came and we began having our meals in good style right off. Indeed we had a nice roast the first day—borrowing knives & forks from the neighbors to eat it with. I forgot to mention that we had the stove put in Friday, the day we cleaned, and two of the carpets down Saturday.

We are perfectly happy.

Everything is beyond our expectations. When *Baby* is here we shall be *too* happy!

The parlor is very recherché. The library is sunny and inviting. The dining-room is a *great* favorite with us. We sit here & write evenings. I am writing here now on our chestnut extension table. Our room is gorgeous. (We have a marble-topped washstand to match) The guest-chamber (straw matting & chestnut set with blk walnut trimmings from Brabrook's) is very elegant. The kitchen (with a Magee stove) is, some say, the pleasantest room in the house and the hall is *not so bad!*

Green furniture in the parlor, piped and no blk walnut, scarcely, visible—got it at L & W's. Katy sleeps up stairs. I let K. go in town Sundays & I get dinner. We breakfast at eight or half past & dine at four. Have tea for dinner. Yesterday Mrs. Farwell & the Angells[7] came out to see us. Prof. Childs & wife (Mrs Norton's sister) called. The Aldrichs called in the morning, also.

Longfellow has inquired *particularly* where we live. The people on the street, when asked where the Howellses live, ask "Literary people?" and say "The third house." So Prof. Childs says he was directed. See what a reputation, already! Mrs F. sent much love— and the Angells. Sweet in her to go to the depot, was n't it? Mrs.

Harris[8] bemoaned you constantly while we remained there. Will must tell the rest. Do write just such another letter! Much love to all

Elinor

We spent the evening at the Winsors'[9] before leaving Boston.
Miss Nourse came out & breakfasted with us Wednesday
The Fields are coming out to see us Tuesday. Went to Prof.
Blot's[10] lecture with Mrs. Fields, Friday.
Dear Aurelia—Tell father that I received his P.O. order for $22, and was duly thankful. I will write soon again. Love to all.

Affectionately,
Will.[11]

Manuscript: Harvard

1. "OUR HOUSE *Sacramento st. Cambridge, May 28ᵗʰ, 1866.* A week ago tonight we were still at the Ainsworths. Now we are fairly settled here. (We moved in Monday the 21st) and Baby is to come down from Brattleboro' with Mother tomorrow. She has been up there a fortnight while we were moving. Her grandfather Howells and Aurelia took her up. Katy McGuire, our servant, is in town today (Sunday) and I got dinner" (EMH, pocket diary). Aurelia had evidently paid a visit to the Meads after staying in Boston with Will and Elinor. Her letter about her stay in Brattleboro is not extant.

2. Whitelaw Reid (1837–1912), formerly with the Cincinnati *Gazette*, later associated with the New York *Tribune* (1868–1905), of which he became editor in 1872. He served as U.S. minister to France from 1889 to 1892 and as ambassador to Great Britain from 1905 to 1912. His sister Chessie married a Mr. Smith of Hanover, New Hampshire, in March 1867.

3. The wife of George S. Hillard (1808–1879), the author of *Six Months in Italy* (1853), a lawyer, journalist, and politician, who lived at 62 Pinckney Street in Boston.

4. The Howellses had met Frank Johnson and Moses Grant in Italy. The former was probably the Rev. Frank H. Johnson, of 24 Pemberton Square in Boston. Moses P. Grant was a partner of Grant, Warren & Co., paper manufacturers.

5. Annie Adams Fields's mother was Sarah May (Holland) Adams, the wife of Dr. Zabdiel Boylston Adams, a noted Boston physician.

6. Lawrence, Wilde & Hull, a store in Boston. E. H. Brabrook, mentioned later in the letter, was a furniture store.

7. Mrs. A. G. Farwell, the widow of a Boston merchant, made her home with the Angells. Mrs. Francis James Child, née Elizabeth Ellery Sedgwick, was the daughter of Robert Sedgwick, of New York. Mrs. Charles Eliot Norton, née Susan Ridley Sedgwick, was the daughter of Theodore Sedgwick, III, of Stockbridge, Mass., and New York. They were cousins. For Howells' reminiscent portrait of Child, "One of the first and truest of our Cambridge friends," see *Literary Friends and Acquaintance*, pp. 211–214. Charles Eliot Norton (1827–1908), coeditor with J. R. Lowell of the *North American Review*, and one of the founders in 1865 of the *Nation*, had found the Sacramento Street house, after house-hunting with Howells. He also helped him to obtain a mortgage (*LinL*, 1:106; *SL*, 1:254n.3). Thomas Bailey Aldrich (1836–1907), then editor of *Every Saturday*, and his wife, Lilian Woodman Aldrich.

8. Mrs. Harris, who is otherwise unidentified, may have lived with the Ainsworths. See 22 August.

9. Justin Winsor (1831–1897), an author, became librarian of the Boston Public Library in 1868, and of the Harvard University Library in 1877.

10. Pierre Blot (c. 1818–1874), an instructor in cookery, author of *What To Eat, and How to Cook It; Containing over One Thousand Receipts* . . . A founder of cooking schools, he was giving a series of lectures on cookery in Boston at the time.

11. The last four lines are in WDH's hand.

To Anne T. Howells

Cambridge, June 11ᵗʰ, 1866.

Dear Annie:

If you'll overlook and forgive my neglect so far since I came to Boston, I will promise to hold up my end of the line for the future. This country life has made us all healthier and happier and better natured already, and it seems as though we had never half lived before, since we were married. Will enjoys writing, in this quiet place where he is never interrupted—and now we may look for great things from him![1] Baby plays in the dirt along with the neighbors' children mornings, amuses herself with her dishes and books afternoons, and takes a walk with papa & mamma about sunset. Tonight her papa has taken her with him, in the horse-cars, down to the printing-office. I see them walking down the street from the corner now—coming back!

Everything goes smoothly about housekeeping. I've got a treasure in my Katy—but she's too good & pretty to stay with me long. The house is so nice & new and convenient, and there are so many arrangements to help off work, that everything goes like clockwork. We've not finished furnishing the house yet by any means, and still it is very comfortable.[2] We expect to put up curtains next week, before flies come—but there's not one up yet. We have had a great deal of company, in the way of callers, and sometimes to tea. Young Sedgewick,[3] Mrs. Norton's brother, and Church of the Galaxy were here a week ago yesterday—Mr & Mrs Clarke, two children & nurse* (of the Ticknor & Fields firm) took tea here last night and Friday Swetzer of the Round Table dines here. We have

*We dont invite people Sunday, but they come

been once to the Angells' to dine—to meet Kate Field,[4] who writes the Recollections of Walter Savage Landor—and to Prof. Childs'* to tea since we were in this house, and tomorrow we are going to the Roundabout Club (a private-house affair) where Emmerson is to read a paper.

We dont care for so much society as this and shall be glad when everybody goes off to the seaside.[5] We've had the yard put in order, so that it looks like a different place and now Will is going to work in it a good deal. I suppose your place is looking beautifully, fresh after all this rain. You'll have to look out, though, if you want to surpass us on grapes, currants, rhubarb, or Lawton blackberries!

I hope your pineapples arrived in good condition, Annie, and that you feasted on them to your hearts' content. Papa occasionally brings home a bananna to Baby—but they are too dear eating for the rest of us 20, 15, & 10 cents apiece!

Much love to all, and tell Vic I am aware of my obligations and will write to her next. Tell Aurelia the Angells always send much love to her

Your loving sister
Elinor

We are anxious to hear if Mother has made up her mind to come— though I dare say Will will have to help her when he comes. Direct to us at *Cambridge* after this[6]

Manuscript: Harvard

1. On 25 May, WDH had written Norton: "We are safely housed here in Cottage Quiet, and have commenced the long-deferred process of feeling at home, and of growing old.... After the life which we have hitherto led in cities, this is singularly free from tumult. Everything is so tranquil about us that I find the agitation of a cow in the pasture across the street very stimulating, and am quite satisfied with it. This morning, however a large dog appeared at the corner of the fence. Presently two men walked up Oxford street, and I was greatly excited. A few minutes later a man drove by in a trotting-buggy: this appeared incredible" (*SL*, 1:253). Cf. also *Suburban Sketches* (1871), p. 13.

2. Elinor had entered in the pocket diary (typescript, p. 11) "May, 1866. Used up every cent of our $800.00 furnishing our house in Cambridge."

3. Arthur G. Sedgwick (1844–1915), a Boston lawyer, and later assistant editor of the *Nation* (1872–1884). William C. Church (1836–1917), editor, or his brother

*He married Mrs. Norton's cousin—also a Miss Sedgewick

Francis P. Church, cofounders in 1866 of the *Galaxy*, which merged with the *Atlantic* twelve years later. John Spencer Clark (b. 1835), an author of textbooks on art education, was in Boston publishing businesses throughout his career. Charles Humphreys Sweetser (1841–1871), journalist and publisher, or, more likely, Henry Edward Sweetser (1837–1870), former reporter for the *Times* and editor of *World*, and cofounder with his brother of the *Round Table*, from which he withdrew in 1866.

4. Kate Field (1838–1896), pseudonym for Mary Katherine Keemle, the actress, lecturer, and journalist. An eccentric and enthusiastic reformer, she was a friend of the Brownings, the Trollopes ("my most chosen friend" [Anthony Trollope, *Autobiography*]), and George Eliot. Her "Last Days of Walter Savage Landor" appeared in the *Atlantic*, April–June, 1866. Francis James Child (1825–1896) taught at Harvard.

5. In *Literary Friends and Acquaintance*, Howells later recounted his impressions of their Cambridge house and neighbors as follows: "Positive beauty we could not have honestly said we thought our cottage had as a whole. . . . But we were richly content with it; and with life in Cambridge, as it began to open itself to us, we were infinitely more than content. This life, so refined, so intelligent, so gracefully simple, I do not suppose has anywhere else had its parallel" (p. 152).

6. On a blank page Howells added a short letter to his sister Annie in which he announced the publication of *Venetian Life*, telling her of the very flattering reviews his book had elicited so far in the *London Review* and the *London Examiner*. He also mentioned his own mother's coming visit in the fall.

To Annie A. Fields

Cambridge, Aug. 2ⁿᵈ, 1866.

Dear Mrs. Fields:

I cannot allow my sketches to be wasted if Mr. Howells' inspiration does fail sometimes; so I write upon a sheet laboriously prepared for a brilliant poem Mr Howells *was* to have written acknowledging Mr. Fields' dozen of claret.[1]

The festive design is not particularly appropriate for this note however, as I write to say that we are very sorry that we cannot avail ourselves of your kind invitation which has just reached us. The baby is not the hindrance but Mr. Howells is to blame for our failure in this case as well as in that of the unwritten verses. He has been prevented by indisposition from working for several days past, and now that he is well he has still his book notices to write, and some other things to do besides.

I only returned from Brattleboro Tuesday. I did not see your

Mother & sister, but Mr Grant of Boston told me they were well
and that he took Miss Adams driving Monday evening.[2]
With *deep* regret that we cannot come,

Affectionately yours
Elinor M. Howells

Mr. Shepard had some trout *all caught* to send you when I got to
Brattleboro—but when I told him you were at the sea side he
feared to send them. I suppose they would hardly be good by the
time they reached you at Manchester!

Manuscript: Huntington

1. Elinor refers to her "festive design" of branches with leaves and wine bottles, on
the left upper corner of the first page.
2. See 22 August.

To Aurelia H. Howells

Cambridge, Aug. 22ⁿᵈ [1866]

Dear Aurelia:

I had scarcely got home from Brattleboro' when Mrs. Fields
wrote up for us to come down to Manchester. At first we excused
ourselves, as Will had a great deal to do on the Magazine, but she
would not let us off, so a little later we had to go.[1] Manchester is
about forty miles up—or as they say here, *down* the coast, and the
Fields, you know go there every Summer. They stay at a farm house
right on the shore, and, as it were on an island, for the salt water
sets up behind it in a sort of "back bay"—making it the dampest
kind of a place. To me it was the most disagreeable place in the
world—especially when the tide was out, and all the dank seagrass
and muddy shore of the back bay lay exposed to the sun, smelling
so strong of salt! I imagine the vegetables which grew on the farm
tasted differently from inland vegetables, and indeed, there was
rather too much of a salt savor to the whole thing for *me*. But Mrs.
Fields likes it and thrives on it immensely. You should have seen
her in her costume—for she & Lizzie Bartol[2] wear one something
like the Bloomer. It looked very droll to see Mrs. Fields in such a
dress—when I had always before seen her in the fussiest kind of

attire—but she is devoted all Summer to getting up her health and looks for the Winter [sketch]. She was as brown as bun and her nose was dreadfully burned—but she was fatter than in the Winter. Will and I both got sick down there, and we were *heart* sick all the time because we had left baby behind—all alone with Katy.³ As it proved she did not suffer at all, but was so contented and happy that she did not even inquire for us—but we got so wrought up that by the time we got to the house that we hardly dared inquire for her. We were only gone three days, and left everything arranged for telegraphing if she should not be well—but we will never leave her with only a stranger again.

The time between Brattleboro and Manchester I spent doing justice to my friends. I had Dr. & Mrs. Ainsworth⁴ & Mrs. Harris, and Miss Sewall all to tea together—and another night Prof. Child and Miss Ireland. I made calls on all the Cambridgians who had been to see me, and sacrificed myself in every way to society. So you see I have had no time to write letters for some weeks? Have I told you that I met Moses Grant in Brattleboro? Yes, he was there and called to see me. After I left he took Molly, Joanna and Carrie Hodges⁵ out driving—and that is the last I have heard of him— for ever since then Molly has been sick—threatened with a fever. I heard of his taking the lame Miss Adams, Mrs. Fields' sister, who is at B., to ride also. He was very agreeable, and we spoke again of what a pity it was that you left before he could take you to ride. Driving out with ladies seems to be his occupation in life. I hav'n't seen Frank Johnson for an age. He is off rusticating. William Bradley⁶ & Josey Hyde both spoke of you at Brattleboro, and wished to be remembered. Willie Mead⁷ dropped in here for one night last week. He has been down visiting a classmate at West Cambridge for some time, and finally saw fit to call here, ("So New Englandish!" Will says) and afterwards came here to stay awhile, but, the very next day after he came, another classmate came over from Danvers to get him to go and camp out—and off he went. Father wrote he was at West Cambridge before he called, and I thought we were getting decidedly *cut* by our brothers—if one could come to New York without coming near us and another could visit in the neighborhood without calling on us. However Jo's explanation is perfectly satisfactory, though we were a good deal disappointed not to see him, and have him see our house.

Will is getting so toasted about his book that I can scarcely manage him. I cant take him down in any way, when the *Westminster*[8] praises him so—and Lowell & Dr. Holmes and people round here flatter him so immoderately. Mr. *Fields* says he is *sure* of fame. But what is fame when you have a *house* of your own?

We have already made arrangements to have our stoves put up within a fortnight—although people prophesy that Sept. will be a hot month. But it is as well to have it done before Will goes. I want the house warm when your Mother arrives—for I still think she will come. I am sorry to hear about her toe. If you will give the size I *think* we can find her some comfortable shoes in Boston—fleece-lined or fur-lined, and very soft. I wish we had brought some from Venice. They have the very kind. They are necessary to keep out the penetrating cold of their stone floors.

Secretary or Chief Ghotic Chase[9] was down visiting at a house near the Fields—a Mr. Haven's. Mr. H. is president of a bank in Boston. Warren the actor was down there too. Baby is over at the Richardsons' playing with Georgie[10] this morning. We like to have her play with other children, but she generally prefers her papa & mamma's company. Does anybody want me to do any shopping for them before Will goes.

Do you ever wear your silk dress, Aurelia? Does the new*er* doctor like it?

Much love to all
Elinor

Manuscript: Harvard

1. Will and Elinor arrived in Manchester on Wednesday, 8 August. They left on Saturday.

2. Probably Elizabeth Howard Bartol, the wife of Cyrus Augustus Bartol (1813–1900), a Unitarian clergyman and pastor of the West Church of Boston. Her home at 17 Chestnut Street was a center for transcendental writers and thinkers.

3. Katy McGuire, their Irish girl. Annie Fields gave her impression of the Howellses' visit in her diary (at the Massachusetts Historical Society): "Wednesday—Morning cool and breezy with a forecast of autumn, afternoon hazy and warm. The boat came over in the sunset holding an unusual load and we strained our eyes to see who the load might be as they rounded the point moving slowly through the glassy water. Presently we discerned Mr & Mrs Howells. The evening was damp and we did not go out after

a somewhat late tea but sat in the parlor hearing tales of Venitian life. Mr Howells lived in Venice 4 years and is deeply imbued with the spirit of the place. He has written a good book about it but I feel as if the work of his life would very probably be a history of Venice. Mrs H. is a pallid, wide-eyed little woman not without great sweetness and purity in her thin face which just escapes being what we call spiritual looking, by a kind of childishness which is a charm, at the same time that it suggests this world and enjoyment of it rather than another. Mr Howells is a thorough student in appearance, not a poet exactly—[and?] I think he is an historian by nature, but we shall see—. . . . Saturday morning. . . . Walked on the beach and swung in the hammock in company with the Howells until our early dinner, then immediately after we started to take the cars for Boston. . . . I soon bade them 'good bye' and felt a little sigh of relief as the train bore them away for they were neither of them strong or well and I feared they would be actually ill if they stayed longer. Their five years sojourn in Venice has helped to enervate constitutions by no means very strong in themselves and their pallid faces were in strange contrast with our healthy sunburnt countenances. Beside they have no rapturous love of country sights and sounds—they love the sea because it reminds them of Venice and their days of early married life specially when it laps up softly on the sands but they did not seem to hear this lonely voice in the nighttime when it brings us news from the far misty deep."

4. The Howellses had boarded at the Ainsworths' in Boston for a month before moving to Cambridge. In a letter to her mother of 28 February 1867 (Alfred), Annie mentions seeing Aurelia's friend, Kitty Ireland, "a wonderful girl," at a musical party at the Childs'. The rest are unidentified.

5. Carrie Hodges was probably a relative of Walter Hodges of New York, mentioned as being in a shipping office by William R. Mead in a letter to his brother Fred of 16 October 1871 (Amherst).

6. William C. Bradley II (1831–1908), son of the Hon. Jonathan Dorr Bradley, an old friend of the Meads, had been Elinor's brother John's childhood friend and Harvard roommate (John N. Mead's journals, Harvard). Josephine Hyde (1831–1899) was the sister of Elinor's close friend, Angelina Hyde Buckingham.

7. William R. Mead, the future architect, then a senior at Amherst.

8. *Westminster Review,* July 1866.

9. Salmon P. Chase (1808–1873), governor of Ohio (1855–1859), Secretary of the Treasury under Lincoln, and Chief Justice ("Ghotic") since 1864. Franklin Haven, president of the Merchants National Bank, lived at 97 Mount Vernon Street, not far from the Fieldses at 148 Charles Street. William Warren (1812–1888), an actor with the stock company of the Boston Museum, was a friend of Fields, Oliver Wendell Holmes, and H. W. Longfellow, and one of Boston's leading citizens.

10. William F. Richardson, their neighbor at Sacramento Street. Georgie was probably the "small boy next door, who is her partner in the mud-pastry line," as Howells wrote Victoria (17 June, *SL,* 1:260).

Pocket Diary

Cambridge, Oct. 7$^{\underline{th}}$, 1866.

Mr. Howells has been in Ohio nearly three weeks[1]—comes Oct. 10$^{\underline{th}}$. he has been to see Laura, and been reminded of old times. I long to hear how Columbus struck him.[2]

Manuscript: Harvard

1. Back in July, WDH had announced his projected visit to Ohio as follows to his friend Comly: "About the middle of September I'm going to my father's in Jefferson, and then I expect to visit Columbus, if only for a day or two. Last year, in the uncertainty I felt about the future, I had no heart to go anywhere, but since prospects have brightened, I find myself homesick for the west, and particularly for Columbus." (8 July, *SL*, 1:264). He left for Ohio on 17 September.

2. Also on 7 October, Elinor entered, on a different page of this diary: "Yesterday I picked all our pears but those on two trees. (Flemish beauties & middle tree next very tall one) I had to do it as Mr. Howells is in Ohio. The grapes are ripe but I am waiting for him to come before picking them."

To Mary D. Howells

[Cambridge] Sunday p.m.
[5 May? 1867]

My dear Mother:

It is nearly three months since I last wrote you—and I have hardly been well a day all that time. As late as the beginning of last week I was discouragingly weak,[1] but the fine weather, "ferrated calisaya bark," and some good claret wine have at last had an effect, and I am quite myself again. Saturday Will & I went in to Boston to buy an entry oil-cloth and stair-carpet—the first time I have been in since I was sick. My sickness has been long and tedious, but such good care has been taken of me that no bad effects remain, and you may think of me as fully as strong as can, and with a little more color in my face than when I was in Jefferson.

We are sorry to hear from Joe that you have a bad pain in your side—or had when he left. A letter from father Howells this morning said nothing about it, so we hope you are getting over it. If you are quite well, and there is no particular need of Annie at home, we are thinking a little of keeping her over Class Day—the 21'st of June—that is if we can arrange—before Joe leaves—some satisfactory way for her to get home—and I think we can: Annie thinks Cambridge rather too pleasant to leave this weather, especially as the winter has been very dismal—both in and out of the house, sometimes.[2]

She has enjoyed her visit greatly, take it all in all, and has made a great many friends here. She has several *admiring* gentlemen friends—why do'n't they insist upon her remaining here? I hope th[e]y wont prove themselves *savages*! Young Wendall Holmes[3] called here today. What do you think of your modest Annie with her hair put up in this manner [sketch]?[4]

Please tell everybody I owe letters that I will write as soon as I possibly can. I did not attempt to write letters during my sickness I was so nervous and low-spirited. Much love to you all from your aff. daughter

Elinor

Manuscript: Alfred

1. Elinor was back in Cambridge since 29 April after her three weeks in Brattleboro where she had been recovering from her miscarriage in late February.

2. On 18 April, Mary Dean Howells had written Annie: "you must not let [Victoria's] letter that she sent a few days ago hurry you home till you are entirely ready—much as we all want to see you. We do not want you to come till you are entirely ready" (Alfred). Annie left Cambridge the last week in May.

3. Oliver Wendell Holmes, Jr. (1841–1935), future Associate Justice of the U.S. Supreme Court, and son of the famous poet and essayist, was then twenty-six.

4. EMH's larger pen and color profile sketch of Annie dated 12 April 1867 is at the Howells Memorial.

To Anne T. Howells

Cambridge, Sunday, June 2ⁿᵈ, 1867.

Dear Annie:

It is but natural that I should write to you exclusively for a while[1]—there is so much to tell you, more than the rest—but you'll be good and read whatever is interesting in the letter to them, wo'n't you? We were surprised to get a letter from Joe yesterday morning. How very quick it seemed! A letter came from Jefferson for you some days ago, which I enclose. I forgot at the last to tell you must deduct those three dollars I owe you from your account. I sha'n't like it Annie, if you do'n't. It is for making over your blue silk which I spoiled.

The principal events of last week—think, you've only been away a week!—were a little party I gave for Winnie on Thursday, and the organ-concert last night. I made the party to please Mrs Richardson[2] and to dedicate the tea-set which cleaning the China-closet brought to my mind—partly, also, to glorify Esther in the white children's eyes by letting her preside. The way it pleased Mrs R. was: that Minnie Hobart, her little niece from Boston was the principal guest, there being only two others—Lily and Iris Gifford invited. I intended asking Kate Pearl, but she was a little old for Winnie Will thought. They had boiled custards, cake, candy & fruit, sandwiches, and weak tea in the little cups. Winnie was rather unmanageable—pitching into her custard the moment she was seated, not even waiting for her bib. After tea they played games, and then took to romping so over my new oil-cloth that I thought there'd be no paint left on it. It is astonishing how much noise five children can make!

Friday we received an invitation from Dr Upham of Boston[3]— the man who projected and performed most of the work of getting the great organ to an organ-concert "at twilight" Saturday evening—Eugene Thayer, organist. I dressed up my prettiest: [sketch] new bonnet, (I stopped at the Square & got narrow back strings and put them on in the milliner's shop) blk silk walking dress & silk sack) and we started. First we called at the Angells'. Mrs Farwell is perhaps a little better. I saw Mrs Angell & excused your not calling. Then we went to Debois & Johann's to see Lark's "Listening" but it was closed. T & F's[4] was closed, so we went to the hall at once. It was early & only Dr Upham was there. He was on the platform and received us. There were seats for perhaps forty placed upon the platform with an exquisitely-printed programme on each—which in the twilight were of no use, and after all the names of the pieces had to be announced. Next came Mr Longfellow, his brother, sister & oldest daughter.[5] He talked to us, and asked for you. Was very sorry to hear you had gone, as he has had a volume of his poems done up for you a long time which he did not send, thinking you were to be here longer. We will send it as soon as it comes.[6] He said it was a pity you could not be at the concert, and we thought so, too. The Fieldses, Holmeses and Whipples[7] soon followed, and then came some others whom we did not know. The music was very fine and sounded well at that *nearness*—"the volume of sound," as Dr U. explained, "passing over us." The organ is blowed by a caloric engine. The vox-humana stop is fearfully real, and very painful to me. Tell Joe the Aldriches were not there, and were not invited, I imagine. The Fieldses procured our invitation for us. It was quite an exquisite affair. The Fieldses go to the seaside this week, but come home for a while in July. Will has been reading Mad. Récamier's Memoirs[8] aloud to me, and I have been sewing. My gabrielle is nearly done all but a few rosettes, and it is very pretty. Esther's calico is finished, waist and all. Mrs. Pearl is better. Harry James is coming here to dinner Wednesday. Miss Grace Ashburner brought me four nicely sprouted gladioli the other day. Willie Mead has sent us an invitation to his Class Day.[9] Mrs Chas. Pierce (Zena Fay)[10] has invited us to the Spring concert of the St. Cecilia Choir (of a hundred and twenty five children) June 5, here in Cambridge. You would enjoy *that*. I wish you could hear it! Mrs Pierce teaches the children, you know?

I asked Winnie to write a little to her aunty, but I dont think she has done it very well and she has blotted a good deal. However, you'll excuse *Crump*. After I hear from you I will write again. Much love to all. Will & Mrs Lewis send much love. Mrs L. tried to work upon my feelings the other day by saying "Now what if Miss Annie *should* get married?" Will met Prof. Child down town, today, who ran half across the green to tell him Mrs Child is threatened with blindness. Poor Mrs C.! I want to go and see her but the Prof. said she could see no one. I wonder if she will have to lie abed three months as she did before!

Manuscript: Harvard

1. Annie was back in Jefferson after her four-month stay in Cambridge.

2. Mrs. Richardson, the Giffords and the Pearls were neighbors on Sacramento Street. For Esther, Mrs. Lewis' eleven-year-old daughter who helped around the house, see 16 June.

3. J. Baxter Upham, M.D., was active in Boston musical circles and published occasional essays on music and medicine. Whitney Eugene Thayer (1838–1889), an American organist and composer, became also an editor and conductor. He gave recitals in America and abroad in 1868, and later opened a private organ studio in Boston. At the time, he was organist in several Boston churches. The concert organized by Dr. Upham was a private, invitational affair for some fifty persons, as distinguished from the usual noon concerts with regular organists at the Boston Music Hall.

4. T & F stands for Ticknor & Fields.

5. It was probably Samuel Longfellow (1819–1892), the poet and clergyman, and

his brother's biographer, who attended the concert with his sister Anne and Henry Wadsworth Longfellow's daughter Alice.

6. Longfellow's note accompanying his "Courtship of Miles Standish" is preserved at Alfred, and reads as follows: "Cambridge June 10 1867 ¶My Dear Miss Howells, ¶I was very sorry to find that you had escaped from Cambridge before I had an opportunity of bringing you the book I promised, and as you see, have not forgotten. ¶Be kind enough to accept it in return for the volume you so generously surrendered, and as a mark of friendly regard[s?], believe me ¶Yours truly ¶Henry W. Longfellow." (The "volume so graciously surrendered" by Annie was a Canadian edition of Longfellow given to her by her brother, who suggested Longfellow would like to have it.) When Longfellow's note arrived with the book at Sacramento Street, WDH wrote Annie: "The book Elinor and I have looked at. Only think how much it must have cost us not to look into your letter! Elinor, yielding to the impulses of her sex said 'Open it,' but I declared that you should not be cheated out of the smallest part of your sensation." A week later he wrote again to urge his sister to "get ready a nice letter *at once,* and have father see that the capitals and punctuations are right" (18 June; letter at Harvard). Annie's thank-you note of 23 June to Longfellow (as well as her 16 May letter of transmittal for the book) are also at Harvard. After expressing delight at having met him, Annie ended: "I shall always prize my 'Miles Standish' beyond everything, as a delightful memento, of my acquaintance with you. ¶I am sorry I could not have been in Cambridge, to have told you at once, how proud I am to receive this—my favorite poem—from its author."

7. Oliver Wendell Holmes (1809–1894), the poet, and his wife, the former Amelia Lee Jackson, daughter of Charles Jackson, Justice of the Massachusetts Supreme Court. Edwin P. Whipple (1819–1886), author and lecturer, whose wife, née Charlotte B. Hastings, was a friend of the Holmeses and their circle.

8. *Memoirs and Correspondence of Madame Récamier (1777–1849),* translated and edited by Isaphene M. Luyster (1867).

9. W. R. Mead was graduating from Amherst.

10. Harriet Melusina Fay, author of *Co-operative Housekeeping,* first serialized in the *Atlantic,* November 1868–March 1869. Founder of the Cambridge Cooperative Society in 1869, she was the wife of Charles Sanders Peirce, the philosopher and scientist, who had lectured on the philosophy of science at Harvard. They separated in 1876 and were divorced in 1883.

To Anne T. Howells

Cambridge, Sunday evening.[1]
[16 June 1867]

Dear Annie!

I enjoyed your account of the journey home very much, and your Brattleboro letter which came yesterday is charming.[2] I hope you will send copies to the Higginsons—both in Brat. and Deerfield—and to my family. You ought to send one to Rev. Frederick Frothingham, Buffalo,[3] and one to *William C. Bradley,* by all means—or will Miss Anna show the paper to him? I'll give the

Childs your messages "in tuto." Will went over there with Winnie this afternoon, and forgot what the messages were, and so said you sent your love, and nothing more. He carried back the books. Mrs. C. is still confined to her room and has Miss Nash there nursing her. The Prof. begged Will to let Esther bring Winnie around often to play with Helen[4] as she has no playmates. Your Mother's sweet letter to Winnie came today. Winnie opened it herself, discovered the flower & the money (put with the rest to buy her silver knife & fork—and we shall call them "Grandma Howells' and Mr. Longfellow's gift") and then insisted on reading it herself, walking 'round the room and improvising aloud thus "My dear Winnie, I love you very much. You must come and see us in two days. Aunt Annie must bring you &c"—but at last she consented to let her papa read it, and it delighted her very much—especially about the peacock.

We should like the peacock of all things if it could be sent. What do they eat, and where should we put it nights & in Winter? Where did you keep it in the cold weather? We have been trying so long to think what pet we could get Winnie, and it seems to me we might manage a peacock, even if we had to build a little pen for it. Do give full instructions if you send it. Could we attach a string to its leg—or would dogs worry it? I know nothing about such *animals.*[5] Tomorrow Will & I go into town—he to see Mr Fields, & I to learn about setting needles at the Howe agency—and we will get your Bedouin pattern. I hope you have got your Miles Standish by this time. Is n't it splendid? I called on Kitty Wells when last in town, and reaped a harvest of compliments for you from her— and she did not add "I shall never let my brother marry."[6]

The dinner with Harry James for a guest was very pleasant. The dinner was "rekirk" [?]: boiled salmon with butter sauce roast sirloin, asparagus and "duck's back potatoe," radishes, lettuce (extra amount of oil, and wooden spoon & fork) *Mrs L's*[7] bread pudding & sauce & coffee. Winnie wore her blue gingham with edging in the neck and Mr. James admired her very much. I wore my gabrille (the rosettes are lovely) and a purple ribbon on my hair. (*I've* braided in yarn, after all, Annie) a pretty way to do is: double a ribbon [sketch] width across, I mean so wide to make it narrower

& firmer, and in the end that comes up on top (just where a comb would be) put in some loops & ends an inch long like [sketch] It looks very cunning. Put in enough to stick up in the air.

The day Harry J. was here I went out into the dining-room after dinner and found the blackest kind of a negro, in green spectacles, sitting there. Mrs. Lewis immediately introduced him to me as "Proffessor Johnson" her son-in-law. He is Prof. of Phrenology, having studied with Fowler.[8] You must send long messages to Mrs L. as she is always very particular to know just what you say. I went over to the Palfreys' the other day & the B. B.[9] entertained me—smi[ling] all the time and looking as though she'd eat me up. Wed. Arthur Sedgewick is coming to dinner, and Thursday is Mrs Washburn's wedding.[10] I shall probably not go to the reception— not but what I'm well enough, but I can be excused Will says and I'd much rather. The "W"[11] does n't come up very well—but most of our things are doing nicely. The Virginia creeper is luxuriant on the sides of the steps, making a bank which covers them and our rose at the side of the porch has literally hundreds of roses & buds. The house is in *perfect order,* (can you believe it?) Will has got his elk-horns up, set on a splendid blk walnut shield, (what if that natural history society *should* claim them?)[12] Whittier has at last put shelves into the closet up there for books. Winnie wears white stockings and Esther looks clean, and everything goes well.

Poor Eddy Mead[13] has had typhoid fever and we never knew it till he had been sick ten days, and Will did n't go in. His sister was here to take care of him, and has taken him to Milford now.

This is a gossippy enough letter! We have heard from Przemysl. Her "go[od but] unhealthy" husband is well again, and she likes Dalmatia.[14] Much love to all

from

Elinor

I *hope* your Father'll be Lieut. Gov.! Rutherford should be Gov.![15] Will forgot to say in his letter that he had seen Howard Ticknor who said he should print you[r] story as soon as he could get an illustration for it. He said nothing about the *pay*! but you'll get the money when the story's printed.[16] Send another. I've been over to East Cambridge today to hear & pronounce on a minister they

think of having at Brat. Charley wanted me to. I liked the man exceedingly[17]
Arthur Sedgewick often hopes you will come back "*soon*"!!!

Manuscript: Harvard

1. Dating established from WDH's letter of 29 June (*SL*, 1:279) about the arrival of the peacock, and the *Ashtabula Sentinel* of 12 June.

2. Annie's "Brattleboro" letter appeared in the 12 June *Sentinel*.

3. Frothingham was associated with the Unitarian church in Brattleboro (1864–1867). Anna Storrow Higginson (1809–1892) was the sister of Francis John Higginson, the physician who practiced in Brattleboro (1842–1867). For Bradley, see 22 August 1866.

4. Helen Child (1863–1903) became a lifelong friend. In *Literary Friends and Acquaintance* (p. 212), Howells tells how her father "imagined a home-school, to which our little one was asked, and she had her first lessons with his own daughter under his roof."

5. In an amusing letter to his mother (30 June, *SL*, 1:279–280), Howells reported: "The peacock came yesterday just as the last course was put on the table; and we all ran down with our mouths full of hot pudding, and welcomed him. Fifteen minutes later we wished him to Jefferson or Jericho. . . . I was expected to do everything. P. thrashed about with his wings and kicked. Winny cried, Elinor danced up and down with fury at my awkwardness. . . . He waked us at dawn by a wild shriek which seemed to come from under our bed and made Elinor think I was beating Winnie to death. . . . We are very grateful to you for sending him, but I don't conceal that he is as yet a blessing in disguise which might be easily mistaken for a calamity. If you have any thoughts of giving us Charles, the horse, please consider our unprepared state."

6. Kitty Gannett Wells's brother was William Channing Gannett (1840–1923), a Unitarian clergyman and author. Back from Europe in 1867, he had written for the *North American Review.* He was ordained in 1868 and married Mary Thorn Lewis of Philadelphia in 1887. Kitty Wells was a girlhood friend of Elinor.

7. Mrs. Lewis, their cook.

8. Orson S. Fowler (1809–1887), lecturer on phrenology and author of many semi- or pseudoscientific publications. Mrs. Lewis is the original for "Mrs. Johnson" (*Atlantic,* January 1868; reprinted in *Suburban Sketches*). Howells used this episode in his book, when he depicted the independent Mrs. Johnson who felt free to invite guests without her mistress's permission. "In this spirit she once invited her son-in-law Professor Jones of Providence, to dine with her; and her defied mistress, on entering the dining-room, found the Professor at pudding and tea there,—an impressively respectable figure in black clothes, with a black face rendered yet more effective by a pair of green goggles. It appeared that this dark professor was a light of phrenology in Rhode Island, and that he was believed to have uncommon virtue in his science by reason of being blind as well as black" (p. 24). Howells also used Esther as a model for Mrs. Johnson's daughter Naomi (pp. 27–28).

9. Probably Mary Palfrey (1838–1917), a daughter of John Gorham Palfrey (1796–1881), the Unitarian clergyman and historian.

10. Marianne G. Washburn, daughter of Emory Washburn (1800–1877), married Samuel Batchelder, Jr., on 20 June. Washburn was a professor of law at Harvard and politically active.

11. Perhaps a plant in the shape of a W (for William?).

12. WDH had asked his father for the elk-horn, which they wanted to use as a hat-rack (*SL*, 1:272). Joseph W. Whittier was a Cambridge carpenter.

13. Elinor's cousin Edwin Doak Mead (1849–1937). The son of Bradley Mead, of Lexington, one of Larkin G. Mead's brothers, he was with Ticknor & Fields. His sister was Anna Maria Mead (1847–1932).

14. For Miss Przemysl, see 24 June 1863 and 26 November 1865. In his letter of 25 February 1867 to Elinor (Harvard), after giving her the latest news of Venice and of some of their Venetian friends, Eugenio Brunetta has this to say of her: "*Miss. Przemizl.* has married Count Capograsso but it was al very unhappy marriage because Mr. Capogrosso a few days after fell sick and was oblized to go to Dalmatia with his wife, and there he is in bed very dangerously ill. The poor bride is now converted in a *Soeur de Charité* and the bridal bed is converted in a bed of thorns. Her uncle is very sorry fort that, but he told me that he had advised her not to marry that good but unhealthy gentleman." WDH quotes this last phrase in his caption under Capograssi's photograph in the album at Harvard.

15. William C. Howells lost the nomination to Samuel Galloway (1811–1872). Rutherford B. Hayes was elected governor of Ohio and inaugurated in January 1868.

16. "Frightened Eyes" finally appeared in *Our Young Folks*, May 1868.

17. This was the Rev. Nathaniel Mighill. Called from his East Cambridge church, he served as pastor of the Centre Congregational Church in Brattleboro from 1867 to 1875. Elinor's brother, Charles Levi Mead, was an active member of the church, where he served as Superintendent of the Sunday School, as clerk, and later as deacon.

To Aurelia H. Howells

Brattleboro', Sept. 7th/67.

Dear Aurelia:

We were quite set at rest about Jo by hearing from himself a few days ago. I hope this fever will so change and renovate his system (I know fevers do often leave people in better health) that he will be quite a new man after it. Of course his sickness has put your Mother back somewhat about her preparations for her visit. We are expecting a letter every day, fixing the time of your Father & Mother's visit. I hope it wont be put off very long. . . .¹ I have enjoyed Brattleboro very much this time. Have seen all my relations nearly, and all my old friends. Heretofore when I have been home since my marriage I have always been so busy with my own affairs that I could not enter into Brattleboro' society at all—but this time we have been to tea-parties and pic-nics and on walks and drives a plenty—to be sure mostly in our own family circle. But that is large this Summer. Joanna & her child & nurse board with Mother at the Water Cure, and Molly is home,² & Willie, too, just at present—though he goes off next week to be usher in a school at West Chester, New York. Gus Shepard comes up frequently & tonight has brought Elliott up with him. We have had a host of cousins boarding at the Water Cure, and Joanna's friends the Leverichs, a party of seven visited her here last week. (Did n't you meet two Mrs Leverichs at Joanna's?[3] They were speaking of seeing Marietta the first night she arrived, so you may have met them that same evening?) Arthur Sedgewick stopped here three days on his way to Ashfield, and was "made to hop," as Winnie says, while in town. He was taken out to tea the evening of his arrival, and dragged over to Chesterfield[4] to a picnic and made to go through all the Brattleboro figures—which of course he did with proper dignity— and was thought by some to be a trifle stiff—but *Annie* knows better than that. He had his face sketched, while here by me, and by that eminent young artist* Fanny Channing—the young lady Willie James wrote so sarcastically about to his sister Alice[5] in a letter read aloud to us one day in Cambridge.** She is daughter of W. H. Channing,[6] who has been preaching so long in England, and has come back to America *so* English! Her father is this radical Unitarian, but her mother is a devoted Episcopalian and so are all the children. Fanny is a beauty & has been much admired in England and has twice been engaged, there. She is also very accomplished, and last Winter made great havoc among the Boston beaux. But Willie James having had it announced to him, by the lady who was to introduce him, that "men just *fell* at her feet" kept rather shy of her it seems. We invited Harry James to come up here, but he does not like to be long away from home—so refused.

Winnie is demoralized by life at a watering-place, and expects to be taken driving or over to "New York" (as she has persisted in calling the Water Cure ever since we came—probably because I have always told her aunt Joanna lived in New York & now she finds her there) all the time. She is in fine health; and is constantly saying queer things, so that she has a reputation for it all over town.

I am feeling quite well today, but have been ailing for a week past from *cutting a wisdom tooth*! At my age to be teething! But

*She is going out to Italy to study this Winter
**Annie must interpret

really the last one has but just come through, and the doctor pronounced it the whole cause [remainder lost].

Manuscript: Harvard

1. Five lines or about sixteen words are undeciphered because of crosswriting on the transparent paper.
2. Mary N. Mead had returned from New Haven, where she had been treated for her eyes.
3. See 24 December 1865 for Charles and Matilda Leverich. The Shepards' daughter had been named after Joanna's friend Mrs. Leverich. Marietta and Larkin Mead had arrived from Europe on 20 March 1866.
4. Chesterfield, N.H., where the Meads had lived before settling in Brattleboro in 1839.
5. Alice James (1848–1892).
6. The Rev. William Henry Channing (1810–1884), Unitarian clergyman, reformer, author, and editor. A preacher in Liverpool, he lived mostly in England, with occasional visits to America. Fanny G. Channing, one of his daughters and a cousin of Lizzie Higginson, later married Sir Edwin Arnold.

✣ IV ✣

Vivacious and Amusing
as Always
1868–1872

*T*hese *were busy and mostly happy years for Elinor. On 14 August 1868 John Mead Howells was born, and on 26 September 1872 the third and last child, Mildred, was added to the family. By May 1870 the Howellses had moved into a rented house on Berkeley Street, near the center of Cambridge (they were unable to buy it). They began to build a new house on nearby Concord Avenue in September 1872. At Berkeley Street they gave their first big party, an occasion when their neighbor John Fiske described Elinor as "vivacious and amusing as always." But she lost her father in 1869, and Will's mother had already died the year before that.*

Though Mrs. Mead and Mary lived with the Howellses frequently after Mr. Mead's death, their most regular visitor outside the family was Henry James, who for a while came every other Sunday night for supper and by 1870 came every Sunday. Elinor found it a little odd that in spite of their long intimacy with him, "we do not know him a bit better than the first time we met him." Still, she read his stories as they appeared, nearly all in the Atlantic *at this time, and mentioned them more frequently than those of any other author in her letters. She also singled out James's review of Will's* Italian Journeys *as the really important one.*

Will himself was turning to fiction. The Wedding Journey *began its* Atlantic *appearance in July 1871, the same month that he became editor of the magazine. Two of their summer trips on the St. Lawrence River in 1870 and 1871 had refreshed the background for this and for* A Chance Acquaintance, *finished while they were vacationing in Princeton, Massachusetts, in 1872. This novel made even more use of his sister Annie than the first one had. In writing to Annie about her impending trip to Boston, Elinor promised her that though the story would not come out until fall, "you can hear it read when you get here, and see how you like yourself." Significantly, in contrast with this identification, in none of her extant letters does Elinor even hint that Isabel March, the wife in* Their Wedding Journey *and several other novels, is a fictional portrait of herself. Her not having done so at least brings into question the widespread assumption that awards her this dubious honor.*

To Anne T. Howells

Cambridge
Sunday, Jan. 19ᵗʰ/68

Dear Annie:

In the first place please thank Grandma Howells from Winnie for the gift she sent, and give her the enclosed letter, which expresses the same thing in another way. The enclosed bit of cloth Winnie got up and put in herself without my knowing it, and when I just discovered it she explained that it was to show her sewing to her grandma. You see there are four or five stitches on it? When her grandma Mead was here¹ she praised her sewing a great deal, and I suppose she thinks Grandma Howells would like to see some of it. She wrote the letter this morning, did it up, sewed the cloth & cut it off and enclosed it and all without our asking her to, or without knowing we were going to write today. Now (though it is Sunday) I have given her one of several little bags which her grandma Mead basted for her while she was here, and a needle and thread, and you shall see the result. She sews over & over very well, and is very persevering about it. She has not yet decided what to do with the money her grandma sent, but she shall buy something useful and lasting when she spends it.

I am glad you like the photo. of me, (I thought it *might* horrify you) and that the few little things in the box were liked.

We like the idea of your going to Cleveland, better than to Columbus, to teach. I should think you might enjoy yourself teaching kindergarten in that pleasant city. Shall I tell you Miss Robbins'² prices? Two gentlemen, to induce her to come out & teach here, pay her rent (a room under the Cong'al church) and provide fuel. She asks $30.00 a year—or $8.00 for a single term (though she says she only takes by the year—but took Winnie a term on trial). She has now eighteen scholars, though she only began with twelve. She has a piano in the school room and for each child a little wooden chair [sketch] with a pocket or slide for books at the left side. These the children bring up and range aroung her in a semi-circle when they recite in their classes.* The gymnastics have to be done slowly and with precision and force of muscle (do you understand?) or they are worthless. Miss Elizabeth Peabody³ is now in Europe studying the kindergarten system there in order to bring it to perfection in this country. Mrs Mason⁴ did not talk of it as a common infant school by any means but as a very deep and original system. It was quite amusing to hear her talk.

I called on Annie Higginson last week. She is soon going to make a visit to her friend Mrs Gordon⁵ (*Mr.* G. is "Vieux Moustache" who writes in the Young Folks and Riverside Magazines) who lives at Newberg on the Hudson. She will stay about two months She got your letter, and spoke to Will of it the day after, but he forgot to mention it when writing home I suppose. She was delighted to get it. When she goes to Ohio, (which is uncertain still) she will certainly visit you she says. She will write you of course—but her Mother has been very sick so she could not yet.

Winnie has given out on the bag—in fact it is growing rather dark—but I will send it as it is. I tied the knot, but she set the first

*The rest of the time they sit in a row along one side of the school-room.

stitch, and once when she broke the thread started again without saying anything to me. I have not touched it myself.

Tuesday

We are in the midst of a great snow storm today—not so bad yet as that one last year, but if it keeps on tomorrow it will be.

Harry James' Galaxy heroine reminds me of a great many persons in general but of no one in particular.[6] Who does it remind you of? Is n't his story in the February Atlantic splendid? It was Harry James you know who wrote that notice of Will in the North-American. Is n't *that* splendid?

Mrs Child & Helen came in a carriage to call on me the other day. Helen is as beautiful as ever, but she is the Cinderella of the family, for the two others are not striking.[7] When they called here Helen had on a black velvet cloak, a white felt hat trimmed with cherry velvet with real lace tabs &c &c. She was rather stunning as you may imagine.

Have we told you that the whole Norton family is going abroad for four or five years next Summer?

My German ivy now reaches from the side of the front window of Will's room up over the top of the book-case, and from there on a string to the horns which it decorates most beautifully. We have had the cherry-tree nearest the Richardsons cut down, so my flowers will have a nice sunny bed next year—and we mean to buy that lot behind us besides, so there will be a great chance for horticulture.

This is an *awfully* stupid letter—but it must go.

Much love to all

Elinor

Will is very much driven or he would write. He is having a hard time writing a notice of the Agassiz' dull book.[8]

Manuscript: Harvard

1. Mrs. Mead had visited them in December (cf. *SL,* 1:289).
2. Either a daughter of the Rev. Chandler Robbins, successor of Emerson at Second Church in Boston, or Abba or Cordelia Robbins, daughters of his brother the Rev. Samuel Dowse Robbins of Waltham, who later conducted a private school in the father's house in Belmont (Belmont Historical Society). For Annie's interest in kindergarten teaching, see WDH's letters of 25 November and 15 December 1867 (*SL,* 1:287, 289–290).

3. Elizabeth Palmer Peabody (1804–1894), a sister-in-law of Hawthorne and a noted educator, who introduced the kindergarten system into America.

4. Possibly the wife of Luther Whiting Mason, a teacher, who was in charge of musical education in the Boston schools from 1865.

5. Clarence Gordon (1835–1920), author of juvenile books and of magazine articles, lived in Boston at the time. His wife was the former Frances Gore Fessenden, of Boston. In his letter of 15 December to Annie, Howells had suggested that she write Annie Higginson to invite her to Jefferson.

6. Marian Everett was the heroine of James's "The Story of A Masterpiece," *Galaxy*, January–February 1868. His story in the February *Atlantic* was "The Romance of Certain Old Clothes." James's unsigned review of WDH's *Italian Journeys* appeared in the January *North American Review*.

7. Helen Child's younger sisters were Susan and Henrietta.

8. The review of Mr. and Mrs. Louis Agassiz's *A Journey in Brazil* appeared in the *Atlantic*, March 1868.

To Aurelia H. Howells

[Cambridge] March 1ˢᵗ, 1868.
Will's thirty first birthday.

Dear Aurelia;

I believe we used to be acquainted, but it is so long since we have exchanged letters that I feel as though I hardly knew you. Is it because I owed the letter? I'll try not to be so careless again. But one always expects an *unmarried miss* to write twice to a *matron's* once. However I know it is different in your case, you all have so much to take up your time, with Henry and the store.[1] You will write a little oftener though, wont you?

We got back to Cambridge Wednesday night and feel now as though we had never been away—though we enjoyed our stay in Boston exceedingly and are feeling brighter for the change.[2] We took our vacation at this time, and mean to stay here all Summer. Sometime this month we are going down to Providence to see my numerous relations there. It is only forty miles, and we have never been yet. My sister Belle is coming to make me a short visit before moving to New Britian, Conn, the first of April. I told you my brother Charley was going there to be connected with the Stanley Rule Co., did n't I?[3] So poor Father & Mother will be left alone in Brattleboro without a child except Molly, who is usually off visiting. At present she is with Joanna, having been down there during

Elliott Shepard's wedding festivities.⁴ Have you seen any notices of the affair? The Boston papers all had them, so I suppose the New York papers did, and you saw them. At any rate it seems too absurd to write about. Maggie has a suite of rooms at home this Winter & next Summer they go to Europe.

Will's room has been greatly improved during our absence by having the white paint made of the same color as the wallpaper. It looks really like a library now. We have put the book-case under the horns, and arranged the room so it looks much better. The front stairs have been painted, and places here and there all over the house. I have fitted up the room out of the kitchen for a servant, putting the furniture of the girl's room down there, and our little Howe [?] bedstead into that room upstairs for a second girl, when I have one.

Winnie did have the measles in Boston we are sure. Drs Angell & Ainsworth both said she looked as though she had had it (you know we had a homeopathic doctor attend her, Dr Clark?⁵) and at the very time she was sick half the kindergarten out here were having it. She is in splendid health now, and good from morning till night. This morning she remarked that she should like to see Uncle Jo. Is nt he going to make his Spring tour before many weeks, and wo'n't he bring Eliza and little Marnie when he comes? I hope he will. We are going to have an early Spring, and Cambridge will be lovely in April. The robbins* have been here a fortnight.

Annie's "Jaunty" is very pretty. She must send us what she writes for the Chronicle. But Whitelaw Reid has declined being editor I see. I want very much to hear about Nell Holman's wedding.⁶ A friend of ours, Mrs. Pierce, was with us at our boarding-house a part of the time. She is writing a book on the woman question which Will thinks good. You shall see it when it is published. She is the Zena Fay I talk so much about.

*I said to Will "There's only one b in robins?" and he replied "Two" so I wrote it so, when he remarked "Of course she is n't a *bird*" showing he thought I meant Miss Robbins, Winnie's teacher. Served me right for asking such a question.

I shopped (bought cotton cloth and household stuff mostly) and made calls and went out to dinner and tea to any amount while in Boston. It seemed like the old times when you were there—only, it was a great satisfaction to know that boarding was only temporary.

> Much love to all
> Elinor.

Manuscript: Harvard

1. The office of the *Ashtabula Sentinel.*
2. In a letter of 26 January to his father (Harvard), Howells had written of their boarding in Boston for a few weeks, explaining that Elinor needed a change (she was expecting her second child in August) and a "rest from housekeeping," and that this would give them a chance to get rid of their servants, who had become a nuisance. In her pocket diary (MS, p. 30), Elinor wrote on the same day: "Mr. Howells' 31'st birthday. Life insurance pending. Have been boarding in Boston at 13 Boylston Place for the last five weeks—dining at the La Grange restaurant. Zena Pierce with us the last week, writing "Coöperative Housekeeping." We sent Bridget away when we left, and took Mrs Davenport, an American woman when we returned last Wednesday (the 26th Feb.) but she wont do [she was an opium-eater, Elinor indicated elsewhere], and leaves tomorrow. T. & F. are going to raise Mr Howells' salary from $2500 to 3500.00."
3. Charles was a director and treasurer of the Stanley Rule & Level Co.
4. Elliott Shepard and Maggie Vanderbilt were married on 18 February 1868.
5. The homeopathic doctor was most likely Dr. Luther Clark, of Pinckney Street.
6. Rejected by *Young Folks* (*SL,* 1:292), "Jaunty" was published in the children's magazine *The Little Corporal* in 1868 (Doyle, p. 21). For Whitelaw Reid, see 28 May 1866. Nell Holman was probably a girl from Jefferson.

To Lilian W. Aldrich

> Cambridge
> Thursday, Sept. 17th [1868]

Dear Mrs. Aldrich:

We seem to be utterly separated by circumstances, but I think of you constantly and long to hear something about you. Possibly I may get in to see you (if you see anybody these days) before your confinement, but if not I suppose we shall meet some time in the Winter! and then we can compare notes.[1]

Will you accept these little shirts, (which I had knit at the same time with my baby's, by a woman here in Cambridge) and believe that the gift though simple carries a great deal of love and sympathy with it.

Do name your baby beforehand and not leave him or her in the exceedingly awkward position our boy has been in for five weeks of having no name. Mr Howells calls him "Brother" for want of anything better.

My nurse leaves on Monday and then I assume the whole care of the infant.

An Irish woman who lives near by comes in to nurse him morning noon & night and I feed him arrowroot &c the rest of the time.

Hoping to hear good news of you

Affectionately

Elinor M. Howells

Manuscript: Harvard

1. Mrs. Aldrich gave birth to twins, Talbot and Charles, on that very day. To WDH's note announcing "I have a fine boy," Thomas Bailey Aldrich replied "I have TWO fine boys, born yesterday morning! (18 September; quoted in Ferris Greenslet, *The Life of Thomas Bailey Aldrich* (1908), p. 89). Of John Mead Howells' birth, WDH had written to Lowell on 15 August, "we had a son born yesterday afternoon about five o'clock— an immense boy with an hexameter voice, and a highly poetical disposition to use it. His mother is very well." And, to Charles Eliot Norton in Europe: "I am wracked by too great good fortune—I am dashed to pieces on the Happy Isles. . . . We have a boy" (SL, 1:297).

To Aurelia H. Howells

Cambridge, Dec. 5ᵗʰ, 1868.

Dear Aurelia:

How I wish you and all the rest could have been present at the baby's christening this morning! Mary Mead was here of our family—and now I wish we could have had the christening whilst Joe was with us—though baby was sick then, and besides we did not then think of having it done for a long time. But we've concluded to go in to Boston to board for six months you know, and I did not like to wait for it till we came back.[1] So we invited the Ashburners, Childs, Palfreys and a few others to be present at the ceremony, and it took place this noon in our parlor.[2] The font or silver vase, was brought from the Church, (my church) and placed on the little table Joe gave us, before the mantlepiece. We had a few flowers from the hot house behind us, and the room looked very pleasant. Mrs Mc Laughlin the nurse brought the baby (dressed in a *Paris* robe, all embroidered on the skirt which my aunt Patton³ gave me, Mrs J. T. Fields' elegant little shirt & gold armlets) to the door, then Mr Howells took him and held him while Mr Mc Kensie prayed, and then the minister took him and baptized him by the name of John Mead. While Will held him he looked about and smiled at the people, and during the whole affair he did not cry a bit. He has splendid fat arms and shoulders and a thick head of hair which he wears parted on one side, and one dimple in the left cheek, just as Will has—and bright blue eyes and a fair skin—and is altogether a nice-looking baby, and everybody of course praised him, and we were very proud. He weighs over fifteen pounds, and Will says his arms ached holding him.

After the ceremony we had coffee & cake and grapes served, and a pleasant chat all around, and then they went home. I took leave of them all as I am not able to go around and make farewell calls. I don't believe my sickness is chronic and I expect when I have been to Brattleboro' I shall be as well as ever. We now intend going to Ohio next Spring, and shall save money for it all Winter. Perhaps I shall leave the baby with Mother if she will assume such a care. I do not think there would be much danger in leaving him early in the Summer, with so good a nurse, and Mother to look after him—but how could I take a nurse way out there—and how could you have her there! I cant tell now how I shall feel by Spring about it—but we shall certainly try in every way to visit you in May—& Will & Winnie will go even if I cannot. Molly sends much love and is going to write to you whilst she is here. She is in good health this Winter. Tell Annie Dr James⁴ inquired for her the first time Will saw him.

I cannot write any more now as I am making an inventory of everything in the house for the man who comes in. Much love to all

Elinor

Manuscript: Harvard

1. Joe had visited them briefly in November, shortly after Mary Dean Howells' death on 10 October (*SL*, 1:302–303, 306). In his letter of 29 November (Harvard), Howells

had told his father of their intention to rent their Cambridge home and board again in Boston for the winter, "Elinor's health is quite broken up, and she must have absolute freedom from care for a while," he had written, adding that it would be more economical than keeping two girls in addition to John's nurse. The plans were for Elinor to go to Brattleboro for a few weeks to recuperate, while her sister Mary took care of the children in Boston. Three days after they moved to Boston on 8 December, Elinor went to Vermont, where she stayed until January 4. They did not go back to their Cambridge house until mid-March. While they were in Boston, Elinor took drawing lessons to illustrate a children's book about Venice based on Winny's first year. This was a project of Howells that he never realized (WDH to his father, 7 February 1869; *SL*, 1:316). On 28 March, Howells wrote his father: "Elinor does not gain strength rapidly, but she is better than she has been since her miscarriage" (*SL*, 1:319).

2. In her pocket diary (5 December, typescript, p. 29), Elinor gives the list of those present at John's christening, among them Mrs. Palfrey and her daughters Anna and Mary, Mrs. Child, Miss Ashburner, and the new minister of the First Church in Cambridge, Congregational, Dr. Alexander McKenzie (1830–1914).

3. Aunt Emily Hayes Patton. See 23 June 1864.

4. Annie Howells and William James had met and become friends during her visit to Cambridge in the spring of 1867.

WDH to EMH

Cambridge, June 25, 1869

Dear Elinor:

The satchel came to-night, and I had an awful pang of wife-sickness when I took out your dear old drab dress. I've got such ideas of your thinness lately that perhaps I half expected to find you in it and was disappointed. However, it really comforts me more to think of your being with your father than it would be to have you here; and what I want is to be with you there. I've rejoiced that in spite of care and anxiety you keep well, and even think Brattleboro' air is doing you good.—To-night I have no report from you though your letter of Thursday morning came through by evening. I hope your father still improves. He has battled harder for life than many a younger man could have done. His survival is wonderful. Was he sensible enough to know of our visiting the old Lexington home?[1] You must write me all particulars. How glad you must all be to have Joanna with you; and how glad she must be. In many things I can see even at this moment that you are a greatly favored family. If this should be your father's last sickness, it is of such gradual course that you can almost accompany him to the verge of the other world. I hope you think of the singularly peaceful and happy frame of mind he is in, and take comfort from his tranquillity whatever are your expectations as to his recovery or his loss.[2] Don't think my preaching too odd—'tisn't often I do preach. Aurelia and father keep saying about Johnny, "How much he looks like his grandfather Mead!" and the likeness is easily seen. I hope he'll have the same sunny disposition.

Winny sends you another letter. It is all of her own composing, of course, and the first clause—which is hard to read—is particularly droll: "Winny sent some flowers to Mamma, hoping you will thank me."

To-day Johnny had an envelope which he threw on the floor, and then he wriggled off my knee in order to stamp on it. Winny and he play a good deal together.—Mrs. Mac's Willy[3]—a very large, handsome boy—was brought by Mrs. Callahan today, and staid till after dinner. I think she loves Johnny best, almost.

—The weather's getting very warm. May it be good for your father.—Father and Aurelia went to Salem to-day, and Winny and I are alone; she kneels down before me waiting till I finish this letter when I'm to read to her.

Your W. D. H.

Manuscript: William White Howells

1. On 23 June WDH had visited Mr. Mead's birthplace in Lexington; he sent flowers gathered there by Winny.
2. Mr. Mead died on 6 July.
3. Mrs. Mac's baby lived in Charlestown, possibly in the care of Mrs. Callahan. Elinor was still in Brattleboro when Willy died unexpectedly (cf. WDH to Henry James, 18 July, *SL*, 1:330). Howells went to the baby's wake, "a wonderful contrast to the scene, I had just witnessed at Brattleboro', where ages of Puritanism had strengthened and restrained the mourners from every display of their grief." Of his father-in-law, WDH said in the same letter (p. 329): "I have been to Brattleboro' to see laid in the ground all that was left of the kind, cheerful, simple old man. He was one who felt so friendly toward the whole world that he imagined it a good one, and led the very happiest life here."

To Victoria M. Howells

Cambridge
Sept. 11<u>th</u>, 1870.

Dear Vic:

Today we are in a torn-up state generally—half our carpets being up and curtains down and nothing in its place—but next Sunday we hope to be peacefully settled in our new home on Berkeley St. We made this sudden move because a house which we had come near taking once before, when it was snapped up, came by chance into the market again. We have taken it for three years with a half-promise from the lady who owns it that we may have it at the end of that time or before, if we like it.

I have drawn a very imperfect sketch of the lower floor by which you will see there is an upstairs kitchen and all is very convenient. There is a pretty door in to the garden out of the dining-room and a large bow-window front of each parlor. There are only wire fences between our yard and our neighbors' yards on either side— one of whom is Mrs (Mattie Griffiths) Brown and the other Ed. Everett's son. Opposite is Richard Dana's house so you see we will have good neighbors?[1]

Mother & Mary are to consider our house their home after this— though they will be away a great part of the time. Mary is coming first, and when she goes back to Jo's Mother will come. We have *two* guest chambers in this house so you see we are more ready for you than ever.

I have one of your silver spoons which I forgot to send by Annie—and why didn't we send Mary Smalley's book by her?[2] Tell Henry Elgner[3] thinks a great deal of him and praises him to everyone. He must not forget her and if he will come and see her she'll keep all the dogs away.

Love to all
Elinor.

Manuscript: Harvard

1. Mattie Griffith was the wife of Albert Gallatin Browne (b. 1835), a lawyer. He had lived in the Boston area since his marriage in 1867, later becoming editor of the New York *Post* and *Herald*. William A. Everett also lived on Berkeley Street, at the corner of

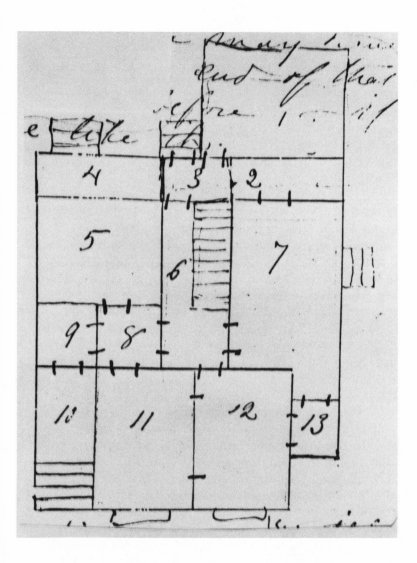

1 kitchen
2 chimney cupboards &c
3 back entry
4 piazza
5 library
6 hall
7 dining-room

8 vestibule
9 outside-vestibule
10 piazza
11 & 12 parlors
13 glass room for plants

Craigie. Richard H. Dana, Jr. (1815–1882), attorney and author of *Two Years before The Mast*, was a nephew by marriage of H. W. Longfellow, whose house was also nearby.

2. A Jefferson friend, Mary Smalley is mentioned in a letter of 4 April 1867 (Alfred). After the Howellses' summer visit to Ohio, Annie had accompanied Will and Elinor and their children on a trip from Niagara Falls to Lake Ontario and down the St. Lawrence. Howells was collecting material for his new novel, *Their Wedding Journey* (*Atlantic*, July–December 1871).

3. Henry's name for Elinor.

To the Howells Family

[Cambridge, 20 November 1870]

[Beginning lost]

his imprudent way of reading their letters about town.[1]

Will is somewhat sick of lecturing.[2] Mr Lowell thinks he will be invited to lecture at Cornell university next May. There are only four more lectures to be given in the Lowell Institute course. The lectures are scarcely of the kind to interest the kind of people who usually go there and Will feels that they are not popular. Prof. Child said "Now these lectures are just the thing for the Lowell Institute;" and it was he who got him the place—and if they are not popular it's not Will's fault. They would be just right to deliver at Cornell.

Brett Harte has sent Will a nice photo of himself and a nicer letter with it.[3] The greatest compliment Will ever got has just come. Lucia Dwinell,[4] a friend of mine, is teacher in the Connecticut Normal School. The headmaster told his pupils the other day, that for style they should read Addison and——*Howells*!!!!

Did I ever tell Annie that Annie Higginson is at times insane—and is more and more a care to her family every year—I never found it out till since her father's death when she has been worse.[5]

What *do* you all think of Harry James "Travelling Companions?"[6] Doesnt he like to touch on delicate subjects? He comes here nearly every Sunday night to tea—and is really quite polite to Molly & me at times—but poor fellow, I excuse him everything, now I know how sick he is! he told Will all about it lately, and it is enough to make your heart ache. Mr McKenzie's baby was christened in church today: Kenneth McKenzie.

Arthur Sedgewick has just been out to Chicago and to Pomeroy to visit Dana Horton.[7] He likes Ohio ever so much and the West generally. He said he noticed when he went through Ashtabula.

Dont you like "Flitting"?[8] Aurelia is having a grand time, isn't she? Give her my love and tell her Moses P. Grant is engaged to a Miss Greeley, whom I know, of Cambridge. How *could* she![9]

Annie, as you see how worldly all Harry James' heroines are, and how he evidently likes that sort of girl, dont you change your opinion of Miss Blunt?[10]

How are Alice and Lizzie[11] getting on? Was it Lizzie's beau who gave Clara Dean that elegant present? How's Pickett? Johnny still shows how he will pinch him. Do tell me all the news. Winnie was greatly obliged for the dish cloth and the samples of Henry's sewing. Tell him we shall be always glad to have him send us writing, sewing or anything to show what an industrious boy he. How strange it will be to have Sam gone again.[12] Do give Florence my love and say goodbye—though I dont know that I shall be any further off from them where they are going. Will sends love and will write soon. Booah[13] is talking more and more. Winnie enjoys her school[14]

Love to all

Elinor

It was a pity Joe did n't give us a call this time just to see the new house.

Manuscript: Harvard

1. This letter is dated from the baptism of Kenneth McKenzie. We are indebted to Joseph Chamberlain, President, Shepard Congregational Society, First Church in Cambridge, Congregational, for assistance in this matter.

2. Having previously lectured at Harvard, WDH began his twelve lectures at the Lowell Institute on 26 October. Reporting on them on 30 October, he had written, "My audience is about two hundred and fifty, which is much larger than the usual audience: in fact, many courses are delivered to twenty-five or thirty people.—Perhaps the girls will care to know that I appear in evening dress, and Elinor at least thinks I'm very 'pretty-looking'" (*SL*, 1:358).

3. In his letter of 5 November 1870 (*LinL*, 1:158–159), Harte addressed WDH as "most excellent writer of excellent English" and promised to visit him in Cambridge later in the winter.

4. Lucia H. Dwinell was the daughter of Deacon Aaron E. Dwinell of the Congregational church in Brattleboro. She later married Edward Chase of New Britain, Connecticut.

5. For Annie Higginson, see 19 November 1856. Her father, Stephen Higginson, had died 11 August 1870.

6. In the November–December 1870 *Atlantic*. Henry James had returned from Europe in the spring.

7. Samuel Dana Horton (1844–1895), an economist and author. Born in Pomeroy, he was a Harvard classmate of Sedgwick's and like him a lawyer.

8. First published in the December 1870 *Atlantic,* "Flitting" is the last chapter in Howells' *Suburban Sketches,* which came out on 19 December.

9. For Moses P. Grant, see 28 May 1866. Miss Greeley was perhaps the daughter of Mrs. Myra A., or of William Henry Greeley, a commercial merchant, who lived on Harvard Street, near the Square.

10. Annie disapproved of the way James had treated Esther Blunt in "A Landscape Painter," as she wrote her mother and sisters on 17 February 1867, when she was visiting in Cambridge.

11. Alice and Lizzie, as well as Pickett (perhaps a beau of Aurelia) are unidentified. Clara Dean was most probably the daughter of Will's uncle William Dean of Pittsburgh. For information on the Deans we are indebted to Dahlia Armon, the Bancroft Library, University of California, Berkeley.

12. Sam and Florence were about to move to a new home.

13. Nickname for John Mead Howells.

14. A home-school meeting at the F. J. Childs', it was taught by Miss Mary Olmstead. Besides Helen Child and Winnie, there were half a dozen more pupils, including two little boys (Elizabeth Ellery Sedgwick Child's journal, Harvard).

To Aurelia H. Howells

Cambridge
January 29ᵗʰ [1871]

Dear Aurelia:

I think your photograph is *lovely,* and we are ever so much obliged for it. What a wonderful artist the Pittsburg man is—even beyond the Salem man it seems to me![1] If Will only had one taken by him to give Mr Clark[2] now—to be engraved for Every Saturday! The only one we have—that which you have—is thought to be too poor to engrave; and Will's beard is in a state at present to prevent another's being taken. Otherwise within a few weeks you might have seen your distinguished brother's face on the first page of Every Saturday. Even now Linton, the engraver has the photo for further consideration, so there is *a* little hope. Did you like Brett Harte's face? There is a perfect furore in cultivated society now about Bret Harte.[3] All the young ladies are in love with him—but it is of no use—he is married.

You are always writing Aurelia and saying you'll write again soon and *tell all about it.* Now I am not going to let you off this time.

You *must* write more about what you did in Pittsburg—the parties you went to, and the conquests you made. Or did you remain faithful to the youth who saves on cravats? I never hear one word of him now. Did things come to a crisis, or is n't he there this winter? I suppose Lucia is fairly off by this time.[4] Did the velvet match as well as you thought it would in the little piece. Please tell me—for I am afraid it was not a *perfect* match? We got Annie's letter enclosing the money order. I thought Puck[5] would suit Henry. I am *delighted* to hear that Henry always wears his slippers, and if you ever write that he goes barefooted I shall turn at once into a statue of Niobe and weep forever.

Booah is getting too cunning for anything, Aurelia. Will makes himself perfectly silly over him—perhaps because he has turned out much prettier and brighter than we expected. His mind is very clear for a child of his age, and he never makes a mistake in speaking without stopping to correct it. This morning he was trying to say "In the baker comes and snickers, when he sees the sugar-lickers" but he said "In the bakers;" whereupon he stopped, gave a little laugh of derision and corrected himself. Just now he pointed at something and said "O, funny!" I said "I dont think its very funny, do you Winnie?" Booah said instantly "Booah no say *berry* funny." His speaking of himself in the third person and not being able to use the negative rightly are the only mistakes he makes in talking now. Will thinks him a remarkable talkish. Winnie goes to school regularly, and knows more about geography already than her parents. Just now I am suffering as much from teething as Booah was last Summer. Actually at thirty three I am cutting a wisdom tooth; and have almost constant head ache and back ache on that account. Dr. Angell says I shall live to be very old I *develope so slowly!* Last night we were in at the Angells' to tea—and what do you think?—Mrs Farwell was out at Arlington, this cold weather, on a visit! I had never seen her except in bed for years and it was *startling* to go into her room and see the bed made up and nobody there. She is very much better, and really enjoys life a little now.

Perabo[6] and his mother were there by accident. Mrs P. is a plain, uneducated woman who speaks broken English. Perabo is more original than ever; and I heard him play for the first time. He played variations for a quartette from Schubert, and then what is called a minuet in A minor from the same—certainly the most melancholy music in the world. He said "These two are my adopted children. Beethoven is grander but these go deeper than anything else" and then he compared Beethoven's music to God and Shubert's to Christ to explain the difference. He went off upstairs in to his room (they have given him a room to use when he pleases, though he lives out at Mattapan with his parents) and Mrs Angell going past asked him why he did not come down. "O you are all so happy—you dont care for the unhappy" said he in the most childish way. Dr Angell still buys pictures, those same three girls still live there, and nothing is changed from five years ago.⁷ Count Rumford, whose life is about to be published, married a great-aunt of Mrs Angell's. I believe I have told all the gossip I can think of.

We enjoy the house as much as ever—and I am quite successful with my plants. Tell Annie Will did not laugh at all at the Rochester entreprise, and as he will be sole editor of the A. by the time her article is finished I dont see how he can help accepting it.⁸ I think it is an excellent idea. There is a full explanation of Vigée Le Brun the artist in the last Galaxy.⁹

Much love to all

Elinor

Manuscript: Harvard

1. Aurelia had gone to Salem with her father in June 1869. She had just returned from Pittsburgh.

2. For John Spencer Clark of *Every Saturday*, see 11 June 1866. WDH's picture was not used. William J. Linton (1812–1897), the printer, was also a poet and an illustrator.

3. Francis Bret Harte (1839–1902).

4. Lucia is unidentified. The "youth who saves on cravats" was perhaps Pickett.

5. Perhaps Elinor's nickname for Henry.

6. Johann Ernest Perabo (1845–1920), concert pianist and composer, who had made his debut in Boston in 1866.

7. A professor of ophthalmology and president of the Boston Philharmonic Society, Dr. Angell was also an art collector, amateur painter, and author of papers on art, literature, and technical subjects. Sir Benjamin Thompson, Count von Rumford (1753–1814), physicist, essayist, and philanthropist.

8. Annie's article on the Rochester enterprise was apparently not published. Howells was to become editor of the *Atlantic* as of 1 July 1871.

9. Elisabeth Vigée, dame Lebrun (1755–1842), the portrait painter, author of *Souvenirs*. J. Durand's "Madame Vigée Le Brun," appeared in *Galaxy*, January 1871.

To *"Dear Girls"*

Cambridge,[1]
Friday [17 March 1871].
Dear Girls:
I have been trying to get time, ever since the party, to tell you about it—but Mother has been sick, and Johnny has been sick and I have been very busy taking care of them. Now they are both well, though not out of doors yet.[2]

You know how we happened to have the Hartes here? Will had a pleasant correspondence with Bret, and when he said he was coming east—last Summer, some time—invited him to visit us when he was on, knowing nothing of his family. Mr Harte did not come then, but later wrote that he was coming to Boston with his wife and two children, and, of course, Will repeated his invitation including the family. Strange that Mrs Harte should have liked to go into a family in that way, without even a personal invitation from me!—but they are easy-going people, and she did n't even attempt any apology or make any remarks on the subject after they got here. Her children were two boys of eight and seven—remarkably well-behaved, or I dont know what we should have done. We liked Mr Harte exceedingly, and Mrs Harte too. He is bland, modest and good natured—quite elegant in his manners, and Mrs Harte is stylish, dignified and sensible to a degree. Of course they are not quite au fait in every thing, but they give you the idea of polished, cultivated people. Her family I should say was superior to his—but there were fourteen children of them (her father was teaching in a girls' boarding school in San Francisco when he [Bret] married her). She has a fine contralto voice and always sang in Starr King's Church.[3] Mr Starrington, her brother-in-law was Starr King's predecessor. Mr Harte is not as good looking as the pictures make him out, and Mrs Harte is positively plain.

There had been a perfect furore over Bret Harte's writing among nice people here, and he was received with open arms. One day he dined at Lowells', the next at Longfellows' and the next they both dined at Agassizs'. One evening we went in to the Fieldses'—but they had to refuse invitations from every quarter—especially he, from clubs and associations.

Do you admire his writings greatly? I have never heard any of you say a word about them. It is the opinion of Lowell, Howells & others that there is a great deal in Bret Harte and he will do even better things. The visit went off splendidly—but *the party!* How shall I do justice to it? You know we've been here five years accepting civilities and never done much in return, and this gave us a grand opportunity to really give our friends a treat—for everybody was curious to see Bret Harte. This is the way I wrote my invitations:

Mrs. Howells requests	It took everybody
the pleasure of	by surprise
Mr Longfellow's company,	as I had
on Monday evening Feb. 28ᵗʰ4	my party
at 8 o'clock, to meet	Monday and
Mr & Mrs Bret Harte.	he only arrived
Berkeley St.	in Boston Saturday.
Feb. 23ᵈ	Sarah Sedgewick[5]

wrote "I cannot write a formal reply to your *thrilling* invitation for Monday." and all were pleased—and nearly everybody we asked came. Mrs Fiske[6] got up off a sick bed, and the high-church people all made an exception of this occasion. Was n't it rash though, to send out the invitations Friday morning before I really knew if they would be here? As it was their engine broke down and they had to stay over night on the way, instead of being here Friday evening as they intended. But there were a great many parties on the [word] and I wanted to secure my guests.

The reason the party was a success was, that we being new people could bring together different sets, making more variety than there is generally in Cambridge parties. (We introduced very generally, but didn't dare present the Fields to the Sedgewicks, for instance—people are so snobbish here![7]) I believe we did not give offense anywhere, and the supper was very nice. Smith the caterer provided it at *a dollar and a half a head!* What do you think of that for William? It is as cheap as we could have got it up in any other way, buying extra dishes &c. The man brought linen, silver, dishes, coffee, chocolate ice-cream, salad bread and cake. Afterwards he & another man washed up the dishes and took them off—and at twelve o'clock all was quiet—or perhaps a little later than that. The company was all off by twelve. The conservatory looked very pretty as my German ivy is all over it, and I got great praise for my *plants.* Is n't that rich? Mrs Ernest Longfellow wanted to know *what I did* to make them grow so, when she came to call. Mrs E. Longfellow is a beauty, and she was dressed in an embroided tunic over pink that night.[8] She & Alice James sat down on a cricket and on a bowseat in the conservatory at supper time and Perry and Sedgewick waited on them. Finally S. stretched his stiff length out at Mrs. L's feet, and Dennett seeing it, said to Mr Howells "Look at the pastoral Sedgewick!"

The bashful Perabo came—flattered because we have no piano. Osgood seemed to enjoy himself greatly and was introduced to all the young ladies: Miss Longfellow, Sara Sedgewick Mary Palfrey &c The "Middy Bear"[9] was, pr—*prev*ented from coming by a cold. The Hartes said they had a pleasant evening and seemed to like Cambridge people. Harte has accepted Osgood's offer,[10] and it is quite possible they may come to Cambridge to live—but I trust they dont expect this ovation to be kept up, for a duller place than Cambridge usually is I dont think they could well find.

Last week arrived by express a big boarded-up thing which contained a rustic basket about five feet high—a present to me, for my conservatory [sketch] It is to be filled with earth. There is a little basket at the top of the arch and a hook underneath for a hanging-basket. I shall not fit it up this Spring—so have put pots of plants in it and it looks very pretty. Mr Harte's note with it was charming. He said he had sent the basket "which you will greatly please my wife and myself by taking into your house as trustfully* as you did us. When it is planted with the tender eglantine and lush** rose let its rough angularities and rude sinuosities remind you of the California strangers around whom your courtesies

*This is the first suggestion on their part that there was anything strange in their coming as they did.
**a joke between him and Will I dont understand

twined so gracefully. The plain prose of which is my dear Madam that we would like you to associate some of your pleasant life at Cambridge ever so remotely with your friends,

Frank and Anna.

Bret Harte"[11]

Don't think because I say to *you* that I think it was queer Mrs. H. could come into a strange family so easily I say it to anyone else. No indeed. You are the only ones I have gossipped to about the party or the Hartes—but I was sure it would interest you to know all particulars. I will enclose a piece of a dress I wore on the occasion. I got it last Fall. It is trimmed with fringe & moss trimming. Lucia's dress, by the way, was *magnificent,* and I should think she would look magnificent in it. After writing all I have I have n't said half.

Will called on young Ranney,[12] but he never returned it. We are going to have him here to dinner or tea one of these days. Vic, your jam is appreciated though little has been said about it. We had it today in this way: Rice cooked in cups, then *cored* and jam put in. With sugar & cream it is a nice fresh dessert The potatoes we had from Ohio have lasted till now, and they are the best potatoes I ever ate.

Aurelia a *part* of this letter is an answer to your letter—so you must write again—and why so mum on Pickett?

Annie you notice Harry James has acted on your advice and left out the objectionable heroine altogether in A Passionate Pilgrim?[13]

I wish you girls one and all could have been at the party—but that's of no use. I'll make a party for you sometime. This house suggests party all the time—and indeed we have entertained a great deal more since we came here. I think of cutting Winnie's hair off. Have you any objection? Booah is so good and cunning these days that we feel like eating him up all the time. I hope Henry is as good as ever, and keeps his shoes on.

Love to all

Elinor

List of People invited to Bret Harte Party.[14]
*those who didnt come

Mr Longfellow
Miss Alice "
Mr Sam[1] " (brother of the poet)
Mr & Mrs Ernest Longfellow
Mr J. R. Lowell
*Mrs. " (out of town)
Miss Mabel " (engaged)
Mr and Mrs Mc Kenzie
Mrs Hopkinson
Mr & Mrs Saml Batchelder
Prof. & Mrs Shaler

" " " " " " Cooke
Mr Dennett (Nation critic)
Mr Adams (Charles Francis A's son)
Mr Perry
*Mr Henry James (sick)
Mrs " "
Miss Alice "
Mr H. James jr.
Mr Wilkie "
*Mr Willie " (sick)
*Prof Agassiz (sick)
Mrs "

Mr & Mrs Alex Agassiz
Mrs Wm. Furness
Mr & Mrs John Fiske
*Mr & Mrs John Brooks (she sick)
Mr James Brooks
Miss Martha Brooks

*Miss Palfrey (sick)
*Miss Anna " (sick)
Henry [?] Palfrey

Arthur Sedgewick
Sara "
Miss Ann Ashburner
" Grace "
*Miss Sedgewick (sick)
*Prof Child (faculty meeting)
Mrs "
*Mr & Mrs Gurney (in New York)
Mr Ralph Keeler

Boston
Mr & Mrs Fields
" " " Aldrich
" " " Clark
" " " Anthony
Mr Osgood
Mr Bugbee (Aldriches' boarder)
*Miss Woodman (out of town)
Perabo
*Mr & Mrs Sam Wells (sick & engaged)
Willie Gannett

64 invited
49 came

Manuscript: Harvard

1. Dating is from Bret Harte's letter of 7 March to EMH.

2. The Bret Hartes, who had arrived on February 25 in Cambridge, had left on March 4, after spending the week at the Howells house on Berkeley street. Parts of Elinor's letter about the party are printed in *LinL,* 1:160–161, with WDH's letter of 5 March to his father on the same subject (also *SL,* 1:365–366). Mrs. Harte was the former Anna Griswold of New York.

3. Thomas Starr King (1824–1864), Unitarian clergyman, lecturer, and writer. An explorer and mountain climber who had extolled the West in his letters to the Boston *Transcript,* he had died of diphtheria at the age of forty.

4. Elinor's revised date is illegible. The party took place on 27 February.

5. Sara Sedgwick (1864–1922), who later married a son of Darwin, was the sister of Arthur G. Sedgwick and of Mrs. C. E. Norton.

6. Abby Morgan Brooks, who had married John Fiske (1842–1901), the historian, in 1864. In his letter of 2 March to his mother (Ethel Fisk, ed., *Letters of John Fiske* [1940], pp. 199–200), Fiske had written: "Monday evening the Howellses gave a very brilliant party for Bret Harte and his wife who with two children are visiting them. Abby and I attended, Abby getting out of bed where she had been for several days, in her enthusiasm over the occasion. . . . Everyone wore his best bib and tucker, the house is well arranged for entertaining, and the supper was delicious—provided by the caterer Smith. Mrs. Howells was very pretty and charming; vivacious and amusing as always. We were among the last to leave shortly before midnight." Joshua B. Smith, of 13 Bulfinch Street in Boston, was the caterer.

7. Across the text, in the middle of the fourth page, is the following, written vertically in WDH's hand: "This is quite untrue. They were not introduced, because I didn't think of it. Will."

8. Harriet Spelman, who in 1868 had married Ernest W. Longfellow (1845–1921), the painter son of the poet. For Alice James, see 7 September 1867. Thomas Sergeant Perry (1845–1928) was a close friend of Henry James and of Howells, who was to help him get a position on the staff of the *North American Review* in 1872. John R. Dennett (1837–1874) was the *Nation's* literary critic.

9. Most of these have been identified earlier. The "Middy Bear" was perhaps Hannah R. Palfrey. In her guest list (see below), Elinor indicated that both she and her sister (probably Sarah H., the eldest of the trio), could not attend because of sickness. Mary Palfrey was the youngest. Elinor had written "prov" then crossed out "ov," adding "*pre-vented.*"

10. In his letter of 5 March to his father (see above), Howells had mentioned that James R. Osgood & Co. might engage Harte to write exclusively for them.

11. The letter is at Harvard.

12. Ranney, a student at Harvard, was the son of Judge Rufus P. Ranney (1813–1891), an Ohio jurist and politician. The father had studied law with Joshua R. Giddings in Jefferson and was a former partner of Benjamin F. Wade.

13. *Atlantic,* March–April 1871.

14. A list of invited guests was included on a separate page. The following have not previously been identified: J. R. Lowell's second wife, the former Frances Dunlap; Mabel, Lowell's daughter by his first marriage to Maria White, who was engaged to Edward Burnett; Mrs. Hopkinson, either Mrs. John P., who lived on Berkeley near Phillips Place, or Mrs. Thomas Hopkinson, widow of the Cambridge judge, and Harvard President Charles W. Eliot's future mother-in-law, who also lived on Phillips Place; Nathaniel S.

Shaler (1841–1906), the geologist, a former assistant to Agassiz, was professor of paleontology at Harvard; his wife, the former Sophia Penn Page; Josiah Parsons Cooke (1827–1894), chemist and author, was professor of chemistry and mineralogy and director of the Harvard Chemistry Laboratory; his wife, the former Mary Hinckley Huntington; Henry Adams (1838–1918), recently appointed professor of history at Harvard; Alexander Agassiz (1835–1910), zoologist, coal mine operator, oceanographer, was the son of Louis Agassiz; his wife, the former Anna Russell; Hannah K. Furness, the widow of William Henry Furness, the portrait painter (1828–1867), lived on Arrow Street, not far from the Howellses; James W. Brooks, a counsellor, John Brooks, and a Mrs. Martha A. Brooks all lived on Oxford Street, near the WDHs' former home on Sacramento Street; Ephraim Whitman Gurney (1829–1886), a scholar of history, law, philosophy, and classics, had taught philosophy at Harvard since 1867 and was the first dean of the faculty; the former Ellen Sturgis Hooper, his wife, whom he married in 1868; Ralph Keeler (1840–1873), the bohemian author, is portrayed in Howells' *Literary Friends and Acquaintance*, pp. 231–234; Andrew V. S. Anthony (1835–1906), the wood engraver, was also in charge of the publication of fine editions with Osgood & Co. and was later associated with Harper & Bros; he had married in 1858 Mrs. James M. Warner, née Mary Aurelia Walker. James M. Bugbee (1837–1913), an occasional *Atlantic* contributor and author of "Boston under the Mayors, 1822–1880)" (*SL*, 2:222); Miss Woodman was probably the sister of Lilian Woodman Aldrich; William Channing Gannett (1840–1923) was the brother of Elinor's friend, Kitty Gannett Wells, and the author of a biography (published in 1875) of his father, the Rev. Dr. Ezra Stiles Gannett (1801–1871). Howells reviewed his book in the *Atlantic*, May 1875.

To Anne T. Howells

Cambridge,
3 Berkeley St.,
Nov. 23ᵈ, 1871.

Dear Annie:

Your father complained in his letter of a lack of news in Will's letters, so I am going now to write you all the small gossip I can think of for his benefit. Let me say first that I have sympathized deeply with you all in the loss of your Cousin Howard. What a sudden unforeseen calamity!¹ I am sorry—*very* sorry—too about the failure of Every Saturday and the consequent failure of the Washington scheme.² It did seem as though an opening had been made for you at last, and I saw you well on the road to fame. I'll tell you a few particulars of the ending of Every Saturday—though I dont know certainly whether Will will like it. Dont tell anybody! Mr Clark undertook to manage the paper much to Aldrich's disgust, and managed it very extravagantly—paying enormous sums for original illustrations, many of which he threw aside because they did not suit his fastidious taste—and besides the pictures he paid for a great deal of original writing for the paper—when the idea it was started on was that the plates of the English Graphic were to be used, and the reading was to be selected matter. Aldrich is thrown out of his place, his salary reduced one half, and he is now after January 1st. to edit something like the old Every Saturday for Osgood and Co. Of course the Chicago crash was what finally killed the paper— but if it had been in good health it need n't have died. Will's account of Quebec which he wrote for E. S., and Shepard's illustrations are on the firm's hands, and they do not know what to do with them.³ Everybody is disgusted with Mr. Clark. Anthony, the engraver, who received a large salary, and with whom Osgood boards, has also lost his place. Aldrich is now editing the paper alone until the first of January which closes its existence. Will's place is as secure and good as ever, but he has to work tremendously now. By & by he is to have more help—but just at present he must do it all himself, to be sure it is well done. He is getting up new departments for the magazine— on art music & literature. He has to go in town to meet contributors twice a week, which I do not like at all, as we have to take dinner without him or have a very late dinner when he comes out. I am afraid it will give us all dyspepsia—and if he has to be in town a great deal I think it will end by our moving in to Boston. One thing is very pleasant this Winter, and that is his open woodfire in his library. The children are in fine health this Winter. Johnny looks very manly in his new little cape overcoat and cap. Winnie still goes to school at the Childs's. Johnny is going to the barber's today to have his hair cut. [sketch]

Hoppin is illustrating Their Wedding Journey capitally.⁴ The book is to be out the first of December. My *plants* are my great excitement this Winter. I do wish you could see my conservatory—it looks so bright and fresh. A German ivy runs all around the top—a white azalea is just bursting in to bloom—the rustic-stand is filled with earth and a variety of delicate little plants: money, German ivy, Colisseum ivy, running nasturtion smilax, a mossy plant, a white running-plant, and in the middle a sort of lily with red leaves. It looks lovely. Big geraniums in pots stand around on the floor, which is white oil cloth, and other pots are on shelves and standing in the windows. The glass doors stand open in to the parlor all the time and everybody says "How delightful!" Today I have planted mignonette & sweet alyssium in wooden boxes. Will has some hyacinth bulbs in his library, I have a big heliotrope in one parlor window and my big English ivy running over a false, looking-glass window in the parlor. Maggie[5] has a shelf of plants in the kitchen, Winnie some little pots in the bath-room, and altogether the house would remind you of Mrs. Cadwells'.

We have had a good deal of company since we came home—not to stay, so much as in the way of meals. I have my splendid girls who make it easy for me to have company. Joanna & Gus came here in October to stay a week—but Chicago took Gus back after two days & Jo would go too.[6] But we made one dinner-party for them, inviting the Aldriches and Prof. & Mrs Shaler. Gus Hoppin came to lunch when he was in Boston.[7] David Neal, an artist whom we met in Italy, and who has sent us big photographs of his paintings from time to time, was here to dinner. Dr. Martin (met abroad) who is here to join Agassiz's expedition has dined here. John Hay dined here awhile ago.[8] I cant remember all, but it has been one person after another straight along. Nicholay and his wife were here awhile ago. Harry James comes every Sunday night to tea. Willie James is much better in health. One queer thing about Harry James is that though we've been *intimate* with him so long we do not know him a bit better than the first time we met him. Occasionally some one like Darwin's sons or Mons. Coquerel come along and the Ashburners make a party for them.[9] The Palfreys have just got back from a six months' tour in Europe. They went to Venice. The "Middy Bear" has more to talk about than ever.

Mabel Lowell is to be married in April, and in June Mr & Mrs Lowell go abroad. The Nortons are to spend the Winter in Dresden. The dreadful Mrs Valerio[10] is to marry a young man, ten years younger than herself, with whom she flirted before her husband died. Fanny Platt has married Tom Fullerton's brother. We quite fell in love with Robert Dale Owen who has been visiting a neighbor.[11] His new book will be "nuts" to you folks who believe in apparitions. It frightened me nearly out of my senses the other evening when I was reading it.

Kitty Wells has moved into a new house down on the Back Bay. She feels her father's death very much.[12] My brother Charley has gone to New York to live—agent for the same company. Mother is coming here to stay three months. Our tennant has just left our house on Sacramento Street, so it is empty. We shall try to let it again, as this is no time to sell. There have four houses been built—mean-looking little cottages—on the corner lot next ours, which you remember as empty. "Bully Bradford" is now chief-of-police in Cambridge.[13] Mr Fields is to deliver his lecture "A Plea for Cheerfulness" in Cambridge next week. Longfellow is to give him a supper after it, and came today to invite Mr Howells. Mark Twain has been in Boston, and Mr Howells saw a good deal of him, and liked him very much. They say his lecture is a great failure.[14] Bret Harte has gone to New York to live. Osgood & Co. (private) paid him $10000. this year—but I guess they will not pay him so much next year. I have been very busy sewing this Fall, but am getting done. Made my bonnet yesterday.—plum-colored velvet lined with light blue. Had a letter from Mary Thatcher the other day. She says Henry Howells & wife have "put in" to her boarding-house in New York, (9th St.) as Georgie was taken suddenly sick in New York.[15] Perhaps I am rather spinning things out.

I must tell you the most brilliant thing Arthur Sedgewick ever did. Down at the sea-side where a lot of Cambridge literati were staying, they used to read Carlyle on the piazza o' mornings. Gov. Andrews' daughter did n't like it,—but was ridiculed for saying so—so she got Arthur to write a nonsensical sentence and Mr Boott, who was chief reader, inserted it in to the French Revolution as he read—and *imagine* it was not detected! So Arthur called attention to it and it was re-read. Old Mr James said it was *a fine sentence,* Grace Ashburner explained it and all approved it, so that the reading went on. At dinner Arthur said he had made it out of the first words he happened to come across in different places—and old Mr James *did n't like it!* The joke around Cambridge is that it is too much like Mr James' own style. Here is the sentence: "Word-spluttering organisms in whatever place; not as now with Plutarchean comparison—apologies; nay, rather without any such apologies—antiphonal too in the main—but born into the world to say the thought that is in them—butchers,

bakers candlestickmakers—men, women pedants—verily with you,
too, is it now or never."[16]
Your father & one of you girls must come on here this Winter or
in the Spring before we go off in to the country. Bear it in mind!

Love to all

Elinor

Manuscript: Harvard

1. A son of Henry C. Howells, Howard Howells (1843–1871), who had died of small-pox on 24 October, was reportedly engaged to WDH's sister Aurelia (Ilka Howells Dufner, "Genealogical Materials," typescript, p. 15 (Howells Memorial).

2. In his letter of 5 October to Annie (*SL*, 1:378–379), Howells, who had acted as intermediary in her behalf with Garfield and John Spencer Clark, the editor of *Every Saturday,* had advised her to go to Washington and write "two or three letters describing the aspect of things during adjournment, and just before Congress meets. . . . write from your own impressions and then send me the result. . . . If you make a hit your future is sure. . . . you have a chance offered to one young writer in a hundred, and a prize worth working very hard for."

3. Howells used the Quebec material he had collected the previous summer in *A Chance Acquaintance,* which William L. Sheppard (d. 1912) was to illustrate (1874 ed.).

4. Augustus Hoppin (1828–1896) illustrated the first edition of Howells' novel, using a sketch and description supplied by Elinor for one of his illustrations, the sketch of the nun's habit (*Their Wedding Journey,* HE, p. xxvi).

5. Maggie was the cook or the second girl. Mrs. Caldwell was probably the wife of Darius W. Cadwell (b. 1821), a noted Jefferson lawyer, former state senator and general in the Ohio Volunteer Army.

6. Gus and Joanna Shepard had returned to New York because of the Chicago crash, A. D. Shepard being with the National Bank Note Company. Just what Shepard's Chicago interests were is not known, but the extensive financial and social dislocations caused by the Chicago fire of 8–9 October 1871 had widespread repercussions. The Howellses were doing an unusual amount of entertaining indeed. In addition to her two "splendid girls," Elinor appears to have enjoyed particularly good health that winter: cf. WDH to his father, 3 December (*SL*, 1:386): "Elinor's health seems on a firmer basis than it has been for a long time," and, to Grace Norton, on 4 February (Schlesinger-Radcliffe): "Mrs. Howells was so well that she had shot up to ninety pounds in weight, and had a faint crepuscular suggestion of not impossible color in either cheek."

7. The WDHs had met David D. Neal (1837–1915) in Italy, as Elinor says. They had also met Benjamin Ellis Martin (d. 1909), the traveler and author, there (pocket diary, typescript, pp. 19–20, 46).

8. John Hay (1838–1905), the future Secretary of State, and John G. Nicolay (1832–1901) had been instrumental in getting Howells his Venice consulship in 1861, when they were both Lincoln's private secretaries. John Hay became a lifelong friend of Howells. For their correspondence, see *John Hay–Howells Letters.*

9. Athanase Josué Coquerel fils (1820–1875), French Protestant minister, orator, and scholar, noted for his sermons and for his studies on the history of Christianity.

10. Mrs. Valerio is unidentified. Laura Platt Mitchell's younger sister, Fanny Hayes Platt (1847–1896) had married Erskine Boies Fullerton (1842–1909) on 19 October.

11. Robert Dale Owen (1801–1877), the social reformer, was the author of *The Debatable Land between This World and the Next* (1872). Born in Scotland, he had become a spiritualist while in Italy, where he had been appointed minister in 1855. He lived in America most

of his life and contributed articles to the *Atlantic* from 1873 to 1875. Owen was visiting Howells' neighbor Albert G. Browne, the attorney.

12. The Samuel Wellses had moved from 10 Boylston Place to 155 Boylston Street. Her father Ezra Stiles Gannett, the Unitarian clergyman, had died on 26 August in a railroad wreck.

13. Isaac Bradford was also the Howellses' former neighbor on Sacramento Street. In his letter of 30 November 1871 to Grace Norton, Henry James, Jr., wrote, among other gossip: "Also Mr. J. T. Fields lectured here on *Cheerfulness* lately (as who should say: I know I'm a humbug and a fountain of depression, but grin and bear it)" (*Henry James Letters*, 1:265).

14. Early lecture topics had been unsatisfactory to both Clemens and his audiences.

15. Molly Rockwell Thatcher, a longtime friend of Elinor's, was the daughter of William H. Rockwell, Superintendent of the Brattleboro Asylum, and the wife of Thomas F. Thatcher, of Philadelphia.

16. With minor punctuation variants and the use of "women, peasants" instead of the "women pedants" in Elinor's version, this "bit of nonsense" is printed in *LinL*, 1:134, after an introduction by Mildred Howells to her father's letter of 12 November 1868 to Charles Eliot Norton. Francis Boott (1813–1904), a musician noted for his songs and for his friendship with Henry James, Jr. Grace Ashburner was an aunt of Mrs. Norton.

To Victoria M. Howells

[Cambridge] February 8ᵗʰ [1872]

Dear Vic:

I had quite forgotten that I owed you a letter. Our correspondence is so irregular that it is hard to remember which is debtor—but pray excuse me this time and I'll not forget again. Will and your father are so unfailing in their letters that it really leaves the "women folks" very little chance to tell a bit of news. And then you know so little of Cambridge—Annie knows the most, and that is why I write to her oftener—that I never have anything interesting to write you about it! But I hope some time you will see us under more favorable circumstances than two years ago at this time when you came. This house is so different! However we expect to be ousted from here as soon as the lease is out, as Miss Donnison positively refuses to sell the house.¹ But Will feels so established on the Atlantic now that we may think it worth while to buy a nice house in Cambridge and settle for good. We have given up all ideas of moving into Boston. It is too expensive. Sam Wells (Kitty Gannett's Sam) has just bought a house for twenty-five thousand dollars and it is not any nicer than we should want—not any nicer than this house—and we feel that we could never afford to live or buy in the city. We find we have got a good deal attached to the place too in the six years we have lived here—and the last year and a half has been so much pleasanter on account of this house and this street that even you would like Cambridge if you visited us now. Ar n't you going to visit us this Spring with your father? You seem to have made no plans yet, but we are fully expecting a visit from Father and one of the girls as early as the middle of April or the first of May certainly. Do write and give some hint of what you are intending to do. Joe writes that he and Eliza are coming east this Summer. One set must come early and the other afterwards. Which shall it be to come first?

We were greatly amused at your father's description of the ceremony of driving spikes on the new railroad as performed by you girls. Annie's account of the railroad men—the Swedes—too, was very funny. The carpet waistcoat must have been a sight. Why does n't she make a sketch of our New Railroad; there seem to be so many amusing things connected with it—at least you people *see* so many droll features in the subject.[2] It is a regular Howells gift—that faculty of seeing more than other people.

Feb. 9th

Tell Annie we had Mr Ranney around to tea last Winter and begged him to come freely to the house—but he has never been near the house since. However we'll try again to see what can be done with the young man.[3] A young Mr Dyer son of Judge Dyer of Chicago who is here in college comes around here quite often and is very pleasant. Your father will know who he is—as his father is a prominent Swedenborgian. We have come across another Sw.g.n.— Mr Chandler of Brookline a grandson of Michael Slaughter—a young man who studied architecture in the same office with my brother Willie in New York.[4] I see more of the Parsonses since we live near them! They are very pleasant people.

I am glad Henry likes the blocks so much. By and by perhaps he will be able to combine them. Dont you think so?

Will went to a dinner-party at Pres. Eliot's[5] Wednesday given for the new Governor. There were twenty gentlemen there mostly old men. Will was the youngest one present.

Mother has been staying with me all Winter, but she dreads the East winds of March so is going soon to stay with Joanna. She wishes to be remembered to your father and sisters, whom she has seen. The children are well. John is getting to be a big noisy boy—but he is a good boy always.

Much love to all.
Affectionately
Elinor

Manuscript: Harvard

1. Probably Miss Catherine L. Donnison, who boarded at 1 Waterhouse Street, corner of Garden Street, nearby.
2. For the probably unpublished series of letters, see WDH to Annie, 20 April 1872 (*SL*, 1:395–396).
3. Louis Dyer (1851–1908) was then a sophomore at Harvard, where he later taught classics before moving permanently to Oxford, of which he was also a graduate. His father was Charles Volney Dyer, M.D., who practiced medicine in Newark, N.J., and in Chicago. An abolitionist active in the underground railroad, the elder Dyer had been appointed judge in the Anglo-American mixed court at Sierra Leone by Lincoln in 1862.
4. This was Theophilus P. Chandler, Jr., the architect whose Boston office was at 14 Devonshire Street. Mead had studied architecture with Russell Sturgis in New York. Michael Slaughter is unidentified. Theophilus Parsons (1797–1882), professor of law at Harvard and a Swedenborgian, lived with his family at 54 Garden Street, near the WDHs on Berkeley Street. Parsons sold the lot on Concord Avenue to the Howellses, and they built their house there in 1872–1873.
5. Charles W. Eliot (1834–1926), president of Harvard (1869–1909). The new governor was probably Cadwallader Colden Washburn (1818–1882), soldier, industrialist, and Republican congressman, who had been elected governor of Wisconsin in January 1872. He may have been a relative of President Eliot's wife.

To Anne T. Howells

Cambridge, April 19ᵗʰ [1872][1]
Dear Annie:

I should have written before to say how glad I was that you were coming only I have been sick and—what is worse—have had a seamstress in the house for a week. I told Will to say I was delighted, and to urge you to come as early as possible, as I and both the children are going into the country early in June. Not but what you could stay here after that, but I wish to see as much as possible of you. The Spring, though it promised to be late, is rushing in now and I think you would find it pleasant here by the first of May—indeed it is delightful now—and now is just the time for society in Cambridge, as people are very lively before getting ready to break up for the Summer. I would not go away till July, only I must be back the last of July.[2] Do hurry up! Dress need not detain you—one street dress and one dress-up will do, and you can find both these in the family I dare say, as your clothes fit one the other. We have already begun to make appointments for you: one is to go to the College Library with Mr Dyer, who has engaged the curator of the Grey collection of Engravings to show them to us. We told him we would wait for our sister.[3]

You will find Cambridge much the same as when you were last here. The Palfreys have returned from Europe—the Childs go on as usual—with a boy Child whom you have never seen[4]—Mrs Cooke still carries on sewing-societies and the Episcopal church—Henry James & Alice and their aunt Mrs Walsh start for Europe next month, but Willie is at home now—&c. Still there are some sad changes when I think of it: the Ashburners & Sedgewicks are in great grief since Mrs. Norton's death—the Higginsons have moved to Brookline, Annie never comes here, Robt. is in New York, and I dont know where Lewis is—Dr Angell is still very sick and they stay out of town most of the time—Dr Gannet is dead and that household is different.

But we have made some new acquaintances who are pleasant in this neighborhood, and the *house* is such an improvement on the other that I think you will enjoy the change. I am getting my flowers in to as good trim as possible on your account. I will have the piano tuned too, before your arrival. Just think of my having a grand piano in the house! My opposite neighbor has moved and gone to boarding; her daughter is in Europe, and she has begged me to keep her piano till Fall when her daughter will return—so we have the use of a piano rent free.

I have just been making myself a *round* dress from your pattern and I like it exceedingly.

We have been negotiating lately between Henry Howells and the youngest of those Lane sisters we boarded with in Quebec, for her to be governess to Henry's children.[5] Of course he will torment her— but she is very anxious to come to the States, and is patient to a degree, and is used to brusque English manners anyway, so I think they will get on—and then there is lovely Georgie with whom she will spend most of her time—they are both French scholars, and will like each other in many ways.

There is no use in my writing you every sort of little thing when I am going to see you so soon. Will's story is not to be published till Fall, but you can hear it read when you get here, and see how you like yourself.[6]

The children are more agreeable than when you last saw them. Winnie has become quite lady-like—she is a laughable mixture of you and Aurelia, both in looks and ways—and Booah—what shall I say for Booah—He is simply *fascinating*. His papa is going to take him in town this afternoon to the store. You know Will now goes in town all day Tuesdays and afternoons Wednesdays? Henry Howells has been on here and to see us twice lately, and John Mitchell twice. Ruddy Platt took tea with us week before last. He is in Yale College. Laura is in a very weak state. She is coming north to the sea-side this Summer. John is building a splendid house right in Mr Platt's lot—on the plan of Prof. Child's house which I took them to see.[7]

Judge Ranney of Cleveland and his son[8] who is in college called here last week—but Will was out and he missed them again. I urged them to stay to tea but they would not. Judge R. had come on to take his son home as he is out of health. He has had something like slow fever. Tell Henry Elgner wants one of his beautiful photographs in the blue suit. Much love to all. I am glad little Beatrice has got up so well.

Affectionately
Elinor

Manuscript: Alfred

1. Year established by Henry James's trip to Europe (*Henry James Letters*, 1:281).

2. Elinor was expecting her third child. Mildred Howells was born on 26 September 1872.

3. This was a collection of some 4,000 engravings bequeathed to Harvard in 1857 by Francis C. Gray, of the Class of 1809. Noted especially for Marcantonio Raimondi's transcriptions of Raphael's compositions, it constitutes a pictorial record of the history of art done by engravers before photographs or color reproductions.

4. The Palfreys and most of those mentioned here have already been identified. The boy was probably Frank Child, who was born on 12 June 1869, after Annie's 1867 visit to Cambridge. Catherine Walsh (1808?–1889), the sister of Mrs. Henry James, Sr. Henry James, Jr., was to escort Aunt Kate and his sister Alice to Europe that summer. Susan S.

Norton had died in Dresden in February during the Nortons' five-year stay in Europe. Dr. Francis Higginson (1806–1872), the Brattleboro physician, had moved to Brookline with his family in 1867. Annie, his niece, was Stephen Higginson's daughter, and Robert and Louis were her brothers. Their father had died in Cambridge in August 1870.

5. The Lane sisters kept a boarding-house at 44½ Anne Street in Quebec, where William C. Howells and his family stayed in 1873–1874, after he had become U.S. consul in Quebec (*SL*, 2:35).

6. After this word, "A Chance Acquaintance" is written in a different hand. First printed in the *Atlantic,* January–June 1873, the book was published on 27 May 1873.

7. Rutherford Hayes Platt (1853–1928) was Laura Platt Mitchell's younger brother. William A. Platt, their father, resided on Broad Street in Columbus, Ohio.

8. Young Ranney did not graduate from Harvard. Three of Judge Ranney's children are said to have died young.

To Anne T. Howells

Mountain House,
Princeton, July 9ᵗʰ [1872]

Dear Annie:

How I wish you were here this bright, clear afternoon to sit on the piazza and take in the beautiful view—a clear sweep way to Waltham Hills, with the Blue Hills of Milton beyond. But you would probably be off with Will and the children on the Mountain picking blueberries just at this time. Later in the evening all the boarders will be out on the piazza, and then human nature would be quite as attractive a study. We have just had an addition to our circle—Mrs. Rev. Guild, a great beauty, who absorbs all attention.[1] Her husband, a Unitarian Minister is Cousin of Chaˢ Norton's and Pres. Elliott's. She was a Miss Cadwallader of Zanesville, Ohio—afterwards of Marietta, where Mr. Guild wooed and won her. She is an Episcopalian herself, and consequently unpopular in the parish wherever they are I hear. (By the way Mrs. G. was saying this noon that after Mr Guild had gone through the Cambridge Divinity School he went to Andover and went through the Course there to see if he could n't become Orthodox— but he could n't it seems.) Mrs Guild looks very much like you in the face, Annie, but her hair beats yours "all holler". It is nearly two yards long! She is very tall and wears it in two long braids behind, which sweep the ground as she sits. [profile sketch] Her front hair she twists in a coil at the back of her head—as big as most people's coils—and yet each braid is as thick as an ordinary head of hair, and even way to the ends. Did you ever hear of her? The house is nearly full now and only pleasant people have come so far. Osgood and the Anthonys want to come but we fear there will be no room for them.[2] Mr Howells wrote for Osgood to come up and see for himself. The Greens, whom your father met, are still here and many of their relations have joined them. Mr G. was telling a gentleman one night that it was a pity he did not meet your father "such a genial agreeable gentleman—and a born naturalist &c" He said he greatly enjoyed those walks he took with your father.

Will is enjoying it very much up here, and is writing as well on his story as if he were at home. Two days will finish it he says. It is *great—* and tell your father I think it grows better and better. I know *you* will be interested in it, Annie. O Annie, how you did strike out in your love-story![3] It seems to me to be immensely ahead of anything you ever did before; and now it will be *Miss* Howells as well as *Mr* Howells in the Magazines. I am provoked to see there is a Fanny Howell already in the field.[4] I hope there'll be no confusion. I have just been reading a wonderful story: "Smoke" by a Russian writer.[5] Will thinks it one of the best novels he ever read. He must contrive to have you girls see it. I'll speak to him. I found, when I went down to Cambridge, that your father had left Romola, and scorned the ivy that I gave him. How he did shorten up his visit![6] I supposed of course he would see the Jubilee through.[7] We enjoyed every moment. [remainder lost]

Manuscript: Howells Memorial, Kittery Point, Maine

1. Emma M. Cadwallader, daughter of Dr. John Cadwallader, had married the Rev. Edward Chipman Guild (1832–1899), pastor in Baltimore (1869–72) and later in Waltham (19873–1880). Elinor was so impressed with Mrs. Guild that she did another color sketch profile of her sitting with her braided hair reaching to the floor. This sketch now hangs in the Howells Memorial Library. According to a note on the back of the sketch, Abby White Howells once remarked "This is 'Mrs. Farrell,'" the heroine of "Private Theatricals" (*Atlantic*, November 1875–May 1876), much later published as *Mrs. Farrell* (1921).

2. J. R. Osgood boarded with the A. V. S. Anthonys on Pinckney Street in Boston. The Greens may have been the family of Samuel A. Green (1830–1918), city physician (1871–1880) and later mayor of Boston, also librarian of the Massachusetts Historical Society from 1868.

3. Annie's story was "Fireworks," published in the *Galaxy*, September 1872.

4. Probably EMH had just read "Woman as a Smuggler, and Woman as a Detective" in *Scribner's Monthly*, July 1872. Later essays by Fanny Howell have not been located.

5. By Ivan Turgenev (1818–1883). An English translation from the author's French

version was published by Holt & Williams in New York in 1872 and reviewed by WDH, *Atlantic,* August 1872.

6. William C. Howells had visited the Howellses in Cambridge and Princeton recently. George Eliot's *Romola* may not have appealed to the senior Howells.

7. Together with T. B. Aldrich, Howells wrote and edited *Jubilee Days,* with illustrations by Hoppin, 1872.

❖ V ❖

A Great Deal of Company
1873–June 1877

A letter of January 1876 to an acquaintance in Bethlehem, Pennsylvania, nicely catches life in Cambridge for Elinor at the time. With the weather very mild and with "no end of gayety," Elinor described the many parties she and Will attended, including an evening of private theatricals at Longfellow's in which several common friends had participated. They had had "a great deal of company in the house" and so many evening affairs that Elinor could hardly catch up with her social obligations. Also, as her husband worked at home, he brought in "a great deal of men's society to the house." Many of the men were authors, some viewed enthusiastically, some skeptically by Elinor.

Best known today among the authors was Mark Twain, who by the end of 1871 had become such a close friend of Will's that he very likely had also met Elinor. But the first recorded meeting was the visit by the Howellses to Atlantic contributor Charles Dudley Warner and his wife in Hartford in early 1874, which also included a dinner with the Clemenses and subsequent visits back and forth between both couples. Elinor corresponded with Susan Warner and Olivia Clemens from this time onward, especially with Susan, who appears to have been Elinor's favorite in spite of, or because of, their divergent tastes in novels.

In the summer of 1876, Elinor's cousin Rutherford B. Hayes received the Republican nomination for the presidency. At a hint from Houghton,

Howells' publisher, Will proposed writing a campaign biography, a project welcomed by Hayes. Elinor also participated in this hastily written book, particularly in regard to Hayes's ancestry, details from diaries and scrapbooks, and portrait. While Hayes's election remained in doubt until two days before he assumed office, Elinor and the other members of the family on both the Mead and Howells sides never seem to have entertained serious doubts about the Republicans' victory under Hayes.

It is likely that Annie's courtship, engagement, and wedding occupied Elinor's attention more than the disputed election of Hayes. Though her letters to Annie appear less earnest than jesting, they reveal much about her attitude toward herself and her own marriage.

To Anne T. Howells

Cambridge,
March 30ᵗʰ/73.

Dear Annie:

You ask why Elinor does not write—as though a *new house* and a *new baby* (the new house is the most care) were not a sufficient excuse for my neglecting all other things under the sun![1] (I do not mean to call you or the baby *things*—try to make sense of it) But I read your letters and letters about you from Jefferson with the keenest interest and am awfully proud of you. I confess I do not read all your book-notices any more than I do those in the Atlantic, but I take Will's word for it that they are good.[2] What I am proud of of course is to see a girl of your age doing such hard work and winning so much credit for yourself without making any fuss about it. I have never yet got it through my head how you came to take such a responsible position at first—it was all done so suddenly and so quietly!

I imagine you must enjoy the society in Chicago, and what opportunities you do have in the way of opera! I should like to see you in the blue silk of two shades immensely.

Will went into Boston to lunch with John Hay yesterday and today he has been lunching with us. He described your Mr Keenan as the youngest-looking man he ever saw—about five years old in appearance he says.[3] He says Mr Scammon occasionally makes overtures to him (Hay) in regard to editing the Inter-Ocean* He likes Miss Bross very much. He says she has become completely enamored of Boston. She has just been in New York for a short time. As she was leaving she met Mrs. Pullman of Chicago. Mrs. P. asked her where she was going and on her replying "To Boston thank Heaven"—Mrs. P. said severely "If I felt so Miss Bross I would not say so!"

Mr Hay says Mr Greeley's two daughters are the loveliest girls in the world—in appearance and in many ways.[4] The youngest I believe he considers somewhat too wilful yet—but she is young. He says it was all a made up story about Ida's being engaged to the young man who was lost on the Boston. Mr Hay came on here to lecture to the Young Ladies' Club of Boston. Tomorrow he is going up to Amesbury to visit Whittier,[5] and Will is going to accompany him.

Mary Mead is with me for a few weeks. She has to stay with Mother now that Mother requires so much care. Mildred is six months old and is quite an intelligent little creature. She knows us all and refuses to go to strangers, comes to the table during dessert and plays with toys. She looks like her papa only she is going to be light like Johnny. But, dear me, how do you know how Johnny looks—not having seen him for so long! He is in pantaloons for good now and is quite a manly little fellow. Mr Lothrop (Rose Hawthorne's husband) was in here today and said they were going to have him to dine with them before long.[6] He is their prime favorite. Today when I asked him if something suited him he said "yes, perfectly." and when I asked him if he had a nice walk he said "Yes, a lovely walk"—which will give you some idea of his style of talk.

Laura Mitchell & her little Lily come out to see us very often.[7] Lily is learning what is called "visible speech" in Boston. She speaks a little and understands her Mother, at least, by the motion of her lips.

*Will says I may have misunderstood this *and not to mention* it.

My cousin Eddy, who nearly bored you to death one day, is going to be an Episcopal clergyman.[8] My brother Willy has stuck up his shingle in New York as an architect. I have been quite miserable this Winter, so have scarcely been anywhere. The first place I went to was, a few weeks ago, to the Palfreys' Golden wedding. It was a beautiful affair—loads of presents and heaps of flowers. The dear old doctor's head was a little turned. He first shook hands with me cordially and then demanded of Will where Mrs Howells was. He carried in his hand the gloves he wore at his wedding. A sister of Mrs Palfrey's was present who stood up with her. The feature of the occasion was a copy of the Milton Shield in gilt metal—a large shield with designs from Paradise Lost embossed on it, with a beautiful presentation letter signed by Charles Francis Adams (who was present), Charles Sumner, Longfellow &c &c. Mrs Palfrey was dressed in purple silk with a showy cap and looked only about fifty with her thick black hair her fine teeth and nice plump figure. Mary wore a black silk with canary-colored crape overdress. Anna wore light blue, and Sarah only black.[9]

I never see Mrs Child as the walking has been so bad and we live so far apart—but Winnie goes there to school every day so I keep posted as to their doings. There is nothing new to tell. Mr Norton and his six children are coming home in May. Arthur Sedgewick is in New York writing on the Nation. Mr Hay says he heard that Sedgewick had said of him to a mutual lady-friend "O but he is only a varnished savage!" because he came from Missouri. Perhaps Arthur's snobbishness never struck you? And let me add that Hay is one of the most elegant of men—though you must know that.

Boysen is considered a great genius in these parts.[10] I think I have a *huge* compliment for him from Mrs Gade, our next neighbor, who is married to a Norwegian and has lived many years in Norway. She thinks his translation of Bjornsen into such noble blank verse perfectly wonderful. I allude to his review in the N.A. Review. If you want to tell him this you may, Annie. He knows Mrs Gade and admires her.

Writing you so seldom I do gossip so! The pen is sometimes

unruly. Write to Will on pure literature, but write me society gossip. I will leave a bit of space for him to add a line.

Affectionately

Elinor

Dear Annie:[11]

I've nothing to add except that a Rev. Mr. Chamberlain of Chicago called on us to-day, and I gave him a card to you. If you don't find him at all extraordinary, I shall not be disappointed.—Elinor has gabbed a good deal, I see. You'll make the usual allowance for a powerful fancy.

Your affte. brother

Will.

Manuscript: Alfred

1. Mildred Howells, their last child, was born on 26 September 1872. The house at 37 Concord Avenue was still under construction.

2. Annie was working as a literary editor of the Chicago *Inter-Ocean*. When WDH learned that she wrote "at the newspaper office among all those editors," he expressed displeasure at her lack of decorum (*SL*, 2:15). While he had apologized for his offensive prudishness on 28 February, Elinor's praise of Annie in this letter appears related to the incident.

3. Possibly Henry Francis Keenan (b. 1849), journalist and novelist. Associated since 1868 with the Rochester *Chronicle*, he also worked for newspapers in New York and elsewhere. A Swedenborgian, Jonathan Young Scammon (b. 1812), was also a lawyer, philanthropist, and leading citizen of Chicago. He had established *Inter-Ocean*, a Republican paper, in 1872, and was noted for his active role in public school education and railroads. Miss Bross was a friend of Annie in Chicago. Mrs. Pullman was Harriet Sanger, the wife of George M. Pullman (1831–1897), the famous inventor and designer of luxury trains and president since 1867 of the Pullman Palace Co. in Chicago.

4. Horace Greeley (1811–1872), founder and editor of the New York *Tribune* and political leader. His two surviving daughters were Ida L., who had converted to Catholicism in Rome in 1871, and Gabrielle, the younger, who was about sixteen years old.

5. John Greenleaf Whittier (1807–1892).

6. Nathaniel Hawthorne's son-in-law, George Parsons Lathrop (1851–1898), author and journalist, was to become Howells' assistant at the *Atlantic* in 1875. John Mead Howells was almost five.

7. Laura's daughter Lilly was probably a pupil at the Boston School for deaf-mutes at 11 Pemberton Square founded in 1869 by Miss Sarah Fuller, its principal. Miss Fuller had been trained by Alexander Graham Bell (1847–1922) to use the "visible method."

8. Edwin Doak Mead (1849–1937), author, lecturer, and reformer. Howells had procured him a place in the counting-room of Ticknor & Fields, where he had been working since 1866. He was to go to Europe for five years in 1875 to prepare himself for the ministry, but he withdrew from the church in 1876 and turned to the writing of books and articles. In 1889, he became associate editor, with E. E. Hale, of the *New England Magazine*.

9. For the Palfrey family, see 16 June 1867. Charles Francis Adams (1807–1886) was the congressman and former minister to England, a friend of Sen. Charles Sumner (1811–1874), the abolitionist.

10. Hjalmar H. Boyesen (1848–1895), the Norwegian-born author, had become a protégé of Howells and a friend of the family (see *Literary Friends and Acquaintance*, pp. 215–223). At this time a tutor of Greek and Latin at Urbana University, a Swedenborgian institution in Ohio, Boyesen later became a professor of German at Cornell and Columbia. Some of his letters to Annie are in the Howells/Fréchette Papers at Alfred University. Entitled "Björnstjerne Björnson as a Dramatist," Boyesen's review had appeared in the January 1873 *North American Review*. Helen R. Gade was the translator of Björnson's *The Happy Boy* (Boston, 1870), reviewed by WDH in the April 1870 *Atlantic*.

11. In WDH's hand.

To Annie T. Howells

[Cambridge] Wednesday morn.¹
[11 February 1874]

Dear Annie:²

After our graceful parting the conductor's elbow smashed my bonnet and I entered the car to find Mrs. Whipple and some stuck-up Cambridge people awaiting me with a smile—of amusement I felt sure.³ In the Cambridge car I found a Cambridge gentleman who, of *course,* insisted on carrying some of my crockery for me—so I arrived home in triumph, to find that another leg of my standard had come off in the various changes of my journey and was lost forever. When I went in I had a standard worth two dollars and no plate. Now I have a plate worth three and no standard. Will thinks he can whittle a leg for it.

Supposing you send the ticket to the concert of Friday *out here*—and then if I go in I shall not have to go way down to Boylston Place for it.⁴

I will either go in or send Will to the concert with Winnie. He would have to go in with her and he might as well go to the concert—and it would do him good. But I should very much like to go with Winnie and will if the weather is fit for her to go.

She says she *must* be home Valentine's morning early to see about valentines—so I guess she'll have to give up spending the night with you this time—and it will be too late to go to see you after the concert. Could n't you come as far as Fera's after the concert (you can calculate better than I about what time that will be) and have a petit souper with us? Write if you will.[5]

You are coming out Saturday I hope? Dont forget to show those braces to Mrs Blake.[6]

Will says he will not go to the concert—and he does nt want to take Winnie in to your house. *I* am going then at any rate, and you must meet us at Fera's.

I hope I dont deprive you or Dora of an *extra* nice concert.[7] I have looked at the programme. It is nice, but not extra.

　Adieu.

We meet at Fera's!

Elinor

Manuscript: Harvard

1. Dating from St. Valentine's Day.
2. After shopping in town with Annie, Elinor had returned home to Cambridge, leaving her sister-in-law, who was boarding at 9 Boylston Place in Boston. Annie returned to Jefferson about the first of May (*SL*, 2:38, 43).
3. For Mrs. Whipple, see 2 June 1867.
4. Howells had procured for Annie two press tickets for a series of symphony concerts.
5. Fera's was a confectioner's and "dining-saloon" on Washington Street in Boston.
6. Probably Mrs. Samuel Parkman Blake, the former Mary Lee Higginson (see 19 November 1856), who lived in Boston.
7. Either Maria Theodora Sedgwick (1851–1916), the sister of Arthur G. Sedgwick, whom Annie had befriended during her Cambridge visit in 1867, or her cousin Dora Howells, a daughter of Henry Craik Howells (EMH, pocket diary, typescript, p. 22).

To Susan L. Warner

37 Concord Avenue,
Cambridge,
March 4th 1874.

Dear Mrs. Warner:[1]

I did think of going to Hartford a little—just to spite Mr Warner—who clearly doesn't want me to come—but it really is quite impossible for Mr Howells and myself both to leave home at this time on account of the children.

I thank you most heartily for your invitation, which was kindness itself, and I live in the hope that we shall some day meet, either at Cambridge or Hartford.[2]

Very sincerely yours
Elinor M. Howells

Manuscript: Watkinson

1. Née Susan Lee, Mrs. Warner was the wife of Charles Dudley Warner (1829–1900), the novelist and essayist, editor since 1861 of the *Evening Press* and since 1867 of the Hartford *Courant*, with which it had merged. Warner's travel sketches, *Saunterings*, had appeared as a book in 1872, followed by *The Gilded Age*, in collaboration with Mark Twain, his neighbor, in 1873.

2. In answer to the Warners' invitation to visit them in Hartford, WDH had written on 17 February that Elinor, whose coming was "out of the question," had allowed him to ask J. R. Osgood to accompany him to Hartford instead. Finally, Howells, Osgood, and Elinor all came to the train on 6 March, together with Mark Twain, who had been lecturing in Boston on the 5th. The Howellses spent the weekend with the Warners, and WDH was much impressed with what seemed to him "quite an ideal life," as he described their "charming visit" with Warner and Mark Twain at Nook Farm to his friend Comly on 21 March (*SL*, 2:56–57). "They live very near each other, in a sort of suburban grove, and their neighbors are the Stowes and Hookers, and a great many delightful people. They go in and out of each other's houses without ringing . . . they call their minister *Joe* Twitchell. I staid with Warner, but of course I saw a great deal of Twain, and he's a thoroughly good fellow. His wife is a delicate little beauty, the very flower and perfume of *ladylikeness*, who simply adores him—but this leaves no word to describe his love for her. As for Warner and his wife they are all that you could desire them." The Howellses, the Warners, and the Clemenses began to visit one another a great deal at this time, either in Hartford or in Cambridge.

Olivia L. Clemens to EMH

My dear Mrs Howells[1]

Don't dream for one instant that my not getting a letter from you kept me from Boston. I am too anxious to go to let such a thing as that keep me. A wet nurse that is tractable and good when I am in the house but who gets drunk when I go away, together with other irresponsible doings by this same nurse when I am not present, lead me to feel that I had better stay closely with my baby until she is weaned, which will not be until next October.[2]

I do wish you and Mr Howells would come down for the next Sunday after this reaches you. *Do come* if possible, remember that I am tied now and cannot go to Cambridge. Mr Clemens did have such a good time with you and Mr Howells, he evidently has no regret that he did not get to the centenial. I was driven nearly distracted by his long account of Mr Howells and his wanderings. I would keep asking if they ever got there, he would never answer but made me listen to a very minute account of every thing that they did. At last I found them back where they started from.[3]

If you find misspelled words in this note, you will remember my infirmity and not hold me responsible.

Hoping you and Mr Howells will come and with kind regards to him.

I am affectionately yours
Livy L. Clemens

[Hartford] Friday Evening
[23 April 1875]

Manuscript: Harvard

1. Dating established by Mark Twain's invitation of the same day. Punctuation minimally regularized by replacing dashes with periods.
2. The baby (Clara L., the Clemenses' second daughter) was ten months old. For Mark Twain's amusing description of the wet nurse, see *Twain–Howells Letters*, 1:71–72. The Howellses and the Clemenses had enjoyed one another immensely when the Howellses had been at Hartford in March, on the first of many highly successful visits. To her mother (14 March), Livy had portrayed Elinor as *"exceedingly* bright—very intellectual—sensible and nice." Clemens had just returned from visiting the Howellses in Cambridge on the occasion of the centennial festivities in Concord and Lexington.
3. The fullest account of the failure to get to the centennial ceremonies is by WDH in *My Mark Twain* (in *Literary Friends and Acquaintance,* pp. 280–282). Having returned "uncentennialed" to Cambridge, the two friends tried to persuade Elinor that they had attended the festivities, but she was not fooled and ridiculed them.

To Miss Webster

37 Concord Avenue, (we wish it could be called,
as it used, Concord *Road*)
Cambridge, January 20th, 1876.

Dear Miss Webster:[1]

We shall never meet again I feel sure, for you will always come North in Summer when we are away and there's no telling if we'll ever get to Bethlehem again. It is possible though that we may meet in Philadelphia at the Centennial, Mr Howells, who just came in says "Why not go and stay at Bethlehem and run down to the exhibition by the day?" So we may meet sometime, after all!

Meanwhile I will solace myself by writing. And I have started another means of communication by sending a friend of ours to call on you the next time he goes to Bethlehem. We took the liberty of giving Mr Stuart Wood of Philadelphia, who sometimes visits your town, a letter of introduction to you—& your mother. He is a pleasant fellow who took the postgraduate course of lectures here, and was a good deal in Boston & Cambridge society and can talk *Boston* any amount.² What must the weather be in Bethlehem this Winter when it is "*dogdays*" here! Getting about is so easy in such weather that there has been no end of gayety, and I heartily wish for a cold snap to end it, for I am getting entirely worn out. You must have heard of the little play Miss Una Farley took the principal part in, through her sister. Do ask her about it if you hav'n't. I will enclose the programme. The play which represents card characters liberated from the pasteboard by a fairy, and cutting up tremendously, was written by Prof. Greenough of Cambridge and was acted by the beauties of the place, and some Cambridge gentlemen and two students from other parts.³ Miss Farley was divinely beautiful in her part—far prettier than the others—and acted perfectly. I think you must have seen her before this as she was going to visit her sister last Winter. Is she not lovely? She has had the goodness to call on me, but I have not had the time to return it yet. When you have young children, and a great deal of company in the house and go out to evening affairs there is very little time for calling—and people here do not wait for you to call before inviting you a second time, so I have got in to a way of shirking calls which will prove my ruin.⁴ Mr Howells brings a great deal of men's society to the house through having his study or office here.

Last week he had a man's supper at which ten sat down. Next day young W. W. Astor the millionaire called & Mr H. in a fit of enthusiasm invited him to lunch next day. Imagine the consternation it spread in one humble household! But we found him charming—cultivated modest and an artist. He modelled two statues in Launt Thompson's studio in New York.⁵ Tomorrow night we are to dine at Mrs Ernest Longfellow's, and have three more engagements ahead. Speaking of *engagements* Miss Marnie Storer is engaged to be married—to a Cambridge lawyer, Mr Warner, a graduate of Harvard—a fine young man.[6] Mrs Storer and her daughter are spending the Winter in Quebec. They have tried nearly every climate for Miss Lizzie, and she finds she feels well in a dry climate no matter how cold. Mr Howells's father lives in Quebec and through him we learn that she is remarkably well there. January 22ⁿᵈ. Perhaps all that about the S's may interest Mrs. de Schweinitz more than you. Perhaps you'll tell her about them and give her my love?[7] Mr Longfellow said last evening that Miss Farley had been quite a long time in Bethlehem—so you must write me what you think of her looks as compared with her sister's.

There is a new rage in Boston: Duvaneck, a young Cincinnati artist. Mr Vinton, a fellow student with him in Munich, brought some of his pictures—life sized portraits & studies—to Boston & exhibited them at Doll's. They are very unfinished, but fine in color and masterly in execution. They are as good [as] the old masters, really. He had never been recognized anywhere before, but Boston went crazy about his pictures. They sold at enormous prices, and now to have a Duvaneck is to be the fashion. He not only paints roughly but chooses rough subjects—such as dirty old German professors, his German landlady &c—but this does not prevent his pictures from being in the most elegant drawing-rooms. William Hunt sanctions them. Duvaneck went back to Munich with more orders than he can execute in years. All this you may have heard before though, as you keep posted in art matters. But twice we have been invited to houses in town to lunch & "see a Duvaneck." We missed the exhibition.[8]

What are you doing this Winter. All sorts of things to get money for the Centennial? That seems to be the object of everything this Winter—balls, theatricals and fairs. The Cambridge people have got up a book among themselves, to sell for the cause. It is called the Cambridge of 1776, and describes things in the form of a Diary, pretending to have been written by Dorothy Dudley in 1776. Lowell, Holmes and the rest. Mr Howells contributed a sonnet "To Dorothy Dudley."[9] That is about all we've done as far as I can recollect—except go to things. How is art flourishing with you? My sole attempt this Winter has been to decorate a white bureau with flowers—and those only conventional, furniture flowers, and *copied at that*! So you see where *I* am! I must end this letter with this sheet and this week. So, with love & regards (properly distributed) to you & yours from me and mine

Very truly yours
Elinor M. Howells.

Please remember me to the de Schweinitzses and the Goundies[10]

I have concluded to add, what I wrote the letter to say—that we were *awfully* sorry to miss you last Summer. We hope you will visit Boston in the Winter the next time.

Manuscript: Collection of American Literature, The Beinecke Rare Book and Manuscript Library, Yale University.

1. The envelope is addressed to "Miss Lizzie Webster, Nisky Hill Bethlehem Pa. Care of the President of the Zinc Works." She was probably the daughter of Benjamin C. Webster, superintendent of the Lehigh Zinc Co. (where the largest steam-engine and pump in the country were put in operation). The Howellses had visited Bethlehem the preceding spring.

2. Stuart Wood (1853–1914), a Philadelphia manufacturer, received a Ph.D. from Harvard in 1875.

3. Una Farley and her sister were perhaps the daughters of Gustavus Farley, a Boston broker who had a house in Cambridge. James B. Greenough (1833–1901), the noted philologist, then assistant professor of Latin at Harvard, was also an author who had a talent for acting. Interested in the education of women, he helped found Radcliffe College. A one-page printed cast of characters for "The Queen of Hearts. / A Fantasia on / a Well-known Theme / in three acts" is enclosed with Elinor's letter. The play, which had been performed at H. W. Longfellow's house on Wednesday 5 January, was reviewed four months later by Howells (May 1876 *Atlantic*; *SL*, 2:38).

4. While Elinor had been in generally poor health through the preceding summer, she now seemed much stronger, and she and Will had an unusually busy social life this winter. "I was out four times and Elinor twice, and we had invitations to parties or dinner every night in the week, I believe," Howells had just reported to his father on 9 January (*SL*, 2:87).

5. William Waldorf Astor (1858–1919). Sole heir to John Jacob Astor's fortune, he was practicing law in New York at the time and had been introduced to Howells by John Hay in his letter of 10 December 1875 (*SL*, 2:113) as "a very fine young fellow" who had "a decided taste for art." Astor, who was later appointed minister to Italy and wrote two books, eventually moved to England and became owner of the *Pall Mall Gazette*. He was made a viscount in 1919 after becoming a British citizen. Howells reported to Hay: "I never told you how very much we liked your Astor whom you sent to me. I asked him to lunch,—to Mrs. Howells' despair. 'Never mind,' I said, 'I'll have Smith send the lunch out from Boston. (Smith is the old colored caterer, friend of Sumner; character; sayer of things. . . .) Smith named over a lot of things for my lunch. 'Oh, good gracious, that won't do,' said I, beginning to rend my garments and looking round for ashes to strew upon my head. 'I'm to have the richest man in America to

lunch. Now, what?' 'My dear sir,' said Smith, 'you want the simplest lunch that can be got.' It was a success" (2 September 1877, *SL*, 2:172). Launt Thompson (1833–1894) was a well-known portrait sculptor. Born in Ireland, he had settled in New York in 1857.

6. Joseph Bangs Warner (1848–1923), the lawyer, was a friend of the Jameses. Margaret (Marnie) Woodbury Storer (1845–1922), his fiancée, who had become a close friend of Annie Howells, was the daughter of Sarah Sherman Hoar Storer (1817–1907), the widow of Robert B. Storer, a Boston merchant. Mrs. Storer was the daughter of "Squire" Samuel Hoar (1778–1856), abolitionist, Whig congressman, and Concord lawyer. Elizabeth, or Lizzie, Hoar Storer (1841–1919) was another daughter of Mrs. Storer. They all lived on Garden Street, not far from the Howellses.

7. Isabel Allison Boggs, the second wife of Edmund A. de Schweinitz (1825–1887), the Moravian bishop and historian.

8. Frank Duveneck (1848–1919) had displayed half a dozen portraits at the spring Exhibit of the Boston Art Club. Three additional portraits had been later shown at the Doll and Richard's gallery on Tremont Street (June and September 1875 *Atlantic*). Painted in Munich, *The Old Professor*, now at the Boston Museum of Fine Arts, had been purchased by the Howellses' longstanding friend, the art collector Dr. Henry C. Angell, in whose house Elinor had probably seen it. According to Dr. Angell's *Records of William M. Hunt* (Boston, 1881), William Morris Hunt (1824–1879), the popular Boston painter, had greatly admired this portrait (Carol Troyen, *The Boston Tradition*, 1980). Duveneck attracted several followers, "the Duveneck boys," who appear briefly as the Inglehart boys in Howells' *Indian Summer* (1886). Frederick P. Vinton (1846–1911), a pupil of Hunt and Rimmer in Boston, who had also studied in Europe, was a noted portraitist of jurists, statesmen, and writers. Howells sat for him in 1881.

9. Edited by Arthur Gilman, *The Cambridge of 1776* was published in 1876, with Howells' poem, "Dorothy Dudley," on the first page.

10. Two Goundies are listed among the members of the Bethlehem Moravian community in Joseph Mortimer Levering's *A History of Bethlehem, Pennsylvania . . . 1741–1892* (1903).

To Anne T. Howells

[Cambridge] Thursday morning[1]
[late February–early March 1876]

Annie dear!

Thank you for confiding this delicious affair to me.[2] So I exclaim at first, but it all depends whether it *is* delicious or not. It's a nasty thing to refuse a man I'm sure. Do stop the affair before it comes to that when you've made up your mind.

Strange how one's feelings change! Now you've really got a chance to matrimonize it does n't seem such a desirable thing— especially now you've a career just opening before you. But that all depends again. Where do the family live? Has Achille expectations or will he have to make his money? And, *is* he nice? You did n't say one word about that. You said Louis F. (why does he write his name with a small f.?) was as elegant as his note.³ Is the brother? *Do* send me one of his letters to read.

I should want to have it all right with papa Frechette & the family. Louis evidently favors the thing and helps it along as far as possible. Is Achille as nice as Ouimet?⁴

I dont believe religion would make any trouble between you. Will did not inspect anything though I opened the letter before him. I told him it was a matter of etiquette—as it is, in one way.

I think it quite according to foreign notions to have a go-between—and it is far better to have Louis than anyone else. Probably he will propose for his brother—if indeed this is not a proposal

I believe in marrying. I believe you would be happy. But you can be happy without as a distinguished authoress.

I dont believe you'd write if you got married. Would he encourage your literary aspirations? Where would you live? How does he look? You *must* tell just as soon as possible. I dont promise to keep another letter from Will. And things must come to a point soon.

Warner says all the best men at that French wedding lately (Mr Richard Greenough showed us a description in a French paper which nearly killed us) came around, or were brought around to the Albion hotel dead drunk after the wedding breakfast and a *frechette* was of them I saw.⁵ Does *Achille* drink? What was his opinion in the Guibord case?⁶

But if you are going to marry him it's none of my business and of course you needn't answer all these questions. It seems so strange to me that all this has happened while you've been so busy with all sorts of things—while you've constantly corresponded with us and kept up to things in all respects. Would it be thought a good match in Quebec? How old is Achille?

Foreigners always have short engagements. O Annie!

I am dying to show the letter to Will. Do give me leave on a postal card. Does Achille write like Louis—and on such elegant paper? They certainly must have good taste! And to think you did not dream of this last October! Did *seeing* Warner & Marnie together make you want to be engaged?

I should think you'd like all the surroundings of a marriage with Achille—but after all it's the man. You will be left alone with him in Canada when your sisters leave or get married elsewhere (dont put the best side out and hope they too will both marry there, for I put the case purposely as severely as possible) he may be paralysed early (I always warn girls with that) and you have to take care of him all your life—he may be jealous if you talk with more brilliant men—he may have french ideas about woman being submissive (though I believe that *is n't* a french idea) &c &c. Is he rich enough to take you to Europe? I'd make that a condition.

I am so excited about this I cant write any thing else now. We are all well. How nice that Joe went to see you! Warner told us all about you.

Affectionately

Elinor

What do Vic & Aurelia say—& your father if you have told him.

Manuscript: Alfred

1. Dating derives from Elinor's mention of Louis Fréchette's letter of mid-February on behalf of his brother. See the fifth paragraph of this letter.

2. Over the Christmas holidays, Annie, who had gone to Quebec in 1874 when her father became U.S. consul there, had met Antoine Léonard Achille Fréchette (1847–1927), a young French-Canadian journalist and official translator in the House of Commons at Ottawa. They had soon fallen in love, and by the end of February, Achille was to declare himself. Their correspondence is at Alfred. For the story of their courtship, see Doyle, pp. 48–52.

3. Achille's elder brother, Louis Honoré Fréchette (1839–1908), was a noted poet also engaged in journalism and politics as a member of the Canadian House of Commons. He had introduced Achille to the Howellses in Quebec and had first written Annie on his brother's behalf. Louis-Marthe Fréchette (1811–1882), their father, was a building contractor.

4. The Ouimets were a prominent family in Montreal and Quebec, where several of them were active in politics, education, law and journalism. The one referred to here has not been identified.

5. Joseph B. Warner had just visited Quebec with Marnie Storer, his fiancée. Mr. Greenough was perhaps Richard Saltonstall Greenough (1819–1904), the portrait sculptor, or possibly his son, a recent Harvard graduate.

6. Probably a reference to Joseph Guibord (d. 1869), the printer, whose burial, which had been refused by the church because of his excommunication, was forced after his widow's court action.

W. D. Howells in the 1870s.

Winifred Howells, 1871.

John Mead Howells.

Mrs. Reverend Guild. Sketch by EMH, 1872.

The library at 37 Concord Avenue.

Mildred Howells.

Winnie, Pilla, and Johnny.
Sketch by EMH, 1 March 1874.

The W. D. Howells family, c. 1875.

Rutherford B. Hayes. WDH,
Sketch of . . . Rutherford B. Hayes, *1876.*

Lucy Webb Hayes.

Winifred Howells, 1877.

Winifred as "Columbia."

Anne Howells Fréchette.

To Anne T. Howells

[Cambridge] Thursday morning[1]
[Spring 1876]

Dear Annie:

Your very satisfactory* letter arrived at breakfast time. I showed it to Will and he was sufficiently astonished—and also amused at your philosophic and funny remarks on the subject. We talked and talked and I think the conclusion we came to was that we were in no haste to have you marry—though before the subject was presented solemnly we would have both said it was the one thing desirable—our theory being that marriage is *the* thing. Of course there is a feeling of losing you a little if you marry a frenchman.

There is hardly anything to say at this critical moment. We'll *support* you either way you decide!

But what I want most is to see some of Mons. Fréchette's letters. Do send at least one. Marmette[2] was nice—but when it came to marrying—I think I should prefer one of my own race. But the Fréchettes are less Frenchy are they not? Would A. be devoted if you were a miserable invalid all your life? An American would, you know. We both think this offer a great honor to you. You've certainly "fetched" one of the best Canadians.

We are in suspense till the thing is decided—but think slowly and surely—you've known him so little

Elinor

Manuscript: Alfred

1. Elinor's letter was probably written in the spring of 1876, after Achille's proposal, but before his week-long visit to Annie and her family in Quebec in mid-July.
2. Probably Joseph Marmette (b. 1844), a French-Canadian author whose name appears on Annie's list of wedding guests (at Alfred) the following year.

*in the way of answering questions I mean

To Lucy Webb Hayes

Cambridge
Sept. 1ˢᵗ [1876]

Dear Cousin Lucy:

I made the same criticism on Cousin Rutherford's picture that you did and *insisted* on Mr Houghton's getting Baker to throw a deeper shade across the eyes.¹ They had to bring him in a carriage from an inebriate asylum to the Riverside Press to do it. He is drunk most of the time. But I think it is slightly improved. He tried to make the nose less sharp & thin but did not dare alter it much. It is not the photo I should have chosen anyway

Mr Howells has been sick but is on his feet again today. Uncle Horatio Noyes is a great help to him on the book—reading ahead and making notes.²

I hope Birchard has recovered from his fever

Much love to all
Affectionately
Elinor M. Howells

Manuscript: Hayes Presidential Center

1. On 30? August, EMH wrote Francis J. Garrison (Princeton) to "insist that the nose should be heavier—not so sharp & thin—the nostril as high & strong as possible, and a heavier shadow thrown across the eyes." (Published with permission of Princeton University Library. William Dean Howells Collection.) Garrison (1858–1916), son of William Lloyd Garrison, the abolitionist, was working at the Riverside Press. Joseph E. Baker, noted for his caricatures of the Civil War, was a lithographer and pencil portraitist. His portrait of Hayes faces the title-page of WDH's *Sketch of the Life and Character of Rutherford B. Hayes.*

2. Horatio S. Noyes (1814?–1883), editor and businessman, was a cousin of Hayes as well as an uncle of Elinor.

To Victoria M. Howells

Pepperell, Sept. 16ᵗʰ 1876

Dear Vic & Aurelia:

Annie is writing you down in the library about Achille—and I am going to write a few words about him myself to tuck into her letter. We have had a nice pleasant visit from him here in the country, and a much better opportunity to get acquainted with him than we possibly could have had in Cambridge.¹ But I felt acquainted with him the first evening he came, for he is one of our sort: quiet, cultivated and perfectly natural in his manners. I was so afraid Annie (I still tremble for Aurelia) might marry one of those incomprehensible foreigners I saw in Quebec, and it is such a relief to find her marrying someone who is *human!*

Both Will & myself are perfectly satisfied with Mr Frechette. I admire his calmness & strength of purpose—the children take wonderfully to him, Winnie saying: "He is just like papa,"—and Will's comment is: "He is no fool" this on just seeing him, but of course he will write you his own opinion presently. I dont wonder Annie likes him, but I do think it is rather hard for the trio to be broken up now it has come to it: I do sincerely offer you my heartfelt sympathy while I offer Annie congratulations. But do try to bear up for poor Annie's sake—who seems to think she is almost committing a crime in getting engaged and revenge yourselves by going and doing likewise.

I thank you all very much for your kind expressions about my loss.² There is not a day that I do not think "I wish I could tell Mother this—" She took an interest in everything concerning us—and every Sunday I used to write her a long letter. It is such a pang to think "She is gone." Winnie thought she *"could n't* die before the *election"*—that was her expression. She took such an interest in her Cousin—and as it happened her last letter was to him. Annie will tell you about it.

We are looking forward with great pleasure to the visit from "Grandpa." We got his letter today saying he was coming to see us before going to the Centennial. One of you girls certainly ought to go along.

Affectionately
Elinor

Manuscript: Harvard

1. After coming to Townsend Harbor to see Annie and to meet her brother and sister-in-law, Achille, who was on his way to the Centennial Exhibition in Philadelphia, had gone back to Cambridge with Howells. Annie and Achille had just become engaged.
2. Elinor's mother had died on 24 August.

To Susan L. Warner

Cambridge,
Dec. 14ᵗʰ 1876.

Dear Mrs. Warner:

I dont know how I shall ever thank you enough for that beautiful scarf and for remembering me in that far-off land! Do those wild Arabs really wear such beautiful head-dresses as that? I never saw one before except in a photograph—and even now I dont know the name of it. You must write me a full description of it, as Mr Warner did of the Scarabaeus.[1] Meanwhile I am content to admire its beautiful texture and color, and I keep it on exhibition in my parlor for the benefit of my friends.

I've a great longing to talk over Venice with you. I am sorry you wernt in better health there. Are you quite well and strong now? I have been sick for two months past with toothache—till I had the tooth out a week ago, and now I am getting back my strength.

Cant you and Mr Warner pay us a visit this Winter so we can talk over Venice and all your wonderful adventures since we met— and persons & things generally? We are seeing no company this Winter—on account of my Mother's death last Summer—but we could have a nice quiet time.

I see Mr Warner has been lecturing in Hartford. Perhaps lecturing will bring him to Boston before long—and then you must come too. We heard of Mr Twitchell at Cornell through Mr Boyeson the other day.[2] What do you think of my letting Winnie be away all Winter?[3] Are you all as gloomy about politics as we? Please remember me to Mr Warner.

Affectionately
Elinor M. Howells

Manuscript: Watkinson

1. In his letter of 22 November (Watkinson), Howells had already thanked the Charles Dudley Warners for his book and for the gifts they had brought back from their trip to Europe and the Orient. Mildred Howells' note at Harvard accompanying the typescript of that letter specifies that "The book Warner dedicated to him was *In the Levant,* and the scarabaeus was set as a scarf-pin, and was one of the very few pieces of jewelry he ever wore." In a note in *LinL,* 1:223, she adds: "When he was summoned to meet the Crown Princess Frederick of Prussia in Venice, Mrs. Howells, in the ex-

citement of the moment, cleaned the scarabaeus so thoroughly that it looked like a piece of green soap, but the son of the family hastily restored it to antiquity with a burnt match." As for Warner's description of the scarab, Elinor may be referring here to a note that had come with the gifts. If so, this note has not been located, though the green scarab mounted on a gold pin has been preserved and is at the Howells Memorial. In his *Mummies and Moslems* (1876), pp. 98–99, reviewed by WDH in the July 1876 *Atlantic*, Warner describes at length the scarabaeus and its significance, and he relates how he asked their guide Ali to select for him half a dozen specimens to take back as gifts.

2. Joseph Hopkins Twichell (1838–1918), pastor of the new Asylum Hill Congregational Church in Hartford, was a very close friend of Mark Twain.

3. Thirteen-year-old Winnie was spending the winter with her grandfather and aunts in Quebec. As for the gloom about politics, see 21 January 1877.

To Victoria M. Howells

[Cambridge] Sunday p.m.
January 21ˢᵗ [1877]

Dear Vic:

We have been feeling Winnie's absence much more lately than the first of the Winter when I was sick with my tooth and we had more to do—and a feeling has been growing on us too that she wasn't doing very well in her French—so when your letter came we at once said we would not wait for any more experiments, but would have her home.[1] She has made a nice long visit and seen a Quebec Winter and improved in health and now I think she can bear a little close studying at home—so she will not be entirely behind her set of girls next Winter.

There is only one consideration which would make it advisable for her to stay—that she should go to the Convent for the rest of the time. If she will do this I renew all the offers & promises I made her before in case she should go. The days are longer now and it wont seem so early to get off in the morning and she would soon have a nice long daylight after school to play in. The English day at the Convent would be just as good for her, as the girls would all jabber French still and perhaps the sisters would make an exception of her and address her in French. I would supply all the aprons she wanted for the dish-washing—and I refuse to believe that the matter of food could make her give up such advantages as she could get at the Convent.

At any rate this is the way the matter stands: Either she must begin to go to the Convent at once or begin to pack up her things to come home.

We will send money directly we hear her decision. But please dont let her lose any opportunity to come—as we would return any sum you gave her to come at once.

I think she will have to borrow a trunk or a bag to bring home all her accumulated riches in—which we would return whenever you said.

I wont begin here to thank you for all your kindness to her as she *may* conclude to go to the Convent. *Nothing* else would satisfy us. It is our final decision. It would be something to remember all her life—a really useful experience to her I think—as they would have no chance to convert her in those few hours and she is quite able to judge of humbug. But they would n't bring any stress to bear on her I feel sure—and you could keep the run of what they were doing. But perhaps they dont take scholars for less than a year—and that would end the matter.

Perhaps some of the Seymour party will be going to New York and would take her.² Will would go to Springfield to get her.

Tell Annie Boyeson asked about Mr Frechette & said he was invited to the wedding. Also that he "must answer her letter"—in his usual patronizing way. Will thinks he is hardening as he grows older—and I find him much more priggish.³ You were right in your surmises about the letter. It was given him by mistake—and I confess I felt very much ashamed that he should have been asked to mail *that* letter, and also fearful—which was very wrong I know.

If Winnie has not bought a white sack tell her I saw a pile of white fur sacks big enough for her for sale at Chandler's for *five* dollars.⁴ Isnt that as cheap as in Quebec?

Mr Frechette has not sent us any photograph yet. I should like one so much. We will try to call on Christine soon.⁵

I sympathize with you in your change of servants. Hope things are going better now. We are utterly in the dark about politics—though of course we dread the counting of the votes—the Dems.

seem to be so satisfied with it.[6] This is a *business* letter. Do excuse it, as it was written directly after dinner & in haste

Elinor

Will got back the letter he sent to Charley Wade today but will now try his business address.

Manuscript: Harvard

1. Returning home after her visit to the United States, Annie had brought thirteen-year-old Winifred Howells with her to Quebec, where they had arrived on 27 September. Her parents had sent her to spend the winter with her grandfather and aunts, and "for the sake of the French," as WDH put it in his letter to Conway (5 June 1877, *SL*, 2:166). Winny did not return to Cambridge until 7 April.

2. General Silas Seymour (1817–1890) was an eminent civil engineer. Appointed in 1872 engineer-in-chief of the North Shore Railway, he settled in Quebec until 1878, when he became president of the Massachusetts Central Railroad Co. and moved to Boston. He and his family are frequently mentioned in the Howells letters at Alfred.

3. Elinor and Will had become increasingly critical of Boyesen's egocentricity (*SL*, 2:22). As Howells had written to his father, "Elinor paints his defects somewhat boldly, but he's undoubtedly a sponge—a finer sort of sponge, but a sponge" (31 January 1875, *SL*, 2:89).

4. A dry-goods store on Winter Street in Boston.

5. Most probably their friend Christine Chaplin (Bush) (1842–1892), an author and artist, who specialized in watercolors of wildflowers and illustrated several little books of her own verses.

6. In the disputed election between Hayes and Samuel J. Tilden (1814–1886), the governor of New York, who had been nominated at the National Democratic Convention in St. Louis, Tilden had received 184 votes against Hayes's 165, with twenty contested votes claimed by both parties finally being awarded to Hayes.

To Anne T. Howells

Elijah Anthony's Farm
Conanicut Island, June 12th [1877]

Dear Annie:

Your father's invitation to your wedding came tonight—[1] but even that does'nt make us realize the important event: When I first direct a letter to Mr Achille Fréchette for Mrs Fréchette—then I suppose I shall take in the astounding fact that you have really begun a new mode of life, and that it isn't all a joke. There are people to this day who consider it a comical thing that Nelly Mead is married—somehow I did n't seem the *kind* to get married! Well you know either of your sisters seem more *the kind* than you—but all I have to say is—if you become as domestic as *I* have become—but perhaps I am saying too much! This air from the gulfstream is affecting all our brains. I *could* n't write for several days past—and now I write as in a dream—but I feel that the time is *so* short in which I can say saucy things to Annie—soon she will have a husband who wont permit his wife to be insulted!

The invitations are *very* pretty. Whose writing is it? I see the paper is from Boston. I feel that a great deal is being kept from me. You are all too busy to tell me now—and afterwards there will be other things! I shall never know. Is Achille to wear morning dress? What? What colored kid gloves shall Will get? Yellow, & a blue necktie?

We have been having a rainy week and are all in the doleful dumps. So far we dislike the sea side very much. We are damp, sticky & uncomfortable. The sea water spots our clothing, and our heads are so heavy! You should see a fire we've got in our chamber in a big old fireplace with a crane in it. Two bricks for andirons & the fire made of corncobs & drift-wood & some green branches of trees. Wood is scarce on this island. We have a nice beach of our own where the children have begun a grotto of shells and pebbles. John made a fire there while it was pleasant this morning. Mr Anthony is to build us a bathinghouse.

Will went over to Newport today. He met Col. Higginson[2] at the boat with an armful of laurel blossoms which he had got for Miss Woolsey's (sister of "Susan Coolidge") wedding tomorrow. She marries Gilman, Pres^t. of St John's College, Baltimore.[3] I hope you will have a less dismal week for your wedding. Will bought a railway guide today and has decided to go by the Vt. Central. He leaves here early Monday morning. He gets back to Cambridge the 23^d and brings Birchard Hayes down here for Sunday & then Tuesday Will & I go up to Cambridge with him to see his father & mother who will be there the 27th.[4] We have asked them to stay with us. The President writes he will if he stays under any private roof. I shall ply Will with questions about the wedding, but it is of Christine Chaplin, when she comes back, that I learn about it. Will *cant* describe anything unless it is on paper. He was weighed since we came here & the scales said 150 pounds—the largest yet! But our landlady, a handsome woman of forty five, weighs two hundred and seventy pounds! She has weighed more.

I dont believe you've the time to read all this nonsense. I send back the scrap of your blue silk which could n't be matched. Well, at any rate you'll be "Ann Taunce"[5] all the same to the children! Pilla spoke of something's being "the reverse" the other day—and of her "usual voice." She beats us all on talking. I wonder what Mr Fréchette would think of *this* boarding-house?[6] The outside is shingled and was never painted—and never had blinds till this Summer. The Eustises (father of young man you met, & his family) have a house on this island & yacht of their own. They stay late at parties in Newport & then come home in it. The last ferry-boat comes at seven. Old Mr James & his family are at Newport this Summer. When my head gets clear I'll write a *farewell* letter! Where are you going to stand to be married? Will Mrs Louis be there?[7] Did the Count get Winnie's picture?[8] Love to all from all

Elinor

Direct to Box 160 Newport R.I.

Manuscript: Harvard

1. Annie's wedding was to take place on 20 June. The plans were for WDH to go to Quebec alone while Elinor stayed with the children on the island (WDH to EMH, 19 June, *SL*, 2:167).

2. A son of Louisa Storrow and Stephen Higginson, and the uncle of Annie Higginson and Agnes H. Fuller, Thomas Wentworth Higginson (1823–1911) was a Unitarian clergyman with a variety of literary and reform interests.

3. Susan Coolidge was the pen name of Sarah Chauncey Woolsey (1845–1905), who wrote poems and prose sketches for newspapers and magazines. Elizabeth Dwight Woolsey, her sister, was the second wife of Daniel Coit Gilman (1831–1908), the first president of Johns Hopkins (1875–1901).

4. For Will's and Elinor's earlier invitation of 28 May to the Hayeses, see *SL*, 2:165. The Howellses went up to Cambridge to join the President and his wife in some of the Commencement festivities, seeing them again subsequently in Newport, where they introduced their two younger children, John and Mildred, to them (2 July, WDH to his father, *SL*, 2:168).

5. For "Anne Thomas."

6. As compared with the cottage in Townsend Harbor, Mass., where he had met the W. D. Howellses and proposed to Annie the summer before. The Eustises are not identified. Of Newport, Howells wrote to his father at the time: "Newport is quite a revelation, and if I had plenty of money I should like nothing better than to see it thoroughly for one season—and then never see it again" (*SL*, 2:170).

7. Achille's sister-in-law, Emma Beaudry Fréchette, did attend the wedding with her husband.

8. The "Count" was José Antonio de Lavelle y Romero, Count Premio Real (d. 1883), the Spanish consul general in Quebec and a close friend of the Howells family there. Under the pseudonyms Fieldat and Aitiache, Premio Real and Annie had published in January *Popular Sayings from Iberia,* an edition of Spanish proverbs in an English translation, that Howells reviewed in the September *Atlantic (SL,* 2:170–171). In a photograph taken during her recent stay in Quebec, Winnie is dressed in American flags as Columbia for a masquerade ball at the skating rink (Aurelia Howells' diary, 30–31 January 1877, Huntington).

To Anne H. Fréchette

On the Island of Conanicut,
R.I.,
7 miles long & not a mile wide.
June 22ⁿᵈ 1877.

Dear Annie:

I had a letter from Will written at Montreal on his way to Quebec (in which he said something *so* ridiculous: that he was going to dress on board the boat as he feared there would not be houseroom enough at his father's!) and in it he said he would write me all about the wedding next day.¹ I have not got his letter yet and am almost dying of curiosity; Perhaps you would like to know how we spent your wedding day at Conanicut? We have a private beach down at the foot of the pasture—a lovely, sandy shore where we go every day, and where there is a bath house and we shall bathe there when it is warmer. Thither we repaired towards noon—and at twelve precisely—reckoning the difference in time here & in Quebec—we stood in a row on the shore: Mamma, Winnie, John & Pilla [sketch] with three stones in each our right hands—a red a yellow & a white & after wishing health & happiness to the newly-married pair each threw in first a red then a yellow & then a white saying

"Here's red for health
and yellow for wealth
And white for peace
May it never cease"

And as the water of Narrangansett Bay where we threw it in mingles in a very roundabout way—perhaps I should say connects—with the water of the Ottawa river—that is the nearest way we could reach you with our wishes. Afterwards, we gave three cheers for our new brother and Uncle. I send you a stone of each like those we threw in—only smaller by fifty times. And I send you some flowers we picked coming home through the pasture after we had performed our mystic rites.

I fancy you saw very little of Will that day? He was to send me Mr Frechette's address. I shall have to guess at it. When you get time do write & sign Annie T. Frechette so I can realize that you've "gone, went and done it." Much love from all to you & our new relative.

Affectionately

Elinor

Manuscript: Harvard

1. Howells' letter of 19 June to Elinor is printed in *SL*, 2:167.

A Country Retreat
August 1877–1880

*W*hen *Elinor's brother William vis-
ited the Howellses on Conanicut Island for a few days in early July, they
may well have discussed plans for a house. For some time Will and Elinor
had longed for a country retreat not far from the city, to escape the "storm
of visitors" and the distractions generally of their Cambridge life. As the
letters of 1877–1878 indicate, further plans and the actual building developed rapidly, so that by midsummer the Howellses had moved into Redtop,
as the house came to be called. A shingle-style Queen Anne cottage, it had
a long sloping red roof, the first story in brick and the upper floors finished
in California redwood shingles. With its gables and windows of different
sizes, and all the woodwork painted "dark green, almost black," it was
something unique. As Howells wrote his father in the spring of 1878, the
place was "blossoming out into a very quaint and peculiar beauty." More
than anything the Howellses were thrilled with the "SUPERB" site "on the top
of a very high hill—from which all Boston is overlooked & the adjoining towns,
and one sees way to the sea," as Elinor wrote at an early stage in the planning.*

*In the immediately ensuing years, Belmont also offered the close family life
that the parents had anticipated. Needless to say, there were visits, most notably*

to the White House in 1880. And plays and novels continued to make their appearances, among the latter The Lady of the Aroostook (1879) and The Undiscovered Country (1880), a novel about the Shakers and spiritualism, named by Elinor and considered by her the "best yet."

To W. R. Mead

[Cambridge] Tuesday Oct. 16th [1877]

Dear Will:

(I hope you dont work much over that transparent paper you sent the plan on—that alone is enough to make you sick—it smells so.[1] I cant keep it near me while I write) We sent you a telegram this morning asking you to come up tonight—but lest you are so busy you cant I write. We think now we will build this fall on Mr Fairchild's land—but we cant afford to pay much as we may not be able to sell this house for some time & will have to pay Mr Fairchild ten per cent rent—so we would like to put all the money on the outside—the shell, and finish when we are able.[2] At present we would only live out there Summers. We would n't put a furnace in, nor paint more than was necessary—& leave everything out that could be done later. We dont want a range. In this way we hope we could lay out quite a good house. Could we?

Your plan was wrong because you did not know the lay of the land. House faces east. Road runs by it on the north. The magnificent view is towards Boston—directly east. The situation is SUPERB. Mr Fairchild's place is a very extravagant one. He gives us this knoll to build on, right in his grounds & almost for nothing.

I should want the library in south east corner—dining room north east kitchen west.

It appears to me like this [sketch]

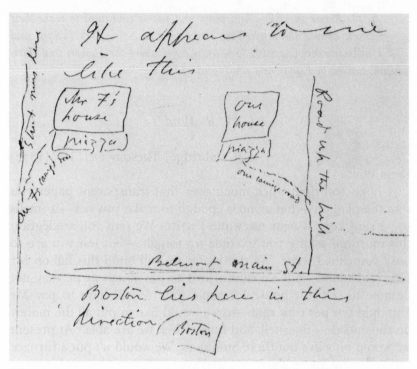

We want you to see a brick & wood house that we like.

You'll have to come soon to build it this Fall—& we must build now or never.

You never saw a finer view that from Mr F's place. His wife was Lily Nelson of Medford Mr McKim may know of her—a beauty & a reformer. Taught down South.

Elinor

Manuscript: Amherst

1. For some time, Will and Elinor had been thinking of moving to the country to escape from the social pressures of Cambridge. On 23 January, for instance, Howells had written to his father, "we both feel the disadvantages of our present uncontrollable way of living so much that we talk very seriously of going into the country for two or three years, now when it w'd do the children so much good, & w'd give us the sort of repose & retirement that we both need" (*SL,* 2:119). Elinor's brother, W. R. Mead, the architect, had been working in New York with Charles Follen McKim (1847–1909), whose brother-in-law, William B. Bigelow (b. 1853) had just joined them as partner in the newly-created firm of McKim, Mead & Bigelow, at 57 Broadway.

2. An art collector and a friend and patron of many writers and artists, Charles Fairchild (1838–1910) was then associated with S. D. Warren & Co., paper manufac-

turers in Boston. In 1880, he was to become a broker with the Boston firm of Lee, Higginson & Co. A set of some thirty letters to Mead, most of them by Elinor, written from October 1877 to July 1878, during the construction of the Belmont house, are in the William Rutherford Mead Papers in the Amherst College Archives. For further details about the building of the house, see Ginette de B. Merrill, "Redtop and the Belmont Years of W. D. Howells and His Family," *Harvard Library Bulletin*, January 1980.

To W. R. Mead

[Cambridge] Wednesday morning[1]
[17 October 1877]

Dear Will:

Of course it will be as well—if not better—to see Mr McKim himself, and it is very fortunate for us that he should happen to be coming on just at this time.

Now what we want is that he should spend all his spare time on us—and to that end why can not he come out here to stay Friday night—or to dinner Friday if he lectures in the morning. Please ask him to do us this favor.

Tell him to take a Garden St. car at the Revere House & go till he comes to Cragie St. We live just where the cars leave Concord Avenue for Cragie St.

We have no idea what builder we should like. Bugbee is too dear.[2] How will Mr McKim find out about a builder while he is here? And the cellar must be dug & the foundation laid *immediately*—as we dont want to build unless it can be begun this fall. The place is two miles out from Cambridge. There is a brickyard half way between. We want the lower story brick. We saw a cheap house over in Winchester six miles from here—with brick below—a five thousand dollar house—& we know the name of the man who built that. Norman.

But we must first talk with Mr McKim. We will take him out to see the situation. Please drop a line—just to keep the pot a biling.[3]

Affectionately
Elinor

Manuscript: Amherst

1. The letter is dated by the postmark.

2. The contractor-builder for their 37 Concord Avenue house, he was probably William A. Bugbee, a carpenter on Brattle Street (1877 *Cambridge Directory*).

3. McKim apparently could not come, as Elinor indicates in her next letter to her brother, two days later. Her brother visited the site with plans at the end of the month.

To Susan L. Warner

[Cambridge] Tuesday afternoon¹
[23 October 1877]

My dear Mrs. Warner:

I feel that it is a perfect imposition for such a sick creature as I am to visit anybody—much more to appear among the Hartford circle as a kind of spectre at the feast—but I *cant* let Mr Howells go to your house again without me and madden me, as he did before, with a recital of pleasant people, pleasant adventures and *that consertavory*² (shows the weak state my sickness has left my brain in) when he comes home—so I am coming!!³

Ten o'clock

Thursday morning is the time Mr. Howells has fixed upon for the start. If that is not convenient pray say so, for we can just as well come later—and if you should put us off it would seem to balance things because we put our visit off in such a cool way.

But if we do not hear from you expect us Thursday morning.

Very truly yours
Elinor M. Howells

Manuscript: Watkinson

1. Dating is from the Howellses' visit to the Warners, 25–27 October. See Elinor's letter of 26 October. On 17 October, WDH wrote Warner that his wife's sickness the preceding week might delay their visit a day or two after the 25th. On 19 October, Warner replied that he would look for them on the 25th, the next Thursday. Howells' letter of 21 October (see note below) anticipated Elinor's, which was sent two days later.

2. Sketch of a hand with finger pointing to her misspelling.

3. After his visit in March, Howells had written Warner: "It was a real joy to meet you and your wife again. I had to give a very full report of what you said, and how you looked, when I got home, for Mrs. Howells and I will never be divorced on account of the Warners. It's our united love that we send them now" (1 April 1877, *SL*, 2:161). A visit to Hartford had been set and then postponed till late October by Howells (20 September). Here Elinor is answering an invitation from Mrs. Warner (WDH to Warner, 21 October, *SL*, 2:176).

To W. R. Mead

Nook Farm,
Hartford, Conn.
Friday morning [26 October 1877]
Dear Will:
We are at Charles Dudley Warner's for a three day's visit—going home Saturday night. Two or three things have occurred to me about the house which I will jot down. Could you slip in a seat halfway up the stairs. The Warners have a conservatory out of their parlor with brick walls going around it which go down below the freezing point and so their plants grow right in the ground. Cant our wall be carried around such a little place—or could the wall be built afterwards? We *must* have room for a sofa in the reception-room. And I think nothing was said about a downstairs water closet. I dare say these things are not important & wont affect the walls of the house—but perhaps the conservatory might. And a blank place ought to be left behind to attach a washroom & shed to, unless we afterwards build a separate establishment like Gus's ice-house, workshop &c.[1] But still the hot water could not be carried to that, so we shall have to have a washroom—and perhaps it would be better to build it at once.

This is all I have to say. Mrs Fairchild has written me a beautiful note.[2] Mr Howells talks of nothing now but our Belmont Home—so I think the thing will go through. We will let our Cambridge house if we cant sell it, and so manage to pull through.

Barrett telegraphs from Cleveland "Play a greater sucess than ever.[3] Crowded Houses. Sure to be a permanent success"—or words to that effect. We are having a capital time here. Mark Twain, Louise Bushnell, Mrs Perkins & Mrs. Stowe (tell McKim as he is a "reformer") in last evening. Going to a party to meet Yung Wing & wife to night—& over to Manchester to dine with the Cheneys tomorrow.[4] I have told Warner that if anybody wants anything *Choice* in Hartford to employ McKim—& have suppressed the fact that my brother is with him. The Clemenses are greatly dissatisfied with their house. Potter architect.[5] It is too much like a church.

Elinor

Uncle Horatio does write at the R. Press every day.[6]

Manuscript: Amherst

1. In A. D. Shepard's New Jersey house.
2. Mrs. Fairchild's note has not been preserved.
3. Lawrence Barrett (1838–1891), the producer-actor, had opened Howells' *Counterfeit Presentment* in Cleveland on 22 October. Barrett's letters to Howells are at Harvard. On the production of this and other plays by WDH, see Meserve. Clippings of reviews of the plays are collected in a blue scrapbook now at Harvard.
4. Mary Beecher Perkins (Mrs. Thomas C.), the lawyer's wife, and her sister, Harriet Beecher Stowe (1811–1896), the famous novelist. Louise Bushnell was most probably a relative (likely a daughter) of Horace Bushnell (1802–1876), the author, preacher, and theologian, and a former Congregational minister in Hartford. In his letter to WDH of 19 October (Harvard), Warner had written that he wanted to take him and Elinor to South Manchester, "the ideal factory village." Of the several Cheney families there, these were probably Colonel Frank W. Cheney (1832–1909), a soldier, businessman, and civic leader, then secretary-treasurer of the Cheney Silk Mills in South Manchester, of which he was to become president. His wife was the former Mary Bushnell, who was to edit the *Life and Letters of Horace Bushnell*, her father. Close neighbors at Nook Farm, the Bushnells, Warners, Frank Cheneys, and Twichells usually summered together in the Adirondacks (Kenneth R. Andrews, *Nook Farm* [1950]). A Yale graduate, friend of Joe Twichell, and the first Chinese to earn a diploma from an American college, Yung Wing (1828–1912) was the director of the Chinese Educational Mission, established in Hartford in 1872. The reception for the Wings had taken place at the Mark Twains' (*Twain-Howells Letters*, 1:207).
5. Edward Tuckerman Potter (b. 1831) was the son of Alonzo Potter (1800–1865), Episcopal bishop of Pennsylvania, and the brother of Henry Codman Potter (1835–1908), a liberal clergyman in New York, also an Episcopal bishop.
6. Horatio S. Noyes.

To Susan L. Warner

Cambridge,
Monday evening [29 October 1877].

Dear Mrs Warner:

Before we got a chance to gush Mr. Warner's sentimental note came.[1] How sweet of you to sit up with us that night!—and how useless!—but we should have done just the same with you. I am sorry Mr Warner is not coming this week—but a longer visit from you both later will be more satisfactory. Barrett is coming to play a night at Worcester—in December I believe—and Osgood wants him to play A Counterfeit Presentment & have a carload go on from here to see it. If this goes off you will surely be there?

Isn't it queer that while I was absent Katy Burbank *called here?*[2] I was sorrier to have missed her here than in Hartford. The Hays are coming to visit us the 6th of November. I have never seen Mrs. Hay.[3] We call it a *perfect* visit that we had at your house—and the vision of your parlor is always before us, making our rooms seem close and stiff & pokey—but at that Belmont châlet we shall have a room like it—only it wont be the same unless the Warners' presence is associated with it—which it will be I hope through numerous visits. (Mr Howells has concluded the purchase of the land since we came home and we are going to build immediately.)

Yes, that was a *brilliant* visit: Mrs Stowe, Yung Wing and Manchester were sensations—but *"of course"* better than all was the sweet home circle and the day it rained. It did us both good. The ride home at night was nothing at all, and we did n't feel tired next day. The children never would have forgiven us if we had not come Saturday. John is delighted with the cocoons, and Winnie at the prospect of going to Hartford.

Winnie has been at the Angells in Boston this afternoon where they have *two* tailless cats—one gray & one black & white. They say they sit up on their hind legs exactly like rabbits.

I hope you'll enjoy your New York visit. With much love & thanks to you all—including Miss Lee & Miss Hess⁴—for making us have so good a time at your house

Affectionately yours
Elinor M. Howells

Manuscript: Watkinson

1. In his note of 28 October (Harvard), Warner had told WDH of their sitting up until eleven P.M. until they thought their friends safely home. He had also exclaimed over their charming visit.

2. Katy Burbank was an old friend of Elinor's from Cincinnati days (7 December 1862).

3. Clara Louise Stone (1849–1914) had married John Hay in 1874.

4. Fanny C. Hesse was Warner's sister-in-law and an occasional secretary of Mark Twain during the 1870s (*Mark Twain's Notes and Journals*, 1979, 3:217). Miss Lee is not identified.

To W. R. Mead

[Cambridge] Friday evening,
Nov. 9ᵗʰ 1877.

Dear Will:

At last the estimates of both carpenters have been submitted to Mr Fairchild; & this morning he telegraphed "Let Myers begin"—

so Meyers is now looking up his workmen.¹ We should have taken him out to see the place this afternoon if it had not been rainy—but are going tomorrow. Mr Emery's estimate was twelve hundred dollars more than Meyers's.

He was in tonight—very pleasant, and hoped we would give him our work after we settled in Belmont.

Will you now send on the contract—specifying the time of payments &c?

We had fifty-three people at our Lunch. The Hays left Thursday morning, & Bayard Taylor spent that (last) night here.²

Barrett's hundred-dollar check has come. Why does n't Joanna write—or Mary?

We've had a *lovely* letter from the President asking us to visit them—naming the exact time—but we cant go. Mr Howells wrote how much obliged we were—but said it was quite beyond our "little possibilities." The Pres. said they were good-natured—no business on hand only a message—that Emily Platt was there &c &c.³

Affectionately

Elinor

Manuscript: Amherst

1. George Myers, a contractor-builder-carpenter from Somerville. While the Howellses were in Hartford, Charles Fairchild had had the land surveyed, and Mead had come subsequently from New York with plans in order to meet with Will and Elinor and Fairchild to decide on the shape and size of the house to be built (EMH to Mead, 28 October). Fairchild's note of 6 November (at Harvard) in answer to Howells' (now lost) letter of 3 November, specifies the terms on which Fairchild was to build the Howells house on his land, the rent Howells was to pay, and his option to buy.

2. The Hays were enthusiastic about WDH's play and Barrett's acting. Hay's letter of thanks after their stay in Cambridge concluded, "Mrs. Hay and I can never forget our delightful visit with you—how you and Mrs. Howells made us feel so perfectly at home with you and your friends—and we shall never be quite contented until we have a chance to show you Cleveland & make you know how you are loved and appreciated there" (*Hay–Howells Letters*, pp. 29–30). For Bayard Taylor, see 20 January 1866 and 14 April 1878.

3. In his letter to WDH dated 5 November (Harvard), the President had invited them to "visit us soon—say from 15 Nov to Dec 1, with wife, chicks, and traps." Emily Platt (1850–1922) was the President's niece and Laura Mitchell's youngest sister, whose wedding was to take place at the White House in 1878.

To Aurelia H. Howells

[Cambridge] Friday morning,
Nov. 16ᵗʰ [1877]

Dear Aurelia:

I was glad to see the cheerful tone of your letter—but sorry that Henry is in such a nervous state—both for his sake & yours. I hope it is only a tempory thing. The *things* that have prevented my writing for so long have been first sickness, then a visit to Hartford, then the Hays's visit here* and, more than all, *getting our house started!* This surprises you I know—but has not Annie told you we were going to make a change as soon as we could, and have'n't I often written that I could not live where we had to go off Summers. (Every *soul* in Cambridge, nearly, leaves in Summer) Well, at last we have accomplished our purpose in this way: Mr Fairchild, brother of Gov. Fairchild¹ of Wisconsin & our Consul at Liverpool, a very rich young man owns a magnificent place just three miles out from here (on the Fitchburg railroad), on the top of a very high hill—from which all Boston is overlooked & the adjoining towns, and one sees way out to sea. He only goes out there Summers—living in a flat in Boston in the Winter. Well, he, knowing our fix, has offered us land at a merely nominal price, and is building a lovely little house on it, (pretty near his house but distant enough for us to be quite independent,) which we are to move in to next Summer, paying low interest on it (about 300) until this house is sold, and then buying it. It is so near Cambridge that we can be here in ten minutes on the cars. We can stay there Summers—and when this is sold, even for a low price, we can *own* it, which we have never been able to say of this house. This Winter Will is going to use every means to sell this house. We go in to the other next July—and, what we do not want Cambridge to know yet, do not come back. We will go to Belmont as if for the Summer (not a very terrible deceit is it?) in order to escape farewells. For we hope not to lose our best friends here—especially if we keep a horse.

*and writing to Annie Fréchette on *business!* sending patterns &c

I should think you would enjoy your little upstairs kitchen very much. It is so pleasant to cuddle up as Winter approaches! We are immensely curious about Daisy's engagement, and also as to the cause of Jenny Seymour's[2] being broken off. I should think you would enjoy the Count's visits greatly.[3] Tell him John Hay, who was at the court of Spain you know, recognized the names of the literary men that he pasted into Winnie's album. Winnie is so well this Winter—almost as tall as I too—and she enjoys her school very much. Yesterday both she and Johnny went to dancing school. They go every Thursday. Winnie is in an advanced class in French—so that my ambition is satisfied. Carol Dean[4] was to come here Thanksgiving—the 29ᵗʰ—but yesterday we got a letter from her saying she had been ill and would like to come today. So we expect her this afternoon.

Marnie Warner[5] was at the lunch we gave for the Hays. The first place she had been to. John Hay did a great deal for the play in Cleveland. But it has also succeeded in Pittsburg—so that we are really counting on some remneration from it. Wont that be splendid? Was n't it droll that Bertha[6] should have seen it—so few have. And she seems to have been delighted. If it comes near here—say to Providence—we shall go down to see it. Barrett is delighted with Bartlett's character; but he is a nervous creature, and has nearly worn Will out with telegraphs & complaining letters and suggestions &c

At last Will has finished a whole new scene—Barrett[7] teaching Constance to paint—and sent it off. It will reach Barrett Monday at Indianapolis, and then our fate is decided. If it pleases him we are *made*! Will is *so* busy—with the whole Atlantic on his hands— no assistant[8] & the autobiographies. He ought not to have had the worry of this play—but there's *money* in it, and it was a great temptation. He is very well, only tired. He goes to a dinner on Beacon St. today—given for Godkin who is about returning to New York. We have been invited to visit at the White House, but have declined. Mary Mead is going. I hope her being a Cousin wont prevent her getting a beau there![9]

The weather here is lovely. Not a sign of snow yet. I think I have told *all* the news—& will keep you posted hereafter. Could you send Annie the part of this letter about the house? Much love to all.

Affectionately
Elinor

These handkerchiefs are from your sisters, Sissy & Lealie.[10]

Manuscript: Harvard

1. Lucius Fairchild (1831–1896) later served as consul general in Paris and succeeded Lowell as minister to Spain in 1880.

2. Probably Daisy Bourinot, the daughter of John G. Bourinot (1837–1902), chief clerk of the Canadian House of Commons, who wrote *The Intellectual Development of the Canadian People* (1882) and *Our Intellectual Strength* (1893). Jenny Seymour was another friend of the Howellses in Canada.

3. Premio Real. Winnie's album of autographs is at Harvard.

4. The daughter of William B. Dean (1838–1922), who went from Pittsburgh to St. Paul, Minnesota, in 1856, where he established a hardware business and served as a director of banks and of a railroad. He was a son of WDH's steamboat captain uncle William Dean, and Will had known each other since boyhood (*LinL*, 2:338). A student that year at Wellesley, Carol stayed with the Howellses during a scarlet fever epidemic at the college, much to Howells' exasperation (WDH to his father, 24 November, *SL*, 1:179–80).

5. Elinor had first written "Marnie Storer." She crossed out "Storer" and added "Warner" instead. Marnie Storer had married Joseph B. Warner in September 1876.

6. Probably Bertha Fieser Krauss, a friend of Aurelia, whose recent marriage is mentioned in Aurelia's diary (Alfred). Barrett brought the play to Providence on 28 December and to the Boston Museum on 1 April 1878.

7. Elinor means Bartlett, the painter, who was in love with Constance Wyatt.

8. George Parsons Lathrop (1851–1898), Hawthorne's son-in-law and Howells' assistant at the *Atlantic* since 1875, had resigned as of 1 September after a bitter exchange of letters (*SL*, 2:213–214). Howells was publishing several "Choice Autobiographies"; a few of his introductory essays had already appeared in the *Atlantic* (Gibson-Arms, pp. 104–105).

9. While the Howellses did not go to Washington until 1880, Elinor's sister Mary did. "Mary writes me of her delightful visit at the Executive Mansion and her letters to me since her return home show that in memory she is living over and over the happy period," her brother Fred wrote the President from Vienna on 15 February 1878 (Hayes Center).

10. Written on the left margin of the first page, this sentence is in a different hand, and seems to be a note to Annie, to whom the letter was evidently sent at Elinor's request.

To W. R. Mead

[Cambridge] Thursday evening
Nov. 22$\underline{nd}$ [1877]

Dear Will:

I ought to have written you so a letter would have got off by tonight's mail, but couldn't. Mr Myers was in this morning—the first time for two or three days.

They have struck that big rock, which extends way out under the house and are having an awful time. It may take a fortnight yet to get the cellar out. But meanwhile the walls are being laid. Mr Myers thinks next Wednesday will be as good a time as any for Mr McKim to come as he thinks he could follow his directions and there are many things he does not quite understand now. He seems to agree to everything you ask. We shall be much pleased to have a visit from Mr McKim next week, and I hope he will come.

I will ask a few questions, and say a few things beforehand that occurred to Mr Myers & myself. If mullion windows how are blinds to be managed?

The window in bath-room is bad—but cant be placed elsewhere that I see. We have concluded to build a little *"picturesque"* shed against the dead wall, back of the library—just to hold double windows, garden tools, carpenter's bench, wood &c. Please make design. And the cellar window will have to be moved in to corner beyond the middle back-door: Mr Myers thinks. (Mr Myers is going to put some extra timbers [?] (long ones) across the house—as a necessary support to the roof—without charge)

The *principal* thing I write about is that I would like to have the pantry built out from the house (Mr M. says it would not cost a hundred dollars) instead of being taken out of the kitchen—*if* it will not make that side of the house ridiculous. It should extend from end along the north side nearly to mullion window in kitchen—8 ft—and have a window in it. Will it be *too* absurd look-ing, or could it, with a shingle roof, be made to look like a bow window a conservatory or anything attractive? [sketch] You see it will have to be passed on the driveway. Of course Mr Myers

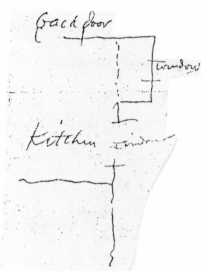

would like to know about it as soon as possible as he can lay the foundations in that part right away. Mr Howells is agreed to the plan, as the pantry cuts in to the kitchen too much. Could you send the dimensions for the thing? Mr Myers thinks eight feet long. How broad then? It could be put nowhere else and that would be very convenient if it *would look well.* I would nt think of having it of *wood.*

I thought the conservatory roof would be *shingled.* We will make it square & not octogon. Mr H. has given in.[1]

I think the new plans are perfectly bewitching. Monday Mr H & I are going up to see the cellar. Mr Myers wants us to, to see how much rock we will consent to leave in the cellar.

We shall expect Mr Mc Kim if we hear nothing further from you. Excuse this haremscarem letter. I am tired after making calls—& just going to bed.

Elinor

Manuscript: Amherst

1. Of the various changes proposed by Elinor, some were carried out, some were not. The Howellses later abandoned the *"picturesque"* shed. Plans were modified to provide for a larger bathroom and conservatory. The architects finally gave in to Elinor about her pantry, a convenient if unsightly feature of the house.

To W. R. Mead

Cambridge,
Jany 16th, 1878.

Dear Will:

I have let you have a good rest and now I must bother you a little again. In the first place, Mr Bigelow was so very satisfactory about everything that there is no need of asking anything till you come up—only I want to make a change in the arrangement of roof in order to gain more room—if you will consent to it. Let me say it is something Mr H. & I very much desire and I *may* do if you dont consent and you need n't include the alteration when you publish your design in the Architect & Builder.[1]

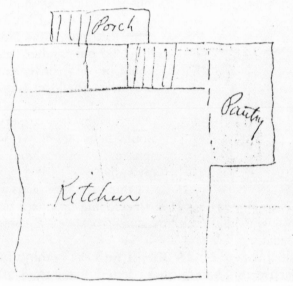

We want the main or the second-main roof to cover the stairs, entry & water-closet and the same amount of room in the story above, & also a second wooden story over the pantry. I am aware you crushed the second story down for the profile effect of the house towards the Fairchilds'—but if you will carry it back as I want it you may build a sloping roof over the steps, which will help it off. The space to me would be very "valuable" in a different sense from what you architects use it. Mr Myers has got on splendidly with the brickwork, notwithstanding the hindrance of the window-sills, and today it will be all finished. But he says he must roof now & board the outside before he puts in any partitions—so you will not need to come for two or three weeks—but please be ready to come then as we've a hundred things in the way of arrangement, ornament &c to ask you.

Was'nt that big window on the stairs to be leaded? (I would prefer to have it, so it could n't be seen through very clearly—& I dont want a curtain over it. Would a corner here & there of yellow glass [sketch] look well? And can the leading be done in Boston? As there will be but two windows leaded it would be scarcely worth while to have them done in New York I suppose?

We enjoyed having Mr Bigelow here and liked him very much— but he had no rest at all. He really had to work every moment. Myers is still worrying over the windows in the second story which have to slide up. He will probably slide the one in my room inside— & perhaps cut the beam some elsewhere. He wants to know if the rafters should be more than two inches broad—those which are carved on the end. His are two. Your measurements would seem to indicate three. I hope you'll allow the change in the roof. Of course you have a plan of the house to look at?

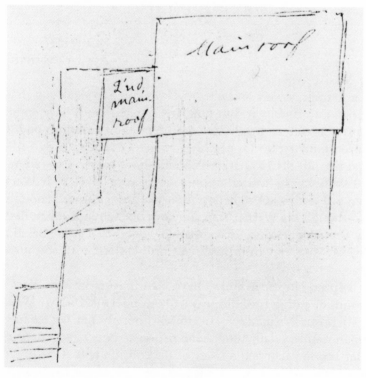

The blue is what we want added

You see the pantry cant have a second story unless the roof is changed—but that is not important as the other reason why we want the roof raised behind.[2]

Have got a nice letter from Fred

Elinor

Manuscript: Amherst

1. The architects did not publish any plans of the Belmont house. No plans have been located as yet.

2. The architects acceded to Elinor's request for a small sloping roof over the back stairs and porch, entry, water-closet, and also the pantry. But they refused to go along with a second story above any of these. The big window on the stairs does not appear to have been leaded, though a stained-glass window is alleged to have been there originally.

To Susan L. Warner

Cambridge
April 14ᵗʰ [1878]

My dear Mrs. Warner:

I spoke so impressively of the letter I was going to write you that I have never dared write it. But now that our husbands have come to be viewed as one and the same by the public I think we ought to be more intimate.[1] So I'll begin.

If I had dreamt Mr Howell's play would have gone off so triumphantly I should have insisted upon your coming on with Mr Warner to see it. I dont want to believe that you were not well enough to come—though Mr Warner says you have been dreadfully pulled down by that sickness,—and company I added that last—he did not really say that, but I do think you tire yourself with so much company.

What a spree our "men folks" have been having in New York! The amount of eating they did was truly wonderful. Did Mr Warner see Mr Boyeson's fiancée? Mr Howells thought her not *too* fine for "human nature's daily food." She proves to be a connection of mine. Her brother married a sister of my cousin's wife. One great thing was accomplished by the visit to New York. Dr Holland & Mr Howells became affectionate friends.[2]

We drive out to the house very often this pleasant weather, and the house begins to look very attractive to us. We hope that the first visitors who cross our threshold will be the Warners. You'll come on to see it before you go to the Adirondacks wont you? We hope to get in by the last of June. I have never liked my shawl since I saw yours—and Mr Howells actually took it to New York to perhaps exchange it for something at Stewart's not quite so striking. But the quiet shawls of a tone some like yours were enormously expensive. They professed to admire my shawl very much, though, & said there was nothing *like* it in the country. I hope there are not many. Ah well, when we go to Constantinople I can get one!

I received Mr. Warner's little note making fun of me about the supper—but I bore it bravely tell him—more bravely perhaps than I did that supper.[3] But a person must have nourishing food when they are exhausted & not raw oysters & salad—and no coffee!

Affectionately
Elinor M. H.

Mr Howells & Winnie send remembrances.

Manuscript: Watkinson

1. Elinor pasted on the letter the following clipping from the Boston *Transcript,* drawing a circle around the sentence about Howells. "The Atlantic for May will have, very appropriately, extracts from Thoreau's journals under the general title of 'May Days.' Professor Boyesen will contribute five sonnets on 'Evolution'; H. James, Jr., a paper on 'Recent Florence'; B. F. Taylor, a Revolutionary ballad, entitled 'The Captain's Drum,' and Mr. Howells another of his charming Adirondack sketches. Professor N. S. Shaler, also, will have a valuable article on the gold and silver question." In a postscript to his letter to Howells of 29 April (Harvard), Warner wrote: "The last thing that happened to me in NYork was to be introduced in a club, as the Editor of the Atlantic."

2. Howells gave his father a detailed account of the "continuous junketing" and round of parties with friends, publishers, and several Shepards and Howellses during his ten-day stay in New York (14 April, *SL,* 2:195–196). The visit had included a dinner at Delmonico's in honor of Bayard Taylor, who had just been appointed U.S. minister in Berlin, and a dinner at the Union League Club, "meeting all the New York sages in politics, literature and finance, among them Tilden, [W. C.] Bryant, and John Jacob Astor." Of this Howells commented: "I enjoyed it all, for the novelty and excitement, and was glad to have it over. I met and made up all old sorrows with Dr. Holland, which I was glad to do." Dr. Josiah G. Holland (1819–1881) was the author of sentimental novels and owner-editor of *Scribner's,* whose work Howells had criticized in earlier reviews as "heavy and trite," and lacking in realism. Boyesen's fiancée was Elizabeth M. Keen, the daughter of a Chicago bookstore owner.

3. Warner's note to Elinor has not been located. But, in the postscript to his letter of 29 April (Harvard) Warner was to acknowledge Elinor's invitation: "Mrs Warner is not well enough to go any where, but was well enough to enjoy Mrs Howells's delightful note."

To W. R. Mead

[Cambridge] Apr 14ᵗʰ [1878]
We have been out to the house today. Can it be that the piazza posts are upside down—or were they meant to diminish toward the bottom

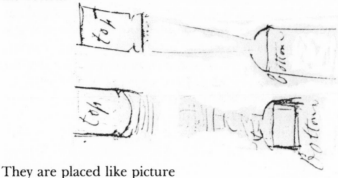

They are placed like picture

E.M.H.

Manuscript: Amherst

To W. R. Mead

Cambridge,
May 22ⁿᵈ [1878]

Dear Will:
We have just returned from a visit to the house, & I will at once report on its progress. It was the first time I had seen the dark paint on the outside, and it exceeded my expectations—though Mr H. had praised it very much. Now I can hardly wait till the brick is scoured & the blinds on, so as to get the full effect.[1] I wrote you what Prof. Shaler said. One lady says it looks *haunted*. Everybody is talking about it now, & a great many go up from Cambridge to see it. The land has been nicely graded around it. All it needs now are the trellises.

Did you mean the turned-work in gable of upper piazza to be

left in light paint? It looks so well so that we think you did. I hope so. The shelves, corner cupboard & window-side of library are nearly done. Hearth to be laid tomorrow.

Myers is impatient for mantles, so as to get hearths for them laid all at once. All are to be brick excepting reception-room which will be tile or blk. marble—brick being too coarse. I hope the mantle to [?] reception-room is being made as well as those for the chambers? Kitchen & china-closet done. Dining-room hardly begun. More men are to be put on.

We like the tiles Mr Bigelow sent. We laid them all out while we were up there. These are broken: (Can you bring some to replace them?)

The Last Supper
John Baptist baptizing Christ
Thomas's Doubt

Some horseman tumbling off his horse

So many [hand points up] are split in two

Corners are broken off these: The Blind Man by the Wayside (a man sitting—with a boy & a dog)
The angel appearing to Hagar in the Wilderness
(Both standing near a well)
The man falling from his horse has not a circle around it like the others. That one is very badly broken. The others could be matched—but will not the people who packed them badly replace them?

Fortunately the big pictures are uninjured. Now are the big ones to be set side by side above the mantle-piece in the dining-room & the others around them—or are the small ones to go around the fireplace. Mr Bigelow will know.

Mr Myers would perhaps like you to come the last of next week. Could you?

I see Mr Mc Kim has come out as an illustrator in Scribner. *Do* ask Russell Sturgis where the book—the "New Bath Guide" published in 1830 & illustrated by Cruikshank can be got. It must be very amusing I think.[2] How unlucky that you should have to be on jury when *our* house was just getting done. Not that Myers has really been hindered a moment!

I know all about Bishop now. *You* did n't know he turned Catholic for love! Molly will tell you what I wrote her. Willie (Dr.) James is engaged to a Miss Gibbens of Boston who teaches. Young Denton who was here in the Scientific School is married to Henry Trowbridge's youngest daughter—Mrs Cam Powell's sister. He lived three years on Cragie St. & I used to see him. Some Cambridge people went up to New Haven to the wedding.[3]

I like the Waring plumbing *very* much. Those [sketch] things hav'n't got to show above the roof. Must go to bed

Elinor

I was going to cut this off and then found I could nt. *So* sleepy! The landscape is *so* lovely about the house now.

Manuscript: Amherst

1. The blinds and other wood ornamentation were painted in a very dark green.

2. McKim's two illustrations, *Death* and *Misery*, appeared in the June 1878 *Scribner's Monthly* in an article entitled "Misery's Pear Tree. A Legend from Flanders," translated by W. Nichols. In the same issue was an article by Sturgis on George Cruikshank (1792–1878), the English draughtsman, political cartoonist, and book illustrator. Elinor is referring to Christopher Anstey's (1724–1805) *The New Bath Guide; or Memoirs of the B-r-d- [Blunderhead] Family; in a Series of Poetical Epistles* (1766).

3. William Henry Bishop (1847–1928), the novelist, later professor of English and Spanish at Yale, and U.S. consul in Italy. His *Detmold: A Romance* had started to appear in the December 1877 *Atlantic*, to which Bishop now contributed "Open Letters from New York" under a pseudonym, reporting on social, literary, musical, and artistic events (WDH to Bishop, 11 September 1877, *SL*, 2:174). Elinor's letter to her sister Molly about his conversation has not been preserved. William James married Alice Howe Gibbens (1849–1922) on 10 July. Huntington Denton, the son of Colonel Gabriel W. Denton, lived at 14 Craigie Street. His wife was probably the former Ellen Eugenie Trowbridge (b. 1856), the youngest daughter by his first marriage of Henry Trowbridge, of Henry Trowbridge & Sons, a well-known shipping house engaged in the West India trade. Henrietta King Trowbridge, another daughter, had married Stephen Camberleng Powell in 1871.

To Susan L. Warner

[Cambridge]
Friday eve., June 14ᵗʰ [1878]

Dear Mrs. Warner:

Our house approaches its ending—the carpenter, who was in to night, hopes to get his "gang" out in about a week and then the painters will riot there awhile—and you know you are to be our first visitors, (a sort of house-Warnering you know) so I write just to let you know that it is about time for you to begin to pack your trunks.

Are you going to the Adirondacks this year—& when? Dont dare to go without coming to see us!

I saw Gertrude Parker lately who said she had seen you in Hartford—but we hav'n't been hearing much from you.¹

Mr Howells is dining at the Nortons tonight. There seems to be an intention of *eating* us out of house & home here quite in the Taylor fashion.² Several of our friends have made dinners for us. The Fairchilds have been living at Belmont for three weeks.

We have fixed on the 3'd of July as the day for our moving. Dont you think you had better begin your packing?

Affectionately
Elinor M. Howells

Manuscript: Watkinson

1. Gertrude Parker was probably a relative of Edwin Pond Parker (1836–1920), a Congregational clergyman in Hartford.
2. A reference to the parties in honor of Bayard Taylor upon his departure in April as U.S. minister to Germany.

To W. R. Mead

[Cambridge] Saturday morning
June 15ᵗʰ [1878]

Dear Willie:

I got no letter from you this morning so I conclude you are too indignant to write. Yours containing Mr McKim's memorandum came yesterday, and I handed it to Myers when he called last night. *Now* I think I *should* like the red paint in the dining-room, so it can be put on, all ready for him when he comes—but if he wants the *wall* painted, as he says, he will have to send Twombly the color of the yellow. Perhaps he has done so.

The library we *really* do not want white. Some one suggested a "silver brown"—whatever that is. That would give it the sobriety we would like, but if the sage-green is thought to be the thing please tell Twombly at once—and the ceiling must not be so dark that one kerosene (study) lamp will not light it.[1]

Myers says the "gang" of carpenters will be out of the house after next week. The dining-room is ready for paint now & the library will be all done today. . . .[2]

We shall be delighted to see Mr McKim.

Affectionately
Elinor

Manuscript: Amherst

1. On 25 March (Amherst), WDH had written Mead: "we are afraid that, even with the third window added by Mr. McKim, the library will be too dark if paneled in redwood. Can you think of some other wood?" Twombly was the head painter.

2. In the rest of the letter, Elinor requested designs for the glass panes for the conservatory windows and for a trellis to be built by the library on the south side of the house.

To W. R. Mead

[Cambridge] July 3ᵈ [1878]

Dear Will:

Myers wants *you*[1] to come up & look over the house the last of next week & verify extras. Can you come? We move in Monday next & shall be all settled by that time. Come over Sunday. Mr McKim's letter was delightful. The vellum suggeston seemed to throw light. *White* it shall be![2]

Trellises bins [?] everything is done but library & stone wall. Winnies window is a success Twombly will be glad to see Bigelow Monday.

We are charmed with the fine work around the dining-room

chimney piece. *We* think the house perfection. The dining-room looks 200 years old.[3]

Elinor

Manuscript: Amherst

1. In his capacity as business manager for the firm and supervisor of the construction of the Howells house.

2. McKim's letter is unlocated. After her violent opposition to his suggestion that they paint the paneling of the library white (see 15 June), Elinor had apparently changed her mind. Cf. also *The Rise of Silas Lapham* (chap. 3), in which Seymour, the young architect, persuades Lapham to paint the woodwork of his drawing-room "white, or a little off white."

3. With its dining-room and its long, sloping red roof, "the most colonial feature of all" (WDH to Bigelow, *SL*, 2:190), "the Howells house was intended by McKim and his partners to be a modernized version of the Colonial" (Richard Guy Wilson, "American Architecture and the Search for a National Style in the 1870s," *Nineteenth Century*, Autumn 1977).

To William C. Howells

Shingleside,
Belmont, Aug. 11ᵗʰ [1878].

Dear Father Howells:

Today we have our first quiet Sunday, for only last night our last carpenter, who had staid to do odd jobs, left. He was not much trouble—sleeping towards the last in a barn nearby & getting his meals in some mysterious way: He always made the kitchen fire, so we suspected him of having a warm breakfast at least, but no traces of it were ever visible. When he first came we had no front door and he slept on a lounge in the dining-room for protection. He was a Maine Yankee with a drawl like Mark Twain and no end of yarns to spin—having been to sea, in the war and almost everywhere. We got quite attached to Gove, and he remarked when he left that he felt kinder sorry to go, for he liked the house and he liked the family.

You dont know in what an excited state we've been living for a month past. The painters staid with us a fortnight after we moved in, and I wonder we did not all die of the smell of the paint. But I've described our discomfort in a letter to Annie, and wont repeat

it here lest you see her letter. Why *doesn't* she answer my last letter!
She owes me two. And did you ever receive a pair of corsets I sent
to your care by mail for her? I am afraid you never did as you
have never spoken of them, and that that is the reason why she
has not written.

Our house is beautiful and comfortable & healthy beyond anything I ever saw. The view is a constant surprise & pleasure—and
the children are in a state of wild delight from morning till night.
There are constant demands—now that Grandpa should come and
see this house, and I think this Fall you must come—and bring
one of the girls. Toronto cant show anything to equal this house—
so I think the attraction is this way just at present—though Annie's
baby may be a piece of architecture even better worth visiting than
this.¹ But I am afraid we wont get that way till next Spring or
Summer—but then we certainly shall. How is little Mary getting
on? You have not mentioned her lately. Annie Hobart came out to
see us the other day solely to talk over the Quebec Howellses. Your
fatherly care, Vic's devotion—especially inviting her to meet the
Garfields—have won eternal gratitude from her.²

I think Fanny Eaton must have been a bright little girl from her
account of her. Think of her having earned the money for the
journey and only fourteen! I think she & Winnie should know each
other. We are going to have the house and Winnie photographed
soon by a traveling-cart man—and will send them to you. How
does Toronto go? I want to hear the girls' opinion. Is everybody
out of town or is it a slow place? We liked Mr Stewart very much.³
Henry's description of the runaway horse was very graphic and
amused our children very much. John is making a fine collection
of butte[r]flies & moths & knows all about them. We are going to
celebrate his birthday next Wednesday by fireworks & a little supper to the four Fairchild children. Winnie is in her glory devising
mottoes and prizes for the archery contest. They & our children
all have bows & shoot very well. This life is going to be a benefit
to our children—& may make a good naturalist out of John. He
has the Howells natural gift for it, and of course can have the best
of advantages for study at Cambridge. They will see plenty of so-

ciety here, and not be distracted by it as they were in Cambridge. We think we have made a wise move. We are puzzled for a name for our house. It is red and on top of a hill. Redtop is Will's choice but no one seems to like it. Sub-Hub, Monte Rose, The Parlor Car & The Spindles have been suggested. The last referring to a good deal of turned work in the ornamentation. The Fairchilds are very attentive & kind—giving us all our vegetables, asking us in to tea to meet pleasant people (Gen. Armstrong the other night) taking us to drive to Concord—and doing all they can to make Belmont seem pleasant to us.[4]

Will goes in town two afternoons in the week, and the rest of the time has perfect quiet to write in. We have the horse of one of Mr Fairchild's friends to use temporarily. It is sixteen years old & safe. Our phaeton has been made as good as new.[5] I shall soon get back in to living order and then will write to the girls and keep you posted as to our life on Wellington Hill. You do not write so often as usual—because you have been getting settled I suppose.[6] Much love to all

Affectionately
Elinor

Manuscript: Harvard

1. Marie-Marguerite (Vevie) Fréchette was born on 16 April 1878.
2. Annie Hobart was probably a relative of Minnie Hobart of Boston. Garfield and his wife had visited the Howellses in Quebec 12–16 August 1877 (*Diary of James A. Garfield*, ed. Harry J. Brown and Frederick D. Williams, 1967–1981, 3:506–507; see also Victoria Howells' invitation to Mrs. Garfield giving directions to 19 Hamel Street, 23 July 1877 [Library of Congress]). Fanny Eaton is not identified.
3. George Stewart (1848–1906), author and journalist, was founder and former editor of *Stewart's Literary Quarterly Magazine*. He was at the time editor-in-chief of *Rose-Belford's Canadian Monthly*.
4. Samuel Chapman Armstrong (1839–1893) had been colonel of the 9th Regiment of colored troops and later an agent of the Freedman's Bureau. In charge of the negroes in and around Hampton, Virginia, since 1866, in 1868 he founded the Hampton Normal and Industrial Institute, of which Charles L. Mead was a trustee and where Mary N. Mead served as librarian.
5. In his letter of 21 July (*SL*, 2:200–201), WDH had told his father how the pony horse he had just bought had run away, smashing the phaeton. This episode found its way into his amusing sketch, "Buying A Horse," *Atlantic*, June 1879.
6. William C. Howells had arrived in Toronto with his family on 10 June, shortly before the WDHs' move to Belmont on 8 July. Wellington Hill is now called Belmont Hill.

Henry James, Jr., to EMH

Wenlock Abbey,
Shropshire.
Aug. 14ᵗʰ [1878]

Dear Mrs. Howells—

I must thank you very tenderly for your generous little despatch on the subject of "Daisy Miller."¹ I am charmed to think that she struck a sympathetic chord in your imagination, & that having been, in fact, so harshly treated by fate & public opinion, she has had it made up to her in posthumous honors. She appears to have made something of a hit; for people appear to have found time to talk of her a little even in this busy & not particularly nimble-witted England. I thank you too for giving me a pretext to send you a reminder of the friendly memory I have of you and of the good wishes I always bear you. I have been hearing from your husband of your transmigration to Belmont & of the beauty of your house & situation there—a picture that fills me with envy even in the midst of English meadows.² May your days be long in your new home, & all your occupations pleasant. One of these days—before it is a very old home—I shall come and knock at its door.

I am spending a few days in one of the most curious and romantic old houses in England—an old rambling mediaeval priory, intermingled with the ivied ruins of a once-splendid Abbey, dissolved by Henry VIII. The place is full of ghosts & monkish relics and is, in every way, delightfully picturesque; I wish that for an hour I might be a well-bred British young lady, so that I might make you a sketch of it. But the lunch-bell tolls, & I can't even stay to make word-pictures. Believe me, dear Mrs. Howells, with every benediction on your house & family,

Very faithfully & gratefully yours
H. James jr

Manuscript: Harvard

1. James's story had appeared in the *Cornhill Magazine*, June–July 1878. "By mid-

1878 'Daisy Miller' had made him the most talked-of American writer in England,"
(*Henry James Letters*, 2:81). Elinor's dispatch is not extant.
    2. Howells' letter to James has not been preserved.

To William C. Howells

Belmont Oct 6ᵗʰ [1878]
Dear "Grandpa:"
    I send you photographs of all the children which we have just
had taken in a cart which stands in the village near our house.
They are pretty good for cheap photos.—and I like them because
the children were not posed. Winnie stared rather hard at the
instrument and so looks cross—& rather cross-eyed. They all have
bangs you perceive—but that is only natural as their father always
wears one. We have had the same man photograph the house also
and will send you one of the pictures as soon as they are printed.
Winnie is getting very tall and quite erect and will I hope be a
strong woman yet. She & John go to Cambridge to school but does
not start until quarter past nine and is home before three when
we give them a hearty dinner and then they amuse themselves the
rest of the day, except for a very little studying in the evening.[1]
John is under W's care on the railroad with the understanding that
when he does anything careless he is to be put to school in Belmont
& not be allowed to go to C. When Winnie cannot go he is to walk
over to Arlington & go by horsecars.
    Two hours later.
    My letter was interrupted by calls from Mr & Mrs Darwin (you
know Sara Sedgewick married Darwin's eldest son?) Miss Sedgewick & Sallie Norton, and, in another party, John Fiske & two
children & his friend Mr Hutton.[2] This is the way we are invaded,
even on Sunday. A show house is getting to be a burden and we
look eagerly for the drifts of Winter to keep people away. I suppose
your Winter will not be so hard in Toronto? Everybody says it is a

lovely city. By & by I mean to write oftener. I have not written any letters this Summer.

Much love to all

Elinor

Manuscript: Harvard

1. Winnie was enrolled at Mr. Justin E. Gale's Young Ladies' School at 17 Berkeley Street. John went to a private school at 27 Everett Street run by Miss Sarah H. Page.

2. Sara Sedgwick (see 17 March 1871) had married William E. Darwin. Miss Sedgwick was her sister Theodora (Dora). Sara Norton (1864–1922) was one of C. E. Norton's daughters. For John Fiske, who had five children, see 17 March 1871. Laurence Hutton (1843–1904), an editor, bibliophile, and author, is noted for his books on the theater and on his literary pilgrimages to cities in Europe and Asia.

To Aurelia H. Howells

Belmont Oct 19ᵗʰ [1878]

Dear Aurelia

It seems some time since we heard from you. I send pictures of the house which have been taken in order that a picture for the Architect may be made from them. Please send half the stereoscopic one to Annie—that she may get an idea of it, as we only have a very few pictures—half a dozen of each.[1]

You may not recognize Will & me on an upper balcony which opens out of my room. The room in front, with two little pointed gables above it & two queer windows in it, is Winnie's. Under it is the little parlor. The room under the piazza is the dining-room—with kitchen—& pantry sticking out—behind it. The window over the front door is in Pill's room—which opens into Winnie's. Above the dining-room is the guest-chamber. Around at the left—behind the conservatory is the big library—18 × 18. John's room is over the kitchen. This—with a big hall & little back entry is all. I shall try to send you some inside views—but only one of each was taken—just for the architects.[2] If you could only see the view in front reaching way to the sea! On the lower piazza are McKim & Willie, Mrs Fairchild & Mary Mead.

Oh but the *color* of the house is half! All the trimings which look so light are dark green—almost black & the shingling is all red.

Winnie has a window looking on to that upper balcony. There is nothing in front of us between us and the shore of Ireland. We are free of company now and getting rested.

Affectionately
Elinor

Manuscript: Harvard

1. This picture was taken on or just before 3 October, when Mary Mead, her brother W. R., and McKim, who had just returned from Europe, had dinner at Redtop (WDH to his father, 6 October, *SL*, 2:209).
2. A page from the McKim, Mead & White scrapbook, 1870–1888, at the New-York Historical Society, New York City, shows the outside picture and one inside view of Redtop: a detail of the library inglenook.

To Lucy Webb Hayes

Dear Cousin Lucy:

Let me introduce Mr Clarence King of New York, whose beautiful book: Mountaineering in the Sierra Nevada you may have read. He is the first young man I have found worthy to be sent with a letter to you—and let me whisper to you that both Mr Howells & I think him the most accomplished man of his age in the country.[1]

Yours affectionately
Elinor M. Howells[2]

Belmont
Jany 4th [1879]

Manuscript: Hayes Presidential Center

1. For Howells' letter of introduction to President Hayes, see *SL*, 2:216–217. Clarence King (1842–1901), the famous geologist, had contributed several articles to the *Atlantic*, where his book had been serialized, May–August 1871, and reviewed by Howells, April 1872. Four months later, King became the first director of the U.S. Geological Survey.
2. EMH's only other extant introductions to Mrs. Hayes in the White House are those of 31 March 1880 introducing George Parsons Lathrop (see 30 March 1873) and of 3 April [1879 or 1880?] (both Hayes Presidential Center), introducing her friends Samuel and Kitty Wells.

To Anne H. Fréchette

Belmont,
Feb 18ᵗʰ [1879]

Dear Annie:

As usual we enjoyed your letter greatly—especially your accounts of "Blossom"—as you seem to call her now. (I knew a Blossom Hobart once—a pretty name it is). We went over to West Medford to tea the other night, at Mrˢ Hallowell's, and there I met a young Miss Hoar of Concord who asked particularly for you, and said she had heard all about the baby through letters you wrote the Storers.[1] Was n't it odd that at the same out of the way house I met a Mrs. Chas. Sedgewick of Syracuse at whose house my brother Fred. had visited. They had travelled together in Europe. She was once a Miss Gannett & Kitty Wells had been out to lunch with her that day—that was Thursday. Saturday after Will & I went in & staid two nights at the Wellses'. She made a dinner-party Sat. night for us, & a tea Sunday. At dinner Tom Appleton, Mrs Tyson, & Rose Lamb. The latter paints finely & took a prize at the N.Y. Academy last year.[2] We met Tom Appleton on the street Sunday & he inquired who she was. I said Rose Lamb. "Roast lamb! said he: O I like that very much." You know he called Nahant (speaking of meat) "Cold, roast Boston"?

We have been dining out a little—have met Helen Bell & Mary Lodge (both widows & brilliant talkers) at the Fieldses' and are going tomorrow to the Fairchilds' to meet Mrs Hodgson Burnett.[3] The Fairchilds are just home from Europe. At Paris, at his brother's the consul's, they met Gen. Grant—& afterwards breakfasted with him at his hotel. Like royalty he was always served first. At first, ladies being present, he made a little motion to change the order of proceedings, but they protested and he submitted. I am not going to California at any rate—but I would like Will to go, and he may go if it is not for too long a time. It would be a good thing for him in the way of stories I think.[4] The Aroostook seems to find great favor with the public.[5] Winnie has written an "*Address to a Skeleton"* for a poem for her father's birthday! She said she thought she would not write "occasional poems" any more. I should think *not!* I will send it to you by & by. She has seen a real skeleton lately at her teacher's, in connection with her physiology. The children are *so* busy these days! all day long with Jenny Moore.[6] Much love to all your family from all of us

Elinor

Manuscript: Harvard

1. Of the several Hallowells who had settled in West Medford, this was probably Mrs. Richard Price Hallowell, the former Anna Coffin Davis (b. 1838). A grand-daughter of James and Lucretia Mott, the abolitionists, whose *Life and Letters* she edited in 1884, she was also an artist. Richard Price Hallowell (1835–1904), her husband, a Boston wool broker and merchant, was an abolitionist, historian, and also vice-president of the New England Woman Suffrage Association. Miss Hoar was probably a daughter of Judge E. R. Hoar of Concord and a niece of Mrs. Robert B. Storer (see 20 January 1876). Charles Baldwin Sedgwick (1815–1883) was a lawyer who practiced in Syracuse and a member of Congress from New York from 1859–1863. Kitty Gannett Wells and her husband lived in Boston.

2. Thomas Gold Appleton (1812–1884), an author and an art patron, and Longfellow's brother-in-law, was noted for his wit. Boston-born Rose Lamb, the portrait painter and pupil of William M. Hunt, had received two bronze medals in Boston exhibits, 1878 and 1879 (Clara Waters, *Women in the Fine Arts from the Seventh Century B.C. to the Twentieth Century A.D.* [1904], pp. 204–205). In Robert Grant's *Fourscore* (1934), pp. 282–283, Mrs. George Tyson, née Emily Davis (1846–1922), is named among the five "highlights of the aesthetic group" in Boston at the turn of the century. She gave readings in her house and was a "domestic sibyl."

3. A daughter of Rufus Choate, Helen Olcott Bell (1830–1918) was the widow of Joseph Mills Bell, law partner of her father at the time of their marriage. "Boston's greatest talker, the most famous Society wit in the city's history," she was noted for her bons mots (Cleveland Amory, *The Proper Bostonians* [1947] pp. 126–129). Mary Greenwood Lodge (Mrs. James) was an editor friend of Annie Fields. Frances Eliza Hodgson Burnett (1849–1924) was the popular and prolific author of children's books, whose *Haworth's* was being serialized in *Scribner's*, November 1878–October 1879.

4. President Hayes had invited Howells and Elinor to be his guests during a trip to California that he was planning (4 January 1879, Hayes, *Diary*, 3:518). Canceled in 1879, the two-month trip to California took place in the fall of 1880, without the Howellses.

5. Serialized in the *Atlantic*, November 1878–March 1879, *The Lady of the Aroostook* appeared as a book on 27 February 1879.

6. In his letter of 23 April 1889 to his father (Harvard), WDH mentions "our old sempstress, Jenny Moore." In 1907 she visited the Howellses when she was lecturing on socialism at Greenacre, a health resort near Kittery Point (WDH to Mildred, 27 August 1907, Harvard).

To Susan L. Warner

Redtop [Belmont]
March 18<u>th</u> [1879]

My dear Mr<u>s</u> Warner:

At last we are to have the pleasure of seeing you both at Redtop! Mr Howells says he has written Mr Warner to express our joy* at the prospect, and also that Mr Osgood has written to ask Mr Warner to a lunch immediately after the lecture.[1] O these *men's lunches*! Of course you are not invited—so I am coming in town to meet you—either before or after the lecture and bring you out to Belmont—on the half past eleven train if you do not go to the lecture, or on the next train, which is not till 2.30, if you do go to the lecture.

Please write how it shall be—and how strong you are—and where we shall meet—whether you want to see anything in Boston, and all that.

Maybe you'll find yourself so comfortable at Redtop that you wont go away on Monday.

Affectionately yours
Elinor M. Howells

*he knew I *would* be joyful

Manuscript: Watkinson

1. Because of travel or sickness, the Warners' visit to Redtop had been postponed several times. As Warner was to talk to the young women of the Saturday Morning Club on 29 March, Howells had invited him and his wife again (7 March, Watkinson). On 18 March (Harvard) Warner had replied that they proposed to stay with the Howellses Saturday and Sunday (29–30 March) and, not knowing whether they would meet in Boston, he had asked for directions. Howells answered that he would meet him at Osgood's after the lecture at the club, and that they would go to Belmont after fetching Mrs. Warner (19 March, Watkinson). This first visit was a success, as Clemens reported to Howells from Paris upon opening a letter to Livy from Susan Warner: "my goodness, how she raves over the exquisiteness of Belmont; & the wonderful view; & Mrs. Howells's brilliancy . . . & the attractiveness of the children; & your own 'sweetness' (Why, do they call *you* that?—that is what they generally call *me*); & the indescribably good time which she & Charley had" (15 April, *Twain–Howells Letters*, 1:259).

C. D. Warner to EMH

Hartford April 9 1879

Dear Mrs Howells,

I would not say that a person who appreciates Philomène could not like a character so ideally pure as the niece of S. M. That would not be true; for she who can appreciate Philomène can appreciate anything.[1] I think I can understand the fascination for you of certainly the most disagreeable woman I ever heard of. Once when we were in Pennsylvania there was killed and laid by the door-step the most disgusting reptile in that disagreeable state, called a "blowing viper." Mrs. Warner looked at it, and turned away with a shiver of disgust and terror all through her frame. But she couldn't resist going back again and again to renew the experience of that shiver of repulsion, until we took the snake away. Do you wish us to keep Philomène away from you?

That *was* a lovely poem, although Mr Howells said you said it was not.[2] We all thought so. What lowly newspapers you have in Boston that never tried to get it, and never mentioned it.

Yours sincerely
C. D. Warner

Manuscript: Harvard

1. Published in Paris by Plon & Cie, 1879, *Philomène's Marriages* was translated from the French of Madame Henry Gréville and published by T. B. Peterson & Bros. of Philadelphia in 1879. Henri Gréville was a pseudonym for Mme. Alice Marie Céleste (Fleury) Durand (1842–1902). Written by Mary Agnes Tincker (1831–1907), *Signor Monaldini's Niece* was published also in 1879 in the No Name series by Robert Brothers in Boston. This fantastic tale ends with the niece, Camilla, drowning either by accident or from despair because the insane wife of the Count who wants to marry her is alive. But happily, between Camilla's supposed death and impending funeral, the Count's wife dies; the Count discovers that Camilla is alive as he keeps vigil over her body, and all is well. For further remarks on this novel, see 20 April.

2. In his letter of 31 March to Mrs. Annie Fields, Howells had mentioned that he was to read a poem "which Heaven is yet to send me" the following Saturday evening at Tremont Temple in Boston, "for the benefit of the Gloucester sufferers." On 3 April, he wrote her again: "I have done a little poem which my wife calls—perhaps too flatteringly—atrocious, and which she would not hear read on any account" (*SL*, 2:222–223). Entitled "Poem. Written for the Occasion," with his name on the programme, Howells' piece was to be a part of a benefit performance (that included also songs, impersonations, and a reading by C. D. Warner) for the bereaved widows and children

of the Gloucester fishing fleet lost at Georges Bank during a storm on 20 February (Boston *Transcript*, 5 April; New York *Times*, 12 March).

To Susan L. Warner

Redtop [Belmont]
April 20ᵗʰ [1879]

Dear Mrˢ Warner:

Philomène arrived safely—& I put her amongst our books as one quite safe for Winnie to read, while I could not rest easy with that vile S.M.s S. around until I had put my protest on its worst pages in pencil—and it is almost as though I had re-written the book![1] There is a difference between coarseness & impurity! I close the subject by sending you a criticism on your favorite by a woman, more besotted by it than yourself. The "friend" she mentions I take to be myself. Not that *I* think the *workmanship* of S. M's N. at all equal to Philomène. The number of times Miss Tinker jumps from the sublime to the ridiculous—and the number of comparisons which she begins and cant carry out—notably that of the Palace to the Prodigal son—are quite too much for a person of my nerves.[2]

I cant resist also enclosing a poem Winnie sent to the Transcript the other day under a nom de plume.[3] You see she has struck out for herself?

I have been thinking of you as sick—& I have been down myself for a week past. That is why I didn't write.

Affectionately yours
E. M. Howells

Manuscript: Watkinson

1. S.M.s S. apparently stands for "Sister Mary's Story," by Jane Silsbee, a pseudonym of Helen Hunt Jackson (1830–1885), best known as the author of *Ramona* (1884). The story was published in the November 1879 *Atlantic*.

2. Mary Agnes Tincker's *Signor Monaldini's Niece* was favorably reviewed in *Scribner's*, May 1879. The book opens with this so-called comparison.

3. Signed Grace Baldwin, Winnie's poem "The Day" had appeared in the Boston *Evening Transcript*, 15 April 1879. An undated clipping of this is pasted in Elinor's scrapbook of Winifred's published poems (Harvard), as follows:

[For the Transcript.]
THE DAY.

Laughing, the Day in innocence awoke,
Bright with new resolutions, unconfined;
And flinging from her shoulders Night's dull cloak,
Descended, like an angel, on mankind.

Blushing, she sank behind the hills at last,
Hiding in fear and shame her glowing face,
And wondering how God's wisdom, unsurpassed,
Should give so fair a world to such a race.

GRACE BALDWIN.

Belmont, Mass.

To Susan L. Warner

Redtop [Belmont]
Thursday [24 April 1879]

Dear Mrs Warner:

It is of no use trying to quarrel with *you*—so I give it up—though of the same opinion still.[1] *But*, we've seen Modjeska—yes, met her off the stage. The Fieldses invited us in to meet her at lunch on Tuesday—and, although we somehow felt that we had a "previous engagement" to meet her at your house—we could n't resist and we went.[2] She certainly is a charmer—much more so off the stage than on, even. She wore a dark green & gold brocade-velvet with a white crape scarf around her shoulders & twisted together around her waist & "sort of " tucked in to her pocket—and a hat which baffles description—rough brown with a red bow on it, and a sweeping purple or plum colored feather. Mr Longfellow was at the lunch & Mrs Thaxter who wrote a sonnet to her, & several others and the whole party went together to see her in Juliet in the evening. O how different she was all painted and in a yellow wig! And Shakespeare is not for her—(though of course we dont think with that wretch Clapp in the Advertiser).[3] She was perfectly bewitching in the scene where she coaxes the nurse to tell her what Romeo said in reply to her messages—but in the tragic parts she overdid terribly.

We spoke together of Mr Warner, whom she thought "very nice" but I did not like to say anything about her visiting you for fear of meddling. And I had not the face to speak of meeting her again when all were adulating her to such an extent. But she was very lovely & ran up & threw her arms around a Miss Cochrane who played Chopin to her & kissed her in *such a way!*

Count Bozenta was very agreeable & the two seemed very fond of each other. She had a picture of her son in a ring under a flat diamond; and she thinks Philomène the most delightful book she ever read in all her life.

Affectionately
EM Howells

Manuscript: Watkinson

1. Probably a reference to Susan Warner's answer, now lost, concerning the novels they had discussed in previous letters.
2. Helena Modjeska (born in Cracow, 1840; died in California, 1909). A widow for some years, she had married Count Charles Chlapowski Bozenta, a patriot and journalist, in 1868. A leading actress with the Imperial Theater in Warsaw, she had first come to the United States in 1876 with Bozenta and her son Ralph during the Centennial Exhibition at Philadelphia and had settled in California. Resuming her acting career, she had gone on a tour of the country. This was her second visit to Boston, where she had played Juliet on April 22. In "Modjeska's Memoirs" (*Century*, March 1910), she recalls the lunch at Mrs. Fields' and quotes the sonnet inspired by her listening to Chopin. Miss Cochrane is not identified.
3. The unsigned review in the Boston *Daily Advertiser* (23 April 1879) criticized Modjeska's performance as inadequate and unsatisfactory, and her tragic interpretation of Juliet (showing nothing of the freshness and joyousness of fourteen-year old Juliet) as a misinterpretation by a foreigner. While praising her sincerity, originality, strength, and the occasional fine touches in her performance, the drama critic stressed her poor delivery and elocution, which he called monotonous and at times unintelligible. The author of the piece was perhaps William Warland Clapp (1826–1891), Boston author and journalist and editor of the Boston *Journal*.

To the Howells Family

Redtop [Belmont],
July 13ᵗʰ [1879]

Dear All:

Will has suddenly been called off to see a horse—with a view of buying it—and so I take his place and write the Sunday letter. (Groans from "All") I didn't say I *filled* it! The gentleman (for he is a gentleman) who is looking up a horse for us, and who came to drive him over to Chelsea is a young farmer of this neighborhood who has a natural eye for horses and has no object in cheating us—so we hope to come out well at or from his hands. He is a Cousin of Anne Whitney the sculptor, who is making a statue of Harriet Martineau.¹ Did Will tell you that Christine Chaplin sent him a little picture of dandelions the other day? I have not read the Colonel's Opera Cloak but hear it well spoken of.²

Will's new story that he is writing will be a success I feel sure.³ It touches lightly on spiritualism & the Shakers. Yesterday Will & Winnie drove down to call on Mʳ Longfellow. The Duke of Argylle had dined with him the day before I believe. The Duke was devoted to Mrˢ T. B. Aldrich on the steamer coming over. It was for her that he bought the American edition of his book immediately on landing.⁴ We enjoy being in our own house so much this Summer! It seems too good to be true. We all went down to Nantasket beach for a day's pleasure last week. The first person I met on board the steamer was my *dentist*: which somewhat marred my pleasure. It is the Coney Island of Boston, and it was great fun to see the people who go down there. Two sisters in purple muslin with Greek caps of the [small sketch] same material waltzed to slow music for the benefit of the crowd. Also a tall thin lady in blk. silk, who looked, Will said, like a closed umbrella solemnly moved around with a man shorter than herself. Mostly we are very quiet this Summer and do not have half the calls we did last year when the house *was* a novelty. Mc Kim has got "Willie" Astor's house—40 ft front on Fifth Avenue—to buil[d], with money ad libitum.⁵ Dr. Bendir of Quebec sent the queerest woman to see us—Mrˢ Harington (one r) Brown.⁶ You will hear of her in Canada I am sure if you inquire. She was there a year. The children are enjoying their vacation extremely. They really make a business of study during the school year and greatly need the relaxation they get now. They amuse themselves from morning till night—John, at present, with bugs & archery, Pilla in picking berries and Winnie in day dreaming. She draws a good deal & very well. But Pill does *everything* well. We are beginning to respect Pill a good deal. She is both clever and good, and seems to have a great deal of character.

We expect to see you the last of Sept. or first of Oct. in Canada. Please write *just when* we shall come. I dont believe we will bring the children this time. You ought one & all to come here and see them and I feel sure you will before long. All send love

Elinor

Manuscript: Harvard

1. Born in Watertown, Anne Whitney (1821–1915), the sculptor, who formerly had her studio in Belmont where her family had moved, now lived in Boston. She is noted for her statues of Charles Sumner, William Lloyd Garrison, Samuel Adams, and Leif Ericsson. Her Belmont cousin is not identified.

2. For their artist friend Christine Chaplin, see 21 January 1877. Her book *The Colonel's Opera Cloak* was published anonymously by Robert Brothers (1879) and was briefly reviewed in the September 1879 *Atlantic*.

3. *The Undiscovered Country*, serialized in the *Atlantic*, January–July 1880, appeared in book form in June 1880. Elinor is reported to have considered it "[his] best yet" (Clemens to WDH, *Twain–Howells Letters*, 1:277).

4. George Douglas Campbell, eighth Duke of Argyll (1823–1900), lord privy seal, postmaster general, university chancellor. Interested in scientific research, he was also author of many books on legal and political questions. The Duke and his daughters had dined with Longfellow, C. E. Norton, and Richard H. Dana on 11 July (*Life of Henry Wadsworth Longfellow*, ed. Samuel Longfellow [1886], 3:300). For an amusing account of Mrs. Aldrich's and the Duke's encounter on board the *Scythia*, see her *Crowded Memories* (1920), pp. 239–245).

5. W. W. Astor's house at 374 Fifth Avenue was completed in 1881 and cost $82,686 (Leland M. Roth, *The Architecture of McKim, Mead & White, 1870–1920: A Building List*, 1978).

6. Louis Prosper Bender (1844–1917), a Quebec physician and author of books on Canada, was to move to Boston in 1884. Mrs. Harington Brown has not been identified.

To Anne H. Fréchette

Redtop [Belmont],
August 15ᵗʰ [1879]

Annie dear!

We are all in a state of excitement over the good news which has just arrived. We highly approve of its being a boy, and of the name you have given him.[1] Well you have done your work up quikly and got a lovely little family—and now I hope you'll take a little rest.

And it all fell to Aurelia again. She must be getting to be a most experienced nurse—but it is to be hoped no tears were shed this time. Do you know that your boy is exactly eleven years younger than John—yesterday having been John's birthday.

And yesterday a letter came from Joe bragging that his baby is the prettiest boy in the world—so we are all satisfied.[2] My sister

Joanna has a new boy about a fortnight old. Her two oldest boys are coming here for a visit next week. They are about John's age & we expect to have lively times.[3] Buttercup Cottage is a great relief however as the children spend most of their time there. It is a little edifice 12 × 8 [sketch] which we have built for Pilla out of some lumber we had left—at the foot of our grounds—almost hidden in a locust grove. There & at the grove—Merryman's Grove—because of archery—our children & the four Fairchildren spend most of their time. We have the new Harper with your story in it. I have not read it closely but it has a brilliant *look!* You see by enclosed slip that the Transcript makes your husband Italian?[4]

We enjoyed Laura Mitchell's visit greatly, and Lily is a lovely looking girl and her voice is pleasant and she makes herself easily understood—so that her deafness does not seem the terrible affliction that it did.[5] Will wishes to add a line. All the children send greeting to their boy cousin I must just go over the page to thank you for Mole's second photo which is even more "knowing" than the first.[6]

Elinor.

Dear Annie and Fréchette:[7]

We are all happy in your happiness, and full of prophetic delight in the little man, whom we hope soon to see. Just think of Pinafore having been written in time for him!

With many congratulations,

Your affte brother
Will.

Manuscript: Alfred

1. Born on 13 August 1879, Annie's and Achille's second child was named Louis Premio-Real William Cooper Howells Fréchette, after his two grandfathers and Count Pemio Real, his godfather (genealogical sheet at the Massachusetts Historical Society).

2. Probably a reference to his youngest son, Joseph Alexander Howells, Jr., born 22 August 1878.

3. Frederick Mead Shepard was born on 30 July 1879. Joanna's two oldest boys were Gus and Burritt, respectively ten and nine years old. Commenting on their stay during some rainy days, WDH wrote his father: "This makes it rather dismal for the boys, and rather lively for those who have to be shut up with them" (18 August, Harvard). The four Fairchild children were Sally (b. 1869), Lucia (b. 1870), Charles Nelson (b. 1872) and John Cummings (b. 1874). Blair, the baby, was only two years old.

4. Annie's story "How That Cup Slipped" appeared in the September *Harper's*. The *Transcript* slip has not been located.

5. For Lilly Mitchell, see 30 March 1873. She and her mother had just left after spending a week at Redtop (WDH to Hayes, 4 August, *LinL,* 1:274). To his father, on 10 August (Harvard), Howells had written: "The Mitchells are still with us, and we are having a very pleasant visit. Lilly, you know, is the child who lost her hearing; but she uses the visible speech in the most wonderful way. She and Winny and [i.e., are] constantly together, and they are very young ladylike."

6. Mole was Howells Fréchette's nickname.

7. In WDH's hand. W. S. Gilbert's and Arthur Sullivan's "H.M.S. Pinafore" had opened in Boston on 25 November 1878.

To Anne H. Fréchette

Redtop [Belmont],
Nov. 8ᵗʰ [1879]

Dear Annie:¹

After a tiresome day I have come down in to the library to quiet down—& John has just handed me your letter—which I will not delay answering a moment, though I am too tired to go up, even our easy stairs to get some decent paper. No, you did not show me any chain—& though I fumbled over the things up in your room & found the earrings &c I saw no chain. Of course my memory may be at fault, but I *remember* no chain. I do *hope* it is not lost. You, or Vic, are lucky at finding things I believe. May your luck not desert you now!

Will has gone in to a Press Dinner. Winnie to Cambridge to her Sewing society—& will remain all night with Edith Young. What has tired me so is a call or visit, from Mrˢ Prof. Child & her Niece Miss Butler.² The call included dinner with us—& as I had let my cook go in town to have a tooth pulled Katy & I had both of us to work pretty hard to bring things about nicely. Mrˢ Child is just as oldfashioned & proper as usual—and gives me good advice. But I enjoy her ever so much. Since our return we have called on Marnie Warner & told her all about you. Just think her Roger is a bouncing boy,—& inquired of us "if we had seen Robert."³ He is modelling himself on that fearful rogue. Robert lately spoke to Marnie of the new Chinese professor's interpreter as the "interrupter." The C. prof.⁴ is a great sensation in Cambridge, which must always have some excitement: He walks about with his little boys—all in full C. costume. His wife cant walk—her feet being only 4 inches long.

She goes in a carriage to receptions and has to be helped up steps. And, fancy, they are cramping their girls's feet in the same way. Winnie has today had a present of two brass candlesticks from Mrs Fairchild as a reward for teaching the children drawing this Summer. Will insisted she should only have a little present.[5] The lamps have rings to fasten on to the candle & support a delicate little shade. Win isn't particularly pleased. Says it seems as though they were more for me. She is reading Tasso's Jerusalem Delivered with great delight. She has read Froude's Caesar and Bacon's Essays. What is she coming to! John had a lovely letter from Mrs Carter of Columbus today saying she was sending him some Ohio butterflies—and asking him to send her some Belmont ones next Summer for Prof. Landis who gave her these.[6] Is n't she a nice little woman? We enjoyed her as much as anybody in Columbus! My journey tired me dreadfully—either jouncing on the cars or visiting very hard. Of course this was after we left Toronto. I think of the sail up the Ottawa as the pleasantest part of our journey. What a pleasant wedding journey it must have been! Say, do you want me to send [remainder lost].

Manuscript: Harvard

1. Leaving the children at home, Howells and Elinor had taken a three-week trip to Canada. After shopping in Montreal, they had gone by boat to Ottawa to visit Annie's family and then to Toronto to see WDH's father and sisters. They had returned on 18 October, by way of Columbus and Cleveland, where they had visited the John Hays (WDH to Clemens, 6 and 24 October, *Twain-Howells Letters* 1:272, 276; WDH to his father, 14 and 26 September, Harvard).

2. Edith Young and Miss Butler have not been identified.

3. Thirteen-year-old Robert Boyd Storer (1866–1885) was Roger's cousin. A grandson of Mrs. Robert B. Storer of Cambridge, he was the son of Marnie's brother, William Brandt Storer (1838–1884), a merchant with Robert B. Storer & Co., and Russian vice-consul in Boston.

4. This was Ko Kun-Hua, an Instructor in Chinese at Harvard (Harvard University Archives).

5. Winny had spent the latter part of the summer at Magnolia, Mass., studying charcoal drawing with Miss Knowlton, a Boston teacher, who thought her highly talented (WDH to his father, 14 September, Harvard).

6. Isabel E. Carter was the wife of Dr. Francis Carter (1814–1881), a distinguished physician and professor of obstetrics at the Starling Medical College in Columbus. The Carters had befriended Howells during his years in Columbus as a journalist, and he often would call on her and her stepdaughters (Howells, *Years of My Youth*, pp. 150–151). Professor Landis is unidentified.

To Rutherford B. and Lucy W. Hayes

Belmont, Apr. 29ᵗʰ [1880]

Dear Cousin Rutherford:

By a happy coincidence your letter came by mail last night just before our expressman brought the books—the édition de luxe of Mr. Howells's writings which it has pleased the President of the United States to bestow upon me.[1] Such a combination of honors upon one small person has nearly turned her head! Nothing in the world could have pleased me so much.

I always longed to see W. D. in a rich dress, but it would have been too conceited for us to put him in it.

The volumes are so pleasant to handle that I am tempted to really begin and *read* his books.

Accept the united thanks of the Howells family. The rest of the letter is to Lucy

Dear cousin Lucy:

Mr Howells and I are just starting for Joanna's where we shall spend a few days. If it is at all a convenient time for you we would like very much to visit you at Washington for a few days arriving Wednesday or Thursday of next week.[2] If it is not perfectly convenient—at such short notice—let us come again—when you are re-elected!

As there will not be time for us to receive a letter from you here, will you kindly write us Care A. D. Shepard at Scotch Plains?

Affectionately
Elinor M. Howells

Manuscript: Hayes Presidential Center

1. The President had given Elinor a set of her husband's writings, handsomely bound upon his order at the Riverside Press. In his letter to EMH of 26 April (Howells Memorial), tipped in *Venetian Life,* one of the twelve volumes, he also asked the Howellses to pay their promised visit.

2. On 1 May (Harvard), the President replied that it would be convenient for them to come the following Wednesday or Thursday. On May 4, he sent them a telegram c/o A. D. Shepard urging them to come, and mentioning Lucy Hayes's return from the funeral in Cincinnati of her brother, Dr. Joseph T. Webb, who had died of a stroke on 27 April.

To Susan L. Warner

Plainfield, N.J.
May 5ᵗʰ [1880]

My dear Mʳˢ Warner:

I was much obliged for your note and at that time was full of going to Hartford on our way back from here—but no less a person than the President of the United States has stepped in and interfered with our plans to such a degree that we may—I might as well say it right out—cannot—go to Hartford to see you, as we expected to, now, but must put off that pleasure yet a little longer. This is how it all happened: We have been repeatedly invited to the White House but never could go. But just before starting for Joanna's there came a letter to me from Cousin Rutherford which seemed to give us but this one chance more ever to visit them in Washington—and as this peremptory letter was accompanied by the gift of all Mʳ Howells's books splendidly bound in morocco, we at once decided to yield to fate and go. Which will use up every moment of time Mʳ Howells can give this month—and we shall be obliged to fly straight home next Wednesday, we shall have a very quiet visit at Washington as Cousin Lucy's brother has just died. We are greatly disappointed not to see you all—but you see the situation? Please give my love to Mʳˢ Clemens and explain and say she shall yet have that hair tonic.²

Ask Mʳ Warner if *he* would n't go to Washington.

Affectionately
E. M. Howells.

Manuscript: Watkinson

1. Will and Elinor were visiting her sister Joanna Mead Shapard, on whose monographed stationery this letter is written.
2. For Livy Clemens' hair tonic, see 16 August.

To Winifred Howells

Executive Mansion
Washington.
Sunday morning, May 9ᵗʰ 1880.

Dear Winnie:

We have only had one letter from you as yet, but know all is well or you would telegraph. I am writing in one of the front bed-rooms of the White House, looking out upon the Smithsonian Institute. An old servant of the Hayeses' named Winnie takes care of our room an[d] us in the kindest way. You used to hear Birchard speak of her?¹ As Cousin Lucy has just lost her brother Joseph—(she is now the only remaining one of her family) she is receiving no company except old friends and it is very quiet here. She has for a companion a lovely Miss Cook of Chillicothe, Ohio. Mⁿ Kilburn, Cousin Rutherford's cousin about sixty years old is here visiting, also. Fanny Hayes is one of the nicest girls I ever saw. I wish you could know her. (The barber of the White House just came to the door & offered his services to Papa. He is a negro, as are all the servants. The doorkeepers & watchmen are white). The President is devoted to us. He went with us to Mt. Vernon yesterday. He never went with any of the visitors before on excursions. So Cousin Lucy says.

Every evening² at half past nine he takes papa to a walk, & raps on our door & asks him to go with him for a turn before breakfast.³ I gave him papa's photographs this morning, which seemed to please him, one taken 1880, one 1872 & one 1855. We are going to church this morning.——Afternoon. The President gave me his arm, Mʳ Howells walked with Mrs. Kilborn & Fanny Hayes & Miss Cook walked together to a little Methodist Church nearby (not the big Methodist church that Grant went to) very near here. After church we had lunch and some very nice talk down in the Blue Room, an oval room which opens on to a balcony from which is a view of the Washington Monument & the Potomac. The President & Papa are going to take a drive now, and a few friends are coming to dinner: Sec. Harlan & wife Carl Schurz & daughter, Mr. Rogers.⁴ I am to go out with Mʳ Schurz, Papa with Miss S.

Neither Burchard nor Webb nor Rudd Hayes are at home.[5]
Rudd has gone to New Mexico for plants. He is an enthusiastic
botanist. Last night Secretary Sherman & Gen. Killpatrick called
& also Miss Rachel Sherman, daught. of Gen. Sherman.[6]
I'd like you to keep this letter Winnie as a souvenir of our visit
to the White House. I was surprised to find Agnes Devens' photograph in the President's photograph book.[7] I did not know she
had been in Washington. M⁰ de Hegemann gave lessons in music
to Fanny Hayes. You & Aggy Suzanne ought to come on here
together next Winter. Tell John we will get him some Executive
stamps. Much love to him & Pilla & the girls from Papa & Mamma.
We expect to be at home Friday.[8]

Mamma

Manuscript: Massachusetts Historical Society

1. Birchard was the Hayeses' oldest son. Lucy Cook, a young cousin of Lucy Hayes
(on her mother's side), who had accompanied her to her brother's funeral in Cincinnati,
had just returned with her. A first cousin of Hayes, Mrs. Sarah Moody Kilbourn
(b. 1812), of Delaware, Ohio, had arrived in Washington during their absence (Hayes,
Diary, 3:599). Fanny Hayes, the President's only daughter, was thirteen years old.
   2. Elinor evidently means "morning."
   3. "Walked with Howells to the new building for the Bureau of Engraving and Printing, to the new Museum building and thence home. Two miles or more between 6 and
7 A.M.—before breakfast," the President wrote in his diary, 11 May 1880.
   4. Secretary John Marshall Harlan (1833–1911), Associate Justice of the Supreme
Court since 1877, and his wife, the former Malvina F. Shanklin, of Indiana. A later
editor of the N.Y. *Evening Post,* Carl Schurz (1829–1906) had come from Germany in
1852. Serving as U.S. senator from Missouri (1869–1875), he had been appointed
Secretary of the Interior by Hayes in 1877. His wife had died in 1876. His two daughters, one of whom was named Agatha, are mentioned in Hayes, *Diary* (5:76). For
William K. Rogers, see 7 December 1862.
   5. The three older sons of the President. Rutherford Platt Hayes or Rud was a student
at Cornell.
   6. General Hugh Judson Kilpatrick (1836–1881). A West Point graduate, he had
been in command of Sherman's cavalry forces during the march to the sea (Garfield,
Diary, 3:355). After the Civil War, he had become minister to Chile (1865–1868), a post
to which he was appointed again in 1881. U.S. Representative (1855–1861) and former
senator from Ohio, 1861–1877, John Sherman (1823–1900) was Hayes's Secretary of
the Treasury. William Tecumseh Sherman (1820–1891), the general, was his brother.
   7. Aggie Devens of Cambridge, a schoolgirl, most probably a relative (perhaps a
niece) of Charles Devens, Attorney-General in Hayes's cabinet, is mentioned along with
Lucy Cook among the seven young ladies who spent the Christmas holidays with the
Hayeses and their children at the White House (2 January 1881), Hayes, *Diary,* 3:633).
Winnie, who had been invited, was unable to go (see 18 November). Mrs. de Hegemann
has not been identified.
   8. "Mr. and Mrs. Howells left us Thursday evening [13 May]. We had a delightful
visit from them," the President wrote in his diary (3:599).

To Anne H. Fréchette

[Belmont] 16ᵗʰ May [1880]

Dear Annie

We arrived home yesterday after a fortnight's junketing in N. York & Washington too tired for anything—the last two nights on cars & boat. We found Achille's drawing—which Will will attend to to-morrow—& your letter.¹ Will is *devoured* with proofs—so I write—but only to make the unsatisfactory statement that we know of *no* favorable comment by foreigners—unless it was by Laboulaye, Dicey, Rose (W. D. thinks that was the name who was here 2 years ago) or others that you know about as well as we.² I can tell you what Ole Bull said at a lunch at James T Fields's just before we left. "When I am in England I like to say savage things to the English. When they are here it is *punishment enough*." Ole Bull's opinion is worth having because he is a friend of all the princes in the world.³ Will says you may quote it. I am *so* stuck up & *so* patriotic since my visit to Washington. I think it will be the most fascinating city to live in in the world sometime. We sa[w] everybody. Gen. Sherman, Alex Stephens, Madame wife of the Marquis de Chambrun, Lafayette's granddaughter & niece of the present Madame de Remusat (had tea there), Indians, Garfields, Hays, Bancroft, Evertses &c &c all personally.⁴

Madame Chambrun has the very forehead [hand points left] [sketch of profile] of Lafayette & so has her brother (his name began with C.) Her husband is lawyer for the French legation.

We were at the White House of course. We will describe our visit later.⁵

Elinor

Manuscript: Harvard

1. Annie's letter has not been located.
2. Edouard René Lefebvre de Laboulaye (1811–1883), the French jurist, also an author, was a political economist specializing in Anglo-Saxon countries, and the originator of the idea of the Statue of Liberty. Elinor probably refers to Albert Dicey (1835–1922), literary journalist and essayist, and Fellow of All Soul's (*Henry James Letters* 1:345). Rose is not identified.

3. Ole Bull (1810–1880) was a Norwegian composer and violinist whose wife was an American. WDH had written his father of taking Winny to the Fieldses on the 16th for

"her first grown-up lunch." "We had a sit-down lunch, and uproarious, story-telling gayety, and after lunch Ole Bull made his fiddle sing to us. It was wonderful: the fiddle did everything but walk around the room" (17 April, *SL*, 2:247–248). Also there were the Clemenses, who were visiting the Howellses at Redtop together for the first time. (The house had been under construction when they had left for Europe in April 1878, and Mark Twain had come alone on December 2, on his first visit to Belmont after his return to America.) "We had a most elegant time in Boston," Clemens wrote back to Howells from Hartford, and "Mrs. C. seems settled in her mind that Mrs. Howells is a perfectly wonderful woman—in fact this poor girl has come home dazed. Boston has been too many for her" (19 April, *Twain-Howells Letters*, 1:299–300).

4. Those not previously identified are: Alexander Hamilton Stephens (1812–1883), former Confederate vice-president and a U.S. congressman (1843–1859 and 1873–1882); George Bancroft (1800–1891), historian, diplomat, and author of a ten-volume *History of the United States* (1834–1874); William M. Evarts (1818–1901), a lawyer for the defense in the notorious Tilton-Beecher trial in 1875 and Hayes's Secretary of State (1877–1881), whose wife was the former Helen Minerva Wardner. The Marquise de Chambrun was the former Marie-Hélène-Marthe Tircuy de Corcelle, the daughter of the French ambassador. Her husband, Charles-Adolphe de Chambrun, the "lawyer for the French legation," was the author of works on constitutional law. The first volume of the *Memoirs of Madame de Rémusat* (1780–1821), the lady-in-waiting of Empress Joséphine, edited and prefaced by her grandson Paul de Rémusat, a French senator, had been recently published in translation in Boston and reviewed in the March 1880 *Atlantic*.

5. See WDH's letter of 17 May to his father (*SL*, 2:253). On the same day, Howells wrote his host from Redtop: "Dear Mr. President: Elinor and I reached home on Saturday morning, and are trying our best to sober down to the realities of workday life, and to reduce our respective dimensions to the scale of our domestic architecture. You can imagine which has the greater difficulty" (*SL*, 2:254).

To Olivia L. Clemens

[Belmont] Aug. 16 [1880]

My dear Mʳˢ Clemens:

Your husband wrote that you were not yet in a condition to use the tonic¹—so I have yet time to add one exceedingly important item:—that the brown tonic must be well shaken and then a little poured into a saucer and applied with a sponge. The white tonic has a cork with a fixture in it which can be unscrewed so you can sprinkle the liquid on through it—after which it must be screwed down again very tightly. Mʳ Howells & Winnie started for Ashfield this morning.² This is royal weather for you to get well in. Mʳˢ Fairchild carried some wine whey to an Irish woman the other day when her baby was three days old—or perhaps five—and found her up doing her washing.

Affectionately
Elinor M. Howells

Manuscript: The Mark Twain Papers, The Bancroft Library, University of California—Berkeley

1. Mark Twain to Howells, (9 August, *Twain–Howells Letters*, 1:319). On 13 August, Elinor had explained about the hair tonic, and she had congratulated Livy on her new baby, born 26 July: "What is her name, who is she most like, and why didn't you *tell* me! I am always in the wrong as to your intentions, it seems." Which prompted in return from Mark Twain the following answer quoted here in part: "REPLIES. 1. Jane Lampton Clemens, (after my mother.) She (the child,) is addressed as Jean". . . . 2. (She is 'most like'—well, say an orange that is a little mildewed in spots.) No—I discover you don't mean complexion, but *who* is she most like? That is easily answered: Mrs. Crane says, Livy; Livy says, my mother; Bay Clemens says, me; Susie Clemens says, Bay; *I* think she looks most like a successful attempt to resemble nobody. Take your choice. ¶Expunged, by order. ¶3. ⟨Didn't 'tell' you,' because we weren't certain; thought it was flesh; supposed it would wear off. But we changed, as time rolled on, toward the last we estimated it at twins, at the very outside.⟩ ¶4. No—O, no, I don't think she had any really definite 'intentions' at the time. ('Intentions' is mighty good!)."

Mrs. Theodore Crane was Livy's foster sister; "Bay" was Clara L. Clemens, their second daughter, and Susie, the oldest. As Elinor had complained about the spine injury incurred the preceding November that still prevented her from car riding, Mark Twain continued: "O dear, I never imagined you were drifting into invalidity as a settled thing. I think we both always looked upon you as a sort of Leyden jar, or Rumkoff coil, or Voltaic battery, or whatever that thing is which holds lightning & mighty forces captive in a vessel which is apparently much too frail for its office, & yet after all isn't. But the spine!—come, this is a surprise, & anything but a pleasant one" (p. 323). To which WDH was to retort (p. 328) "I have been wanting to tell you how wonderfully good that comparison of her to a Leyden-jar was. She never *does* quite go to pieces, but it always looks like a thing that might happen." Clemens' letter concluded with "Love to Winny, & John, & Howells. And speaking of Howells, he ought to use the stylographic pen, the best fountain-pen yet invented—he *ought* to but of course he *won't*—a blamed old sodden-headed conservative—" (pp. 321–324).

2. "I have been away again for a little while, spending three or four days with Norton at his summer place in Ashfield. Winny went with me, and visited his girls," WDH wrote his father on 22 August (*SL*, 2:263–264). "The occasion was a fair which they were holding in benefit of the village academy, and in which Norton took a great interest. It ended with a dinner, and then we had speeches, in which Winny thought I bore my part 'very nicely,'" he continued. Ashfield is in the northwestern part of Massachusetts, at the foot of the Berkshires.

To Thomas Bailey Aldrich

Redtop
Belmont, Oct. 9ᵗʰ 1880.

Dear Mͬ Aldrich:

I have seldom been more pleased than I was when Mͬ Howells handed me the lovely little volume of poems you sent me.¹

It is certainly the most exquisite cover I ever saw and most appropriate for the gems it contains. There was s[om]e company, including Miss Whitney and Dan French over at the big house the other night,² and Mͬ Howells by request read aloud several of the poems. You ought to have over heard (that's not *exactly* the word) the hush that followed the reading of Sleep and Identity. Mͬ Weis's reading did not produce half so much effect on his audience.³

Sometime—on this desolate wind-swept hill—we are going to have a cosy visit from you & Madame. I fully intended it should have been when the F.s⁴ were here, but *steady* company all Summer long has prevented, and I find myself a good deal used up by the excitement as yet. My love to your wife & Mother & to Mrs & Miss Woodman.⁵

Yours sincerely
Elinor M Howells

Manuscript: Harvard

1. Most probably *XXXVI Lyrics and XII Sonnets Selected from Cloth of Gold and Flower and Thorn* (1881). The little book is bound in vellum, with red lettering and two ornamental cherubs on the front cover.

2. Daniel Chester French (1850–1931), the Concord sculptor, like Anne Whitney a pupil of Dr. Rimmer, and a friend of Emerson, whose bust he had sculpted in 1879. His "Minute Man" had won the competition in Concord in 1875. He is noted also for his seated figure of John Harvard (1882), his colossal personification of the Republic for the Chicago Exposition of 1893, and his marble seated Lincoln in Washington.

3. This was perhaps John Weiss (1818–1879), the author, lecturer, preacher, and translator of Goethe and Schiller, whose famous wit and eloquence amazed, puzzled, and finally exasperated his congregation.

4. The Fairchilds.

5. Lilian Aldrich's mother and sister (or possibly her aunt).

To Susan L. Warner

Redtop [Belmont],
Nov. 18ᵗʰ [1880]

My dear Mᵣˢ Warner:

Your very sweet note arrived, and the then lovely water-pot. The last was so very lovely that it made us ashamed of ever having joked about it, and it more than justifies your enthusiasm on the subject. I am perfectly "gone" on it and think I never saw anything in pottery more simple & elegant in form and more satisfactory in color & design. All those exquisite curves, and that nice sharp edge at the top! [sketch] It is really perfection. Its newness & its association with you add to its charm, and with it I intend to captivate my friends & confuse my enemies.

Last week was bright and ended triumphantly with your visit.¹ This week is dull and we are all sick: Mᵣ Howells with a boil on his face, Winnie with headache &c &c. Mᵣˢ Wyman says you overdid Sunday.² You ought to have staid quiet after junketing over here—but it will not interfere with your general gain I trust. I was sorry—as was Mᵣ Howells that it prevented your coming in to see him yesterday—but if you stay a week longer you must surely go in—and perhaps I can meet you there.

Winnie has been invited to go to the Symphony concert by some friends—and is going though she is not well.³ I feel sure you'll be there too, sick or well. I told her to look out for you.

Yes, you have a good taste in crockery, if not in books.⁴

With much love from us all.

Sincerely & devotedly yours
E. M. Howells.

We got Mrs. Fields's book of poems from her today.⁵

Manuscript: Watkinson

1. In her letter of 10 November (Watkinson), Elinor had invited Susan Warner, who was staying with friends in Cambridge, to come over to Belmont for a short visit that week.
2. Née Elizabeth Aspinwall Pulsifer, Mrs. Wyman was the wife of Dr. Morrill Wyman (1812–1903), a distinguished physician who had practiced in Cambridge for over sixty years and was a founder of the Cambridge Hospital. Mrs. Warner had been visiting

them at their 81 Sparks Street home, and Warner had written Elinor on 9 November to give her his wife's address.

3. Reporting Winny's "triumphs at school" to his father on 29 February, Howells had added: "I think she is growing up a strong, tranquil nature" (*SL*, 2:245). As she was "fagged from her school" (20 June, *SL*, 2:257), Winny had spent part of the summer in Magnolia again, with their friends the Shalers, sketching and writing. She had just had the joy of seeing her "beautiful" poem "Magnolia" published in the *Youth's Companion* on 14 October and praised by Longfellow (3 and 17 October, Harvard). On 28 November (Harvard), Howells told his father of the headaches and vertigo that she had started to suffer from, and he had to decline the Hayeses' invitation to Winny to visit them at the White House during the holiday week (27 November, Harvard, 1 December, Hayes Presidential Center). To Mark Twain, on 13 February 1881, he was to write: "You know perhaps that Winny is quite broken down. She has not been in school for five months, and for a while she could not cross the room alone" (*Twain-Howells Letters*, 1:348). This was the beginning of her long and baffling illness.

4. A joking reference to their continuing disagreement about *Philomène* and other stories.

5. Annie Fields, *Under the Olive* (1881).

To Susan L. Warner

[Belmont]
Friday [10 December 1880]

My dear M̲r̲s̲ Warner:

I left you to thank M̲r̲ Warner for his photograph and I never acknowledged the Thanksgiving in the Courant—which we all enjoyed—but today I must write about his letter in the Childhood's Appeal—and it is so much easier writing to you that I hope he will consider it all one—which M̲r̲ Howells stood up by the register in my room and read aloud before I was up, Winnie reclining on the bed in demi-toilette to listen.[1] It is perfectly delicious—the kind of thing that makes you feel good! And it had such a good effect that my lazy man started right off to write something for the Abused Children himself—giving up his forenoon to it too. He had attempted to do something once before one evening, but failed—but he will succeed this time with such inspiration I feel sure. It is an actual adventure of John's that he is going to write.[2]

We have been to see Bernhardt twice this week—in FrouFrou and the Sphinx—but we had the moral courage to stay away from her reception. She is certainly fascinating—but *great, never*! She ought nt to be compared with Modjeska as a tragedienne.[3]

We have had Björnsen out to lunch. He is a lovely soul. We are going to make him stop over in Hartford when he goes through sometime.[4]

Last night came some photographs of Beaver Brook and the mill-wheel, which you did not see, from Mᴿ Underwood. They are to illustrate an article on Lowell in Harper, so you will yet see that wheel.[5]

Affectionately

E M Howells

We have Mᴿ Warner's photograph framed nicely. It is very satisfactory

Manuscript: Watkinson

1. In his letter to Elinor of 9 November (Harvard), Warner had written: "I am going to mail you a photograph, just come from Venice, to take the place of the one you do not like and I detest—as I promised you." An account of the holiday in the hills of Western Massachusetts when he was a boy, "The Old-Fashioned Thanksgiving," had appeared in the Hartford *Courant* on 25 November. Warner's letter of 8 December had been published in *Childhood's Appeal,* "an occasional newspaper issued for the benefit of the Massachusetts Society for the Prevention of Cruelty to Children" during the State Fair for Abused Children held in Boston in December.

2. Entitled "A Perfect Success," the piece appeared in *Childhood's Appeal* on 17 December (reprinted in the *Harvard Library Bulletin,* April 1980). It is a charmingly told adventure of twelve-year-old John Howells in the woods near Redtop.

3. Sarah Bernhardt (1844–1923), who had left the Théâtre Français in Paris after being fined, had recently arrived with her own troupe in the United States. The Howellses had seen her at the Globe Theatre on Tuesday in Meilhac and Caillavet's *Froufrou,* and in Octave Feuillet's *Sphinx* on Thursday.

4. Björnson (see 30 March 1873), the Norwegian author, who was touring the United States that year, came to Redtop for lunch on 3 December. Howells described the occasion in a letter to his father, two days later: "On Friday we had to lunch here the Norwegian poet Björnson, who is spending the winter at Mrs. Ole Bull's, in Cambridge. I don't know whether you've ever heard me speak of his books; but he is a great genius. Personally he is huge, and very fair. . . . I like him extremely. He is a hot Republican, and just now is in disgrace at home for having spoken disrespectfully of the king: I think he called him a donkey" (*SL,* 2:270). The Ole Bulls were living in Lowell's Elmwood house while he was in England as minister to the Court of St. James, a post for which Howells had recommended him, and to which he had been nominated in January (WDH to Hayes, 7 August; Hayes to WDH, 8 August 1879; *SL,* 2:235–236).

5. F. H. Underwood's "James Russell Lowell," *Harper's Monthly,* January 1881, contained among other illustrations that of "Beaver Brook" and of the "Wheel of the old Mill on Beaver Brook" as well as one of "The Waverley Oaks," celebrated in Lowell's poem. A favorite haunt of the poet, Beaver Brook and its valley in the Waverley part of Belmont were a few miles from his and the Howells houses.

✤ VII ✤

From Belmont as Far as Venice

1881–1883

$\mathcal{S}$omething has already been said in the introduction about the beginnings of Winny's illness, her father's breakdown in the fall of *1881*, and the European trip of *1882–1883*. On all these subjects, especially on the year in Europe, Elinor's letters offer significant though not complete details.

The letter to Olivia Clemens that opens this section reveals the deep concern for Winny, now nearing the end of the first year of her illness and undergoing a rest cure. She has improved and now can attend school, Elinor observes in the letter to her father-in-law that announces Will's sickness. In this difficult time Elinor exhibits considerable managerial ability, as shown by her handling of her husband's literary affairs (the uncompleted Modern Instance had just begun serialization) and coping with the family's future in the Belmont house.

Will's slow recovery delayed finishing the novel and the trip to Europe, one taken partly in the hope of restoring Winny to full health. After visiting his father and sisters in Toronto and seeing the "Annie family" in Montreal, the Howellses sailed from Quebec for Liverpool on 22 July and arrived in London on the 31st. From London they went to Switzerland in mid-September, then to Italy at the beginning of December, where they stayed until the end of May, returning home by way of Paris and London. They reached Boston on 16 July. The year was a fabulous one for the whole family, even

Winny, in spite of not providing the hoped-for-cure—as Elinor's letters record with the special glow that radiates while she is traveling. Her artist's eye, not limited to seeing places, also appreciates people, such as the Bulgarian "Daisy Miller" at Le Clos and the priest at Will's birthday celebration in Siena.

On returning home, the Howellses, who had given up Belmont before they left, briefly lived on Louisburg Square in Boston. Among the quotations lettered by Elinor on the library walls of Redtop was the Shakespearean one, "From Venice as Far as Belmont." Settling in Belmont had apparently been regarded as the final move, and in a way the leave-taking marks a sad reversal in the Howellses' expectations. Yet as one views the full pattern of their various moves, the just conclusion may be that in their apparent restlessness they found the greater satisfaction.

To Olivia L. Clemens

Redtop [Belmont],
August 26ᵗʰ [1881]

My dear Mͬˢ Clemens:

It always seems like an "opportunity" to write you when Mͬ Clemens is here.—though I know there is nothing to prevent me at other times. You will smile when I tell you that I *was* just going to write, and that I have hardly been myself enough to write since I left Hartford—owing first to my having been sick and then to my great anxiety about Winnie.

Now, although she has not left her bed a moment for more than a month and will be in bed some months longer, I feel light-hearted, for I am sure we are at last acting in the right direction and that she is really gaining in flesh and strength. We changed from the old treatment—of keeping her lively and stirred up to one of perfect rest, some like Weir Mitchell's—under the direction of Dͬ· Putnam, a nervous specialist.[1] She actually takes eight meals or lunches a day—so that we have named her The Lunch Fiend.

Of course she has a nurse—so that I am not at all worn out: but it has been a very wearisome Summer.

I hardly think we will get off to Europe before next Summer, and in that case we shall go in to Boston for the Winter. Winnie will be quite well by that time, and you will make us that visit you did not make last Spring. You will soon be looking forward to getting settled in your own house after all those repairs. M͞r Clemens speaks enthusiastically of the improvements—but it never occured to me that the outside *could* be improved.[2]

Branford seems to have agreed with M͞r Clemens. I hope you are all in as good condition as he. M͞r Howells is going down to Manchester next week to pay his respects to M͞rs Fields whom we have not seen since her husband's death. She could not see people at first.

Today Osgood has a lunch—*without* ladies. I think it will hardly be as jolly as usual on account of the state of the President.[3]

Much love to Susy & Ba

Affectionately
Elinor M. Howells

Manuscript: The Mark Twain Papers, The Bancroft Library, University of California—Berkeley

1. S. Weir Mitchell (1829–1914) was the best known nervous specialist of his time in the United States, though today his theory and methods are in poor repute. James Jackson Putnam (1846–1918) also was using the rest cure of Mitchell, but late in his life he became at least a Freudian sympathizer (for his correspondence with Freud, see *James Jackson Putnam and Psychoanalysis,* ed. Nathan G. Hale, Jr. [1971]).

2. Perhaps EMH would have liked to have underlined "outside" instead of "*could.*" See her remarks about the Clemens' house looking too much like a church in her letter to her brother, 26 October 1877. The house had recently been expanded, with interior decoration by Louis Tiffany.

3. Garfield was to die 19 September. On the luncheon WDH commented to Warner, 3 September: "We had a perfectly lurid time when Clemens was here: Garfield was at his lowest, and we did everything but shed tears; it was the saddest lunch I ever saw" (*SL,* 2:295).

To William Bradley Mead

Belmont,
Nov. 9͞th '81

Dear Cousin William:[1]

I have been sick abed or I should have written sooner. The Brattleboro' man[2]—I am sorry for your sake to say—is right. Lark came home from H. K. Brown's & set up a studio & soon after made his Ethan Allen—which is now at Montpelier.[3] I imagine the snow statue was made about 1854. I think Ethan Allen was modeled about 1858. The Snow statue was colossal & very classic.[4] An untaught boy could not have done it—though the newspapers always said such a one did. As for teaching: H. K. Brown never did anything but criticise the young fellows who worked in his big barn of a studio—but they saw him work and had access to casts and statues from which they modeled.

The thing which Lark did all by himself, & which attracted the attention of an artist who was staying at the Water Cure, was to cut a pig in marble.[5] This artist sent him to study with Brown—who liked him and took him in to his family to board. On referring to your letter I see Lark was at H. K. Brown's in 1855. As he was born in '35 he would have been twenty when he made it—but I remember he was nineteen—and the papers kept him nineteen for 6 or 8 years afterwards.

We laughed a good deal over what you wrote about Dr̲ Breen.[6] Excuse me for scratching, but I must not let out the secret until the Dec. Magazine is published. I believe I did know that Will was building in Providence[7]—but Winnie has been sick in bed all Summer and I have not known much outside the house. Now we are all going to Europe the first of January—partly on her account.

I want very much to hear about Upham's Corner, if anything about Gabriel Mead has been discovered.[8] Eddy & Anna were out here to tea last week.[9] Eddy seems really very sick and in an unhappy state of mind. We should like much to see you & your wife before we go abroad. Will you not call upon us? I cant propose anything more in the state we are in but we always give people something to eat. Please remember us to your wife. Mr̲ Howells joins me.

Your affate̲ Cousin
Elinor

Manuscript: William White Howells

1. William B. Mead (1823–1909), a dentist in Providence, R.I., was the son of Elinor's uncle Bradley Mead (c. 1792–1871) and of Charlotte Hastings Mead (1796–1841).
2. Probably Henry Burnham, author of *Brattleboro* (1880).
3. Larkin G. Mead, Jr., worked in the studio of Henry Kirke Brown (1814–1886) in

1855–1856 and boarded with the Brown family in Brooklyn. At that time Brown was designing the equestrian statue of Washington for Union Square, unveiled 4 July 1856, now considered his highest achievement. During Larkin's stay in Brooklyn he wrote frequently to the family in Brattleboro; typed copies of some two dozen letters, eleven of them addressed to Elinor, are in the collection of Mrs. Katherine Lois Mead Hasbrouck, Larkin's great-niece. *Ethan Allen*, completed for the capitol at Montpelier in 1861, was later moved to the National Statuary Hall in the capitol at Washington.

4. *The Recording Angel*, as the sculpture is called, had been modeled on New Year's Eve, 1856. See *SL*, 1:343.

5. The pig was sculptured in 1854.

6. *Dr. Breen's Practice* appeared in the *Atlantic*, August–December. The December chapters were 11 and 12, in which Grace Breen rejects Dr. Mulbridge and marries the mill owner Libby. The "scratching" partially obliterates "And you who have said Rochester."

7. The William C. Chapin house was begun July 1881 and built 1882–1883.

8. Gabriel Mead (1587–1666), who came to America in 1635 and settled in Dorchester, was an ancestor of EMH. Uphams Corner is in Dorchester, south of Boston, and is now undergoing restoration.

9. Edwin D. Mead and his sister Anna.

To William C. Howells

Belmont,

Nov. 17ᵗʰ [1881]

Dear Father:

Will has been threatened with gastric fever, but is better today.¹ He is greatly disappointed that his sickness will interfere with the long-planned visit to Toronto—but it may only postpone it a short time. I wrote to Joe this morning to say Will could not go at this time, so he could arrange his affairs, perhaps, to go later. It is very unfortunate, also, that the banquet to Louis Fréchette takes place today—as Will was counting greatly on going to it.² What hastened Will's attack I think was a visit from J. J. Piatt & his wife & two children—one of whom cried all night.³

Winnie is wonderfully improved and has gone in to Boston to attend Miss Wesselhoeft's school—taking at first only just what lessons she wants to. She is with girls of her own age, which is what she needs most of all. Tomorrow evening she is going to a little dance. She says she has not felt so strong for two years as she does now. If she continues well and is happy in Boston, probably our trip abroad will be given up until May.

We were glad to get Henry's picture. And Annie sent the laughing one of her boy with a long letter the other day. Mᵣ Moore has been out here again to tea.⁴ We think him very pleasant. Harry

James has become a fussy old bachelor—very advisatory & instructive in his conversation.
    Much love to all

Elinor

Manuscript: Harvard

1. The onset of WDH's long illness.
2. The Institut Canadien-Français gave a banquet for Louis Fréchette at Young's Hotel, preceded by the gift of a gold pen and pencil and followed by many toasts (Boston *Transcript*, 18 November).
3. John J. Piatt (1835–1917) and Sarah M. B. Piatt (1836–1919) were both poets. Piatt had been the other author of WDH's first book, *Poems of Two Friends* (1860).
4. Moore's first visit is noted by WDH in a letter to his father, 23 October 1881. Except for the description of Moore as "a modest, nice young fellow in every way" (*SL*, 2:299) and as a friend of William C. Howells, nothing is known of him.

To William C. Howells

[Belmont] Sunday,
Nov. 20ᵗʰ [1881]

Dear Father:
    Will's fever ran on and took the rheumatic form, so that he was full of aches and pains—but he feels so relieved of pain this morning that his doctor (who is an acknowledged pessimist)[1] says that if he continues to improve as much for the next twenty-four hours he will get up from his sickness very soon—but he says there is a possibility (and he always mentions the very worst possibilities) that the fever may run on and be a regular low, congestive, or rheumatic fever—lasting three or four weeks. But he said that at ten o'clock this morning—and as Will, after sleeping well all night, has slept from then till now (after 12 o'clock) and is still sleeping, I hope the sickness has taken a favorable turn. I wrote this morning because I got the chance, but shall add a line just before sending this to the mail, to give you the latest news.

Evening, 7 o'clock
    Will has slept most of the day—his fever is still going down—the pain has left his heart and is now lodged in the right shoulder & elbow—but the pain is not severe. The doctor said yesterday that

after leaving the heart it would not go back there again. We shall be very careful not to do anything to drive it there again. Will sends his love & says tell you he is much better.

Affectionately
Elinor

Manuscript: Harvard

1. The doctor was Walter Wesselhoeft (1838–1920), son of the founder of the Water-Cure at Brattleboro. At least at this stage in WDH's illness, Elinor's description of it is the fullest that is extant. In a letter to W. C. Howells that probably follows this one, "Noon Tuesday" (22 November? Harvard), Elinor reports that her husband has made "a *decided change* for the better" and adds the detail that "the abdomen has become quite soft."

To Henry James, Jr.

Belmont,
Sunday [27 November 1881].

Dear Mͬ James:

I am glad to be able to write that Mͬ Howells is decidedly better today—and if no further complications arise he will be speedily well[1]

Meanwhile I am so tired and distracted that I fear I couldn't talk straight to you if you came out & might not even be able to see you—so dont make the long journey for nothing—nor let us lose your visit—but wait a few days yet

I told Mͬ Howells of your kind inquire (though I keep a great deal from him and he reads no letters) and he joins me in thanks & rembrance

Yours very sincerely
Elinor M Howells

Manuscript: Colby. Special Collections.

1. EMH's note replies to one from James of 25 November in which he asked for "three lines" about WDH's illness and expressed the hope that he might visit in a few days. Then on 28 November (Harvard) he wrote again: "I will spare you, most tenderly; especially as you had my mother & sister yesterday. But the 1ˢᵗ moment I *may*, I will come."

To D. C. Heath

Belmont,

Dec. 8ᵗʰ [1881]

Dear Mʳ Heath:¹

Mʳ Howells is much better;—so that he walked into an adjoining chamber his own today—but still our plans are so undecided that I must still ask for two week's grace and then I will write something positive. I know enough of the future already, however, to beg you not to lose any other house for this; for it is quite uncertain whether it is to be let, after all, and whether furnished if let at all.² I am not at liberty to say more at present—but please do not count on this house, as things have taken a new turn and everything is very uncertain.

Please excuse the mystery of this letter. It will all be very simple when explained.

Yours very truly

Elinor M. Howells

Mʳ Howells sends his regards and thanks you for your kind sympathy.³

Manuscript: American Antiquarian Society, D. C. Heath correspondence.

1. Daniel C. Heath (1843–1908) was a junior member of Ginn & Heath until he established his own publishing house in 1886. On 21 November (American Antiquarian Society) Elinor had written Heath about Howells' illness, noting that if the Howellses did not leave for Europe by 1 January they would not leave until May, "and that would not answer Mrs. Heath's purpose, I fear. . . . you had better not count at all on our house."

2. In a draft of a letter to Fairchild of 6 October WDH had spoken of the possibility that he might "give up Redtop to you finally, and store my furniture" (*SL*, 2:298–299). Since the house remained empty until 1883, presumably Fairchild did not ask for rent during this period. On 13 March 1883 WDH wrote Fairchild from Florence (Harvard): "After the first pang was past, Elinor and I rejoiced that you were to have the Walkers in Redtop. The empty house has been heavy on my heart every time I thought of it; and I hope the Walkers won't mind our guilty ghosts flitting about the glimpses[?] of their hearth fire in the cool evenings. Ah, that library!—. . . . But if I begin to recognize your kindness, where should I stop?" Francis A. Walker (1840–1897), a political economist, was president of the Massachusetts Institute of Technology. Elsewhere in the letter WDH writes, "I have no more respect for bronchial tubes than I have for the plumbing that Myers put into Redtop."

3. In another hand on the fourth page appears this note: "Please return with any comments or suggestions you think proper ¶W[?]."

Winifred Howells to Samuel L. Clemens

Boston, Jan 18th, 1882.

Dear Mr. Clemens

Thank you very much for that beautiful book you sent me.[1] The binding is so exquisite that I hardly dare handle it, and I have to keep looking back to the first page, on which you have written to assure myself that it is really mine and "from the Author". It seems almost too good to believe.

I have not finished it yet, but I think the loveliest place I have read is where the Prince finds the calf in the barn and is so glad to cuddle up to something warm and alive.[2] After his horror at first feeling something beside him it was such a relief to find what it really was that I was almost angry with the picture for letting me know a little too soon.

I was very sorry to be obliged to leave the Prince and Miles Hendon in prison today when I finished reading for, though I know it is all coming out right, I cannot bear to have such a splendid fellow as Miles suffering such indignities.[3]

I never knew or realized before how in old times the Laws hindered instead of helping justice; and now they seem to me much worse than if there had been none.

But most of all the book makes me glad to think that I live in these times.

Please let me thank you again for remembering me in such a lovely way.

Very Sincerely Yours,
Winifred Howells.

Manuscript: The Mark Twain Papers, The Bancroft Library, University of California—Berkeley

1. One of a few copies (estimates vary from eight to fourteen) of *The Prince and the Pauper*, printed on China paper and bound in white linen, stamped in gold. Winifred's book is at the Howells Memorial.

2. Perhaps implicit in this remark is Winifred's own felt need to "cuddle up to something warm and alive."

3. Winifred is at chapter 27 or 28, about four-fifths through the short book. Along with the childlike tone of the letter, her slowness in reading suggests the fluctuating nature of her illness. Or did she feel that this "Tale for Young People of All Ages" was not quite for her? A few weeks later, 19 February, WDH reported to his father: "Winny

is doing a great deal of gayety this winter, and is enjoying it immensely. Of course, she enjoys it in a very Winnyish way, though. She reported that she had such a delightful time at her last party because a charming young man had talked to her about—J. Stuart Mill!" (*SL*, 3:9).

James Russell Lowell to EMH

Legation of the United States[1]
London
24 August, 1882.

Dear M⁼ Howells,

Smoothe your anxious brow once more. He *is* a Count since 1782, & his name is Don José Antonio de Lavelle y Romero which means that his mother's name was Romero. I humiliate myself & eat whatever amount of umblepie you choose to prescribe. With love to your husband who has done even better than I prophesied (& that is saying a good deal).

Me pongo a ll pp enanos de V⁼²
Santiago de Lowell y Spence[3]

Manuscript: Harvard

1. For EMH's explanation of this letter, see her own to William C. Howells, 30 August–1 September.
2. Lowell varies a customary complimentary close, "Me pongo a los pies de usted" (I throw myself at your feet), by writing, "Me pongo a los pies enanos de Vueseñoría" (I throw myself at the tiny [lit. dwarfish] feet of your Ladyship). Raymond R. MacCurdy, of the University of New Mexico, has helpfully provided the translations.
3. The signature adds an amusing self-canonization (St. James) and the family name of Lowell's mother, Harriet Traill Spence.

To William C. Howells

[London, 30 August–1 September 1882]

. . . Charles D Warner lunched with us.[1] Monday we go to M⁼ Anna Lea Merritt's to meet Holman Hunt (Willie Bright son of John B. is invited to meet Winnie) and this Saturday we all lunch with Green the historian. He is far gone in consumption. I have just refused an invitation to the Keegan-Paul's as our time is so taken up that we cant go to see sights.[2] We want to go to Hampton Court, Winsor &c & never get the chance. We now expect to leave here the 1ˢᵗ of October.[3] As malaria is said to be in Venice we think we shall stay a month in Switzerland on the way. We read "Hare's Walks about London" or some book on it all the time and have maps of London spread out over all our tables—and we dont mean to miss a great deal. Will has already been taken by Hutton—who is writing a "Literary Guide to London"—to many choice places.[4] Winnie is eagerly studying up details of English history so as to understand all she reads about.

We have lived faster here the past month than we could have possibly done at home—for this climate helps you out so. You sleep heavily and you dont wake up feeling sore. At present M. D. Conway's two youngest children are visiting our children and they are all playing a game. They have once been in America & rave about it. I gave them ice cream & (French) banannas for dinner, and they were delighted. Both things are unusual here. Mrs Conway has invited us for the 12ᵗʰ to meet Mʳˢ William Morris at dinner—a thing I greatly desire.[5] Lowell, who is awfully schoolmasterish, would have it that Premio Real must be a Count of very new creation—because he was stationed in Canada as Consul, as far as I could judge—and though rather puzzled when he heard his name was De Lavalle which he admitted was about the highest name in Spain, he stuck it out. But afterwards finding his credentials in some Spanish official guide he wrote me a charming note in which he said he would eat any amount of humble pie I would prescribe [letter breaks off].

Manuscript: Harvard

1. The four pages of the letter that remain begin on the fifth page. The letter is dated as having been written after the Warner lunch (29 August) and before the Green lunch (2 September).

2. Anna Lea Merritt (1844–1930), an English painter and engraver; William Holman Hunt (1827–1910), one of the founders of the Pre-Raphaelite Brotherhood; John Richard Green (1837–1883), the author of the *Short History of the English People* (1874); and Charles Kegan-Paul (1828–1902), publisher and author. John Bright (1811–1890) was associated with Richard Cobden as a leading representative of the manufacturing class in England; his son William Latham Bright was a member of parliament (1885–1890).

3. In a letter to Lucretia P. Hale, 25 September, Mrs. Hale wrote that the Howellses left London on 18 September (corroborated by WDH in *SL,* 3:32) and that they had been "very jolly, delighted with London, themselves, and the world in general" (*Letters*

of Susan Hale [1919], p.137). The WDHs had left London earlier than at first planned so that Howells could catch up on his writing.

4. Augustus J. C. Hare (1834–1903) was the author of many travel books including *Walks in London* (1878). Laurence Hutton published *Literary Landmarks of London* (1885).

5. Moncure D. Conway's two youngest children were Dana (b. c. 1865) and Mildred (b. 1869). In his *Autobiography* he describes Mrs. William Morris as the Beatrice to D. G. Rossetti's Dante.

To Victoria M. Howells

Pelham Crescent
So. Kensington [London]
Aug. 31ˢᵗ '82.

So you see the country is safe!¹

Dear Vic:

You ask to have Elinor write—whence I conclude Will's letters are not devoted to gossip. One reason why I dont write oftener (I wrote a long letter to Annie which she was to show you) is the hopelessness of telling you the hundredth part of the queer things we hear see & experience here. I dont know where to begin. Will & I are even more astonished at things than the children—for to them the whole world is strange & new—but with our ripe judgement comparing England and the Untited States—our ways and theirs—is rich fun I assure you. And we have had most ample chances for observation of society—for although, as they told us, everybody was out of town, we have been to no end of dinner parties and entertainments. I must say it has been chiefly amongst artists who seem to have staid late in to the season to finish their pictures by the Summer light—for they dont have much in Winter.² Indeed Tadema & most of them have to go to Rome in Winter to paint. If not what they do in the fog has to be rubbed out & done over if a bright day comes. Harry James told of one evening when the fog was so thick that when he had left a house he could see nothing—not even the house he had left—and had to wander about with his arms stuck out for fear of running in to somebody. Cabs could n't drive and he was two hours groping his way home— only a short distance off. One lady described the atmosphere to me as "pea-soup." Since I wrote we have dined in great style at Mᶜ Lowells'.³ Carpet across the sidewalk & three lackeys to receive us.

Sir George Bonnard took me down to the table.[4] M⌃ʳˢ L. said on introducing us when I first arrived—"And will you take Mrs Howells to dinner." Mrs L. was in real Spanish lace (Why does nt Aurelia acknowledge the *imitation* Spanish scarf I so munificently bestowed on her?) over white satin—tied with red ribbons at the elbow & waist—a dress that would have been becoming to Aurelia who looks some like her. M⌃ʳˢ L. came for us to drive the other day. only W. & Pilla went. She took them down Constitution Hill[5]—a place only the ministers & privileged people can go. Today Will has gone with Osgood to Oxford.[6] Yesterday [remainder lost].[7]

There is a picture of Hanlan for sale in a shop just across the Crescent.[8]

Dont stop writing. We feel a gap so Will got quite worried at not hearing from you

Manuscript: Alfred

1. Above this line a newspaper clipping appears: "THE COURT. / OSBOURNE, AUGUST 28. / The Queen went out this morning. / Princess Beatrice walked, attended by Miss Bauer."

2. Lawrence Alma Tadema (1836–1913) had married the sister of Edmund Gosse's wife, who on "Sunday, October" (Harvard) was to write Elinor thanking her for the illustrated edition of *A Chance Acquaintance*, Cable's *Grandissimes*, and Björnson's *Arne*. Gosse, who was the English representative of the *Century*, had introduced the Howellses and Tadema. See Paul F. Mattheisen and Michael Milgate, eds., *Transatlantic Dialogue: Selected Correspondence of Edmund Gosse* (1965), pp. 92–94.

3. A letter from Lowell to WDH, "X⌃ᵗʰ August—1882" (Harvard), explains Lowell had said Friday, presumably for a dinner or party, and adds that he will take Mrs. Howells and Winny for a drive tomorrow. But the date is probably 10 August and refers to an earlier occasion.

4. Sir George Bonnard is unidentified.

5. Constitution Hill is between Green Park and Buckingham Palace.

6. On WDH's trip with Osgood, see his letter to Clemens, 1 September (*Twain–Howells Letters*, 1:413–414, and *SL*, 3:27–28).

7. The two postscripts appear respectively on the second and first pages, written across the text.

8. George A. Hanlon (fl. 1840–1860) was a wood engraver of the Irish school who also worked in London.

To Victoria M. Howells

Le Clos,
Nov. 28ᵗʰ 1882.

Dear Vic:

I did get your letter which I proceed to answer now. We have fully made up our minds to go to Florence next Monday or Tuesday, and I am getting things ready for the great move. The Grenier family, with whom we have had a very pleasant intimacy (and who have been the principal instruments toward our learning to understand and to speak French) leave before we do. Mlle Reuss, the Alsatian lady also leaves Monday—and our young Bulgarian has— I believe—left already so that Mlle. Columb will be quite alone.[1] She has seemed out of place & homesick here, alone in this remote place—speaking French & English hardly at all—(she speaks modern Greek, Bulgarian & German) and today she has gone alone to Basle where she says she is to meet her brother, an officer in the Prussian army who cannot come further to see her than there. She is only sixteen and has gone quite alone. Could Daisy Miller do more than that? And she arrives at Basle at half past seven without bag or baggage! Mlle Columb could not hinder her. I think it was a ruse to get away—that she will send for her trunk & never return. She is rich & cultivated & has charming manners. Her name is Elizabeth Keskari. Her father was a trader of furs (I think) in Leipsic where her mother, a widow lives now. Mlle. K. came to this region for her health—& by accident got into this house. I think she expected to be in some gay hotel or pension & is dissatisfied. Her Mother cannot read—but was not "sold," Mlle Keskari proudly says. Most women in Bulgaria are *sold* for wives! At one time Mlle K. lived in Constantinople. She belongs to the Greek Church. There is a beautiful little Greek church at Montreux, where many Russians spend the Winter. Yesterday Will took all the children to see Vevey. They visited the church where Ludlow, the judge who condemmed Charles I, is buried—and John Phelps his clerk, near him. A tablet is put up which says Phelps was so convinced of the righteousness of his act that he signed his name on every page of the condemnation—which I think Ludlow read to the king and adds that this memorial to him is put up by William Walter Phelps & another Phelps—his descendents beyond the seas.[2] Will knows Mr W. Phelps very well and is going to write to him that he has seen it. The children have worked very diligently here, and we hope they will have a little fun in Florence—especially Winnie. The mother & sister—quite Winnie's age—of Clarence King are spending the Winter there; and with them their Cousin M^{rs} McKaye of

Cambridge, who is very fond of Winnie.³ Then Stillman the artist has a daughter just Winnie's age—whose mother was a Miss Mack of Belmont. You know perhaps that she committed suicide in Athens by taking prussic acid—& Stillman afterwards was married—greatly against her father's will—by the beautiful Miss Spartalee of London—daughter of a rich Greek banker—the toast of the artists, and known to have refused several noblemen. We knew the first M͞r͞s Stillman in Rome when he was consul there—and we know her sister Bella Mack very well. Your father may have heard of David Mack the abolitionist—her father.⁴ I see that M͞r͞s John Brown (the martyr's wife) has had a reception in Boston at D͞r Talbot's.⁵ I saw M͞r͞s David Mack's name among those present. All the old abolitionists met to do her honor. Winnie just remarked, looking at Scribner of November, that she thinks the verse Longfellow wrote in her album as pretty as the one published. Hers were written after those were—in 1876 or 7.⁶

Will is "getting it" for his rather unqualified expression about Thackary—but he did n't say he did n't like Thackary—& as his conscience is clear he just works on—not minding much either blame or praise. He is in Montreux this afternoon. He often walks over there to keep down his "sitz fleisch" as he calls it. He has an immense correspondence and has to write many letters besides working on his story.⁷ We were very glad to have Annie's letter. I wont return it until I get home—when we will exchange letters. I am going to write next to Aurelia—and then to mon beau-pére. *Do* tell us more about Roche.⁸ Where is he now? And did Annie give him up? Will has just come in bringing the letter to Winnie & Pilla from your father—dated Nov. 9͞th and Nov 14͞th just after Joe left. It was begun the 9͞th & finished the 14͞th. I dont believe we have lost any of your letters—and how often it happens that one arrives before one is quite finished! Pilla is quite gratified at her grandpa's praise and will write again. John will describe the vintage, the birds &c. Wine is made mostly of white grapes (thin skinned ones) and they look very nasty being pounded—much as clothes used to be pounded in the wash with a wooden pounder. I make a shape of the pounder & the tub they are pounded in—[sketch] but John will tell all. We shall have been here eleven weeks.

We were seven in London. It was cheerful to hear that you were *all* well this time. Is Aurelia really going to Quebec? Please remember me to Mary Hurley. There may be a little break in the correspondence while getting settled in Florence—but perhaps not. Will is getting lots of material for the future, though working very hard at present. His story is dragging a little. Edmund Gosse of London & Will have struck up a great friendship. He is a lovely character. Do read his Life of Gray.[9]

Glad the Frechette children seem to be getting over whooping cough. Glad always to hear that Henry is quieter. Do send us an Annie letter occasionally as she does not write to us—nor we to her—I must own. But it is all the same thing if you will exchange our letters. Much love to all from all

Elinor

Manuscript: Harvard

1. In the fall, on their way to Italy, the Howellses stayed at the pension of Mlle. E. Colomb in Le Clos for more than two months. Mme. Grenier, the widowed cousin of Mlle. Colomb, had also been staying there with her two children; she is described by WDH as "one of those exquisite persons, who keep a pretty gentleness and a young habit of blushing, far into her gray hairs" (*SL*, 3:34). WDH briefly mentions the "young Bulgarian" in a letter to his father, 5 November 1882 (Harvard)—"We have had an addition to our queer household in the person of a young Bulgarian girl, who speaks modern Greek, as well as German and some slight English and French." He does not comment on Mlle. Reuss.

2. The grave of Edmund Ludlow and the tablet commemorating John Phelps are in St. Martin's Church. William Walter Phelps (1839–1894) was a U.S. congressman (1873–1875 and 1883–1889) and a diplomat.

3. Florence Little King, whose second marriage was to George S. Howland, a New York merchant; and their daughter, Marian Howland, a half-sister of Clarence King's. Marian Treat McKay, married from 1878 to 1890 to Gordon McKay (1821–1903), of Boston, wealthy inventor, manufacturer, and philanthropist. After her divorce, she married a count in the German embassy at Washington.

4. Marie Spartali Stillman (1844–1927) had sat for portraits by Burne-Jones, Whistler, and D. G. Rossetti, and was a painter herself. Stillman was married to Laura Mack in 1860. The Howellses had met both of them in Italy in the 1860s. She died in 1869, and her oldest daughter, probably referred to, was named Bella after her sister. The father David Mack (1803–1878), a graduate of Yale who had been associated with a number of Transcendentalist circles, was known best in Belmont for his Orchard Hill Family Boarding School for Young Ladies, though he was also active in the Unitarian church and the library.

5. Probably Israel T. Talbot (d. 1899), for many years dean of faculty at the Boston University School of Medicine.

6. The poem, in Longfellow's hand, appears in the *Century* (until recently *Scribner's Monthly*), November 1882:

> She who comes to me and pleadeth
> In the lovely name of Edith,

Will not fail of what was wanted.
Edith means the "Blessed"; therefore
All that she may wish or care for
Will, when best for her, be granted!
Henry W. Longfellow
Jan 1. 1873.

7. Toward the end of his November *Century* article on Henry James, WDH had remarked that the new fiction had become "a finer art in our day than it was with Dickens and Thackeray" in a paragraph that antagonized the British to such an extent that even his new friend Gosse took exception. The story was *A Woman's Reason.*
8. Roche and Mary Hurley (mentioned a few lines later) are unidentified.
9. *Life of Gray* (1882) was a favorite of the Howellses. See *SL,* 3:31 and passim.

To Anne H. Fréchette

Pensione Togarazzi,
via Salustio Bandini,
Siena, Tuscany, Italy.
February 17ᵗʰ, 1883.

Dear Annie:

That two of your letters should have gone unanswered so long! But if you had ever been in Florence you would understand that no one *could* write letters there. When I tell you that our life there was more exciting even than in London you will understand what it must be. And it was all a mistake. We had been quiet so long that we had promised Winnie that she should have a little dancing & fun when we got to Florence—and therefore we rather courted society at first—and I assure you we *got* it! It used Winnie up directly and prevented our sight-seeing—but we kept staying on & on hoping we could snub people off and have a little quiet—but it grew worse & worse, so that we were finally obliged literally to flee before the pursueing crowd to this quiet retreat. Will is obliged to go back to make some more studies in Florence, but we are going quite incognito.¹ We shall only stay a fortnight at most there. As for Rome, we are now so near it that I hate to turn back without seeing it—but we are afraid to take Winnie there in her rather low state of health, for fear of her contracting the fever; so possibly John & I shall run down alone for a few days. Will does n't care in the least to see it again. He is going to write about all the cities

in Tuscany, so that he will be obliged to see Arezzo, Pisa, Leghorn
& Lucca. And then we shall go on to Venice—thence to Paris for
a week & then to England. Will may run up to Edinburg but none
of the family will accompany him—unless Winnie should. So much
for our plans.—Winnie is cutting a wisdom tooth—but still she is
improving since we came here. We are staying in an old palace on
the edge of the city, and our windows overlook the plains as far as
the Appenines, and a fresh wind blows constantly from the hills.
I think we shall stay here a month, as Will wants some time to
write, & it is cheap & healthy. John Hay & his wife are coming up
to spend Sunday at the hotel here.[2] They came on from Cannes,
where they have left their children, especially to meet us at Florence—& then we ran off & left them after three days. I suppose
we had two hundred calls in Florence. I counted 111 long before
we left, just from cards. We had a private parlor to receive in. I
had the worst of it—for I had to make calls *every* afternoon—&
then perhaps go out to dinner afterwards. We had letters from
our English friends in London which got us in for no end of things
then all the Americans called—& we met many Italians and even
Germans & Russians—for instance: Carl Hillebrand & D͞r
Homberger[3] were very polite to Will—& Winnie had quite a flir-
tation with a young Russian. We saw a good deal of the Stillmans.
You know she was the beautiful Miss Spartalee of London—a
Greek? At the hotel we met Lady Sarah Spencer (sister of the L. L.
of Ireland) & Lady Maude Seymour who had read Will's books &
were very polite. Also M͞r Edward Stanhope, Disraeli's secretary.[4]

We bought John a mandolin & he has learned to play it very
well. Larkin modelled a medallion of Will—which I enclose. He is
modelling M͞r and M͞rs Hay.[5] John Hay is "down" on a Grecian
temple for a monument to Garfield. John is in long pantaloons for
good. He studied Latin with an Irish clergyman the whole time in
Florence—& Pilla had a French teacher. Of course we did see *most*
of the Florentine sights—especially the outdoor ones—& we shall
do the rest of the galleries & churches when we go back. Will at
last got desperate and *hired a historian* (the poverty stricken Italians
will do anything for money) to take him about the city and explain
its history. He paid the cavalier (he was a noble) ten dollars for a

week's service. I will end this mere *poster*, by thanking you for your delightful letters. We are impatient to see your articles. We thought Pauvre Elise a very deep & good study.[6] *I* like that sort of subjective writing so much better than any other. Will thinks his new story a sort of failure.[7] I wonder what you all will think. We hope Sam's accident is not going to prove a serious injury. It is a long while since we have heard from you—nearly three weeks—but storms in the ocean may have delayed the letters. I end hastily. All send much love to all

Elinor

Manuscript: Harvard

1. The studies eventuated in *Tuscan Cities* (1886), which describes the pension at which the Howellses stayed in "Panforte di Siena," part 7.

2. In a letter to John Hay of 18 March 1882 WDH had suggested that the Hays might come to Florence the following winter. They met the Howellses in London the summer before and then briefly in Florence and Siena. See *Hay–Howells Letters*, pp. 56–67 passim.

3. Karl Hillebrand and Heinrich Homberger were German writers, the latter of whom had written on WDH in German magazines.

4. Lady Sarah Spencer, the sister of John Poyntz (1835–1910), Earl Spencer, Lord Lieutenant of Ireland (1868–1874 and 1882–1885). Though his wife was a Seymour, no Lady Maude appears in Burke's *Peerage*. Edward Stanhope (1840–1893), formerly undersecretary of state for India, was at this time leader of the opposition against Gladstone.

5. The Howells medallion is in the John Hay Library at Brown University. For a picture of it, see *Hays–Howells Letters*, p. 46. The Hay medallion (presumably Mrs. Hay does not appear with him) has not been located.

6. "Pauvre Elise" first appeared in *Our Continent*, a small New York weekly, and then was reprinted in the *Ashtabula Sentinel* (Doyle, p. 54). The story is that of an attractive habitant, the servant of the narrator, who has an unhappy love affair, attempts to become a seamstress, and finally enters a convent, inviting her former mistress to attend the ceremony at which she takes her final vows. It fully merits EMH's praise as "a very deep and good study" and WDH's comment that "there is not a false note in it" that he made to his father in a letter of 31 December 1882 (Alfred).

7. The first installment of *A Woman's Reason* was in the *Century*, February 1883.

To William C. Howells

Siena, March 4ᵗʰ/ '83.

Dear Father:

. . . It is of no use for me to begin describing our adventures— as Will is going to write them all out, and Pennell the artist, who is with us, is to illustrate them.¹ But is it not amusing that we should have a priest for a landlord? I offered him some chocolate drops the other morning, and he most gracefully refused saying "It is not permitted." In Lent & at that time of day I suppose. I must tell you about our celebration of Will's 46ᵗʰ birthday. All the other birthdays have been most religiously kept by the children—and they had no idea of letting us off a bit this time. So I went to a shop & ordered: a big round cake—the kind they call Bocca di Dama or Lady's Mouth, because it is very sweet—to be made, which Pilla decorated and inscribed with candy—and I asked Maddalena, the priest's niece, to buy a boquet for me to give him for Mʳˢ Angell,² (who always presents him one at home and had written me to do so for her here) I bought him a copper bucket—beaten in ornamental figures [sketch] and bound with brass—such as the peasants use here—we bought the priest's Roman lamp, [sketch] which he desired to have, for him, and some pictures of Siena. On the morning of his birthday the cake arrived all right but the flowers were late. At eleven o'clock the boquet arrived—and imagine our surprise at seeing this thing: a pyramidical structure a foot high & a foot & a half across entirely made of camellias—fifty-one in all—the lower row red—then white with migionette between, then variegated with pansies between, surmounted with a big white camellia surrounded with a fringe of goldengrass & yellow flowers. The whole was in a lace-edged paper and was really a splendor to behold. Beside this came five hand boquets each containing camellias—so that we counted sixty three elegant camellias in the room at once. Which is not so surprising when I tell you that there are bushes loaded with them right under our windows here. And it seems that all these flowers were a present from Maddelina who got them—for nothing I suppose—of a "gardiner she knew." By the way day before yesterday a young girl in the house brought Winnie lovely primroses & crocuses & anemones & snow drops which she had found growing in a meadow outside the city. But lest you get false ideas of things here, let me add that it snowed last night—and that the mountains in the distance have snow on their tops. Still the grass is green everywhere—the olive keep their leaves and many vegetables of last year are green in the gardens.

I have seen peas six inches high. Last week they were planting potatoes. We have an open wood fire all the time. It is much colder here than at Florence—as Siena is on a high hill.

To go on with the birthday: we opened a bottle of champagne (which M⸰ Alexander an American lady in Florence gave us for our lunch on the cars coming here) and invited the priest & his niece and a M⸰ Wathen & daughter (English) to join in our festivities.[3] The priest added two bottles of his own wine, and all went off very merrily—until the lunch bell (12½) interrupted the celebration. I say lunch, but we have nothing until then except a cup of coffee & bread & butter—and they call it breakfast. The Wathens gave Will a pretty paper weight. We have had a talk about Rome since I began, and it is about decided that I shall go along too—leaving Winnie & Pill & M⸰ Pennell here.[4] Will seems to wish me to go principally so I shall take John in to the Catacombs— which he seems to dread. The fare is half price, now on account of an exhibition in Rome—so I may as well go. And Winnie is rather pleased at the prospect of a quiet time. We may not do this—so dont be alarmed till you hear again. . . .

Affectionately
Elinor

Manuscript: Harvard

1. In the omitted three paragraphs, about a third of the whole letter, EMH mostly comments on the activities of the Ohio Howellses, though she also mentions the possibility of taking John to Rome, a trip referred to later. Joseph Pennell (1857–1926) was commissioned by R. W. Gilder to illustrate the articles on the Tuscan cities.

2. Mrs. Henry C. Angell.

3. Francesca (Esther Frances) Alexander (1837–1917), author and artist, had gone to Florence in 1853 with her parents and was the author of several books on Tuscany. Mrs. Wathen and her daughter have not been identified.

4. Evidently on the same day the family had another talk, for in a letter of this date to C. D. Warner (*SL*, 3:56) Howells announced that Pennell was going to Rome with John—"the rest of us joyfully refrain."

Winifred Howells, Belmont, October 1878.

John Mead Howells, Belmont, October 1878.

Mildred Howells, Belmont, October 1878.

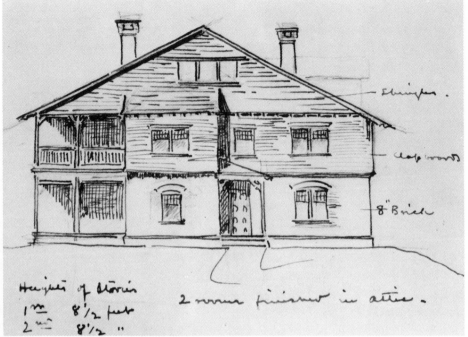

Sketch of the Belmont house. McKim, Mead & Bigelow, 1877.

Redtop, Belmont. Exterior view with EMH, WDH, McKim, and Mead, October 1878.

Redtop, Belmont. Detail of the library inglenook, October 1878.

Redtop's library inglenook. From Harper's Monthly, *February 1881.*

Library inscription designed by EMH. From the Boston Evening Transcript, 22 May 1920.

Redtop today.

John, Mildred, and Winifred, Boston, c. 18

Mrs. W. D. Howells.

Beacon Street.

Winifred Howells at twenty.

✤ VIII ✤

Beacon Street and Other Hazards of New Fortunes

1884–1891

For most of the summer of *1884* the
*Howells family stayed at Kennebunkport, Maine, while except for occasional
visits there Howells remained in Boston, renovating the newly purchased
house on the waterside of Beacon Street. Their house turned out to be one
"hazard of new fortunes," as Will was to entitle a novel six years later,
though neither Elinor nor he recognized it at the time. Here was to be the
fashionable home where Winny would be introduced to society, but in 1887
with her further decline they sold it. Until Winny's death on 2 March 1889,
nearly all the places where they were to live depended on what was hoped
to be for her well-being: Auburndale, Lake George, Dansville, Buffalo,
New York City, and Little Nahant. After her death they were drawn back
to the Boston area for a while, partly because John was at Harvard and
most of Mildred's friends lived there. Here the Howellses were mostly in an
apartment on Commonwealth Avenue, though they spent their first summer
at the Brooks Place in Belmont. At the end of 1891, after summers in the
Adirondacks and Saratoga and Intervale, they returned to New York on as
permanent a basis as was possible for them.*

*Because of the scarcity of Elinor's letters during this period, we know
less of her response to what Will was doing in fiction than at other times.*

But from such a letter as Henry Harland wrote her at the end of November
1888, almost exactly at the time Winny was placed under the care of
Mitchell, we may assume that her interest in her husband's work and in
the literary politics of the time remained active and informed. From letters
of others to her it is evident that she recognized his great accomplishments
of these years in such novels as The Rise of Silas Lapham (1885) and
A Hazard of New Fortunes (1890). Finally, she appears to have fully
sympathized with his developing social awareness at the time of the An-
archist trials, for as he remarked in a letter to Annie from Buffalo in
November 1887, Elinor and he wished to be settled very humbly where they
could be identified with the struggling masses. Of course, neither of them
ever realized this Tolstoyan dream. In her letter to Annie a couple of months
later, Elinor mentioned the speech of Laurence Gronlund, which first interested her husband in socialism.

To Lilian W. Aldrich

[Boston] Sunday [23 March 1884]

Dear M͏ͬ͢ Aldrich:

Will you lend me the loan of the two double candlesticks [sketch]
standing on top of the bookcase in your front parlor? I think they
would be just the things for the two ends of the very broad, or
long, desk M͏ͬ͢ Abbott is to use to read on?[1] Dont you? I have single
candle-sticks, & those for three candles, but very heavily hung with
glass.

I will send for them on Tuesday if you please. We are going out
to Brookline to pay the last tribute to M͏ͬ Fuller this afternoon.[2]

M͏ͬ͢ Abbott came in yesterday. She is counting on M͏ͬ Aldrich's &
M͏ͬ Bugbee's[3] being present at her reading. M͏ͬ Howells hopes Aldrich will come. He would certainly have been at the reading had
he been an *old* friend of M͏ͬ͢ A's.

M͏ͬ͢ Abbott suggested something "Turkish*" over the desk; I
should think it better plain myself—and should have to borrow the

*or said M͏ͬ͢ Mifflin had it[4]

article—which might not look well in conjunction with the deep yellow hanging behind her.

Come & see.

Manuscript: Harvard

1. As reported in the *Saturday Evening Gazette* of 8 March 1884, "Mrs. C. P. Abbott, the daughter of the late Stephen B. Ives, of Salem, is to give a series of five readings descriptive of 'Life and Experiences in India,' the first taking place at Mrs. T. B. Aldrich's, 57 Mt. Vernon street, next Tuesday afternoon." Mary Perkins Ives Abbott (1857–1904) had married a Calcutta merchant; later she published two novels, *Alexis* (1889) and *The Beverleys* (1890).

2. George Fuller died on 21 March and his funeral was on the 23rd.

3. For James M. Bugbee, see 17 March 1871.

4. Probably Jane Appleton Phillips Mifflin, wife of George H. Mifflin, of Houghton, Mifflin and Co.

To Lilian W. Aldrich

[Boston] Monday evening
[14 April 1884]

Dear Mrs. Aldrich:

I think I ought to write at once to say that M⸬ Howells would have to leave your house by a quarter past two on Friday as the tickets to the Wendell Phillips Services say people must be in their places by 2.45[1]—which if the breakfast or lunch is fashionably delayed by some Matthew Arnold might deprive him of his full share of victuals[2]—

EMH

I dare say your "men folks" are going—so you would have arranged accordingly, but I believe we agreed to communicate ideas on the subject?

Manuscript: Harvard

1. A memorial service for Wendell Phillips, who had died in February, was held on Friday, 18 April, at the Tremont Temple.

2. Matthew Arnold was in Boston in November and December during his American lecture tour. Because WDH was out of town when Arnold first arrived, Elinor "is greatly excited about having to call on Matthew Arnold alone," as Howells wrote his father on 11 November 1883 (Harvard). In *My Mark Twain (Literary Friends and Acquaintance, HE,* pp. 272–273), he reports the conversation when Elinor told Arnold that he was visiting Mark Twain: "'Oh, but he doesn't like *that* sort of thing, does he?' 'He likes Mr. Clemens very much,' my representative answered, 'and he thinks him one of the greatest men he ever knew.'"

To Lilian W. Aldrich

302 Beacon St [Boston],
Tuesday eve¹ [4 November 1884].

Dear M͇ Aldrich:

Your proposal is *most* kind—but there is no *rest* for me until everything is ready for the Gosses—and then I mean to lie abed all day for a week.²

Today I have been out—shopping to be sure, but in the air a good deal & feel much better.

I had the shops all to myself being Election Day.

Affectionately
E M Howells

Manuscript: Harvard

1. The letter is dated from mention of Election Day.
2. The Gosses were to stay with the Howellses beginning 1 December during the husband's lectures at the Lowell Institute (see *Transatlantic Dialogue: Selected American Correspondence of Edmund Gosse*, 1965, p. 154).

To Lilian W. Aldrich

302 Beacon St. [Boston]
Jan͇ 26ᵗʰ [1885]¹

Dear M͇ Aldrich:

I would like to come so much—especially if we were to be by ourselves—but have already made one engagement for Wednesday and *I* never try to do two things the same day. I have not been well of late—and am saving myself up for a week of dentistry when I shall go nowhere—it uses me up so. I took a fearful cold at a dinner-party where they had had the windows of the dining-room open to keep the flowers fresh, and had such a huge wood fire in the drawing-room that the guests were all in a perspiration before going in. One lady was in low neck too.

I shall have to give up society altogether I find. *My* constitution cant stand this sort of thing. And you cant carry a shawl to a dinner party you know.

I am going to call upon you—though it is you who owe the call—very soon.

How much difference neighborhood does make even in a city! We *never* see the Angells any more than you.

Affectionately yours
Elinor M. Howells

Manuscript: Harvard

1. The only other January when the Howellses were at Beacon Street was 1887. Of the two years 1885 appears more likely.

To Horace E. Scudder

[Boston] May, 1885[1]

Dear Mʳ Scudder:

Do receive my humble thanks for your review of Emilia. I hav'n't enjoyed anything so much for a long time.[2] It more than repays one for wading through that absurd book. I thought it the wildest book since Signor Monaldini's Niece,[3] but had no idea of what opportunities for criticism it offered until I read your article. And I did n't suppose you would *dare* to be so funny over Harvard's professor of literature. I only hope he is capable of being taught that it is but one step from high romance to comedy and next time he will not step quite so far. I hear he wrote a society novel first which was a failure.[4] But he is young. All the affected fashionables are praising the book. Your review will create a sensation.

Sincerely yours
E M Howells

Mʳ Howells has neither read the book nor this note. He will read your review & then the book. We go to Great Barrington to board for June leaving June 1ˢᵗ

Manuscript: Harvard

1. The "1885" may be in another hand.
2. The review of *The Duchess Emilia* by Barrett Wendell (published by James R. Osgood & Co.) appeared in the June 1885 *Atlantic*. With its action in Italy and New England, the novel's major motif is the transmigration of Emilia's soul. Both in his summary of the story and in his judgment, Scudder derides the book: "The whole

manner of the author is fatal to that deep reality which is essential to genuine romance. He is not himself wholly possessed by the story. . . . The result is that from beginning to end the reader perceives himself in the presence of an affectation." At the end, however, the reviewer shows some mercy: "We would really much rather not scoff, and if Mr. Wendell will cultivate his natural voice, and sing us a song with as much purity of feeling as goes to this book of his,—for that is its one redeeming feature,—we promise to applaud with the heartiest."

3. On *Signor Monaldini's Niece*, see C. D. Warner to EMH, 9 April 1879.

4. At this time Barrett Wendell (1855–1921) was an instructor of English at Harvard. The first novel, to which EMH refers, was "Plaster of Paris," written in 1881–1882 and not published, though probably seen by WDH since Osgood, whom he often advised, published *Emilia*. The Wendells and Howellses were on good terms, for Mrs. Barrett Wendell has recorded in her journal (at Harvard) that on 24 March 1885 she took tea at Mrs. Howells' to meet Mr. and Mrs. Mark Twain. Of the Clemenses' visit generally Twain wrote WDH on 26 March, "We did have a most noble good time in Boston" (*Twain-Howells Letters*, 2:524).

To Anne H. Fréchette

Buffalo,
Jan. 6ᵗʰ, 1888.

Dear Annie:

Of all our Silver Wedding presents the old letter you sent was the most appropriate and had the most sentiment in it. It took us straight back to Casa Falier and the little parlor in which it was written.[1] Our only regret on reading it was that we had put all those stamps on the envelope when it could have been sent for nothing, as we found out only after we had spent fifty dollars in stamps.

Yes, one other regret we had; we found Padre Giaccomo (who sent the figs) spoiled by the flattery* when we were last in Venice. But Will did much towards that by putting him in to his book.[2]

I came across this cutting in a Buffalo paper the other day.[3] Howells Fréchette will have a distinguished name to shoulder when he goes out into the world. Mʳ T. S. Perry, who is in Paris, had sent us Le Figaro with the notice of the Poète Canadien.[4]

We see a good many Canadians here.[5] Mrs. Buel who boards at the Niagara is from Ottawa—daughter of an Irish M.P. whose name I dont know and who is now dead. She puts on many airs.

*of Americans & English

Her husband is a lumber merchant. Then M⁫ Jebb here is a particular friend of the daughter of Agnes Strickland whom we met in Toronto &c &c. We shall be glad to see M⁫ Peck whom I think Will remembers.[6]

Last night the Fortnightly Club here discussed socialism, having invited Grönlund who wrote "Ça Ira" and the "Coöperative Commonwealth" to speak.[7]

There is lots of fashion and money in Buffalo, but very little literature. For creature comfort we would like to stay here, but sometimes we pine for the "culture" of Boston. *Your* friend M⁫ Keenan, who is staying in Scranton, is coming here soon to visit M⁫ Jebb in this hotel. Have you any message?

I dont believe we shall remain here long. Winnie is just as well without us now that she is in full "treatment." Will has a scheme for looking up certain phases of society in New York. I dont think we will ever live in Boston again, Winters, though we may be near the city Summers. Will has got all he can out of society there, for literary purposes, and John will have to go to New York finally.[8]

So you have *bought* a house?[9] Well, that is better than building. But we have sworn never to do either again. I think yours must be very nice and I know it is pleasantly situated if in New Edinboro.

Have you heard that my sister Mary is assistant librarian at Hampton Institute at present? She got the place through Charley who is a trustee. With much love to Achille and the chicks.

Affectionately
Elinor

Manuscript: Alfred

1. Probably the reference is to Elinor's first letter to her father-in-law, 4 January 1863. The first extant letter to Annie is 14 November 1863.

2. Padre Giacomo Issaverdanz is described in "The Armenians," *Venetian Life* (2d ed., 1867), pp. 195–197 and passim.

3. The clipping, attached to the letter, mentions Louis Fréchette's new book *La Légende d'un peuple* and his commission to translate *King Lear* for performance at the Comédie Française. It concludes: "A master of the English language and literature and an ardent Shakespearean student, M. Frechette will approach his work with every qualification."

4. The article in *Le Figaro*, by Jules Clarette, is noted in the clipping, but the letter from Perry has not been located.

5. The Howellses stayed at the Niagara Hotel from November 1887 through January.

6. Probably EMH is recalling a niece (rather than a daughter) of Agnes Strickland (1796–1874), best known for her "Lives of the Queens of England"; for an unnamed niece was mentioned in a letter from Annie to Elinor of 13 May 1883 (Alfred). The other people mentioned in this paragraph have not been identified.

7. For WDH's interest in Laurence Gronlund (1846–1899), the prominent socialist writer, see *SL*, 3:216. His review of both books appeared in *Harper's Monthly*, April 1888.

8. After Winifred's death in 1889, the Howellses did return for a time to the Boston area, though with the fall of 1891 they went back to New York.

9. The reference may be to the actual purchase of the house at 87 MacKay Street, into which the Fréchettes had moved in 1886.

Henry Harland to EMH

35 Beekman Place [New York],
November 27 [1888][1]

Dear M⟨rs⟩ Howells:

Many thanks for the "Confessions", which I return to-day by post.[2] In parts I found it rather amusing, though the author's constant straining after originality and his consuming desire to reveal his familiarity with recondite French literature were rather tiresome. As for his measure of M⟨r⟩ Howells, was it not simply silly and puerile? I don't believe he has read M⟨r⟩ Howells at all. His sort of cynicism and originality is after all very cheap and shallow, and I should think it would be very easy. Let us decry all that is sweet, natural, wholesome, and go in for all that is malarial, artificial, extravagant. For a pet we will keep a Python. We will love all that the world hates, even if it is bad and deserves hatred. We will hate all that the world loves, even if it is good and deserves to be loved. A sort of conventionalized unconventionality, cut-and-dried, made by rule.

I don't know when I have enjoyed anything as I did my visit at your home the other evening.

Always Yours
H. Harland

Manuscript: Harvard

1. The year is established by the appearance of George Moore's *Confessions of a Young Man* (London, 1888) and Harland's sailing for Europe in July of the following year. This was the period of Harland's being an "ardent convert to realism," as Howells described him to Perry (14 April 1888, *SL,* 3:224). The friendship continued through Harland's espousal of aestheticism and editorship of the *Yellow Book.* Again writing to Perry, this time from San Remo, 9 January 1905, Howells remarked, "We can at least talk literature together, and leaving out our great difference, we hate the same things specifically" (*SL,* 5:118). Also, from San Remo Elinor writes of seeing the Harlands (to John, 18 January 1905). At this time Harland was suffering from tuberculosis, which caused his death on 20 December 1905.

2. Of Howells, Moore observed in part: "He is vulgar, is refined as Henry James; he is more domestic. . . . Henry James went to France and read Tourgueneff. W. D. Howells stayed home and read Henry James" (pp. 254–255). The famous witticism did not permanently disconcert Howells, who reviewed Moore's *Celibates* and *Esther Waters* favorably in *Harper's Weekly,* 27 July 1895, along with James's *Terminations.* In *Heroines of Fiction,* 2:212, Howells tells how he had long shied off from Moore as an imitator of Zola, "but his 'Esther Waters' showed me how mistaken I was. That is a great book. . . .'"

WDH to William C. Howells

New York, March 17, 1889.

Dear father—

We are wearing the time away, and as we get farther from the moment of our loss I suppose it will be less terrible. I will not needlessly dwell upon it; but I must tell you how bravely Elinor has borne it, with always a first thought for me in her own anguish. This is the more generous in her because she yielded to me in all the details of this last attempt to restore Winny to health. It is forever too late, now, to undo any part of it, and I can only console myself by thinking of the absolute confidence I had in what was known to me as the highest skill and greatest experience. I am sure that Dr. Mitchell did all he could for Winny, and that he suffers almost next to ourselves at the result.

John went back to Boston on Monday, and he will look up a house for us in that vicinity. We scarcely know what plans to make beyond something that shall keep us together, and be best and cheerfullest for the children.

Pilla is very well, and is having belladonna put in her eyes previously to having glasses fitted to them, as one has been found near-sighted.

All join me in love to you all.

> Your aff'te son,
> Will.[1]

Manuscript: Harvard

1. In his next letter to his father, 22 March, Howells continued: "You know that there can be but one thing in my mind. Elinor and I talk it over continually, and when I am alone I recall and reclaim all I did, and reconstruct the past from this point or that, and dramatize a different course of events in which our dearest one still lives. It is anguish, anguish that rends the heart and brain: but you can understand how inevitable it is. It must go on, I suppose, till the futile impulse wears itself out" (*SL,* 3:248). Winny had died on 2 March. John was a sophomore at Harvard.

Samuel L. Clemens to EMH and WDH

> Hartford, November 14/89

Dear Mrs. Howells:

It was a most refreshing visit to me, & brought back my youth by bringing back the friends of my youth, with voice & ways & aspects remembered out of that ancient time.[1] (Thirty-five is Youth to 54.) Mrs. Clemens was hoping to hear that the whole three of you could come down to us for a visit this winter, & I have as good as promised her that it will happen.[2] I based this on the fact that you were very much your old self, yesterday, & might justly be expected to continue to improve in strength until the necessary capital of it should be acquired for the journey. So mote it be. And it is time the new generation of us were getting to know each other.

Dear Howells: Stedman decided with you; so I shall make it read—

"Dismember me this animal, & return him in a basket to the base-born knave his master, &c".[3]

> Yrs Ever
> Mark

How the priest haunts me![4] How sweet he was; he was embodied music. And don't you suppose he leaves that rare & clinging impression with everybody? That is *success*.

P.S. Later. My facts are astray. We return home *Wednesday.* So, whatever day you come, we shall be here & expecting you.

S L C

Manuscript: Harvard

1. Clemens had seen *A Foregone Conclusion* in the Boston performance of 12 November at the Tremont Theatre and stayed overnight with the WDHs.
2. The WDHs visited the Clemenses on 11 March 1891. "Mrs. Howells enjoyed herself immensely," Howells wrote on 15 March (*Twain–Howells Letters,* 2:639).
3. Both Stedman and Howells were reading *A Connecticut Yankee in King Arthur's Court* in proof. WDH was to review it ("Editor's Study," *Harper's Monthly,* January 1890). The Iowa-California edition of *Connecticut Yankee* (1979) has the text as in the letter except that "&" appears as "and"; in the first edition (1889) "his master" was changed to "who sent him."
4. Alessandro Salvini (1851–1896) played the priest, Don Ippolito, in the play.

WDH to William C. Howells

Boston, March 30, 1890[1]

Dear Father:

I want to tell a dream that I had about Winny, yesterday morning. I have dreamed several times about her since she died: directly after, when I was wild to have some assurance that she still existed, and when she seemed to come to prove it. At one of these times, I took her cheeks between my hands, and said, Have you come back to poor Papa? and cried over her.

A few months ago I seemed to find her standing in her mother's room, in a gray dressing gown she often wore. Elinor said, "Here's Winny," and I took her in my arms and held her close and long. When I let her go far enough to speak, she said in a glad way, "Then you *do* love me?" and I choked out, "O, *love* you!" That was all.

But this last time the dream was longer. The night before, I had to go back from the room where I sleep to my study, and I lit a match in front of a large photograph of Winny, that stands on a desk. It is a sad face, and frowning from the strong light thrown upon her by the photographer, when it was taken. But as I saw it by the match light the melancholy face seemed to smile at me; and I had the thought of this in my mind, I suppose when I fell asleep, for when I saw her in my dream, just before I awoke in the morning, she had a smile on her face. We seemed to be in some sort of hotel; and Elinor and I were sitting at a small table set for four, when Pilla and Winny came in together, as if to breakfast. I said to Elinor, "Why here's Winny!" and I took her hand, which was slight but not thin, and had a warmth which I recognized as that of health. Neither Elinor nor Pilla seemed surprised, though I kept saying to them, "Here's Winny! Don't you see? It's Winny!" She was dressed in a dark blue, which she sometimes wore, and she was girlish and slender, but looked strong; this impressed me; and she had a humorous, half-teasing gayety, which was peculiar to her when she was with Pilla, and that was always so pretty in her. I knew that it was her spirit, and I began at once with what is so much in my mind about her. Throughout our talk, she had a surprised air, mixed with something puzzled and amused, as if she wondered at my curiosity about something which was to her very plain and simple, and yet that she was not wholly free to speak of. "Are you happy there, Winny?" I asked. "Yes," she said. "At first I was lonesome, and cried." Then she laughed and said, as if joking at the form of the verb she used, "but I haven't *weeped* since." I was holding her hand, glad of its warmth and strength after so much sorrowful sense as I used to have of her sickness; and now I looked very earnestly into her face and said, "Is it interesting, there?" I was aware that I asked this because all the accounts of the other life have made it seem dull. She answered, "Yes," in a way as if she would like to explain. "But not like here?" I went on. "No," she said, and she still kept that look, so like her, and so pertinent to the moment; and then the dream changed. She and Pilla were sitting on a window sill, somewhere. I introduced some young man to Winny. "This my daughter who is dead; this is her spirit." She slightly acknowledged the introduction, as if there were nothing strange in the affair; and then we were going up hill, as if to our house at Belmont, and I put my hand on her waist behind, to help her, and said, "I don't suppose you need that now." "No," she said "I am well," or something like that; and then the dream ended.

It was fantastic in its changes; but otherwise it was perfectly real; and it was true to Winny, in the sort of wise reticence it showed in her. I felt all the time that she could have told me fully every thing I wished to know, if it had not been against counsel that she knew the good sense of, but which she was free to disregard if she would. She apparently hesitated between her wish to gratify me, and her knowledge of what was best for me.

This dream has somehow given me courage and consolation. I could not say why, for I can account for it in a perfectly natural way. I feel as if I might dream again of her, as if I might continue that very dream. But of course I have no warrant for such a feeling. However, I have written fully out for you.

W. D. H.

Manuscript: Alfred

1. The location of the original is unknown; it was returned to WDH by his father (*SL,* 3:278), who made the copy from which this text is taken. At the end the senior Howells added: "I have copied this dream for myself. It seems most natural to me, and the most probable view of the world of spirits and what its appearance would be to us, and just what would occur at a casual meeting of those concerned in this dream ¶Wm. C. H."

WDH also sent a copy (or the original) to Howard Pyle on 22 December 1890, remarking: "I do not always feel sure that I shall live again, but when I wake at night the room seems dense with spirits. Since this dream which I wrote out for my father I have had others about my daughter, fantastic and hideous, as if to punish me for my unbelief " (*SL,* 3:299).

Kate Douglas Wiggin to EMH

[New York]
Hotel Albert Dec. 24ᵗʰ [1891][1]

Dear Mrs. Howells:

Now that your revered husband has ceased to read books for a time, I dare to send you one of mine, which however I pray you wont leave within reach of the lion's maw, lest the old habit & the thirst for blood overtake him. I have had the leaves rubricated at fabulous expense, as I wanted it as red as possible *out*side, having my private doubts of its *in*side ever being read.[2]

I miss you all "more than tongue can tell" as the children say—
With Merry Xmas & Happy New Year I am yours really
Kate Douglas Wiggin

Manuscript: Harvard

1. Kate Douglas Wiggin (1856–1923), best known as the author of children's books, especially *Rebecca of Sunnybrook Farm* (1903), knew the Howellses well and apparently was a favorite of theirs after they became acquainted with her in 1890. On 25 March 1895 WDH wrote her an affectionate letter (Wiggin, *My Garden of Memory*, 1923, p. 275) sending her a copy of the 1891 *Venetian Life*, illustrated by Childe Hassam and others, as a wedding gift, and five days later he and Mildred were present at her wedding to George C. Riggs. Later that year, on 30 November, he reviewed one of her books in *Harper's Weekly.* The year of the letter to EMH is fairly well established, for though the WDHs were at 241 East Seventeenth Street (cf. "I miss you all"), they had been at the Albert in late November. Mrs. Wiggin's opening remarks on WDH's ceasing to read books points to his editorship of the *Cosmopolitan,* to which he committed himself on 8 December.

2. *Timothy's Quest* (1890) has red covers, and while not technically a rubricated book, does have the color on the "*out*side," as Mrs. Wiggin observes. The quest of the title is that of a ten-year-old orphan boy who searches for a private home where he and his three-year-old sister will be adopted. The youngsters succeed in overcoming the caution of the well-to-do spinster to whom they apply. Mrs. Wiggin develops the story with liveliness and with less sentimentality than might be expected. EMH may have found the juvenile novel mildly enjoyable without positively rejoicing in it; and she probably would have liked the "local color" of the new England country types.

Wild-Goose Chase

1892–1901

$\mathscr{I}n$ New York City Elinor and Will mostly lived in an apartment on the south side of Central Park, though now and then they tried hotels or apartments or even a house elsewhere. Outside the city, they frequently made brief stays at Atlantic City (Elinor was especially fond of its vaudeville shows); they tried Garden City several times; and they once made an ancestral pilgrimage to old Newbury, Massachusetts, early home of the Noyeses. In 1894 Will went with Pilla to spend the summer with John in France, but Elinor stayed home because of the shortness of the planned trip, which turned out even shorter when Will was called home because of his father's stroke; and Elinor, Will, and Pilla went to Carlsbad in the summer of 1897 for Will to take the cure; in their following trip, mostly in Germany, they had a reunion with John, who had just graduated from the Ecole des Beaux-Arts.

Over most of these years they also were on the lookout for a country place and tried many places in their "wild-goose chase for a summer home" (WDH to Aurelia, 28 July 1895, Harvard). They experimented with Magnolia in Massachusetts, Far Rockaway on Long Island, York Harbor and Kittery Point in Maine, and then Annisquam in Massachusetts, all with varying degrees of disappointment except for Kittery Point, where in the fall of 1902 they bought the cottage that they had rented that summer. "We have at last a home for our old age," Will wrote Aurelia. "Elinor simply dotes on it, and Pilla is in love with the beauty of the place; John always liked it, and for once we seem agreed about a sojourn."

To John Mead Howells

[Intervale]
Wed. p.m. July 27 [1892][1]

Well, John, you always did "play possum"; but this time you completely took us in![2]

I no more thought you could get in! Such a short time to prepare, mathematics not being your forte, & that awful trouble coming on just before you left & all! I only hope you wont collapse now. I guess joy will cure you. We all hope you will go to Carlsbad first of all & get yourself up, then you can go where you please. Spend a little money & have a good time. You have *distinguished yourself* and it will be said of you all your life "He went right in to the Ecole with scarcely any preparation." Papa really did *not* think you would get in, n d[3] Pilla. I guessed papa hoped you would come home & go back again. But dont take that awful voyage twice. Travel in Europe rather.

Papa is going to write to Sam.[4] First Uncle Gus's telegram came— then uncle Will's letter & then your cablegram from Cosmopolitan.[5] I am really glad it came just in the way it did. We thought Gus must be mistaken & then it gradually dawned upon us. It was exciting enough I assure you. Pilla will write later. I enclose (a part of) Gus's telegram. It is a happy day for us. Pill & Papa will be photo'd for you (by Percy Greig) reading your cablegram. 3 for 25 Hurrah! Rah! Rah! Rah! 91! 91! 91![6] Bully for you. Papa has sent you a cablegram.

Emma[7]

If you go to Interlacken get me both a pin & small earrings

Manuscript: William White Howells

1. Though the envelope is postmarked "New York," the letter was written at Intervale, in view of Shepard's telegram and WDH's letter to John of the same day, which has an Intervale dateline. The postmark may be explained by the letter's having been put on board ship in New York.

2. John had gone to Paris in March and had just passed the final entrance examination to the Ecole des Beaux-Arts.

3. The "n d" appears to stand for "nor did."

4. Sam has not been identified.

5. Augustus D. Shepard had telegraphed WDH at Intervale: "Heartiest congratulations upon John's—great success we are—as happy over it—as possible can be" (William White Howells). Evidently William R. Mead had written and John had cabled his father at the *Cosmopolitan*.

6. Percy Greig remains unidentified. The 91's refer to John's class at Harvard.

7. EMH's family called her "Emma" or "Emmer."

Hamlin Garland to EMH

Dear Mrs. Howells:

Being western and very rural, I said to Mr. Howells as I put him into the cab, "Now I wish you'd tell Mrs. Howells and your daughter to call upon me if I can be of any service during your absence."[1]

He promised he would but he may forget it. I believe in being neighborly and if I can drop in occasionally and thaw out the pump or kill a chicken for you, it will give me pleasure.

I'm coming in soon and bring "Prairie Folks."[2] I want you to read it.

Very sincerely

Hamlin Garland

Jan 9/94.[3]

107. W. 105 [New York].

Manuscript: Harvard

1. WDH wrote to Mildred from Jefferson on 12 January 1894 (Harvard): "I guess Garland would have gone before, if he hadn't understood I was to leave the house at six, instead of the station. He charged me, if either of you needed him while I was gone, to let him know! After all, a good heart is a good deal."

2. On 13 January Mildred's engagement book (Polly Howells Werthman) records that Garland was at lunch and brought *Prairie Folks*.

3. As Howells left New York on 10 January, Garland has evidently misdated the letter. It seems likely that he wrote it later in the evening of Howells' departure.

To Mildred Howells

[Saratoga, c. 18 July 1894][1]

It is perfectly delightful to think of Mͬͨ͞ Whistler's calling out "Johnnie Howells." Is Miss Philips young? No, I think John spoke of Purvis de Chavannes as being attentive to her—so she cant be.[2] I wish Lucia could read your description of M͟r͟s͟ Whitman at the Whistler's.[3] But I dont know where she is. Off on her car somewhere. Do you write to her care Charles Fairchild, Newport? If you ever get time do tell her about that Sunday at Whistlers'. Still you could never write it so well again. You *can* describe a scene, Pilla! I seem to have *seen* the "plain but charming young man who always seems to be at the Whistlers'." Papa says your pink silk dress is *perfect.* Do thank the Lusks from me for all their kindness to you—and thank M͟r͟s͟ Clemens too.[4] The strike seems to be over.[5] Pullman refused to arbitrate—like a czar. *Every* paper is down on *him,* many are down on Att. Gen. Olney for telling the Government to interfere. That was the first time it has ever done it—and *that* was also czar-like. Olney was a corporation lawyer when he was chosen by Cleveland for his cabinet—which was not liked at the time.

Mamma.

Manuscript: Harvard

1. The letter appears on the inside leaves of a letter from WDH to EMH, 16 July 1894, reporting on his father's improvement and saying that he expects to leave Jefferson on Friday (20 July). This is the fullest account of Whistler's garden party by a member of the WDH family, Mildred's letter having disappeared (though in her engagement book she records "The Salon and the Whistelers" on Sunday 24 June). In a letter to John, after noting that "the poison of Europe was getting into my soul," WDH cryptically exclaimed, "When I think of the Whistler Garden!" (*LinL*, 2:52). According to James, Howells had said to Jonathan Sturges on that occasion something like what Strether told Little Bilham in *The Ambassadors:* "Oh, you are young, you are young—be glad of it; be glad and *live.* Live all you can; it's a mistake not to" (*Notebooks of Henry James,* ed. E. O. Matthiessen and Kenneth B. Murdock [1947], p. 226); as James added, he amplified and improved a little. Six years later he wrote WDH that Sturges had quoted only five words by Howells (*Henry James Letters,* 4:199).

2. The Miss Philips was one of the sisters of Mrs. Whistler, Ethel Birnie or Rosalind Birnie. Pierre Puvis de Chavannes (1824–1898) was an academic painter, though highly regarded by the avant-garde of his day. He founded the society that sponsored the annual exhibit at Champ de Mars in which Mildred later had watercolors.

3. From the context it is not clear whether Lucia Fairchild Fuller was touring out from Saratoga, where EMH was, as her base, or from elsewhere. Mildred had been at her wedding in 1893 to Henry Brown Fuller (*SL*, 4:51). Sarah de St. Prix Whitman (1842–1904), a painter, had designed the covers of a book by WDH, most likely the Franklin Square Library edition of *A Shadow of a Dream* (1890). In 1886 and 1899–1904 she was a trustee at Radcliffe, where a dormitory and window are named after her.

4. William T. Lusk (1838–1897), professor of obstetrics at Bellevue Hospital Medical College, was in Europe with his daughter Mary Elizabeth (a friend of Mildred's), prob-

ably her stepmother, and possibly other members of the family. The Clemenses also
went to Etretat.

5. The Pullman Strike, beginning as a boycott by the American Railway Union of
Pullman cars, resulted in Cleveland's sending Federal troops to protect the mail and
interstate commerce. Richard Olney, the Attorney General, swore in a number of dep-
uties and secured an injunction.

To Mildred Howells

[Magnolia] Tuesday [7 August 1894][1]

Dear Pill.

I wrote you a long letter yesterday—& today your letter of July
26ᵗʰ has come. It *is* pretty hard to have to wait for John so long.
You dont mention the climate of Etretat as compared with Magnolia, nor say you have once been in bathing. Mˢ Gratwick says
Dick Norton is engaged to Edith White of Cambridge, daught. of
Prof. White who has been in Athens for a year past.[2] Is she your
age? Jack Gade was here in rough clothes last night, dancing with
Grace Maynadier whom he brought—& looking like fun.[3] He held
her arm up as though he was feeling her pulse [sketch]. Everybody
laughed.

Mamma.

Manuscript: Harvard

1. This note is included in a letter of 7 August 1894 from WDH to his daughter.
The Howellses, coming from Saratoga, had arrived at The Oceanside in Magnolia on
1 August. John was to join his sister at Etretat after the end of the term at the Beaux-
Arts.

2. Richard Norton (1872–1918), son of Charles Eliot Norton, later became director
of the American School of Classical Studies in Rome. He and Edith White, daughter
of John Williams White (1849–1917), professor of Greek at Harvard, were married in
1896. Martha Penelope Weare Gratwick, the wife of William Henry Gratwick (1839–
1899), lumberman and ship operator, had probably heard the news of Norton's engagement from her son, also William Henry Gratwick (1870–1934), Harvard 1892, a
friend of Norton and later an industrialist and philanthropist.

3. John A. Gade (1875–1955) graduated from Harvard in 1896. Later he began his
architectural practice in the office of William R. Mead. His mother, Helen R. Gade (née
Allyne) had met the WDHs in Venice and lived near them in Cambridge. See 30 March
1873.

To Anne H. Fréchette

40 West 59ᵗʰ street [New York],
Feb. 16 [1895].

My dear Annie:

I am very sorry, but on many accounts we are obliged to put Vevie's visit off till Easter, and then I can only ask her for a fortnight. But if she should come April 8ᵗʰ, could she not go from here to Ohio, instead of returning to Canada—and thus escape the worst part of the Canadian Spring?¹ Of course, in asking her, Will expected that the journey should be no expense to you—and he will see that she gets to Ohio safely, also at his expense. Easter is the most charming time in New York—the Park will be in its glory and all the ladies—and poor people too, for that matter, will be in straw hats. Still Vevie will need one light weight woolen dress—or perhaps a thick one—though I have a woolen dress to give her— too small for Pilla—that will *fit* her without alteration. I think it is even short enough—this one—though I could quickly alter it if it is not. And we have no end of wraps, hats & gloves that she could use—so dont bother about her dress. Just one thin dinner-dress & a pretty thicker dress (woolen perhaps) for lunches, will do. She can take off her hat at the theatre (Pill *always* does) so she will only need whatever hat you think proper for her to make calls in. (Nobody is ever in in New York.) Vevie shall sleep late mornings and not get overtired I promise you.²

I am glad you are back in your house, which will be much better for Achille. Your description of life at the Convent was delightful.³ Pilla is working very hard drawing from life and modelling in a studio—and Will is working hard, as usual. I may as well tell you that every magazine, except Harper, has refused a story or even a paper from him saying they are full for two years to come, and unless Harper orders one, which they are considering till March 1ˢᵗ we shall have to step down and out.⁴ Of course he is an old story. Harry James—whose play has just been hissed in London— is in a worse case. I am going to enclose his last letter—and ask you to return it quickly—.⁵ He does not allude to the failure of his play. That is too bitter to speak of. John & Mildred lunched with him last Summer in London. Will is considering many new schemes for the future. We are not worried as both John & Pill can easily support themselves and of course we have laid up something. But what we always supposed would happen has happened.

With much love to Achille Vevie and Howells

Affectionately

Elinor

Manuscript: Harvard

1. Vevie arrived in New York on 10 April. She returned to Ottawa on 29 April by way of Montreal, where she was met by her uncle Louis Fréchette (26 April, Vevie to her family, Alfred).

2. Evidently Vevie had not entirely recovered from an illness which Howells regarded as much like Winny's and for which he prescribed the S. Weir Mitchell rest cure in a letter to Annie of 17 November 1893 (Harvard; see also *SL*, 4:53).

3. The Fréchettes had bought the MacKay Street house, New Edinburgh, in the 1880s, but recently they had lived at the "Convent" in Ottawa.

4. Though Howells' career at this point hardly seems as faltering as Elinor suggests (a major problem was spacing out his novels so that serialization in several magazines did not overlap), he himself wrote to Aurelia 21 April 1895, when his "Life and Letters" department in *Harper's Weekly* was beginning: "It has been trying for me of late to place my work, and though I shall now have to do more work, I shall be less anxious" (*SL*, 4:103).

5. The unfriendly response of the gallery to *Guy Domville* and the hissing of the author, who was persuaded to appear on stage after the performance on 5 January 1895, are examined in Edel, *Treacherous Years*, pp. 61–89. The "worse case" letter from James, 22 January 1895, appears in *Henry James Letters*, 3:511–513.

To Anne H. Fréchette

[New York] Thursday evening [18 April 1895].

Dear Annie:

Vevie has gone with her uncle Will to Brooklyn this evening, to see young Salvini in Hamlet, and to go behind the scenes and be introduced to the principal actors.[1] Salvini is just beginning to play Hamlet, and commissioned our little (Watkins) Cousin Paul Kester to come over and offer Will a box. Neither Pilla nor I could go, so Will, Vevie & Pauly (as we call him—or sometimes "Pretty Polly") will have a whole box to themselves. While she is gone I will talk behind Vevie's back—and tell you that we think her, to begin with, exceedingly pretty—with her ivory skin her beautiful teeth her dimples, red lips and pretty hair. She always had a pretty figure tiny feet and was graceful. Now she is very graceful in speech and manners and full of (French?) tact. Although she looks young she is quite able to take care of herself, and she shows she is used to society in every motion. She meets all sorts of people here and they all like her and she gets on perfectly with them. We think her very sweet, and very clever. Today she met Olga.² (I have always thought I should like to hear *Achille* speak French with Olga. Olga studied at the Paris conservatory—and admitted to me, once, that she had thoughts of the stage.)

Tonight Vevie's tact was put a little to the test, as Joe Evans (late President of the Art League) stayed to dinner and sat beside her.³ He is a dreadful humpback (with a *beautiful* face) and his chin came scarcely above the table. But Vevie behaved beautifully. He is a Harvard graduate and has a brilliant mind. I took Vevie to the New Art League this afternoon and she greatly enjoyed the pictures. Yesterday we saw Midsummers Night Dream at Daly's—with Ada Rehan in it.⁴

You must not be frightened at Vevie's doing so much. We take her mostly to matinées instead of evening entertainments and she never gets up till late. She is looking finely. Saturday the Hasty Pudding Club of Harvard is to act in New York and Pilla will take Vevie to see the play, which is Proserpina.⁵ This is the great treat of Easter to the followers of Harvard. Vevie's dress seems to come out all right on every occasion and she has shown great judgement in the few things she has bought. We are glad to hear that Achille thinks of taking up miniature painting which seems to be all the rage just now. Lucia Fuller fairly supports herself by painting them.

Affectionately
Elinor

Just the end of a little note from Lucia Fairchild Fuller with a compliment for Vevie in it⁶

Manuscript: Alfred

1. On 25 April Alessandro Salvini also appeared in *Hamlet* at a special matinée in New York. Paul Kester (1870–1933), a playwright, had known WDH since 1893, and beginning in 1896 they were to collaborate on a play version of *The Rise of Silas Lapham*

(see *SL,* 4:117, 177, and Meserve, pp. 481–83). His mother was the granddaughter of WDH's great-aunt.

2. Olga Kilyeni Mead, the wife of William R. Mead.

3. Joe Evans (1857–1898), a painter, studied at the National Academy of Design, at the Art Students' League, and in Paris at the Ecole des Beaux-Arts under Jean Léon Gérôme. He had also recently become vice-president of the American Fine Arts Society.

4. *A Midsummer Night's Dream* was reviewed in the New York *Times* on 14 April: "Miss Rehan's Helena is as beautiful and sympathetic as ever. . . ." Ada Crehan Rehan (1860–1916) was the leading actress in Augustin Daly's company from 1879 to 1899.

5. In a letter to her family, 22 April (Alfred), Vevie describes the play as "the funniest play imaginable": "there were three principle characters in the play who were girls & the 'Havards' as they are called looked *so* like girls with wigs & classic dresses except when they moved." She adds that yesterday "Venus (in Proserpina) called." The cast (Harvard Archives, for 25 April) identifies "Mrs. Venus, an Immortal Professional Beauty" as Edwin Godfrey Merrill (1873–1950), a future bank executive.

6. The following fragment is enclosed with the letter: "Come soon to see me & let me see your beautiful little cousin again. Yours ever, Lucia." EMH's postscript ("Just the end . . . in it") appears on the verso.

To John Mead Howells

[New York] Sunday, 12ᵗʰ [January 1896]

Dear John:

I know it is wicked but I cant but be glad that the English are getting insults right and left. It is amusing to see how they have backed down about Venezuela.¹ I wish Russia Germany & France *would* unite & whip England—but they wont.

When you & Pilla were at Redding in Berkshire Co. England, you were only a few miles from Cholderston, Wilts where Rector Noyes lived, whose son James, on graduating from Brazenose Oxford found that he could not conscienciously preach in the English Church and so started for the wilderness of America—in 1634— accompanied by his brother Nicholas (from whom *we* descend) and his Cousin Mʳ Parker—also a university graduate. They settled in Newburyport, and James Noyes was the first Minister there.² Nicholas married either a Little or a daughter of Tristram Coffin (the same family of Sir Isaac of Cape Cod that Wᵐ Amory brags of being related to) and you can yet see the Noyes house, the Little house and the Coffin house—in Newbury, the old part of Newburyport.³ When you come home we will go up & see it. My great-grandmother Hayes's ancestor Russell of Hadley Mass. kept Goff

& Walley, the judges of Charles I st hid in his cellar two months.[4] I have the Noyes ancestry entire from Choulderstone down—& I guess it can be traced back of the Rector in England.[5]

Manuscript: William White Howells

1. The Venezuelan Boundary Dispute threatened war between England and the United States when WDH wrote his "Life and Letters" column in *Harper's Weekly*, 4 January 1896, objecting to Cleveland's recent message to Congress. By the time of Elinor's letter, however, England had acquiesced and the meetings of the Venezuela boundary commission had begun. A treaty followed early in the next year.

2. Nicholas Noyes (1614?–1701), ancestor of Elinor, came to New England in 1634 and settled in Newbury. James Noyes (1608–1656), whom he accompanied, attended Oxford but did not graduate. They were sons of William Noyes, the rector at Cholderton, whom another son, Nathan, succeeded. Nicholas served as deputy of the General Court at Boston on several occasions and became deacon of the first parish in Newbury (1683–1684). James Noyes and Thomas Parker (1595–1677), who attended Trinity (Dublin), Magdalen, and Leyden and received his M.Phil. from the University of Franeker, were joint pastors of the church in Newbury.

3. EMH later learned from *Noyes Descendants*, a copy of which, with her annotations, is in the Howells Memorial, that Nicholas married Mary Cutting, c. 1640. His grandson James, son of Timothy, married Sarah Coffin, the great-granddaughter of Tristram Coffin, in 1713. Another grandson named James, son of James, married Sarah Little in 1729. Tristram Coffin (1605–1681) was at Newbury for a time but in 1659 purchased Nantucket in association with others. Admiral Sir Isaac Coffin (1759–1839) remained a British subject but retained his interest in America. William Amory (1869–1954), manufacturer and Harvard classmate of John, was a great-great-grandson of Elizabeth Coffin (1741–1823), aunt of Sir Isaac and wife of Thomas Amory. All three houses at Newbury are still standing and are illustrated in John Mead Howells, *The Architectural Heritage of the Merrimack* (1941); the Little and Coffin houses belong to the Society for the Preservation of New England Antiquities.

4. The Reverend John Russell (1626/27–1692), who left Connecticut because of religious dissension, was a founder of Hadley. The regicides William Goffe (d. 1679?) and his father-in-law Edward Whalley (d. 1675?) came to New England in 1660 and in 1664 went into hiding under the protection of Russell.

5. *Noyes Descendants* speculates that William Noyes may descend from William des Noyers, one of William the Conqueror's barons.

Though Elinor's letter is unsigned, it appears complete.

To Anne H. Fréchette

115 East 16th st [New York],
Nov. 28th [1897]

Dear Annie:

Hamlin Garland has just been here hoping to get a letter of introduction to you in Ottawa, whither he goes next week. Will happens to be out, but will write one for him this evening.

Meanwhile I write to explain that he is coming & that, though he is a rough one, he is to us a diamond both as a writer and as a man. I am sorry he has taken up the Indians & now Klondyke so violent—as those subjects dont interest me. But his stories are sometimes perfect, "The Little Norsk"—"Main-travelled Roads" &c[1]

We were greatly interested in your & Vevie's Court Costumes that you sent.[2] Vevie's is so graceful!

I enjoyed the trip abroad more than all the others.[3] I became young again & forgot death and disease & sickness & money. But it is all the harder for me to get back into the traces again.* Will & Pilla settle right down to work again and dont seem at all rattled. John's being here is quite an excitement all the time, as he is anxious about work of course. And that University Building cant be started until the funds are raised—probably they will be by next Spring.[4] We dont like so public a life any more and are going back into the Dalhousie Flats.[5] Not into the same one though—in to a lower down & cheaper one. And now that Pilla is twenty five and has had her fling and is going in for serious illustrating &c we need not entertain as we have done, and I hope our life will be more serious & simple. You see I am an old woman?

One of our most delightful experiences were the three (foggy) days in London, when we were almost constantly with Harry James (now old & much softened) hearing his wonderful talk. I loved the sentiment & the Castles of Germany (though we hated the militarism) the pictures of Holland (though not the smells) & I enjoyed looking up the old parts of Paris (though I hate the damp climate). You will get all this in Will's story.[6]

A swell Hon. Drummond & wife of Montreal were at our table coming back—She like Mrs. Bourinot, much younger than he. But a half French half Irish sugar planter of Barbadoes—M⸢r⸣ Louis Bert de la Marre—kept us laughing all the time at table.[7] We met *every* kind—& saw lots of royalties. One thing I found out that took down my pride considerably: they dont, any of them, care a straw for Americans over there—& dont know much about us. Even in Paris

*so that I would rather never go again. It is too upsetting.

they would not miss the *huge* American Colony a day if it all left. They are quite sufficient for themselves and interested in their own affairs. The Germans really *dislike* Americans. The English had a little time of being interested in Americans—but now they are dreadfully sick of us—Even the Prince of Wales is I fancy.

I almost wish I had staid at home and kept my conceit! New York seems *so* incomplete after Paris & London! Berlin is crude enough—& not so impressive as New York—& *such* a vile climate! We have the *climate*! You cant imagine the sunless damp dark days in Europe unless you go there. I could not live without the *sun*. Of course there is Italy.[8] But the houses have no means of being heated and the winds in Nice & Rome are very treacherous. And Florence in a valley, with the Arno slowly running through, is *so* chilly. I never had so bad a cold in my life as there. Will & I got frightened at the prospect & so came back. Then too we heard of the great poverty & suffering in Italy & hated to witness it. Think of it: Italy could have made herself comfortable after she got free and not have been molested any more than Switzerland. But she wanted to be a Great Power—and the army & navy cost so much that the people are nearly starving.

Here I will end my lecture. I hope you understand the situation!

Affectionately
Elinor

Manuscript: Harvard

1. In 1896 Garland published a half-dozen Indian sketches, the first of them "In the Happy Hunting Ground of the Utes," *Harper's Weekly,* 11 April 1896. Though *The Trail of the Goldseekers* did not appear until 1899, and though the first Klondike article did not come out until March 1898, the WDHs were aware of his projected work. *Main-Travelled Roads* had been published in 1891, with WDH writing an introduction for the 1893 edition; *A Little Norsk: Ol' Pap's Flaxen* was published in 1892.

2. The court costumes had been worn by Vevie and her mother at the main social event during the session of Parliament in the spring, when about six hundred guests were presented to Lord and Lady Aberdeen in the Senate Chamber. In a letter to Aurelia, 28 March 1897 (Alfred), Annie wrote: "Last night we went to the Drawing Room and I'm sure you would have thought the little woman looked dainty and sweet in her glistening white (china) silk, with her bare white arms and neck—and her trim head decked with three white ostrich tips and a tulle veil hanging down her back. She carried a bouquet of pale pink carnations. She made her two courtesies very gracefully, and enjoyed the whole show greatly." Evidently Annie did not send the studio photograph of herself and Vevie until the Howellses returned from Europe.

3. From mid-July until the end of October the Howellses were in Europe, with WDH

first taking the cure in Carlsbad. In his travel novel, *Their Silver Wedding Journey* (1899), he closely follows their itinerary through his characters, omitting however their going to Holland, Belgium, Paris, and London after the tour of Germany.

4. John and his future partner, Isaac Newton Phelps Stokes (1867–1944), had won the competition for the University Settlement, at Eldredge and Rivington Streets, while still at the Ecole des Beaux-Arts.

5. The Dalhousie was at 40 West 59th Street.

6. In a letter of 11 October James had offered to go to Paris or even Italy to see the WDHs. Instead, the WDHs left Paris on 26 October and saw James before sailing from Liverpool on the 30th. About the London meeting James wrote on 27 November (both letters at Harvard): "It was a joy to me to be reminded by your wife (please tell her so) of a thousand tender things of the old sweet time. And I love your daughter—I break it to you thus" (also *Henry James Letters*, 4:60–61).

7. George A. Drummond (1829–1910), Montreal banker and art collector, and his wife Grace Julia Hamilton Drummond. Emily Alden Bourinot, the wife of John G. Bourinot (see 16 November 1877). Louis Bert de la Marre is unidentified.

8. Though Elinor seems to suggest that she and her husband had gone to Italy, they had only considered it (cf. James's reference to that possibility above). Later in the paragraph, the remark on having "so bad a cold in my life as *there*" may refer to the 1883 visit to Italy.

To Anne H. Fréchette

Dec. 7 [1897]

The Westminster [New York]

Dear Annie:

I am so tired from moving that I must write lying on my back. But I want to tell you first: how nice we think your children's stories are, and how pretty a book they make, with the illustrations. Those little cats are charming. Who did them?[1] But, most of all, we like Howells's letter. It ought to be utterly satisfactory to your parent hearts, for it is both sweet & manly. How charmingly he laments the boredom of having *two* distinguished uncles! It is a precious letter, ever to be kept, & I return it carefully.[2]

I live only in my family now. I never go out anywhere, and I am old & sick. But Will seems in better health. He and Mildred have just been lunching at the Albemarle with Alice & Ernest Longfellow.[3] (*We* have given up "keeping tavern"—which really our housekeeping seemed to ammount to & never invite anybody now. But Will & Mildred go a great deal.) Only Will never goes out twice in the evening in succession since his illness. Mildred works at writing

& drawing & got 65 dollars for an illustrated poem today from St. Nicholas.[4] It is about a Dutch mother who scrubbed all the features off her children's faces. John is *hard* at work—with good prospects. But the University Settlement Building wont be started till Spring—though he has little jobs meanwhile. But New York eats up so much money! And the children wont live anywhere else—& now Will likes to be near his *doctor*—for he dreads those attacks.[5] I wish *I* had not gone abroad, almost. For it used me up both physically & mentally. When we got there we were *eager* (after leaving Carlsbad) and I did too much. I did not feel it then. I felt 10 years younger—from the excitement. But I feel it now. I hope to get into my niche again.

I am glad you liked Hamlin—though everybody does. He is a pure, noble soul. Brander Matthews says "I *love* Garland!"[6] I have said so much about Howells. But I am greatly interested in Vevie's work—both artistic & social. Will she have the strength to draw & go to parties, both. It is what Mildred has always tried to do—But the parties would interfere with her work awfully. Now it is less parties & more work. She read a paper at the "Little Sisters of the Quill"'s meeting the other night.[7]

Elinor

Manuscript: Alfred

1. *The Farm's Little People* and *On Grandfather's Farm* were published by the American Baptist Society in Philadelphia, 1897, but did not attract buyers. See Doyle, p. 85, who approvingly quotes WDH's letter to Annie that "It is a perfect picture of child character and child thought . . . charmingly simple and natural."

2. The letter has not been located.

3. Both Longfellows were children of the poet. Alice (1850–1928) was a founder of Radcliffe. Ernest (1845–1921), portraitist and landscapist, left his art collection and an endowment to the Boston Museum of Fine Arts. The Albemarle was close to Madison Square.

4. "Going Too Far" appeared in the *St. Nicholas*, March 1898.

5. Homer I. Ostrom (1852–1925) had advised WDH's trip to Carlsbad (*SL*, 4:151). His office was on 47th Street, while The Westminster was on 16th Street.

6. Brander Matthews (1852–1928), novelist and professor of dramatic literature at Columbia, was a friend of the Howellses.

7. Mildred's engagement book (Polly Howells Werthman) indicates a meeting of the "Little Sisters of the Quill" on 4 December. To judge from this and another entry, the group met at various members' homes on Saturdays.

To Sylvester Baxter

40 West 59ᵗʰ St.,
New York, April 22ⁿᵈ [1900]

Dear Mr. Baxter:[1]

Mr Howells is in such a state of mind that I am emboldened to write you and see if you can do anything for him. He is set on having a country house with a garden where he can raise vegetables. We have searched up & down the coast from Boston to York and cannot find a house within our means. Everything seems to be taken—even in Ipswich, Essex, Newburyport—and those least desirable places.

So now I think we must go inland. And I wondered if you would know of any old (or new) house without all the improvements in any of the places near you: either in Malden or Melrose or Wakefield or Saugus. We rather prefer a village house, if it has a little privacy.

I see there are many places advertized in Bridgewater—but I suppose that is a very hot place in Summer? We prefer above Boston of course. Mr Howells would buy outright if he liked a place very much. But we would like to hire with a view of buying. Or if we liked a locality we would hire a house there hoping something would eventually turn up.

I suppose that "Messenger Place" in your vicinity is awfully dear?[2] Now please dont go looking about and bothering yourself; but write me a little advice about places & prices.

It is my own scheme, writing to you. But Mr Howells has come back from Ipswich—where we hoped to find something—so discouraged that he is losing sleep.

We would come up to Boston to see anything that seemed promising. We want very little land & would like a good many rooms—but we could add to a small, desirable house.

I dont believe you can help us but I must send this dispairing cry. Please give my love to your wife—and dont drag her in to this miserable business.

You understand that Mr H. is getting old and would rather stay mostly in the country and perhaps only board when he comes to New York?

Sincerely yours
Elinor M. Howells.

I take another sheet to thank you for sending me that precious thing about Fred Mead that Gurnsey wrote.[3] I am keeping it to show Minnie when she returns from Europe.

M͞r Howells has lately been reading aloud the Joy of Captain Ribot in English (the good translation with your preface) and we have all enjoyed it so much![4] We dont like the ending, though.

M͞r Bishop of Yale has a photo. of Valdes's new wife.[5] Valdes wrote something like: "She is quite young—but she says she loves me" to Bishop.

Manuscript: Huntington

1. Sylvester Baxter (1850–1927), newspaper man and civic reformer, was a longtime friend of the WDHs. The year of the letter is established by internal criteria, including the oblique reference to WDH's return from a house-hunting week in New England (cf. *SL,* 4:234).

2. The Boston *Transcript* of 14 April 1900 has a real estate advertisement of the Messenger homestead, formerly owned by Charles A. Messenger. The seventeen-room residence in Ashland Street, Melrose, with a billiard room, stable, and carriage house, located on an acre of land (well-graded lawns, shade and fruit trees) was offered for sale or rent. Baxter lived in Malden; Melrose is the next town to the north.

3. Probably what Gurnsey wrote about Frederick G. Mead was a letter rather than an article. The Gurnsey or Guernsey has not been identified. Minnie was Marie Louise (Myers) Mead (1860–1948), Frederick's widow.

4. WDH reviewed *La alegría del capitán Ribot* in *Literature,* 12 May 1899. A translation with a preface by Baxter was published by Brentano's the same year.

5. William H. Bishop, novelist and former contributor to the *Atlantic,* was at this time teaching at Yale. Palacio Valdés married Manuela Vela y Gil on 8 November 1899.

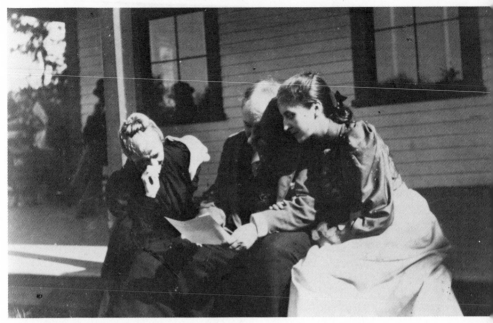

EMH, WDH, and Mildred at Intervale, 1892.

EMH and Mildred at the Dalhousie, c. 1895.

lred Howells. Lithograph by J. A. McNeill Whistler.

ie-Marguerite Fréchette. Sketch by Mildred Howells,
_5._

ie and Marie-Marguerite Fréchette, Ottawa, 1897.

The Howells summer house, Kittery Point, Maine, exterior view.

The Howells summer house, garden steps "where E. used to sit."

The Howells summer house, library.

EMH, WDH, and Mildred on the steps leading to the library.

EMH, Larkin and Marietta Mead, WDH and Mary N. Mead, Genoa, 1904–1905.

EMH, WDH, and Laura Platt Mitchell at Kittery Point during the Portsmouth Conference, 1905.

John Mead Howells in his New York office, c. 1908.

Mildred in EMH's last New York apartment at 130 West 57th Street.

✤ X ✤

The Laugh We Had Together
1902–May 1910

T̶hough Elinor had regarded the 1897 European trip as the last she would take, she went abroad on two more occasions: England and Italy in 1904–1905 (as it turned out to witness Will's receiving a degree at Oxford, though she had already planned to join him before the news came) and Rome in 1908 for three months (with Naples, Leghorn, Genoa, Monte Carlo, Paris, and London for shorter stays). From Kittery Point she and Will also made brief trips in the summer to other parts of New England—notably a tour to visit her relatives in September 1905—and in 1907 she and Will took a trip through the Great Lakes, accompanying John on his way to Yellowstone. "E. considers this the greatest trip of her life," he wrote to his brother Joe.

With all the travel, the Howellses were also caught up in house-building again, this time a studio apartment on 57th Street, described by a reform-minded friend as "cooperative, if not quite Altrurian." The usual delays in construction that prevented the family from moving into it until January 1909 occasioned Elinor's last extant letter, a delightful outburst on contemporary taste in decoration.

In the introduction the story of Elinor Mead Howells' final illness has been told. What gives this last section its title is a part of a sentence in a letter from Will to Grace Norton shortly after Elinor's death: "I wished to

286

go first, and in a day of deceitful respite I told her so, with the laugh we had together."

To Grace Norton

[Kittery Point, ca. 24 August 1903][1]

Dear Miss Norton:

I thank you so much for remembering me—& the beautiful book—& the beautifully written preface. I was glad to read Loti— as I thought I never could. And now there is at least one of his exquisitely written studies that I cant finish—nor being able to read all the details of the "remains" of a child in a grave. He is rather *charnelly* minded I think. But I have enjoyed most of the book.[2]

M⸢r⸣ Howells & John have a grudge against him because the was taken into the Académie and Zola was not.[3] John who keeps up in French literature has with him, here, a book "Eve Victorieuse" the author crowned by the French Académie.[4] Have you read it? It is about America. We all like Henry James more than anyone else and read him aloud—both his old & his new things—constantly. The Sacred Font is the only book that baffles us. The first half can be understood—but, then, are those people trying to put the writer off the scent—or what? M⸢r⸣ James wrote that he only intended it for a short story.[5] Then, I think he became fascinated with the characters & went on & on. What do you think?

Manuscript: Schlesinger Library, Radcliffe College: Grace Norton Papers, John Graham Brooks Collection.

1. Principal evidence for establishing the date is WDH's remark in his letter to Grace Norton of 24 August 1903 (Schlesinger-Radcliffe) that he has "not yet read James's introduction" but has glanced at it. His note begins: "Mrs. Howells had begun a letter to you, but was not able to finish it to her liking, because it was 'too gossippy.'"

2. *Impressions* (a translation of *Figures et choses qui passaient*), by Pierre Loti, with an introduction by Henry James. The London edition had appeared in 1898 and the New York edition (from the same plates and with the same handsome covers) in 1900. The first sketch, "The Passing of a Child," concerns the death of a two-year old son of a servant and visits to the new grave. Though it is less charnel than EMH indicates, subsequent stories fulfill her description.

3. Loti was elected to the Academy in 1891 and accepted in 1892.

4. *Eve Victorieuse* (1900), by Pierre de Coulevain, pseudonym of Hélène Favre de Coulevain (1871–1913).

5. On 11 December 1902 (*Henry James Letters*, 4:250–251) James wrote to WDH: "I am melted at your reading *en famille The Sacred Fount*, which you will, I fear, have found chaff in the mouth & which is one of several things of mine, in these last years, that have paid the penalty of having been conceived only as the 'short story' that (alone, apparently,) I could hope to work off somewhere (which I mainly failed of,) & then *grew* by a rank force of its own into something of which the idea had, modestly, never been to be a book."

To John Mead Howells

[San Remo] Sat. morn. Oct. 29 [1904]

Dean John:

I might as well describe our housekeeping & you can keep it, with papa's for notes.¹ After getting in we found the place had not been swept or dusted since last April. They expect tennants to do the cleaning—so everything is inch deep in dust. The old woman cooks a little too richly & cant see flies or dirt in the food and blows dust off our plates with her mouth. Otherwise she is perfect & charming. I shall get a young woman in for cleaning after I am rested. In the meantime I am warm enough & fed enough and dont have to dress so am resting and improving. The flies nearly eat us up, but we˙ *cant* shut out the sun as the Italians do. The chambermaid at Paradiso said they & the mosquitoes lasted *all Winter.*² No mosquitoes here yet. Gill turned up and called twice the first day.³ Quite a bore. Pilla has never responded to what we wrote about those girls. She likes the Pension but not Florence.⁴ She wouldnt like it here—except for Gill—unless pleasant English turn up.

I dont want any English! The Anglo-americans agents were rude to us, so papa wont go in their room anymore.⁵ They have found out who he is & yesterday the old one came out. Papa just goes & talks with Rubino & says he shant talk to them any more Thats the way to treat them. They came out to talk to us in the outside room & wouldn't let us go in their room one day. I pushed in through— as it was a private matter—and said "I cant talk about it out here" The older man sat down & began writing. We only staid a moment. So Banker, Gill & dust are my only grievances now. To be sure I did not sleep a wink the first night as the bed was narrow & the spring cosi: [sketch] But we took that off & put it under the bed & put two mattresses on about 3 tiny pieces of iron. But I slept well, all the bed clothes slipped off the night before. O, for the beds the eating and the heating of New York! Dont go out to dinners in cold rooms and you will get on. Go to see Evelyn. Ask all about H. James & write it.[6] He promised he'd write—but he hasnt. The sun is glorious today. Hope your coat is allright. By March put it away in a big pillow case with plenty of moth balls. Papa's overcoat was always the greatest care to me. The moths even got in in February and he had to have it mended in Gunther's and they never really got out & it cost him a lot.[7] If you only had a *camphor* box to *always* keep it in! *Never* leave it out. or a tin trunk. Do get a box to keep it in. Your little French trunk, without the tray would do. They couldnt get in there if it was shut tight. Charles would brush it & hunt for moths in the Spring. Moths are the curse of our climate. But they must have them in Italy as they do in England. And you know the Bible speaks of them "where moth and rust doth corrupt." Do read the 5th & 6th & 7th chapters of Matthew—the finest chapters in the Bible The 6th is the best.

We shall get through the Winter perhaps. I *may* go *back* with Mary in December & let Pill take my place a little while, if worse comes to worst.

Poor Lark Mead! Do find out what he is doing. Uncle Will keeps track of him. Tell Olga all about me & give her my love.[8]

A man made papa 2 willow waste baskets in one day. The smell of the heliotropes comes in the windows. I envy you so going home that I cant even say "Bon voyage." I would stand even a poor one. And you arnt seasick.[9]

Mamma

I shall think of you all the time on the ocean and be glad to hear you are *safe* home

A bad spirit seems to pervade the Canopic.

Manuscript: William White Howells

1. WDH did not publish anything on this place until much later—"A Memory of San Remo," *Harper's Monthly,* February 1920. A letter to Norton a few days later is

typical of his own response (4 November, *SL*, 5:110): "I am as far out as Italy, you see, with the bluest of skies over my head, and the bluest of seas at my feet; but I would give them both for an hour in your painted woods, and my wife would give them for much less. She is openly and I secretly homesick, as in fact we have been all summer in spite of the intense interest of our English sojourn."

2. The Howellses had been at the Hôtel Paradis et de Russie before moving to an apartment.

3. William Arthur Gill (1865–1834), journalist, came to the United States in 1900, when he and the Howellses became friends. As a former fellow of Oriel College, he may well have brought WDH's name to the attention of a former colleague and thus been instrumental in the awarding of the Litt.D. Oxford. Generally EMH was fond of Gill; as her husband wrote him on 20 November (Bristol County Reference Library): "My wife was more than pleased, she was toucht, by your writing to her, but she is not well enough to write today (which is letter-day), and she wishes me to write and say how glad we both are that you are well, and that the climate of Venice is behaving so hospitably."

4. With her Aunt Mary N. Mead, Mildred was visiting in Florence, where the Larkin G. Meads lived.

5. The Anglo-American Agency and Bank had been used by Howells, but on 24 October he wrote Harper & Brothers (American Antiquarian Society) to address him in San Remo at Villa Lamberti.

6. Evelyn Garnaut Smalley was the daughter of George Washburn Smalley (1833–1916), European correspondent for the New York *Tribune* and friend of Henry James, who at this time was lecturing in America. On 4 November WDH wrote Norton: "James offered to write me 'an early letter' from America, but I do not make him even the most tacit reproach, for I know how these things go." Indeed, the first letter James wrote Howells was on 1 March 1905 (Harvard), which he sent through John (also 1 March, Howells Memorial). In the letter to John he remarked, "I really hope my letter will catch him before he does face homeward; his sailing without it will have made my long delay of it so much more horrible." To WDH he apologized for his "hideous long silence."

7. C. G. Gunther's Sons, furriers, were at 184 Fifth Avenue. Charles Lynch was the Howells servant—"our colored Charles, a creature built of mauve velvet," as Howells describes him in a letter of 22 December 1907 (*SL*, 5:235).

8. William R. and Olga Kilenyi Mead.

9. John was sailing from Genoa to Boston on the *Canopic*, 30 October. Thus the letter, probably directed to Genoa, was a "bon voyage" itself.

To John Mead Howells

[San Remo, late December 1904 or early January 1905][1]
I have got some pretty new envelopes in which to send letters to you in. They are named the "extra strong." I have used some very rotten ones heretofore. Pilla is much worried because she doesnt here from Sylvia.[2] She got S's letter to you but has never heard from S. Letter lost, probably. She will write to Sylvia, and she does not hear from Mary. How long it takes to get an answer! I guess

we will not put an ocean between us again. And yet Pilla likes the Eng. people at the hotel. So much better than Americans. It is such a pity she cant *live* in England itself on account of climate. Otherwise she would. She says she doesnt care for the English she meets at Bermuda. What *is* she to do!! We came over so she could place herself congenially—and now it is the *climate* that interferes. So we are no nearer settling than ever. She says she wont stay at Kittery Point next Summer. She'll probab[l]y be there early & late though. Then she wants to go with Mary, to a hotel somewhere. It's pretty expensive—but papa will have to stand it.

This doctor here has greatly worried her by saying the *membrane* is *gone* in that sore place & he cant restore it—That is what Pilla thinks he has said. I am going to ask him today. Meanwhile his treatment is soothing it. The swelling of the glands in her neck he thinks may be maleria from Venice or Florence—or from the grip she had in England. I think she'll be better when she goes back— but if she is she'll proceed instantly to use herself up & get sick. That's her temperament. I may as well tell you all this so you will know how to act when she gets back. Dont count on her strength or resistance *at all*. She *cant* toughen—& rest is her only hope. The doctor here says she cant even go out evenings.³ With that throat she can never do much. But an even steady life would probably finally cure it. I look for much help from the American Summer. With the first Autumn cold she'll have to go off to Bermuda I suppose. But even *that* may not agree with her *now*. I want her to go to Dᵣ Dench. His treatment helped my deafness much. Papa sometimes acts discouraged between Pilla & me. But not often. He pegs away like a little brick sometimes wishing he could give up the Easy Chair—but when he adds and write *novels* I always say "Your essays are so good—and you have written so many novels!"

Now this dismal letter is to be *answered* by a gay one.⁴

Manuscript: William White Howells

1. Exact dating is hard to establish. The letter appears to antedate that of 18 January. Mildred's throat trouble had begun in Florence.

2. Sylvia Scudder Bowditch, the daughter of Mrs. Horace Scudder and the wife of Ingersoll Bowditch, was a close friend of Mildred's.

3. Dr. M. G. Foster practiced at San Remo during the winter season and lived at the Villa Lamberti, where the WDHs also had their apartment. He is probably the "young

Scotch doctor" mentioned in the letter of 20 November to Gill, who in spite of his driving hard bargains with the residents of San Remo may well be "the lightest spirit" in the town. Edward B. Dench (1864–1936) practiced medicine in New York City and was a professor of otology at the Bellevue Hospital Medical College.

4. Most of the San Remo letters are unsigned, though all those printed appear to be complete.

To John Mead Howells

[San Remo] Jan 18ᵗʰ [1905]

Well, your letter by the Deutchland arrived this morning. I have sent it to Pilla, but it seems to me it was dated Jan. 6. By the way I wish you would take up writing "Yours of Jan——&c" arrived. I get mixed up about letters—it takes so long to get an answer. Your letter was delightful—about Mark Twain's house & Mark.¹ Papa roared. But now dont forget to tell me the number of his house on 5ᵗʰ ave. or between what streets that gothic edifice stands. We shall be so glad to hear about H. James. He is *comic* looking & talking now. Pilla giggles every time she thinks of his looks. Those [sketch] tufts of hair at the side & his queer figure. He *used* to be tragic & beautiful. I hope you made out between you that we see the Harlands here.² James *must* know how greatly "London Films" are praised. All England is in raptures over it—the first one on London.³ Papa read us a bit about Wells near Bath this morning which was awfully good. His Bath will be splendid. Yes the English book *is* a success already. It is to be big & expensive but it will be in demand in England. Papa thinks Duneka does not treat him very well—is jealous or something. O if only the Harpers had not failed! With bother about Pilla & lots of other bother Papa has not written you. He says tell you he is sorry—& do *you* write often. Papa told Maria today that *you* were dining at Maria's *here*.⁴

Last week all the lining broke down in the parlor-stove. Now its broken in the dining-room. The man just came, too late, as Maria had built the fire for dinner. It's just "inching along" here, with great risks to health—from which Pilla & I suffer sometimes. Papa & Mary are tough as knots. The English at the hotel call Pilla the "angel" and make much of her, & she is very happy. She likes these English so much that she would rather be here next Winter than in Bermuda—though she sees the drawbacks & knows she couldnt be here without us. She is pretty sick just now. But this doctor is doing her lots of good & giving her good advice for the future. She'll be up & rushing about perhaps by the time you next write her. She has had to leave a pretty room because a woman had asthma above her & people walking about all night. Now she is next aunt Mary however with door open between & better air. The storm seems to be over.[5] In February they *pretend* it will grow warmer—but we have been *so* deceived! Maria says there are strawberries by March. Pilla is pleased at the idea of going back on the Moltke April 5th (if we can get passage) She has rather turned against Bermuda—the people are *so* stupid there. But what can she do? I think she would go to live forever in England if she could stand the climate.

Manuscript: William White Howells

1. Mark Twain was renting 21 Fifth Avenue, on the southeast corner of 9th Street.
2. WDH often mentions Henry Harland in his letters from San Remo, as in those of 27 December 1904 and 9 January 1905 (*SL*, 5:115, 117–118).
3. "London Films," the first twenty-six pages of what was to become the book, appeared in the December *Harper's Monthly*, the same month as another section in the *North American Review*. On the English response WDH wrote to T. S. Perry, 9 January (*SL*, 5:118): "Some of the London papers quote it from the magazine, picking out plums which are sugared, but leaving the bitter almonds which flavor the whole. It is droll." Meanwhile Howells was working on the Wells and Bath essays that later became a part of *Certain Delightful English Towns*.
4. The "*here*" is part of the joke. Maria was the housekeeper at San Remo; Maria's, on 22nd Street near Sixth Avenue, where John often ate, was a restaurant famous for its "spaghetti hour" and chicken dinners. Among its patrons were Oliver Herford, friend of the Howellses, Charles S. Reinhart, an illustrator of WDH's books, and George Luks, the painter.
5. As WDH observes in several letters and in "A Memory of San Remo," an intense two-week cold wave began on 31 December.

To John Mead Howells

[San Remo] Sat Jan 28 [1905]

Dear John

It is a glorious day & I shall go out after 2 days in. Pill will surely come up here. Her hotel is up a road just beyond the Paradiso— a sunny clean road very steep, & the hotel much higher. She has got a letter from Sylvia at last.¹ S. adores the necklace & had worn it. Thinks you very kind. Papa has had a suit of clothes made here nothing extra—& the other day he brought in these *green* (in the light) samples to have an overcoat made of! I scouted the idea so that he gave it up. He buys awful neckties.

If weather like this would *hold* I would be off to Nice or Mentone—but the nights have an *awful* chill & the English dont promise that there wont be winds & storms soon.

Mary said to Lady Eliott²—at the hotel—I suppose March will be beautiful? Lady E said "Have you ever been here? Well, perhaps we have had so much wind now it will be less windy" and March is the month for the Battle of Flowers & all that.³ *What* a humbug! The Paris Herald made such fun of the weather record from the hotels on the Riviera that they have come down several degrees

The Ormonds live in the *greatest* magnificence here—Violet Sargent is spoken of in the papers as Mrs Sargent Ormond.⁴ They associate only with Marquises & Dukes & Princesses. So when Violet told M꞊ Ameglio that *you* were her "chum" it raised us to the skies in her eyes

But old Madame Taine⁵ was dreadfully bored by the company she met at the Ormonds'. She came in their grand carriage—coachman & footman & called on papa for a little change. He gave her his book.

Duneka writes that old Clemens is poorly & that he thinks it will do him good to see papa. Tonight at *our* restaurant Maria (Pill has wonderful Kodaks of her) sole fricassee chicken fresh green peas (grown outdoors in San Remo) and pears stewed in red wine

Nocera to drink—the best tablewater in the world Yet I am putting in wood two sticks at a time—and sleeping under piles of clothing. Thermometer only 60 & less in the daytime & south & I read it is 61 sometimes now in New York. I am almost mad on the subject of climate—& shall spend all next year studying ours. How *different* Europe is! Ours must be the *dryest* climate in the world

Manuscript: William White Howells

1. Sylvia Scudder Bowditch. Mildred's engagement book for 15 October 1904 (Polly Howells Werthman) notes "Got Sylvia's pendant" in Venice.

2. Harriette Emily Elliot, daughter of the first earl of Ravensworth and of Isabelle Horatia, daughter of Lord George Seymour, was the widow of Sir Charles Elliot (1801–1895), an admiral.

3. In "A Memory of San Remo," *Harper's Monthly,* February 1920, WDH pronounced the carnival as having a procession "better at San Remo than I ever saw elsewhere," though he concludes, "I have the impression that the affair was largely under English management."

4. Violet Sargent Ormond (b. 1870), was the wife of Francis Ormond and the sister of John Singer Sargent (1856–1925), the painter. Mrs. Ameglio was probably the wife of Alberto Ameglio, the U.S. vice-consul in San Remo.

5. The former Mlle Denuelle, daughter of an architect and patron of the arts, who married the critic Hippolyte Taine (1828–1893) in 1868. She was coeditor of *H. Taine: Sa Vie et sa correspondance* (1902–1907).

To John Mead Howells

[San Remo] Friday Feb. 3ᵈ [1905]

Enfin!

The Spring has come—the sun rises way over in the east so that we have twice as much—wild flowers abound—& the English are in *Summer* hats & white dresses! We have waited & waited for a letter from you—but now we are both going to write. Pilla was here to lunch with me yesterday, as papa went up to Taggia to see the house where Ruffini died.¹ She was apparently perfectly well—though she does not go out to late teas nor evenings yet. Today she & Mary take tea with her doctor's wife (the man who has treated her *so* wisely) and tomorrow Sir George King (really the biggest—because most learned) man here, gives a tea for her & papa & aunt Mary & Mrs. Wayte & Miss Perks (an authoress & Pilla's bosom friend) in his parlor at the Anglais.² Pilla says her throat is as well as it is all Winter in New York. I think it will all heal up here. The doctor thinks her recovery marvellous. She still says Bermuda is better for her throat.

You may read this letter to Kate Mead if you wish. Daisy & her family & *two* automobiles (I found it in the Society paper of the Riviera) one 60 horsepower the other 28 are over at Cannes.³ Mʳˢ Grote the historian's wife dressed *awfully*—in stamped velvet &c— Sidney Smith called her Grota & said she was the origin of the word grotesque.⁴ Papa says your place was Trafoi. Pilla & Mary are very happy at their hotel. Lady Eliott (Earl of Ravensworth's daught—name Liddell & probably related to Alice in Wonderland) came up to Pilla & said she would like her to know her daughter— this after looking Pilla over for a whole month.[5] The *English* girls in the hotel are very loud & fast. They talk nothing but hockey & golf. Isnt it strange about Lady Katherine Mead?[6] Give it to Kate. I have somewhere got a Mead of the Horse Guards for you. Mead is a good name in England—or Ireland. Mead*e* is the name of the Earl of Fitzwilliam.[7] Eben Jordan has taken the Duke of Argylle's sacred castle—where Victoria Tennyson &c planted trees.[8]

Papa & I are going off to Nice or Mentone for a change. I will put this letter in with papa's—but send a letter that I wrote *before* it along, too. Keep them for notes (I make a lot in the Year Book). Papa will sometime write delightfully of San Remo—though he has been so disgusted that he has *said* he wouldnt. Maria, alone, is worth writing up. The San Remoites have such filthy habits that you cant believe they are much advanced. In his book Ruffini was constantly lecturing them about it—but they havent improved. Cramer,[9] of the Paradiso has taken a Villa nearby & filled it with Russians. There are two grand Dukes over at Bordighera: Cyril & Michael, at present.[10] Cannes is *full* of wounded Russian officers. Dr. Antonio in Tauchnitz, is *quite* worth reading. Buy it. And then buy Carlino which contains also San Remo revisited—the most humorous piece of writing that ever—Papa adores it.[11]

Mamma

Manuscript: William White Howells

1. In a letter to John of the same day (Howells Memorial) and in "A Memory of San Remo," WDH mentions his trip to Taggia. Giovanni D. Ruffini (1807–1881), Italian-English novelist, wrote *Doctor Antonio* (1861), which WDH called "one of the most delightful novels ever written in English."

2. Mary Antoinette Lovett Cameron was the widow of William Wayte (1829–1898), who taught at Eton beginning in 1853 and for several years was professor of Greek at the University College, London. Sir George King (1840–1909), botanist, was prominent for his collections and experiments, especially with quinine, in India. He had retired in 1898 and was to die in San Remo. Miss Perks has not been identified.

3. Probably Joanna Hayes Shepard, called Daisy (1876–1952). She had married Osborne W. B. Bright (1870–1933) in 1900; her first child was born 11 July 1903, and her second child, born 23 October 1905, was named Elinor.

4. Harriet Lewin (1792–1878), a biographer, married the historian George Grote (1794–1871) in 1820. Sydney Smith (1771–1845), the celebrated wit, knew her well and

mostly admired her, though in a letter of 9 October 1843 he remarked "She is very clever and very odd." Elinor may have heard her anecdote from the W. Sydney Smiths (otherwise unidentified) who were in nearby Cannes.

5. Henry George Liddell (1811–1898), dean of Christ Church and coauthor of *Liddell and Scott's Greek Lexicon,* was the eldest son of the brother of Thomas Liddell, baronet and later baron of Ravensworth, the father of the first earl.

6. Lady Katherine Mead has not been identified.

7. Richard James Meade (1832–1907) was the fourth earl of Clanwilliam in the Irish peerage. In one of EMH's scrapbooks (Harvard) is a clipping announcing the death of the third earl, which occurred 7 October 1879; and on 29 April, 12 May, and 11 October 1907 (Massachusetts Historical Society), William B. Mead wrote Elinor about family history and genealogy, mentioning the Clanwilliam family.

8. Eben Jordan (1857–1916), president of Jordan, Marsh & Co., the Boston department store, rented Inverary Castle in Argylshire in 1905. A journal entry of 27 September 1875 (*Victoria in the Highlands,* ed. David Duff, 1968, p. 312) records the tree planting but not Tennyson's presence; after breakfast Victoria "met the Duke and the rest in the pleasure-grounds, where I planted a small cedar of Lebanon. . . . Then went on a little farther to where the road turns near the river, and planted a small silver fir, opposite to a magnificent one which my beloved Albert had admired in 1847."

9. S. or L. Krahmer was the owner of the Hotel Paradiso.

10. Cyril (1876–1938), a cousin of Nicholas II, and Michael (1832–1909), a brother of Alexander II, were staying at the town about six miles west of San Remo. The officers were from the Russo-Japanese war.

11. *Dr. Antonio* had first appeared as a Tauchnitz volume in 1861; *Carlino and Other Stories,* in 1872. Among the "other stories" is "San Remo Revisited."

To John Mead Howells

Villa Lamberti [San Remo]
Wed, March [29, 1905][1]

You've arrived! And *how* you have worked! And to outdo *those* two firms which generally suit everybody & all kinds is TREMENDOUS. More to me almost than if it was McK M & W because they are not so practical.

I'm tired from packing. Can only say your letter has made the whole Howells family proud & happy. Think! I actually had said I wished you had nt gone into another big competition. Which shows I didnt *know.* I m glad about Deville.[2] Glad about everything. Pilla was pleased, with your invitation to share you're flat. She got *malaria* at that Pension in Florence—that's what is the trouble. But she's all right & Saturday we leave this place for Genoa. I dont say half as bad things about S.R. as most people. Lots die here. You say N Y is healthy. How about spotted fever—40 deaths a day? But the snow will be all gone when we arrive April 16. The Southe[r]n voyage has already cured John Hay and on the Moltke it will be a *treat*, till we meet[3]

Emma

Now I lie back!
I am a bit tired from packing
Lark & M. are to be at Smith's we at Continental[4]

Manuscript: William White Howells

1. The letter is dated by one to John from WDH that was enclosed with it. In a letter to Mildred of c. 17 March (Harvard) John had told her that Howells & Stokes had won the competition for the Title and Guarantee Trust building at 176 Broadway. The two competitors EMH mentions were the firms of Carrère & Hastings (New York Public Library and others) and Clinton & Russell (many business buildings).

2. Henry Deville, a friend of John's at the Beaux Arts, was an architect in the office of Howells & Stokes for many years and also an etcher, who in 1914 had an exhibit in Paris.

3. The Howellses called on the Hays in Genoa but missed them (*Hay–Howells Letters*, pp. 126–127). They had left San Remo for Genoa on 1 April and were to sail 5 April on the *Moltke*. The crossing turned out to be stormy, with the ship arriving on 17 April in New York, a day behind schedule.

4. The Hôtel Continental des Etrangers and the Hotel Smith (with an English landlord) were near each other in Genoa.

Henry James to EMH

Lamb House, Rye, Sussex.
August 14th 1905.

Dear, dear Mrs. Howells.

Infinitely touched & delighted am I at your sweet letter just received, which ministering angels prompted you to write. I like to think—I *revel* in thinking—of you 2 intelligent twain sitting there in the summer nights & sipping the Golden Bowl in that delicate gustatory fashion. Well, I'll be brazen & say, that as it is, I think, it should be read. Never, also, I believe, was a Subject, a Situation, pumped so dry as that one—striking you as so pumped & left so thirsty for any remaining shade of mystery—by the time the last page is reached. It is *over-treated*—but that is my ruinous way & why I have never made my fortune. No matter if I have made yours! We are thinking of you all, in your neighborhood, immensely, in Europe just now, as Accessories to Portsmouth drama, & I should be able to swagger, immensely, also, before a larger gallery than this little place, were I disposed, by saying that I had *almost* seen the Wentworth Hotel, while with you, & the Navy Yard, & the rest of the Scene of Action.[1] But I keep all this romantic consciousness quite tenderly to myself, as part of my memory of those 2 wondrous days at Kittery Point and at Sara Jewett's place— so wondrous has everything become to me since my return here— in such iridescence does it all shine as seen from the re-entered, contracted tent of the Pilgrim at rest.[2] The Pilgrim is going to immortalize you in his own particular ruinous way however—overtreat & pump you dry till you will gasp for a drop of neglect. I find a beautiful summer here & the sense of the return to private life & the resumption of literary labour most sustaining. But all the things I didn't say to Howells, while I was with you, still whirl about the room like importunate bats—tormenting me overmuch. Tell him, please, that I just love his new London chapters & wish they wd. go on & on.[3] Also please that I love *him* & constantly hang about him, & am firm in the faith that we shall have you here again. There is distinctly room for you, always. And believe, again, that your letter was a joy, dear Mrs. Howells, to your affectionate old friend

Henry James[4]

Manuscript: Harvard

1. The Portsmouth Peace Conference, 9 August–5 September, took place across the river from Kittery Point. In a letter to Aurelia of 13 August WDH remarked that he and Mildred had gone to the admiral's breakfast for the peace commissioners, and as he wrote John St. Loe Strachey a month later he "saw all the envoys several times" (*SL,* 5:129). The Wentworth Hotel was on New Castle Island, off Kittery Point.

2. In June before James left America (on 5 July), he visited the Howellses at Kittery Point and Sara Orne Jewett in South Berwick.

3. "American Origins, London Films" had appeared in the July and August issues of *Harper's Monthly.*

4. A brief note from James to EMH dated 20 September 1905 (Harvard) thanks her for the "Conference photos" and adds: "My Peace for your Passion!"

To Lucy E. Keeler

the St. Charles Hotel,
Atlantic City, N.J.
February 7ᵗʰ [1906]

Dear Cousin Lucy:¹

I am glad you liked the photo. of Uncle William. I have seen
Mary Ann Bigelow's big one, which must be the same as yours,
and I prefer the small one, it is so much more distinct and sharp.
I gave one to my brother William Rutherford, who was named for
Uncle William. You see my Mother was just the age of her uncle
and they were great friends. So she named her son for him. Will's
wife who is a Hungarian was very much impressed by the Consular
uniform. I should have said that it was taken in Portsmouth from
the daguerreotype Annie Wright possesses.² She insists I shall have
it. But I prefer the photographs.

Mᵣ Howells & John were so glad to have those notes of your
mother's. They were very entertaining, and I thank you very much
for copying them for me. John is the collector of family history
and traditions, and is glad of any contribution to it. I have no end
of Noyes history, beginning with the grave and Church of the rector whose son came to Newburyport—both graduates of Oxford.
But, as neither of my children show the least sign of marrying I
am wondering what will become of all the accumulations which are
now in a store house, I suppose they will naturally go to some
Howells in Ohio who will make nothing of them.

We live in a rough way in Kittery Point now but I cant keep any
treasures there as the house is left to itself in Winter. So all our
paintings and relics are stored in New York and we never see them.
My son lives in a hotel in New York, and my daughter elects to
spend her Winters in Bermuda. So Mᵣ Howells & I have little encouragement to set up housekeeping for a few months in the Winter.

I will enclose a characteristic letter of Henry James's to show you
his horror of the way we live.³ He visited us in Kittery, but he
evidently considers *that* much too informal to be a "home." He is
very fussy and has quite an elegant house, with a walled garden

and five servants, in Rye—where M⁑ Howells & Mildred visited him. He thinks so much of *things*. But his family were always travelling when he was young, and he has told us how he hated it. So this is the result. Please send back the letter as I treasure all he writes so much.

John Mitchell lunched with us the other day, when he was here acting in Francis Wilson's Company.[4] Mildred is meeting the Hastingses in Bermuda. She admires the daughters and finds Mrs. Hastings very like Cousin Laura only "more determined."[5]

I remember your Mother perfectly. I wonder if you are like her.

I am not much of a letter writer—as it exhausts me too much— and besides I have a rheumatic hand. But the "Pamphlets" have never reached us!

Yours sincerely

Elinor M. Howells

M⁑ Howells has gone up to New York (3 hours) to attend a dinner to Mr. Mu[nr]o editor of the N A Review[6]

We expect to be here all this month.

Manuscript: Hayes Presidential Center

1. Lucy E. Keeler (1864–1930) was the daughter of Isaac M. and Janette Elliott Keeler of Fremont. William Rutherford Hayes (1804–1852) was a favorite uncle of Mary J. Mead, after whom her son W. R. Mead was named. Another cousin, Mary Ann Bigelow (b. 1838) was the daughter of Martha Billings Hayes and Russell Hayes, who was a brother of Lucy's grandmother Linda Hayes Pease Elliott.

2. See 12 January 1896 on the Noyes ancestors. On 21–27 September 1905 the Howellses made a "tour of relationing." Among places visited were Epping, Nashua, Keene, Brattleboro, Putney, Greenfield, and Old Hadley in search of details about the Hayses, Meads, Noyeses, Russells, and Smiths. At the end of the trip EMH stayed with Mary N. Mead at Northampton when WDH returned to Kittery Point.

3. Probably EMH enclosed James's letter of 14 August 1905, though "the horror of the way we live" is only implicit in it.

4. Francis Wilson (1854–1935), actor and manager as well as author.

5. Emily Hayes Platt (1850–1922), Laura Platt Mitchell's sister, who in a White House ceremony of 1878 had married Russell Hastings (1835–1919), breveted a brigadier general in 1865. Her two daughters were Lucy W. (b. 1880) and Fanny (b. 1882). Laura Platt Mitchell had visited the Howellses at Kittery Point for two days, as WDH reported to Aurelia on 3 September 1905: "Elinor and she had a great talk-up of old times. She is sixty years old, but without a wrinkle or a grey hair" (*SL*, 5:131). Possibly Lucy E. Keeler herself had written several of the pamphlets; one of her publications of about this time was *A Guide to the Local History of Fremont, Ohio* (1905). In addition to her work in history, she also wrote on other subjects, most notably nine essays about gardening that appeared in the *Atlantic* between 1913 and 1924.

6. David A. Munro (1848–1910) was editor of the *North American Review* (1895–1899),

and after reorganization of Harpers in effect the editor, though George Harvey became nominal editor.

To Eliza W. Howells

[Kittery Point] Oct. 11ᵗʰ
1907.[1]

Dear Eliza:

What a delightful visit you must have had at the Harriotts! I have never written you what treasure the old letter, Joe printed in the Sentinel, is for me.[2] For my mother & a sister or two went way down from above Brattleboro' (*how* I cant imagine—though grandfather Noyes took my mother to Boston in a *chaise* to get her wedding finery) i.e. Putney Vt. to Miss Pierce's school in Litchfield Conn.[3]

They boarded in Rev. Reeves's family. Mʳˢ R. was a sister or something of Aaron Burr's.

I have often heard my mother talk of the school—but in those days took little interest in old things.

We have got Pilla back. She saw the Hay & the Mills there.

I send a little cutting from the New York Herald that may interest you—though you probably see the Herald[4]

We are pulling up to go to New York—& perhaps to Rome.[5] Will will tell you all about it.

Affectionately
Elinor

Manuscript: Hayes Presidential Center

1. "1907" may be in a different hand.
2. In "Notes by the Editor," *Ashtabula Sentinel*, 3 October 1907, Joe wrote about going to Salt Cay to visit the family of Alexis W. Harriott, the British official there, who with his brothers produced salt from the ponds on the island. On 15 August Joe had published a letter of 23 September 1817 written by Eliza F. Wadsworth, a student at Miss Pierce's school, to the mother of a Jefferson resident, Helen C. Burgess, who had sent it to him.
3. For the trip of Elinor's mother to Litchfield in 1820, there is a manuscript in EMH's hand entitled "My Mother's School Experience" (Polly Howells Werthman). As noted there, Tapping Reeve was a judge, and his first wife was a sister of Aaron Burr.
4. The clipping from the *Herald* is no longer with the letter.
5. In the letter of 20 October to Joe (Harvard), WDH gave full plans, noting the

proposed date of sailing as 30 November, with the return home on 25 June. Because of John's wedding, the sailing date was later changed to 4 January 1908. Travel letters would be published serially in the New York *Sun* and then appear as *Roman Holidays and Others* (1908).

To Abby MacDougall White

Hotel Bellevue,
Boston Nov. 12<u>th</u> [1907]

My dear Abby:[1]

I had hoped to see you before this, but as it is impossible for me to leave Boston for some days yet I have to tell you in this way how happy you have made me for my son.

I hope soon to see you and talk the whole joyous affair over with you.

Yours affectionately
Elinor M. Howells.

Manuscript: Howells Memorial

1. Abby MacDougall White (1880–1975) became engaged to John Mead Howells in early November. She was the daughter of Amelia Jane MacDougall (1850–1885) and Horace White (1834–1916), editor and owner of the New York *Evening Post* and earlier chief proprietor and editor of the Chicago *Tribune*. The marriage was to take place on 21 December 1907 in St. Paul's Chapel, designed by John and completed earlier in the year, at Columbia University.

To Mildred Howells

[New York] Sat. p.m. [21 November 1908][1]

Is it possible the paint in the flat dining-room is *white*. Well then there is no use putting on dingy paper of course.

I dont like the peacock paper—but the room cant be pretty anyway, with our old blk. walnut chairs with their shabby dark leather seats. They are not mahogony color—just imitation blk. walnut with dark marroon seats almost black now

The peacock paper sets my teeth on edge—in spite of Whistler's room example. It looks so painty.[2]

Now the doors are *mahogony* I don't see that anything can be done—just leave it white

Get any paper but peacock   I am so sick of peacock

All New York is in Peacock

Manuscript: Harvard

1. The letter is dated from one by WDH to Mildred on the same day. The Howellses were preparing to move into their studio apartment at 130 West 57th Street, as they finally did on 15 January.

2. When Whistler's original peacock room was exhibited in London before being shipped to the United States in 1904 (now at the Freer Gallery), much comment appeared about it.

WDH to Aurelia H. Howells

Stratford House [New York],
11 East 32 st.
Nov. 28, 1908.

Dear Aurelia:

You will think I move as often as Sam, but it is for a different reason. Elinor thought she was not well at the Burlington; and we came here for the very few weeks, as we hope that we shall now be in our flat.¹ It will be papered this coming week, and then we shall move in, and arrange things afterwards.—I am glad your hotel is so pleasant, and that you are happy and comfortable. John's wife bore him a fine boy, at 6.45 yesterday morning: weight 7½ lbs. It is a peaceful little fellow, with black hair and at present he looks like me. Abby is very well, and John grins from ear to ear.² Of course Elinor is tremendously excited and we share the parents' joy in greater measure than I could have dreamed. There is a grandfatherly rivalry between Abby's father and me. He saw the boy first, but I profess to have had an interview with him in which he apologized for Mr. White's precipitancy as the foible of an older man. . . .³

Your affte brother,
Will.

Manuscript: Harvard

1. Within a few days, however, they moved back to the Burlington at 10 West 30th Street, where they stayed until at last getting into their apartment.

2. In a letter of 6 December to Annie (*SL*, 5:265), WDH adds a few details: "I can't tell you what a joy he is to granny and me. I suppose his mother and father like him too, but that is mere nature. Still, we allow them a small interest in him. Abby is getting on very well: we had great fears for her, though her health remained so good throughout; all her mother's children were seven months babies, and she died giving birth to the last. However the whole affair has gone splendidly; and John is as happy as the day is long."

3. Two short paragraphs on household matters in Jefferson have been omitted.

WDH to Henry James

130 West 57th st. [New York],
Oct. 24, 1909.

My dear James:

I hope you'll care to know that we prospered home over seas that laughed at the old notion of a rolling deep. But our gayety ended with a smooth passage of the customs, for I found my wife in bed after a severe surgical operation in a Boston hospital.[1] It was to rid herself of a danger that had haunted her for 20 years, but which is now absolutely at an end. She had, with Mary Mead's help, pluckily kept the event from me, knowing that I could not get home in time, and to keep me in my ignorance had sent all the letters from Mary to be posted at Kittery Point. Of course she is still very nervous from the "shock," but is otherwise well, and I found her greatly taken with "Crapy Cornelia."[2] She thinks it one of your masterpieces, and so do the Franklin Square people, especially dear old Alden and Harry Harper. Now we are hurrying up the November "Putnam" in order to get at your story in it.[3]— I haven't Smalleyed yet, but the dovelike Badger has sent me a very sweet letter about the Year Book which he has definitively accepted and is getting out.[4] I don't know just when he's to get it out.

Pilla is down at K. P. "amid the yellowing bowers"[5] of her garden. She too liked "C. C."

The John Howellses are back from Colorado, and twice a day we glory in our grandson, who has a way of clawing my moustache which I can't less than call inspired. He also shows a remarkable genius in the inspection of my watch.

My wife joins me in affectionate wishes.

Yours ever
W. D. Howells.

Manuscript: Harvard

1. As WDH wrote to Clemens, 5 November 1909, "She had some swollen glands removed" (*SL*, 5:291), also recalling that in 1889 she had been examined for the same condition and been "divinely reassured" by the specialist. Another letter notes that Elinor's life was feared for after the operation and "nothing but morphine could save her from the pain" (*SL*, 5:290).

2. "Crapy Cornelia," the story of a renewed friendship by a middle-aged man and woman, came out in the October *Harper's Monthly.* In his response, James thanked WDH for the "pat on poor Cornelia's crapy little back" (*SL*, 5:290). On 25 December WDH also reported that he was reading *The Tragic Muse* (1890) aloud with Elinor (*SL*, 5:294–295).

3. "The Bench of Desolation" appeared in *Putnam's*, October 1909–December 1910. In the same letter, James depreciated the lack of proofreading and the way the story was divided. Henry M. Alden (1839–1919) was editor of *Harper's Monthly* from 1869 until his death. J. Henry Harper (1850–1938) was head of the literary department of the firm.

4. *The Henry James Year Book,* ed. Evelyn G. Smalley, was published by Richard Badger in 1912. WDH contributed "One of the Public to the Author."

5. The phrase is from Tennyson's "Song" ("A spirit haunts the year's last hours / Dwelling amid these yellowing bowers").

WDH to Aurelia H. Howells and Anne H. Fréchette

. . . 130 West 57th Street
[New York] May 8, 1910.

Dear Aurelia and Annie:

I telegraphed last night. You will know how I feel.

Last Sunday was her birthday, and when I told her she smiled. She smiled a day or two later when I told her something about Billy; she loved him so; but after that first brightening on her birthday, there was after 24 hours, only a peaceful oblivion, which last night at ten minutes to eight became the peace everlasting.

Pity me.

> Your aff'te brother
> Will.

Manuscript: Harvard

WDH *to Grace Norton*

> 130 West 57th st. [New York],
> May 13, 1910.

My dear Miss Grace:

It is kind of you to write me, and it was sweet to see you that other day. My wife had been an invalid for years, and last September, to save herself from the cruelest of maladies, she underwent an operation which she kept a secret from me while I was in Carlsbad taking the cure, and let me know of it only when I came into her room here. The "shock" ended in the complete nervous breakdown from which she could not rally because her vivid life was worn out. At first she suffered cruelly from neuritis, but the morphine, blessed be God, saved her from that.—She was a great intelligence, a clearer critical mind than I have ever known, and she devoted herself to my wretched ambition as wholly as if it had been a religion. No more generous creature ever breathed, and whilst she breathed I did not know how selfish I was.—I wish I could feel, as you do, thankful to be *left*, but I wished to go first, and in a day of deceitful respite I told her so, with the laugh we had together. She would have survived with dignity, and the beauty with which women adorn widowhood.—I hope sometime to see you. Pilla and I will probably go to England for the summer.

> Yours affectionately
> W. D. Howells.

Manuscript: Schlesinger-Radcliffe, Grace Norton Papers, John Graham Brooks Collection

Index

Designed by
Sally Harris
Summer Hill Books
Madison, Connecticut

Typography by
Brevis Press
Bethany, Connecticut

Ginnette de B. Merrill has taught French literature at
Vassar College and Wellesley College. George
Arms is Professor Emeritus of English at the
University of New Mexico.